THE
VORTEX

THE VORTEX

AN ENVIRONMENTAL HISTORY OF THE MODERN WORLD

FRANK UEKÖTTER

UNIVERSITY OF PITTSBURGH PRESS

This project has received funding from the European Research Council (ERC) under the European Union's Horizon 2020 research and innovation programme (grant agreement no. 101019086).

This book reflects the views of the author only. The funding body is not responsible for any use that may be made of the information contained therein.

Published by the University of Pittsburgh Press, Pittsburgh, Pa., 15260
Copyright © 2023, University of Pittsburgh Press
All rights reserved
Manufactured in the United States of America
Printed on acid-free paper
10 9 8 7 6 5 4 3 2 1

Cataloging-in-Publication data is available from the Library of Congress

ISBN 13: 978-0-8229-4756-1
ISBN 10: 0-8229-4756-0

Cover art: AdobeStock
Cover design: Joel W. Coggins

If you know exactly what you are going to do,
what is the point of doing it?

Pablo Picasso

CONTENTS

Acknowledgments ix
Introduction | *A New History for a Young Century* 3

PART I: ESSENTIALS

1. Potosí | *Rich in Metals* 39
2. Sugar | *The New Organic* 50
3. The Canal du Midi | *The Great Mobilization* 64
4. Sustainable Forestry | *The State of the Woodlands* 75
5. Shipbreaking in Chittagong | *Leftovers* 87

PART II: APPROPRIATIONS

6. The Land Title | *To Own a Place* 107
7. Breadfruit | *Food Choices* 118
8. Guano | *The Fertility Business* 129
9. Whaling | *Resources at Sea* 142
10. United Fruit | *The Great Corporate Banana* 154

PART III: IRREVERSIBLE

11. The Dodo | *Species That Perish* 173
12. The Boll Weevil | *The Nemesis of Monoculture* 186
13. The Little Grand Canyon | *The Perils of Erosion* 199
14. Cane Toads | *Immigrants on the March* 212
15. Saudi Arabia | *The State of Resources* 225

PART IV: TECHNOLOGY TAKES COMMAND

16. London Smog | *In the Age of Coal* 245
17. The Water Closet | *Producing Cleanliness* 258
18. Chicago's Slaughterhouses | *Animals on the Line* 269
19. Synthetic Nitrogen | *Farming on Steroids* 280
20. Air-Conditioning | *Engineering the Climate* 294

Interlude | *Opium*   303

PART V: RUPTURES

21.	Cholera	*The Nature of Disease*	319
22.	Baedeker	*Guidance for Seekers*	333
23.	Gandhi's Salt	*To Change the World*	346
24.	The 1970 Tokyo Resolution	*International Conventions*	359
25.	The 1976 Tangshan Earthquake	*Somewhat Natural Disasters*	370

PART VI: THE FINAL RESERVES

26.	Kruger National Park	*Reserved Nature*	391
27.	Eucalyptus	*Supertrees*	405
28.	Hybrid Corn	*Breeding Ambitions*	417
29.	Aswan Dam	*Damming and Developing*	429
30.	The Rice-Eating Rubber Tree	*Speaking of Dependence*	447

PART VII: THE AGE OF CATASTROPHE

31.	Holodomor	*The Politics of Hunger*	469
32.	The Pontine Marshes	*Fighting for Space*	482
33.	The Chemurgy Movement	*The Business of Biofuels*	494
34.	Autobahn	*The* Endsieg *of Automobilism*	506
35.	The Pine Roots Campaign	*The Totality of War*	519

PART VIII: THE GREAT ENTRENCHMENT

36.	Battery Chicken	*The Industrialized Animal*	543
37.	*Lucky Dragon No. 5*	*Atoms without Limits*	558
38.	DDT	*Learning from a Book*	572
39.	*Torrey Canyon*	*Coping with Technological Failure*	584
40.	Plastic Bags	*Ephemeralia*	595

Coda	*The Pandemic*	605
The Mess We're In	*An Inconclusive Conclusion*	611
Appendix	*Making Choices*	639
Notes	651
Index	783

ACKNOWLEDGMENTS

EVERY BOOK IS A JOURNEY. THIS BOOK WAS AN ODYSSEY. MORE PRE-cisely, it was an odyssey in a post-heroic age, where certainties and allegiances fall apart under the eyes of the despairing protagonist. The voyage began when I was deputy director at Munich's Rachel Carson Center for Environment and Society and thought about ways to bring a diverse group of academics into a conversation. It took a number of years until the endeavor turned into a book project, a few more until I had a draft, and yet a few more until the manuscript finally went to press. I encountered plenty of challenges and siren calls along the way, and I certainly had more than one episode of self-doubt. But, as Homer might have said, you really learn something on a journey.

While my esteemed 700 BC colleague drew on the blessing of Gods, twenty-first century historians enjoy more worldly support. This book would not exist without the Volkswagen Foundation, which supported me through a generous Dilthey fellowship, the environment and memory project at the Rachel Carson Center, and the University of Birmingham, which became my academic home when writing started in earnest. Along the way, I received critical feedback at academic events in Berlin, Bielefeld, Birmingham, Bochum, Bozeman, Cambridge, Chapecó, Chemnitz, Chicago, Düsseldorf, Eckartsau, Florianopolis, Frankfurt, Geneva, Gießen, Guimarães, Halle, Hanover, Hay on Wye, London, Lucerne, Minneapolis, Moscow, Munich, Nagoya, Oberhausen, Osaka, Oulu, Paris, Philadelphia, Potsdam, Rome, Saint Peter, Seattle, Seoul, Starkville, Stuttgart, Toronto, Tübingen, Versailles, Visselhövede, Washington, Wilmington, Wrocław, and Zagreb. A special word of thanks goes to the students of my Munich years who made the "environment and memory" project a truly memorable experience. I also wish to thank the staff at seventeen archives in Germany, the United States, and the United Kingdom, whose holdings are cited in this book.

Corey Ross was usually among the first readers, provided plenty of helpful feedback, and was a partner in some crucial discussions. Comments or inputs on individual chapters came from Laura Bradshaw, Courtney Campbell, Jane Carruthers, Reginald Cline-Cole, Matthew Francis, Jim Giesen, Simona Grothues, Keith Jones, Jonathan Harwood, Matthew Hilton, Simon Jackson, Karena Kalmbach, Tait Keller, Claas Kirchhelle, Rob MacKenzie, John McNeill, Mo Moulton, Kazuki Okauchi, Jotham Parsons, Jeremy Pritchard, James Pugh, Sadiah Qureshi, Joachim Radkau, Jonathan Reinarz, Klaus Richter, Manu Sehgal, Sarah Waltenberger, Jim Webb, Edward Wilson, Jennifer Wunn, Shirley Ye, and Amir Zelinger. While Odysseus, as captain, was entitled to instruct his sailors, support from these readers was completely voluntary, and I really appreciate their efforts. It helped in so many ways big and small.

It bears recognition that fifteen of these readers are current or former Birmingham colleagues. However, I also wish to thank the faculty member who choked on his lunch sandwich when I told him about this project: for those who write a world history, it is crucial to maintain a sense of humility. I learned a lot from working with Birmingham's Ikon Gallery and the Birmingham Institute of Forest Research (BIFoR), as I did from Maryum and Eesa and their parents and grandparents next door—the best neighbors we have ever had. All that makes this book very much a Birmingham project, down to my office in a university building that, while not otherwise known as inspiring, offers two of my features of global modernity, separate toilets for staff and students and an air-conditioning unit. (It would be three features if asbestos had made the list.)

I am reluctant to mention more names, as the attempt at a complete account would soon run out of control while remaining exactly that: an attempt. It is also opportune to note that acknowledgments are subject to the law of diminishing marginal utility, and twenty-first century academia has produced plenty of lengthy prefaces to make that point. Joachim Radkau and John McNeill deserve a word of thanks because they supported this project knowing that it would be a critique of their respective world histories. Klaus Tenfelde helped and inspired me from my student days to his project on the history of German mining that brought me to rethink my work on agriculture in resource history terms. He died in 2011, and it just so happened that my first presentation on "the vortex" took place in Bielefeld, in a seminar that he had once chaired, on the day of his funeral. I also wish to mention Joel Tarr, who said in conversation with an aspiring graduate student more than twenty years ago that "there is nothing wrong with being ambitious." I guess he could not have seen this coming.

It was probably illusionary to think that a monograph on the circuitous routes of modernity would have a straight path toward the press, but it was smooth sailing once Abby Collier took the project under her wings. I also wish to thank my copyeditor, Therese Malhame, as well as Amy Sherman, Alex Wolfe, Joel W. Coggins, and Sandy Crooms at the University of Pittsburgh Press, and Jürgen Hotz at Campus, a fabulously open-minded German publisher that had the German translation in press before the English original. I appreciate the feedback from five anonymous reviewers and more than a dozen reluctant agents and editors, whose comments made this a better book. The manuscript was finished in August 2018, and I made various updates to the draft until November 2021. Nonetheless, references remain imperfect, and so does the book, which is the inevitable price of writing on a huge canvas. And in a way, a smooth, near-perfect narrative would run counter to my main argument. Nobody should expect a smooth ride when navigating a vortex.

I reserve my deepest thanks for my family. Simona was a formidable partner throughout this journey (which makes for another Odysseus mismatch) and shared the trials and tribulations of world history writing. And then there was Johanna, who was born while this project was under way and became a source of inspiration in her own right. I mean that in more than a metaphorical sense. Among the many risks of writing this book, one of the greatest was to turn into a cynic. The project brought me face to face with more suffering, violence, and misery than any sane human would want to stomach. The temptation was huge to turn this book into a scathing indictment of the dark, predatory soul of the human race: *homo homini lupus*. But cynicism does not work with children. Maybe it should not work with adults either.

THE VORTEX

INTRODUCTION

A New History for a Young Century

EVERY HISTORY REFLECTS THE EXPERIENCES OF ITS AGE, AND SYNTHESES are particularly susceptible to the tides of the times. Among the experiences that have shaped this book, the most obvious is the bundle of processes commonly known as globalization. The need for planetary perspectives is probably beyond debate in an age where the world is more interconnected in more ways than ever before. It shows in a boom of scholarship on environmental issues in the Global South. Environmental historians have moved beyond their traditional mainstay in Western Europe and the United States, and regions such as Russia and the Middle East, which John McNeill described as "near black holes" of environmental history as recently as 2011, have gained more attention.[1] As result, today we are in a much better position to write environmental history in a global context. No longer does a synthesis need to start in Europe or North America because research on the rest of the world is simply too scarce.

Second, we have learned in recent decades that we should be gravely concerned about the state of our planet. We do not lack achievements on specific issues and in certain places, but the overall trend in the interplay between humans and the natural environment is clearly negative: anthropogenic climate change, loss of biological diversity, pollution, resource woes—we know about the environmental toll of global modernity in greater detail and with more certainty than ever. We also know that these challenges will not be met in satisfactory fashion anytime soon, and they will not go away miraculously either. Only a fool would deny that environmental challenges will define the twenty-first century—and it should leave us unimpressed that the number of fools remains at disheartening levels.

Neither of these points is new. The severity of the environmental crisis and the interconnectedness of the globe have inspired environmental histories from the early 1970s to John McNeill's *Something New under the Sun* and Joachim Radkau's *Nature and Power*, the two magistral syntheses that greeted the new millennium for the purpose of environmental history. I am greatly indebted to this intellectual tradition, certainly more than the notes in this volume suggest, and yet this book departs from a third observation that marks a break with existing scholarship. The environmental challenges of the twenty-first century look different from the challenges that earlier generations of scholars were writing about.

Looking back at McNeill's and Radkau's books after two decades, it is odd to realize how these authors confidently structured their narratives along spatial and thematic lines. McNeill discussed the lithosphere and pedosphere, the atmosphere and the hydrosphere, with subchapters looking into forests, whaling, and so on, and Radkau gave much attention to nation-states from the United States to Bhutan.[2] However, global historians have become wary about the container theory of statehood, and the same holds true for biological spheres. Entanglements figure prominently in the recent literature: humans, ideas, commodities, and species traveled the world in all sorts of directions, and scholars have made the case that interconnections matter in many different ways.[3] Convenient compartmentalizations, be they geological, biological, or political in nature, look increasingly dubious in a global age.

These doubts were historiographic as well as political. When the world entered a new millennium, there was no way to deny that the agenda of environmentalism was a wild mix of issues: pollution, garbage, exhaustion of resources, protection of landscapes, endangered species, and so forth. The severity of problems also differed around the globe. London smog was a thing of the past in the British capital by the 1970s while other industrial areas were still wrestling with coal smoke, and particulates remain one of the world's major killers. Or maybe the real problem about London smog was the use of fossil fuels. It also became clear that responses to environmental problems had repercussions in other realms. Numerous nature reserves, dearly needed to preserve biological diversity, went at the expense of Indigenous populations. And what about environmental solutions that created new problems, such as when sewage treatment led to an accumulation of toxic sludge? Despite plenty of expertise, environmental challenges defy big comprehensive blueprints—as seen in 2020, when the fiftieth anniversary of the first Earth Day coincided with the global spread of COVID-19.

Except for a lunatic fringe, everyone knows about the environmental toll of industrial civilization today. What is recognized less often is the growing uncertainty on how to approach this toll. Priorities, approaches, institutions, even the definition of problems—once upon a time, we thought that we had clear ideas and settled routines on all these things, but received wisdoms have come into doubt. Global environmentalism is not like a Beethoven symphony, where you know after the first few measures what is coming. It is more akin to an orchestra with many instruments and even more pieces of music, and more often than not, the sound that prevails is not so much due to a heroic conductor as to the powerful lungs of some of the players. We live in an interconnected world, and we can no longer define priorities and externalities in the traditional manner, simply as givens. Priorities are matters of perspectives, and what qualifies as an externality for some may be someone else's life.

In a global twenty-first century, we can no longer write environmental history with a predefined set of concepts and a fixed hierarchy of issues. Agents, nation-states, tools, the political and cultural framing of issues—categories that were self-evident a generation ago—have become blurry and contested. Some authors, notably those under the influence of the Anthropocene thesis, have tried to cut through these ambiguities and imposed categories by force, borrowing on the presumably ironclad findings from the natural sciences, but this suggests a global consensus on environmental matters that does not exist: there is no Archimedian point in today's world. In fact, it has dawned on us that the idea of a global consensus has always been more fiction than reality, the result of a Western intellectual hegemony that looks irretrievably gone. In our age of globalization, an environmental history of the modern world needs to view the past from the perspectives of the industrialized West and the Global South, the policymaker and the consumer, the expert and the layperson, the urbanite and the farmer, and so on.

At first glance, this approach might smack of relativism, but it actually comes down to the exact opposite. Environmental historians have long made the point that we face limits in our interaction with the natural environment, but these limits are not just material and ecological: they are also about institutions, about laws and government bodies, about mindsets and interests, about technologies and expertise, and they are about the shadow of history. As humans have engaged with environmental challenges over time, they have accumulated a legacy that this book seeks to dissect. Materialities and technologies, laws and experts, institutions and cultural tropes—many things frame our engagement with the nonhuman world, and we can only understand them as products of a long and complicated history. Once we start to untangle this history and take stock of our environmental legacy in its full complexity, it becomes clear that there is no room on our planet for an environmental "anything goes." In an interconnected age, where past experiences are entrenched in technologies, materialities, institutions, cultural tropes and ecological conditions, very few things "go."

While this book seeks to complicate environmental history in some respects, it also aims to make it more transparent in others. A global scope can bring out something that is typically lost in case studies on specific times and places: environmental challenges are remarkably similar in otherwise different parts of the world. The laws of nature hold force everywhere, and this makes for a baseline of similarities that this book seeks to tease out. In other words, environmental history provides an opportunity to make world history *simpler*—completely unlike cultural or political history, where going global quickly leads to a cataclysm of complications. Invasive species, pests, dams, sewers, and other phenomena have triggered similar responses around the globe, and if we compare individual trajectories, we can map recurring paths in the human engagement with environmental challenges. In fact, similarities are so strong that one can understand global challenges by departing from a single example—assuming, of course, that this example is chosen judiciously.

The conclusion will take it from here and summarize the stories in

INTRODUCTION

this book as individual threads in a global entanglement of five big trends: the Great Need, the Great Externalization, the Great Reckoning, the Great Regulation, and the Great Narrowing. This introduction seeks to outline the methodological essentials of this endeavor, and the most basic is about what this synthesis aims to achieve. The book seeks to map the environmental trajectory of the modern world and analyzes the environmentalisms of our time as products of history, but it refuses to deliver the classic synthetic overview. More specifically, this volume does not offer the kind of intellectual order that syntheses typically do, and it is not about indecisiveness on the part of the author: it is what the state of the planet demands. While writing was under way, a German professor inquired, "Wie das alles zusammenkommt"—how does it all come together? There was a simple answer to that question: it does not come together at all, and that is precisely the point. The chapters in this book end with open questions, moral dilemmas, or a realization that solutions merely bought time, and I will not hand out solutions ex cathedra. We do not live in a world where things simply add up: where we are accumulating wisdoms, technologies and solutions and are getting better all the time. If things do not make intuitive sense, if moral and practical inconsistencies linger, and if we are stuck with unsatisfactory or paradoxical outcomes, that is just where we are. This book is a world history for an age where things seem to fall apart—where we see what is coming, where we have the tools, technological ones and others, and plenty of experience, and yet somehow fail to get our act together.

In other words, this world history seeks to turn complication into an object of study. Every chapter moves up to the present and traces material, technological, institutional, and cultural developments along with the making of ambiguities and fault lines. Complication was made in concrete places, by real people driven by interests, mindsets and experiences, in ways that are open to scrutiny: we can arrive at a much more sophisticated understanding of our own predicament, the material, political, and intellectual mess that we are wrestling with, if we view it as a product of history. Challenges and responses coevolved over time, and they were shaped by vested interests, artifacts, power relations, institutions, and cultural tropes. We can arrrive at a better understanding of our concepts, our definitions of problems, our priorities, and our taboos if we take a closer look at how we came to them.

This approach has consequences that go down to the basics of history writing. There is no way to tell this history in traditional linear fashion: we need a nonlinear, entangled mode of historical narration that does not reproduce the intellectual hierarchies of the Western world but rather exposes them to scrutiny while being scholarly rigorous, reasonably specific, and sufficiently precise. Nonlinear narratives need a combination of top-down and bottom-up perspectives, and I hope to convince readers that this is best achieved by departing from one exemplary case. In order to bind the individual narrative threads together, this project needs a conceptual metaphor that highlights the interconnectedness and the dynamism of the modern world, and the vortex is the best metaphor that I could think of. But first of all, this world history departs from the realization that the environment, obviously the most basic category of any environmental history, is not just a given: it has a history that matters, like everything else. We need a new history in order to write narratives that make sense in a global twenty-first century, but this history is not without precedent. In some respects, this book argues for a return to an environmental history that we once had.

THE FLOW OF HISTORY

As concepts go, the environment is one of the younger ones. According to Paul Warde, Libby Robin, and Sverker Sörlin, it was a product of the postwar years, when hopes for the future and fears about humanity's destructive potential mixed in a way that made for an intellectual watershed of the first order. The environment was global, it was all around us, and it was under threat, and the details of the diagnosis were evolving within these general parameters.[4] Most people treat the environment as a self-evident concept today, and the word serves to define ambitions, policies, and academic disciplines, including a historical one. However, the environment was not invented out of the blue, and neither was environmental history. Historiographic surveys typically focus on the upswing of research since the 1970s, but there was such a thing as an environmental history avant la lettre, though not in the sense of a self-conscious academic field. It was merely a hodgepodge of scholars and books, held together not so much by a moral cause as by a realization that the nonhuman world mattered somehow, but it can be inspiring to read books from the time before the environment became a paradigmatic concept. It serves to challenge conventional wisdoms about the place of nature in human history, and it brings out the hidden assumptions in our modes of historical narration. After 1945, the environment was first and foremost a bundle of problems, but one could also see it as the inevitable base of all human life.

Warde and his coauthors did not mention Fernand Braudel and his expansive *La Méditerranée*. Mostly written during his time as a prisoner of war in Nazi Germany, it discusses the Mediterranean world in the age of Philip II with great attention to the natural environment. Braudel famously opened his lengthy treatise—the original edition contained some 600,000 words, which is double the size of this book—with remarks on the different speeds of history. He distinguished three fields, each with a rhythm of its own.[5] In his opinion, the fastest but also the most superficial one was *l'histoire événementielle*: he emphatically confined the history of events to "surface disturbances, crests of foam that the tides of history carry on their strong backs." The second field was about societies and economies, "the history of groups and grouping," where change took place at a more measured pace. Finally, there was "a history whose passage is almost imperceptible" and where change was about "constant repetition, ever-recurring cycles." This history was about the natural environment.[6]

On first glance, Braudel's *La Méditerranée* seems of limited value for a study of modern history. His argument about cyclical change probably worked for the sixteenth-century world, but it is a far cry from our modern predicament, where acceleration and accumulation of harm seem to be the norm. While natural environments are typically vulnerable and under threat in modern history, they seem rather static in Braudel's narrative—though he did note in prescient manner that "everything changes, even the climate."[7] For the most part, Braudel discusses how the natural environment created challenges for humans, not the other way round. Against this background, it might seem that the only benefit of invoking Braudel for the present endeavor is that he provides an excuse for writing a really big book.

However, three points deserve closer scrutiny. The first concerns Braudel's broad understanding of the environment. He talked about natural conditions, but he also covered transhumance and nomadism, agriculture in the mountains and in the plains, droughts and famines, shipping and urban life, epidemics and the seasons. There were plenty of problems in Braudel's environment, but problems were not what defined the natural world: it was the setting where people tried to make a living. It was something to work with, a realm that was entangled with humans on so many levels that the entire distinction between humans and the environment became dubious: "Human life responds to the commands of the environment, but also seeks to evade and overcome them, only to be caught in other toils."[8] For Braudel, the environment was not in trouble. It was just there.

Second, Braudel's narrative showed how natural environments created baselines of similarities. He was certainly not oblivious of the diversity of his geographic realm. He devoted individual sections to the Tyrrhenian Sea, the Adriatic and other parts of the Mediterranean, and he looked extensively at the Sahara, Europe, and the Atlantic Ocean as well. But he did not leave it at a mere account of peculiar geographies. He looked at famines and how cities reacted by creating grain storages and regulating food markets. In his remarks on epidemics, he identified general urban patterns such as "using disinfectants based on aromatic herbs, destroying by fire the belongings of plague victims, enforcing quarantine on persons and goods . . . , recruiting doctors, introducing health certificates"—though Braudel did not fail to note that notwithstanding all these efforts, "the rich had always sought their salvation in flight."[9] Some of these responses will resurface in the cholera chapter.

Third, Braudel was onto something when he couched his concept in hydrological terms. A watery metaphor was probably no coincidence for a scholar who pursued a vision of "total history." A pool of water is indeed total for those who are floating in it, and water in motion has a fearsome dynamism that can dwarf human agency. Braudel's waves of history were massive, they crushed humans and exposed them to the elements, and the natural environment has not lost its ruthlessness since the sixteenth century. We now have better technology and better knowledge about many pertinent challenges, but the nonhuman world is still beyond our control. It is just that waves are a rather poor metaphor for the flow of history in modern times.

The natural world continues to operate in cycles, but humans have created new material flows that are more unidirectional. Huge resource streams flow into the urban centers of the modern world, and equally huge streams of waste emanate from them. Global trade networks channel billions of tons of oil each year, with many millions of tons of everything from wheat to oranges to boot, and the pollutants that they produce have changed the global climate, among other things. In fact, the human footprint has grown to such an extent that scientists talk about a new geological epoch, the Anthropocene, where humans stand on a par with the forces of nature.[10] The modern world has an unprecedented hunger for stuff, and it has an equally unprecedented ability to put other stuff into motion, but movement no longer comes in the form of waves. It is not a deluge either, though it is devastating beyond measure. A deluge peters out after a while, and it comes mostly from a single direction, but neither is the case in our time. In the modern era, wanted and unwanted flows have increased relentlessly, they are the cumulative product of many individual streams, and there is plenty of turbulence.

In light of these features, it seems more appropriate to view the flow of history in the modern era as a giant vortex. The emphasis is on the singular: we cannot understand global flows if we study them one by one. They have mixed and interacted in an eminently haphazard manner, and that has created plenty of crosscurrents and secondary and tertiary vortices. The vortex also features a lot of intended and accidental obstacles that shape the flow in one way or another. There is a lot of movement inside a vortex, there are moments of calm, and it can be smooth sailing for a while, but there is ultimately no escape from the material heft of masses in motion. A vortex is about dynamism and momentum, and it is about humans who tend to underestimate the forces that they are wrestling with. A vortex can look harmless, and it can be fun to drift along in the outer fringes. But when you move toward the core, things are getting serious.

VORTEXIAN HISTORY

As understood in this book, the vortex is a metaphor, and it should be read accordingly. The masses that humans have set into motion are terrifying, but weight is ultimately just one of numerous properties that the nonhuman players on the stage of history can claim. Momentum serves as a descriptor for forces of nature that are just there, whether humans like it or not. It can be about toxicity, or an awkward smell or taste, or the ability of a plant, animal, or pathogen to multiply rapidly, or something else. The key is that the vortex highlights the powers of the nonhuman world and their unrelenting inclination to push humans around.

As with every conceptual metaphor, the case for the vortex rests on analytical merits. The vortex evokes a sense of dynamism and forces beyond control, particularly if we think of it on a planetary scale. The metaphor also evokes a sense of turbulence and chaos. There are rough waters inside a vortex, and while some people may profit from beneficial currents, they should never be too confident about their good fortune. Humans are at the mercy of the elements inside a vortex, but they are not passive victims. They can navigate the water, they can create barriers or channel flows to their advantage, but it takes time, energy, knowledge, and a decent understanding of human means.

The vortex is about the power of the nonhuman world, but the metaphor is ultimately anthropocentric. It would be a mere feature of the natural world, perhaps equivalent in significance to the Great Red Spot on Jupiter, if it were not for its frightening ability to harm and swallow humans. At the risk of stating the obvious, the vortex is also a product of human inputs, particularly the voracious hunger for huge amounts of stuff that runs through modern history. At the same time, modern history includes plenty of efforts to tame, control, or even stop the flow, and assessing these efforts is a major part of the following text. In short, the vortex is not a raw force of nature. It is a highly regulated endeavor, albeit in an uncoordinated, if not chaotic fashion. There is no control center (though the dream of building one remains alive), but regulations matter even if they ultimately come down to a patchwork.

Regulations can take many forms: law, codes of practice, policies enacted by states or corporations, expert knowledge, cultural tropes. The vortex metaphor suggests a tepid primacy of the material, but it is a soft determinism, if it is a determinism at all. Scholars have played parlor games between environmental determinism and cultural relativism, but that is a type of academic ping-pong that has outgrown its usefulness, if it had any at all: many pertinent discussions bore all the marks of academic turf wars. In this volume, the guiding question is not whether institutions are more significant than cultural tropes or technologies, material properties, or ecological interactions—the focus is on how they mattered in specific contexts. Sometimes a book made history, as in the case of Rachel Carson's volume vis-à-vis the insecticide DDT. Sometimes it was a technology with "killer app" quality, such as the slaughterhouses of Chicago. And sometimes institutions win out, as in the case of the land title—one of the most successful inventions of Western modernity, still on a victory run in spite of a broad array of problems.

Many flows have clashed and converged, and it was the same with human responses to these flows. Institutions interlocked, and so did cultural tropes, business practices, materialities, and so forth. These interactions are a defining issue in this book, and they follow no discernible pattern: the web of interlinkages has grown over time and continues to grow, and we can only understand it as a product of history. Cross-references highlight these interconnections in the following narrative, and readers are encouraged to follow them even if it has an erratic feel. Doing so provides an idea of the fragmentary nature of the project of modernity.

It is important to recognize that these entanglements are stabilizing and constraining in equal measure, and hence a bit different from the entanglements that global historians have frequently recorded with a sense of enthusiasm. In an age of globalization, it is tempting to view entanglements as a mind-broadening force, and scholars have produced ample evidence that this is the case for cultural entanglements. Six decades ago, William H. McNeill wrote *The Rise of the West*, a sweeping overview of human history where civilizations thrived on interactions, and that was long before postmodernism celebrated exuberant possibilities.[11] But more often than not, environmental entanglements, and particularly those that include technologies and materialities, end up *constraining* the range of options. If several trends overlap in the human engagement with environmental challenges, people may come to a point where few options remain. It can happen with amazing speed. In the case of COVID-19, it took just a few weeks to get governments to a point where the only remaining choice was about which kind of lockdown they should enact.

The vortex can be brutal, but it is also hard to figure out. Currents are diverse, turbulences are hard to predict, and that is not the least among the attractions of this conceptual metaphor. A vortex features resilience as well as fragility, and one should never be too certain about what will prevail. After all, a multitude of rationales were in play in the modern world, from food systems to transport routines and modes of waste removal, and the general trend of the modern world is ultimately a composite product of these different rationales. The world is not a bathtub, where events are predictable until everything has gone down the drain. As understood in this book, the vortex is out at sea where things are constantly in flux, where a broad range of actors navigate more or less at will, where the seabed is diverse and widely unknown, and where a big wave can make a difference.

Water lacks buffer quality, and this means that movements can resonate in unexpected places. And yet interactions may surprise those schooled in the first law of ecology, which suggests, in the words of Barry Commoner, that "everything is connected to everything else."[12] The following narrative suggests an important addendum: some interconnections matter far more than others. The following stories feature a multitude of rationales and a range of entanglements, but neither is unlimited in numbers. As this book will show, it is possible to highlight the crucial rationales and the critical interconnections in any given field on a few dozen pages—with the notable exception of climate change.[13] It is not the sheer number of interconnections that make for the tensions and dilemmas that lie at the heart of the present discussion. It is that rationales and interconnections operate with different degrees of autonomy and force. Some things are tightly connected to some other things, but other connections are loose or insignificant in the grand scheme of things. The web of modern life is not a network of countless atomized entities with an infinite number of linkages. It is more akin to a set of tightly interwoven clusters: food chains, energy systems, expert communities, and so forth. Each of these clusters has an iron pulse of its own, and there is no guarantee whatsoever that they work in sync.

Most people associate a vortex with the threat of drowning, and a pessimistic undertone may not be entirely inappropriate for an environmental history of the modern world. In the twenty-first century, few people are confident that humanity's engagement with the natural environment is bound for a happy ending. But drowning is no foregone conclusion in a vortex, and a bird's-eye view is only one of numerous perspectives on a vortex. Much of the following discussion is about how the vortex looks for those caught in the flow, and that makes for a dizzying array of perspectives. Drifting along in a vortex, one can observe how things move around, how they change in shape and substance, how things come together in new arrangements, and how humans are at the mercy of the elements, or at least feel that way. This makes for a rough ride in the following pages, and that is by all means intentional. This book grew out of a sense of disaffection with world histories where the narrative was oozing order.

For all the material heft of masses in motion, humans retain agency inside a vortex, albeit to widely different extents. Some command powerboats that rule the waves (and make a few more) while others are condemned to swim on their own. Some people have access to the levers that regulate flows while others need to work with the currents as they are. By and large, agency is about resources and power relations rather than smartness. Idiots can run powerboats, too, and more than one captain has proudly steered his boat into the shallows, a reef, or the abyss. Disorientation and delusion are common experiences inside the vortex because so many things are constantly in motion.

INTRODUCTION

The one thing that unites everything inside the vortex is that it is exceedingly difficult to look beyond the roaring waters. A lot of things are open to many different perspectives, and vantage points and experiences are constantly in flux, but taken as a whole, the vortex is cognitively closed: a distinct universe that offers no certainties beyond its purview. There is no island inside a vortex, no terra firma where alternative projects can take root. As seen in the Gandhi chapter, alternative projects were merely another thing that was drifting along, and even an otherworldly figure like the mahatma was riding with the waves at times. And why would he not? There are many ways to stay above water in a vortex, but apathy is rarely a winning strategy.

It is an open question whether people, if given a choice, would wish to swim in rough waters. But for those who live in the twenty-first century, the question is ultimately pointless: we are condemned to engineer and navigate the vortex for better or worse. There was a world with humans before the vortex, and there may be one after the vortex has run its course, but engaging with either provides limited guidance, if any, as we wrestle with the mess that we're in. If there is a silver lining, it is that we do not need to learn how to navigate and engineer the vortex from scratch. This would be an obvious point if it were not for a few people who suggest otherwise.

THE ANTHROPOCENE DELUSION

When I started writing this book, the Anthropocene was the buzzword du jour. It was already around for more than a decade—I will talk about the birth of the Anthropocene in the chapter on the 1970 Tokyo Resolution[14]—but the "Big Bang" of the Anthropocene discourse, a meeting at the Geological Society of London and a subsequent cover story of the *Economist* in May 2011, was still fresh. A decade later, the Anthropocene has entered the environmental literature and the environmental imagination, but popularity has taken its toll. The concept is cited prolifically, sometimes ad nauseam, but its precise meaning is more diffuse than ever. Even textbooks on the Anthropocene have abandoned the idea of a coherent concept and they note by way of introduction that they "do not cover all the modes of understanding that might be brought to bear on the Anthropocene."[15] There is also a mushrooming array of sister terms: capitalocene, carbonocene, technocene, plantationocene, wasteocene, and so forth.

The blame goes mostly to a chatty academic field that is named the humanities. The Anthropocene is a pretty straightforward concept in natural-science terms, thanks in large part to rigorous processes of scientific consensus building. The International Commission on Stratigraphy set up an Anthropocene Working Group in 2009 and gave it the task "to assess whether the Anthropocene could be considered a potential chronostratigraphic/geochronologic unit." The Working Group replied in the affirmative seven years later and proposed 1950 as the threshold year for a new epoch.[16] The official proclamation was pending when this book went to press. The Anthropocene is also a familiar term in earth system science, a multidisciplinary endeavor that aims to model material and energy flows on a planetary scale. Insights from earth system research are cited throughout this book.

It is open to debate whether historians should embrace a chronological framework where everything is about one single threshold.[17] But when it comes to the significance of humans for planetary processes, the evidence is clear. Humans have played an outsized role in the environments on this globe, and it is impossible to understand the present state of this planet if we do not account for the impact of humans, and none of the following remarks shall suggest otherwise. In the words of the Working Group on the Anthropocene, "It is clear that human beings are now operating as a major geological agent at the planetary scale, and that their activities have already changed the trajectory of many key Earth processes, some of them irreversibly, and in doing so have imprinted an indelible mark on the planet."[18] No other species comes close in its significance for material and energy flows on planet earth, and there is no serious debate whether something like the Anthropocene exists in scientific terms. The question is how it matters for our view of history and the way forward.

We have entered a new age in terms of planetary materialities and ecologies. But does this mean that we have new *people*? Humans launched what we retrospectively call the "project of modernity," and they have entered new ages of lesser scales and scopes all the time: the age of automobility, the age of industrial chemistry, the age of fascism, and so forth. But as this volume shows, entering a new age did not mean that humans shed their baggage and started from scratch. Quite the contrary, the global environmental legacy that this book seeks to dissect is the cumulative product of numerous processes that involved an even greater number of groups all over the world, and we keep adding to this legacy on a daily basis. There is no delete button in the modern mind, and as the following narrative shows, humans are far less forgetful than they usually think. A lot of history resonates in today's environmental thought and action, except that we rarely recognize this history as such.

The Anthropocene literature is huge, but it seems that these divergent views are impossible to reconcile. We have entered a new epoch in environmental terms, but we do not have new mindsets, cultures, or societies. We enter the Anthropocene with a lot of baggage: culturally, politically, economically, socially, materially, and technologically. In fact, the key challenge is to find new paths that are compatible with the minds, the routines, the institutions, the artifacts, and the expectations that we have acquired over decades and centuries. The Anthropocene does not mean that we write on a blank canvas. We merely add another layer of notes on a densely scribbled parchment with many different colors and all sorts of handwriting.

If we stick to this metaphor, the task of this book is to make that parchment legible in full and to tell the stories it records, at least in some broad outlines. The next task, far beyond the abilities of the present author, will be to bring these insights to bear on a circle of experts who seem to get overexcited about the earth in their hands. We see it in glaring form in the recent debate over climate modification. The Anthropocene debate has boosted the standing of the geoengineering community, as it provides their megalomaniacal schemes with a semblance of normality. If we are going to tamper with the earth system anyway, why not do it along the lines of sophisticated knowledge from a world-leading class of experts?[19] One might shrug it off if it were not so characteristic of an obvious strand of expertocratic self-aggrandizing that runs through the Anthropocene discourse. For all its diversity, the Anthropocene discourse has operated in top-down mode since its inception, and it shows. A few years back, earth system governance might have appeared as the pinnacle of intellectual arrogance, but there is actually a journal with that name now.

It is not wrong to have big plans for a planet in trouble. If anything, climate policy has been too timid in recent decades, and that is not the only environmental challenge where humans have fallen behind planetary needs. The delusion starts with the idea that expertocratic blueprints can actually be engines of change—they are mere lubricants at best. Readers can see the outcomes of expertocratic delusions scattered through this book, and this is a topic where history speaks with exceptional clarity: there is a difference between good intentions and good policies, and seeing like a state is not a good idea, particularly when it becomes the only way that people see the world.[20] We really should not go there another time.

NARRATING THE NONLINEAR

This book is about a paradox. Never have more people agreed that the environmental crisis is global, and never has it been less clear what this means. History cannot solve this puzzle, but it can provide an idea of how we came to this point. More precisely, it can provide a multitude of ideas to that effect. These ideas are large and small, obvious and hidden, material and cultural, and no, they do not add up. Insights do not even come in clear packages. In this book, narratives work on different levels and in many different directions, all in an effort to sow a healthy distrust in the stories that we have become accustomed to: linear narratives with a start, an ending, and a clear moral message. These stories provide a sense of order and orientation, or rather the semblance thereof. But as this book shows, linear narratives do not really work inside the vortex.

The essay style of this book is not just a fancy extra that serves to keep readers in thrall throughout a long text. It is a logical extension of the conceptual framework, and perhaps the only way to explore the vortex in adequate fashion. This is a book about dynamism, and that calls for a narrative that allows readers to *experience* dynamism. The essay style is a means to evoke the sense of dizziness that life in the vortex tends to produce, and readers are advised to brace themselves for a bumpy ride. One chapter connects the battery chicken with neoliberalism, sea turtles, o.b. tampons, and the Falklands War (yes, there is a connection) while another links a women's cooperative in Bangladesh with the space shuttle. Materialities can create bizarre connections, and they are more than anecdotal: they are what modern life is about. Modernization operates through a global web with many nodes, weaving continues in our time, and that should instill a sense of humility in every conversation about the flow of history. As Heraclitus might have said, you never step into the same vortex twice.

Nonlinear narration faces limits at the printing press. The complexity of the planet is fundamentally at odds with the linearity of writing. Texts are about one word after another, one sentence after another, and one paragraph after another, and our collective obsession with mobility and connectivity has not changed what one might call the natural order of textuality. At some point, authors need to decide on a certain sequence of chapters, if only because publishers wish to have pages in order before binding, but readers are encouraged to subvert this given structure and follow interlinkages or one of the following paths through the book. Jumping chapters is a common sin among readers, but in the book at hand, it is actually encouraged. You may learn more from this volume if you do not read it from cover to cover.

Interconnections can be about many different things, and this book pursues them in many different forms: spatial, temporal, causal, material, and imagined. The case of guano may serve as an example. It relates to synthetic nitrogen because both substances were used as fertilizers. Guano was a commodity from Latin America, just like Potosí's silver and the bananas of United Fruit. Farmers used fertilizer on fields for which they had a land title, and probably on land that was previously a swamp and had been drained with the help of eucalyptus trees. Guano was one of the first auxiliary substances that farmers were buying, thus starting a tradition that would continue with calcium arsenate (used lavishly in the fight against the boll weevil), hybrid seeds, and DDT, and advice on guano use came from a new cadre of chemically trained experts. Seen in this way, guano is no separate chapter of environmental history. It is a node in a web that has evolved over time.

There is no given hierarchy among these nodes, the linkages, the paths, or the rationales that they imply. Such an approach subverts the convenient compartmentalizations of traditional case studies, but more significantly, it challenges the self-perceptions of powerful stakeholders. Generations of agriculturalists have argued that feeding the world is the cardinal challenge, and utilities and oil companies have made a similar case about energy, but in the twenty-first century, it should be clear that these fundamentalisms serve specific interests. Feeding the world is obviously important, but so are fertile soils, and food safety, and labor relations, and all the other things that are at play in global food chains. A hierarchy of issues, or a conviction that only one of these aspects really matters, is at odds with the experience of recent years that all these issues matter—somehow. The mess we're in calls for some kind of balance rather than a hierarchy of issues.

With that, the chapters of this book are an effort at decentering narratives. They present different storylines that intersect, clash, or come together in expected and unexpected ways. To be sure, this does not mean that all interconnections have equal weight. Interconnections are in play on many levels and in all sorts of forms, they vary in intensity and resilience, and they do not always make a difference. Some connections are superficial or anecdotal, others are matters of life and death, and many are somewhere in between, and we can assess their significance more clearly if we stay close to real-world events. While other global histories look at large countries or general problems, this book prefers specific places with specific issues: silver mining in Potosí, the dodo on Mauritius, Chicago's slaughterhouses, and London smog.

Individual chapters sketch a story that revolves around a certain place, an artifact, a commodity, and so forth, but case studies serve a higher purpose in this book. They are exemplary explorations of general challenges—mining, the extinction of species, the industrialization of the meat commodity chain, and particulate emissions—and they mirror places that exist all over the world: mining regions, islands, hubs, and industrial cities. In a modern world where everything is in flux, it seemed to be the least intrusive approach to organize the book around challenges, and the ultimate goal of each chapter is to explore one or more challenges that humans have met all over the world. They may not capture all the experiences that have shaped the environmental history of the modern world: the rough waters of the vortex do not allow for comprehensiveness in the real sense of the word. But they capture the challenges that matter the most if you want to understand the interplay of humans and the natural environment in the world that we live in.

The chosen examples serve as nubs for similar experiences from other parts of the world, but always in a tepid fashion. That leaves plenty of room for individual experiences, and the essay style invites readers to add their own trains of thought. If you have a nature reserve that you care about, or if you care particularly about an endangered species, or if you have a favorite food, it may be a worthwhile attempt to map your own views in light of the chapters on Kruger National Park, the dodo, or a selection of chapters from the agricultural path. As an added benefit, this helps to exorcise the pretense of comprehensiveness that weighs heavily on any world history book. Every synthesis is selective, and global ones are selective in the extreme, but that becomes less of an intellectual problem if the narrative fails to convey an air that this is the full story. In each individual chapter, the goal is to identify the defining threads that, in my humble judgment, run

INTRODUCTION

through every story of its kind on the globe, and readers are encouraged to put that claim to the test in light of their own experiences.

An exemplary approach hinges on the judicious choice of these case studies. I elaborate about the quandaries of making choices in the appendix, and it shall suffice to say at this point that these choices were about significance (many were about a global first, or a defining act in material or metaphorical terms), about covering the full range of environmental challenges, and about capturing the diversity of the world. The narrative seems to digress at times, but looking at other places and artifacts helps tease out the common themes that resonated around the globe. For example, the chapter on the Canal du Midi makes a point of bringing in other canals and other traffic links, including the railroad and the motorway that run parallel to the Canal du Midi today. This does not mean that all canals around the world are essentially the same. We have plenty of case studies about specific traffic links, and I draw on some of them, but we are underplaying the value of case studies when we treat each canal as immediate to God. The Canal du Midi, along with many other projects, is also about what I call the Great Mobilization.

The narrative offers two different ways to engage with the interconnectedness of the modern world. The text includes cross-references to other chapters whenever issues are touched upon that are discussed in greater depth elsewhere. These cross-references operate in the manner of hyperlinks, and today's readers do not need further instructions as to their use. The second option is the following list of paths. They highlight a range of topics or key challenges that run through this book, and readers should see these paths as virtual sections on a par with the eight official ones. The chapters of this book are nodes that make sense in different narrative threads, and these threads go into different directions, intertwine and reinforce each other, or stand at odds. It matters beyond individual themes. If we read chapters as treatises on specific topics as well as crossroads of various big trends, we gain a deeper understanding of the improvised, patchwork nature of our modern existence. The "project of modernity" was (and is) more akin to a large construction site with a multitude of blueprints, work crews, and professional skills, where communication and coordination were often fragile at best.

In other words, do not let the table of contents fool you. This is not a book with forty chapters in eight parts plus one interlude and a coda as bonus material. It is a book with twenty-nine parts, and on average, individual chapters come up in eight of them. And if you find that a confusing overabundance of roadmaps, or if you are not into roadmaps at all, let me remind you that environmental historians have an innate sympathy for ramblers. Readers are invited to subvert the prelaminated structure and chart their own individual path through this book. Chapters cover different issues, different time periods, and different places on the globe, but they all give an impression of the dynamisms, the interconnections, and the many choices, some irreversible, some open to change, that people have made as the flow of modern history has run its course. Experiences inside the vortex have never been uniform, and will never be. Neither should the experiences of those who read this book.

PATHS THROUGH THIS BOOK
THE AGRICULTURAL PATH

Food is essential to human life, but the daily bread changed dramatically in modern times. It was produced with new methods in distant places, it was judged with new categories like calories, new corporations supplied seeds and fertilizers while other companies took care of slaughtering and marketing, and then there were the unexpected events: pests and diseases, soil erosion, and slaves who refused to eat their designated food. A lot of things are waiting to be disentangled along the path of modern agriculture.

Sugar (chapter 2), the Land Title (chapter 6), Breadfruit (chapter 7), Guano (chapter 8), United Fruit (chapter 10), the Boll Weevil (chapter 12), the Little Grand Canyon (chapter 13), Cane Toads (chapter 14), Chicago's Slaughterhouses (chapter 18), Synthetic Nitrogen (chapter 19), Opium (Interlude), Hybrid Corn (chapter 28), the Rice-Eating Rubber Tree (chapter 30), Holodomor (chapter 31), the Pontine Marshes (chapter 32), the Chemurgy Movement (chapter 33), Battery Chicken (chapter 36), and DDT (chapter 38).

THE PATH OF INDUSTRY

Just like agriculture, industry was about doing familiar things in a different way: with new technologies, new resources, new institutions and expert groups, a growing distance between producers and consumers, and way more stuff. The path of industry is about innovation, about risks and unexpected side effects, about fabulous wealth and who gained a share of it—and about changing ideas concerning what we really need.

Potosí (chapter 1), Shipbreaking in Chittagong (chapter 5), Guano (chapter 8), Whaling (chapter 9), United Fruit (chapter 10), Saudi Arabia (chapter 15), London Smog (chapter 16), Chicago's Slaughterhouses (chapter 18), Synthetic Nitrogen (chapter 19), Air-Conditioning (chapter 20), Hybrid Corn (chapter 28), Aswan Dam (chapter 29), the Chemurgy Movement (chapter 33), the Pine Roots Campaign (chapter 35), Battery Chicken (chapter 36), *Lucky Dragon No. 5* (chapter 37), DDT (chapter 38), *Torrey Canyon* (chapter 39), and Plastic Bags (chapter 40).

THE WOODEN PATH

Forests are the world's most complicated terrestrian ecosystem. They are also about plants that grow rather slowly, at least until modern science invented the turbocharged eucalyptus. Most of all, the forest history of the modern era is about the rise of state power and the incorporation of woodlands into resource-hungry economies. The wooden path also looks into the many other uses of trees and forests that fell by the wayside over the course of modern history.

Sustainable Forestry (chapter 4), Breadfruit (chapter 7), Kruger National Park (chapter 26), Eucalyptus (chapter 27), the Rice-Eating Rubber Tree (chapter 30), and the Pine Roots Campaign (chapter 35).

THE MINING PATH

The underground was full of dangers—and full of precious stuff. The mining path traces the way from silver mines to bulk commodities like coal and bauxite, from mountains with tunnels to giant holes, from resource boomtowns to resource states, and it inquires about pollutants and leftovers. Do you want to have a piece of Potosí's Cerro Rico? Just take a deep breath.

Potosí (chapter 1), Guano (chapter 8), Saudi Arabia (chapter 15), and London Smog (chapter 16).

THE PATH OF THE ANIMALS

It was people who built the modern world, but they had plenty of nonhuman companions. They were, among other things, hunted, slaughtered and eaten, watched and adored, bred and confined, feared and killed, and pushed into oblivion. Plenty of reason to look them in the face.

Guano (chapter 8), Whaling (chapter 9), the Dodo (chapter 11), the Boll Weevil (chapter 12), Cane Toads (chapter 14), Chicago's Slaughterhouses (chapter 18), Kruger National Park (chapter 26), Battery Chicken (chapter 36), and DDT (chapter 38).

THE INFRASTRUCTURE PATH

Infrastructure is a neologism. People were content to talk about roads, canals, and other concrete things until the rise of new systemic technologies that became the backbone of modern life. Infrastructures served transportation, moved and delivered water, and satisfied food and energy needs and the longing for cool air. They also created plenty of problems. But who can imagine a modern world without infrastructures?

The Canal du Midi (chapter 3), the Water Closet (chapter 17), Chicago's Slaughterhouses (chapter 18), Air-Conditioning (chapter 20), Cholera (chapter 21), Aswan Dam (chapter 29), the Pontine Marshes (chapter 32), Autobahn (chapter 34), and *Torrey Canyon* (chapter 39).

THE PATH OF ENERGY

Energy is a physical quantity—and a myth of modernity. It was available in many forms, open to transformation, ephemeral and omnipresent, and it came at a price. It was not necessarily the end user who was paying it.

Sugar (chapter 2), Sustainable Forestry (chapter 4), Breadfruit (chapter 7), Whaling (chapter 9), Saudi Arabia (chapter 15), London Smog (chapter 16), Synthetic Nitrogen (chapter 19), Air-Conditioning (chapter 20), Opium (Interlude), Aswan Dam (chapter 29), the Chemurgy Movement (chapter 33), Autobahn (chapter 34), the Pine Roots Campaign (chapter 35), *Lucky Dragon No. 5* (chapter 37), *Torrey Canyon* (chapter 39), and Plastic Bags (chapter 40).

THE PATH OF POLLUTION

Dirt is matter in the wrong place, but good luck explaining that to workers in Bangladeshi shipbreaking. Pollutants mirrored and reaffirmed the fissures that ran through societies. They also became subject to control programs—sometimes. They have consequences from urban decay to global warming. They also provide a showcase on how environmental problems are social constructs. If you follow the path of pollution, you will not call it a problem again without inquiring for whom, and in which way.

Potosí (chapter 1), Shipbreaking in Chittagong (chapter 5), London Smog (chapter 16), the Water Closet (chapter 17), Synthetic Nitrogen (chapter 19), Cholera (chapter 21), the 1970 Tokyo Resolution (chapter 24), *Lucky Dragon No. 5* (chapter 37), DDT (chapter 38), *Torrey Canyon* (chapter 39), and Plastic Bags (chapter 40).

INTRODUCTION

THE COLONIAL PATH

Colonialism has many faces, and none is pretty. The colonial path explores spatial inequality in its full breadth: the new order of the globe, the tools, the profiteers and their bounty, the sources of power, and the resistance. The focus is on the consequences of colonial relations and how they survive through daily routines. Colonialism is about doing.

Potosí (chapter 1), Sugar (chapter 2), Sustainable Forestry (chapter 4), Shipbreaking in Chittagong (chapter 5), the Land Title (chapter 6), Breadfruit (chapter 7), Guano (chapter 8), Whaling (chapter 9), United Fruit (chapter 10), the Dodo (chapter 11), the Boll Weevil (chapter 12), Cane Toads (chapter 14), Saudi Arabia (chapter 15), Opium (Interlude), Gandhi's Salt (chapter 23), the 1970 Tokyo Resolution (chapter 24), Kruger National Park (chapter 26), Eucalyptus (chapter 27), Aswan Dam (chapter 29), the Rice-Eating Rubber Tree (chapter 30), Holodomor (chapter 31), the Pontine Marshes (chapter 32), DDT (chapter 38), *Torrey Canyon* (chapter 39), and Plastic Bags (chapter 40).

THE WATERWAY

Look at the earth from space, and you understand the rationale for this path: more than two-thirds of our planet is underwater. The waterway looks at the open sea, at canals and rivers (free-flowing and dammed), at transportation and maritime resources, and at clean and dirty water and all the stuff that goes with the flow.

The Canal du Midi (chapter 3), Shipbreaking in Chittagong (chapter 5), Guano (chapter 8), Whaling (chapter 9), the Water Closet (chapter 17), Cholera (chapter 21), Aswan Dam (chapter 29), the Pontine Marshes (chapter 32), *Torrey Canyon* (chapter 39), and Plastic Bags (chapter 40).

THE PATH OF CHEMISTRY

Three centuries ago, chemists were sorcerers of sorts who believed in things like phlogiston. Today they make up one of the most powerful branches of the natural sciences, and they underpin one of the greatest industries in the world. This path looks at the chemists, their work, their products, their power (or lack thereof), and their blunders. Yes, they did kill small children. And they got away with it.

Potosí (chapter 1), Guano (chapter 8), Whaling (chapter 9), Synthetic Nitrogen (chapter 19), Opium (Interlude), the Chemurgy Movement (chapter 33), the Pine Roots Campaign (chapter 35), DDT (chapter 38), and Plastic Bags (chapter 40).

BOTANICAL TRANSFERS

Eucalyptus in California, breadfruit in the Caribbean, cane toads in Australia—in the modern era, organisms traveled like never before. This path is about how it happened and why it matters. (Spoiler: Kew is overrated.)

Sugar (chapter 2), Breadfruit (chapter 7), the Boll Weevil (chapter 12), Cane Toads (chapter 14), Eucalyptus (chapter 27), and Hybrid Corn (chapter 28).

BUILDING THE STATE

Police states, welfare states, failing states, l'état, c'est moi: the power of the state is a defining feature of global modernity. It mattered for environmental challenges—and, in turn, state power was built through the environment. Do you think that there should be a law? Maybe read these chapters first.

Potosí (chapter 1), Sugar (chapter 2), the Canal du Midi (chapter 3), Sustainable Forestry (chapter 4), the Land Title (chapter 6), Whaling (chapter 9), United Fruit (chapter 10), Saudi Arabia (chapter 15), London Smog (chapter 16), Opium (Interlude), Gandhi's Salt (chapter 23), the 1970 Tokyo Resolution (chapter 24), the 1976 Tangshan Earthquake (chapter 25), Kruger National Park (chapter 26), Aswan Dam (chapter 29), the Rice-Eating Rubber Tree (chapter 30), Holodomor (chapter 31), the Pontine Marshes (chapter 32), the Chemurgy Movement (chapter 33), Autobahn (chapter 34), the Pine Roots Campaign (chapter 35), Battery Chicken (chapter 36), *Lucky Dragon No. 5* (chapter 37), DDT (chapter 38), *Torrey Canyon* (chapter 39), and the Pandemic (Coda).

BUILDING PROFESSIONS

They were the priests of a new epoch—and the people that nobody knew. There is a profession for every environmental challenge today (and sometimes more than one), but that is the result of a long history. This path traces the making of expert groups, their quest for resources and power, their struggles, their resilience—and the stories that they tell. Experts may not save the planet. But we will not save the planet without experts either.

Sustainable Forestry (chapter 4), Breadfruit (chapter 7), Guano (chapter 8), the Dodo (chapter 11), the Boll Weevil (chapter 12), the Little Grand Canyon (chapter 13), Cane Toads (chapter 14), Synthetic Nitrogen (chapter 19), Air-Conditioning (chapter 20), Cholera (chapter

INTRODUCTION

21), Baedeker (chapter 22), Eucalyptus (chapter 27), Hybrid Corn (chapter 28), Aswan Dam (chapter 29), the Chemurgy Movement (chapter 33), DDT (chapter 38), *Torrey Canyon* (chapter 39), and the Pandemic (Coda).

ECONOMICS 101

Sounds like a pathway for nerds? You may not be interested in the economy, but the economy is interested in you. This path provides a tour d'horizon of economic concepts that were in play in the modern era. It is a rather male-heavy path, but it includes a female Nobel laureate who criticized the discipline's penchant for models and the man who penned the theory of interstellar trade. The toolbox of economics deserves scrutiny even if you will never buy antimatter futures.

Potosí (chapter 1), Sugar (chapter 2), the Canal du Midi (chapter 3), Shipbreaking in Chittagong (chapter 5), Whaling (chapter 9), United Fruit (chapter 10), Saudi Arabia (chapter 15), Chicago's Slaughterhouses (chapter 18), Gandhi's Salt (chapter 23), the Rice-Eating Rubber Tree (chapter 30), and Battery Chicken (chapter 36).

THE DEVELOPMENTALIST PATH

Say what you will about development, but it is complicated. So maybe there was that one thing that could change everything and catapult a place, a region, or a country into modern times? This path is about a dream that refuses to die.

The Canal du Midi (chapter 3), Guano (chapter 8), United Fruit (chapter 10), Saudi Arabia (chapter 15), Eucalyptus (chapter 27), Hybrid Corn (chapter 28), Aswan Dam (chapter 29), the Rice-Eating Rubber Tree (chapter 30), the Pontine Marshes (chapter 32), and DDT (chapter 38).

THE CARNIVOROUS PATH

Gandhi ate meat—though only to give it a try. Others were less restrained. This path traces the full range of consequences.

Guano (chapter 8), Whaling (chapter 9), the Dodo (chapter 11), Cane Toads (chapter 14), Chicago's Slaughterhouses (chapter 18), Gandhi's Salt (chapter 23), and Battery Chicken (chapter 36).

THE PATH OF DISASTER

Sometimes things do not go as planned. And sometimes things go disastrously wrong. This path is about the latter, the sudden events, the cata-

clysms—*and about what happens when the dust settles. Disasters matter, but not in the way most people think.*

The Boll Weevil (chapter 12), the Little Grand Canyon (chapter 13), Cane Toads (chapter 14), London Smog (chapter 16), Cholera (chapter 21), the 1976 Tangshan Earthquake (chapter 25), Holodomor (chapter 31), *Lucky Dragon No. 5* (chapter 37), *Torrey Canyon* (chapter 39), and the Pandemic (Coda).

THE PATH OF WAR

War is no longer the father of everything, as Heraclitus surmised in ancient Greece. But in sixteen of the following chapters, the imprint of war was strong enough to warrant inclusion in this path. Chapters look at preparations for war, resource allocation in war, wars for oil (and quite a few other resources), and at the mental worlds that this has produced. Fighting for change? Campaigning for justice? That is where it starts.

Potosí (chapter 1), Sugar (chapter 2), the Canal du Midi (chapter 3), Whaling (chapter 9), United Fruit (chapter 10), the Boll Weevil (chapter 12), Synthetic Nitrogen (chapter 19), Opium (Interlude), Cholera (chapter 21), the 1976 Tangshan Earthquake (chapter 25), Holodomor (chapter 31), the Pontine Marshes (chapter 32), Autobahn (chapter 34), the Pine Roots Campaign (chapter 35), *Lucky Dragon No. 5* (chapter 37), and DDT (chapter 38).

THE MOBILITY PATH

We live on a small planet, but it is large enough to make mobility a challenge of its own. This path looks at the technologies that facilitate mobility in one form or another, at the resources in play, and at the consequences for humans and environments. Please choose your speed judiciously.

The Canal du Midi (chapter 3), Whaling (chapter 9), Cane Toads (chapter 14), the Water Closet (chapter 17), Chicago's Slaughterhouses (chapter 18), Cholera (chapter 21), Baedeker (chapter 22), Aswan Dam (chapter 29), the Chemurgy Movement (chapter 33), Autobahn (chapter 34), the Pine Roots Campaign (chapter 35), *Torrey Canyon* (chapter 39), Plastic Bags (chapter 40), and the Pandemic (Coda).

THE PATH TOWARD THE MODERN CITY

For the first time in human history, the majority of people live in cities. This path is about how we made it happen—and what it means for the world.

Potosí (chapter 1), Shipbreaking in Chittagong (chapter 5), London

INTRODUCTION

Smog (chapter 16), the Water Closet (chapter 17), Chicago's Slaughterhouses (chapter 18), Air-Conditioning (chapter 20), Cholera (chapter 21), Baedeker (chapter 22), the 1976 Tangshan Earthquake (chapter 25), Aswan Dam (chapter 29), and Autobahn (chapter 34).

PART I

Essentials

IN THE BEGINNING WAS THE STUFF.

As Europeans moved beyond the confines of their continent, a number of grand narratives came into play that bestowed the endeavor with a sense of glory. In these readings, Europeans sought to spread the Christian gospel, escape inhospitable conditions, learn about the world, liberate unfree people, or bring the achievements of Western modernity to those who seemed unable to develop them on their own. Historians continue to disagree on how to view the ideas that underpinned Europe's global outreach, or whether to take them seriously at all, but they are generally unanimous that there was also something else: material benefits. Take the commodities out of the last five hundred years of world history, and it does not make sense anymore. It is unlikely that Europeans would have embarked on such a long and complicated foray beyond their ancestral homes if it had not been for stuff: food to store and eat, metals to make coins or weapons, timber and fibers for factories, and trinkets to show and impress. When all was said and done, there was always something to bring back home.

Some of the stuff was actually new, like corn and potatoes. Other stuff was probably known but unfamiliar in most of Europe, like sugar: some Mediterranean regions were cultivating sugar cane before the rise of transatlantic colonialism, but most people of medieval Europe satisfied their sweet tooth, to the extent that they had one, with honey. And then there was the stuff that everybody knew, like silver, where supplies suddenly exploded. The materials of modernity were not always new, but they were produced in new ways and available in unprecedented amounts, and a different world evolved as new modes of production took shape.

The following five chapters explore this transformation through two commodities (silver and sugar), one transport link (the Canal du Midi), one political doctrine (sustainable forestry), and one activity (shipbreaking in Chittagong). In other words, this section is about how the people of the modern era obtained metals, food, and timber, how they mobilized these commodities and some other things, and how they got rid of waste. People had done all these things for ages, but modernity offered new ways to deal with these issues, and these new ways were what today's software developers call a "killer application": the modern ways were never the only ones at hand, and certainly not the best ones for everyone involved, but sooner or later, people all over the world found these new ways irresistible. The transformation was, after all, about essentials: about how people eat, build, move around, and how they deal with leftovers. It's not everything that life is about, but few people make it through life without these things.

The new ways drew on existing practices, but that should not nourish doubts about their novelty. Late medieval mining in Central Europe paved the way for Potosí, but in the same sense that the Inca roads paved the way for the autobahn. The new ways were of a different quality in their mode of operation and in their consequences for peoples and environments. More specifically, they were different in scope, scale, and speed, and had notable similarities as to when these differences came into play over the course of the modern era. By and large, it was scope that came first, then scale, then speed.

Potosí's silver provides a case in point. The precious metal traveled between America, Europe, India, and China and turned a transcontinental exchange that had previously been a trickle into a defining current of the emerging world economy. Potosí's Cerro Rico also produced far more silver than any mine in Europe. Economies of speed never really came to the Andean highlands, as the deposits were long past their prime when steam engines and other industrial technologies transformed mining and shifted the focus of primary resource extraction from precious metals to bulk commodities. These innovations also brought another shift in scale, and the mines of the age of industry had outputs that dwarfed even the rich veins of Potosí. They also produced toxic legacies, and they are not always buried in the ground: Potosí's legacy shows in the background concentration of mercury in the global atmosphere. Containing toxic substances in abandoned mines is an ongoing challenge, and the material leftovers will almost certainly outlast cultural memory despite the best efforts of pertinent authorities. Potosí's silver mountain survives in somewhat diminished form to this day and enjoys protection under the UNESCO World Heritage program. Mass destruction mining, a twentieth-century technology, would have literally eaten up the Cerro Rico.

It is no accident that it was the city rather than the mountain that became famous all over the world. Resource extraction hinged on urban hubs that provided technological and commercial services, plus services for men who did a tough and dangerous job. A good part of the new towns of the modern era owed their existence to primary resources. It could be anything from a shabby gold-rush town to a world-class city like Potosí. The town was oozing richness in its prime, but the splendor always had an air of fragility, and not just because the fate of mining towns was invariably tied to a finite mineral resource base. Order was a relative thing in bustling mining towns, and the sources of trouble ranged from perennial labor conflicts to illicit trade. An extra dose of testosterone did not help, though mining was not always the province of men: a good part of Potosí's workforce was female.

Potosí produced a precious resource, but it also consumed resources like food and mercury for ore processing. The same held true for the sugar plantations in the Americas: they were nodes in multiple networks, or that is how the world economy categorized them, for one of the crucial resources, slaves, was arguably more than an economic unit. Just like Andean silver mining, sugar production drew on established practices from agricultural technologies to the use of unfree labor, but it catapulted them into a new orbit. Large technological artifacts for processing, division of labor, externalization of human and environmental costs, and the categorical need to make it all run as smoothly and quickly as possible—sugar plantations were harbingers of industrial production regimes. They were also ephemeral on the ground. For those who owned a plantation, sugar was about getting rich quickly, another novelty of modern production methods. Unlike other commodity chains before the dawn of modernity, the network of sugar was a centripetal system where wealth accumulated at certain points.

Sugar, initially a luxury item for European elites, turned into a mass-produced essential. More precisely, it was a commodity that changed the meaning of essentials: there was no tradition of sugar consumption in most of Europe, and yet people in many different countries fell for the sweet life—a transcultural convergence of tastes that critics of the global food system have targeted more than once. Consumers have replied with everything from low-carbohydrate diets to apathy, but few have embarked on a return to premodern modes where people drew a significant share of their food from working the land with their own hands. Most consumers rely on faceless food systems to provide them with their daily bread or whatever tastes and pockets demand. In the Western world, private gardens are usually hobbies rather than necessities. On Caribbean plantations, slave gardens were means of survival, and slaves were not alone in seeking foodways that offered a buffer against disaster. The modern food system was a gamble, but those with money could act as if it did not matter to them.

The Caribbean lost its pivotal place in the global sugar economy in the nineteenth century, as competition from beet sugar and substitutes like saccharine entered the market. Just like cane sugar, sugar beets brought an intensification of land use, and production of organic resources became increasingly concentrated in specific regions. Substitutes mirrored a growing role of scientific knowledge in food production. But there was more to the world of sugar. Protectionist policies and subsidies were also part of the picture, and they were intrinsically connected to the new production regime. It was about the new capital intensity of food production: advanced producers shouldered huge investments that called for safeguarding from whatever authority was willing to listen or get bribed. David Ricardo suggested that free trade would allow entrepreneurial people to exploit comparative advantages, but the new sugar economy was also breeding corporate interests that were more interested in profits than principles. The modern food system has the ability to feed an unprecedented number of

mouths, but it has also created stark inequalities on different levels, and whether it is the best of all possible worlds is open to debate.

The globality of silver, sugar, and many other commodities relied on the transformation of world transport, but the corresponding infrastructures were more than mere service providers. The Canal du Midi was cutting-edge technology, seventeenth-century style, but it was also an act of state, an instrument of royal power that brought absolutist power more forcefully into a peripheral region, a military project, and a transformative agent for the Languedoc. Like many infrastructure projects, the canal brought a multitude of mobilizations from goods and information to political hierarchies and pathogens. The economics of transport projects depended strongly on the specific circumstances and on what counts as an economic benefit, but projects were more than tools of business. Connecting places could take place in many different forms and with many different results, and the outcomes of modern infrastructure projects range from Egypt's Muslim Brotherhood to the state of Panama.

A tax farmer, Pierre-Paul Riquet, built the Canal du Midi, and his descendants ran it for profit until the French state took over in 1897. But construction was really a group project: Riquet's success was due to support from vernacular knowledge and a loyal workforce that Riquet gained by treating workers exceptionally well. Construction was also ongoing. Vauban rebuilt major stretches of the Canal du Midi after Riquet's death, and repairs have been a necessity ever since: infrastructures collapse in the absence of maintenance. Traffic arteries have their resilience, but like human arteries, they are not static, and the same holds true for their social, economic, and biological environments. Today's boom in pleasure boating in the southwest of France is only the latest outcome of the perennial reinvention of the Canal du Midi.

Absolutism framed the Canal du Midi as evidence of human power over nature, but this was more about royal representation than the environment. The powers of early modern states held greater material significance in another realm: the doctrine of sustainable forestry turned woodlands from a habitat with a multitude of uses into assets that served the financial interests of the state and other owners. The ensuing conflicts over control of the woods raged for generations, and they only petered out—to the extent that they did—with growing urbanization. Forest conflicts continue to rage in the Global South, and the legacy of state control is an obstacle rather than an opportunity in regions such as the Sahel, but the fate of forestry in post-Socialist

Eastern Europe also shows the fragility of state power closer to the place where sustainable forestry was born. Early modern states forged a deep connection between statehood and forestry, but the alliance also worked in reverse.

A growing profession of foresters underpinned the states' power grab, but academic authority showed the same combination of iron-clad authority and fragility. Sustainable forestry moved from control over forestry to managing and improving its biology, which created plenty of follow-up problems that disciplines like applied entomology tried to keep in check. To a significant extent, forest research was repair science, an attempt to manage problems that would never have emerged without human meddling. The reductionism of specialist expertise was invariably at odds with the biological complexity of woodlands, and academics did not fail to notice, but it was one thing to envision holistic expertise and another to actually foster it. The father of applied entomology in Germany, Karl Escherich, dreamed about mixed forestry, but when it came to appointing a successor for his chair, he was eager to recruit another specialist. The committee found one among Escherich's assistants.

Monoculture was practice rather than dogma in the woods, and to the extent that academic expertise provided guidance, it offered improvised makeshifts rather than ultimate solutions. Hans Carl von Carlowitz did not mean to build an academic profession when he cited the wood scarcity trope, but the concept was crucial for the rise of modern forestry. It drew attention away from the significant cognitive and conceptual problems of academic expertise, and it galvanized attention in a most helpful way: Why quibble about details in the face of a horror scenario? Sustainable forestry was a shock doctrine (see chapter 25.2, Crisis Mode), except that it thrived on an enduring myth rather than a momentary disaster.

All modes of resource extraction entailed waste, as did subsequent stages of production, and that was a growing challenge in a supercharged global economy. There were plenty of preexisting routines concerning recycling and reuse, but they were overwhelmed by the sheer quantities of stuff that industrial societies spewed out since the nineteenth century. At the same time, the heft of materiality called for efforts to deal with the leftovers, and the final chapter traces the changing constellations of municipalities, war economies, and private entrepreneurship over the past 150 years. Shipbreaking in Chittagong is the latest, globalized stage in the perennial search for an ultimate sink, and surely not the last. Together with derelict mines, waste heaps may one day rank among the most enduring legacies of modern civilization.

But recycling and waste management were not just about materialities and technologies. They were also about people, which is important to stress in light of a long and transcultural tradition to marginalize the human dimension of the waste business. It would be an understatement to say that social and ethnic discrimination was a part of the business—it was crucial for how things worked. The dismal fate of migrant workers on the Bangladeshi coast is only the latest incarnation of an old phenomenon.

The chapter shows that shipbreaking is subject to ongoing national and international efforts at regulation, but it is crucial to recognize that these efforts are up against powerful tides of history. Wastes have their own rules, material and other, and political reform was rarely a match for the raw silent power of needs and technologies. If there is a tradition among the tremendous changes of practices and waste flows in modern times, it is about the transgression of limits in virtually every dimension: material, social, spatial, and temporal. We may never find that ultimate sink (at least that is what Joel Tarr told us in a landmark article), but the search has energized modern societies for generations—if energizing is a good word for the waste business. Few people have gotten excited about leftovers during the course of the modern age. They have dealt with it because they had to or were paid for it.

Shipbreaking in Chittagong has acquired a measure of fame due to dramatic visuals, but the true drama at play escapes the naked eye. Lives and livelihoods are fragile on the Bangladeshi coasts, as is the institutional scaffolding that underpins the business model, a topic that will be explored in greater depth in part II. It is no coincidence that the following chapters deal with illicit activities in many different forms: modern modes of production have defied political control more than once, and their global hegemony may not end anytime soon—and neither will the effects and costs that they have brought to people all over the world. Just like the scrap metal workers in Chittagong, we are stuck with our "killer apps" for the foreseeable future.

1

Potosí

Rich in Metals

I. GOING UNDERGROUND

Local histories typically display a sense of pride. The exuberance of the baroque period did not encourage understatement either. And so we are told that the mountain to the south of Potosí was really a "perfect and permanent marvel of the world," a "singular work of the power of God," the "emperor of mountains" and a "clarion that resonates in the whole wide world."[1] For Bartolomé Arzáns de Orsúa y Vela, the author of the *Historia de la Villa Imperial de Potosí,* no words were too great when it came to his hometown, and he opened his book with lavish praise along these lines. It seemed like a long shot for an author who had never left the Andean highlands, but Arzáns, who lived from 1676 to 1736, was not all that exceptional. Potosí and its silver had triggered enthusiasm since the earliest days of colonial mining. The Habsburg emperor Charles V bestowed it with a widely cited phrase for its coat of arms: "I am rich Potosí, the treasure of the world, and the envy of kings."[2] Adam Smith mentioned Potosí's mines as "the most fertile in all America" in *The Wealth of Nations.*[3] Potosí lingered as a synonym for riches even when its mines were long in decline. Reflecting on real estate and capitalist accumulation in *Das Kapital*, Karl Marx wrote that "the mines of misery are exploited by house speculators with more profit or less cost than ever were the mines of Potosi."[4] To this day, the Spanish phrase "vale un Potosí," literally "worth a Potosí," translates as "priceless."

When the Spanish conquistadores learned about Potosi's silver deposits in 1545, mining had been flourishing in central Europe for a full century. New technologies, new modes of corporate organization, and the discovery of new deposits underpinned a long boom that changed geographies and economies.[5] Thanks to its silver mines, a sleepy Tyrolian village, Schwaz, turned into Austria's second-largest urban agglomeration after Vienna.[6] The copper mines of Falun in Sweden, the Wieliczka salt mine near Kraków and the tin mines of Cornwall gained a reputation far beyond their respective regions for centuries.[7] Around 1520, the Bohemian town of Joachimsthal began to strike a heavy silver coin that is remembered to this day because the *Joachimsthaler* became the etymological ancestor of the word "dollar."[8] Joachimsthal's town physician, Georg Agricola, wrote a monumental treatise on mining technology, *De Re Metallica*, that summarized the age's technological achievements. It was published posthumously in 1556.[9]

Mining was a technology unlike any other in the late medieval world. Large mines embraced the division of labor long before it became a hallmark of industrialism: a successful mine hinged on axes and sledgehammers as well as cutting-edge chemical knowledge. Turning solid rock into gravel was barely half the job. Fresh air had to find its way into the mine while ore and water had to get out. Roadways and shafts needed timber frames. Humans and draft animals had to be fed. Waterwheels needed enlightened operators, maintenance, and a reliable supply of water. No other production regime of the late medieval period had a similar size and complexity, and no occupation offered such a broad range of mortal threats. In his *Technics and Civilization*, Lewis Mumford pointed out that the mine was "the first completely inorganic environment to be created and lived in by man."[10] But for all its challenges, mining also brought tremendous wealth and, in the case of Habsburg's Charles V, the crown of the Holy Roman Empire. It was the silver of Schwaz in combination with the deep pockets of the House of Fugger that got him elected in 1519.[11]

In short, silver mining was well established as a source of economic and political power by 1545, but Potosí brought it all to a new level.

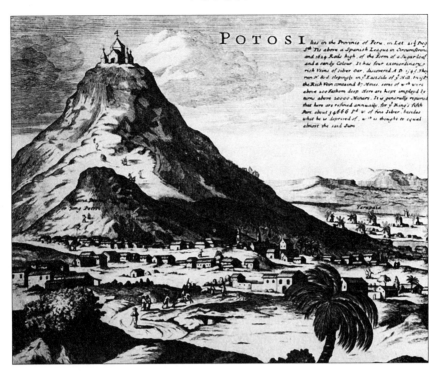

1.1 View of the Cerro Rico and Potosí by Bernard Lens, in Herman Moll, *Map of South America* (London, 1715). Image, Wikimedia Commons.

The name of its mountain, Cerro Rico (Spanish for "rich mountain"), said it all: the spot had wealth written all over it. When output peaked in the last quarter of the sixteenth century, Potosí alone produced half of all silver in the Americas, the world's dominant producer for centuries.[12] Around 1600, the transatlantic flow of silver exceeded Europe's domestic production by a factor of eight.[13] Potosí was "the motor of the Spanish economy between the first strike in 1545 and the 1660s," and the repercussions were ultimately global.[14] Dennis O. Flynn and Arturo Giráldez have argued that "the singular product most responsible for the birth of world trade was silver."[15]

A booming commodity needs grateful buyers. As the economist Erich Zimmermann declared, "Natural resources *are* not, they *become*."[16] European elites embraced silver to show off their wealth, the silver peso entered monetary vocabularies across Europe, India consumed a significant amount, but China was the most important buyer.[17] Domestic production in China had fallen in the late fifteen and early sixteenth centuries, and a growing population with a silver-based currency generated almost insatiable demand.[18] It did not save the late Ming dynasty from collapsing and maybe even hastened its demise, but the complementary interests, helped by a direct shipping link from Mexico's Acapulco to Manila on the Philippines, made for a thriving business. "The combination of low supply-side production costs in Spanish America and Chinese-led demand-side elevation in silver's value in Asia generated probably the most spectacular mining boom in human history."[19]

Resource flows also depend on a robust institutional framework. While Central European mines organized themselves along comprehensive *Bergordnungen*, business in Potosí went down a more disorderly path, as hundreds of corporations worked on the Cerro Rico at the same time.[20] When they sold their silver, institutions from the local mint to the traders came into play, and a combination of contracts, transport networks, and the momentum of ingrained routines channeled resources along certain ways. Immanuel Wallerstein made resource flows a cornerstone of his commanding synthesis of the modern world-system and put the mines of Hispanic America on his map as one of two key peripheral areas in the sixteenth century, but the general direction of resource flows allowed different paths within and beyond the letters of the law.[21] Smuggling was a part of Andean mining from its inception, as some of the output bypassed the Potosí mint and went to French, English, and Dutch smugglers along the Pacific coast and to the Portuguese in Brazil.[22] Illicit trade was always part of the resource business (see Interlude, Opium), and its exact share remains anyone's guess.

The stream of silver was only one of several resource flows that intersected in the Andean highlands. Processing Potosí's ores depended on an amalgamation process since the 1570s, and the mines bought copious amounts of mercury from Huancavelica in Peru, Almadén in Spain, and Idrija in Slovenia.[23] Mines needed firewood to heat the mixture of ore and mercury and timber to shore up shafts and seams.[24] Feeding an urban population at an altitude of four thousand meters was another challenge, all the more as workers needed calories (see chapter 7.2, Numbers Games) as well as stimulants in the form of coca leaves (see Interlude, Opium). Potosí was not the only place where hungry miners transformed agricultural systems. When copper mining brought a quarter of a million migrant workers into the sparsely populated Yunnan Province in southwest China between 1750 and 1800, rice production flourished in neighboring Burma.[25]

Spanish colonial mining built on technological achievements from late medieval Europe. A group of German miners even made the trip across the Atlantic and brought new smelting techniques to New Spain in the 1530s.[26] However, David Brading and Harry Cross have argued that the transatlantic parallels are weaker than the similarities with another production regime of the New World: "American silver mining offers fewer comparisons with its German antecedents than with that other great colonial industry, the manufacture of sugar" (see chapter 2, Sugar).[27] Just like sugar plantations, silver mining produced a high-value commodity for distant consumers in a specific geographic realm of limited size, and the prospect of huge profits seemed to justify extreme conditions for workers and environments. Potosí even had an equivalent to plantation slavery in the *mita*, a forced labor regime inherited from the Inca that became just as infamous as the transatlantic slave trade.[28] "The *mita* labor system was a machine for crushing Indians," Eduardo Galeano wrote in *Open Veins of Latin America*, a classic in the vein of dependency theory (see chapter 30, Rice-Eating Rubber Tree).[29] And just as the Georgian houses in Bristol failed to reveal the conditions on the sugar plantations that provided the money for their construction, it was easy to forget about the miners' hardships in the streets of Potosí.

2. BOOMTOWNS

Arzáns's *Historia de la Villa Imperial de Potosí* was an expansive work. It ran to some 1,500 large folio pages, each of them filled with close handwriting.[30] But capturing the city's splendor was truly a challenge, as countless buildings and lavish ceremonies were clamoring for the author's attention, and Potosí was not the only resource town that had a lot to show. Incorporated as a mining camp in 1881, Aspen had electric light after four years, piped water (see chapter 17, Water Closet) after five, and the third-largest opera house in the state of Colorado after eight.[31] There was also an opera house in the Brazilian city of Manaus, some nine hundred miles upstream from the mouth of the Amazon River. Built from imported materials between 1891 and 1896, it provided a sphere for high culture in the middle of the rain forest, paid for by rubber revenues during the heydays of tapping (see chapter 30, Rice-Eating Rubber Tree). The gold-domed structure was not the only extravaganza that Manaus granted to itself. Manaus also had piped gas and water, electric light and a telephone network, an artificial harbor, and the first electric streetcar line in South America.[32]

But mining cities did not always show the wealth that they generated. "The architectural magnificence of Latin American mining cities, like Guanajuato, Zacatecas, San Luis Potosí, Ouro Preto and Potosí where the magnates vied with each other in their magnificent patronage of church building, could not provide a greater contrast to the makeshift, sleazy air of the mining towns of California, the Canadian Northwest, Victoria and New South Wales," the Warwick professor Alistair Hennessy wrote.[33] Gold-rush towns were notorious for the various manifestations of uninhibited masculinity, and, when the frenzy flamed out, the biggest money was often with those who handled the supplies. It could be the start of a global career. Levi Strauss made a fortune selling denim pants in the wake of the California gold rush, and one of the giants of beer brewing, SABMiller, grew out of a company that sold Castle beer to thirsty gold miners in South Africa.[34]

Gold rushes thrived on the mobility of independent men, but many of them ended up joining a workforce. California had crushing mills with large workforces within four years of the first discovery of gold.[35] The nineteenth century was when industrialism came of age, and a good part of the proletariat was sweating underground, particularly in the coal mines that fed the new hunger of Western societies for fossil fuels (see chapter 16, London Smog). While up to a third of Potosí's workers were female, the miners of the industrial age were an eminently male community, and tender souls were well advised to keep their distance.[36] When Arzáns went down the shafts of the Cerro Rico, the trip left him traumatized for life. The lights of his party went out underground, and they sat in the dark for hours until another miner happened to pass by. He also was not into chewing coca leaves.[37]

The miners of the industrial age won a special place in labor history. As a recent handbook article declared, "In the twentieth century miners were at the forefront of radical movements and policies in many countries."[38] They offered a charismatic combination of numbers and symbolic power, and when they went for industrial action, it sent shockwaves through societies. More than one miners' strike entered national histories: the US anthracite coal strikes of 1877 and 1902, the strikes of German coal miners along the Ruhr in 1889 and 1905, Britain's general strike of 1926, and the National Union of Mineworkers strike in South Africa in 1987.[39] And then there was the blood that was spilled. The 1914 Ludlow massacre and the ensuing ten-day war in the southern Colorado coalfields hold "a key place in the martyrology of the American labor movement."[40]

The mythology of labor suggested a united army of miners, an argument that lingers in some recent publications.[41] But in reality, differences in age, status, and skill level ran through the workforce, and clever capitalists exploited these divisions as best they could. Potosí's *mita* certainly did not create a homogeneous group, and not only because some workers ran away or paid for relief from service. The colonial workforce was a hybrid of forced and nominally free labor.[42] Even when the Bolivian government formed a new state-owned corporation, the Corporación Minera de Bolivia (COMIBOL), after the revolution of 1952, only a fraction of the workforce had access to the benefits of nationalization. COMIBOL's labor pyramid included a salaried labor aristocracy, workers in cooperatives who were paid by the pound, peasants who leased tailings for reworking, peasant women scouring rock dumps, and destitute marginals whom COMIBOL officials allowed into the tunnels at night for a fee.[43]

Working in a mine was always a danger to limbs and lungs, and conditions in Potosí amplified the risks: silicosis was a peculiar challenge at an altitude of four thousand meters. In line with modernity's trust in numbers (see chapter 7.2, Numbers Games), writers have offered staggering estimates of the human toll. It remains open to debate whether the Cerro Rico really "consumed 8 million lives" over three centuries, as Galeano has claimed, but only because of a brutal fact: nobody was counting.[44] The precise fate of most workers remains a mystery for lack of sources, but nobody doubts that Potosí's mines produced plenty of misery and dislocation along with the silver. Work was terrible, and so was the outlook for life after the mine. As the anthropologist Michael Taussig wrote, "The mines spewed forth a class of homeless and masterless people—a colonial lumpenproletariat—whose presence and energy were to become very noticeable in swelling the mass of discontent and rebellion, particularly in the great Tupac Amaru Indian nationalist uprising of 1780."[45]

The splendor of Potosí looked more ambiguous with knowledge of the workers' plight, and that is not just the wisdom of hindsight. Arzáns was fully aware that the glitter was tied to a lot of Indian sweat and blood, and he wrote about it in his history. He did not make his remarks in a revolutionary mood. He was a traditionalist in the days before the Enlightenment, a "timid, retiring scholar" who wrote with "a sense of resignation" and shared "that peculiar Baroque quality of disenchantment with the world."[46] And if you were living in Potosí in the early eighteenth century, melancholia was a perfectly appropriate state of mind.

3. ABANDONMENT

The opera house in Manaus closed its doors when the Amazon rubber boom ran out. For all their splendor, resource boomtowns inevitably declined when the material flows that underpinned their existence dried up, and while there were many paths toward insignificance, they were all painful. In the case of Potosí, it was a long decline rather than sudden collapse. Mining continued throughout the colonial period, but the town was long past its prime when Arzáns was writing his voluminous chronicle in the early eighteenth century: Potosí's all-time record year was 1592.[47] Simón Bolívar climbed the Cerro Rico and planted the flags of Colombia, Peru, and Argentina on the summit in the fall of 1825 when Latin American independence was all but secure, but hopes for a postindependence boom fell apart when a London-based Potosí, La Paz and Peruvian Mining Association collapsed in a stock market crash later that year.[48]

Conditions within newly independent Bolivia did not look more favorable. As Paul Gootenberg has argued, Latin America was saddled with a hamstrung economic elite, a "free-trade 'lumpenbourgeoisie'" of entrepreneurs that "had never stood up for themselves."[49] With a bow to the dependency school (see chapter 30.2, In Their Theories), Gootenberg noted that, as "nationless appendages of world economic currents, unfit for nation-building," Latin American elites followed up on what they had learned over centuries on the periphery of the modern world-system: "It was the region's peculiar 'colonial' role in the world economy—as purveyors of exports—that truncated the historic role played by nationalist entrepreneurial elites elsewhere."[50] Mining in Potosí has been sputtering on in Bolivia ever since independence, too strong to die and too weak to create the wealth of a nation.

Mining in Potosí shifted from silver to tin in the late nineteenth century, and the transition was about the quality of the remaining ores as well as the changing preferences of the global economy. The precious metals lost their former preeminence in mining as industrial societies showed a growing hunger for bulkier commodities: iron, copper, tin, bauxite, and coal. The new mines of the nineteenth century were bigger in every respect, and they featured the latest in contemporary engineering from steam engines, originally invented to keep British mines dry, to electric light. By the early twentieth century, technological progress also changed mining as it had been known for ages when a young American engineer named Daniel Jackling invented opencast copper mining. Instead of following the richest veins into the underground, miners dug up all the ore indiscrimately and then used chemical means to extract the precious metals. It was an expensive technology because it required giant shovels and huge facilities for processing, but by the end of 1905, Jackling had convinced the Guggenheim Exploration Company to pay for a trial run at the Bingham Canyon Pit in northern Utah.[51] It became a watershed in the global history of mining. Instead of moving miners into the mountain, mass destruction mining moved mountains into the factory.

The new mines were larger than anything from the days of Potosí, but that did not raise their profile in collective memory. Few people know about opencast copper mines in Chile or iron-ore mining in the Australian outback nowadays, and they do not *have* to know. Today's resource flows are anonymous, or rather *made* anonymous. "Not only do modern science and technology backed by wants and needs create resources; they also destroy them and reconvert them into 'neutral stuff,'" Erich Zimmermann wrote.[52] Only insiders are familiar with the geography of commodities, and company names no longer reveal places of origin: the Anglo-Iranian Oil Company changed its name to British Petroleum in 1954.[53] Even precarious types of mining fly under the radar. Niger, Gabon, Madagascar, and Namibia were all producing uranium ore, but they barely figured on the mental map of nuclear power (see chapter 37, *Lucky Dragon No. 5*) until concerns over proliferation emerged in the post–9/11 world.[54]

Bingham Canyon Pit has been a National Historic Landmark since 1966, but it is also one of America's worst sources of toxic waste according to an assessment of the US Environmental Protection Agency in 1994.[55] Mining waste may be among the most enduring legacies of the age of industries, to be outlasted only by the holes themselves. In his best-selling book *The World Without Us*, Alan Weisman credited coal mining in Appalachia with creating scars in the land that are "good to endure a few more million years."[56] Lead, cadmium, arsenic, and other metals are well-known toxics, and modern mining has dislodged them in tremendous amounts. In Canada's Northwest Territories, the Yellowknife gold mine blew 237,000 tons of arsenic trioxide dust, a waste product (see chapter 5, *Shipbreaking in Chittagong*) from ore processing, into underground chambers, enough to kill each human on planet earth one hundred times. The company went bankrupt in 1999, and taxpayer-funded cleanup efforts include refrigeration technology to freeze the dust in its place.[57]

Abandoned mines are not the only places where the ravages of colonial resource use are etched into the land. Writing on the Caribbean sugar island of Nevis (see chapter 2, Sugar), Marco Meniketti observed that "the sugar plantation landscape is . . . similar to relic mining districts once resource extraction ends."[58] But leftovers from mining can spread in insidious and largely unrecognized ways. Arsenic contamination of drinking water has been labeled "one of the worst and most widespread environmental problems currently facing humanity."[59] Mining is not the single culprit, as some of the pollution has natural causes, and yet few people know that "more than 100 million people may be at risk from utilizing arsenic-contaminated groundwater."[60] Latin American silver has left a permanent mark in the environment through the mercury that went into the amalgamation process. Cumulative losses in South and Central America are estimated at 196,000 tons, and "very little is currently known about the fate and effects of the unprecedented quantities of mercury discharged in the silver and gold mining areas."[61] What we do know is that mine wastes can exhale mercury into the atmosphere, and the residents of Potosí "continue to breathe toxic air, ingest mercury-laced dust, and are otherwise exposed to the myriad risks of mercury intoxication."[62]

Mercury can stay in the air for months, which makes it a double-edged pollution problem: it combines high exposure in former or active mining regions with lower but pervasive exposure for everyone living on this planet (see chapter 37.1, Global Pollution). In other words, the legacy of Potosí lives on in a significant contribution to the background concentration of mercury in the global atmosphere. The other part of the legacy is a cluster of old buildings in the Andean highlands. "Potosí had no raison d'être apart from silver," Alistair Hennessy wrote, and what once was one of the largest cities in the world is now "a museum town set in a forbidding treeless landscape."[63] (The absence of trees triggered an aforestation campaign with thousands of eucalyptus trees [see chapter 27, Eucalyptus] in the 1990s.[64]) Potosí has been inscribed on the UNESCO World Heritage List since 1987 and on the List of World Heritage in Danger since 2014.[65] Restoration work is ongoing in the city and on the Cerro Rico, where a government project is filling sinkholes to keep the summit from collapsing, all while mining continues on that very mountain.[66] Protected areas have always been places of compromise (see chapter 26, Kruger National Park), and yet one can read a deep symbolism into a project that pours cement, polyethylene, and sand into a mountain that humans have honeycombed for almost half a millennium. It seems as if, in a horribly belated act of repentance, humans are now going out of their way to provide stability in a town that has never had any.

2

SUGAR

THE NEW ORGANIC

I. SMALL ISLANDS, GLOBAL NETWORKS

In 1759 British forces occupied the Caribbean island of Guadeloupe. As the Seven Years' War continued during the following four years, Britain had a lively discussion over the island's value. More precisely, it had a discussion over its value as compared to another territory that France and Great Britain were fighting over at the time: Canada. Sure enough, Canada was much bigger and made more of an impression on a world map, particularly when mapmakers used the Mercator projection that inflates countries far away from the equator. But size was not everything when it came to the business of conquest. Small islands were easier to defend, and this island was very much worth defending because it offered prime conditions for the production of sugar. Could Canada offer anything similar by way of export products? The citizens of London fought a veritable pamphlet war over the question, and they were not alone in their infatuation with Caribbean sugar islands. When France signed the Treaty of Paris in 1763, it gave up its claim to Canada in return for Guadeloupe and neighboring Martinique.[1]

Booming commodities were a key part of modernity, but few careers were as spectacular as sugar.[2] While medieval Europeans had satisfied their sweet tooth through honey and tree saps, cane sugar claimed the market in the early modern era. Like many other commodities, sugar was originally a luxury item that turned into a mass product with growing volume and declining costs. It was about prestige: sugar was an object of what Thorstein Veblen has called conspicuous consumption, not least because it mixed well with other luxury products like tea, coffee, and chocolate.[3] And it was about the energy boost: sugar offered plenty of calories (see chapter 7.2, Numbers Games) that went straight into the blood, a perfect match for the bodily needs of factory workers in the age of industry. As Sidney Mintz has argued in his seminal *Sweetness and Power*, calories from Caribbean sugar underpinned the Industrial Revolution in England. By 1900, sugar provided almost a fifth of the calories in the English diet.[4]

Sugarcane was not native to the Caribbean, nor did the region pioneer its agricultural use. Humans first began to squeeze a sweet juice from cane in southern Asia several thousand years ago, and sugar production remained in the hands of peasants in India and China into the twentieth century.[5] Sugarcane spread in the Mediterranean between 700 and 1100 as part of what Andrew Watson has called the Arab Agricultural Revolution.[6] Its journey across the Atlantic followed the path of Portuguese and Spanish overseas expansion, and as Madeira, the Canaries, Cape Verde, São Tomé, and Brazil all fell for the new crop, planters grew accustomed to using African slaves on sugar estates.[7] The combination of sugarcane, long-distance trade, and unfree labor was well established before it moved to the Caribbean, and yet it was here, starting in the 1640s, that its transformative power played out in its most dramatic form. Largely depopulated after the genocidal Hispanic conquest, the islands were devoid of customs and traditions, agricultural or otherwise, that could stand in the way of a new type of agribusiness with global connections and scant regard for the needs of the place. In his synthesis of the modern world-system, Wallerstein argued that sugar transformed the Caribbean, barely under European control by 1600, into a part of the European world-economy.[8]

Barbados is commonly credited as the first Caribbean island that experienced a "sugar revolution," and other islands of the eastern Caribbean followed swiftly: "Guadeloupe in the 1650s, Martinique in the 1660s, and St Kitts, Nevis, Antigua, and Montserrat in the 1670s."[9] It was a type of farming that differed enormously from European traditions, a "bonanza agriculture" wherein planters could make a fortune within a few frantic years.[10] But for all the wealth that sugar produced, its distribution along the commodity chain was highly unequal, quite in line with what Wallerstein suggests in his world-systems theory, whose inspirations include dependency theory (see chapter 30.2, In Their Theories).[11] Sugar was not the only commodity where scholars have observed a difference between modern and non-European patterns in the distribution of benefits. Giorgio Riello found a crucial difference between cotton production in Asia and Europe. While the Asian system, operating at its peak from around 1000 to 1500, was "a centrifugal system based on the diffusion of resources, technologies, knowledge and the sharing of profits," the European system (dated 1750–2000) was "a centripetal system, one based on the capacity of the centre to 'exploit' resources and profits towards its productive and commercial core."[12]

However, the resilience of the European world-economy must not distract from the enormous volatility in what Wallerstein has called the periphery. Supply lines changed enormously in the greater Caribbean, and so did the relative importance of individual producers. Barbados, Jamaica, Saint-Domingue, and Cuba were all at sometime the leading sugar island. Labor became a particular source of unrest. In addition to a warm climate and plenty of water, sugarcane needed a lot of helping hands, and people were scarce in the Caribbean. Even more, sugarcane called for a tightly organized labor regime. The sugar content declines from the moment when cane is cut, and that forced plantation managers to coordinate harvesting and processing in sugar mills in ways reminiscent of the industrial factory. Barbados, which was planned as a settlement for white immigrant labor in the 1620s, went for African slaves when the mainstay of the colony shifted from tobacco to sugar, and the island was what Philip Curtin has called "a mature plantation colony" by the 1680s: it had three Black people for every white one and one slave for every two acres of arable land.[13]

Resistance runs through the history of African slavery in the New World, and it took many forms from the refusal to eat breadfruit (see chapter 7, Breadfruit) to large revolts. In Jamaica, the struggle between planters and Maroon communities of runaway slaves escalated in a prolonged war that ended with a treaty in 1739. In exchange for liberty and 1,500 acres of land, the Maroons pledged to return future runaways to their masters.[14] Half a century later, the French Revolution triggered an uprising on Saint-Domingue, by far the largest sugar island at the time. When the guns fell silent after thirteen years of atrocious war, the survivors formed Haiti, the first nonwhite republic in the Americas. It was a milestone in the abolition of slavery, a powerful symbol for Black people, and a shock to plantation owners in the rest of the Atlantic world.[15]

Over the course of the nineteenth century, labor regimes grew more diverse. Some countries tried to stick to the old ways, and the prime Caribbean sugar producer, Cuba, was Latin America's next to last country to abolish slavery in 1886.[16] Peru stuck to slavery until the windfall from guano sales (see chapter 8, Guano) allowed the country to pay off slave-owning planters in 1854.[17] In the Dutch East Indies, sugar factories without plantations flourished on Java, supplied by mandatory cultivation of sugarcane in neighboring villages and a forced labor regime.[18] In Queensland, a self-governing British colony that would later merge into the Commonwealth of Australia, sugar production began with plantations in the 1860s, but small-scale European farmers and cooperative central mills took over after the government imposed a ban on the recruitment of Melanesian workers in 1885.[19] Around the same time, a Melanesian archipelago, Fiji, had the Australia-based Colonial Sugar Refining Company build its sugarcane industry with indentured labor from India.[20] In the early 1900s, vertically integrated US-based corporations (see chapter 10, United Fruit) drove the expansion of sugar production in Cuba, Puerto Rico, and the Dominican Republic.[21] All the while, European sugar beets were growing into a formidable competitor to cane over the course of the nineteenth century.

However, sugar regions were far from homogeneous. Slave populations, often portrayed as faceless in traditional histories, were remarkably diverse in terms of background and status, and the same held true for the producers.[22] Commodities had their requirements, but they did not dictate modes of production. When it came to sugar, Caribbean estates were competing with European peasants, and other commodities showed similar diversity. In spite of the name, the "banana republics" of Central America were actually quite diverse. United Fruit (see chapter 10, United Fruit) controlled banana production in Guatemala, competed with other multinationals in Honduras and Nicaragua, failed in Ecuador due to a government policy favoring independent farmers, and El Salvador focused on coffee rather than bananas.[23] Some regions could even accommodate several modes of production, though coexistence bore the seeds of conflict. In 1937, Mauritius saw the worst riots in its history when the sugar factories, which were controlled by estate owners, decided to cut the purchasing price by 15 percent for Uba cane, a variety with a lower sucrose content that was favored among the island's small farmers.[24]

The natural environment provided another set of complications. The introduction of new plants produced what John McNeill has called a "creole ecology": sugarcane, bananas, and citrus fruits stood at the center of "a motley assemblage of indigenous and invading species, jostling one another in unstable ecosystems."[25] Caribbean ecologies experienced another round of biological shocks when the introduction of new cane varieties from other parts of the world resulted in the unintended transfer of new pathogens, and planters have been struggling with transnational epidemics ever since.[26] And then there were the effects on soil fertility that commonly resulted from an excessive reliance on a single plant. Planters did recognize the perils of monoculture. Barbados introduced cattle in order to recycle nutrients back into the soil and developed cane-hole agriculture, where sugarcane grew up surrounded by a protective circle of traditional food crops like yams, corn, and peas, and Saint-Domingue was also exporting indigo, cotton, cocoa, and coffee, but at the end of the day, it was sugar that ran the show.[27] On St. Kitts, sugar and rum claimed no less than 97 percent of exports to the British Isles on the eve of the American Revolution, "indicating the extent to which monoculture had been pushed."[28] And then, environmental learning was not a one-way street. Caribbean sugar planters had learned a lot about the benefits of forests, the combustion of cane pulp, intercropping, and crop rotations by the nineteenth century, and yet much of the acquired wisdom was lost on Cuba, where sugarcane thrived on the use of virgin land and fuelwood from shrinking forests.[29]

While Britain was pushing monoculture in its Caribbean colonies, it pursued a different approach back home. The result was another agricultural revolution. While the traditional European three-field system left the land to lie fallow for a third of the time, the introduction of turnips and clover allowed for the construction of elaborate crop rotations such as the Norfolk four-course system.[30] The interplay of different plants and the combination of livestock and crop husbandry provided an effective way to boost soil fertility before the age of commercial fertilizers (see chapter 8, Guano), and yet the difference to Caribbean monoculture was one of technology rather than principle: both aimed for a more intensive use of the land. Subsistence production gave way to the sale of commodities on distant markets, and homegrown food from garden plots and marginal land was merely a kind of insurance against bad times. Innovations such as hybrid seeds (see chapter 28, Hybrid Corn) and battery cages (see chapter 36, Battery Chicken) were matters of commercial prospects, and the traditional ways ended up in museums if they no longer brought a decent return on investment. When synthetic nitrogen (see chapter 19, Synthetic Nitrogen), DDT (see chapter 38, DDT), and other little helpers allowed European farmers to work with narrower crop rotations down to monocultures in the postwar years, they abandoned their textbook ideas of proper farming and never looked back.[31]

In the new world of organic production, crop rotations and monocultures were negotiable, but the quest for profit was not. It was money, rather than homegrown food, that kept the modern farmer alive, and if market access hinged on a transcontinental commodity chain with all its inherent uncertainties, that was just a matter of transaction costs. The global exchange of agricultural commodities was standing practice long before David Ricardo penned the corresponding theory of comparative advantage. Ricardo argued that in a world of free trade, countries could maximize the wealth of nations by focusing on what they could produce with the lowest relative costs.[32] It was a brilliant idea that earned Ricardo a place in the pantheon of modern economics, and yet there was one fundamental problem: for people on the ground, the wealth of nations was a pretty abstract idea.

2. POWER GAMES

In September 1913 a funeral procession was approaching the Swiss border to Germany. The customs officials on duty had seen a number of these processions recently, and as common decency commanded, they had let them pass quietly. However, the growing incidence of these processions looked suspicious. Why did people from Switzerland suddenly seek to get buried in German soil? The officials felt that it was time to take a closer look. They stopped the people in mourning, took a deep breath, opened the coffin, and found plenty of white powder instead of a corpse. The procession's real purpose was smuggling saccharin, an artificial sweetener that was produced in Switzerland and illegal in Germany. The funeral party ended up under arrest.[33]

As the twentieth century progressed, smuggling white powder became a global business model, and color is not the only thing that connects saccharin smuggling along the Swiss–German border with the worldwide trade in illicit drugs (see Interlude, Opium). Like heroin, originally a trademark of Bayer, saccharin was a product of industrial chemistry. Constantin Fahlberg, a Baltic German with a doctorate in chemistry, discovered saccharin during a stint as a research fellow at the Johns Hopkins University in Baltimore in 1878. Saccharin was much cheaper than sugar and filled a market niche as the sweetener of the poor. With low production costs and a high value density, it invited smuggling after Germany imposed its ban in 1903. Authorities estimated that more than one thousand people were living from the illicit saccharin trade in Zurich alone. And like the ban on narcotics, the rationale for banning saccharin appeared dubious when one took a closer look. After coming to market in 1887, saccharin had triggered a brief medical debate in France, but concerns had long subsided by the time German parliamentarians were casting their votes. The ban on saccharin had very little do with health and consumer protection—apart from tobacco, perhaps no other food product has been studied as intensively as saccharin—and everything with the power of big sugar.[34]

While the origins of sugarcane are lost in the fog of prehistory, sugar beets were a project of science from the beginning. Andreas Sigismund Marggraf, a chemist with the Berlin Academy of Sciences, discovered in 1747 that some beets contain sugar, and his successor in that post, Franz Karl Achard, set up the world's first beet-sugar factory in 1801.[35] France's continental blockade helped the nascent industry, all the more as the French state continued to protect it after Napoleon's fall.[36] Systematic breeding (see chapter 28, Hybrid Corn) increased the sugar content, which stood at between 2 and 3 percent in 1800, to 14 percent around 1900 while the demanding plant taught peasants the merits of better farming skills.[37] No other crop offered German farmers a higher return per acre, and while production remained confined to regions with excellent soils, farmers in these regions could make a killing with sugar beets. The boom turned into a frenzy after 1880, and beet sugar surpassed cane sugar on the world market around 1890.[38] Sugar became Germany's most important export product for a while, and the country was even the world's leading sugar exporter from 1895 to 1900.[39] Sugar beets are one of the great success stories of nineteenth-century European agriculture, and yet it all hinged on governments that supported sugar beets generously and unwaveringly over decades. Banning an artificial sweetener was just one of many favors.

After launching the modern agribusiness, sugar blazed the way for another feature of modern agriculture: subsidies. To be sure, sugar was never a textbook case for functioning markets. David Ricardo complained in 1817 about the "monopoly price" for sugar, where market prices had "lost all connexion with the original costs."[40] But during the nineteenth century, price manipulation became a matter of government policy. When Germany asked the agricultural experts at its embassies about the state of sugar production in selected countries toward the end of the nineteenth century, the reports showed that lavish government support was firmly entrenched around the globe. The German emissary in Buenos Aires wrote that Argentina had the world's highest production costs and the lowest yield of sugar per acre, and yet sugar producers would only need to point to their enormous outlay of capital to win another round of export subsidies. "It seems almost unthinkable that any government or parliament would ever consider sacrificing this industry for the love of the Manchester doctrine [of free trade]."[41] Writing on Bulgaria, another German expert found a sugar industry "akin to a greenhouse plant": near Sofia a Belgium consortium had built a sugar factory whose fate was completely dependent on subsidies.[42] In Mexico, sugar production was in the hands of a syndicate that just about satisfied domestic demand, but the German emissary saw overproduction on the horizon, and generous export support would be likely: the sugar producers were "very influential people" that included the son-in-law of Porfirio Díaz, Mexico's president of thirty years.[43] And it was not just governments who were paying the price. Commenting on the recent expansion of Russian sugar production, the German emissary in Saint Petersburg found that sugar beets were bound to become a threat to the health of the soil (see chapter 13, Little Grand Canyon).[44]

The report on Bulgaria blamed the penchant for sugar beets on "national vanity," a quest for autarky and agricultural advancement, and the magic word "industry."[45] But subsidies were ultimately about corporate power: sugar producers had powerful lobbies, formal or otherwise, and it took a particularly courageous government to put a thriving business sector at risk. Sugar production had become a capital-intensive business over the course of the nineteenth century, and the shiny factories with expensive equipment and many well-paid jobs were bound to collapse if the government did not contribute its share.[46] With that, sugar foreshadowed what would become a powerful rationale for agricultural subsidies in the twentieth century: the most advanced producers were usually the ones with the highest capital investments, which also made them the producers who were the most exposed to price fluctuation and thus most deserving of government support. And then, sugar subsidies were not just about farmers. When the United States went for protectionism in the Great Depression, Coca-Cola, the world's largest industrial sugar consumer since the mid-1910s, launched a massive lobbying effort against sugar tariffs.[47]

Subsidies were fought over perennially throughout the twentieth century, and sugar producers were not always the winners. In fact, sugar was subject to a landmark free-trade agreement, the Brussels Convention of 1902 that banned cartels, tariff walls, and export supports. The Brussels Convention on sugar even had an unprecedented Permanent Commission for arbitration and enforcement.[48] Unlike what neoliberal mythology suggests, free trade is not a state of nature that miraculously materializes in the absence of government interventions. Quite the contrary, open markets are the creations of effective nation-states. The Brussels Convention collapsed during World War I, and for all its merits, it was never a complete success. After all, there were other ways to give domestic sugar industries a helping hand. The ban on saccharin was a direct response to the Brussels Convention that German sugar producers saw as a mortal threat.[49] A few years later, sugar producers launched an association that sought to boost domestic sugar consumption through the elimination of lingering sentiments "that sugar is a luxury product."[50] In the twenty-first century, the World Bank argued that sugar was the second-most-protected commodity after rice.[51]

Farmers and corporations had bargaining power. The situation was more complicated with a view to the workers. When the Western powers abolished slavery in the Caribbean, they expected the former slaves to morph into a proper working class, with liberals extolling the virtues of wage labor.[52] However, the rural proletariat became a wildly heterogeneous mix of ethnicities and contractual obligations, and not just in Latin America. Sometimes rural workers came together in powerful movements. César Chávez and the United Farm Workers fought for better wages and working conditions in California's agribusiness.[53] Brazil's Landless Workers Movement occupied sugar estates in a push for land reform (see chapter 6, Land Title) from below.[54] But more often than not, plantations offered miserable jobs, and the plight of the workers was at the same time evident and invisible. Recalling his pioneering anthropological work in the 1950s, Sidney Mintz put it as follows: "Everybody knew that there were millions of people in the world, nearly all of them people of color, working at ghastly jobs producing basic commodities, mostly for consumers in the West, but hardly anybody had thought about it—including myself, until I was sent to Puerto Rico."[55]

Plantations flourished in colonial and postcolonial environments, but their rationale did not remain confined to the periphery of Wallerstein's world-system. Chicago's slaughterhouses (see chapter 18, Chicago's Slaughterhouses) introduced the logic of industry into the meat business in the late nineteenth century. In the postwar years, innovations like battery chicken (see chapter 36, Battery Chicken) spread the guiding principles—large technological systems, division of labor, and a disregard for human and environmental costs—to the other segments of the meat commodity chain, and the industrial style of production held the world of agriculture firmly in its grasp by the new millennium. We can even find the legacy of agribusiness in the east of England, where the Norfolk rotation once opened the door to a new world of agriculture.[56] In Lincolnshire, companies like Staples Vegetables produced kale and other crops for supermarket chains, used the free movement of workers within the European Union to bring in labor from Poland and other Eastern European countries, and paid them the minimum wage. Housed in a fenced camp on company grounds, their presence made Lincolnshire a hotbed of anti-immigrant sentiment.[57] When the United Kingdom chose to leave the European Union in a referendum in 2016, it was Boston, the seat of Staples Vegetables, that recorded the highest anti-European vote.[58]

3. THE SWEET LIFE

In the fall of 1890, Joseph Chamberlain paid a visit to the United States. The British politician had married the daughter of the US secretary of war two years earlier, but he found this stay less than satisfying. When he grew bored with New England society, he went to Montreal, where he found more interesting conversation. Chamberlain happened to meet the governor of the Bahamas, who gave him a glowing description of his archipelago. He was particularly excited about sisal, a plant that grew as a weed on the island and would make for high-quality hemp.[59] The governor's enthusiasm proved contagious. A sisal plantation was just to Chamberlain's taste. He had recently incurred heavy losses with South American securities and sought ways to restore his fortune. Sisal was exactly the kind of development project that Chamberlain, a fervent imperialist, was advocating at the time.

And best of all, he had a son, Neville, then working in his first job as an accountant back home in Birmingham, who was ready to cut his teeth in an imperial challenge. After all, plantations were a place where real men were made: from Francis Willoughby, governor of Barbados in the 1660s and a key figure in the migration of Barbadians to Suriname, Jamaica, and other Caribbean islands, to Minor Keith, one of the founders of United Fruit (see chapter 10, United Fruit).[60] Unfortunately, young Neville did not make the cut.

The family bought twenty thousand acres in the Bahamas in 1891, but the project was beset with problems from the outset. Clearing the land took longer than anticipated. Labor conflicts were simmering, particularly after Neville Chamberlain lost his right-hand man to drinking after the death of the man's wife. The world price for sisal collapsed. The sisal plants that he had bought for a cheap price in Mexico turned out to be of inferior quality. When the plantation finally produced some sisal in 1896, a fire consumed the baling shed. Worst of all, Neville Chamberlain did not have a genuine interest in horticulture or even a proper understanding of the needs of plants: he was deeply offended when his sisal plants failed to grow in unison. After seven lonely years for Neville and no hope for any improvement, the family cut its losses, which ultimately ran to £50,000. Neville Chamberlain returned to Britain, embarked on a career as a politician, and eventually made it to 10 Downing Street in 1937, where he tried to placate Hitler with an appeasement policy. It did not fare any better than the sisal project.[61]

Plantations remained a cornerstone of global commodity chains throughout the twentieth century, and yet their hegemony was an embattled one. Pests like the boll weevil (see chapter 12, Boll Weevil) continued to rock plantation ecologies. In 1930, the League of Nations launched an investigation into labor conditions at Firestone's rubber plantations in Liberia.[62] The intensive use of limited space had a price, ecological and otherwise, but for the men who could overcome these obstacles (and yes, it was a job for men), it was a winning formula. As food production grew in quantity, it clustered more and more in specific regions. To mention one of the more drastic examples, California claims two-thirds of the global production of almonds.[63] In the twenty-first century, agricultural intensification even won friends among environmentalists who argued that concentration of production allowed "more room for non-human species" (see chapter 26, Kruger National Park).[64]

2.1 "Greetings from Jamaica." A postcard from the British West Indies depicts work in the cane fields. Image, The Tichnor Brothers Collection, Boston Public Library.

The trouble in the land was not of much concern for urban consumers. To the extent that they recognized a food problem, it was that they ate too much of it: today, for the first time in human history, more people are obese than underweight.[65] Sugar was a popular lightning rod for critics of the global food system. In 1972, the British nutritionist John Yudkin published a book titled *Pure, White and Deadly*, which identified sugar as a cause of heart disease, diabetes, and other health problems, and countless authors have followed his path with widely divergent levels of sophistication and style.[66] Artificial sweeteners like saccharin were booming, but fighting obesity took more than substitutes. Low-carb diets have long emerged as a discussion point among overweight people, and conversations about dieting are now so much a part of the Western lifestyle that it is easy to forget that the health toll of sugar is not just an issue in the affluent world. When the leaders of fifteen Caribbean nations made the case for slavery reparations in 2014, type 2 diabetes was part of the indictment.[67]

In short, the sweet life of the new organic was already an ambiguous achievement in times of peace. Warfare jeopardized global commodity networks in ways that defied preparations and planning, and more than one agricultural producer cursed these dependencies in times of crisis. Many residents of Mount Lebanon starved to death during World War I (see chapter 31.3, Blame Games) after converting their slopes to mulberry cultivation in previous decades.[68] Unlike silk producers, sugarcane planters could at least eat their commodity, and any worries that dental surgeons might have had paled in times of war, but war could throw food systems into disarray for many years. For example, scholars have argued that the French Revolution "damaged agriculture, or at least condemned it to a generation of stasis."[69]

But for all the legitimate concerns, the global food system was a modern marvel. It had an unprecedented ability to deliver cheap and reliable supplies over long distances, and yet this was probably not just an achievement in its own right. The resilience of the modern food system was also due to hidden reserves (see part VI, Final Reserves) that provided crucial support in case of trouble. Looking across the numerous contestations and the endless conflicts on plantations and other sites of production, it is hard to understand the system's permanence without some inconspicuous buffers against disaster. For many peasants, the right balance between commercial and subsistence crops (see chapter 30.1, In Their Dreams) was one of the most vexing issues, and their concerns could resonate widely. In early twentieth-century Java, the feasibility of a crop rotation with sugarcane and rice triggered a decadelong controversy that the Dutch colonial authorities struggled to contain.[70] There was even space for other crops on Caribbean sugar plantations, which tended to increase their resilience. Planters grew some of their foodstuff on marginal land or gave small plots to their slaves, who tended to their own gardens and enjoyed a little corner of freedom. It was an insurance against starvation, and maybe more. Judith Carney has argued that slave gardens also served as "botanical gardens of the dispossessed" (see chapter 27.2, Botanical Exchange).[71]

The modern food system was a gamble, and not just for embattled workers and their more or less successful masters. It took a leap of faith to abandon subsistence modes and become dependent on the monetary economy, all the more as the supply of food relied not just on faceless producers but also on a functioning state. In the modern era, matters of food were also matters of authority, as the new meaning of hunger (see chapter 31, Holodomor) serves to attest. The large-scale Sahel famines sharpened global awareness for the precariousness of food supplies, and yet one does not need to go to the ends of the earth to learn about the risks of the modern food gamble.[72] The promise of reliable supplies wore thin even in the heart of twentieth-century Europe. Norman Davies has argued that few people did as well in Communist Poland as the "peasant-workers" who combined small-scale agriculture with a factory job. They had "the best of both worlds—a high cash income all the year round, a cheap supply of home-grown food, and an independent base."[73]

3

THE CANAL DU MIDI

THE GREAT MOBILIZATION

I. THE STATE OF TRANSPORT

A ditch was the place to be in Toulouse on November 17, 1667. The town's notables mixed with the clergy, the Parliament of Languedoc, and some six thousand workers to watch the start of a new construction project. The archbishop of Toulouse took two stones in his hands, blessed them, and had them put into the ground as part of the foundation for a lock. Drums were beating, guns were firing, and commemorative medals were thrown into the crowd amid cries of joy and "Vive le Roy." Even the weather was unseasonably nice. It left a deep impression on the author of the *Annales de Toulouse*: "God was present."[1] And if you think about it, divine support was arguably a good idea for a project that was pushing the limits of contemporary technology.

When the Canal du Midi opened some fourteen years later, it stretched 240 kilometers across the continental divide. It connected Sète, a port on the Mediterranean Sea, with Toulouse, where ships could enter the Garonne River toward the Atlantic Ocean. The canal had 101 locks, some 130 bridges, and a tunnel with a length of 165 meters that workers dug in a weeklong frenzy in defiance of an order from the governor of Languedoc to halt construction. The project even included a reservoir at Saint-Ferréol (see chapter 29, Aswan Dam), as the canal ran through a dry watershed region. With a crest length of 780 meters and a thickness of 140 meters at its base, the dam was the largest man-made barrage of its age. Numerous other problems were lurking along the way, and builders were improvising with the course of the canal, the design of locks, and many other things. The project also faced hostility among the region's nobility. But that was part of the plan.[2]

Complaints about roads were legion in seventeenth-century France, but the Canal du Midi was about more than solving a transport problem.[3] It was literally an act of state: the project was meant to showcase and consolidate the power of the absolutist regime. Construction began pursuant to an edict of Louis XIV, and his minister of finance, Jean Baptiste Colbert, monitored progress as best he could. The Canal du Midi tightened the grip of the center in a peripheral region, and control over water held symbolic power. As Chandra Mukerji writes, "Bending such stuff to the will of the Sun King and changing the geography of the continent to do it was a project worthy enough to compete with war in ambition and symbolic possibility."[4] Few narratives of the Canal du Midi fail to mention how rulers through the ages had dreamed of linking the two oceans: Augustus, Charlemagne, François I, and Henry IV.[5]

The Canal du Midi mirrored a political style that drew power and prestige from knowledge of and control over the natural environment. While workers were digging their way through Languedoc, Colbert set out to create new academic institutions. He established the Académie des Sciences in 1666 and the Observatory in 1667. He took over administration of the Jardin du Roy and made it the hub of a network of botanical gardens (see chapter 27.2, Botanical Exchange) that served France's growing colonial empire.[6] In 1669, Colbert commissioned a survey of France's forests that is nowadays credited as a milestone in the making of sustainable forestry (see chapter 4, Sustainable Forestry).[7] All the while, Colbert built a clientelist network with appointees selected for their loyalty, efficiency, and good stewardship and used them to establish a new government structure parallel to the existing one, effectively bypassing the great nobles and provincial elites.[8] It came down to a new type of territorial stewardship, depersonalized and technocratic, based on knowledge and a desire to remake, "improve" the land. Absolutism was as much about the display of royal glory in Versailles as about its approach to the land, and the Canal du Midi was a case in point.[9]

New roads and canals were not just about moving earth and laying stones. They were also about information: building new transport links called for familiarity with the topography and knowledge about local assets and obstacles. And they were about establishing contacts between heretofore distant places. For people with a progressive bent, supporting better transportation seemed like the most natural thing. "Of all inventions, the alphabet and the printing press alone excepted, those inventions which abridge distance have done most for the civilisation of our species," wrote the British historian and Whig politician Thomas Macaulay in his *History of England*. "Every improvement of the means of locomotion benefits mankind morally and intellectually as well as materially, and not only facilitates the interchange of the various productions of nature and art, but tends to remove national and provincial antipathies, and to bind together all the branches of the great human family."[10] But as so often, progress was a matter of perspective, as new transport links were also about power. The Canal du Midi was not unique in facing local opposition, as infrastructure projects challenged the status quo: they cut through existing land holdings (see chapter 6, Land Title), they brought obstacles for those who wanted to cross, and they created new entitlements. And this was a peacetime project.

Military rationales had shaped transport networks ever since the Roman and Inca Empires built their famous roads to entrench imperial power. The Canal du Midi provided a convenient link between the French naval base at Toulon on the Mediterranean and the arsenals of Rochefort and Brest on the Atlantic Coast.[11] Other military planners of the time were more ambitious. The Spanish Empire sought a canal from the Rhine to the Meuse in order to divert traffic from the unruly Low Countries, and some blueprints aimed for a diversion of the Rhine itself. Construction started near Rheinberg in 1626, and a few miles were actually built before the project stalled. It received a temporary hit when Spain ran out of money after a Dutch West India Company force captured the Mexican silver fleet in 1628, and a terminal one when the Low Countries conquered Rheinberg in 1633.[12]

Clever generals favored investments in mobility throughout the ages. Between 1804 and 1812, Napoleon spent twice as much on road construction as on fortifications.[13] Railroads played a crucial role in military planning ever since quick movement of troops helped suppress the revolutionary movements of 1848/1849.[14] Built between 1887 and 1895, the Kiel Canal allowed the ascendant German navy to transfer ships between the Baltic Sea and the North Sea without going through international waters.[15] The Suez Canal facilitated the movement of European warships and troops toward colonies in Asia and East Africa.[16] In short, transport links figured prominently in military planning, and they figured prominently in the public imagination. Historians of Nazi Germany have tried for many years to convince the general public against all odds that autobahn construction (see chapter 34, Autobahn) was not driven by military interests.[17]

The connection was not just about military strategy but also about mindsets. It was a general, Guillaume-Henri Dufour, who led a comprehensive topographic survey of Switzerland from 1833 to 1865, and the resulting "Dufour map," the country's first map that was based on accurate measurements, became a focal point of Swiss nation-building.[18] A unit of the US military, the Army Corps of Engineers, has played a crucial role in the construction and maintenance of navigable waterways since 1824.[19] The quest for control was particularly important in colonial settings, and railroads were an important part of what Daniel Headrick called "the tools of empire." They helped consolidate European power once the machine gun had done its part.[20]

Like many subsequent projects, the Canal du Midi was literally about state-building. The guiding spirit was Pierre-Paul Riquet, a tax farmer who submitted a project outline to Colbert in 1662.[21] The plan was attractive for a minister who sought to boost state revenue, as Riquet offered to build the canal on his own account in return for income from tolls and other rights. The plan was less impressive in terms of technical expertise, but Riquet won Colbert's favor by completing an experimental trench on time and under budget. Colbert did not have much in the way of alternatives: in the absence of eighteenth-century creations like the Corps des Ingénieurs des Ponts et Chaussées, expertise was generally scant.[22] In any case, Riquet rose to the challenge and found innovative solutions like locks with curved walls, which proved more stable than the original rectangular design. Riquet also drew on the knowledge and skills of local peasants and artisans, some of them female.[23] Later generations would celebrate Riquet as an eminent man of Gaul, and he received his share of monuments, including one paid for by his descendants.²⁴ But if these monuments were to reflect the true genius of the project, the pedestals would need to be far more crowded.

Riquet died in the fall of 1680. The canal was finished a few months later and swiftly included into the decoration of the Hall of Mirrors in Versailles.²⁵ But completion was a relative term at the Canal du Midi. In 1686, the country's foremost builder of fortifications, Vauban, went to Languedoc to deal with numerous unresolved issues. He had some fifty viaducts added over the following eight years and rerouted the canal in numerous places, and Vauban's work provided an object lesson on the fluid boundaries between construction and maintenance.²⁶ Repairs rarely make headlines, all the more so as they usually take place long after ribbon-cutting ceremonies and invocations of divinity, but they are just as important for the circulation of traffic.²⁷ Thus, in a strict sense, canals, streets, and railroads are never really finished. It's just that the rate of construction goes down at some point.

2. CONNECTED

Infrastructures built connections between different people, but they also built fortunes. Some of America's leading universities grew from the donations of rich men who made their money in the nineteenth-century railroad business, and several of these men are immortalized in the names of institutions like Stanford University and Johns Hopkins University to the present day. Some tycoons had enough money for more than one university: the oil magnate John D. Rockefeller, who made a part of his fortune through pipelines and secret rebates from railroads, helped launch the University of Chicago, the Rockefeller Institute for Medical Research, and Central Philippine University in the Philippines, a US colony at the time.²⁸ Riquet was less fortunate in his lifetime and actually sold his home in Béziers a year after the start of construction in order to keep the project afloat. He was financially ruined at the time of his death, and it took his descendants until 1724 to pay back his debts, but they could eventually live comfortably from canal revenues.²⁹ The French state did not buy the Canal du Midi from Riquet's heirs until 1897.³⁰

Infrastructure projects were expensive, and their economics were a gamble both for the builders and the regions that they crossed. The notables of the Languedoc eventually made their peace with the Canal du Midi, as it provided an object lesson on the benefits of mercantilism, but a boom was no foregone conclusion.³¹ In his book *The Peas-*

ants of Languedoc, Emmanuel Le Roy Ladurie challenges "a certain canal mythology popular with some historians" by pointing to the decline of the region's viticulture under Colbert and Louis XIV: wine from Agde and Béziers "had never been so mediocre as in the half century following the completion of Riquet's great undertaking."[32]

Outcomes took many different forms. Some projects were ahead of their time. The transcontinental railroads would have made more sense, economic and otherwise, if the United States had built them in the 1890s instead of the 1860s.[33] Some projects limped along over decades. The Freedom Railway connecting Zambia with the Tanzanian port city Dar es Salaam, built with Chinese support from 1970 to 1975, became a source of endless problems, partly because its rationale was exceedingly political: it gave landlocked Zambia an outlet for its crucial copper exports without going through colonial Angola or white-controlled Rhodesia and South Africa.[34] Economy was also a secondary concern when Russia built the Trans-Siberian Railroad, as "the image and prestige of the Russian government in the nation and the world were always more important considerations."[35] And some projects were complete failures. New Zealand built a steel-reinforced concrete bridge across the Whanganui River in 1936 when settlers sought to develop the remote Mangapurua Valley on the North Island. The last families left six years later, reducing the bridge's function to a scenic one. The "Bridge to Nowhere" is nowadays listed as a Category 1 historic place and a tourist attraction (see chapter 22, Baedeker) in Whanganui National Park (see chapter 26, Kruger National Park), accessible only by walking a forty-minute trail.[36]

Individual projects could fail, but quantitative studies have stressed the general benefits of efficient transportation. Challenging received notions of the French countryside as a *société immobile*, Philip Hoffman has argued that falling transport costs "seem to explain much of early modern productivity growth, not just in France, but in Germany and England as well."[37] The case for better transportation looked like a truism of economics until Robert Fogel published *Railroads and American Economic Growth* in 1964.[38] Fogel argued that the American economy would have grown just as well if railroads had never been invented: canals could have satisfied the country's transport needs up to 1890 with only marginally higher costs. His book became a classic of econometrics, and it did not prevent him from becoming a Nobel laureate in 1993, but Fogel's argument may say more about economics' infatuation with numbers (see chapter 7.2, Numbers Games) than about transport history.³⁹ No country ever eschewed railroad construction in favor of canals, leaving a definitive answer adrift in the shallows of counterfactual history. But we do know that canals would have made for a different economy.

Canals were probably competitive in terms of transport costs, but they were also slower, and speed was important for the modern economy. The Chicago slaughterhouse (see chapter 18, Chicago's Slaughterhouses) would not have thrived with canal boats: it was the speed of rail and the refrigerated railroad car for chilled meat that accounted for the system's efficiency. Furthermore, ships would have produced a seasonal economy: canals froze during the winter while railroads could operate all year.⁴⁰ Unlike canals, where locks imposed strict limits on growth, railroads could easily realize economies of scale. In fact, it was scalability that turned America's railroads into engines of growth: since the operating costs of trains rose only marginally with additional cars and passengers, railroad companies tried everything to boost traffic from experimental farms for eucalyptus (see chapter 27.2, Botanical Exchange) to speaking tours for dry-farming apostles like Hardy Webster Campbell (see chapter 13.2, Saviors of the Soil). And then it remains anyone's guess how canal operators would have used a transport monopoly. In Britain, where canals thrived during the Industrial Revolution, competition with ships was an important part of railroad development. There were even plans for new canal projects in the late nineteenth century when railroad companies imposed excessive freight rates.⁴¹

The economic case was beyond doubt from the viewpoint of individual cities. Riquet certainly understood the importance of transport links for communities. He chose a longer and more difficult route towards the Mediterranean in order to lead the Canal du Midi through his hometown of Béziers.⁴² The citizens of Telgte, a small town in Westphalia, were aghast in 1812 when they learned that Napoleon had commissioned a new road from Wesel to Hamburg that skirted the town. The military was worried that troops would get stuck in Telgte's medieval streets, though a French engineer was also concerned about isolating the town: if the road were built, people would need "to leave town to see what is going on in the great wide world."⁴³ Many nineteenth-century towns struggled to get railroad access and cursed builders when they chose a different route. The matter did not become moot in the age of flight. The rise of Atlanta Airport began in 1926 when the city successfully outbid Birmingham, Alabama, for a stop on the New York–Miami airmail route.⁴⁴ Infrastructures continue to reflect the inequalities of the world. A book titled *Europe's Infrastructure Transition* pointed out that in 2011, there was only one bridge across the Danube along the 470 kilometers where it forms the border between Bulgaria and Romania whereas Hungary's capital, Budapest, has nine bridges across the Danube alone.⁴⁵

Infrastructure has its own laws. It is often vulnerable in economic terms because of the enormous amount of fixed capital, but it has shown remarkable resilience. It tends to interweave with societies and economies on so many levels that it becomes difficult to imagine life without it. Farmers and industrialists rely on it for selling goods, people make it a part of their mental maps and their identity, and sometimes the technical requirements shape entire regions: the Panama Canal's water supply hinges on the protection of large swaths of forest.⁴⁶ In other words, the Canal du Midi was not just a ditch across a continental divide but "a brute fact in the countryside," as Mukerji has written. "It was something to work with and work around like a mountain, not something to debate or query about its history."⁴⁷ It was usually a sign of trouble when infrastructures fell into disuse. When China failed to keep the Grand Canal functioning after the Second Opium War (see Interlude, Opium), it abandoned a traffic link that had served to unite the country politically and economically since it was built in the early 600s.⁴⁸

The Canal du Midi certainly showed a remarkable ability to survive. A parallel railroad was built in the nineteenth century and a divided highway (see chapter 34, Autobahn) in the twentieth, and yet the canal continued to operate. Its role was obviously diminished, but the canal remained full of water, and lockkeepers continued to do their job. Barge operators even dreamed about modernization in the 1970s, but preservation gained the upper hand.⁴⁹ The Canal du Midi became a UNESCO World Heritage Site in 1996, and today it is firmly in the hands of recreational pleasure boating (see chapter 22, Baedeker).⁵⁰

3. UNSETTLED

Situated in the heart of the Suez Canal Zone, Ismailia was a great place to watch ships. However, Hassan al-Banna was more concerned about humans and their souls. The son of a watchmaker with a degree from Cairo University, al-Banna became a teacher at an elementary school in Ismailia at the age of twenty-one. He saw a society adrift. Ismailia was the seat of the Suez Canal Company and a British military base,

3.1 The Canal du Midi in Carcassonne. Tourists are the defining users today. Image, Krzysztof Golik / Wikimedia Commons.

and that gave al-Banna a chance to observe how white men dominated the country.[51] He founded a new organization, the Muslim Brotherhood of Egypt, in 1928. It became "the largest, most effective, and most influential Islamic political party for the next half century."[52]

Ferdinand de Lesseps built the Suez Canal with a view to the interests of white men, but once it was open, the canal forged links between many different people: officials and workers, soldiers and prostitutes, cosmopolitan businessmen and Muslim pilgrims on the hajj to Mecca. Large ships carried passengers, goods, and guns of war through the canal while a ferry service shuffled Bedouins and their camels across, as the canal cut through a traditional caravan route at al-Qantara.[53] Like many other traffic links, the Suez Canal was a place of multiple mobilizations: intended and unintended, consequential and superficial, enduring and ephemeral. The web of transfers continued below the waves, as the canal gave fish and plants from the Red Sea a chance to migrate to the Mediterranean. Marine biologists speak of "Lessepsian migrants," and some of these migrants have become so abundant that they support new fisheries.[54]

Infrastructure projects had a mobilizing effect already when they were being built, as they typically called for legions of workers. It was not necessarily a free choice: construction of the Suez Canal depended heavily on forced corvée labor, in spite of concerns on the part of Lesseps about the contemporary antislavery movement.[55] His next project, the Panama Canal, was even more disastrous for the workers, as malaria and yellow fever brought the death toll to up to forty a day. The project collapsed in 1889.[56] In the subsequent US-led project, mobilization occurred even before construction resumed: when Colombia did not ratify a treaty for the canal's completion, the United States supported a separatist movement that staged a successful coup in 1903. The new Republic of Panama signed the contract for the canal only fifteen days after its creation.[57] Yet workers were not necessarily on the losing end. Riquet's project thrived on good labor relations, as up to twelve thousand workers, including six hundred women, enjoyed decent pay and unheard-of benefits such as sick leave and payment on rainy days. Colbert had suggested corvée labor and requisitions, but Riquet wanted none of this and gained a loyal workforce in a hostile region.[58] Even the timing of the Toulouse ceremony showed his concern about labor. In a rural region like Languedoc, it was easier to recruit workers when the harvest was done.

Of all the mobilizations that occurred in the wake of new infrastructures, perhaps the most dreaded were about germs, as traffic links were also disease vectors. Cholera (see chapter 21, Cholera) only made it to Europe in the nineteenth century because of the increasing speed of transport. The Suez Canal had a comprehensive disease-prevention policy with officials, hospitals, and cordons sanitaires. Contagions were color-blind, but the same could not be said of disease policies, as the authorities devoted particular scrutiny toward Muslim pilgrims. At times pilgrim ships had armed guards on board and a Canal Company boat in their wake, with standing orders to shoot pilgrims who tried to jump ship.[59] Transport history is as much about growing speed, volume, and safety as about the efforts of bureaucrats and experts to deal with the repercussions of mobility. There were different classes of travelers with different means, needs, and rights.

In short, the great mobilization was messier than traditional tales of technological progress have suggested. It was about intended and unintended mobility, about people, goods, information, and pathogens, about displays of power and forced migration: the mobility history of France includes the glorious Canal du Midi as well as Napoleon's exiles on Elba and Saint Helena and Alfred Dreyfus's condemnation to Devil's Island off the coast of French Guiana.[60] Even the dichotomy between growing mobility under Western auspices and Indigenous stasis is misleading: a significant strand of modernity is about the control of mobile people—from gypsies to nomads.[61] Mobilization could take many different forms, and the consequences were as diverse as the underlying causes. Emmanuel Le Roy Ladurie has argued that the great losers of the Canal du Midi were the Huguenots of Languedoc. Waterways do not have a religion, but the people who pay for them do, and the region's episcopate gave the canal their blessing in return for the destruction of the reformed churches.[62]

In ecological terms, the Canal du Midi was rather benign. It did not trigger a Lessepsian migration because the freshwater canal served as an effective barrier between the seas. It could also cope with riparian use. Women used it to wash their clothes and farmers siphoned off water for their fields, but neither became a contentious issue. The waters have stayed calm along the canal since the days of Riquet and Vauban, and most tourists cherish the leisurely pace nowadays when they tug along with their boats. But the sense of quiet can be deceiving: the *Ceratocystis platani* fungus, first identified in New Jersey in 1929, came to France in 1945 (see chapter 14, Cane Toads), perhaps quite literally on the heels of American soldiers, and it is particularly fond of traveling along the Canal du Midi, where the trees are conveniently standing in line.[63] The Canal du Midi may be in for yet another round of mobilization.

4

Sustainable Forestry

The State of the Woodlands

I. TO OWN A FOREST

It was a case of writer's block. The topic at hand was forestry, and the author approached it with admirable breadth in his book-length treatise. He looked into all the contemporary issues from seeds and soils to the use of the axe, and the remarks mirrored the experience of a lifetime. But when it came to the guiding principle, the author was stuck: there was simply no word for the idea that he had in mind. He talked about "continuiren," the continuous use of woodlands, he used the Latin word "conservation," he quoted Cicero and the Old Testament, but it just did not feel right. And so it happened that, on page 105, the author coined a new phrase: "nachhaltende Nutzung"—sustainable use.[1]

The birth of sustainability had the air of a forceps delivery, and the author, Hans Carl von Carlowitz, did not terribly like the result: "sustainable use" came up only once on the 432 pages of his *Sylvicultura oeconomica*. Carlowitz, who published his book in 1713 and died a year later, would surely be surprised to learn that "sustainability" turned into a global buzzword in the wake of the 1987 Brundtland report (see chapter 24.2, Buzzwords).[2] As Paul Warde has pointed out, "Carlowitz has been hailed as a prophet of sustainability today, but he was not arguing in the modern mode."[3] Yet terminologies were arguably a minor issue for a man who had spent his life working toward sustainable forest use. Words were negotiable, but the need for action was not: forests were "among the greatest treasures of a country," and they were under threat.[4]

A concern about forests had an existential dimension in a society that relied on its resources in many different ways. Timber was crucial construction material and the stuff that built powerful navies. Firewood kept houses warm during the winter and meals prepared all year round. Noblemen went to the woods to chase powerful stags while ordinary folks were hunting for berries and mushrooms. Domestic pigs grew fat from fallen acorns and chestnuts, which gave them distinct biological characteristics that set them apart from the waste-fed (see chapter 5, Shipbreaking in Chittagong) pigs of China (see chapter 36, Battery Chicken).[5] Some resources even served as the foundation of premodern chemistry.[6] The people of early modern Europe went into forests on a regular basis, though some trees saw more humans than others. *Grimm's Fairy Tales* and Dante Alighieri's *Divina Commedia* provide different accounts of what could happen if you got lost in the woods.

Carlowitz looked at these different uses from a particular angle. At the age of thirty-two, he became Vice Berg-Hauptmann, the second-highest position in the mining administration of the early modern state of Saxony, and made it to the top job shortly before his death.[7] Mines (see chapter 1, Potosí) were a coveted source of revenue for early modern rulers and harbingers of an industrial-style division of labor. They were also prime consumers of wood, and the supply was potentially a limiting factor for production. That made reflections on the economic use of woodlands a natural endeavor, and yet it would be naive to read the *Sylvicultura oeconomica* as a mere collection of ideas. The real issues were money and power.

Sustainable forestry was born in mining regions, but it gradually advanced toward a general principle of forest use in Central Europe. States enacted forest ordinances and set up schools to train a new cadre of officials who gave teeth to the new rules. It was about revenues: claiming wood as state property and selling it on the market brought money into the state's coffers. It was about the Enlightenment, which encouraged reflections on the rational use of resources. And it was about authority: the claim to the forests taught respect for the state, its officials, and the written law. It all came down to a powerful amalgam of knowledge and sovereignty: forestry made the state, and the state made forestry.[8]

Of course, competing uses were not irrational but merely different. For one, they were different in geographic scope: while states sought to stake a claim to all trees within their jurisdiction, most people were focusing on woodlands in their immediate surroundings. Competing uses were often nonmonetary: they drew on customs and entitlements that had grown over time. Forest use was also variable. People consumed more firewood in a severe winter, and the demand for building material changed in the aftermath of a disastrous fire (see chapter 25, 1976 Tangshan Earthquake), a type of event that premodern cities had to reckon with.[9] In fact, forests were a treasured resource precisely because they could tolerate fluctuations in the intensity of use: they were a kind of insurance, a buffer against all sorts of disasters. Many people cherished the forests precisely because they did *not* know what they would need the forests for.[10]

The claim of the state was up against pressing needs and long traditions, and as enforcement goes, forests were a difficult case. Officials could not guard every tree, and policing looked arbitrary without a powerful rationale. The legitimacy of sustainable forestry hinged on a veritable horror scenario, the prospect of an imminent scarcity of wood that only rational management under the aegis of the state could forestall. The specter of a "timber famine" struck a nerve in a society based on forest resources, and it allowed the state to occupy the moral high ground: it took a long view of people's needs while individuals focused narrowly on short-term gains. Officials never grew tired of warning about a coming "timber famine." Carlowitz's *Sylvicultura oeconomica* referred to it almost sixty times.[11]

The trope became the shibboleth for foresters all over the world. From Sweden to South Africa, officials depicted very different forests as uniformly under threat, with the only hope being farsighted management by the state.[12] The ensuing struggles could last for generations: in the northern Indian state of Uttarakhand, which won global fame when the Chipko movement pitted tree-hugging women against commercial forestry in the 1970s, conflicts were already simmering after the government asserted proprietary rights over "district protected forests" in 1893.[13] While the flight to the cities took much of the heat out of forest conflicts in Central Europe, struggles in the colonial world grew in intensity in the late nineteenth century when colonial forestry turned from a mere idea into an institutional reality. Authorities passed new laws and created academic institutions such as Dehradun in India, all with a view to commercial as well as political goals: they sought revenue and control of unruly people who used the woods as a place of refuge.[14] One does not have to go as far as Robert Harrison, who depicted forests as the antithesis of civilization.[15] But forestry was clearly about the state of the woodlands in more than one sense. It was about rationalization, order, and the spread of the monetary economy.

Germany played a key role in the global career of sustainable forestry. The country exemplified the European idea of statehood, and it had academic institutions of transnational fame. Bernhard Fernow, chief of the Division of Forestry within the US Department of Agriculture since 1886, was born in Germany and received a thorough training in forestry before moving to the United States while his better-known successor, Gifford Pinchot, had enrolled in the Ecole Nationale Forestière in Nancy, France, whose directors between 1825 and 1880 were all German-trained.[16] India's inspector general of forestry, Dietrich Brandis, had a German doctoral degree, as did his successor Wilhelm (William) Schlich, whose five-volume *Manual of Forestry*, "the epitome of forest practice in India and the rest of the British Empire," included two German forestry books in translation.[17] Germany launched an afforestation program when it seized the Kiautschou Bay concession from China in 1898.[18] Carlowitz played a marginal role in all of this, as the crucial events happened after the end of the Seven Years' War in 1763: the creation of academic institutions, often in remote places with abundant woodlands such as Tharandt in Saxony and Eberswalde in Prussia, and the prolific writing of a golden generation of scholars in the early nineteenth century, which textbooks now celebrate as the "classics" of German forestry.[19] The *Sylvicultura oeconomica* merely supplied the buzzword.

Timing and tree types differed, as did the real situation out in the woods. Historians have had bitter controversies about whether the timber famine was a hoax or a reality, which is hardly surprising given that the issue at stake was the founding myth of a global profession.[20] State control varied enormously in intensity, and it was little more than a carefully cultivated fiction in some realms.[21] But when the state cracked down on customary forest use, it was typically the poor who were suffering the most. States sought to make money with the forests, and they had little patience with those who did not have any. They did not enjoy critical reporting either. The young Karl Marx ran into trouble with the Prussian censors when he published five articles in the *Rheinische Zeitung* on a law against wood theft in the fall of 1842.[22]

Sustainable forestry is an unfinished project, all the more so as its accord with the state also worked in reverse: a crisis of state authority immediately fell back on the forests. When the legitimacy of Eastern European administrations crumbled after the fall of Socialism, illegal logging became an endemic problem. Sometimes the skeleton of state control remained intact, if only to provide camouflage for the enrichment of corrupt officials, as in Romania. And sometimes control vanished entirely. When Albania slid into chaos and violence in 1997 after the collapse of Ponzi investment schemes, people took to the streets as well as the woods, cut down trees, and sold them for quick money.[23] The technological means were modern, but the idea was not. Once again, forests served as the people's insurance for difficult times.

2. SPECIALIST TREES, SPECIALIST MINDS

German professors are powerful creatures. When they invite their institute to a special event, the underlings will duly oblige. Karl Escherich, a professor of forestry at the University of Munich, brought his folk to watch an up-and-coming local politician whose fiery speeches were the talk of the town in early 1920s Munich. Escherich particularly enjoyed the way the speaker phrased his concerns in simple terms, an underrated skill in German academia.[24] Escherich's infatuation became something of an embarrassment when his favorite politician, whose name was Adolf Hitler, turned from talk to action, launched an amateurish coup in the fall of 1923, and landed in prison. But some blunders can look like clever career moves after a while. When Hitler was the dictator of Germany ten years later, Escherich became the first Nazi-appointed rector of Munich University.[25]

It was common after 1945 to ignore these allegiances and focus on scholarly credentials, and Escherich already had an impressive academic career when Hitler was still lounging around in Vienna.[26] Munich was his third professorship after seven years in Tharandt and a brief stint in Karlsruhe. When he moved to Munich in 1914, he came with the programmatic goal to push entomology, the study of insects, toward applied research.[27] He founded the German Society for Applied Entomology and launched the field's defining journal. His academic work drew on excursions to exotic places, with Ceylon, Brazil, Eritrea, and the Island of Djerba, today part of Tunisia, among the destinations.[28] As befits a German professor, he summarized his wisdom in a four-volume magnum opus on the insects in Central European forests.[29] In short, Escherich was no academic lightweight, and yet his academic work is no less ambiguous in retrospect than his political allegiances. There was a reason that insects were such a hot topic in early twentieth-century forestry.

German forestry did not leave it at securing its claim to the woodlands. It aimed to increase returns and thus intervened in natural regeneration. Foresters compared growth rates of various trees and conducted experiments with imported specimens (see chapter 27, Eucalyptus). In many cases, their decision was to focus on a single tree that promised the greatest returns and plant identical stands of the same age. But as it turned out, these plantations, typically fast-growing conifers, were much more vulnerable to pests. Forest insects usually had a favorite tree, and they multiplied quickly in monocultures. That is where Escherich came in, for that was the meaning of the word "applied" in the name of his academic field: it suggested, in the crudest terms, that entomologists should not just study insects but also find ways to kill them.

Monoculture was never a dogma in German forestry. Another Munich professor, Karl Gayer, wrote an entire book about "the mixed forest" a generation before Escherich, with the argument being ecological as well as economic: since markets and demands were in a state of flux, it seemed like a good idea to bank on more than one tree.[30] Another generation back, Wilhelm Pfeil, the maverick among the founding fathers of German forestry, urged foresters to look carefully at the local environment and "interrogate the trees."[31] In the nineteenth century, the rationale for coniferous plantations was often remarkably ambiguous: they came across as a temporary solution, a remedy for devastated soils that seemed to leave no other option.[32] There were the forests as they were and the forests as they should be, and both had their own set of rules.

It was a state of mind that bordered on the schizophrenic, and Escherich provided a case in point. In a 1935 speech as rector of Munich University, he praised the mixed-forest *Dauerwald* concept, as its biological diversity increased the forest's resilience to biological threats.[33] But his academic field, applied entomology, was a classic example of a reductionist discipline. Pathogenic insects were obviously part of a much bigger ecological community, but reflections on the wider context became obsolete if applied entomology offered effective solutions for insect woes. Specialist disciplines call for specialist skills, and Escherich turned from mixed-forestry dreams to a staunch defense of a distinct academic profile when it came to appointing his successor. He felt that professorships in forest entomology should by all means go to candidates with a thorough training in zoology, and he was shocked that some candidates under consideration for his chair came from a general forestry background.[34] It was the fight of a lifetime, and it focused on the classic issues of academic specialization: crucial skills, career paths, and jobs. To his relief, Escherich's chair ultimately went to one of his former assistants.[35]

Escherich's research won acclaim beyond Germany's borders, and he received awards like the honorary membership of the Swedish Entomologiska Föreningens in 1942.[36] German forestry is best seen as part of a transnational web of knowledge, and inspirations could travel in different directions. Carlowitz went on a Grand Tour of Europe (see chapter 22.1, Manual for a New Age) before settling into his mining job in Saxony, and Escherich experienced nothing short of an epiphany abroad. It was that galvanizing experience of twentieth-century academia, a visit to the United States, that made him a lifelong crusader for the practical application of entomological expertise. Andrew Carnegie paid for an extended study trip along the network of the federal Bureau of Entomology in 1911, and the means and the spirit of the American fight against insects made a lasting impression. Escherich was particularly inspired by the campaign against the boll weevil (see chapter 12, Boll Weevil), which he learned about at the Dallas branch of the Bureau of Entomology: here he found a kind of optimism that was dearly lacking among his German colleagues.[37] The idea of a German Society of Applied Entomology came to him after his departure from Dallas, allegedly while watching the Texas steppe from a dome car.[38] He explored biological methods and aerial spraying campaigns and instructed foresters on how to use them, and he confidently declared upon his retirement that the major pests, heretofore feared among forest owners, were now "pretty much under control."[39] But at the end of the day, these remedies never quite satisfied Escherich's penchant for more holistic solutions. In his memoirs, he described aerial spraying as a makeshift for the next few decades until an ecologically sensitive forestry would have restored "the broken balance" in Germany's woodlands.[40]

Biology never played much of a role in making the case for plantations, but other rationales made them more alluring. Calculations of future yields were much easier in identical stands, and these calculations were dear to the hearts of German foresters. They allowed sustainability to be defined in precise numbers (see chapter 7.2, Numbers Games), and Paul Warde has called sustained-yield theory "the cornerstone of modern forestry."[41] A faith in numbers spread within the profession, and authorities like Georg Ludwig Hartig, another founding father of German forestry, filled hundreds of pages with instructions for the collection and compilation of data.[42] And then there were other motives that Prussian foresters were merely whispering about. Planting conifers cost money for seeds and workers, and Pfeil found these expenses so high that contractors routinely made a nice profit.[43]

The single-plant forest followed on the heels of the single-purpose forest. Timing and circumstances differed, but the hope for economic returns and jobs, combined with the self-interest of state agencies to assert their authority, made forests a political concern long after fears of wood scarcity had lost their existential dimension. In the postwar years, Scotland experienced a government-sponsored "planting bonanza" with Sitka spruce and other American conifers (see chapter 27, Eucalyptus) that increased the forested land from 6 percent in 1960 to 17 percent in the new millennium.[44] The economic benefits for remote areas were not what the sponsors had hoped, though, and Margaret Thatcher pulled the plug on supportive tax breaks for plantation forestry in 1986, but that did not end the allure of forestry for the powers that be. The Forestry Strategy of the devolved Scottish government aims for woodlands on 25 percent of land area in the second half of the twenty-first century, and the sins of the past provide merely another rationale for government intervention: the government recognizes "negative environmental legacies" such as "poorly designed forests that have yet to be 'restructured.'"[45] Plantations provide opportunities even in failure, and applied entomology is not the only endeavor that drew benefits from the need for repairs. The stabilization of frail monocultures is the province of specialized academic research while the management of more resilient mixed stands typically calls for broader minds.

In Scotland, environmentalists and foresters seek to overcome the legacy of monoculture, but their goal is anything but new. As Oliver Rackham has observed, "Objections to plantations are nearly two centuries old."[46] One of the most ambitious attempts to shift course came in Nazi Germany when Hermann Göring appointed an advocate of

the *Dauerwald* doctrine, Walter von Keudell, as head of his national forest service in 1934. However, the rank and file tried its best to stall, the demands of the war economy created more imminent needs, and von Keudell finally made way for a more compliant successor after three hapless years.[47] Aldo Leopold, who came to Germany on a Carl Schurz fellowship in the fall of 1935, was unimpressed by what he saw and wrote a disillusioned article on "deer and *Dauerwald* in Germany."[48] Just like Scotland, postwar Germany saw another push for coniferous monocultures, many of which later succumbed to winter storms and acid rain.[49]

But even where academic wisdom or good luck brought a plantation to maturity, chances were that the intended demand did not exist anymore. Markets were still rather stable when the founding fathers of German forestry were writing their seminal books in the early 1800s, but that was bound to change. The transport revolution of the nineteenth century (see chapter 3, Canal du Midi) broke up local supply networks, and Germany received an increasing amount of timber from the forests of Norway and the Baltic Sea region.[50] Coal provided an alternative to firewood, but underground mines needed huge amounts of high-quality timber.[51] The oak trees in the Forest of Dean in Western England, placed under special protection by an Act of Parliament upon instigation of Admiral Nelson in 1808, matured when the Royal Navy was no longer building wooden ships.[52] In the twentieth century, autarky regimes and pillaging in times of crisis made for additional complications while affluent societies developed yet another layer of perspectives on forest resources from pulp and paper to plastic furnishings. Many conifers were planted in an age of wood-burning stoves and harvested for IKEA bookshelves.

Sustainable forestry thrived on the optimizing spirit that defined modernity's approach to organic resources from hybrid corn (see chapter 28, Hybrid Corn) to battery chicken (see chapter 36, Battery Chicken), but the slow growth of trees gave the experience in the woodlands a peculiar twist: calculations came to naught when parameters changed over the years. Shifts in commercial uses and environmental conditions have made a mockery of many a forester's intentions, and in the twenty-first century, the change of local climates in the wake of global warming will likely emerge as yet another realm of complications. Carefully planned forests turn into living monuments for the visions of the past, and yet it remains an open question whether anyone cares to read them as such. Rackham has pointed out that calculations on the profit margins of British forests suffer from one cardinal problem: landowners who set up plantations rarely bothered to keep the paperwork.[53]

3. TREES WITHOUT FORESTERS

Distance helps if you cannot see the forest for the trees. The African Sahel, the semiarid zone to the south of the Sahara, was firmly etched into the world's conscience for devastating famines since the 1970s (see chapter 31, Holodomor) until satellite pictures offered a different perspective.[54] They showed green, and ever more of it. Rainfall had grown more abundant since the drought years, and scientists were curious as to what kind of green was growing. They looked closely at shades of green and seasonal variation, but it took a combination of satellite observations and perspectives from the ground to get a clear picture.[55] As field research goes, the Sahel is a less than perfect destination, but the results were of interest beyond the region. They provided a glimpse at a possible future of the woodlands: trees without foresters.

The Sahel had departments of forestry since colonial times, and they jumped into action when starving children made global news and international aid poured in. Foresters set up nurseries and launched tree-planting campaigns in order to halt the advancing desert.[56] It was fairly successful in generating work for foresters and their affiliates, and millions were spent on planting, weeding, fencing, and guarding seedlings. It was less successful in its results, as many trees did not grow to maturity.[57] That changed with the entry of farmers, who acted in what might be called the spirit of Pfeil, though it was probably akin to what farmers had been doing ever since the Neolithic Revolution: they looked at the local conditions and the trees at their disposal and worked from there.[58]

Trees offer a number of benefits to agriculturalists in the Sahel. They create favorable microclimates, increase soil fertility and crop yields, enrich local diets with fruits and seeds, and provide firewood and construction material. Furthermore, they do not create fateful dependencies on expensive inputs such as fertilizers (see chapter 8, Guano, and chapter 19, Synthetic Nitrogen) or commercial seeds (see chapter 28, Hybrid Corn).[59] Just like the forests of early modern Europe, trees in the Sahel "serve as a safety net in times of crises": they increase the resilience of farming in places that are only a few steps away from disaster.[60] Farmers do not even need to plant trees: they can grow from live stumps, though pruning and protection demand attention and skill. The practice, commonly called farmer-managed natural regeneration, spread through development initiatives, demonstration projects, and word of mouth. In Niger, where promotion started in the Maradi region in 1983, farmers were nurturing trees in one way or another on millions of hectares twenty years later.[61] Such a success in one of the poorest countries of the world showed that farmer-managed natural regeneration does not claim huge resources, but it does hinge on autonomy over land management decisions.[62] No farmer will attend to trees if a forester can come along and claim them as state property.

Of course, approaches and results vary enormously in a zone that reaches across an entire continent. The overall balance can be ambiguous: a region in Senegal experienced a gain in biomass but a loss in biodiversity.[63] Farmer-managed natural regeneration may suffer from a return of the dry years of the 1970s. Neoliberal policies may intervene in unexpected ways, as cuts in fuel subsidies could stimulate interest in firewood.[64] Pressure on trees in the Sahel also depends on the availability of wood supplies from more abundant stocks farther south.[65] And then there is the wider context of a region fraught with instability: the insurgencies of Boko Haram in Nigeria and Islamist militants in Mali have jeopardized forest management and everything else. Agroforestry is no panacea when it comes to responsible stewardship for trees. But then, neither is state management.

Agroforestry in the Sahel reflects a change of tide in the management of the world's woodlands. Sustainable forestry grew in lockstep with an ascendant state, and while local communities could put up resistance, they were usually on the defensive throughout the modern era. But budding administrations have become rare in recent decades, and that has left foresters with a choice between two options. They can rethink traditional approaches and reach out to other stakeholders, which is what foresters are experimenting with beyond the Sahel. In France, ecologically minded administrators even encouraged woodland pasturing in mountain regions, thus pushing for a revival of a custom that two centuries of French forestry sought to expunge.[66] Or they can choose to ignore the erosion of state power and cling to what they have. Opinions diverge among foresters and even within departments. A young Nigerian official spoke out in support of community participation at a national workshop in 1991, only to be reprimanded by a senior forestry administrator who apologized on his behalf: "He is a young man who does not know that forestry is ultimately about power and control."[67]

The authoritarian spirit of sustainable forestry is not dead, and its future depends on whether it can latch on to a powerful political cause. It could be the quest for energy, which may stimulate interest in renewable supplies like wood. It could be the sequestration of carbon in the fight against global warming: reforestation in the Sahel was barely an established scientific fact when climate researchers were already out calculating the region's potential as a carbon sink.[68] Carbon capture may become big business in the twenty-first century, and given the experience with twentieth-century development projects (see chapter 32.3, Planning Development), it would not be surprising if these projects gravitate toward regions like the Sahel, where people have few means to make themselves heard. Or maybe politicians will just ask foresters to make money and hope that urbanized societies do not care all that much about the state of the woodlands. As it happened, this was the path that the birthplace of sustainable forestry took in the new millennium. Some three hundred years after Carlowitz was struggling to find words, German forestry was firmly in the grasp of neoliberal profit seeking.[69]

5

SHIPBREAKING IN CHITTAGONG

LEFTOVERS

I. THE RECYCLING BUSINESS

It all began with a cyclone. Rough waters in the Bay of Bengal left the *MD Alpine* adrift in 1960, and when the winds calmed down, the ship was stranded in a tidal zone near Chittagong. It sat there for four years, and when the Greek owners abandoned all hopes of getting it back to sea, a local company, the Chittagong Steel House, bought the wreck for scrapping.[1] It proved a business with potential. The nascent Bangladeshi economy could well use a cheap supply of scrap metal, and an empty seagoing vessel is around 90 percent steel.[2] There was also plenty of space on the long, flat beach north of Chittagong. Once under contract for dismantling, ships could steer toward a designated spot along the coast at high tide and speed through the mud until they got stuck. What had been a singular act in the 1960s was a major part of the local economy in the 1980s, and business kept growing. When it was at its peak in 2008, Bangladesh accounted for half of all ships scrapped worldwide.[3]

As founding myths go, shipbreaking in Chittagong was one of the less glamorous industries. But in recycling, glamour was typically the least of all concerns. In a way, an abandoned ship on the beach was the perfect metaphor for the nature of the business: recycling was always about opportunism. Businessmen had to find a resource in demand, an artifact beyond the prime of its use, and a way to turn the latter into the former at competitive costs. Recycling was about the raw creativity of capitalism at work, and sentimental souls did well to look at J. M. W. Turner's painting *The Fighting Temeraire* rather than a real scrap yard.

Melting scrap was usually easier than turning ore into metals, and that made recycling a part of humanity's engagement with metals since prehistoric times. Precious metals were reused throughout the ages with particular care, and those who own silver need to live with the chance that it contains a few atoms from the thirty silver coins that Judas Iscariot received for the betrayal of Jesus. Even in the deeply religious sixteenth century, Catholic kings confiscated ceremonial silverware from churches and had it melted down in order to pay for military campaigns.[4] Nothing was sacred in the recycling business, and certainly not the burial grounds of infidels. In his history of Latin American mining (see chapter 1, Potosí), Kendall Brown has pointed to "a form of macabre pseudo-mining, the robbing of pre-Hispanic graves," which yielded "considerable gold."[5]

Recycling was no distinct branch of business until the nineteenth century. Blacksmiths, jewelers, and other craftsmen were collecting and reusing metals as part of their daily routines until industrialization changed the rules of the game. With the increasing volume of metals in play and the growing distance between owners and users of material, a niche opened for commercially savvy middlemen. It was a business with its own set of challenges, not least its anticyclical nature. Many traders made their best purchases in times of crisis when other businessmen cut capacity and sought a quick dose of cash. On the plus side, the initial investment was low: it took barely more than some wheels, space for storage, and a few tools to go into scrap recycling. It became the province of small- and medium-sized companies, and while large, vertically structured corporations (see chapter 10, United Fruit) eventually gained a foothold in the business over the course of the twentieth century, small businesses have retained a share of the market even when it comes to supersized objects. Bangladeshi shipbreaking is the province of thousands of small and medium-sized firms in Chittagong and Dhaka that form a close-knit community underpinned by family ties.[6]

It was easy to enter the scrap-metal trade, but one had to be flexible to make it. Recycling was also an inevitably dirty business, and that left a stain on its reputation. All this made it attractive for those who carried a stigma anyway. A good part of the scrap-metal trade in France and Germany was in Jewish hands.[7] In the United States, immigrants gravitated toward scrap metals in great numbers during the late nineteenth century, and stereotyping followed suit. Carl Zimring has noted that "by World War I, xenophobia and hygienic concerns produced an image of a stereotypical hook-nosed Jewish junk peddler or swarthy Italian scavenger bent on dirtying the streets and morals of urban America."[8]

Facing discrimination is never fun, but at least clever entrepreneurs had a chance to get rich. The outlook was dimmer for workers: the labor history of the recycling business is an endless tale of misery from the backyards of New York City to Chittagong. Even Bangladeshi workers shunned shipbreaking, and migrants from poor rural regions made up much of the workforce.[9] Some jobs required special qualifications. Those handling the gas torches were typically the best-paid workers, and companies also paid for rhythmic singers who helped casual workers to synchronize steps. But most jobs were about brawn rather than skills, and unskilled workers received meager pay in return for plenty of dirt and all kinds of workplace hazards.[10] Most jobs were also transitory in the recycling business, which did not encourage solidarity. In the United States, unionization of scrap-yard labor did not start until the 1930s, and it is an ongoing process in Chittagong.[11] Being a labor leader in the shipbreaking business is a job with another set of dangers, for some owners "maintain paid gangsters to prevent trade union pursuits."[12]

As the twentieth century progressed, industrialists took a growing interest in efficiency, which had consequences for the contents of waste heaps as well as the structures of the recycling business. In fact, it was the reuse of residues in petroleum refineries that gave birth to the term "recycling" in the 1920s, and the word remained an engineering term until environmentalists appropriated it half a century later.[13] Industrial experts sought ways to use resources more efficiently and closed loops for materials that had previously gone to waste, and the recycling business effectively split into an expert-driven endeavor that large corporations maintained in-house, and the open market where countless independent companies sought to make some money. Some waste reuse endeavors yielded enduring results. In 1940, chemists at the German branch of Coca-Cola cobbled together "left-overs from left-overs" and thus created a new fruit-flavored soft drink named Fanta.[14]

All the while, state authorities remained on the sidelines. They occasionally jumped into action when scrap yards became too smelly or workplace hazards too egregious, but recycling was not a matter that seemed to call for any comprehensive policy. The rise of environmentalism in the 1970s challenged this long-standing negligence, but tighter regulations could play out in different ways. The shipbreaking business went for relocation. Dismantling of ships had traditionally been part of shipbuilding in Western harbor towns, but with shipyards struggling and the prospect of higher costs due to environmental regulation, the industry moved south. What followed was a race to the bottom, as shipbreaking moved to Taiwan and South Korea and then on to India, Bangladesh, and Pakistan.[15] National regulations were essentially there to be escaped from, and the law of the seas presented no headaches at all. An article in *Maritime Policy and Management* noted in 1998 that shipbreaking was "probably the only area in shipping that is unregulated."[16]

Before environmentalism changed the rules of the game, the only times when state governments were interested in recycling were times of war. Nazi Germany sought control over recycling in the 1930s, which nicely dovetailed with its goal to eliminate Jews from Germany's economy.[17] While defending freedom during World War II, Britain's government "made it a crime to discard or destroy ferrous and nonferrous metals, paper, rags, string, rubber, animal bones, and food scraps."[18] Comprehensive resource mobilization was part of total war (see chapter 35, Pine Roots Campaign) while smaller military conflicts stimulated more improvised modes of recycling. During a brief 1940 war between Dubai and Sharjah on the Persian Gulf, soldiers returned fire in the most literal sense by scooping up and reusing enemy bullets.[19]

Some wartime experiences played out in curious ways. The experience of resource scarcity during and after World War II inspired two Dutch women to launch a recycling drive in the early 1970s, which made the Netherlands the first Western European country with a separate glass-collection system.[20] However, most states found it hard to justify recycling after the end of hostilities: France, for one, ended its collection drive in 1947.[21] But while state policies were sporadic into the postwar years, the lower ranks of government, particularly municipal authorities, had gone into the recycling business in their own ways since the mid-nineteenth century.

2. THE WASTE BUSINESS

Nineteenth-century households knew plenty of options when it came to recycling and reuse beyond metals. Stitching and repairing were among the natural duties of housewives. Rags went to collectors who sold them to paper mills. Decaying foodstuff went into thick soups or the trough of domestic animals. Combustible material could go into the stove, and ash was spread on garden plots as fertilizer (see chapter 8, Guano). And for the small amount of stuff that just did not seem to fit anywhere, there was usually a hole in the ground that could take care of things, a time-honored routine that has filled more than one archaeologist with gratitude. The profession has learned a lot from the waste heaps of bygone ages.[22]

The growing cities of the nineteenth century jeopardized some of these routines. A good part of the urban population, and particularly the urban poor, lacked access to garden plots, hungry animals, or a conveniently located hole in the ground, and yet they produced significant amounts of waste. Garbage emerged as a public health issue in the burgeoning cities of the industrial age, and just as with sewer construction (see chapter 17, Water Closet), it was the sheer material heft of foul-smelling masses that pushed urban authorities into action. Sometimes a crisis helped to speed up decision making. Shanghai introduced concrete refuse containers after a cholera epidemic (see chapter 21, Cholera) in 1907.[23]

Cities introduced garbage collection at different speeds. In France, Lyon and Saint-Étienne introduced trash cans in the 1850s while Marseille delayed introduction until 1913.[24] Garbage collection also fell prey to the familiar challenges of municipal politics. In the mid-nineteenth century, New York's city inspector, Alfred White, created a franchise system for waste removal and awarded a lucrative contract to his own dummy company.[25] Half a century later, George Waring turned New York's street-cleaning department into a model force with white uniforms and military-style discipline while fending off attacks from the notorious Tammany Hall political machine.[26] The campaign changed refuse management in the United States and made Waring a national figure, though fame proved an ambiguous blessing for Waring when he won an assignment as special commissioner of the US government to Cuba after the Spanish–American War. He arrived for an investigation of sanitary conditions in Havana in October 1898, contracted yellow fever, and died within a month.[27]

Waring ran a municipal department, but many cities were reluctant to take matters into their own hands and signed contracts with private companies instead. It was not always a matter of principles. In late nineteenth-century France, Socialist city governments refrained from municipalizing trash collection in Lyon and Saint-Étienne while liberal Bordeaux took charge in 1889.[28] But no matter how garbage services were organized, cities invariably discovered one inconvenient truth: trash collection was hugely expensive. Municipal garbage services entered the recycling business in order to recover some of the expenses, which effectively meant a new approach to an existing business. While scrap-metal dealers and rag collectors were cherry-picking wherever trash was produced, garbage services took everything to the outskirts and then looked for ways to use some of the stuff. Waring organized a waste-separation program in New York City and searched for grease, metals, and fertilizer.[29] Oakland, California, used refuse for land reclamation in San Francisco Bay, and Los Angeles maintained a swine-feeding program until 1914.[30] The latter approach faced limits, though, for pig populations grew to astronomical numbers in large cities. When Nazi Germany launched a huge waste for pigs program with centralized feedlots, much of the kitchen garbage spoiled before reaching its destination. Losses ran to 14 million reichsmarks by the end of 1940.[31]

But even with all these efforts, a lot of stuff was waiting to be dealt with, and thus began what Joel Tarr has called "the search for the ultimate sink": authorities sought a final solution for the waste problem and ended up moving from one set of problems to the next, as presumed sinks revealed themselves as merely different ways to circulate stuff (see chapter 19.3, Running Cycles).[32] In the late nineteenth century, Vienna dumped its trash in improvised landfills on the city's outskirts and soon realized that urban growth caught up with these sites.[33] New York City tried ocean dumping, but a lot of garbage washed up on nearby shores, and courts forced the city to end the practice by 1934.[34] Engineers learned to set up sanitary landfills with compaction, stratification of waste, and a protective layer of earth, only to realize that sanitary landfills caused serious groundwater pollution.[35] Incinerators became notorious sources of air pollution.[36] Short-term solutions turned into long-term liabilities, and some innovations even failed on their own terms. In Colombia, an advanced landfill, meant to operate as a bioreactor to accelerate decomposition, collapsed in a major landslide "to some extent because the operators did not appreciate that increasing the liquid content of the waste would drastically decrease the stability of the landfill."[37]

Local authorities initially met with resistance when they entered the waste business. When the prefect of the Seine department, Eugène Poubelle, introduced dustbins to Paris in 1883, protests arose from those who searched trash for a living. But a revised ordinance settled the matter after a few months of conflict, and *la poubelle* became the French word for waste containers.[38] It came down to a division of labor: private collectors took whatever they liked, and garbage companies took care of the rest. However, the quality of municipal services differed enormously from place to place, and the primacy of urban sanitation produced a stark contrast between city and country: what happened beyond city limits was a second-rate issue at best. While urban industries faced restrictions early on, pressure was much weaker farther out, and some industries could operate more or less at will. The shipbreaking industry of Bangladesh, located on the northern outskirts of Chittagong, was one such industry.

3. GETTING ETHICAL

The business of leftovers changed again in the postwar years, and once more, it was about the crude material force of accumulating masses. Affluent societies flourished all over the West, and they changed the nature of the waste problem. Mass consumption increased the volume of garbage, new materials like plastics (see chapter 40, Plastic Bags) entered the waste stream, and low resource prices put pressure on profit margins in the recycling business. Sanitary landfills grew at unprecedented speed, and the search for the ultimate sink entered another round on a new scale. Perhaps worst of all, the challenges were directly related to a new ethic: for the first time in history, throwing things away was supposed to be fun. As Frank Trentmann has argued, "The thesis of the 'throwaway society' was the natural twin of the 'affluent society.'"[39]

Conspicuous consumption had always carried an air of wastefulness, and that was not an inherently negative thing: Veblen's *Theory of the Leisure Class* includes remarks on "conspicuous waste."[40] But when consumerism turned into a mass movement, the critique achieved a new level of urgency. In 1960, the US journalist Vance Packard published *The Waste Makers* and became "the first to popularize a critique

5.1 Dismantlers at work on the beach north of Chittagong. Image, Naquib Hossain / Wikimedia Commons.

of the throwaway society," and waste has been a fixture in the environmental discourse ever since.[41] It made for an additional layer of complications in the management of waste. Gone were the days when one could make recycling decisions solely based on costs and benefits: for many people, recycling routines became habitual commitments to a conservation ethic. Businessmen and policymakers had the technology for all sorts of leftovers, but dealing with sentimental citizens was a challenge in its own right.

The garbage crisis provoked a number of obvious responses: separate bins for different types of waste, fines for illegal dumping, and a plethora of educational campaigns. But it also provoked a new spatial order of trash that escapes the eyes of Western supermarket users. Environmental sentiments in urban areas increased the costs of waste disposal, and trash began to gravitate toward places where opposition was more muted. In other words, trash got on the move. In the United States, thirty-two million tons of municipal solid waste crossed state lines in 2000.[42] Other waste streams crossed national boundaries. Shipbreaking was just one of the more glaring examples.

In other words, affluent societies did more than simply increase the volume of refuse material. They also led to a transgression of spatial boundaries: today's waste business is truly global. It was a silent process, driven by material exigencies rather than concepts, and it defied rationalizations that one could share in public. During his tenure as chief economist at the World Bank, Lawrence Summers signed a memorandum that argued for more dirty industries in less developed countries, on the grounds that African countries were "vastly under-polluted," a phrase that triggered outrage when the memo was leaked.[43] Inequality and discrimination has always been part of the waste business, but it took on a different quality when it played out globally and found a justification, or the attempt thereof, in the ice-cold language of global capitalism. In the United States, waste sites in the proximity of poor African American neighborhoods gave rise to the environmental justice movement.[44]

Diplomacy (see chapter 24.3, Getting Serious) took several shots at regulating the global shipbreaking business. In 2003, the International Maritime Organization adopted guidelines that looked at the full life cycle of ships from construction to dismantling, but as nonbinding recommendations, the guidelines were only a declaration of good intentions.[45] The Hong Kong International Convention for the Safe and Environmentally Sound Recycling of Ships, adopted under the auspices of the International Maritime Organization in 2009, is more ambitious, but it has yet to enter into force.[46] The Basel Convention on the Control of Transboundary Movements of Hazardous Wastes and Their Disposal entered into force in 1992, but unscrupulous traders have found ways around the treaty, and it has not stopped ships from beaching at Chittagong with asbestos and other toxic materials on board.[47]

It was a classic clash between high ideals and material needs. It had more than a whiff of neocolonialism to store the leftovers of Western affluence in poorer countries, and yet it was just so convenient. Governments were torn between moral commitments and realpolitik when negotiations on the Basel Convention were underway, and it was not just about the usual suspects. One of the loopholes was due to lobbying from the Federal Republic of Germany, a self-styled environmental leader since the 1980s.[48] Like other industrialized countries, West Germany suffered from a shortage of landfill capacity in the late 1980s, and the cash-strapped GDR was glad to provide an outlet. It was a thriving business that both sides wanted to retain, and the governments of the two German states secured an exemption for bilateral deals. The Basel Convention was adopted in March 1989. The Berlin

Wall fell eight months later, and within a year, West Germans were reunited with their trash.[49]

The flow of wastes was usually hidden from sight, not least because this served the interests of pertinent companies. But garbage stories can sell on the media market, particularly when waste is really yucky or otherwise spectacular and a cast of seedy characters is at hand, though it usually takes investigative reporting to get to the bottom: the muckrakers did not get their name for nothing. However, exposure of unwholesome conditions is merely a first step toward change. The plight of shipbreaking workers has been in the press since a series in the *Baltimore Sun* in 1997.[50] The authors won the Pulitzer Prize for investigative reporting for revealing "the dangers posed to workers and the environment when discarded ships are dismantled," but political consequences were a different matter.[51]

Regulation of shipbreaking is an ongoing endeavor on different levels. The European Union adopted a regulation on ship recycling in 2013 that includes certification of safe and sound facilities within and beyond the EU, and since 2019, dismantling in a certified facility is mandatory for vessels flying the flag of an EU member state. No Bangladeshi yard had filed an application by November 2021, but yards in Turkey, China, and India had applied, and the list included twenty-six certified facilities in EU countries, eight in Norway, one in Northern Ireland, eight in Turkey, and one in the United States.[52] In Bangladesh, a patchwork of environmental and labor laws has created a system of certifications and permissions, though enforcement by government agencies has been less consequential than public interest litigation from the Bangladesh Environmental Lawyers Association.[53] The association won a landmark ruling from the Bangladeshi High Court in 2009 that "directed closure of all shipyards that operated without clearance from the Department of Environment."[54] The industry replied with a human chain demonstration that brought thousands of protesters to the street.[55] It was more than fearmongering when Chittagong's shipbreaking industry warned of the loss of jobs in the wake of tighter environmental regulation. Bangladesh only became the world leader in shipbreaking after a decision of the Supreme Court of India called for higher standards in 2003.[56]

Shipbreaking in Chittagong remains contested on multiple fronts, but everyone agrees that it provides spectacular pictures. For critics of global capitalism, the beached giants of the sea, eviscerated to the bone by men and children in ragtag clothes, became "icons, circulating a counter discourse of antiglobalization and contesting the logics of capital," but Chittagong's ships also provided the setting for a scene in the superhero movie *Avengers: Age of Ultron*, which ranks among the dozen highest-grossing films of all time.[57] Both readings are apocalyptic in their own peculiar ways, and both struggle to capture the full reality of shipbreaking: from the perennial search for the ultimate sink to the workers maimed and killed and the toxic mix of heavy metals, asbestos, and persistent organic pollutants (see chapter 38, DDT) that percolate into beaches, bodies, and food chains.[58] But then, the business of leftovers has always been about being selective.

PART II

APPROPRIATIONS

PILLAGING BECAME MORE COMPLICATED THAN IT USED TO BE.

The previous chapters have discussed how resource allocation changed over the course of modern history. This section takes a closer look at the institutional scaffolding that underpinned the endeavor. The modern world economy hinged not only on superior scope, scale, and speed but also on an amenable context. In the age of global modernity, appropriation was not just about taking what was needed: it involved a complex patchwork of agencies, organizations, laws, and cultural tropes. Some of them, like property laws, were created consciously. Others, like numbers, merely came into use over time. Some innovations were about the reduction of complexity, like vertical integration in large companies. Others merely increased the unknowns, as the enduring mysteries of human tastes serve to attest. The one thing that the following innovations had in common was their geographic origin. They were all born in the West.

Institutions have received a lot of attention in discussions about the rise of the Western world. Niall Ferguson has even argued that Western power grew from six "killer apps," one of which, property rights, bears a certain resemblance to one of the following chapters.[1] This section tells a less heroic story, and not just because environmental historians have a legitimate interest in the price of progress. For one, it explores more specifically who was winning when "the West" gained the upper hand: from those who held titles to land to the expert groups that flourished in the wake of challenges. It also gives due consideration to competition between the different powers of the West: the breadfruit project thrived on the rivalry between France and Great Britain in the Caribbean. In fact, the superiority of Western approaches was probably to a significant extent about what they *failed* to resolve. Some of the "killer apps" worked because they did *not* deal with certain things. There was a tension between the empowering nature of land ownership and rural inequality that did not lose its political volatility until urbanizing societies lost interest in land reform. The world's oceans remain largely unregulated, and the same holds true for the botanical exchange, fertilizer use in agriculture, and food choices—not to speak of the liberties taken by large corporations such as United Fruit.

The new institutions worked best when people forgot that they were actually new. The land title is so common today that few people in the West recognize how exceptional it really is. For most of human history, people held multiple overlapping claims to land rather than a single registered title that gave the legitimate owner exclusive authority. Land titles had no built-in incentives for long-term management and thus opened the door for environmental abuse, but they offered a simplicity that facilitated commercial transactions. They also meshed with a Western penchant for the written law and a sovereign state. Even Indigenous people and conservationists learned to play by the rules of the land, if only for lack of choice.

Landownership encouraged the individual pursuit of happiness—some argued that it was a crucial requirement—but the egotism of landowners was also a source of problems. They were particularly glaring in the burgeoning cities, and urban authorities acquired a range of legal tools to keep them in check such as compulsory removal of waste (see chapter 5.2, Waste Business) and sewage (see chapter 17, Water Closet). Agricultural land use had environmental repercussions as well, but regulation arrived only in a somewhat haphazard way toward the end of the twentieth century. Out in the countryside, land titles were first and foremost a problem of social justice. Ownership patterns fortified socioeconomic hierarchies, and a more equitable distribution of land was under debate whenever societies forged a new social contract. Few countries made it through the modern era without a serious debate over land reform.

Some countries enacted sweeping reforms while others merely discussed until people grew tired. Several trends took much of the heat out of struggles over land distribution. Landed elites gradually lost their grip on political power even without expropriation. The flight to the cities made landownership a less contentious issue. Those who remained in agriculture could buy or lease land on burgeoning markets. Agricultural technology shifted the focus from ownership of land to sophisticated use. Soviet collectivization tarnished the quest for land reform in the West, and the Cold War delivered the coup de grâce. However, conflicts continue to simmer beyond the Western world, and efforts range from land occupations to state policies endorsed by the World Bank.

Just as land does, food allows for many different systems of classification. One set of categories was about social status. When the British brought breadfruit to the Caribbean, it was conceived as slave food, a means for plantation owners to lower the expenses for feeding their slave populations. Few people thought about the views of the slaves on these matters, but as it happened, Caribbean slaves refused to eat breadfruit, and the tree-borne fruit ended up as pig feed. Breadfruit eventually came into use for human consumption over the course of the nineteenth century, but the stigma lingers: the fruit is still waiting for a culinary fashion that might catapult it into a new orbit. As it stands, the breadfruit remains stuck at the bottom of the food hierarchy, an asset in reserve for regions that lack food security. People continue to eat breadfruit, but they typically abandon it when something else becomes available.

Taste and costs have always mattered for food choices, but numbers did not figure prominently until the modern era. Now every food item comes with precise information on caloric value and amounts of vitamins and minerals, notwithstanding numerous critiques of quantified nutrition. Numbers are an indispensable language of modernity, a transcultural marker of objectivity and neutrality, a claim to authority, and a mode of crisis management, as Theodore Porter has argued. Whether quantification helps to deal with problems is subject to ongoing debate, but some statistical correlations have acquired the status of laws. The correlation between income levels and food expenditures that Ernst Engels identified in the mid-nineteenth century is one of the most robust laws of economics.

The breadfruit story also shows the power of the institutional network that allowed species to travel the world. Plant transfers are probably as old as human civilization, but botanical gardens in combination with potent navies greatly expanded the scale and speed of the exchange. Sometimes things did not go as planned—the mutiny on the *Bounty*, by far the best-researched chapter of the breadfruit story, is part of postimperial lore—but a recurring blunder was due to the botanists' having neglected to transfer cultural scripts along with the specimens. The breadfruit transfer also showed a striking contrast between ironclad commitment to the core mission and freewheeling exchange of other plant species over the course of the long journey. Biological improvements were due to careful planning by key figures like Joseph Banks as well as sheer chance.

Institutions were also crucial for the construction of a commodity chain that brought guano from Peru to European and North American farms. Many things had to fall into place: transport services, credit arrangements, sales networks, and a workforce that dug guano on the tiny Chincha Islands in the Pacific. The flourishing guano trade inspired the colonial powers to search for deposits elsewhere, and Spain, formerly the imperial hegemon of Latin America, tried to seize Peru's islands by force. The United States started its first foray into imperialist acquisition of land in the quest for guano, though material returns were meager and the legal quagmire enormous. Peru's guano was the defining commercial fertilizer for decades, the first mass-produced commodity of its kind and an innovation with repercussions far beyond the commodity chain.

Guano allowed richer harvests, but it also brought material and intellectual dependencies to the farm. Agriculturalists had to use commercial fertilizer on a regular basis, and they needed information on dosage and timing that defied traditional agricultural knowledge. Farmers relied not only on a material supply network that spanned the world but also on input from experts and advisers. It was a step with profound implications: fertilizer decisions were among the first business decisions that agriculturalists placed into the hands of people beyond the farm gate. It fostered networks and liabilities that became a defining part of modern farming. Today's agriculturalists are invariably caught in a web of material supplies and flows of information that is a far cry from yeoman dreams of peasant autonomy.

However, the web of institutions was based on a finite material resource, and the collapse of Peruvian guano production in the 1870s plunged the country into chaos. But unlike other mineral resources, guano came back: a science-based bird conservation regime allowed production of guano as a renewable resource. However, sustainable guano came to compete with fishmeal production in the postwar years, as both were tapping the same resource—the abundant maritime resources of the Humboldt Current. The government favored fishmeal and suspended the bird conservation program in the 1960s, but the shift of policy soon came to look ambiguous. Fishmeal production ran into trouble when the anchoveta population off the coast of Peru collapsed in the El Niño of 1972.

Institutions mattered even when they implied the absence of rules. The open seas were not a state of nature—it was a transnational construct that grew to a significant extent out of whaling conflicts. Policing whaling ships was not particularly difficult in principle, as the size of pelagic whaling made it hard to hide from scrutiny. It took large ships, dedicated crews, and a lot of capital to hunt cetaceans beyond coastal waters, and it all hinged on commercial networks that brought products from the hubs of whaling to end users. Hunting grounds changed, and so did the nations that dominated global whaling. Basques, Americans, Norwegians, and Britons were among the leaders at one time or another, but the legal framework remained remarkably stable into the postwar years. National sovereignty ended three nautical miles from the coastline, and much of the recent change was about pushing that line farther out into the sea, but that did not serve the goal of expanding the protection of maritime resources.

Whale products were diverse, but they had one thing in common: it was not easy to recognize that they were whale products. Oil for illumination and lubrication, baleen for umbrellas and corsets, leather gloves from whale skin—whales were simultaneously omnipresent and invisible in modern societies. When hydrogenation technology turned whale oil into raw material for margarine and soap, whales became virtually impossible to discern as a commodity. Creating invisibility in commodity chains was a major achievement of modernity, a remarkable countertrend to the growing interconnectedness of the globe. Or maybe invisibility made global interconnections easier? Either way, invisibility hinged on the use of modern technology. Invisibility opened the door for overuse, but that was a minor concern in the whaling business until far into the twentieth century. Scholars have taken a different view, and the historical literature has shown an inclination to depict cetaceans on the brink of extinction for many different times and places.

Knowledge about stocks and reproduction cycles has always been fragile, and specifying when global whaling moved beyond sustainable levels remains a matter of speculation. The one certainty is that the whaling business was long beyond its prime when Greenpeace and other environmental groups began to seize on the issue. It led to a worldwide ban that activists defend fiercely, but that says more about human sentiments and the power of iconic pictures than about marine ecologies. The environmental toll from commercial fishing is far greater than that of whaling since the 1980s, if not much earlier, a result of the factory-freezer trawler that revolutionized global fishing in the postwar years in the same way that the whale hunt changed with the introduction of harpoons and stern slipways. Whaling is mostly a thing of the past, but its business model continues to rule the waves.

The final chapter in this section discusses an innovation that shaped resource allocation worldwide: the large, vertically integrated corporation. Companies like United Fruit controlled entire commodity chains, and a fine-tuned administrative machine realized economies of scale, scope, and speed that pushed other firms to the margins or out of business. Large corporations lacked inherent limits to growth: expansion was not a problem as long as corporate structures and hierarchies grew accordingly. Or so it appeared from inside the company, for new corporate giants like United Fruit were open to criticism from more than one side. However, consumers liked cheap bananas, and they typically failed to recognize (or care) that others paid a hefty price for the cornucopia. Open flow of information was crucial inside the large corporation, but it came to the outside world in measured doses with carefully calibrated spin, and consumers bought it. For all the organizational prowess of vertically integrated corporations, it might have been in vain if it had not been for consumer amnesia.

United Fruit became infamous for its business practices in and beyond Latin America, but resistance from humans was only one of the challenges. The banana business also wrestled with the characteristic challenges of monoculture, particularly in the form of soil-based fungi that ultimately forced United Fruit to shift to a new banana cultivar—a change that required significant adjustments in technology as well as a comprehensive consumer education effort. The new banana looked different and tasted different, but the marketing department made sure that it found grateful buyers. It was a feat that only the visible hand of management could achieve.

United Fruit changed its name and now operates as Chiquita Brands International, but it matters more for the company's image than for its mode of operation. Large, vertically integrated corporations continue to rule the banana business, including its organic branch, and while there are opportunities for externals at selected places of the commodity chain, the bigger question is about the resilience of the corporate behemoth in this and many other parts of the resource business. Maybe the strength of United Fruit and numerous other corporations is not so much about size and performance as about a lack of alternatives.

United Fruit no longer operates in the manner of an occupying army, and yet a different type of organization, more transparent and less greedy, is nowhere to be found. The same holds true for commercial fertilizers, the open sea, land titles, and our infatuation with numbers: none of this evokes a sense of pride in the twenty-first century, and yet we are stuck with them for lack of something better. It may be the ultimate irony about the institutional scaffolding that underpinned appropriation in modern times. The modes of appropriation have appropriated us.

6

THE LAND TITLE

TO OWN A PLACE

I. ORDER IN THE LAND

In 1967 Stanford Research Institute submitted a report on the state of agriculture in Ethiopia. Written for the country's government with support from the United States Agency for International Development (see chapter 32.3, Planning Development), it highlighted a broad array of problems that curtailed agricultural productivity. Fertilizer use (see chapter 19, Synthetic Nitrogen) was marginal, seeds were inferior (see chapter 28, Hybrid Corn), land was underutilized, and capital was scarce. But the advisers, who were about to draft a program of agricultural improvement, were also pondering a more fundamental issue: who actually owned the land?

The answer was that there was no answer. Inquiries into Ethiopian property structures uncovered a maze of entitlements, tributes, and tenancy agreements, many of them unwritten; titles were held by the government, the Orthodox church, district chiefs, tax collectors, officials paid in land rather than salaries, and private persons. "The present multiplicity of land tenure systems and diversity of taxes levied on land can, as a rule, be traced back to feudal and military services originally paid to landlords and the Government for varying rights to use land," a primer on Ethiopian land tenure noted.[1] The advisers could not envision a program of agricultural improvement without more clarity on this issue. They called for cadastral surveys to be made "expeditiously," and after that, "steps will probably have to be taken by the Government to simplify the ownership situation by extinguishing many instances of multiple titles, even if this necessitates compensating those who must give up their titles."[2]

As so often in development projects, a clash of civilizations lay at the root of the problem. It would have been easy to resolve the issue in the United States: all it would have taken was a trip to the land registry, where information was on file about the owners of the land. But in Ethiopia, land rights were a patchwork of different titles—rights to use, collect taxes, harvest certain products, and so on—that had accumulated over time, and in that, Ethiopia was arguably closer to the normal across the ages. In fact, most European countries had had such a patchwork in premodern times. The land title—a simple, unique identifier of the property holder, registered in writing in a government institution—is a grandiose act of simplification, and an act of empowerment for the title holder: "Limited and not always saleable rights *in* things were being replaced by virtually unlimited and saleable rights *to* things."[3]

Several long-term trends came together in this modern marvel. The first was the European penchant for written law. Codification could serve a variety of purposes—England's *Domesday Book* of 1086, the first comprehensive survey of landed wealth in medieval Europe, was ultimately about the legal and fiscal basis of the Norman dynasty[4]—but it is hard to imagine the trust in land registers without a centuries-old tradition of legal titles in writing that held force in an independent court of law. The second was the capitalist desire to facilitate commercial transactions. Sales and purchases, tenancy leases, and mortgages were all much easier when there was only one party to go to, and easier still when an independent agency served as a clearinghouse with reliable information.[5] The third was the power of the state, specifically its ability to set up land registers, settle disputes in a court of law, and evict unlawful occupants. The land title is a showcase for the interdependence of capitalist individualism and the sovereignty of the modern state. "Property and law are born together, and die together," Jeremy Bentham wrote.[6]

The connection was particularly clear in the colonial world, where property rights were intrinsically linked with matters of sovereignty. From the viewpoint of the Europeans, land was basically there for the taking, with treaties, concessions, and deeds providing for a thin veneer of legality. Settler societies were particularly greedy, as "free land" was their lifeblood, and they were not shy to offer a rationale. In 1823, the US Supreme Court ruled in *Johnson v. McIntosh* that "all the nations of Europe, who have acquired territory on this continent, have asserted in themselves, and have recognized in others, the exclusive right of the discoverer to appropriate the lands occupied by the Indians."[7] The decision became a cornerstone of American law.

Indigenous people resisted as best they could, but nowhere was their defeat as complete as in terms of legal philosophy. If they were lucky, they could retain some traditional rights and territories for themselves, but these titles were invariably phrased in the language of Western land law. We can see this most clearly when Indigenous people fought colonial intrusions in court. Indigenous people did not necessarily lose, but the decision to go to court already implied an act of surrender: in order to make their case, they had to abandon their own ideas about landownership and embrace a foreign and frighteningly egoistic philosophy. As scholars have shown regarding negotiations between the Maori and British settlers in New Zealand, "Colonial attempts at tenure reform necessarily rendered fixed, certain, and simple what was fluid and complex."[8]

Needless to say, the system of land titles took time to evolve. Since the Land Ordinance of 1785, the United States had been committed to dispensing land in the public domain, but proper implementation was a different matter. Throughout the nineteenth century, surveyors struggled to catch up with the drive toward Western lands, wrestling with all sorts of difficult clients: squatters, speculators, estate owners, and cattlemen. Andrew Jackson's land commissioner, Elijah Hayward, appointed to create order and uniformity in the General Land Office, grew so desperate about his job that he ended up as an alcoholic.[9] And then, America at least had a land office. Great Britain did not introduce compulsory registration of land titles until 1925 and did not complete the work for all of England and Wales until 1989.[10]

The meaning of land titles was not uniform all over the West. For instance, opinions diverged on how to deal with the underground. Whereas the English legal tradition allowed splitting the subsurface into multiple layers with different owners, German law knew only one single title so that German landowners invariably held rights down to the center of the earth.[11] Opinions also differed on whether people were allowed to walk on private property. Nordic countries gave hikers unlimited freedom to roam on private land while English aristocrats liked their "no trespassing" signs.[12] And land titles owed much of their power to technologies of subliminal enforcement like barbed wire, which began its global career when cattlemen on the American Great Plains sought a method of fencing that did not require large amounts of wood.[13] But for all these details, there was a convergence on the fundamental. For capitalists of all nations, the concept of one title to land proved impossible to beat, and it was typically a sign of modernization when overlapping or collective land titles were abolished. As Jürgen Osterhammel wrote in his global history of the nineteenth century, "No state is 'modern' without a land registry and the legal right to dispose freely of real estate."[14]

It attests to the power of the concept that when the nature conservation movement emerged all over the West in the late nineteenth century, it readily embraced the concept. Countries developed widely different ideas about what kind of nature was in need of protection, but there was a remarkable convergence on what protection meant in legal terms: conservationists designated a plot of land, secured a legal title, and sought to fend off intrusions as best they could. Some conservationists thought on a grander scale and aimed at the preservation of entire regions, but at the end of the day, most came to focus their attention on specific tracts with protected status: national parks, nature reserves, and inalienable land holdings (see chapter 26, Kruger National Park). In a world of land titles, conservationists were playing by the rules.

Titles facilitated transfer and use of land, but they also encouraged disrespect for the interests of others. That became particularly evident in cities, where unregulated building could have awkward consequences. The public interest as well as the interests of neighbors called for some restrictions on the free use of urban land, and governments came to develop a whole range of rules and regulations: building codes, restrictive covenants, and zoning laws as well as more unorthodox creations such as the "air rights" that New York invented to regulate the growth of skyscrapers.[15] In the countryside, doubts came from a different direction.

2. AGRARIAN REFORM

Clarity about property rights was good for commerce, but it could also serve other purposes. In 1871, a British aristocrat, Lord Derby, urged an inquiry to disprove allegations about an undue concentration of landownership. After several years of counting and accounting, the results were available in print; David Cannadine has called it "the first comprehensive account of landholding in Britain in nearly a millennium." It was not the kind of testimonial that Lord Derby had hoped for, though. It showed that 710 individuals owned a quarter of all land in England and Wales. For the entire British Isles, less than 5,000 people held nearly three-quarters of the land.[16] Henceforth, land reform was high on the political agenda until 1914.

Britain was hardly exceptional in this respect. All over the world, plantations, estates, and haciendas occupied large swaths of land while the masses tried to make do with little if any land. That became a problem as states abandoned feudalism and increasingly based their legitimacy on the individual pursuit of happiness. The Jeffersonian idea of small independent farmers as the backbone of democracy exemplified a grand Western tradition that suggested a deep link between citizenship and the possession of property.[17] "Since the Enlightenment, few ideas about land have been as durable as those concerning land as a guarantor of individual liberty."[18]

Large landowners found it hard to come up with a convincing rationale for inequality. A German scholar, Theodor von Bernhardi, even wrote a book of more than 650 pages about "reasons that have been put forward for large and small landownership" as part of a bid to join the Russian Academy of Sciences, only to end up in double failure: Bernhardi concluded at the end of his treatise that there was no clear answer, and his academic hopes came to naught through intrigue.[19] Explanations for large landholdings ultimately came down to history, and they often looked similar to the Ethiopian case, where landholdings mirrored feudal and military allegiances of bygone eras inherited over generations. In other words, the rationale of large landholdings was usually that some ancestor had been particularly brave or ruthless, or had simply arrived at the spot earlier than others and staked a claim. It was not a rationale that went down easily with the landless.

Thus, whenever societies forged a new social contract, changes to landownership patterns were on the agenda. Action started during the

French Revolution when rebellious peasants instilled the National Assembly to abolish feudalism in a legendary night session on August 4, 1789, a decision that was greater in symbolism than in substance because a good share of feudal privileges were transformed into titles under civil law.[20] After seceding from the Russian Empire, Estonia passed a Land Law in 1919 that provided for the expropriation of large estates while the country was still fighting its war of independence.[21] Klaus Richter has called it "the only route to survival for the Estonian Republic": with two-thirds of the peasantry landless and 60 percent of the land owned by the Baltic German nobility, the legitimacy of the new nation hinged on the creation of new small farms.[22] Latvia passed a similar bill a year later, and other newly independent countries in East-Central Europe followed the same path with various levels of enthusiasm.[23] In the 1930s, the Mexican president Lárazo Cárdenas distributed forty-four million acres of land, including entire haciendas, to peasants in order to fulfill the promise of the Mexican Revolution.[24] And then there were the botched attempts. In 1959, West Pakistan enacted a Land Reforms Regulation that sought to purge the country from feudalism, but implementation was so bad that some landlords collected rent from peasants for land they had already bought.[25]

The most comprehensive reform took place in Russia. Immediately after seizing power in Saint Petersburg, the Bolsheviks enacted a Land Decree that declared all private titles void without compensation and empowered peasant communities to take charge. The decree secured peasant loyalty for years, unlike the second stage of Soviet land reform: only Stalinist terror assured the creation of kolkhoz collective farms that started on a grand scale in 1929.[26] The reasons for collectivization and its role in the famines of the early 1930s (see chapter 31, Holodomor) are among the most divisive issues of Soviet historiography. However, Socialists were not alone in implementing land redistribution programs. After the Japanese defeat in World War II, a key concern of the American occupying forces was comprehensive land reform.[27] Enacted swiftly under the aegis of General Douglas MacArthur, it effectively ended decades of tenant unrest and won praise for its radicalism from Che Guevara, of all people, when Cuba was enacting an ambitious land reform program in 1959.[28]

However, there were also countervailing trends. While some policies sought to make ownership structures more equitable, others bolstered land concentration. Sometimes it was about the spoils of power. The Somoza family controlled 20 percent of Nicaragua's arable land when the Sandinistas toppled Latin America's longest-running dictatorship in 1979.[29] Sometimes it was about improvement projects that played out to the benefit of rural elites. When Egypt began to regulate the Nile's waters in the nineteenth century, the prelude to the construction of Aswan Dam (see chapter 29, Aswan Dam) that often escaped the environmentalists' critique, large estates "were the best instruments and the first beneficiaries of the hydro-agricultural revolution."[30] And then there were the unintended effects. In the American South, small farmers paid a particularly heavy price for the spread of the boll weevil (see chapter 12, Boll Weevil), and so it went for many other biological crises of monoculture.

The push toward land reform eventually lost steam in Western countries, sometimes with amazing speed. In Britain, the decades-long debate over the aristocracy's landholdings reached the boiling point when Chancellor of the Exchequer David Lloyd George launched a Land Campaign in the fall of 1913.[31] However, when Lloyd George pushed the issue again with his 1926 Land Programme, it failed to excite voters and was "quietly forgotten" by the time of the next election three years later, contributing to the decline of the Liberal Party in British politics.[32] Land reform has been a dead issue in Britain ever since, and so it went sooner or later all over the West.

Several factors contributed to this remarkable turn of affairs. One was the intrinsic link to issues of power. Land reform was also about the power base of old landed elites. Scholars have fought bitter debates over when and how the aristocrats lost their political hegemony, and their increasing marginalization became a key indicator of a country's democratization.[33] However, there is little disagreement that their grip on power weakened sooner or later, and with that, land reform ceased to serve as a political football. All the while, migration to the cities took some of the urgency out of land reform, if only by offering a plan B to the disaffected. In rural areas all over the world, peasants could either wait for the day when land reform would arrive or seek better fortunes in urban areas. For many, the second option sounded more attractive.

The flight to the cities also had a second effect: it lowered the profile of the farming sector. Agriculture's contribution to gross domestic product was shrinking wherever industrialization took hold, as did its share of the working population, and that changed the rationale for land reform. The agricultural sector did not take its marginalization lightly. American farmers rebelled in the late nineteenth century in what has come to be called the Populist Revolt, and most European countries had agrarian parties during the interwar years.[34] There was even a "Green International," an alliance of European peasant parties headquartered in Prague.[35] However, urban societies failed to get excited about rural property structures. As Lloyd George learned the hard way in 1926, land reform was no longer an issue that galvanized an entire nation. When Western people voiced concerns about wealth and social inequality, they increasingly thought of topics other than land.

The change of agricultural technology undercut the drive for land reform in more subliminal ways. Farmers relied on inputs to a growing extent: new machines, new seeds (see chapter 28, Hybrid Corn) and new chemicals (see chapter 38, DDT) boosted yields per acre to an extent unheard of in human history. With that, the question was no longer just who owned the land but also what people made of it, and those who did well could pay more for land than those who did not. As a result, land markets gained an unprecedented dynamism that amounted to land reforms from below: instead of waiting for a government commission, successful farmers signed a lease. In the late eighteenth century, Arthur Young, one of England's leading agricultural reformers, coined a phrase that others would quote ad nauseam: "The magic of property turns sand to gold."[36] As it turned out, synthetic nitrogen (see chapter 19, Synthetic Nitrogen) could do the job just as well.

The productivity revolution in agriculture gained a particular dynamism after 1945. It helped to overcome hunger and food rationing in postwar Europe, and declining prices for farm products, helped by industrial-style inventions such as battery cages (see chapter 36, Battery Chicken), boosted mass consumption by significantly increasing discretionary income. In the Global South, where change was so dramatic that it was dubbed the Green Revolution, the productivity revolution seemed to change the ground rules for rural development. In 1977, a study noted that the Green Revolution "has to a certain extent diverted attention away from land reform, even to the extent that some governments and observers maintain that the green revolution has dispelled the need for land reform."[37]

The Cold War made the issue even more contentious. The Soviet Union brought collectivization to Eastern Europe after World War II and offered it to countries in the Global South after decolonization. Ethiopia was among the countries that followed the Soviet approach, thus fulfilling the wish of the Stanford researchers in a way that they surely did not appreciate. On March 4, 1975, following on the heels of a Declaration of Socialism, Ethiopia nationalized all rural land and placed equitable redistribution into the hands of newly formed Peasant Associations.[38] The architect of Japan's land reform, Wolf Ladejinsky, was facing charges of Communist sympathies during the "Red Scare" of the 1950s, and his personal rehabilitation did not dispel the dark suspicions now hanging over the idea.[39] To be sure, the West never abandoned land reform completely, and the US government warmed up again to the idea in the wake of the Cuban Revolution. Most Latin American countries had land reform laws on the books by 1970.[40] But then, implementation was notably cautious, as overambitious efforts inevitably raised questions about which side a country was on in the Cold War.

3. IN SPITE OF ALL DOUBTS

The Cold War eventually came to an end, and the Russian federal government decreed in December 1991 that all collective and state farms must either disband or turn into joint-stock companies and partnerships with limited responsibility. However, Russia did not allow the sale of agricultural land until 2003, a fact that nicely reflected the sad state of agriculture in the former Soviet Union.[41] Ownership patterns were notoriously opaque, corruption was endemic, and productivity remained below potential. The land title did not come out of the Cold War as an undisputed winner. Some thirty years after the demise of Communism, it looks more akin to a leftover.

Nowhere is this lingering crisis more evident than in the Global South. "Agrarian reform is back at the center of the international debate over rural development," declared a 2006 book of the Land Research Action Network, an international working group of researchers, analysts, nongovernmental organizations, and social activists.[42] The spectrum of proponents and approaches is broad: it ranges from the World Bank, which prefers market-based solutions, to grassroots action groups like Brazil's Landless Workers Movement, which has organized land occupations with tens of thousands of families.[43] The latter groups have received their share of romantic encomia, particularly from critics of neoliberalism.[44] But they have also faced incomprehension. That became particularly clear in Zimbabwe, where Western observers discussed the expropriation of white landowners predominantly in terms of Mugabe's power play. Interest in the case for land distribution remained scant, though realities on the ground were arguably as clear as they get: large estates with ownership skewed along racial lines are hard to justify in a postcolonial world. For one, the World Bank acknowledged the need for land reform and endorsed a blueprint from Zimbabwe's government in 1998.[45]

The failure of Cold War panaceas contributed greatly to the resurgence of the issue. The Green Revolution did not become the miracle cure for landownership problems, and in fact exacerbated rural inequality in significant ways. With its emphasis on high-yield varieties, the Green Revolution package was well suited for large farmers who could afford mineral fertilizer purchases and other investments while small peasants stood on the sidelines.[46] Protests from Indigenous groups, from Mexico's Zapatista movement to Canada's Inuit, have merged claims for land with the quest for autonomy, putting authorities on the defensive.[47] After ignoring Native land claims for two centuries, the High Court of Australia finally acknowledged in 1992 that the Aboriginal peoples could not possibly be "trespassers on the land on which they and their ancestors had lived" for tens of thousands of years.[48] The coexistence of Crown sovereignty and customary Aboriginal claims has since become conventional wisdom.

Environmental concerns have opened yet another line of critique. In the light of erosion (see chapter 13, Little Grand Canyon) and groundwater contamination from industrial-style agriculture, doubts emerged whether one should leave the care of the land solely to the owner. Under the impression of the Dust Bowl, Roy Kimmel, coordinator of government agencies for the United States Department of Agriculture, argued "that private ownership carries obligations as well as privileges." Speaking at a meeting of the Kansas State Board of Agriculture in 1938, Kimmel acknowledged that "this concept is something new in American life," but "in the face of present-day conditions, this assumption must be revised."[49] In the same spirit, the US Soil Conservation Service fantasized about comprehensive land use planning in the early years before settling for more modest policies.[50] In the 1990s, the European Union enacted controls for nitrates (see chapter 19, Synthetic Nitrogen) and pesticides (see chapter 38, DDT) that brought farmland into the sphere of environmental regulation. Gone are the times when European farmers could do as they please on their fields.

Doubts even grew about transfers of land, the very thing that titles had been invented for. When foreign investors started to buy large tracts of land in the Global South in the new millennium, the practice was dubbed "land grabbing." The move coincided with a global rise of commodity prices that made land an attractive investment, but higher commodity prices also exacerbated concerns about food security; more recently, low interest rates have pushed investors into the land business. And then there was that discomforting feeling of déjà vu. Speculation, dispossession and forced migration, subsistence production under threat, the smell of corruption, and the desire to stake out claims in a shrinking world—it felt as if the scramble for land in nineteenth-century settler societies was being reenacted on a global scene.[51]

In a way, land grabbing triggers all the reservations that we have accumulated about the concept of property in land. It raises concerns about feeding the local population, as investors are primarily interested in export production. It reminds people of past experiences about the intrinsic links between large property holdings and political power. It breeds inequality: whatever investors have on their minds, profits will most likely be global while problems will be local. And it is anyone's guess whether investors will show an interest in environmental sustainability. Reservations about land grabbing are huge, and yet investors are under no obligation to care. They do not need public acclaim. They need a land title.

That, after all, is the great paradox of the history of property in land. The concept has received plenty of criticism from numerous sides, and yet it has shown remarkable resilience. Property owners can still go to the land office and assert titles in a court of law. The sense of pride that nineteenth-century capitalists displayed about the modernization of landholdings has long dissipated, but the concept remains alive, and those who challenge it can easily end up in debt or in jail. But where is the alternative? A new understanding of property as land that is more sensitive to neighbors, environments, and future generations, is nowhere in sight. American law schools routinely teach their students to criticize *Johnson v. M'Intosh*, but it remains the law and is quoted in court decisions several times a year.[52] It seems that the land title, once an icon of modernization, has turned into a concept that we are stuck with for lack of something better.

7

BREADFRUIT

FOOD CHOICES

I. IMPROVEMENTS NEEDED

British sugar had seen better days. France had lost the Seven Years' War, but it looked invincible on the plantation front (see chapter 2, Sugar). Sugar exports from the largest French possession in the Caribbean, Saint-Domingue, grew fivefold between 1720 and 1775, and costs of production were so low that British planters struggled to compete. Saint-Domingue had surpassed Jamaica as the Caribbean's leading sugar island in the early eighteenth century, and its output approached that of the entire British West Indies by 1770. Even more, there was still land in reserve on Saint-Domingue, which French planters put into production over the following two decades.[1] Nervousness among British planters grew further when another war broke out, this time over independence for Britain's North American colonies, and when warfare came to an end in 1783, the British defeat increased the planters' economic troubles: American independence meant that the West Indies were "cut off from their principal source of supply as well as from an important market for their molasses."[2] All the while, the climate of the Caribbean exposed planters to the unsettling risk of hurricanes (see chapter 25, 1976 Tangshan Earthquake).[3] Against this background, plantation owners sought ways to cut costs, and one of their ideas revolved around a plant from the other end of the world: breadfruit.

The first European encounters with the breadfruit date to the sixteenth century, and James Cook's South Sea voyages gave British planters a rough idea of the fruit.[4] Breadfruit grew on trees, and it served as a staple food on the islands of the South Pacific. Furthermore, cultivation did not seem to take much labor, a crucial point for Caribbean slaveholders. Planters were delighted to hear from Joseph Banks, who had served as a botanist on James Cook's first voyage, that Pacific islanders procured breadfruit "with no more trouble than that of climbing a tree and pulling it down."[5] And then there was the allure of the South Sea: the abundant breadfruit looked like a fruit from paradise, "a symbol of a simple and idyllic life free from worries about work or property."[6] Caribbean sugar plantations were far removed from Eden, but planters firmly believed that feeding slaves with breadfruit could save them a lot of money.

The Royal Society of Arts offered a prize for the introduction of breadfruit to the West Indies in 1777, but nobody expressed interest in claiming it.[7] Institutions were a better investment: new botanical gardens in Saint Vincent and Jamaica increased the chances for newly arriving plants to survive.[8] All that was missing was an effort to bring breadfruit plants from the South Sea to the Caribbean, and as this project dragged on, Banks emerged as the crucial figure. He had been president of the Royal Society since 1778, he ran the Royal Botanic Gardens at Kew, and he had the ear of King George III.[9] In December 1787, HMS *Bounty* set sail for Tahiti under the command of Captain William Bligh.

What followed became one of the epic stories of British seafaring. Bligh tried to sail around Cape Horn, but headwind forced him to change course and take the longer journey around the Cape of Good Hope.[10] The *Bounty* finally reached Tahiti in October 1788, and Bligh secured the cooperation of local chiefs. Collection began in November and continued for months while his crew mingled with the locals, paying particular attention to the female part of the population. In April 1789, Bligh left Tahiti with 1,015 potted breadfruit plants in a purpose-built section of the ship, but he did not get very far.[11] Some three weeks into the return journey, a mutiny led by the master's mate,

7.1 "The Mutineers turning Lieut. Bligh and part of the Officers and Crew adrift from His Majesty's Ship the Bounty." Painted and engraved by Robert Dodd, 1790. Image, National Maritime Museum, Greenwich, London.

Fletcher Christian, seized control of the *Bounty*. The conspirators set Bligh and loyal crew members adrift in a longboat, dumped the plants into the sea, and returned to Tahiti.[12]

It was not the first time that a ship carried breadfruit trees. Originally from the Malay Archipelago, breadfruit had spread across Micronesia and Polynesia along with human migration.[13] But unlike Oceanian settlement, the voyage of the *Bounty* was recorded in abundant detail, and it invited storytelling. A court-martial ruled on the mutiny back in Britain while Christian and some of his crew members settled on Pitcairn Island, but the real events gave writers and filmmakers plenty of room to project motives and character features onto the protagonists. Was William Bligh really the tyrannical captain that Charles Laughton played in the US movie of 1935? Was it the South Pacific with all its charms, and specifically the loose sexual mores, that drove the crew into revolt? Or was watering a thousand breadfruit plants simply below the dignity of a navy crew? The court-martial took the side of Bligh, whose longboat reached Timor after 3,600 miles at sea, but many narrators found other plots more exciting.[14]

The mutiny on the *Bounty* left its mark in the collective imagination, but it merely delayed the breadfruit transfer. It was institutions and interests in different parts of the world that drove the botanical exchange (see chapter 27.2, Botanical Exchange), and an occasional mishap left these powers utterly unimpressed. Banks lobbied for a second attempt, and the Admiralty commissioned two ships, the *Providence* and the *Assistant*, in 1791. In a remarkable display of British stubbornness, William Bligh was again placed in command, and this time he fulfilled his mission. He assembled more than two thousand breadfruit plants in Tahiti and delivered those that survived to Saint Helena, Saint Vincent, and Jamaica.[15] Back in Britain after two years, the Royal Society of Arts could finally hand out its breadfruit award, and the accolades for Bligh were about a fruit as well as a symbol.[16] Plant transfers of the breadfruit type showed how London "could feed its plantation colonies, add to their wealth, and demonstrate its care for its creoles."[17] Why bother that Britain had actually lost on the breadfruit front against France, which had brought the tree to Saint-Domingue in 1788?[18]

The *Providence* and the *Assistant* made a number of stops on the way to Tahiti and back, accepting and dispensing other plants as they went along. It made for an interesting contrast to its core mission. While Bligh and his crew conducted the breadfruit business with ironclad determination, they dealt with other species in the manner of a speculative broker in a global botanical bazar. They traded nectarine trees for various fruit trees in Cape Town, took a eucalyptus tree (see chapter 27, Eucalyptus) on board in Tasmania in exchange for pomegranates and strawberries, and they planted pineapples, guavas, figs, and firs on Tahiti while potting breadfruit trees.[19] The French acted in the same manner: their breadfruit transfer was the result of a supplement to a shipment of pepper plants, cinnamon, and mango trees from Mauritius.[20] Reciprocity was a cardinal virtue in the botanists' global network, and yet it had a certain air of arbitrariness: the botanical gardens of the world probably filled up more by chance than by design. Even when it came to commemoration in the Linnaean taxonomy, botanists were remarkably flexible. Bligh became immortalized in the Latin name of the ackee fruit (*Blighia sapida*) in spite of the fact that it was a slave ship that brought it from West Africa to Jamaica in 1778.[21] It was Bligh who brought the ackee fruit from Jamaica to England in 1793, but that was arguably a minor achievement. He certainly had plenty of space on his ships after unloading all those breadfruit trees.[22]

As it happened, the Caribbean breadfruit tree looked equally ambiguous. The economic prospects for British planters had changed when the plant was finally at their disposal. Food imports from North America had resumed, and slaves were in revolt on France's Saint-Domingue since 1791. When they emerged victorious after thirteen years of savage war, their new country, Haiti, withdrew from the global sugar market, and Jamaica was again the world's leading sugar exporter.[23] Breadfruit trees spread on the sugar plantations of the West Indies nonetheless, but when they reached maturity and produced the green, balloon-shaped fruits, planters made a stunning discovery: slaves refused to eat it.[24] It lacked the heroism of a mutiny, but slaves had agency on many different levels, and they had more than one reason to demur. Breadfruit was different from their customary food, and it lacked a distinctive taste of its own. A tree-borne fruit implied competition for the coveted slave garden, where slaves could cultivate their own food with a degree of autonomy. A little rebellion at the dinner table could also be a way to gain leverage in the perennial conflicts with slaveholders. But whatever the motives, there was no way to ignore the gap between the culinary status of the breadfruit and a project that Richard Drayton has called "the model for a variety of future British endeavours in economic botany."[25] After two expeditions, one mutiny, and plenty of strained nerves among botanists and gardeners around the world, all the breadfruit gave to the West Indies was just another pig food.

2. NUMBERS GAMES

Statistics is not an obvious career path for those who want to achieve enduring fame. But for Ernst Engel, it was an article in the bulletin of Saxony's statistical bureau of 1857 that made his name.[26] Analyzing family budgets of Belgian workmen, Engels found that there was a statistical correlation between income levels and food expenditures: "The poorer a family, the greater the proportion of its total expenditure that must be devoted to the provision of food."[27] The discovery had profound implications, for Engels argued that the proportion of income spent on food was "an infallible measure for the material welfare of a population in general."[28] The correlation entered the textbooks of economics as Engel's law, and researchers have discussed its validity and significance ever since. The *New Palgrave Dictionary of Economics* declared in its 2008 edition that Engel's law was "one of the earliest empirical regularities in economics and also one of the most robust."[29]

Engel's law mirrors the transformation of Western food systems over the course of the nineteenth century. His Belgian workers only spent so much on food because they could no longer feed their families from their own gardens and fields. A growing number of people bought their food on markets, and these markets were more and more abundant. Contrary to the prophecies of Thomas Robert Malthus, the growth of global food production outpaced global population growth: world output increased by 1.06 percent per year between 1870 and 1913.[30] Famines were henceforth a matter of distribution and prices rather than quantities (see chapter 31, Holodomor), the farming sector declined in importance while industry and services were expanding, and in accordance with Engel's law, food claimed a declining share of expenses in the more affluent parts of the world. Engel's Belgian workers spent about two-thirds of their income on food in the 1850s, which would put them on the borderline between medium and high vulnerability to food insecurity nowadays. In the rich countries of the twenty-first century, food claims less than 15 percent of total expenditures.[31]

However, Engel's law was not just about the long farewell to agriculture's traditional preponderance. It was also about a new way to look at the world: statistics. Numbers hold a special place in modernity. They are the one language that all modern societies understand, they are a cross-cultural marker of objectivity and neutrality, and if we can trust Theodore Porter, they are also a mode of crisis management. In his book *Trust in Numbers*, Porter argued that expert systems typically moved toward quantification not when they were ascendant but when they were in trouble.[32] Be that as it may, numbers are a crucial part of communication about the many problems of modernity, and yet they were always controversial. It was about more than the many things that could go wrong on the long path from data collection to the statisticians' conclusions. It was about the language. And it was about the claims to authority that experts and governments attached to it.

Numbers were a tool of power, and authorities were keen to stress how statistics gave them superior knowledge. "All these remarks might seem like worn-out truisms," Engel wrote in his article of 1857, but he argued that they were really much more: he called them "truths of mathematical validity" and "emanations of a natural law."[33] Numbers never spoke for themselves, and yet those who sought to boost their own standing were inclined to suggest otherwise. But even when numbers and correlations were correct, there was always room to discuss conclusions and whether they mattered. A century and a half after Engel's emphatic proclamation, the *New Palgrave Dictionary of Economics* noted somewhat laconically that Engel's law was an "unsurprising partial correlation with many alternative interpretations."[34]

Statistics on incomes and prices were not the magic keys when it came to food choices, and the same held true for the other numbers that came into play in the food business. The calorie was an established thermal unit in physics before it entered the popular vocabulary in Europe and North America as the quantitative description of the energy content of food in the late nineteenth century.[35] The breakthrough was World War I, when bureaucracies struggled to manage food supplies: Nick Cullather has argued that "military rather than hygienic necessities made the calorie an international standard measure of food."[36] A range of vitamins and other micronutrients were identified over the course of the twentieth century, and the University of Hawaii conducted a pioneering investigation in 1927/1928 that identified cooked breadfruit as "a fair source of vitamins A and C and a good source of vitamin B."[37] Precise numbers accompanied the statement, though findings of this kind were usually a mere start for complex discussions over intended effects, side effects, and recommended consumption.[38] Yet nutrition in numbers survived countless contestations, and that allows us to learn from a can of Tropical Sun Jamaican Breadfruit in a British supermarket that its drained content has 60.84 grams of carbohydrates, 3.6 grams of protein, and 259.2 calories.[39] But no numbers can give you certainty about whether you will like it or not.

Modern societies have statistics for everything from soil erosion (see chapter 13, Little Grand Canyon) to pollution mortality (see chapter 16, London Smog), but they do not settle questions about farm practices and matters of life and death. Food choices reflect prices and availability, cultures and fashions, childhood memories and the momentum of ingrained habits, and numbers are a rather crude way to capture this diversity.[40] They can also reflect government policies, sometimes with unintended consequences. Few Japanese had consumed whale meat (see chapter 9, Whaling) until the Japanese School Lunch Act of 1954 introduced "a policy of stable, daily, long-term use of large quantities of whale meat in all elementary and middle school lunches across Japan, in order to improve the nutrition of Japanese children."[41] A generation later, Japan was more committed to whaling than was any other nation on earth. Food consumption is not necessarily a matter of individual choice, but food choices do always matter for social status, and students of social stratification know about the significance of food ever since Pierre Bourdieu dissected France's class society in his book *Distinction*. Bourdieu analyzed how food preferences and the bodies that they produced revealed people's place in society.[42] And if you have never eaten breadfruit, that says something about you, too.

3. FEEDING MASSES

Caribbean eaters eventually changed their minds about the breadfruit. According to John H. Parry, it was "after Emancipation" that breadfruit "became an important source of food for communities of free peasants."[43] It is open to debate whether that was a statement about causes or chronologies, but breadfruit did find a place in diets and beyond. The fruit is rich in starch, dietary fiber, and a good source of iron, calcium, and potassium, the tree provides shade and firewood, and various parts have a range of uses in folk medicine.[44] However, it also serves as a less than favorable metaphor for men in the French Caribbean, which points to an enduring image problem.[45] Some ninety countries grow breadfruit today, but most of the trees are scattered around backyards in the humid tropical lowlands of Oceania and the Caribbean.[46] Few Europeans are familiar with breadfruit. Banks sent samples to Göttingen University, a French cultivator tried to naturalize breadfruit in the Département Alpes-Maritimes, and Kew Gardens recorded a first flowering of breadfruit in 1995, but import statistics reveal that the West has yet to develop a taste for the fruit.[47] A dozen years ago, annual exports stood at around one hundred tons for both Samoa and Fiji and one thousand tons for the Caribbean.[48]

Food preferences can evolve in mysterious ways. Some 240 years after its arrival from Africa, the ackee is Jamaica's national fruit and a cherished national symbol, quite an achievement for a fruit that is toxic when consumed unripe and caused a poisoning incident as recently as 2011 that left 148 Jamaicans hospitalized and 23 dead.[49] Breadfruit has no risks of note beyond falling off a tree during picking, and yet it lacks popularity. A survey among farmers in Trinidad and Tobago recorded views "that breadfruit was 'hog food' or associated with poverty."[50] In the atoll countries of the South Pacific, breadfruit trees are frequent in private gardens but rare on markets and in shops, and the fruits are consumed only in case of need. Younger people seem to be particularly averse: "In Kiribati and Tuvalu, the younger generation prefers to eat rice instead of breadfruit."[51] When economic sanctions were imposed on Haiti between 1991 and 1994, food choices

7.2 Breadfruit tree on a 1970s stamp from the Comores. Image, Shutterstock.

reflected the going hierarchies when consumers moved down from rice to plantains and on to breadfruit.[52] We may never know whether the mutineers on the *Bounty* really cursed the breadfruit trees, as Charles Nordhoff and James Norman Hall suggested in their novel of 1932, but it would not be surprising if voodoo-loving Haitians were inclined to think that he did.[53]

All this makes for a reality of breadfruit that is very different from the utopian dreams, both botanical and religious, that inspired eighteenth-century improvers. Observing the fruit during James Cook's first voyage, Joseph Banks thought about Eden. "These happy people may almost be said to be exempt from the curse of our forefathers," Banks wrote in his journal.[54] Compared with the toils of European agriculturalists, food production in the South Pacific looked amazingly simple: "If a man should in the course of his life time plant 10 such trees, which if well done might take the labour of an hour or thereabouts, he would as compleatly fulfill his duty to his own as well as future generations."[55] But if Banks had taken a closer look, he might have discovered a more complicated reality. Breadfruits were traditionally eaten along with bananas and taro, and it was a seasonal fruit.[56]

Bank's shortsightedness probably mirrored more than the euphoria of an Englishman in warm weather. It revealed one of the great mysteries of the botanical exchange (see chapter 27.2, Botanical Exchange): painstaking attention to plants went along with negligence regarding the social practices that surrounded them. The breadfruit affair was not the last time that crucial knowledge failed to make the passage along with the biological material. In spite of plenty of sugarcane plantations around the world, canegrowers in Australia were unsure about cultivation methods when they entered the sugar business in the 1860s.[57] When Kew's botanical network successfully transferred Amazon rubber seeds to Southeast Asia (see chapter 28.2, Legal Titles), it failed to include information on harvesting techniques, forcing the director of Singapore Botanic Gardens, Henry Nicholas Ridley, into lengthy experimentation to reinvent the wheel.[58] Botanists were inclined to isolate plants from their human context, only to learn that biological reductionism caused problems and disappointments at the destination. In the case of the breadfruit, the result was a double error, as botanists were ignorant not only about food preparation but also about the nature of slavery. If the breadfruit transfer was ultimately "an attempt to displace a growing abolitionist revolution with a scientific one derived from the new knowledges of tropical botany," as Elizabeth DeLoughrey has argued, the project fell short of its goal.[59]

Food systems are webs of commodities and cultures, and there is no way to exchange one item for another and expect that everything else will stay the same. Even handbooks, another favored classification system of modernity (see chapter 22.2, The Age of Handbooks), have struggled to create order. Gastronomic writing emerged as a distinct genre after the French Revolution, but it remains on shaky ground even after two centuries of literary development. Stephen Mennell has characterized gastronomic writing as "a brew of history, myth, and history serving as myth."[60] As it stands, the breadfruit brew is one of the more difficult ones to stomach.

Recent efforts seem to take the breadfruit as it is and promote the fruit on food security grounds. Researchers market it as "a key component in traditional agroforestry-based cropping systems" and, perhaps with a nod to Banks, "a low-input crop . . . that produces food for many decades with relatively little maintenance, care, fertilization or pest control."[61] America's National Tropical Botanical Garden established a Breadfruit Institute in Hawaii in 2003, and its work includes promotion and distribution of trees as well as the study of more than 120 distinct cultivars, whose properties are still widely unexplored. One of their rays of hope is "anecdotal evidence that breadfruit can help mitigate type II diabetes (see chapter 2.3, The Sweet Life)."[62]

The breadfruit seems stuck in its place at the bottom of the food hierarchy, which makes it an island of stability in the dynamism of modern foodways. Some foods advanced over time from luxuries to everyday treats such as sugar (see chapter 2, Sugar) and chicken (see chapter 36, Battery Chicken). Others fell out of favor, and some have experienced meteoric rises from below, and it remains anyone's guess whether these careers were due to inherent properties or conspicuous consumption. Breadfruit made it on a list of "top 25 superfoods" from the Caribbean Food and Nutrition Institute in 1997, and it would not be the first exotic commodity to profit from a culinary status upgrade, windfall profits along the commodity chain included, but it is probably less than likely.[63] Those who have consumed a can of Tropical Sun Jamaican Breadfruit know rather well that a breadfruit food fad has a long way to go.[64]

8

Guano

The Fertility Business

I. FECES FOR SALE

In the 1850s, German agriculturalists became enchanted with the guano song. Written by Victor von Scheffel, otherwise known for hearty student drinking songs, and sung to the melody of Heinrich Heine's "Die Loreley," it tells the story of a lonesome island in the middle of the ocean where many thousands of birds live undisturbed by humans. "None of them fails to do its duty," Scheffel reported, and that duty was to defecate. By virtue of amassing excrement, the birds are engaged in a collective effort at mountain building. In a moment of avian hyperbole, Scheffel has the birds fantasizing about "covering the entire ocean over the course of history." But humans thwart the building project when they recognize that the mountains make for excellent fertilizer. They turn excrements into a commercial product, and the poem ends with a grateful Swabian peasant blessing the "splendid birds" for their "superb manure."[1]

In the nineteenth century, exotic commodities thrived on a sense of mystery, and Scheffel did his best to provide the somewhat profane process of guano production with an aura. He even managed to insert a whiff of intellectualism by taking a swipe at the German philosopher Georg Wilhelm Friedrich Hegel. According to Scheffel, the birds' metabolic materialism raised doubts about Hegel's case for the supremacy of the spirit.[2] It would seem that familiarity with German idealist philosophy was still a given among the fertilizer salesmen of the nineteenth century. For once, the educated and the economic bourgeoisie, so often at odds in German history, were chanting in sync, as Scheffel's poem skillfully combined intellectual wit with economic sense. But like so many commercials of later years, it was just as remarkable for what it said as for what it left out.

Guano was not the first commercial fertilizer. A book-length "comprehensive overview" of 1800 listed forty-five different types of fertilizer and discussed their merits and pitfalls.[3] In late Tokugawa Japan, where ecological constraints were particularly harsh, peasants even came to use fishmeal as fertilizer.[4] Neither was this the first time that businessmen turned feces into a commercial product. Chinese cities had a flourishing trade in nightsoils—the Victorian euphemism for human excrement (see chapter 17, Water Closet)—since the late Ming dynasty, and city dwellers in Japan and India were equally familiar with the business. But these were local networks, and the supply was inevitably constrained by the number of human bowels.[5] In contrast, guano was available in seemingly unlimited quantities, and it was about big money. Whereas the nightsoil trade mattered only for the cities and their hinterland, guano was a global commodity.

After the first guano shipments arrived in Europe in 1841, the guano business developed with the vigor of a gold rush, except that the government of Peru maintained a firm grip on exploitation. It nationalized the rich deposits (see chapter 15, Saudi Arabia) on the Chincha Islands off the Peruvian coast, and the ensuing export boom singlehandedly turned a nation in default into one of the continent's most reputable debtors.[6] Monopoly prices inspired a quest for alternative supplies. In 1853, the British Admiralty asked navy officers "to discover any unoccupied islands from whence guano may be freely obtained by British merchantmen."[7] Napoleon III gave similar orders as emperor of France, which led to the annexation of Clipperton Island, some eight hundred miles southwest of Acapulco. It was not the proudest moment of La Grande Nation. A rough sea prevented the captain from raising the flag in person, exploitation was delayed after samples indicated bad quality, and when mining finally began decades later, France stood on the sidelines, and the mining company acquired a concession from Mexico after Clipperton Island received a visit from a Mexican gunboat.[8] For all the interest that guano stimulated among the Great Powers, it also held a distinct ability to make them look foolish.

Guano even prompted the United States to embark on its first imperial venture. The Guano Island Act of 1856 gave US citizens the right to take possession of unclaimed and unoccupied islands with guano deposits and put it at the discretion of the president to declare them appertaining to the United States. It was a distinctly American approach: it reflected the country's overabundance of entrepreneurial energies, the lack of a powerful navy, and the innocence of a nation that was still new to the business of colonial acquisitions. The Guano Island Act inspired merchants to file claims by the dozen, and some seventy islands made it onto the list of "appurtenances," including some that did not actually exist.[9] It also inspired a lot of head-scratching as to what "appertaining" actually meant. Even the US State Department, which produced a 969-page typescript on the legal status of guano islands in the 1930s, was clueless as to the meaning of "appertaining": it found that the word "lends itself readily to circumstances and the wishes of those using it." Summarizing seventy-five years of policy, the diplomats declared that "the only conclusion which can fairly be drawn . . . is that no one knew what the Guano Act really did mean."[10]

It would seem that when it came to guano, the legal challenges were bigger than the technological ones. Guano mining was a low-tech endeavor that required little more than picks, shovels, and brawn. The Chincha Islands did not even have a proper harbor. Digging fossilized dung under a searing sun was an unforgiving job that fell mostly to Chinese coolies, which made Peruvian guano part of the ignominious nineteenth-century forced labor system that took hold after the British ban on the slave trade. Edward Melillo has argued that, with ammonia-laden air, strict production quotas, and nowhere to flee on the small islands, "guano excavation ranked among the world's deadliest and least remunerative jobs."[11] Peruvian newspapers regularly featured notes about suicides of guano workers.

In short, Scheffel was probably well advised to keep silent about business practices in the guano trade and focus on the production of excrements and their philosophical implications. He was more accurate when it came to the consumption side: farmers were truly grateful for guano. The United States imported 140,000 tons in 1855, and the most eager buyers were Southern plantation owners who saw the productivity of their fields declining due to monoculture.[12] British farmers were equally enthusiastic, all the more so as the repeal of the Corn Laws in 1846 forced them to brush up their competitive edge. Even Scheffel's Swabian peasant had probably heard of guano and bought some at great expense.[13] Improving agriculture through better crop rotations and more intensive land use was a transnational concern in Europe, and inputs such as guano helped make sure that the ground would tolerate the new methods. For those who read farming journals or went to agricultural assemblies in the mid-nineteenth century, guano was hard to escape.

But for all the enthusiasm, guano use remained the privilege of a minority, though one that grew in importance throughout the modern era. Even in Great Britain, the world's leading importer, only one in four farmers used it between 1840 and 1879.[14] The purchase required command over significant financial resources and an entrepreneurial attitude toward innovations and investments. Many farmers saw guano as a gamble, and if they bought it nonetheless, they waited nervously for harvest time to discover whether the outlay of capital had paid off. And then there were those who, for the time being, managed soil fertility with other means. For example, Brazilian coffee farmers ignored the guano ships that traveled along the country's shore and simply cleared forestland when soils showed signs of exhaustion.[15]

The Chincha Islands were tiny, the length of the largest one being less than a mile, but that has not kept historians from bestowing them with global significance. "Guano . . . thumped a shock wave through the ecology that unifies producers and consumers," Steven Stoll declared in his history of American soils in the nineteenth century. "The biological foundation of American society became a one-way transfer of material from some point of extraction or production to the farm where it went into the crops, and from there to consumers."[16] The one-way stream typically led from colonial and postcolonial countries to the West, and that has prompted scholars such as Gregory Cushman to speak of a form of imperialism.[17] But guano was only one of the more conspicuous ways in which nutrients were now moving around the globe. Nutrients also traveled in the guise of sugar (see chapter 2, Sugar), cotton, and all the other commodities that the burgeoning trade networks of the nineteenth century shuffled around in ever-growing quantities. Guano was peculiar only because for the first time, nutrients went on their commercial journey in the form of feces.

ened the doors for commercial fertilizers, and a growing ucts competed for the farmers' attention from the mid-entury onward. In retrospect, guano was the perfect com-ugurate the business of fertility: it was familiar enough by virtue of its excremental provenance yet sufficiently exotic to generate confidence in other heretofore-unknown substances. When the nutritional value of guano began to decline, a German company began to treat it with sulfuric acid in order to improve solubility, thus blurring the boundary between organic and chemical fertilizers in a helpful manner.[18] Monopoly pricing may have helped, too: the economic historian William Mathew has argued that the high price of guano allowed other products, most notably superphosphates, to gain market share in the 1850s.[19] Western agriculturalists increasingly relied on a material supply network of growing size and complexity, and the practice usually spread quicker than awareness of the ensuing problems. And this was not the only network that fertilizers brought to the farm.

2. FARMERS' LITTLE HELPERS

In the fall of 1845, the Liverpool-based Muspratt works launched a new brand of patent manures. It seemed like a good time to enter the fertilizer business. Guano had proven the merits of fertilizer investments in previous years, and harvests were bad in 1844 and 1845—the latter year now famous as the start of the potato famine (see chapter 31, Holodomor). Most crucially, the product came with the authority of one of the eminent scientists of the day, Justus von Liebig, who had published the first two editions of his landmark book *Agricultural Chemistry* in 1840 and 1842. Liebig sent his faculty colleague and brother-in-law Friedrich Knapp to Liverpool and paid a personal visit to the Muspratt works in September 1845. With a hefty price of £10 per ton and a recommended dosage of one ton per acre, it was an expensive product, but surely one that would pay. The producers were so confident about success that they dispensed with field trials before marketing.

Liebig's patent manure became a disaster. Fused in a furnace to reduce solubility, the stuff "remained on the surface like a glass dressing."[20] Some farmers took on additional expense to plow the material in, only to discover that it had no effect on growth. Muspratt quickly changed the manufacturing process, but the new product performed no better: it lacked nitrogen, a nutrient that Liebig thought plants could obtain from the atmosphere for free. It was a humbling experience for one of the founders of modern chemistry, and his biographer William Brock acidly declared that "agricultural science was a good deal more complicated than Liebig had initially thought."[21] Liebig abandoned fertilizer issues for a while and eventually revised his stance on nitrogen in the seventh edition of *Agricultural Chemistry*, but the fiasco was about more than an ill-fated product. The episode sheds a revealing light on how agrochemistry was a curious blend of Promethean might and persistent ignorance.

Liebig advocated a new understanding of soil fertility. Earlier generations had adhered to what Steven Stoll has called "dunghill doctrines": a blend of experiences and intuition with a humus theory that ultimately went back to Aristotle.[22] Farmers were familiar with a range of practices to maintain soil fertility, but they lacked a proper understanding of the nature of fertile soils until Liebig proposed his "law of the minimum." In Liebig's reading, plant growth hinged on a number of factors, including a range of chemical nutrients, with the scarcest resource limiting growth even if other resources were plentiful. In practical terms, Liebig called for the use of mineral fertilizer in order to make sure that the soil held all chemical nutrients in sufficient amounts.

According to Jacques Hadamard, it is the natural urge of historians of science "to prove that nobody ever discovered anything."[23] They had it rather easy with Liebig, as Carl Sprengel penned the theory before the publication of *Agricultural Chemistry*. The mineral theory of plant nutrition is best seen as the result of a cumulative process of intellectual gestation in early nineteenth-century chemistry, and Liebig stood out mainly as its most visible proponent. He was a passionate preacher of the new gospel and acted as a kind of "chemical gatekeeper": his *Agricultural Chemistry* was a lucid summary of contemporary wisdom about how chemical expertise could help farmers.[24] It was a watershed moment in the history of modern agriculture. The fertile soil was now open to rational analysis, or so it seemed.[25]

In reality, soil fertility was about more than chemistry. It was about a complex interplay of physical, chemical, and biological processes, and geology, temperature, and rainfall were only some of the more important factors. Ecologists sometimes describe the soil as "the poor man's tropical rainforest" since a single spoon of earth holds a biological diversity on a par with Amazonia, and scientists still struggle to understand the full range of processes that create fertile soils.[26] Given this context, agrochemistry offered a starkly reductionist view. It treated the soil as if it was merely a temporary storage for nutrients on their way to the crop.

Furthermore, agrochemistry was much better at retrospective diagnosis than at prescriptions. Scientists could set up fertilizer demonstrations on experimental plots and show that guano worked well because it was rich in nitrogen and phosphorus. However, farmers were more interested in what they should do for the next growing season, and they did not find the law of the minimum very helpful in the search for answers. How should they identify the minimum factor that constrained growth, and what were "sufficient amounts" of nutrients? Analytical methods did not allow for quantifying the concentration of critical nutrients in the soil until far into the twentieth century, and that left scientists somewhat clueless when it came to precise instructions. Max Gerlach, the founding director of Germany's renowned Kaiser Wilhelm Institute for Agriculture, put it bluntly in an article of 1926: "For once we need to be clear that we are unable at this point to tell the farmer, based on any available formula, how much fertilizer he should use on his field in the upcoming year."[27]

It did not help that fertilizer research was about a lot of money. Liebig's patent manure adventure in Liverpool illustrates the blurry borders between academic research and commercial interests, and the debacle did not make him more cautious: Liebig returned to fertilizer speculation in 1857, when he put money into a Bavarian superphosphate factory.[28] The stakes increased with the growth of the fertilizer business, as big corporations invested generously in research institutes, academic journals, and close relations with university professors. The Limburgerhof agricultural research station, set up in 1913 when the BASF chemical company started synthetic nitrogen production on a commercial scale (see chapter 19, Synthetic Nitrogen), was just one of many well-funded initiatives under the auspices of industry.[29] When it came to agrochemistry, it was hard to define where research ended and where salesmanship began—and fertilizer manufacturers had no interest in marking a clear boundary.

In short, agrochemistry called for self-confident, commercially savvy experts who could sell a gospel and the product to go with it. Didn't Liebig's blunder with patent manure show that you could get away with a mistake if you stood on the right side of progress? Was it really necessary to record all the ambiguities of fertilizer experiments? Wasn't it in everybody's interests to tweak things here and there so that input-heavy agriculture could follow its predestined course? Suggestions along those lines reached the ears of a recent graduate of Munich's Technical University who got his first job at Stickstoff-Land GmbH, a fertilizer company in nearby Schleißheim, in 1922. He did not like what he heard, nor did he enjoy his meager salary and his subordinate position as *Hilfssachbearbeiter*, and there were more exciting things going on in contemporary Munich. He quit his job as an assistant clerk and focused on his political career, where his craving for honesty and uprightness had more room to flourish. Fifteen years on, he was the head of Nazi Germany's security apparatus. Another five years on, the *Diplom-Landwirt*, whose name was Heinrich Himmler, was the mastermind of the Holocaust.[30]

In any case, it was clear that farmers had good reasons to have a sense of caution when meeting fertilizer people. It was less clear what would follow from that sense of caution. Where would farmers turn for advice if they did not trust the promises of the friendly fertilizer people? Sometimes states reacted and set up research institutes and extension services, but it often proved that their independence was merely a matter of degrees. And even if a farmer had confidence in a certain fertilizer manufacturer, he could not be sure whether someone would dilute the product along the commodity chain and make a quick profit. It was usually a long way from the factory to the farm, and many people were under no illusion as to what reckless businessmen would do when nobody was watching.

Ironically, the latter worry paved the way for the institutionalization of agrochemistry. Their analytical methods were not good enough for testing soils, but they were good enough to check the nutritional value of fertilizers and identify adulterated products, and that is what chemists did in growing numbers. For example, the Royal Agricultural Society of England hired its first chemist in 1849.[31] Serial testing showed how the quality of Peruvian guano declined over the years, and that taught a lesson on the merits of chemical expertise in agriculture.[32] Fertilizer testing became a routine activity, the first chemical tests that agricultural research institutions conducted en masse, and they allowed chemists to secure a strong position in a nascent field of expertise: on the eve of World War I, chemists ran no fewer than fifty of Germany's sixty-five agricultural experiment stations.[33] It was a success that differed markedly from Liebig's chemical heroism. Agrochemistry flourished not because it could guide farmers with superior authority but because it helped solve problems that people would not have had without mineral fertilizer.

Fertilizer tests were popular, and farmers grew more confident of mineral fertilizer as a result, but they also implied a significant shift of authority: testing was as much about chemistry as it was about power. For the first time, farmers gave command on a matter of farming into the hands of strangers. Henceforth fertilizer decisions relied on outside expertise, and farmers lacked means to check the lab results. It was an important step for people who took pride in being masters in their own house, and it was the first such step down a long flight. Farmers lost control over their seeds with new varieties such as hybrid corn (see chapter 28, Hybrid Corn), they worked with engineers and mechanics on tractors, milking machines, and other new technologies, and they surrendered authority over pest control when DDT (see chapter 38, DDT) and other chemicals promised a panacea straight from the factory. Innovations were gnawing away at the autonomy of the farmer, as a widening range of experts offered advice, services, and products.

It all came down to a new type of farming where everything hinged on cooperation with outsiders: the former masters of their own trade became team players. Cooperation usually went along with cozy rhetoric of companionship and shared identities, and yet it was an open question whether trust really came from deep convictions or from a lack of alternatives. That problem became particularly evident in the Global South after 1945 when the Green Revolution offered high-yield seeds that hinged on lavish doses of mineral fertilizer. "The Green Revolution was essentially a seed-fertilizer package since the new seeds were bred to be high 'consumers' of fertilizers," Vandana Shiva has noted.[34] The package had its price, though, and many farmers faced a veritable leap of faith. They could jump on the bandwagon and confront new financial and institutional dependencies. Or they could stand aside and be on their own.[35]

Of course, the troubles of the Green Revolution were still beyond the horizon in the nineteenth century. A guano purchase was expensive, and it required a degree of trust in salesmen and agrochemists, but it was a far cry from the fateful decisions that later generations would face. Just as with the global cycle of nutrients, farmers were sleepwalking into a new age of shared authority, where expertise was spread more evenly than the accompanying risks. A fertilizer salesman who overplayed his cards risked losing customers, and an overambitious scientist risked a dent in his reputation, as Liebig learned in the aftermath of the patent manure affair. A farmer who listened to the wrong experts risked going into debt for seeds, fertilizer, and other necessities of modern farming. If the gamble did not work out, farmers would lose their land, and perhaps more. Over the past twenty-five years, at least a quarter million Indian farmers have committed suicide.[36]

3. RUNNING EMPTY

Nauru House is an impressive building. Located at 80 Collins Street in Melbourne, it has 52 floors that rise to a height of 190 meters, which made it the tallest building in town when it opened in 1977. It lost that distinction after a few years, but with its octagonal shape, Nauru House continues to stand out among the box-shaped buildings that dominate the central business district. It even earned itself a nickname among the locals, though it wasn't one that looked good in a marketing brochure. Nauru built the skyscraper with revenues from its phosphate mines, and mindful of guano and its origins, Australians called Nauru House the "Birdshit Tower." It was a bit of a misnomer, as Nauru's phosphates were not of avian origin.[37] Unfortunately, that was not the only problem about the building.

Phosphate mining on Nauru was part of the global scramble for mineral fertilizer reserves that engulfed islands around the world in the wake of the guano boom. Some places saw mining in a frenzy. The Ichaboe Island off the coast of today's Namibia won a brief moment of transnational fame in the 1840s when British ships stripped it of guano within less than two years.[38] Other places never saw a boom. The British Navy found the deposits on the Khuriya Muriya Islands off the coast of today's Oman disappointing, but it annexed the islands anyway.[39] Among the disperse assemblage of US guano islands, less than two dozen were actually mined.[40] Some places experienced multiple bouts of extraction. In Australia's Great Barrier Reef, phosphate mining followed on the heels of guano, and when fertilizer people could not find any more raw material on these islands, they focused on the islands themselves: in 1900 coral mining commenced on at least twelve areas of the Great Barrier Reef and continued for over four decades. Sugarcane farmers in Queensland were wrestling with acidic soils, and a cheap supply of lime from nearby coral reefs was just what they needed.[41] These were the same farmers who would introduce the infamous cane toads (see chapter 14, Cane Toads) to Australia in the 1930s.

Nauru had been a German colony since 1886, but during the first years of colonial rule, its main commercial asset was dried coconut meat. It took the Germans more than a dozen years to discover the phosphate deposit and a few more to negotiate a contract. The mining rights went to the London-based Pacific Phosphate Company, and extraction began in 1906.[42] It was just as destructive to the land as mining on other islands, but there was one important difference. The Chincha Islands did not have human residents, nor did the islands that the United States "appertained" through the Guano Islands Act of 1856. But colonial Nauru had a population of about 1,400 people spread over some twenty square kilometers.[43] Even worse, the island comprised little else than phosphate holdings and a small coastal strip. Paul Hambruch, who visited Nauru as part of a German South Sea Expedition in 1910, found former mining areas "infertile" and "completely useless," and while depletion was still "some 500 years" into the future, Hambruch saw the day approaching when the indigenous population would be forced to abandon their native home.[44] Nonetheless, mining continued for almost a century.

There was no concern about landscape restoration, and Nauru was perfectly typical in this respect: the same held true for Potosí (see chapter 1, Potosí), Appalachia, the Soviet Donbass, or England's aptly named Black Country.[45] Simply put, the state of the land after mining was not an issue: the Guano Island Act explicitly declared that the United States would not have any responsibility for the islands after plundering their assets.[46] They were *terrae nullius* in the language of international law, and the future of "nobody's land" was nobody's business.

It was not that people of the nineteenth century were insensitive to global nutrient connections. A book of 1843 ventured that guano brought back nutrients that soil erosion and sewage had washed out into the ocean, mirroring an awareness that nothing could get lost on a finite planet.[47] But people did not see that as much of a problem as long as the global flow of nutrients was still a trickle: arguments about ecological imperialism thrived better against the background of a world fertilizer demand in the range of two hundred million tons per year.[48] In the nineteenth century, guano production was principally about money and power, with both circulating rapidly in the guano years: Peru had fourteen changes in government from 1850 to 1875. The stability of guano's commodity chain went along with political instability, and some scholars have argued that the rapid sequence of governments actually bolstered guano's rule. With power relations in constant flux, decision makers lacked the political capital to tamper with the export regime.[49]

The boom years came to an end in the 1870s. The quality of guano was declining, and speculation ran rampant as to an impending exhaustion of deposits. Peru's debt burden had increased rapidly during the guano years, and creditors were getting nervous. Peru defaulted on its foreign debt in 1875.[50] From 1879 to 1883, Peru and Bolivia fought and lost the War of the Pacific against Chile. Scholars have offered different readings, with dependency theorists (see chapter 30.2, In Their Theories) blaming neocolonialism while others pay more attention to domestic politics, but they generally agree that Peru wasted its resource endowment with few lasting benefits.[51] When Nauru began to acquire assets such as Melbourne's Birdshit Tower after gaining independence in 1968, it seemed to be on a better track.

Curiously enough, Peru performed better when it came to restoring the guano supply *after* the collapse of the export business. Starting in 1909, the government set up a remarkably successful bird conservation program that focused on the species with the highest output of excrements. It was not without environmental ambiguities. In order to boost guano production, marksmen targeted species like the Andean condor that preyed on guano birds and their nests. Several El Niño events reduced the bird population, but the conservation regime survived for half a century, reaching an all-time production record of 332,223 tons of fresh guano in 1956. The lion's share went into Peru's domestic cotton and sugarcane production (see chapter 2, Sugar).[52]

The regime ran into trouble around 1960 when Peru embraced free-market economics and sought to boost exports.[53] Thanks to the nutrient-rich Humboldt Current, the country's terrestrial waters had abundant stocks of anchoveta, and Peru became the world's largest fish-producing nation within two decades, with most of the catch going into fishmeal production. Unfortunately, the guano birds were feeding on the very same marine life, and unlike the domestic guano business, fishmeal brought export revenues. That tilted the scales, and the government suspended its guano bird conservation program in 1966.[54]

The boom of fishmeal exports was closely linked to the expansion of industrialized meat production in the United States (see chapter 18, Chicago's Slaughterhouses). Farmers cherished it as a protein-rich animal feed, and chicken farmers (see chapter 36, Battery Chicken) were the most enthusiastic buyers. That made for an interesting return to the global metabolism of the guano years. In the mid-nineteenth century, the Humboldt Current fed the guano birds, whose excrements were mined, sold, and spread on fields across the West, and a good part of the harvest then went into meat production. The fishmeal industry cut out the birds and the fields and put marine resources directly into the animals' trough. Mindful of the lessons from guano bird conservation, the government set up a scientific monitoring program, and Gregory Cushman has argued that "Peru's anchoveta fishery was among the most carefully supervised and rigorously regulated the world had ever seen."[55] In the late 1960s, Peru's fishing industry claimed one-fifth of worldwide commercial landings by tonnage.[56] But the boom did not last. The anchoveta population collapsed in the El Niño of 1972 (see chapter 9, Whaling).[57]

At first glance, Nauru's investments looked more farsighted by way of comparison. But as it turned out, financial assets were no less fragile than ecologies. When Nauru's phosphate supplies neared exhaustion in the 1990s, a stunned island population discovered that their wealth was gone. Corrupt politicians, shoddy advisers, and dubious investments had claimed it all, with the relative share of the booty being just as unclear as the precise circumstances. Angered creditors seized Nauru House to cover their losses, an investment corporation bought it in 2004, and Nauru is now widely seen as a tragic example of the "resource curse" (see chapter 15, Saudi Arabia).[58] Moral verdicts are always cheap when it comes to resources, but it is crucial to recognize that the parties along the commodity chain—governments, mining companies, merchants, fertilizer salesmen, and farmers—were all acting in what they perceived as their own best interests. But maybe that makes the experience of collapse even harder to stomach.

9

WHALING

RESOURCES AT SEA

I. HUNTING FOR PROFITS

Christian von Rother was a man of energy. Born into a peasant family in Silesia, he rose to prominence in the Prussian state administration after the country lost its war against Napoleon in 1806. He represented Prussia in the negotiations over France's war reparations in 1815, became a member of the Prussian State Council in 1817, and administered the country's debt for decades. He also became head of Prussia's Seehandlung in 1820, a state-owned bank that Rother turned into a powerful development agency. The Seehandlung set up model companies, built a road network (see chapter 3, Canal du Midi), and maintained a merchant navy, all with the goal of boosting commerce and trade and catapulting Prussia into the age of industry.[1] And so it fell to Rother to decide on an interesting business proposal in 1839: Should Prussia go into whaling?

By the nineteenth century, whaling had long since moved beyond the coastal hunts for sustenance that seaside communities had conducted for thousands of years. The Basques had pioneered pelagic whaling in the open sea when they extended their traditional hunting grounds in the Bay of Biscay northward around 1500.[2] Early modern Europe could well use extra amounts of oil for illumination and lubrication, and other seafaring nations entered the whaling business over the course of the following centuries. French, English, Dutch, and Danish ships were heading to the whale-rich waters along the fringe of the Arctic Sea ice in changing numbers, driven by a search for profits as well as national interests. British whaling flourished in the wake of an act of Parliament that created a forty-shilling bounty per ton in 1750, a subsidy that even Adam Smith could approve of.[3] While Smith readily conceded in *The Wealth of Nations* that whaling did "not contribute to the opulence of the nation," it did "contribute to its defence" in that it was a school for hardy sailors that might be of use in times of war.[4]

The United States was the global leader in the whaling business when Rother pondered his decision, and one of the reports at his disposal came from Friedrich Ludwig von Rönne, Prussia's ambassador in Washington. Rönne painted a glowing picture of a thriving industry. According to his account, whaling gave work to some ten thousand sailors while total investments ran to $12 million, and that was only the seafaring part: Rönne put the value of the entire whaling business at $70 million.[5] Whale blubber required processing, and so did the baleen from the mouths of whales, a filtering system of strong yet flexible rods that found a variety of applications from umbrellas to corsets.[6] Whalers also sought markets for other body parts at their disposal, and uses ranged from leather gloves to dog food.[7] Processing technology, division of labor, heavy capital requirements, and a desire to turn leftovers into profit—whaling was an organic industry akin to Chicago's stockyards (see chapter 18, Chicago's Slaughterhouses), and just like the meatpackers of the Midwest, whaling fed markets without natural limits. Unlike coastal whaling, the hunt in the open sea did not come to an end when the stomachs of Indigenous people were full.[8]

As befits a Prussian official, Rother took a diligent look at various aspects of the whaling business, but there was one issue that he did not worry about: access. The sea and its resources were open to anyone who could muster a ship, a crew, and the capital for a long journey. In fact, whaling played a major role in the making of the open seas. When English, Danish, and Dutch whaling interests clashed around Spitsbergen in the early seventeenth century, the Dutch prevailed with the doctrine of mare liberum commonly attributed to Hugo Grotius. National sovereignty henceforth ended three nautical miles from the coastline, a distance that was meant to reflect the contemporary range of canons.[9]

A Prussian whaling expedition would not have infringed on the sovereignty of other nations, but Rother did worry about labor. Contemporary hunters sat in small rowing boats that crews moved into the immediate proximity of whales. The harpooner threw a razor-sharp spear with a line attached once the moment seemed right, and if it stuck, the wounded animal dragged the boat on a journey that could last hours. When the whale finally grew tired, it was killed by another stab that pierced its lungs. A lot of things could go wrong during the hunt: whaling crews fell into icy water, lost track of the mothership, or got hit by an animal many times their own size. It was a masculine job even by the generous standards of the naval world, and loss of life was simply part of business.[10] Of 787 vessels in New Bedford's whaling fleet, 272 were lost at sea.[11]

All this made American whaling towns remarkably similar to mining boomtowns (see chapter 1.2, Boomtowns). Whaling was a business in which those who were lucky could make a fortune, and whaling towns attracted men who were adrift. Herman Melville's father went bankrupt and died when his son was twelve, and young Herman traveled to Liverpool and back on a merchant ship, made a trip to the frontier in Illinois, and then headed to New Bedford. He served on the whaling ship *Acushnet* for a year and a half, jumped ship in Polynesia—a regional tradition, as the breadfruit story showed (see chapter 7, Breadfruit)—and then went on to write *Moby-Dick*, the book that defined the collective imagination of whaling like no other.[12] New Bedford was also a favored destination for fugitive slaves who found work on whaling boats and onshore, no questions asked. In fact, New Bedford was probably "the best city in [antebellum] America for an ambitious young black man." Residents included Frederick Douglass, America's most important Black intellectual in the nineteenth century.[13] Writing skills were not an obstacle in whaling, but physical strength, skills, and perseverance were a must, and Rother found that a Prussian whaling expedition would be hopeless without expertise from abroad. Hiring foreign hands was a time-honored tradition in the whaling business, and in Rother's opinion, foreigners would need to fill most of the jobs.[14]

9.1 Whaling as depicted in the 1861 book *The Arctic Whaleman* by Lewis Holmes. Image, Library of Congress.

It was a bit too much even for the energetic Rother, who ultimately decided against the proposal "for now." He conceded that he was charmed by the prospect of profits, but the many difficulties, the long journey, and the uncertainty of success made this, in Rother's words, "a very daring speculation."[15] However, Prussia discovered another asset in the course of its investigation, and that was the author of the proposal. Louis Bahre was a man of the world, nineteenth-century style: he had traveled all over Mexico, a larger country in 1839 than it is now, he had plenty of contacts in Latin America and a deep knowledge of geographic and nautical matters, but after returning to his native Hamburg, Bahre had found himself confined to an office desk under the tutelage of his father. It is not clear whether Bahre sought a good investment or an adventure when he penned his proposal, but Prussia's representative in Hamburg felt that this might be a man worth remembering. Specifically, he recommended Bahre "for some kind of mission in overseas countries on scientific or mercantile matters," an odd combination from a twenty-first-century perspective, but it made perfect sense in the world of whaling.[16] As the bodies of cetaceans were piling up along the ships and shores of the modern world, so did the body of cetacean knowledge.

2. MAMMALS INTO MARGARINE

The whaling business changed in the decades after Rother's decision. Ships grew bigger, and they ran on steam instead of sail. They also featured the grenade harpoon gun that Norwegian Svend Foyn introduced to whaling in the 1860s.[17] Hunters could now target whales from the relative safety of a ship's bow, and they could target larger and faster species. The use of whale products changed just as dramatically. Whale oil lost its hold on the illumination market to petroleum, but hydrogenation—a chemical technology whose high-pressure catalytic version created synthetic nitrogen (see chapter 19, Synthetic Nitrogen)—gave whale oil a new commodity life. Hydrogenation turned liquid oil into hardened fat, which made it a priced raw material for the burgeoning food and soap industry. As Gordon Jackson wrote, "Whaling became, in effect, an adjunct of the margarine and soap industries by 1914."[18]

Change did not come without its share of discontents. In Norway, the global leader in the late nineteenth-century transformation of whaling technology, fishermen in the country's north attacked whaling as the embodiment of capitalist arrogance toward the fishing proletariat, and tensions escalated to the point where a horde of fishermen destroyed a whaling station in 1903. Socialist agitators helped to whip the fishermen into revolt, but they were less than helpful in stemming the tide of technological change.[19] The food and soap industry was increasingly in the grasp of large multinational corporations (see chapter 10, United Fruit), with corporate consolidation culminating in an Anglo-Dutch merger that created Unilever, and multinationals knew all sorts of ways to deal with protest on the ground.[20] In fact, fishermen faced competition from the business model of whaling in their own trade after World War II when factory-freezer trawlers revolutionized ocean fishing.[21]

Technology was changing, and so was the geography of the whaling business. An increasing number of ships moved into the waters around Antarctica since 1900, a particularly inhospitable part of the oceanic world and the last region that had heretofore escaped the hunters' attention. The first year that whaling catches in the Southern Hemisphere exceeded those in the north was 1909, and so it remained for the following decades: of the 2.9 million whales that humans killed for commercial purposes in the twentieth century, more than 70 percent were caught south of the equator.[22] Antarctic whaling relied on local whaling stations initially, but advancing technology such as the stern slipway, which allowed crews to pull dead whales on board for dismantling, liberated whaling from a regional support network. The new ships shaped the collective imagination of whaling, not least since the cool technological efficiency of processing (see chapter 18, Chicago's Slaughterhouses) made for a startling contrast to the blood that was spilled, and they opened the business for anyone with enough money for a floating factory. Antarctic whaling turned into a free-for-all.[23]

Norway and Great Britain were the leading whaling nations of the interwar years, but they faced growing competition from two newcomers. Japan entered pelagic whaling in the 1930s, which was a double source of irritation. It was part of the country's pelagic imperialism, and the stalwarts of whaling in the West were stunned that Japan was more interested in whale meat than in blubber.[24] While Japan had a long tradition of coastal whaling, Nazi Germany was a complete newcomer when it decided to go into whaling in 1935. Germany was wrestling with a "fat gap" and relied heavily on imports, which was a major drain on the country's currency reserves. Antarctic whales were supposed to come to the rescue, and the Nazis hired Norwegian crews in order to get started.[25] It was the maritime part of the Fascist quest for autarky, which had previously inspired the "battle for grain" and land reclamation in the Pontine Marshes in Italy (see chapter 32, Pontine Marshes), but the capitalist producers in Norway and Great Britain were not interested in moderation either. The result was a dramatic increase in catches. Numbers for the Southern Hemisphere rose from 11,127 whales in 1923 to 34,648 in 1929 and 57,777 in 1937.[26]

But did the oceans really hold enough whales to sustain the boom? The shift to Antarctic waters suggests that whaling exceeded environmental limits elsewhere, and the enormous growth of catches in the twentieth century was certainly alarming. However, the historical literature readily identified precarious scarcities long before whales emerged as an environmental icon, and the rationale often appears dramaturgic rather than scientific.[27] As organic production goes, whaling was a rather volatile industry, a remarkable fact in light of the enormous investments in capital and expertise that it required, and a decline of business can have many causes. For example, strong evidence suggests that "the decline of American whaling [after the Civil War] antedated serious problems of whale numbers."[28]

The reality was that nobody really knew. Whales were slow-breeding creatures, but the oceans of the world were vast and full of unknowns, and one of the prized tools of modern science, the experiment, was obviously useless in the quest for answers. Even the shift to the Southern Hemisphere is ambiguous, for whaling continued in northern latitudes and did not reach its peak until 1966, six years after catches peaked in the south.[29] The global stock of the different varieties of whales was anyone's guess, and nobody had a good model for commercial or physical extinction in the interwar years. Extrapolating from other maritime resources was not an option either. As Carmel Finley has shown, the concept of maximum sustainable yield for fisheries was politically and economically constructed in the postwar years before it became scientific orthodoxy.[30]

While experts and diplomats were negotiating about the future of the global commons, the consuming masses had different problems. An extra dose of fat was an attractive offer in the face of fragile world markets and the experienced reality of famine (see chapter 31, Holodomor). Any remaining concerns met with the creative and manipulative powers of government, science, and the food industry. When Germans were reluctant to eat growing amounts of fish in the 1930s, researchers came up with a fish sausage that the Ministry of the Interior labeled, with an earnestness that only German officials can muster, the "Neptun-Fisch-Bratwurst."[31] It is a matter of perspective, like so many things in this book, whether that was the creativity of the food industry at work or an underreported Nazi crime.

In light of the Neptun-Bratwurst, marketing whale products was a rather easy job. Margarine and soap did not reveal the maritime origin of the raw material, and producers had no incentive to create transparency. The modern food system has a remarkable ability to obscure origins and procedures and get away with it, as sugar (see chapter 2, Sugar) and Chicago's stockyards (see chapter 18, Chicago's Slaughterhouses) serve to attest, and mammals-turned-margarine were just one more step in the industrial creation of invisibility. US manufacturers invented the fish stick in the 1950s when the new factory-freezer trawlers left them with more frozen fillets than consumers were willing to eat.[32] In the twenty-first century, factory-made tomato purée, produced in Italy from Chinese tomatoes, floods street markets in Ghana.[33]

Among the many challenges of the whaling business, ethical qualms were minor into the postwar years. Even former whaling nations did not show signs of remorse. The United States sent its last whaling vessel to sea in 1924, but it had no problem allowing Japanese whaling in the 1946–1947 season "as a solution to the country's food shortages."[34] World War II had provided the whales of the world with a few calm years, but the hunt resumed with vigor in the postwar years, and 1964 became the record year with a global catch of 82,194 whales.[35] Cetaceans were still, in the words of Eric Jay Dolin, "swimming profit centers to be taken advantage of."[36] It probably helped that commodity chains seem to defy the collective imagination in mysterious ways. It was arguably a moral disgrace to turn majestic animals into faceless commodities, but only if one thought in interconnections, and it appears that human memory is more comfortable with pictures and events. In fact, people were struggling to think in terms of systems even when the tide turned against whaling.

3. HUNTERS' REMORSE

According to the King James Bible, "God created great whales."[37] As the largest mammals in the world, playful at sea yet helpless ashore, they touched a nerve among sentimental souls long before the age of ecology. In his 1861 book *La Mer*, the French historian Jules Michelet rhapsodized about "the whale, the grandest animal, the richest life in all creation" that "man madly pursues."[38] Carl Schmitt was in a similar mood when he called the whale "the most humane of all creatures, more humane than man who exterminates it with savage cruelty."[39] Better known as the crown jurist of the Third Reich who infamously defended Hitler during the first years of Nazi rule, Schmitt wrote about "the poor leviathan" who "has almost disappeared from our planet" in a long essay, *Land and Sea*. The booklet was published in 1942, when the Nazis were no longer keen on juridical flank protection, and Schmitt probably failed to realize that at the time of writing, there was a gap between his cetacean sentimentalism and the transcontinental killing spree of his political bedfellows.[40] In any case, Schmitt did not spell out a conservation philosophy and left it at a lament about how "the cannon . . . had turned the poor whale into an easy target."[41] Michelet, writing when harpoons were still thrown by hand, was more outspoken. He blamed whalers for seeking "the enjoyment which executioners and tyrants feel" and argued for "half a century of absolute peace."[42]

It is a romantic idea that these sentiments pushed conservation diplomacy into action, and environmental historians have sometimes nourished this fantasy.[43] The reality was different: "Early efforts to regulate whaling were not aimed at the protection of whales but, rather, at securing a high price for whale oil," Maglosia Fitzmaurice wrote in *Whaling and International Law*.[44] Whaling was first and foremost about commercial and nautical interests, and with huge capital investments and the vagaries of world trade during the interwar years, the overarching goal of international whaling regulation was to gain some stability in a notoriously volatile business. Whaling was not controversial on principle, and did not become so for decades. When the US House Subcommittee on Fisheries and Wildlife Conservation held hearings on marine mammal protection in the early 1970s, speakers from the Sierra Club, the Audubon Society, and the National Wildlife Federation talked about professional management of a renewable resource. The idea of a comprehensive ban for perpetuity was still beyond the horizon.[45]

The international regulation of whaling (see chapter 24.3, Getting Serious) began with a 1931 Geneva Convention and a 1937 London Agreement, and like so much of 1930s diplomacy, they were essentially about good intentions that failed to make a difference.[46] The United States took over after World War II and shepherded the creation of the International Whaling Commission (IWC) in 1946, but even leadership from the Western superpower did not forestall extensive jockeying among the vested interests.[47] Some defied the IWC openly. Aristotle Onassis, a naturalized Argentinian citizen with Greek roots in Anatolia, sent the aptly named *Olympic Challenger* on whaling missions in the 1950s; the ship was owned by a company in Uruguay that flew the flag of Panama and thus escaped IWC regulations in spite of a board of directors full of US citizens located in New York.[48] Others defied the system in clandestine ways. Soviet trawlers conducted illicit whaling between 1948 and 1972, and while other IWC members harbored "strong suspicions," they "chose never to tackle this issue head-on."[49]

However, while whaling increased to its postwar peak, the economic rationale looked increasingly dubious. Thanks to hybrid seeds (see chapter 28, Hybrid Corn) and other improvements, agricultural production grew dramatically, which made the exigencies of the autarky years a fading memory. Baleen lost many of its applications in the wake of the boom of plastics (see chapter 40, Plastic Bags). Other uses of whale products remained in place, and yet whaling increasingly looked like a solution in search of a problem. After the mid-1960s, it was previous investments in equipment, careers, and national prestige that kept whaling going, or rather limping along, for business was clearly on a downward slope. "Into the middle of the 1970s, the central question about whaling was how much to restrict catching to ensure more catching in the future," Kurkpatrick Dorsey has noted.[50] When Greenpeace sent its first ship on an anti-whaling mission in 1975, global whaling catches had decreased 57 percent from their peak eleven years earlier.[51]

For many generations, whaling had profited from a heroic aura. Hunting had a long and noble pedigree in Western culture, and the epic struggle between the biggest and the smartest animals on planet earth was the type of drama that discouraged critical questions on the legitimacy of the endeavor. And then there was always *Moby-Dick*, whose cultural legacy ranges from Hollywood films to a global coffee chain. If Starbucks were to be named today, the marketing department would probably veto a novel character that, as chief mate on the *Pequod*, was complicit in the killing of whales.

However, the whales of the environmentalists were different. To start with, they were all one: environmentalists were reluctant to distinguish between varieties, though threat levels varied dramatically between, say, blue whales and minke whales.[52] The environmentalists' whales were also intelligent and had a sense of community, and they were singing songs that are among "the sounds of earth" on the golden records that the Voyager 1 and 2 spacecrafts have carried outside the solar system.[53] Greenpeace activists felt so confident about the intelligence of whales that crew members tried to communicate by playing musical instruments in zodiacs on the first anti-whaling voyage.[54] Counterculture myths were not to everyone's taste, but iconic pictures of mammoth whales (see chapter 39.1, The World Was Watching) had wide appeal, and whaling became a defining issue of Western environmentalism. Color pictures helped. Gray whales in the gray sea would have made less-than-perfect Kodak moments in the age of black-and-white photography.

Environmental campaigning reached a milestone when the International Whaling Commission imposed a zero-catch limit for all commercial whaling in 1982.[55] Hunting continued on a much-reduced scale, and IWC meetings continue to see their share of diplomatic haggling, but the real threats to whales have recently been of a different nature: noise, marine debris from discarded plastic (see chapter 40, Plastic Bags), and accidental catches in fishing nets. Pollutants accumulate in whales, which pass them on in measurable quantities to the few people who still consume their meat: the traditional Greenland diet, where hunting falls under an exemption for Indigenous whaling, exposes natives to dangerous levels of cadmium, mercury, and PCBs (see chapter 38.3, Banner Slogans) from the consumption of Arctic predator animals.[56] However, whaling remained in the public spotlight, and not just because of continued activism from Greenpeace and Sea Shepherd, its militant offspring. The documentary film *The Cove*, which covers dolphin hunting in Japan, won an Academy Award in 2010. When the team from *The Cove* conducted a covert operation and found whale meat in a top-end sushi restaurant in Santa Monica a few days later, the story became front-page news in the *New York Times*.[57]

The Cove shows a bloodbath, but tuna fishing, for one, is pretty bloody, too. The blood is perfectly visible in Salvador Dalí's iconic painting of 1967, *Tuna Fishing*, and when Michelet called for a "universal code, applicable to every sea" in 1861, he talked about whales and fish alike.[58] Nonetheless, whaling and fishing hold different places in the moral universe of Western environmentalism. It is certainly not due to the absence of problems: it is likely that overfishing was a bigger ecological problem historically than excessive whaling.[59] And while the regulation of whaling continues to hinge on negotiations among diplomats, nongovernmental organizations, and vested interests, much of the world's fishing is under the control of national governments, which sought to push their spheres of sovereignty farther out into the sea in the postwar years. The traditional three-mile limit seemed archaic when nations recognized that any country could seize the rich maritime resources at their doorstep with the right technology.

Fishing played a major role in the expansion of territorial claims. Mexico established a two-hundred-mile zone in 1945, followed by Argentina in 1946 and Chile and Peru in 1947. However, enforcement was delayed for a generation when the United States, citing the concept of maximum sustainable yield, pushed aggressively for open seas.[60] Subsequent conflicts were about more than dubious scientific concepts and international conventions. Shots were fired when two NATO members, Iceland and Great Britain, fought the so-called cod wars from 1958 to 1976. It ended with a humiliating defeat for the erstwhile naval superpower. Iceland successfully enforced a two-hundred-mile zone, which became the agreed global standard with the 1982 UN Convention on the Law of the Sea (see chapter 24.3, Getting Serious).[61]

But claiming national sovereignty and sustainable fisheries management were two very different challenges, and the latter had its share of spectacular blunders. The cod stocks off the coast of Newfoundland and Labrador were a resource under scientific management from the Canadian Department of Fisheries and Oceans, and yet they collapsed in spectacular fashion.[62] The outcome for other stakeholders was no more impressive. Somali fishermen turned to piracy when state authority collapsed and illicit fishing and dumping of hazardous waste in coastal waters (see chapter 5, Shipbreaking in Chittagong) depleted Somalia's fishing grounds.[63] It would not have happened without the collapse of the Somali state, but other African governments do not protect domestic fishing either. Many states have signed treaties with the European Union, which offers handsome payments to African rulers in return for access to domestic waters.[64]

The risks of fishing for the marine environment are legion, but the commercial risks are a different matter. While Christian von Rother shied away from whaling because it looked just too speculative, today's trawlers rely on generous payments from the governments of industrialized countries. Figures vary depending on what qualifies as a subsidy, but one estimate from the UN Food and Agriculture Organization has put the annual amount at $50 billion, and unlike the eighteenth-century subsidies that Adam Smith was willing to defend, there are no military spin-off benefits.[65] Just as with agricultural subsidies (see chapter 2.2, Power Games), government payments stand as an entitlement rather than a lever for environmental policy, and fishing quotas exceed sustainability standards more often than not. When it comes to managing the resources of the sea, the ban on whaling was probably the easy part.

10

UNITED FRUIT

THE GREAT CORPORATE BANANA

I. VERTICAL INTEGRATION

There are many ways into the plantation business. Minor Keith came to it by building a railroad. Costa Rica's coffee elite sought a transport link from the country's Central Valley to Puerto Limón on the Caribbean coast, and Keith landed the job by mediation of his uncle, the US railroad tycoon Henry Meiggs, at the age of twenty-three. Starting in 1871, it was a daunting project, plagued by shaky finances, forbidding terrain, and a murderous climate that killed thousands of workers. By 1883, the Costa Rican government had become so desperate about completing the project that it offered Minor Keith direct control over the railroad's operation. It also gave him a ninety-nine-year lease for 800,000 acres of land adjacent to the railroad, some 6 percent of the country's territory. Eager to recruit and retain labor and to find cargo for his line, Keith encouraged his workers to plant bananas on his newly acquired holdings. By the early 1890s, his Tropical Trading and Transport Company was sending a million stems of bananas per year through Puerto Limón to consumers in the United States.[1]

A growing number of Americans were developing a taste for bananas (see chapter 7, Breadfruit) at that time. During the 1876 Centennial Exposition in Philadelphia, visitors could buy bananas wrapped in shiny tinfoil for ten cents apiece. Ships took them on board as extra cargo in the Caribbean and tried to make it to a US port before the priced commodity turned into brown mush. Bananas were a perishable fruit, and that provided the export business with an air of speculation. More than a hundred companies entered the banana trade in the last three decades of the nineteenth century, but only twenty-two were still in the business by 1899. However, business structures took a more permanent shape that year when Minor Keith merged his holdings with the Boston Fruit Company to form a new corporate player: the United Fruit Company.[2]

The assets of both companies complemented each other nicely. Keith had sold his bananas via New Orleans while Boston Fruit served consumers in the Northeast. Keith's bananas grew in Latin America whereas Boston Fruit focused on Caribbean islands. Keith owned a railroad while Boston Fruit had grown out of a steamship company. And both had learned about the inherent risks of the banana business: events such as hurricanes (see chapter 25, 1976 Tangshan Earthquake) made it advisable to spread production over the region. The new company controlled the entire commodity chain from the plantation to retailers in American cities, a crucial advantage for a perishable good, and losses declined with the spread of refrigeration technology (see chapter 20, Air-Conditioning). The new company was bound to emerge as the new behemoth in the banana trade, all the more so since United Fruit pursued an aggressive expansion policy and purchased other companies. A few disgruntled dealers formed a loose association called "The Anti-Trust Company."[3]

The association mirrored more than the traders' declining fortunes. United Fruit was not the only large corporation that people were grumbling about in late nineteenth-century America. It was an era that came to be known as the Gilded Age, a time when people experienced rapid industrial growth and technological advances but realized that the fruits of progress were distributed in highly uneven fashion. Large companies grew out of nowhere into hegemonic positions. The Standard Oil Company monopolized the petroleum business, Thomas Edison and George Westinghouse built corporate empires in the electric industry, an oligopoly controlled Chicago's stockyards (see chapter 18, Chicago's Slaughterhouses), and in 1901, a merger created United

States Steel, the world's largest steel producer and the first company with a capitalization of more than $1 billion. It was an age of crony capitalism, with endemic corruption and violent labor struggles. In short, it was an age just as Adam Smith had envisioned it, more precisely the Adam Smith that usually gets short shrift in business school curriculums. As Smith so aptly noted in *The Wealth of Nations*, "People of the same trade seldom meet together, even for merriment and diversion, but the conversation ends in a conspiracy against the publick, or in some contrivance to raise prices."[4]

The country had little historical experience with large corporations. The Hudson Bay Company had operated to its north and the Dutch West Indies Company to its south, but the United States had traditionally been a nation of small businessmen. As late as 1840, most companies were limited in capital and technological means, they had few if any employees, and they focused on one specific task. The new corporations were huge in every respect, and they combined different operations under the same roof. United Fruit ran plantations, railroads, a shipping line, and distribution to retailers, and the company would add further branches with growth. Salaried managers were running operations as the owner-operators of yore were fading into the background. Large corporations were fond of innovation, but the secret of their success was coordination between the different parts: by controlling all steps along the commodity chain from the plantation to US markets, United Fruit could speed up operations so that bananas would no longer rot on the dock. Where dealers had previously wasted time by haggling over prices or waiting for ships to arrive, managerial cooperation made sure that all parts of the company were operating in sync.

With that, the members of the Anti-Trust Company had good reasons to be concerned. The formation of United Fruit indicated the end of their business model: there was little room for local banana dealers in a world of large corporations that ran like giant clockworks. When it came to bananas and many other resources, the winning formula was vertical integration: combining production, transportation, and trade under one roof allowed for economies of scale, scope, and speed that smaller businesses could not realize.[5] United Fruit would change its business model several times and delegate some tasks to externals when it seemed opportune, but it never forsook the combined benefits of size, vertical integration, and operation by salaried managers. The large integrated company was the future of business for bananas and many other products, and it has defined the face of global capitalism ever since. The business historian Alfred Chandler summarized it as follows: "Rarely in the history of the world has an institution grown to be so important and so pervasive in so short a period of time."[6]

Big business drew criticism from the start. In the United States, two large political movements of the late nineteenth century, the Populists and the Progressives, pledged to challenge the rules for large corporations.[7] Under pressure from antitrust proceedings, United Fruit divested some of its holding, but it remained the dominant player in the banana business. It claimed half the US market between 1910 and 1951, with no competitor claiming more than a 20 percent market share until the 1950s.[8] It also controlled markets abroad after buying Elders and Fyffes, a company that brought bananas from Jamaica and West Africa to Britain.[9] But all this ultimately hinged on the stunning ability of large managerial companies to deliver cheap goods en masse. Within a generation, the banana advanced from a world's fair curiosity to a staple in the working-class diet.[10] The craving for bananas varied over time, but it never disappeared: consumers enjoyed bananas in milkshakes, in their cereals, or simply as a snack. They were cheap, delicious, and approved by food authorities, and for most buyers, that was all they wanted to know. Bananas became a showcase for consumer amnesia. However, things looked a bit different at the other end of the commodity chain.

2. TROUBLE IN THE LAND

United Fruit entered a region that was notorious for political instability. Most of Latin America had shrugged off colonial rule in the early nineteenth century, but the long wars of independence left a gaping power vacuum. Weak infrastructures made national integration difficult, many borders were contested, and a multitude of stakeholders became entangled in perennial rivalries. Sudden changes of government were the rule rather than the exception: the average term of a Mexican president was eight months between 1833 and 1855.[11] Even family ties did not assure the grip on power, as Costa Rica's president learned in 1859 when he was deposed and replaced by his brother-in-law. He went into exile in El Salvador, returned with an invasion force a year later, and ended up in front of a firing squad.[12] But for all the hustle and bustle, the socioeconomic underpinnings of political power did not change all that much over time. In Costa Rica, a coffee elite ruled without interruption from 1870 to 1948.[13] There were a few political assets that had enduring value throughout the region: landownership (see chapter 6, Land Title), military power, and a light skin color.

In a way, United Fruit fitted squarely into this panorama. It owned large swaths of land, its shareholders were white, and it had the backing of the region's hegemonic military power. However, an amphibious landing of US Marines was merely the most spectacular tool that United Fruit had at its disposal. The key to United Fruit's power was that it was everywhere: the company ran railroads and harbors, controlled the telegraph lines, set up schools and housing, and ran its own stores. It even had political representatives that negotiated with governments as if United Fruit were a sovereign power. "The banana empire is... the expansion of an economic unit to such size and power that in itself it assumes many of the prerogatives and functions usually assumed by political states," Charles David Kepner and Jay Henry Soothill wrote in a landmark study in 1935.[14] Seen from the ground, United Fruit was a corporate giant whose tentacles spread into all parts of society, and that is how it came to be known: *el pulpo*—the octopus.[15]

In light of its omnipresence, it is not surprising that United Fruit entered the literary imagination of Latin America. It figured prominently in the works of Miguel Ángel Asturias and Gabriel García Márquez, both recipients of the Nobel Prize in Literature. Pablo Neruda even wrote a poem titled "The United Fruit Co." that began with how God parceled the earth upon creation to Coca-Cola, Anaconda, Ford Motors, and other US corporations.[16] Passions also ruled in nonfiction literature, which for a long time featured, as one scholar put it in 1993, a mixture of "company-supported apologias, journalistic critiques, and politically inspired attacks by Caribbean nationalists."[17] For scholars in the dependency theory tradition (see chapter 30.2, In Their Theories), United Fruit looked like a textbook case for manufactured underdevelopment at the hand of a multinational corporation.[18] Even scholarly compendiums make no bones of their stance when they discuss the company's role. *The Palgrave Dictionary of Transnational History* of 2009 notes, "United Fruit symbolized the worst aspects of multinational corporate behaviour, a source of the exploitation and poverty common to people who lived in the 'banana republics.'"[19]

Nobody liked United Fruit, and at times it looked as if United Fruit did not even like itself. "I feel guilty about some of the things we did," Sam Zemurray declared after stepping down after a life in the banana business and some twenty years at the helm of United Fruit. "All we cared about was dividends."[20] United Fruit was an embattled company, and that left its mark on minds and spirits. It was no fun to bribe local potentates or otherwise coax them into embracing the Faustian bargain of their countries with the banana empire.[21] And that was only one of multiple fronts.

Labor conflicts were a constant source of trouble. Even a scholarly volume that seeks to bring out the diversity of the banana region came to a sobering conclusion: "Confrontational labor relations seem almost inherent to banana production."[22] Conflicts ran along lines of class as well as ethnicity, particularly when United Fruit hired Black people of West Indian descent rather than Hispanic natives in order to divide the workforce.[23] Banana cultivation hinged on manual labor, and the harsh conditions on the plantation found a reflection in equally harsh labor conflicts. One of the most infamous episodes was the 1928 army massacre of demonstrating banana workers in Ciénaga, Colombia, where estimates of casualties run into the four digits. The event might well be forgotten if Gabriel García Márquez had not made the massacre the climactic event in his novel *One Hundred Years of Solitude*.[24]

The environment was no more friendly than the workers. Since the 1890s, growers had wrestled with a fungus whose wilting symptoms became known as Panama disease. Another fungus, Sigatoka, spread rapidly in the mid-1930s.[25] Growers responded with toxic chemicals and new techniques such as flooding of fields, but the most important strategy was moving on to new land. United Fruit always possessed far more land than it actually used for cultivation, and with a view to disease problems, this was surely a good idea. But in societies full of impoverished landless people, that was also a provocation.

As so often, the call for land reform (see chapter 6, Land Title) went hand in hand with the push for democracy. Guatemala had endured a particularly harsh dictatorship under General Jorge Ubico, which ended in 1944. Seven years later, Jacobo Arbenz Guzmán assumed the presidency with a mandate for change, and agriculture was the crucial issue. According to Guatemala's census of 1950, 70 percent of the country's arable land was in the hands of 2.2 percent of the landowners. United Fruit alone owned more than 550,000 acres, and it had only 15 percent in cultivation. No land reform could possibly have left the company's holdings untouched. The government offered compensation for expropriated land, and 1,700 acres came from the president's own estate, but neither made Arbenz's policies less provocative. For United Fruit, land reform was a red flag.[26]

It was another clash over the meaning of land titles (see chapter 6, Land Title). For Western corporations like United Fruit, property was sacred. For the Guatemalan masses, land was the key to a better life. It did not help that the government based its compensation on United Fruit's tax declaration, as the company suddenly found that it had undervalued its property for tax purposes by a factor of twenty-five. Twenty years earlier, United Fruit had bought the land for about 2 percent of what the U.S. State Department now found a just price.[27] This being Cold War times, the issue brought up charges of Communist inclinations and rumors of infiltration, and that got the US government into action. The conflict ended in June 1954 when a CIA-inspired coup drove Arbenz into exile.

The coup became a symbol of American imperialism. Critics cited it endlessly as definitive proof that for the United States, fighting Communism (or what it defined as such) was really a smokescreen for the pursuit of corporate interests. They would also point to the personal ties between United Fruit and the Eisenhower administration. John Foster Dulles, the secretary of state, had argued cases for United Fruit as a lawyer. His brother Allen, head of the CIA, had been on the company's board, and president Eisenhower's private secretary was married to a PR executive at United Fruit.[28] The company had friends in high places, but the Guatemalan coup was only a success in the short term. As a diplomat at the US embassy in Guatemala mused in retrospect, "Having a revolution is a little like releasing a wheel at the top of a hill. You don't know where it's going to bounce or where it's going to go."[29]

3. MORE THAN ONE OCTOPUS

Among the eyewitnesses of the Guatemalan coup was a young Argentine doctor named Ernesto "Che" Guevara. He had come to the country toward the end of 1953 while drifting around Latin America and stayed because of the allure of Arbenz's reform policy. He left convinced of the need for armed struggle against US imperialism and with a thirst for frontline action. In the wake of the coup, he took refuge in the Argentine embassy for a month before moving on to Mexico City, which had a tradition as a sanctuary for left-wing political exiles. That is where he met Fidel Castro, who had fled Cuba after spending twenty-two months in prison for an insurrection against the regime of Fulgencio Batista. A decade later, United Fruit had lost its Cuban possessions.[30]

Cuba was never important for United Fruit, but the episode reflects the company's declining fortunes in the postwar years. Only five days after Arbenz's resignation, the US Department of Justice filed an antitrust suit against the company—a move that nicely showed how US capitalism was more heterogeneous than has been assumed by dependency scholars (see chapter 30.2, In Their Theories). The case was closed four years later with a consent decree that forced United Fruit to sell some of its holdings.[31] Central American governments displayed a growing readiness to confront United Fruit, bolstered by Cold War fears in the United States that the oppressed masses of their countries were prime fodder for Communist insurgents. Even the CIA was displeased after the Guatamalan operation and noted in an internal directive that the company's policies were detrimental to US interests.[32] All that caused the company's profits to fall dramatically, and the value of United Fruit's shares fell from $70 in 1950 to $15 in 1960.[33]

United Fruit was still able to act in impressive fashion. When disease problems became overwhelming in the late 1950s, it abandoned the traditional Gros Michel banana and shifted to the pathogen-resistant Cavendish cultivar. It was a display of corporate prowess, as change was not limited to the plantation. Cavendish bananas were prone to bruising and required boxes and extra care in handling, and United Fruit diligently retooled its commodity chain accordingly. They also looked and tasted different, and it took a huge marketing and advertising campaign to ensure that consumers would go along. The weapon of choice was Miss Chiquita, a high-heeled cartoon figure that drew on the fruit's exoticism. Originally invented for a radio commercial in 1944, the cheery figure became the widely recognized face of the Cavendish brand.[34] The new bananas also brought higher yields per acre, but that was a decidedly mixed blessing.[35] During the postwar years, improved farming methods and the growing use of mineral fertilizer (see chapter 19, Synthetic Nitrogen) led to a sharp increase of per-acre productivity, which in turn led to chronic overproduction.[36]

Under pressure to regain its profitability, United Fruit did voluntarily what the Guatemalan government had tried to make it do by force: it gave away land (see chapter 6, Land Title). For sober accounts, unused land was simply a drain on the balance sheet. United Fruit reduced its possessions during the 1950s and continued to do so more aggressively throughout the 1960s.[37] But United Fruit did not abandon vertical integration. The contracts it drew up with planters specified quality criteria, delivery dates, and other terms of business, which allowed it to let go of land while maintaining a firm grip on banana production.[38] Whereas Chandler had spotlighted the "visible hand" of management, postwar businesses learned that the invisible hands of intermediaries and dependent producers could reduce all sorts of risks.[39] Wrestling with the challenges of monoculture was now somebody else's problem, and then there were the other risks of doing business in a dangerous world. When the United Self-Defense Forces of Colombia, a right-wing paramilitary umbrella organization, threatened its business operations in Colombia, the company used intermediaries to channel money to these groups, a deal that earned it a US government investigation for supporting terrorism in the post–9/11 world.[40] The banana men learned that a clever octopus did not put its own fingers into everything.

In 1970, the Wall Street investor Eli Black took control of United Fruits and merged it with his other holdings to form a new company, United Brands. His tenure ended on a Monday morning in February 1975, when Black smashed the glass of his office window on the forty-fourth floor of the Pan Am building in Manhattan and jumped.[41] Another investor, Carl Lindner Jr., tried his luck with the company, moved its headquarters to Cincinnati, and took stock of its potential. United Brands was a diversified company that included activities like meatpacking (see chapter 18, Chicago's Slaughterhouses) with no connection to bananas. Vertical integration was not much help here, but speculation skills were, and when the British pound was strong, United Brands sold its British subsidiary Fyffes to Irish investors in 1986.[42] Four years later, the company changed its name again and henceforth operated as Chiquita Brands International, thus using its own marketing creation to wash off an inconvenient past. Yet none of this brought back the thriving company that United Fruit once was. It applied for bankruptcy protection in 2001.[43]

However, the octopus was not dead. It merely looked like there were now several octopuses around that sometimes mingle and sometimes fight. The latter obviously makes for a spectacular show of force, and so it became global news when the United States and the European Union became embroiled in what was soon termed a "banana war" during the 1990s. Europe had maintained a protected market for bananas from former European colonies in the Caribbean, and the United States filed a complaint against the policy with the World Trade Organization. After several years of high-stakes wrangling, the European Union was forced to dismantle most of its preferential trading arrangements. One study summarized the outcome as "a giant step forward for those who promoted neoliberal economic policies in the agricultural sector."[44] However, the United States had not brought the case just because it liked neoliberalism. It went to the World Trade Organization after intensive lobbying from United Fruit in its new Chiquita Brands incarnation.[45]

In the new millennium, the banana business made headlines with a different issue. Plant diseases were back, and they were more terrifying than ever. Chemicals were no match for yet another fungus named Tropical Race Four, and banana planters, knowing well about the shift from Gros Michel to Cavendish bananas half a century before, set out to find a new cultivar that was immune to the disease and palatable to consumers. Once more, it was a suicide that came to symbolize the peril. One morning in 2001, United Fruit's banana breeder Phil Rowe hanged himself from a tree in his experimental field after forty years of trying in vain to find that new banana. Journalists were writing speculative reports about the coming end of bananas, enthused both by the yucky topic (how often does a fungus make the news?) and by the prospect of the banana business finally finding its environmental nemesis.[46] Or is the pandemic more imagined than real, a smokescreen of the boll weevil type (see chapter 12, Boll Weevil) that obscures a more sinister truth—say, an attempt to legitimate a crisis mode (see chapter 25.2, Crisis Mode) that allows for take-no-prisoners action?[47] The banana business breeds paranoia just as well as Tropical Race Four.

For all the changes in the banana business, one thing could generally be taken for granted: consumers liked bananas. But did they like United Fruit? When fair trade bananas (see chapter 23.3, Alternative Projects) reached European consumers in 1996, they found a market niche wide open. Fair trade seeks to give farmers a better deal and invests in communal projects; independent certification makes sure that the terms of trade serve poverty alleviation and sustainable development.[48] It looked as if there was finally an escape from the grip of the octopus, and consumers were willing to pay a premium for these bananas. But time would show that fair trade was not really anathema to the world of big business. Just like the common banana, fair trade varieties need efficient transport and marketing, and both are cheaper with vertical integration and economies of scale. It did not take long for the corporate heavyweights to seize on the opportunity, and so it came that Fyffes grew into the largest supplier of fair trade bananas in Europe. A thriving business typically brings up the specialists from mergers and acquisitions, and as it happened, Fyffes did consider a merger with an American multinational company in 2014. Its name was Chiquita Brands International.[49]

PART III

IRREVERSIBLE

NOTHING IS IRREVERSIBLE IN HUMAN HISTORY, UNLESS IT IS.

When people say that something is irreversible, it is usually a declaration of faith rather than a statement of fact. Narratives of rise and fall run through the history of empires, corporations, and individuals, and scholars have told them in all sorts of modes: heroic, tragic, comic, epic, nemetic. Irreversibility smacks of determinism, and if we can trust Reinhart Koselleck, it violates a deep-seated sense in the modern imagination that history is open-ended.[1] But material realities do not respect human sentiments, and a good share of modern environmental history is about developments that are eminently one-directional. Species have perished. Monocultures became embattled. Soils have washed away. Invasive organisms have spread dramatically. Resource endowments have expired. And we will need to live with the repercussions for the foreseeable future.

Irreversibility is alarming in retrospect, but contemporary perspectives were different. A significant part of the following discussion is about how humans developed concepts for extinction, erosion, and other irreversible processes. Once these concepts gained currency, history looked different, and this section touches on several popular narratives that are retrospective constructions: the dodo, the Dust Bowl, the cane toad blunder, the US–Saudi alliance. These stories were anything but dramatic in their own time, but irreversibility gives them a sting that will not disappear for the imaginable future. Many chapters in this book are about loss and a feeling of remorse, but these are the overwhelming sentiments in this section. Even the Saudi oil state has failed to make people happy.

The extinction of the dodo provides a case in point. The flightless bird perished from Mauritius at a time when there was no scientific concept for extinction. It also perished in a highly inconspicuous way: we do not know the exact time of the dodo's disappearance or the precise cause. It makes for a stark contrast to the event's biological significance: evolution over millions of years was brutally severed when Dutch ships arrived on an island in the Indian Ocean. We did not even acquire a good picture of the bird until it was gone, though modern science has been remarkably good at filling the gaps in our knowledge. The dodo was an early victim of what may be the sixth mass extinction in the history of our planet.

Natural history showed a strong interest in the dodo in early modern times, but it was a highly selective type of interest. If the animal had not had feathers, or if it had not been so endearingly clumsy, the dodo might have produced little more than a note in a forgotten diary. Selectivity was a recurring feature in engagements with endangered species: science in the nineteenth century, conservation movements in the twentieth, and even genetic conservation projects, currently more fantasy than reality, follow a hidden hierarchy defined by visuality and Eurocentric cultural norms. Filled with memories from the Vereenigde Oostindische Compagnie to *Alice in Wonderland*, the dodo stands out.

Conservationists continue to wrestle with this imbalance. Red lists have grown tremendously since the 1960s, but they are still far from comprehensive. Humans were more proactive in the conservation of biological diversity if there was a commercial use. Seedbanks have been at work for a full century collecting and storing the diversity of the world's crops. However, both efforts effectively focus on preserving genes in living organisms, and that is only part of the challenge. Seedbanks know a lot about cold storage, but they are less good at conserving past cultivation practices and uses, and conservation biologists work with depleted stands, impoverished gene pools, and an industrial civilization with a tremendous hunger for space. Few environmental problems are more glaring than extinction (once you have a concept for it), and few have produced more dissatisfactory responses.

While extinction could go unnoticed, few agriculturalists failed to recognize the problems of monocultures. The boll weevil was one of many biological challenges that struck monocultures everywhere on the globe, and it triggered a recurring set of responses: feverish debates, hectic creation of scientific institutions, and quests for technological fixes and resistant varieties. It was a collective process of learning by doing, particularly in cases like that of the boll weevil for which no previous experience was at hand, and it was an irreversible loss of innocence. Many monocultures began as processes of cognitive simplification, and biological contestations were Mother Nature's way of telling farmers that it did not actually work. In the beginning, and only in the beginning, monocultures looked amazingly simple.

The boll weevil changed products and production methods, but that was not just a matter of commerce and technology. The insects' impact resonated in socioeconomic structures and racial hierarchies, and different groups recorded different consequences in different time frames. Some sharecroppers used the boll weevil to gain leverage in negotiations with landowners, but they fared less well in the long run. Like most biological challenges to monocultures, the boll weevil ultimately played out to the advantage of those with money, knowledge, and access to the latest technologies. But the relationship also worked in the opposite direction: a biological challenge was a convenient way to process unpleasant memories. A Southern mythology depicts the boll weevil as the bug that ate the Old South, effectively naturalizing a development that was eminently human. It helped to evade accountability for racism and other inglorious reasons for the demise of King Cotton.

A thick network of narratives surrounded the dodo before modern science looked at the bird with its own rationale. It was similar with soil erosion. Humans had plenty of practices regarding soil conservation when it became a distinct professional field with academic credentials and a grand narrative. As institutionalized in the United States in the 1930s, soil conservation was about nothing short of saving civilization. In this reading, societies have collapsed throughout human history for neglecting the soil, and only determined action could forestall history repeating itself. A leading expert, Walter Lowdermilk, even went to Jerusalem and proposed an Eleventh Commandment for soil conservation. Such an amendment was perhaps the greatest weapon in the arsenal of the Judeo-Christian tradition, and for a generation in which Eurocentrism was a fact rather than a problem, it did not really matter that an Eleventh Commandment was a difficult sell in China, a country where Lowdermilk had previously lived, because the idea of divinely imposed law was foreign to Chinese tradition.[2] Soil conservation expertise also thrived on dramatic pictures. One of these icons depicted a seriously eroded landscape in the US South: Providence Canyon, known informally as Georgia's Little Grand Canyon.

Providence Canyon was young, a result of the cotton boom of the nineteenth century, which in turn was an example of the expansion of agricultural lands in that century. Frontier societies were inherently unstable due to the vagaries of distant markets, scant knowledge about the environment, and frail power structures. Soils were no more stable, and erosion was a major problem, though more in retrospect than in contemporary contexts. Frontier farming was a gamble, and if it did not work out and soils vanished, people simply moved on. It was a convenient solution in the absence of scientific and other authorities, but it was not a sustainable one. People could not move on infinitely on a finite planet.

Soil conservation grew against the backdrop of ignorance and carelessness, but it remained embattled. The world's largest expert body, the US Soil Conservation Service, was less an expression of a remorseful frontier society than a result of two intertwined disasters in the 1930s, the Great Depression and the Dust Bowl. Expertise was also more impressive in ambition than in substance: the grand narrative about soil and civilization also served to distract from the shaky knowledge base during the early years. But even after the hectic early years, soil conservation found it difficult to sell its long-term concerns to farmers struggling to pay their bills, and to the extent that this cooperation worked, it was built on subsidies, technological assistance, and support of more intensive land use. The fight against soil erosion lacks a powerful lobby to this day: in an urban world, vanishing soils do not generate much excitement even among environmentalists. While soil conservationists evoked the Little Grand Canyon as a warning sign, locals were reluctant even to acknowledge that it was the result of human-caused erosion. Providence Canyon became a state park in 1971, but the rationale was scenic rather than environmental.

PART III

The gap between agriculture and the rest of society was also a critical issue in the story of the cane toad. A native of American rain forests, it came to Australia and numerous other places on the Pacific rim through the global expert network of sugar producers. The giant toad was supposed to help Queensland canegrowers in their fight against cane grubs, and agricultural interest ceased once cane toads failed in their assigned task. The rest of society was stuck with a species that multiplied rapidly and moved beyond the sugarcane region. It shocked ecosystems and humans alike. Australia's native fauna did not include toads, let alone toads that gave predators a lethal dose of poison, and people found it difficult to get used to toads that could weigh more than four pounds.

Like many nonnative species in the twentieth century, the spread of the cane toad was framed in the language of total war. Scientists painted a different picture after years of painstaking research: cane toads were disruptive to Australia's ecosystems, but the impact was more complicated and more multidimensional than convenient clichés suggested. Some species were unaffected, others faced local disturbances, and some impacts were actually positive, like the reduced parasite burden for native frogs. But scientists with complicated findings were fighting an uphill battle against people with dramatic pictures. Like other chapters in this section, the cane toad story is about the disconnect between biological and visual history. Few people realized that the results of the scientists were actually scarier than the narrative of invasion. After decades of research, we do not know how the spread of cane toads will play out.

Cane toads are hard to purge from the Australian continent, though science-led eradication efforts have grown in sophistication over recent decades.[3] However, locals took matters into their own hands and started their own extermination drives. It was grotesquely ineffective, but was it really meant to be effective? It was probably more about human needs than about cane toads, a way to vent anger, meet new people, and placate archaic hunting instincts. Maybe it was also about acting out an exuberant masculinity and about reaffirming an embattled human supremacy: in a world with plenty of environmental problems that are the direct result of mankind's irresponsible behavior, it was gratifying to have a problem that spread just by itself. One of Australia's most popular documentaries was about cane toads, or perhaps more about the responses that they triggered from humans. The real story of the cane toads was about biological and cultural coevolution.

The final chapter looks at mineral deposits, which offer a particularly glaring example of irreversibility: modern resource extraction is pillaging a store of assets in a flurry that was built over geological time. Once brought to market, mineral deposits are irretrievably gone, but that insight has not played a significant role in Saudi Arabia or elsewhere. Modern mining rewards a predatory style of business, a get-rich-on-the-cheap mentality where gaining a grip on the prize was the overriding concern, if not the only one. The alliance between Saudi Arabia and the United States was unlikely by any measure except for one: oil.

The deal turned a young kingdom held together by little more than an aging Ibn Saud into a regime that has survived plenty of doomsayers over the past seventy-five years. Oil revenues also built a state apparatus with widely different levels of efficiency and fateful path dependencies: few things have played a more powerful role for state-building in the non-Western world than resource revenues. Saudi Arabia was not the first rentier state, and it is disheartening to observe the insignificance of previous experience. Resource-led development has brought countries down individual paths, and Saudi Arabia had its institutional success stories—for instance, a good central bank and a professional national oil company—and yet observations on other resource states from previous centuries ring strikingly familiar. Writing about Peru's guano years, Shane Hunt argued that "perhaps the most pernicious effect of the rentier economy" was "psychological": "In the rentier economy, wealth is generated merely by ownership, not by effort."[4] More than eighty years after the first discovery of oil, Saudi Arabia's economy remains wedded to oil, for better or worse, and the challenges of decarbonizing the Saudi economy may pale in comparison with future challenges when the wells run dry. Modern history has a body of stories about resource states whose resource base expired, and none of them bodes well. The interaction of state-building and socioeconomic development with a material resource flow makes for a fateful entanglement in boom times as well as times of collapse.

The Saudi Arabian oil state shows the characteristic opaqueness of a resource economy. Assets and modes of decision making are notoriously intransparent and subjects of much speculation. But would those who do not work in the resource business really want to take a closer look? Collective awareness about resource extraction is typically scant, and it is a matter of debate whether this is due to the small number of people working in this field—less than one in five hundred

Saudis works for the national oil company—or because of a dim awareness of unwholesome stories. As the Indian novelist Amitav Ghosh wrote in 1992, "The history of oil is a matter of embarrassment verging on the unspeakable, the pornographic."[5] For all the importance of cheap resources for modern societies, discussions are usually confined to expert and business circles, and if outsiders chime in, the conversation is often framed in moralizing metaphors of addiction and disease for want of a better language. It is a pretty shambolic way to talk about an event that can only happen once in the history of our planet.

All that makes irreversibility a curious blend of hard material facts and a remarkably flexible economic, political, and cultural context. Some things are no longer possible, but people retain a choice among a range of options. In fact, some trends do look reversible to an extent. Humans have stopped invasive species, and maybe future research will even develop options to purge cane toads from the Australian continent. Monocultures can control biological challenges to an extent that evokes the innocence of the early years. Soils do rebuild, evolution can restore biological diversity, and genetic reconstruction might even resurrect the dodo from its grave. But all these things require time and resources to an extent that may prove prohibitive: What are the odds for funding a genetic mega-project to reverse the sixth extinction at a time when we do not even have the money for comprehensive studies on many species that are threatened or invasive? And would it really help us in our ongoing efforts to navigate the vortex if we took irreversibility as just another cultural construct? As it stands, we are struggling to accept irreversibility as a fact of modern history. But it might be better, both for the world and for our own mental universe, if we tried to come to terms with it.

11

THE DODO

SPECIES THAT PERISH

I. NATURAL HISTORY

In September 1598, a fleet of five ships approached an uninhabited island in the Indian Ocean. They had started their journey in the Netherlands twenty weeks earlier, and this was their first stop on the way to the Dutch East Indies. A few men went ashore to explore the island and look for food and freshwater. Portuguese ships had visited the island in 1500, which was henceforth known as Ilha do Cirne, and Arab seamen had been there even earlier, but information about the island's geography and resources was scarce.[1] In fact, the Dutch officers disagreed on whether they were really mooring at Ilha do Cirne or rather at Rodrigues, a smaller island 350 miles to the east. But when the crewmembers returned, they brought good news: the island had a natural harbor and plenty of freshwater. They also brought "eight or nine large birds: dodos."[2]

The ships stayed for two weeks. Officers renamed the island Mauritius in honor of Maurice of Nassau, the contemporary *stadtholder* of the Netherlands, and several expeditions studied the island's flora and fauna.³ Accounts of their discoveries received a grateful audience back home, with German and Latin translations enlarging the realm of readers beyond the Netherlands, and European literati learned that Mauritius had birds "as large as two swans" and that "their flesh is good to eat."⁴ Both statements would subsequently be disputed, as most statements about the dodo were at one point or another, but uncertainties barely diminished the infatuation of Europeans with the flightless bird. Natural history had a new enigma.

Collecting and classifying plants and animals was a transnational endeavor in early modern Europe. Seafaring nations delivered a steady stream of specimens and observations, and scholars exchanged samples as well as opinions about their inner nature. With its taxonomic rigor, its disciplined collaborators and its hierarchies, the emerging network of natural history invited comparisons with the military, down to the ironclad routines and the uniforms that Linnaeus devised for his excursions.⁵ Systems of classifications were oozing order, and yet they were the result of a dynamic and open-ended process of communication and exchange. Species were "not born but made in a process of negotiation between botanists, their patrons, and the expediencies of the marketplace."⁶

Many species were competing for attention in early modern Europe, but the dodo had one crucial advantage: it looked different. The bird was not an obvious variety of a well-known species, challenging taxonomists to find a place in their classifications where it might fit in. There was also something oxymoronic about a bird that could not fly, and a bulky body and an oversized beak gave the dodo an endearing clumsiness. Linnaeus certainly found that the dodo's ineptitude was a defining characteristic, and he selected its binomial name accordingly: *Didus ineptus*.⁷

But while interest in the dodo was strong, the same could not be said of contemporary knowledge. Eyewitness accounts from the non-European world were edited and rewritten in ways that remain a mystery to this day, and biological material was hard to come by: few specimens made it to Europe dead or alive, "perhaps as few as 3–4."⁸ The natural history book market was competitive and did not reward intellectual modesty, and authors plagiarized other publications in ways that would make first-year history students blush. "The dodo of European natural histories was assembled from a scant collection of dismantled parts and travel accounts," Natalie Lawrence wrote.⁹ Scholars continue to wrestle with the ensuing uncertainties, and even

11.1 Picture of a dodo in George Edwards's *Gleanings of Natural History* (1760). The guinea pig serves to illustrate the dodo's size. Image, University of Wisconsin Digital Collections.

coffee-table books come with a warning that "what is actually known of the living, breathing dodo is minimal."[10] A number of prints and paintings survive, including an iconic one by the Flemish painter Roelandt Savery that is on display in London's Natural History Museum, but the degree of artistic license remains anyone's guess. According to Julian Hume, a court artist in India, Ustad Mansur, produced "almost certainly the most accurate and reliable coloured rendition of the Dodo" when he drew a picture of a live specimen in the menagerie of Mughal Emperor Jahangir.[11]

If we can trust this picture, the dodo was not quite as fat as European depictions suggested. Maybe artists had seen overfed dodos in European captivity, or the bird's most distinctive feature, its blatant lack of mobility, had captured their imagination. Or maybe it was about the Dutch East India Company (Vereenigde Oostindische Compagnie, or VOC) that claimed Mauritius in the seventeenth century, a company so infamous that VOC was read as "vergaan onder corruptie" (perish under corruption) after its demise.[12] Lawrence has argued that the dodo was "the VOC's bird," and a gluttonous bird was a powerful symbol of the VOC's insatiable appetite for commodities and profits.[13]

A thick cultural web surrounded the dodo long before the Oxford don Charles Lutwidge Dodgson, better known by his pseudonym Lewis Carroll, made the dodo his alter ego in *Alice in Wonderland*.[14] The culture of the dodo became so powerful that some early nineteenth-century naturalists wondered whether the bird was just an imaginary creature.[15] It was natural history's intellectual and institutional heir, academic science, that brought the real bird back into the myth, and the relationship between the dodo and the scientists was mutually beneficial. When the British Association for the Advancement of Science met in Oxford in 1847, the university commissioned the dissection of a dodo skull that had survived in the collection of the local Ashmolean Museum for some two hundred years. The results were presented in an evening lecture with Prince Albert in the audience, and it helped to raise the profile of science at Oxford.[16]

The dissection came to the conclusion that the dodo was "a very aberrant member of the family *Columbidæ*."[17] In other words, the dodo was a pigeon, albeit a really strange one, a view that was ridiculed initially but eventually became accepted knowledge.[18] In 1865, an excavation found subfossil dodo remains in the Mare aux Songes marsh near the southeastern coast of Mauritius, which greatly increased the bone material at the scientists' disposal and put the rumors about the bird's existence to rest.[19] More recently, a DNA-based study revealed that the dodo's "closest living relative is the monotypic Nicobar pigeon (*Caloenas nicobarica*) from the Nicobar Islands" in the Bay of Bengal.[20] Scientists also analyzed cross-sections of dodo bones and deduced that the bird bred around August, that chicks grew rapidly, and that molting occurred between March and the end of July.[21]

However, exploration and classification were just one part of Europe's engagement with exotic species. The other part was exploitation. In the case of Dutch Mauritius, the greatest biological asset was ebony. The Dutch cut some thirty square kilometers of ebony forest during the sixty-six years of settlement, which amounted to "perhaps 6–7% of the primeval lowland forest."[22] Dutch colonialists did not see a need for something akin to the European project of sustainable forestry (see chapter 4, Sustainable Forestry), and the impact paled in comparison with the toll of sugar cultivation under British rule (see chapter 2, Sugar). Native forests still claimed 70 percent of the island's surface around 1800, but the boom of sugarcane in subsequent decades brought widespread deforestation with severe repercussions for indigenous species.[23] The endangered Mauritius kestrel survives in two populations separated by fifteen kilometers of agricultural land, and ringing programs have not identified a single migration event.[24]

The dodo never stood on a par with ebony, let alone sugar. It had some prestige value, which likely brought the dodo to the Mughal court as a gift from British merchants seeking a favorable trade deal.[25] But unlike eucalyptus (see chapter 27, Eucalyptus) and other objects of the biological exchange, dodos were never brought anywhere to breed, and their greatest practical use was for provisioning. Obtaining the meaty flightless bird was a matter of collecting rather than hunting, and countless exemplars that would have cheered European naturalists were unglamorously consumed by hardy sailors. Whether they enjoyed it remains subject to debate. The dodo's French name, *oiseau de nausée*, suggests that the bird's greasy meat left even gastronomically challenged sailors nauseous.[26] However, *oiseau de nausée* was probably a corruption of *oiseau de Nazare*, which may be the contemporary name of a nearby island, and accounts of the bad taste are disputed.[27] As David Quammen has noted, "The story of the dodo is obscured by a fog of uncertainties."[28]

Exploration and exploitation look like contrarian concepts in retrospect, but contemporary views were different. Both went hand in hand in and beyond the colonial world. Heligoland, a North Sea island that served as a popular resting spot for migrating birds, became a Mecca for bird lovers precisely because it had a long tradition of bird hunting. Heligoland's residents made birds a cornerstone of their diet when their farmland shrank due to coastal erosion, and they accumulated an impressive body of knowledge about the different species and the best ways of catching them.[29] Local lore had it that Heligoland's churchgoers would sometimes leave mass in droves when a flock was approaching.[30]

Hunting did not look problematic as long as environments were teeming with wildlife. "We have generally lost from our collective memory any notion of the scale and size of wildlife populations before intensified human predation," John Richards wrote in his environmental history of the early modern world.[31] But hunting eventually claimed its toll, and islands typically faced limits fairly soon because they lacked hinterlands that could resupply dwindling stocks. It did not escape people's attention. The slaughter of tortoises raised concern on Mauritius and neighboring Réunion, "with orders banning hunting dating from as early 1671."[32] As on Heligoland, an aristocrat from Hanover wrote in 1826 that the large flocks of birds were gone and that quite a few years had passed since the islanders' hunting fever had last caused the suspension of religious service.[33]

No laments marked the fading of the dodo. We are even in the dark about the precise cause: Was it human consumption, or invasive species (see chapter 14, Cane Toads) like ship rats, or something else? We only know that dodos, a popular topic in Dutch accounts until 1620, are largely absent from subsequent reports.[34] Scholars continue to discuss the precise timeline, with some evidence pointing to an extinction date around 1690, but the end is beyond debate.[35] Less than a century after a few Dutch men went ashore in Mauritius, the dodo was no more.

2. THE SIXTH ONE

The dodo went extinct in an age that did not know what extinction was. There was no need for such a concept at a time when the natural world was God's creation and just a few thousand years old. But this received wisdom came into doubt during the eighteenth century when fossil remains and other findings had naturalists scratching their heads. The breakthrough occurred in the Age of Revolutions when geology and paleontology "burst the limits of time," in Georges Cuvier's memorable phrase, by expanding the realm of earth's history and making extinction a plausible concept.[36] Humanity's understanding of its place on planet earth would never be the same.

Like every new scientific paradigm, extinction remained contested for a while.[37] One of the skeptics was Thomas Jefferson, who had personally studied fossil mastodon bones, held a mastodon tooth and a thigh bone in his personal collection, and wrote about the elephant species in his *Notes on the State of Virginia*. When Lewis and Clark embarked on their expedition to the American Northwest in 1804, Jefferson hoped that they would find live exemplars in the unexplored

expanses of the American West.[38] But as extinction became an established concept, another question arose: was there ground for concern? The irreversible loss of a species is poised to trigger nostalgic sentiments, but it was not necessarily a biological problem. In light of Charles Darwin's theory of evolution, extinction was just the inevitable downside of the survival of the fittest. Competition implied that some species could fall by the wayside, and the clumsy dodo looked like the perfect candidate for such a deserved fate. Should people really intervene and obstruct natural selection? But selection was not always an evolutionary process.

According to the theory of evolution, extinction was supposed to be a background phenomenon that occurred at a fairly constant rate. However, the fossil record told a different story: paleontologists identified dramatic changes in the rate of extinction. At certain times, more than 30 percent of plants and animals perished all over the world, and these events captured the imagination of scientists and the public. Paleontologists commonly speak of five mass extinctions over the past 540 million years, and the last of these "big five" events occurred some 65 million years ago: it marked the end of the dinosaurs. Depending on the magnitude of the event, recovery of biotic diversity can take 10 million years or longer.[39] It is more than a matter of paleontological interest, as we may currently live in the midst of the sixth mass extinction event in earth history. In his book *The Future of Life*, the American biologist E. O. Wilson warned that "as many as half [of earth's species] may be gone by the end of this century."[40]

It is beyond doubt that the sixth mass extinction is the result of human agency.[41] But then, human agency can mean very different things when it comes to vanishing species. It can be as big as global climate change, which is moving climate zones with dramatic consequences for ecosystems all over the world. Or it can be as trivial as the rats that escaped from the Dutch ships mooring at Mauritius. Extinction can result from the incidental action of a small group such as the Queensland farmers who introduced cane toads (see chapter 14, Cane Toads) to their sugarcane fields. It can come from the dissipating remains of plastic bags (see chapter 40, Plastic Bags) in the open sea. It can even emerge when a world-class scientist runs amok, as shown in Robert Koch's proposal to eliminate wild game from German East Africa in order to get rid of tsetse infection (the idea sank when conservationists pointed out that extermination was unbecoming for a *Kulturvolk*.)[42] Or extinction can result from hunting for gain or pleasure, and it did not always take a bloodbath akin to pelagic whaling (see chapter 9, Whaling). The Caribbean monk seal was rare and lived in small fragmented populations, and occasional hunting was enough to push it into oblivion.[43]

Observing extinction was one thing; fighting it another. According to Mark Barrow, it was a hodgepodge of seven different motives: naturalists argued for species preservation on the grounds of aesthetics, usefulness, ecological stability, evolutionary precaution, nationalism, science, and ethics.[44] Wilson argued that humans "should judge every scrap of biodiversity as priceless" and "should not knowingly allow any species or race to go extinct," but that was wishful thinking in light of a diverse set of values.[45] Not every species was equally useful, aesthetically pleasant, or prone to nourishing nationalist sentiments, but if a species pressed the right buttons, preservationist sentiments could fuel a dramatic cultural rebirth. The dodo was a case in point.

Species preservation gave the dodo a new meaning that superseded the curiosity status of natural history and the whimsical adjudicator of the caucus race in *Alice in Wonderland*.[46] The dodo henceforth lingered as an icon of conservation, and its narrative revealed a sense of postcolonial guilt: Ursula Heise called it "a recurrent symbol of the destruction of nature wrought by the imperialist expansion of European modernity."[47] The dodo is one of a handful of species of "almost mythical status" that are gone forever. In his *Story of Conservation*, William Adams mentions the dodo along with the quagga, Steller's sea cow, the great auk, and the passenger pigeon.[48] In somewhat ironic fashion, conservationists turned the dodo's evolutionary disadvantage, its helplessness in the face of predators, into a postmortal advantage, as the flightless bird seemed to cry for human help from its grave. Thanks to natural history, there was also a sufficient body of relics and paintings to sustain a modern myth, something that cannot be said of every species that went extinct since the dawn of modernity. The only written record of the Ascension crake, another flightless bird on a lonesome island, is a few lines in the diary of a seventeenth-century English traveler.[49]

Needless to say, there was always something arbitrary about the choice of a handful of species for a problem that was about large numbers. On the Mascarene Islands, which include Mauritius, Rodrigues, and Réunion, "the Dodo was only one of at least 48 endemic species of terrestrial vertebrates that became extinct before 1800," Samuel Turvey and Anthony Cheke have noted.[50] Mammals and birds usually ranked high on the list of conservation sentiments, particularly if they were large, strong, or particularly cute, while animals were out of luck when they were small, ugly, or devoid of fur and feathers. Endangered plants received far less attention, and when it came to parasites or microbes, things were truly dark: Wilson called bacteria "the 'black hole' of biodiversity."[51] Wildlife reserves for large animals can also help to preserve lesser known species that are equally endangered, and some ecosystems rely on "keystone species" such as the elephants that keep forest growth down in Hwange National Park in Western Zimbabwe.[52] Sometimes iconic species can also help save treasures of nature. The Białowieża Forest on the border of Poland and Belarus survived as a unique European wilderness because it is home to the European bison.[53] But for all these benefits that escaped a naive critique, there was something shallow about a conservation discourse that thrived on visuals of charismatic animals. It obviously fit the predilections of late twentieth-century media societies, but the bias goes back to an age before color magazines.

Next to the naturalists, it was hunters who turned extinction into an important concern of conservation in the late nineteenth century. Frontier regions played a pivotal role: the decline of formerly rich stocks of wildlife gave some tough men pause. Theodore Roosevelt saw the dwindling big game population in the Dakota Badlands and founded America's first wildlife conservation organization, the Boone and Crockett Club.[54] A similar reckoning among hunters in the British Empire led to the creation in 1903 of the Society for the Preservation of the Wild Fauna of the Empire, whose elitist founders were soon "lampooned as 'penitent butchers.'"[55] Inspired by the demise of the bison, the National Zoological Park was founded in Washington in 1889 as a preserve for North American animals, and captive breeding programs became an important part of zoo management over the course of the twentieth century.[56] Preservation of species was a major driver behind the dramatic global expansion of protected areas since 1960 (see chapter 26, Kruger National Park), a growing number of red lists sought to capture threats to plants and animals in all their diversity, and the fight against extinction became a fixture of global conservation diplomacy with the Convention on Biological Diversity of 1992 (see chapter 24.3, Getting Serious). But while campaigns were gaining momentum, the conservation community made a disturbing discovery. As resources were growing, so did the doubts about goals and strategies.

3. CONSERVING DIVERSITY

One of the legacies of natural history was an ongoing process of intellectual fragmentation. "During the eighteenth and nineteenth centuries, natural history . . . gradually splintered into a series of more specialized disciplines," Barrow observed.[57] Academic science continued the trend toward compartmentalization in disciplines and subdisciplines, notwithstanding ecology's insights about interconnections on many levels. Experts like Wilson, who was a world-leading expert on ants, have looked beyond their field of specialization, but more often than not, taxonomists have focused on their own academic niche. Even in the eyes of a scientist, the world's plants and animals are rarely a family of equals. It is no coincidence that red lists have long focused on classes of species whereas the idea of a red list for ecosystems only gained traction recently.[58]

However, the red lists have also seen tremendous change in their own right. The first two Red Data Books, compiled under the auspices of the International Union for Conservation of Nature (IUCN), dealt with mammals and birds and were published as loose-leaf editions in 1966; a third loose-leaf volume on reptiles and amphibians came out two years later.[59] The compilation of national and global red lists gained momentum in the 1970s, and the IUCN Global Species Program currently manages data for 138,300 species and aims for a target of 160,000 species, a goal that would cover less than 10 percent of the world's known species.[60] While the first Red Data Books "included only the categories 'rare' and 'endangered,'" with "a three-star designation" for "the most endangered species," the IUCN lists used 4 different categories by the late 1960s, and when a review discussed 151 lists of threatened species with 57 different categories in the 1980s, it found that "the extinct category is the only status category whose definition is unequivocal."[61]

The fight against extinction came a long way in the postwar years. When the International Technical Conference on the Protection of Nature convened for a landmark meeting in Lake Success, New York, in 1949, its conclusions included "a list of 14 mammals and 13 birds in need of action," mostly "large well-known animals" including subspecies and local populations.[62] But in the twenty-first century, conserving biodiversity is big science, and today, few if any experts will recognize the full IUCN Red List. Among the world's animals, whose representation in red lists far exceeds plants, 776 are currently extinct or extinct in the wild, 3,273 are critically endangered, 4,219 are endangered, and 5,949 are vulnerable.[63] And like every attempt to quantify the complexity of the natural world (see chapter 7.2, Numbers Games), these figures need context to understand the full drama at play. Even when numbers improve and species move to a less alarming category, the gene pool may reflect the perils of the past. The Mauritius kestrel has rebounded from only 4 known individuals in the 1970s to between 500 and 800, but inbreeding led to a substantial loss of genetic variation.[64]

The survival of species depends on a population of sufficient size, and defining that size is not just a matter of ecological knowledge. It is also about personal experiences, as marine biologist Daniel Pauly argued in a widely cited article of 1995. Writing about overfishing, he found that estimates of remaining stocks suffered from what he called "shifting baseline syndrome": "Each generation of fisheries scientists accepts as a baseline the stock size and species composition that occurred at the beginning of their careers, and uses this to evaluate changes."[65] In other words, scientists tend to take stocks as their yardsticks that are already depleted and under threat, and the lack of earlier scientific accounts leaves them in the dark as to what a healthy population might look like. Pauly did not offer a solution, though he proposed to take anecdotal evidence more seriously. It might be a good idea if the experience of the dodo did not suggest that this can easily lead scientists into another maze of ambiguities.[66]

However, biological diversity is not only about the number of animals. It is also about genetic diversity among members of a given species, and not only because a diverse gene pool increases the resilience of plant and animal populations. Agricultural plants in particular thrive on intraspecies diversity, as it helps staple crops to adjust to local conditions and provides raw material for breeding (see chapter 28, Hybrid Corn). The latter is a particular matter of concern because high-yield seeds tend to displace landraces with lower productivity while the variety of these landraces serves as a backbone for seed development. In other words, seed development thrives on the very genetic resources that it puts at risk, a paradox that has haunted plant breeders for generations. As Tiago Saraiva has noted, "Plant breeders have been worrying about the increasing genetic uniformity of European fields since the beginning of the twentieth century."[67]

Breeders sought to counter the trend with systematic seed collection and storage. Founded in late nineteenth-century Russia, the Bureau of Applied Botany and Plant Genetics at Saint Petersburg's Botanical Garden (see chapter 27.2, Botanical Exchange) became "the mother of all modern, science-driven seed banks," propelled to worldwide prominence through the work of the geneticist Nikolai Vavilov beginning in the 1920s.[68] The institute famously survived the siege of Leningrad during World War II as well the vagaries of Soviet and post-Soviet rule, which included an attempt to raze it to make way for luxury homes in 2010.[69] A global system of seed banks seeks to conserve agrobiodiversity today, with Spitsbergen's Svalbard Global Seed Vault at its pinnacle, but its achievements may remain a mystery even with the best of monitoring. As it stands, breeding is a success story in terms of yields per acre, but it will never become known whether an extinct landrace, if only it had been crossed with the right complement, would have allowed larger harvests, greater resistance to pests, or other benefits in the perennial struggle of the world's agriculturalists against the many contestations from Mother Nature.

The only certainty is that seed banks are expensive, and this makes them fit into a pattern of costly high-profile endeavors that define the public imagination of the fight against extinction. Wildlife biologists saved the California condor with a comprehensive program that included puppets to feed condor chicks.[70] In the 1960s, rangers and scientists in South Africa developed tranquilizing technology to relocate and save the white rhinoceros.[71] One of the veterinarians celebrated the success with a popular book titled *The Flying Syringe*, and zoological gardens use every birth of an endangered species for a flurry of press releases, but the amount of money and energy that these efforts claim stands in inverse relation to the number of species that they actually affect.[72] David Hancocks, who has served as director of zoos in the United States and Australia, put it as follows: "We cannot save the world's endangered wildlife through the few successful breeding programs in zoos, just as one cannot save a language simply by holding on to a rare document."[73]

As the celebrated father of biodiversity, Wilson gave seed banks, zoological gardens, and other research institutes their due, but he was under no illusions as to their overall impact: they "will save a few species otherwise beyond hope, but the light and the way for the world's biodiversity is the preservation of natural ecosystems."[74] However, nature reserves have always been contested on the ground, and the number and intensity of conflicts have grown dramatically as ever more land has been put under protection since the 1960s (see chapter

26, Kruger National Park). And then, how much land do you need to save the flora and fauna of the world? Wilson seriously suggested aiming for nature reserves on 50 percent of the earth's surface.[75]

So is it perhaps time for a radically different approach? In 1992, the American physicist and science fiction writer Gregory Benford called for a broad program for the collection and freezing of biological specimens from all over the world so that a future generation with better technology might reproduce extinct species from cold storage.[76] The following year, moviegoers learned that microbiological resurrection might not go according to plan when cloned dinosaurs ran amok in Steven Spielberg's *Jurassic Park*. However, cryogenic conservation does exist: Elizabeth Kolbert ends her journey through the world of mass extinction in the Institute for Conservation Research of the San Diego Zoo, where the remains of vanished species rest in tanks of liquid nitrogen.[77] In 2013, *The Futurist* published an article from a master's student who proposed bringing the passenger pigeon back from extinction, but it found that genetic reconstruction would only be the beginning: the comeback would also need facilities for pigeon training and the large forests whose disappearance doomed the species in the late nineteenth century.[78] Nobody knows whether gene-based resurrection will work, but it will likely be expensive. With that, cryogenic conservation may replicate the fundamental imbalance in the human response to extinction: large projects for a selected few and apathy toward the masses.

As it stands, the dodo is lucky in that its cultural legacy is overwhelmingly positive. Nature lovers cherish its picture while scientists continue to subject its remains to the latest methods, and conservationists use it as an icon for the struggle against extinction. Even the dodo of *Alice in Wonderland* is a gentle creature, as shown in adjudicating the caucus race that "*everybody* has won, and *all* must have prizes."[79] The worst that has been said about the dodo concerned the digestive troubles of those who ate them. But people may take a more critical view a few hundred years from now. They may wonder about the mismatch between our intellectual and material investments in one really strange bird and our negligence toward many other species, and they may wonder what we were thinking. Stuck on a planet that has wasted its biological diversity, they will probably cite our infatuation with the dodo as evidence of how modern societies failed to grasp the real problem about extinction.

12

The Boll Weevil

The Nemesis of Monoculture

I. THE RULE OF KING COTTON

The trouble started in Texas. In the early 1890s, Texan cotton growers noticed a previously unrecognized beetle in their fields. It was tiny, only between two and seven millimeters long, and a long dark snout made it particularly awkward looking even by the generous standards of the insect world. It multiplied rapidly, and its favorite food was pollen in the cotton plant's unopened flower buds, precisely the part that grows into bolls with the desired fibers.[1] Over the following decades, cotton planters watched in horror as the beetle ate its way through the US cotton belt and lost none of its vigor on the way. When it finally reached Georgia in 1921, it consumed 45 percent of the season's crop.[2] The insect became notorious throughout the South, and most people were familiar with its name long before it actually arrived: the boll weevil.

Cotton was not the only cash crop of the American South, nor was it the first; that distinction goes to tobacco and rice.[3] But cotton plantations supplied the raw material for the textile mills of England, and

THE BOLL WEEVIL

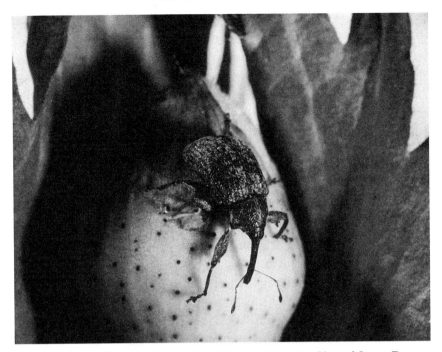

12.1 A boll weevil. Image, Agricultural Research Service, United States Department of Agriculture.

as textile production was driving the Industrial Revolution in England, cotton became the South's defining crop. Fueled by rapidly growing demand, the population swelled from less than two million in the 1790s to twelve million in 1860.[4] In 1860, cotton alone amounted to 61 percent of all US exports.[5] The Civil War suspended trade with Europe and led to an expansion of cotton production from Turkey to Peru, but the South regained its prominence in world markets: it produced ten million bales of cotton in 1900, with two-thirds of the harvest going into export.[6] But all this was at stake once the boll weevil, endemic in Mexico, migrated across the Rio Grande (see chapter 14, Cane Toads).

The weevil's advances were so dramatic that the insect came up in Theodore Roosevelt's 1904 State of the Union address.[7] After all, it was hard to imagine a South without cotton. Trade connections, farm credit, transport, and processing technology were all geared toward cotton production. Perhaps most crucially, cotton was deeply ingrained in the labor system, which in turn played a huge role in stabilizing racial hierarchies after the abolition of slavery. In short, there was no way to abandon cotton, or even to challenge its hegemony, without putting the fundamental pillars of Southern society at risk.

Even a temporary suspension was out of the question: the Texas government ignored proposals in 1894 and 1895 to impose a quarantine zone to stop the weevil's northward crawl.[8] The boll weevil highlighted the fateful interdependence between a society and its hegemonic crop.

The experience of the cotton South was hardly unique. The nineteenth century saw a massive expansion of export-oriented monocultures. All around the globe, societies were betting their economic fate on the production of commodities for industrializing Western countries. And sooner or later, societies learned that this strategy entailed risks beyond the economic dependencies of modern commodity chains (see chapter 30, Rice-Eating Rubber Tree). Unlike factories, the industrial twin of the new agriculture (see chapter 2, Sugar), monocultures were open to biological contestations: the concentration of identical plants in certain regions made them prime targets for pests and pathogens. For example, banana trees had been wilting on Latin American plantations since the 1890s due to Panama disease, which was caused by a soil-based fungus (see chapter 10, United Fruit).[9] Another fungus caused coffee leaf rust on plantations across Africa and Asia in the late nineteenth century, which opened the door for Brazil to become the world's leading coffee producer.[10] European winegrowers were terrified to learn about phylloxera, a tiny insect that has attacked the roots of grapevines across Europe since 1864.[11]

Threats were diverse, but they produced similar responses from growers. They wanted a quick and inexpensive cure so that everything could stay the same, and the authority that was supposed to deliver that cure was modern science. When the first reports about the boll weevil reached Washington in 1894, the US Department of Agriculture sent an entomologist to Texas.[12] Over the following years, growers and state governments hired new experts and created research centers all over the South, and most plantation economies did the same during the nineteenth and twentieth centuries. When it came to scientific expertise, nothing galvanized interest in farming circles like a devastating plague, and pests, fungi, and bacteria became first-rate agents of academic institution-building. Monocultures created their science, not the other way around (see chapter 4.2, Specialist Trees, Specialist Minds).

Scientists were glad to respond. Biological threats offered jobs and a mission: solving an intricate problem was a powerful demonstration of the merits of modern science. And yet for all the enthusiasm that scientists brought to the fight against the boll weevil and other threats, their task was, as academic jobs go, a particularly delicate one: they were supposed to find a cure, and better be quick. The scientists were essentially doing a repair job, and academic freedom was only a faint idea out in the fields. Panicked planters were an inconvenient clientele. Texas fired its first state entomologist in 1902 after his suggested remedy had shown disappointing results in field trials.[13] When the plantation owner Alfred Stone did not appreciate the wisdom of the Delta Branch Experiment Station that he personally had helped to create, he and his neighbor went on their own research trip through Texas, Louisiana, and southern Mississippi in the fall of 1910. After traveling 1,600 miles over fifteen days, they summarized their observations in a report that they titled, without undue modesty, "The Truth about the Boll Weevil."[14]

Solutions were in high demand, and inventors were not shy about offering devices to desperate growers. Mechanical weevil catchers of dubious merit went on sale in suffering regions, and engineers continued trying to build an operational weevil-collecting machine until the 1960s.[15] Some communities had more trust in human hands and sent schoolchildren into cotton fields to pick weevils.[16] Poisons were another option, but they faced a particular challenge with the boll weevil. The beetle spent most of its life nestled inside the cotton plant and was thus hard to reach for substances applied to the outside.[17]

For most planters around 1900 the best defense was to shorten the growing season. Boll weevil populations built up over the year and were at their most devastating in late summer, allowing planters to limit damage through an early harvest. Early maturing seed varieties and proper fertilization (see chapter 8, Guano) helped speed up growth, with planters usually being more enthusiastic about the first option. They dispatched agents to regions where quicker-growing varieties were in use, and Texans even formed a Boll Weevil Convention in 1902 to obtain the right seeds and cheaper freight rates from railroads.[18] Some communities ventured to diversify, which earned the boll weevil a monument of its own—a rare achievement for a devastating pest. Faced with the advancing boll weevil, a county in Alabama went into peanut production and became so successful that it erected a statue to gratefully commemorate the infestation. Featuring a toga-clad woman carrying a supersized weevil like a torch of liberty, it has been standing on Main Street in Enterprise, Alabama, since 1919.

12.2 The Boll Weevil Monument in downtown Enterprise, Alabama. Image, Carol M. Highsmith Archive, Library of Congress.

People could be happy in a South without cotton, but not for long. A few years after the monument was erected, cotton regained its pivotal place in the county.[19]

The shift to new varieties came at a price. The length of the growing season is related to the length of cotton fibers, and fiber length is related to quality. When the boll weevil finally reached the Atlantic around 1920, it effectively terminated the production of the prized Sea Island cotton in South Carolina, and that was only one of some fifty longer-staple varieties that ceased to exist commercially.[20] It was not an unusual outcome for a monoculture in crisis. The modern world of commodities is not just the result of corporate decisions and consumer preferences but also of Mother Nature exercising veto power. Coffee leaf rust induced numerous planters to shift from arabica to robusta beans, which were inferior in taste but resistant to the devastating fungus.[21] United Fruit even reinvented the banana when it abandoned the traditional Gros Michel variety in favor of the disease-resistant Cavendish. It was the birth of what would henceforth be known as the Chiquita banana (see chapter 10, United Fruit).[22]

Researchers eventually found an effective insecticide against boll weevils in the form of calcium arsenate. Arsenic was a well-known poison if ever there was one, and calcium gave the compound the adhesive quality needed to reach its target. Calcium arsenate is known today to be carcinogenic, but it was already ranked an awkward substance at the time because it did not differentiate between cotton plants and agricultural workers in its sticking ability. But it was precisely the kind of panacea that planters had been seeking all along. First advocated for use in 1918, calcium arsenate sold three million pounds in 1919 and ten million pounds in 1920.[23]

2. SIZE MATTERS

The fight against the boll weevil was not just about technologies and profits. It was also about power in a divided society. Like all plantation societies after the abolition of slavery, the South had sought a new arrangement for land and labor. Its solution was sharecropping: white planters remained owners of the land, rented plots to Black tenants and received a share of the harvest as payment. It was essentially a perpetuation of less-than-free plantation labor, a step bigger in legal terms than in substance. Sharecropping allowed the former slaves to work the land without direct, gang-labor-style white supervision, but it ex-

12.3–4 Souvenirs for sale in Enterprise, Alabama. Image, Frank Uekötter.

posed them to a credit regime controlled by their former masters. The land reform (see chapter 6.2, Agrarian Reform) that Black people had hoped for after the Civil War never materialized.

Sharecropping had a profound influence on the response to the boll weevil. One of the most effective control measures, cutting and burning infested stalks after the harvest, was mostly ignored because it ran counter to the logic of sharecropping. As contractual obligations ended with picking and delivery of the harvest, tenants had no incentive to perform the work, all the more so since tenants were free to move on once they had paid their debt.[24] While planters were horrified by the plague, tenants could also see the boll weevil as an opportunity to gain leverage in their perennial negotiations with the owners of the land. Some planters found talk about the boll weevil more dangerous than the plague itself. When a delegation of experts came to Greenwood on an educational tour through the Mississippi delta in 1909, the town's leaders prevented them from giving public lectures and hurried them out of town.[25]

The intervention mirrored the planters' penchant for control as much as their failure to actually achieve it. Black farmers had their own networks of information, and they were keen observers of their local environments.[26] They could even learn about the boll weevil by listening to blues music.[27] However, the impact of the boll weevil on Southern society is an intellectual minefield, as contemporaries tended to blame the insect for the decline of cotton production and the mass migration of Black people to cities in the North. If it had not been for that tiny insect, the myth suggests, King Cotton could have reigned forever. In such a reading, the boll weevil was the bug that ate the Old South.[28]

Scholars have gone to great lengths in discounting the boll weevil's credentials in social engineering. The argument suggests an illusionary level of stability: like most plantation societies, the Cotton South was always in a state of flux.[29] Researchers have also pointed to the persistence of cotton and the expansion of acreage. The peak year for cotton production was 1929, when forty million acres were under cultivation.[30] Highlighting infestation also served to distract from other causes of the Great Migration such as bad credit, racism, and the hardships of farm labor.[31] "This little insect made a good villain, and its depredations could be placed in the 'act of God' category," Arvarh Strickland wrote.[32] Blaming the weevil was much easier than facing up to the stark systemic injustices of Southern society. When it came to avoiding awkward issues in modern societies, naturalization was a popular option (see chapter 25, 1976 Tangshan Earthquake).

But for all these caveats, the short-term impact was huge. A recent article declared that the boll weevil "hit local communities with the force of a tsunami," a metaphor that is probably more appropriate

12.5 A few blocks from the monument, this motel welcomes travelers without entomophobia. Image, Frank Uekötter.

than the authors have realized.³³ Communities can rebuild after a tsunami, but they have changed. Infestation disrupted local economies, reduced land values, and forced planters to negotiate with tenants on new terms. It also caused an enormous degree of migration within the region, with some counties receiving veritable demographic shocks from boll weevil refugees.³⁴ The boll weevil was much better at further destabilizing a society under stress than defining ways to the future.

The Cotton South survived the boll weevil. But as the twentieth century progressed, the crop's grip loosened, and a New South, more urban and industrial, left its mark on the land. Those who travel the South today see more woodlands than cotton fields, the result of the region's last plantation boom: expanding paper mills brought vast pine forests into the region.³⁵ Visitors also see sprawling suburbs, particularly around metropolitan centers such as Houston and Atlanta.³⁶ Farm population in the South decreased 83 percent between 1940 and

1969, and cotton accounted for only 7 percent of agricultural revenues by the early 1970s.[37]

Much of the change occurred in the years after 1945, when mechanical cotton picking reduced the need for human labor and sharecropping disappeared. But while the boll weevil did not terminate the reign of King Cotton, it changed modes of production. Calcium arsenate was expensive, the lower quality of cotton depressed revenues, and production took more land: few planters achieved yields per acre on a par with pre-infestation levels.[38] Marginal land went out of production, and mechanization and farm consolidation received a boost: the new weapons in the fight against the weevil required sufficient acreage and expertise to justify the expenses. In short, the new cotton farmer was bigger and better educated, an inaugural to a century that saw farming become ever more reliant on science and technology. Innovations such as synthetic nitrogen (see chapter 19, Synthetic Nitrogen) and hybrid seeds (see chapter 28, Hybrid Corn) were some of the next steps on a path that eventually made agriculture one of the most capital-intensive industries. Once more, the boll weevil boosted a general trend of modern agriculture. Rarely did a biological threat work out to the advantage of the small farmer.

3. NEVER FORGET

By the mid-1920s, it seemed as if the boll weevil and the cotton South had agreed on a truce. Cultivation methods had changed, investments per acre grew, and white planters were spending more time in their fields. And yet, in spite of the boll weevil's finally being endemic throughout the region, cotton was still an important part of Southern agriculture. Calcium arsenate remained popular among planters and loathed among workers, but the latter's concerns became less significant when aerial dusting came into use. Army pilots made the first experimental flights in the delta region in 1922, and only a few months later, growers could hire pilots and planes from Huff Daland Dusters, Inc. to do the awkward job. The company later started carrying passengers instead of poison, and still does. It also chose a new name, Delta Air Lines, a tribute to its origins in the Mississippi delta.[39]

The boll weevil continued to take its toll, but losses were relatively small during the 1920s and 1930s. In fact, losses were probably a good thing for planters, as they prevented cotton prices from falling even further than they did in the early 1930s.[40] The weevil returned with a vengeance in 1949, which became the worst year since 1923.[41] Planters resorted to newly available pesticides like DDT to fight back, thus opening an age of prolific spraying (see chapter 38, DDT). These pesticides were much cheaper than their predecessors, and farmers grew accustomed to blind use irrespective of infestation.[42] And yet, after two decades of chemical warfare, the fight against the boll weevil still claimed one-third of all pesticides used on farm crops in the United States, and a good part of that went toward controlling follow-up problems, as chemicals made no distinction between bad and beneficial insects. Losses for cotton planters came to between $200 million and $300 million per year.[43]

Would the South have to live with the weevil forever? Dreams of annihilation were as old as the infestation, and they were not confined to the lunatic fringe. The Ford Motor Company proclaimed in 1921 that its new Fordson tractor would finish the boll weevil because it facilitated plowing under infested stalks.[44] The National Cotton Council created a Committee on Boll Weevil Eradication in 1969 and started a pilot eradication project in southern Mississippi in 1971, but skeptics were not hard to find. Dale Newson, the chairman of the entomology department at Louisiana State University, pointed to the fire ants spraying campaign, whose main achievement was to get Rachel Carson interested in DDT (see chapter 38.1, Panaceas). An eradication campaign against the boll weevil would be "long, costly, and futile—one might say an entomological Vietnam."[45]

The challenge was as much about technology as about consensus-building. Researchers found a weevil-eating wasp in Mexico, synthesized a pheromone and used it to lure the beetle to insecticide-coated sticks, and isolated a lethal fungus from dead weevil carcasses. GIS mapping and computer modeling to track infections came into use.[46] And yet, thanks to the weevil's reproductive power, it only took a single neglected field to undercut the entire endeavor. It was not for nothing that eradication campaigns always began with a vote among growers, who paid 70 percent of the program's costs.[47] Texas passed a Boll Weevil Law in 1987 that required growers to shred and plow under cotton stalks by September 1, and set up air patrols to assure compliance.[48]

War metaphors have permeated insecticide talk for decades, and Edmund Russell has shown that this was more than careless rhetoric: tools and mindsets were remarkably similar.[49] And like all good warriors, weevil eradicators liked to boast about impending victory. The first full-scale eradication trials began in southern Virginia and northern North Carolina in 1978, and today the boll weevil is gone from large parts of the cotton South. Growers typically experienced a growth in yield of up to 10 percent, and pesticide use dropped sharply once elimination was achieved in a region.[50] In the wake of eradication, cotton acreage in Virginia, Georgia, Florida, and the Carolinas grew by more than 700 percent from 1980 to 1995.[51]

Yet all this is only a success for the time being. It only takes a major storm to blow weevils over the land to start a new infestation and prompt agriculturalists into frantic action.[52] The eradication campaign, for all its merits, did not achieve its ultimate goal—it only shifted the front lines. In fact, any other outcome would have been surprising, as the fight against pathogens (see chapter 21, Cholera) has produced a similar result: in spite of lavish investments in medical research and sanitary infrastructures (see chapter 17, Water Closet), humans have only achieved global eradication of two diseases, smallpox and rinderpest. For all the others, we have just achieved the ability to more or less keep them at bay.[53]

When it comes to elimination, it would seem that bugs and microbes are still more successful than humans. The globalization of pathogens is an ongoing process, and agriculturalists are facing ever new challenges in addition to the unresolved old ones. After felling more than two million citrus trees, Florida's Department of Agriculture and Consumer Services canceled its citrus canker eradication program in 2006 and accepted that the disease had become endemic.[54] The same officials are currently just as helpless about citrus greening, a bacterial disease that turns fruits bitter, makes them drop before they are ripe, and thus threatens to wipe out Florida's orange industry.[55] In southern Italy, *Xylella fastidiosa* bacteria kill olive trees, helped by the populist Five Star Movement that challenged the European Union containment regulations and the underlying science.[56] Journalists are speculating as to whether a soilborne fungus named Tropical Race Four may mean the end of the banana (see chapter 10, United Fruit). In the twenty-first-century world, one can no longer practice monoculture without a sense of nervousness.

If we phrase it in military terms, the eradication campaign against the boll weevil has not achieved victory but merely a contested truce that is only holding thanks to diligent monitoring. A vast network of field traps allows researchers to identify infestations early, and DNA evidence helps track the origins. The Southern Plains Agricultural Research Center in College Station, Texas, even hired a pollen expert who checks dead weevils for pollen grains that may give away its itinerary.[57] When it comes to the boll weevil, cotton growers are still glancing nervously over their shoulders, and the same holds true for the other pests, diseases, and fungi in the monocultures of the world. In the delta region, extension service pamphlets implore cotton growers to remain on the watch: "Although the boll weevil has been successfully eradicated from Mississippi cotton fields, it must never be forgotten!"[58]

13

THE LITTLE GRAND CANYON

THE PERILS OF EROSION

I. THE VIRGIN LAND

The end is always near in the American Bible Belt. You can even find a touch of the apocalypse in your own backyard. Something along these lines was on the mind of the Lower Chattahoochee Valley Area Planning and Development Commission, a government body in southwestern Georgia, when it made the case for a new state park (see chapter 26, Kruger National Park) in 1969. "More than anything else in the state, the canyons look like the end of the world," the commission declared.[1] The area under discussion, named Providence Canyon after a nineteenth-century Methodist church, offered stunning canyons with a depth of up to 250 feet, and enchanted visitors dubbed the place the "Little Grand Canyon." The comparison was arguably a bit overblown when it came to size—the original Grand Canyon was a mile deep and up to eighteen miles wide. But like its namesake, the walls of Georgia's Little Grand Canyon featured a spectacular array of colors—"divinity pinks, taffy browns and caramels, vanilla creams and fondant pastels," in the words of the commission. "Here for the first time you realize that the earth may be beautiful all the way through."[2]

The invocation did not leave the powers that be unimpressed. The state of Georgia bought some one thousand acres from private owners, declared Providence Canyon a state park in 1971, and obtained a federal grant to develop it for tourism (see chapter 22, Baedeker).[3] Today's visitors find all the amenities that they expect in such a place: designated trails, secure fencing, campsites, picnic areas with restrooms, a visitor center with a small museum and a gift shop, and a large red warning sign that hiking down the canyon "is not recommended for persons with heart problems or in poor physical condition." Scenic beauty is the primary attraction of Providence Canyon, but there is more to the site than trails and multicolor walls. It is testimony to one of the key trends of modern history: the great land rush.[4]

The quest for new land was a crucial part of the global expansion of Europe, and it came to a head in the nineteenth century: no other century has expanded the acreage under cultivation to such an extent.[5] Plows broke the sod from the Russian steppe to the Argentinean pampas, transforming allegedly virgin soils into land for cotton, wheat, and other commodities for world markets. Even in the Caribbean, where the land hunger of sugar plantations was a familiar phenomenon beginning in the seventeenth century, export-oriented agriculture grew to completely new dimensions. Whereas 43,000 tons of sugar made Cuba the world's largest exporter in 1820, production moved beyond the 1-million-ton threshold toward the end of the century (see chapter 2, Sugar).[6]

The virginity of the land was usually more imagined than real. Humans had left their mark on the land long before the arrival of white settlers, and it is a matter of debate whether Indigenous knowledge, or population density, or greed made the difference. In fact, the difference was often a matter of degrees rather than absolutes. "It is important not to make too sharp a distinction . . . between nomadic pastoralism by the indigenous population and settled, arable farming by the incomers," David Moon remarked about the Russian steppe.[7] Frontier mythologies have long obscured the extent to which the colonizers and the colonized exchanged ideas and artifacts in what Richard White has termed "the middle ground."[8] Migrants could also draw on experiences from similar regions or adapt farming practices and crop rotations that they had learned back home.[9] The new agriculture used whatever looked promising, and inputs could come even from those who were lacking the most essential rights. Since Judith Carney's *Black Rice*, the contribution of African slaves to American rice cultivation has been a matter of scholarly debate.[10]

The vagaries of distant markets exacerbated the ecological unknowns. Events in other countries could jeopardize local production routines and ecologies with amazing speed. For example, Russian settlers plowed the steppe with new vigor when the repeal of the British Corn Laws increased the demand for grain in the 1840s. They also reduced their herds of sheep because they could not compete with cheaper wool from Australia.[11] When the American Civil War disrupted the supply of cotton from Southern planters, European textile manufacturers embarked on a frantic search for new supplies, resulting in cotton booms in places as different as India, Egypt, Brazil, Turkmenistan, and Togo.[12] And then there was the notorious instability of frontier societies, where state authorities, codes of law, ethnic hierarchies, and many other things were in a state of flux.

In short, the expansion of agricultural land was a gamble, and nowhere was that more apparent than with a view to the soil. Freshly plowed grasslands typically held a store of nutrients that crops would deplete over time, prompting debates over how to restore fertility. Southern planters had led vigorous discussions over the use of lime, guano (see chapter 8, Guano), and other remedies since the early nineteenth century.[13] Furthermore, while grass and woodlands usually covered the ground well through the seasons, plants like cotton and corn left much of the ground exposed to the elements for months, which greatly increased its vulnerability to wind and water erosion. Soil types and weather conditions would determine whether erosion would become a problem, and settlers were usually less than certain about both.

Soils were deemed so significant that they became part of the historical imagination. In 1926, Avery Craven published a famous book titled *Soil Exhaustion as a Factor in the Agricultural History of Virginia and Maryland*.[14] In North Africa, tales about land degradation and desertification provided a justification for colonial development projects as France sought to revert centuries of Arab mismanagement and restore the ancient granary of Rome.[15] It was a veritable "environmental history before environmental history," though the crude environmental determinism at the heart of the argument led many scholars to ignore environmental issues altogether. Recent discussions about agriculture

13.1 The gullies of Providence Canyon in 1922. Note the cornfield being engulfed. Image, Georgia State Archives.

and the environment usually cite these readings in an act of exorcism, seeking to create as much distance as possible from their own intellectual endeavors.[16]

Yet soils were indeed fragile, and Providence Canyon provided a case in point. Like much of the American South, southwestern Georgia went into cotton in the first half of the nineteenth century, and gullies developed rapidly in the wake of cultivation. The beginnings of Providence Canyon are shrouded in a fog of local lore, but it was already a threat to nearby buildings in 1859, when a local congregation decided to move its church away from the abyss.[17] Once a thin layer of topsoil was gone, surface erosion set in, and subterranean water streams above an impermeable layer of clay washed out underground pipes that eventually caved in. The sands of Providence Canyon eroded both from the surface and from below, and that allowed erosion to advance at dramatic speed.[18] It was obviously frustrating for the owners of the land, though probably not frustrating enough to inspire an adjustment of farming methods. A picture of Providence Canyon from 1922 shows a cornfield right on the brink.[19]

2. SAVIORS OF THE SOIL

The quest for new agricultural land did not end with the dawn of the twentieth century. Fascist Italy drained the Pontine Marshes in the interwar years (see chapter 32, Pontine Marshes), and the Soviet Union launched the Virgin Lands campaign to put eminently nonvirgin land in Kazakhstan to new uses. However, these campaigns thrived on the combined powers of authoritarian states and scientific expertise, and proponents gambled with the land as well as their own careers: Khrushchev's Virgin Lands campaign played a major role in his rise and fall.[20] In the nineteenth century, the search for good farming practices was still firmly in the hands of those who worked the land. Settlers followed widely different learning curves on the ground, and some were remarkably successful; for example, Mennonites, the Germanized descendants of Dutch Anabaptists, won wide acclaim as the best farmers on the Russian steppe.[21] However, many farmers embraced a somewhat less sophisticated solution when they encountered troublesome soils. As long as land was abundant, they simply moved on.

For most of the nineteenth century, scientific experts were standing on the sidelines, hampered by both a lack of resources and a lack of knowledge. When Hardy Webster Campbell invented his own method of dry farming and toured the American Great Plains with eager support from railroad companies and other commercial interests, the government's agricultural science network was unable to approve or disprove the concept.[22] In the Russian steppe, experts advanced over time from their ardent reading of European agrarian reformers toward field studies and experiments. It often took a disaster (see chapter 25, 1976 Tangshan Earthquake) to boost the standing of agricultural expertise. When a serious drought caused the harvest to fail on the Russian steppe in 1833, the government set up a Committee for the Improvement of Agriculture. However, when the Russian soil scientist Vasily Dokuchaev called for two research stations in the wake of disastrous dust storms in 1892, the Department of Agriculture and Rural Industry turned him down.[23]

The American government was more responsive when a similar disaster, the Dust Bowl, struck the Southern Plains four decades later. President Franklin D. Roosevelt made soil conservation a part of his New Deal, and a charismatic chief, Hugh Hammond Bennett, led the ensuing campaign for almost two decades.[24] He created a huge federal agency from scratch: founded in 1933, the US Soil Conservation Service had 13,331 employees only four years later.[25] Appropriations continued to grow after Bennett left the helm, and the Soil Conservation Service won broad acclaim as "one of the more successful agencies spawned by the New Deal."[26] Renamed Natural Resources Conservation Service in 1994, it exists to the present day.

Like most New Deal agencies, it had a turbulent start. It began in 1933 as the Soil Erosion Service within Harold Ickes's Department of the Interior but moved to the Department of Agriculture in 1935 and was renamed Soil Conservation Service. The change was more than a formality, as Ickes sought to build a department of conservation while the Department of Agriculture was traditionally a development agency.[27] Furthermore, the soil conservation staff was competing for the ear of the farmer with county agricultural agents, the cooperative extension branch of the Department of Agriculture since 1914.[28] When the Supreme Court annulled subsidies under the Agricultural Adjustment Act in 1936, the Department of Agriculture set up an Agricultural Conservation Program that paid farmers for planting soil-conserving crops, a program that clashed with the work of the Soil Conservation Service and had about four times their funds.[29] Fieldworkers were also struggling to find work for unemployed men that the Works Progress Administration dumped in their lap.[30]

But the Soil Conservation Service had a powerful sense of mission. For Bennett, soil conservation was a moral crusade, and it was about nothing less than the future of America. As Bennett declared in a memorandum for Ickes in 1934, "Unrestrained soil erosion, if permitted to continue, will result in the virtual elimination of civilization from great areas of the United States."[31] Even in the midst of the Great Depression, Bennett found that soil erosion was one of America's greatest problems, and he waged his campaign with a zeal that bordered on the religious. As Arthur Schlesinger wrote, Bennett preached his gospel "with Old Testament wrath."[32]

For Bennett, the fight for the fertile soil was as old as humanity, and history showed what failure would mean. "Great civilizations and great nations of the past have disappeared because their wealth of soil was washed away by erosion," Bennett declared in a national radio broadcast in 1939.[33] He was not alone in this point of view. The New Deal gave birth to a distinct literary genre that survives to the present day: sweeping global histories that find neglect of the soil throughout human history, along with the terminal result. The tradition includes, among others, *Behold Our Land* by Russell Lord (1938), *Topsoil and Civilization* by Vernon G. Carter and Tom Dale (1955), *Far from Paradise* by John Seymour and Herbert Girardet (1986), and *Dirt* by the University of Washington geologist David Montgomery (2007).[34] Randal Beeman and James Pritchard have called them "soil jeremiads."[35]

Bennett's view of history left its mark on political strategies. If soil conservation was really about the future of human civilization, the mission justified drastic means. As a result, Bennett's crusade initially looked rather authoritarian: his early correspondence talked about "regulations of land use in the public interest" and "zoning laws requiring all property owners to cooperate in erosion control through the proper use of their own land and through financial contributions."[36] Bennett even had no qualms about quoting Mussolini (see chapter 32, Pontine Marshes) in a congressional hearing, as he liked to see the Fascists fighting erosion without petty concerns about landowners and costs: "The Italian Government does not profess to apply an economic yardstick to its program."[37] Dictatorial powers were no problem for Bennett as long as they served the right cause.

Bennett's crusade brought the canyons of southwestern Georgia, heretofore scarcely known beyond the region, into the national spotlight. "For soil conservationists, New Dealers, environmental writers, and even a few southern liberals, Providence Canyon served as the poster child of southern soil abuse," Paul Sutter has noted.[38] Bennett showed a picture on page four of his 1939 book *Soil Conservation*, a massive tome of almost a thousand pages, and other authors were equally enchanted.[39] Interest was less pronounced when it came to doing something about these gullies. When the Soil Conservation Service was scouting Georgia's Coastal Plains in search of the right place for a demonstration project in 1935, it chose Muckalee Creek near Americus, some forty miles to the east of Providence Canyon. The gullies looked spectacular in pictures, but they were a hopeless case even for an agency that was not otherwise lacking ambition.

Like other demonstration projects around the country, Muckalee Creek was officially established "for the purpose of employing and demonstrating all known practical methods of erosion control."[40] In reality, they were sites of institutional learning. Local conditions were waiting to be explored, tools and methods were put to the test, and they served for on-the-job training: soil conservation was one of the few professions where qualified people were in short supply during the Great Depression. Staff at Muckalee Creek did not dare to propose a crop rotation until after a year of observations and learning about the best way to build dams in gullies, and that was a fairly typical experience.[41] Soil conservationists "were doing a type of work that had not been done before," and some made no bones about their initial ignorance in retrospect. A demonstration project in Northern Georgia frankly declared that "in the early days of the project much of the work... was 'hit and miss.'"[42]

All that made for a notable gap between visions and realities. While Bennett and his associates pledged to save American civilization, his staff was frantically trying to understand the realities on the ground. It did not help that many farmers were skeptical of outside interventions, down to fears that "the Government would take over their lands" (see chapter 6.2, Agrarian Reform).[43] And then there was the gap between land use ideals and economic necessities. The Soil Conservation Service proposed to "use each acre for the purpose for which it is best suited," but what if a soil-conserving crop rotation reduced the farm's income?[44] Farmers and soil conservationists could find some common ground in the end, but that took a process of mutual adaptation. It also took money. The Soil Conservation Service was generous not only with words of wisdom but also with equipment, plants, and services, and thus became part of the tradition of farm subsidies (see chapter 2.2, Power Games) that grew out of the New Deal.

All that put Bennett's moral crusade on a slippery slope. Many conservation measures were ambiguous in that they allowed a more intensive use of the soil, and these measures often won the day because they helped to boost per-acre productivity. The work of the Soil Conservation Service increasingly gained a business-friendly touch, and the fight against erosion gradually turned into a quest for intensification. Bennett was reluctant to embrace the trend and left it to his successor Robert M. Salter to draw the conclusion from the experience in the field. "We need to concentrate on increasing yields," Salter declared at the annual meeting of the National Association of Soil Conservation Districts in 1952.[45] His bluntness earned him a spat with Bennett who felt that Salter was "wrecking soil conservation," and discussions have raged ever since as to whether paying farmers for erosion control was good policy, a sellout, or something in between.[46]

The Soil Conservation Service never abandoned the great story of soil and civilization, as it had more appeal than the ambiguities of everyday work. That left its mark on global discussions when soil conservation became part of American development policy during the Cold War. On the surface, it all looked like a matter of moral vigor. When Walter Lowdermilk, the second in command in the Soil Conservation Service until 1947, gave a speech in Jerusalem in 1939, he suggested that it was time for an Eleventh Commandment: "Thou shalt protect thy fields from soil erosion and thy hills from overgrazing by thy herds, so that thy descendants may have abundance forever."[47] Bennett went on a two-month tour through South Africa in 1944, and his disciples followed suit, crisscrossed the globe, and turned American soil conservation into a global endeavor.[48] One federal official resigned from his government job and went to Abu Ghraib College in Baghdad, a place that would see other types of US activities decades later.[49] Soil conservationists abroad talked more about terraces, strip cropping, and plowing on the contour than about the bottom line, and few dared to spell out that the American way of soil conservation effectively hinged on a federal government in spending mode. Donald Worster has calculated that "more than $2 billion was spent by New Deal agencies in the thirties to keep the farmer of the plains region in business."[50]

3. WHO REALLY CARES ABOUT EROSION?

With the Dust Bowl firmly entrenched as a global icon, it has become difficult to recognize the oddity of its rise to prominence. The Southern Plains were a peripheral region by almost any standard, devoid of a metropolis, a particular commodity, or a distinctive landscape. It was also a latecomer in agricultural development. The region figured on nineteenth-century maps as the "Great American Desert," and the great plow-up in the 1920s was but a minor sequel to what had happened in the nineteenth century. The dust storms were certainly not unprecedented in history, nor were they the first to become subject to scientific scrutiny. When drought fell upon the Russian steppe in the early 1890s and dust storms plagued the region, some of the accounts were from field research stations.[51]

But when it came to capturing the public imagination, reports from soil scientists were no match for the spectacular photographs that showed a wall of sand moving across the Great Plains. It was pictures that made the Dust Bowl, as they touched a nerve in society at large. People associate soils with stability and permanence across cultures, and soil on the move mirrored a society in which all certainties were evaporating: it was the perfect symbol for disturbing times. Just like the Great Depression, the dust storms were a menacing and disturbing threat that came out of nowhere and struck helpless victims indiscriminately. Everyone could relate to the experience of Great Plains farmers, if only to put one's personal troubles into perspective. "People beset by depression needed someone white with whom they could identify and who was worse off than they were," the rural historian

David Danbom has written.[52] The only people who had mixed feelings were the plainsmen themselves, who did not like their home being portrayed as a disaster region (see chapter 25, 1976 Tangshan Earthquake). But voices from the ground did not count for much in a mythology that was about so much more: the Great Depression, the future of the American dream, soil conservation, and modernity as such. The story of the Dust Bowl is also the story of the cultural occupation of a peripheral region.[53]

It provided a fitting illustration of what it meant to fight erosion in the twentieth-century world. The frontiers of the nineteenth century were crucial arenas of social development that formed the imagination of entire nations: from Frederick Jackson Turner to South African Afrikaner myths.[54] But as industrialization and urbanization changed societies all over the West, agriculture lost its former preeminence in minds and economies. Everything about the countryside was shrinking: its share of the population, its economic significance, its political pull. Agriculture became a marginal part of modern societies, and fewer and fewer people were in touch with the fragile soils. Pictures took the place of firsthand experience, and pictures were more open to different readings and more easily forgotten.[55]

The change of production methods created further obstacles for soil conservation. The fight against erosion was about the long view: the goal was to look beyond the needs of the day and preserve fertility for future generations. But farmers and policymakers were also interested in the next harvest, and this divergence of chronological horizons hampered many a conservation program. Dokuchaev's research program failed in the wake of the 1892 dust storms because it aimed for a comprehensive investigation of the steppe's biological history while the Russian government sought quick solutions.[56] Short-term thinking spread with particular vigor in the years after World War II, if only for lack of a choice. In the new agriculture, the crucial factors were size, speed, and the sophisticated use of technology. It made farming enormously capital-intensive, and that discouraged reflections on the long haul: while nineteenth-century farmers could simply ride out a few bad years with some reserves (see part VI, Final Reserves), postwar farmers had contracts, bills, and installments that called for immediate attention. "Get bigger, get better, or get out," the US secretary of agriculture Earl Butz told America's farmers in the 1970s, and agriculturalists in other parts of the world heard similar advice.[57] It was good for urban consumers, as food prices fell to historic lows. It was not so good for those who cared about the future of the soils.

Even environmentalists were reluctant to embrace soil conservation. Western environmentalism had its mainstay in urban areas, and if urbanites cared about farmers at all, it was usually in an adversarial mode. Farmers used dangerous chemicals such as DDT (see chapter 38, DDT), they ignored animal welfare (see chapter 36.3, Liberation Movements), and their pervasive use of synthetic nitrogen (see chapter 19, Synthetic Nitrogen) put water supplies in jeopardy. If, in addition to all these sins, the farmers were ruining their own soils, wasn't that really their own problem, if not a well-deserved fate? Soil conservation also lacked the kind of moral clarity that environmentalists were craving. It was about local solutions, about balancing different parameters, and about a compromise between conservation and profit. Most crucially, soil conservation was an eternal struggle, and a single drought or a long period of rain could ruin the achievements of many years. In other words, soil conservation was a difficult issue for environmental campaigning.

Victory is an elusive concept when it comes to saving the soil, and the Soil Conservation Service was never in a position to display "mission accomplished" banners. According to the most recent overview, the 2007 National Resources Inventory, an acre on the Southern Plains is losing 6.2 tons of soil per acre to wind erosion every year. Some 70 years after the Dust Bowl, that amounts to an annual loss of more than more than 200 million tons of topsoil.[58] All over the world, governments, experts, and stakeholders discuss what that means for feeding a global population of 10 billion or more in the twenty-first century, and solutions range from comprehensive land use planning to the invisible hand of the market. The one certainty is that the solution of the nineteenth century is no longer in the cards. We no longer have the land for another great plow-up.

Today's soil jeremiads do not lack staggering statistics (see chapter 7.2, Numbers Games), but their concerns are a tough sell in an urban society. Even the management of Providence Canyon State Park is reluctant to push the issue. The website of Georgia's Department of Natural Resources leaves it at an opaque reference to "poor farming practices during the 1800s," and the displays at the local museum are equally short on remarks about monoculture, or the global web of cotton, or the learning curve of settler societies.[59] And that is already

13.2 Providence Canyon in 2015. Image, Frank Uekötter.

an improvement over earlier states of denial. When the Lower Chattahoochee Valley Area Planning and Development Commission made its case in 1969, it blamed "the ruthless hand of nature."[60]

Nature reserves (see chapter 26, Kruger National Park) are usually meant to last forever, and Providence Canyon State Park was no exception: it was to be preserved "for the benefit of present and future generations," as Georgia's governor, Zell Miller, declared in an executive order in 1998.[61] Human-made landscapes can last a long time. Left over from gold mining in the first century AD (see chapter 1.3, Abandonment), the spectacular cliffs of Las Médulas in northwestern Spain were inscribed on the UNESCO World Heritage List in 1997.[62] But in the case of the Little Grand Canyon, the whiff of eternity quickly dispersed on the ground. Erosion is an ongoing process, easily identified in muddy runoff and exposed roots. Park managers are struggling to contain the spread of kudzu, a fast-growing vine that the Soil Conservation Service brought to the South in the 1930s, which has since achieved a notoriety akin to Australia's cane toads (see chapter 14, Cane Toads). In remote places, tourists (see chapter 22, Baedeker) are etching graffiti into the canyon walls. The designation of Providence Canyon as a state park was not the end of its history, any more so than the creation of the Soil Conservation Service was the end of erosion. When it comes to fertile soil, it does not take a bible reading to experience a touch of the apocalypse.

14

CANE TOADS

IMMIGRANTS ON THE MARCH

I. A GOOD IDEA AT THE TIME

Australia was a latecomer to the sugar business (see chapter 2, Sugar). Sugarcane was first planted near Brisbane in the 1860s, more than two centuries after the sugar revolution on Barbados and several decades after European sugar beets emerged as a competitor. Sugar grew into an agricultural industry along Australia's eastern coast in Queensland and northern New South Wales, far away from thriving southern towns like Sydney, Melbourne, and Adelaide, and it became a pillar of the regional economy. By the 1930s, sugarcane was the defining agricultural crop in Queensland, accounting for two-thirds of the state's produce and a quarter of its income.[1]

Like every monoculture, Queensland sugar faced challenges on multiple fronts. The labor regime changed several times over the first fifty years. Sugar cultivation began with indentured labor from Vanuatu and the Solomon Islands, and Queensland looked poised to become another plantation economy.[2] However, recruitment from

Pacific islands became so difficult that the Queensland colony tried to annex New Guinea in 1883, which caused Britain to declare a protectorate on the southern coast of New Guinea where Queensland blackbirded some six thousand natives in 1884.[3] "Colored" workers were expatriated after the new Commonwealth of Australia adopted its infamous White Australia policy in 1901, and sugar cultivators, now a community of family farmers, received generous government protection in return for the higher costs of white labor. Price protection and production quotas provided sugar producers with a measure of economic stability until ascendant neoliberalism inspired deregulation in the late 1980s and early 1990s, which meant that for most of the twentieth century, the prime challenges for sugarcane farmers were biological in nature: irrigation and drainage (see chapter 29, Aswan Dam), soil erosion (see chapter 13, Little Grand Canyon), and pests and diseases.[4]

Australian sugarcane farmers faced their first biological crisis when a fungus caused severe losses from red rot disease in the 1870s. Some farmers abandoned sugarcane for maize or potatoes, but most survived by switching to less susceptible cane varieties.[5] In order to curb the spread of the sugarcane weevil borer, the Colonial Sugar Refining Company (CSR), a milling company that drove the application of science in Australian sugar, introduced tachinid flies from Fiji in 1914. This did not eliminate the pest, but it reduced its damage, which already qualified as a success in the perennial fight against biological contestations (see chapter 12, Boll Weevil). In the fight against cane grubs, CSR recommended soil fumigation while experts were searching for an animal with an appetite to match.[6]

Biological pest control became fashionable in the wake of excessive pesticide use in the age of DDT (see chapter 38, DDT), but applied entomology was actually older than the promise of better living through chemistry. It had come into its own in the second half of the nineteenth century, a somewhat natural sequel to the botanical exchange (see chapter 27.2, Botanical Exchange) that catapulted species like breadfruit (see chapter 7, Breadfruit) into a global orbit: after scouting the world for the most productive organisms, experts were scouting the world for another set of organisms that would help get a grip on the problems of the new organic production regimes.[7] More than once, biological pest control was essentially about the introduction of natural enemies that globe-trotting species had left behind in their na-

14.1 A cane toad. Image, Matthijs Kuijpers / Alamy Stock Photo.

tive homes, but the botanical exchange produced new challenges at the same time. The sugarcane weevil borer had likely come to Queensland with new cane varieties from New Guinea.[8]

Applied entomology developed earlier in the United States than it did elsewhere, but the growing network was ultimately transnational in nature. A native of Waldkirch, Germany, Albert Koebele was employed by the US Department of Agriculture when he went to Australia and discovered a ladybird that saved California's citrus industry from a devastating pest. He moved on to Hawaii, introduced parasitic insects from the rest of the world by the dozens, and thus laid the foundation for a world-class center of biological pest control.[9] Koebele also looked into Queensland's cane grub problem during a visit in 1891. He urged his hosts to think about the introduction of an animal that Australia did not have naturally: toads.[10]

Four decades later, a delegate from Queensland's Bureau of Sugar Experiment Stations traveled to the Fourth Congress of the International Society of Sugar Cane Technologists in Puerto Rico. The cane grub problem was still unresolved, and as it happened, one of the conference papers was about a toad. *Bufo marinus* had come to Puerto Rico just a few years earlier, though the toad, a native of tropical rain forests in the Americas, had been introduced on Barbados and Martinique in the early nineteenth century. The paper discussed the food habits of *Bufo marinus* and provided a glowing description of its potential for pest control. A delegate from Antigua took issue with the presentation and noted that the toad had become a nuisance on his island, but other delegates were more impressed. A shipment of *Bufo marinus* was on its way to Hawaii within a month, and the animal, soon known in sugar circles as cane toads, went on to other Pacific islands over the following years: Guam, Taiwan, Fiji, New Guinea, and the Philippines. Queensland received 101 specimens in 1935.[11]

The speed was far more impressive than the underlying knowledge, but pest control experts were men of quick decisions. In 1923, the director of Puerto Rico's Insular Experiment Station brought cane toads to the US territory on an impulse after watching a few hungry toads consuming insects under a lamp in Jamaica while waiting for his ship.[12] It was not scientific protocol, but seeing cane toads in action was quite an experience. Cane toads have a voracious appetite, and they are big: females can grow to a weight of more than two kilograms.[13] They also eat everything that fits into their big mouth, and cane toads have been observed eating a rat 20 percent their weight.[14] A powerful toxin from its parotoid glands protects the toad from large predators. It looked like the perfect exterminator for all kinds of pests, a kind of DDT with warts (see chapter 38, DDT), and optimistic scientists were glad to send them to different places with different insect problems. It might have worked if evolution had not given insects the ability to fly.

As it turned out, cane toads were not much help in the fight against cane grubs. They just sat on the ground while grubs multiplied in the soil below or buzzed over the toads' heads. Cane toads were also inclined to go where insects were abundant, and that was not necessarily in a field of sugarcane. On Hawaii, where a scarcity of insects and inhospitable volcanic soils drove cane toads into the vicinity of human settlements, biologists found that they were prolific consumers of cockroaches.[15] In Australia, beekeepers were among the first to complain about the new species, as cane toads gathered around beehives for a reliable stream of insects.[16]

Sugarcane farmers returned to soil fumigation, and chemicals took over after World War II. A new insecticide, benzene hexachloride, brought splendid results when applied in late spring or summer; it is now banned under the 2001 Stockholm Convention on Persistent Organic Pollutants (see chapter 24.3, Getting Serious).[17] Perhaps the last hurrah came in August 1937 when a pair of cane toads was on display at the Brisbane Exhibition, but excitement about the biological curiosity waned when the giant toads showed up on people's doorsteps.[18] Cane toads had come to stay, and they were spreading: they multiplied rapidly, they ate a lot of other species, and when they were eaten themselves, the predator gained a lethal dose of poison.[19] When the International Union for Conservation of Nature (IUCN) published the booklet *100 of the World's Worst Invasive Alien Species* in 2000, cane toads were one of three amphibians on the list.[20]

2. INVASION

The IUCN's booklet sought to raise awareness for "the risks of further harmful invasions," and it offered a number of dramatic examples. Introduced into Lake Victoria in 1954, the Nile perch "contributed to the extinction of more than 200 endemic fish species through predation and competition for food." The damage extended beyond the waterline, as the flesh of Nile perch was oilier than the flesh of native species and thus required more firewood to dry, which contributed to regional deforestation and subsequent erosion. Lake Victoria was also plagued by the water hyacinth, a fast-growing plant that was "found in more than 50 countries on five continents." Other featured stories included the brown tree snake on Guam, the tropical Caulerpa seaweed, which escaped into the Mediterranean from the Monaco Aquarium in 1984, and feral pigs. All in all, the booklet argued that "alien invasion is second only to habitat loss as a cause of species endangerment and extinction" (see chapter 11, Dodo).[21]

The impacts of invasive species varied widely, and so did the causes of their global spread. Trade and botanical exchange were important venues of transfer, but they were barely the only ones. American soldiers brought the brown tree snake to Guam as unintended military cargo during World War II.[22] The Burmese python came to the Everglades because bored or overtaxed pet keepers released them from captivity.[23] Colombia has had wild hippopotamuses since they escaped from the derelict private zoo of drug lord Pablo Escobar (see Interlude, Opium).[24] And sometimes we know only the place but not the cause. The red fire ant spread across North America from the port of Mobile, Alabama, but the precise circumstances remain a mystery. After all, interest in fire ants was limited until 1957 when the United States Department of Agriculture launched a massive and spectacularly unsuccessful eradication campaign (see chapter 38.1, Panaceas).

The cane toad was different in that it was introduced on purpose. Even more, its spread resulted from a cascade of failures within the expert community. It began with the shoddy science behind the paper at the 1932 Puerto Rico conference. The experimental design did not allow statements on the cane toad's potential for pest control, and the decline of cane grubs on Puerto Rico between 1931 and 1936 was likely due to several years when the wet season was exceptionally wet and the dry season exceptionally dry. Delegates also failed to follow up on the sobering experiences in Antigua. Back in Queensland, release of toads was rushed because the next meeting of the International Society of Sugar Cane Technologists was about to convene in Brisbane in August 1935. When Australia's Department of Agriculture imposed a ban on further releases later that year, the Bureau of Sugar Experiment Stations fought to repeal the decision.[25] But for all the individual mistakes, the real blunder was about expertocratic tunnel vision. Stopping an invasive species was not what sugar scientists were programmed to do.

Biological pest control was about delivering solutions for embattled farmers. There was no reward for caution, and certainly not in the years of the Great Depression. Economic imperatives were always in the room when biological experts convened, and not just in a metaphorical sense. At the Puerto Rico conference, the most fervent proponent of cane toads was an English-born businessman named John Waldron, director of the Oahu Sugar Company and chairman of the Hawaiian Sugar Planters' Association.[26] Producers hired and fired many of the experts, and any sense of societal obligations paled in comparison to the scientists' allegiance to their commodity network. And then there was the experience of biological fieldwork: for men who spent their lives in inhospitable environments accumulating plenty of scratches, bruises, and diseases along with the coveted species, it was hard to get worried about an awkward-looking amphibian that was no threat to humans. Scientists only had to look at their own bodies to see that everything came at a price. Albert Koebele was in his mid-fifties when he quit his job on Hawaii in 1908 and returned to his boyhood home of Waldkirch, where he spent the last sixteen years of his life in failing health, the gruesome result of the maladies that he had acquired on his global journey.[27]

Revealingly, cane toad research effectively came to an end when sugarcane farmers discovered that the species failed in its assigned task. Cane toads were spreading, much to the dismay of ordinary Queenslanders, but this did not stimulate scholarly interest. Cane toads figured as a mere curiosity in biological circles, and when they became a scientific object during a field course of the Organization for Tropical Studies in Costa Rica in 1970, it was research of the testosterone-fueled kind. The group discovered that tadpoles of the *Bufo marinus* variety did not hide when threatened and speculated that it might be due to an awkward taste. In order to test their hypothesis, eleven students and faculty members volunteered as "mock predator," chewed tadpoles from eight different species, and recorded palatability on a scale of one to five. The volunteers gained confirmation of their hypothesis and a really bad taste in their mouths. It is the kind of experiment that would have raised eyebrows on the ethics committees of later years, but the scientists made up for moral qualms with beer.[28] The experiment received a measure of fame when the lead researcher won the 2000 Ig Nobel award for biology.[29]

A giant toad looked custom-made for the age of visual media, and the filmmaker Mark Lewis produced one of the most popular documentaries of Australian history, *Cane Toads: An Unnatural History*, in the 1980s. It thrived on numerous allusions to movie history, the Queenslanders' redneck reputation in the rest of Australia, and plenty of jumping-off points for readers of postmodernist literature.[30] The media covered the invasion with sensationalist frontline reports, grateful about an enemy that was neither politically volatile nor prone to canceling advertising. Enraged citizens jumped into action and collected cane toads for euthanization or smashed them on the spot. When it came to invasive species, humans were inclined to draw on the rationales of total war (see chapter 35, Pine Roots Campaign), not least because it suggested moral clarity: invaders were villains, collaborators were traitors, victims deserved sympathy, and the only legitimate response was an all-out fight. But did it really make sense to think in these terms?

The language of war has raised awareness of the perils of biological invasions among governments and experts. If cane toads were discovered for pest control today, they would undergo a number of tests that would preclude another blunder.[31] But fears of invasion are a difficult guide for action if there are too many enemies and too many battlesites. It has always been difficult to keep track of all invasive animals, plants and pathogens, and they have shown up in the most inconspicuous of places. In the postwar years, Queensland canegrowers were struggling with groundsel bush, a type of weed that arrived from the United States and was clogging their drainage channels.[32]

Cane toads were among the invaders that achieved iconic status, but that was always a small group. Few people know about avian malaria, the small Indian mongoose, or the Indian mynah bird, all of which made it onto IUCN's list.[33] Even the cane toad is widely unknown outside Australia, and it is debatable whether awareness within the country has made people sufficiently cautious. When two outback towns in western Queensland found too many insects in local gardens in the 1970s, they brought in cane toads on purpose.[34] As to the men who brought the cane toad to Australia, there was none of the recrimination that one might expect for officials who triggered an invasion out of carelessness. The toad's willing collaborators remained in their posts, and when Australia's delegate to the 1932 Puerto Rico conference, Arthur Bell, died in 1958, the Queensland branch of the Australian Institute of Agricultural Science and Technology voted to name a student research paper award after him.[35]

When biologists finally embarked on scientific studies of the invasive species, their publications painted a nuanced picture. As with all good science, it came with caveats about unknowns and methodological limits, but the results challenged some of the pillars of the battle against cane toads. For one, cane toad damage was far from universal, and some animals were barely affected: the effects on the bird population were minimal.[36] For another, the damage looked different than expected. The spread of cane toads did affect the native frog population, but the overall impact was about rarefication rather than elimination (see chapter 11, Dodo).[37] Even when it came to damage among top predators, which succumbed to the toads' poison in great numbers, the results were ambiguous. A study of mortality among freshwater crocodiles showed that some populations were devastated while others were not.[38] Perhaps most important, studies found that cane toads were fond of disturbed habitats, which was often synonymous with places of human activity.[39] Looking across their well-watered and well-lit backyards, suburbanites were inclined to think that cane toads were everywhere. In reality, cane toads in the backyard were in their favorite place.

It was one of the archetypical conflicts of modern conservation: people with dramatic pictures stood against scientists with more complicated findings. Moreover, people had a powerful narrative, that of invasion, while biologists had none. As a branch of biology, ecology offers quantifiable observations rather than value judgments, and while scientists could show that worries about extinction were overblown, they could not rule it out on principles. In the absence of a better rhetoric, scientists fell back on military metaphors. One of the leading scientists, Rick Shine, wrote in his research memoirs that "invasion-front toads were the amphibian equivalent of the troops sent in to assault vigorously defended castles during the Napoleonic Wars."[40]

There is actually a template for the impact of cane toads, but it runs against hegemonic concepts of culpability. The common reading implies a mistake somewhere in the past that no one would repeat in our own time. According to a powerful cultural script, perhaps enhanced by Enlightenment thinking, it is individuals rather than systems that are to blame for accidents (see chapter 39.2, Drawing Up Lessons). But today most biological invasions happen accidentally, and if we look beyond cultural conventions, there is no way to deny that the effects of cane toads are remarkably similar to the effects of modern societies. They spread continuously and irrevocably and throw environments off balance without much regard for what was there. Humans have even targeted the same group of animals. Just like cane toads, legions of hunters took pride in targeting large predators: wolves, lions, tigers, dingoes, whales (see chapter 9, Whaling). In essence, invasive species are just another disturbance that ecosystems experience in modern times. Humans treat them as distinct moral problems nonetheless, or maybe because of that.

When the IUCN convened a meeting of scientists in Kuala Lumpur in 1999, it faced an unexpected problem. The group was supposed to compile in time for the millennium the aforementioned list of the world's worst invasive species, but for all the disagreements around the table, the experts were unanimous about one point: among the broad range of invasive species, humans were clearly the worst. The IUCN did not allow it and insisted on a list that focused on animals, plants, and pathogens, and the reasons remain anyone's guess.[41] Maybe it was about definition problems or political goals. Maybe it was about disaffection with human self-pity. Or perhaps the decision mirrored a realization that the concept of invasive species satisfies a deep desire of modern people. The public furor about cane toads and other invasive species revealed a sense of gratefulness for an environmental problem where human responsibility was not quite so glaring. Environments do not need invasive species. But maybe humans do.

3. METAMORPHOSES

Australia's nature was isolated for much of its evolutionary history, and that made it particularly vulnerable to new animals and plants. The fight against these invaders runs through the modern history of Australia, and it has seen some spectacular successes. An introduced moth eliminated a plague of prickly pear cacti in the 1920s, an achievement of biological pest control that was surely on the mind of the protagonists in the cane toad affair.[42] After World War II, researchers discovered a virus for feral rabbits that achieved a 99.8 percent fatality rate in laboratory experiments and killed up to 95 percent in the wild.[43] Australians have fought the advancing cane toad physically and scientifically, but it has failed to make much of a difference. The toad's toxicity renders populations immune to natural predators, and their phenomenal reproduction rate means that even severely decimated populations can rebound quickly. When cane toads were found breeding near Sydney Airport in 2009, authorities managed to extinguish the population with a vigorous campaign over several years, but with millions of adult cane toads in the country, it is just a matter of time until the next one jumps off a truck. The only documented case of permanent extermination happened on a tiny island with a single freshwater pond off the coast of Bermuda.[44]

Cane toads will be a part of Australia's nature for the foreseeable future, and they will keep expanding their range. However, these are not the same cane toads that arrived some ninety years ago. Conditions in the rain forest were different from conditions at the invasion front, and researchers have found significant changes in the toad's skeletal morphology. In other words, the cane toad did not arrive as a ready-made "invasion machine" on the Australian continent but rather became one in response to the new conditions. It showed in dramatic change in the speed of invasion, which increased from fifteen to sixty kilometers per year.[45] Evolutionary change also means that cane toads may adjust to drier and colder climates and spread to heretofore inaccessible parts of Australia. The species' range may change even more in response to global warming.

Cane toads were adapting, and so were the environments that they disturbed on their march across the continent. Native Australian frogs learned that eating cane toads was not a good idea.[46] Native mammals had similar experiences and increased their life expectancy by leaving the poisonous toads alone.[47] For species that refused to learn, evolutionary change was the path of adaptation: researchers found a decrease in the jaw size of snakes that prevented them from consuming lethal-sized toads.[48] Some effects were even positive: cane toads killed native parasites and thus reduced the parasite burden for native frogs.[49] To be sure, cane toads remain a disruptive force of the first order, and yet their effect was not as one-dimensional as the imaginary of invasion suggested. Ecosystems invariably change in response to a rapidly multiplying species without natural enemies, but a multitude of adaptations and feedback cycles makes predictions about future developments a gamble. More than eight decades after the arrival of *Bufo marinus* on the Australian continent, Rick Shine put it as follows: "Cane Toads mow down the top predators, and we still don't know where it will all end."[50]

Species and environments changed naturally, but learning was a more complicated thing among humans. Anger about the warty invaders remains strong, and modes of adaptation are not necessarily about making peace. Cane toads have been subjected to various machismo activities from pub races to alcohol-fueled toad busting. Scholars have drawn parallels to Australia's harsh immigration regime, but there may be more to it.[51] Maybe the battle against cane toads comes down to a subliminal reaffirmation of human supremacy. The fight allows people to adopt a heroic posture as unselfish stewards of embattled environments, knowing that cane toads cannot harm humans unless eaten. Or maybe it is just a tale about the superficial lure of visuality (see chapter 39.1, The World Was Watching). One of the iconic scenes in Mark Lewis's *Cane Toads* showed a swerving Volkswagen minibus on a toad-filled road that seeks to smash as many amphibians as possible. It may be the perfect visualization for the fight against cane toads: forceful, futile, and probably fake. According to Rick Shine, the toads were actually melons.[52]

However, *Cane Toads* was not just about fear and loathing. The documentary also showed retirees talking affectionately about the toads in their backyards. It could point toward a looming cultural accommodation, and it would not be the first of its kind. Some migrating species have become natives at the other end of the world, like eucalyptus (see chapter 27, Eucalyptus) in Israel.[53] Some developed a love–hate relationship with locals, such as the boll weevil (see chapter 12, Boll

14.2 A bar in Port Douglas, Queensland, advertises a pub race with cane toads in 2004. Image, LatitudeStock / Alamy Stock Photo.

Weevil) in the American South. Some are stuck with a stigma for centuries, like the breadfruit (see chapter 7, Breadfruit) in the Caribbean. And some species have roots that are all but forgotten: it never mattered for Queensland farmers that they were close to the native home of sugarcane. Societies have found many different ways to deal with migrants, and it is now a given that there is "no universally accepted set of terms to describe an alien species."[54] Even native range can be a remarkably flexible concept. When Rick Shine visited Costa Rica, a local biologist told him that cane toads were an invasive species.[55]

All the while, scientific research continues, but it struggles to make itself heard—a remarkably defensive posture in a modern age that bred more than one self-confident profession. But biological research will never deliver the kind of moral clarity that the mythology of invasion suggests, and it may not even answer all the important questions. Research has flourished since the making of *Cane Toads*, and yet knowledge shows some glaring gaps. For example, research has focused overwhelmingly on northern Australia even though migration into the more populous south is bound to become controversial. According to an article of 2015, no study had been published on the impact of cane toads in southern Australia since 1993. It was a surprising result in light of the cultural vibrancy of cane toads, and it cast a light on the fragility of expert networks in conservation: "We have failed to recognize a major ecological problem unfolding in a place close to major cities where logistics are straightforward, and robust experimental designs are possible."[56] After decades of feverish debates and plenty of activities, Australia's engagement with cane toads comes down to a sobering conclusion, one that we can observe more often in conversations about migrants: we don't really know them.

15

SAUDI ARABIA

THE STATE OF RESOURCES

I. THE GREATEST PRIZE

In February 1945, King Ibn Saud boarded an American warship in the port of Jeddah. It brought him to the Great Bitter Lake in the Suez Canal, where he met the American president Franklin D. Roosevelt. It was barely a meeting of equals. In the Kingdom of Saudi Arabia, Ibn Saud ruled some three million people who mostly lived in poverty in a harsh desert environment, and the trip to the Suez Canal zone was only the second during his long reign that had taken him beyond the borders of his country. Roosevelt had just started his fourth term as president of the United States and had come straight from the Yalta Conference with Joseph Stalin and Winston Churchill. Roosevelt and Ibn Saud talked for five hours and touched on many issues, including their bodily ailments as two aging men. When Roosevelt learned that Ibn Saud envied his wheelchair, he left him with his spare one as a memento. It helped forge one of the most unlikely alliances of modern history, which also proved to be one of the most resilient.[1] The deal between the leading democracy of the West and the royal fiefdom—

Saudi Arabia is the only country named after its royal family—is still on after seventy-five years in spite of numerous predictions that it could not last.[2] It has survived, among other things, Nasser's Pan-Arabism, the 1967 Arab–Israeli War, the Islamic Revolution in Iran, two American-led wars against Iraq, the terrorist attacks of 9/11, various other acts of terrorism, and the Arab Spring. The incumbents of 2022, King Salman and Joe Biden, are the seventh Saudi king and the fifteenth American president to honor the agreement.

One can summarize the essence of the agreement in a single word: oil. Saudi Arabia holds the world's largest petroleum reserves, and the United States has always been the world's leading consumer. The deal was a perfect match in material terms, but the rationale was rather young when Roosevelt and Ibn Saud met. As recently as 1941, the United States had shown its lack of interest in Saudi Arabia's oil by refusing Lend Lease aid for the country.[3] However, World War II catapulted the country into a global leadership role and the US secretary of the interior Harold Ickes warned of an upcoming scarcity of oil in 1943, and these concerns meshed with news from geologists who had explored the potential of the Persian Gulf region.[4] Oil had been discovered in Kuwait and Saudi Arabia five years earlier, and the geologists found that reserves were huge. In fact, they were so huge that, in the explorers' opinion, the Middle East was bound to dethrone the United States as the traditional world leader in oil production. One member of the mission put it to the US State Department as follows: "The oil in this region is the greatest single prize in all history."[5]

By the mid-twentieth century, the modern world had plenty of experience with resource-led development. Sugar (see chapter 2, Sugar) had made the Caribbean a part of the world economy, guano (see chapter 8, Guano) had turned a nation in default into one of Latin America's most reputable debtors, and revenues from rubber tapping had built an opera house in Manaus (see chapter 1.2, Boomtowns). But mineral wealth inevitably ran out after a while, and the built relics provided stark reminders of bygone boom times. "Condemned to nostalgia, tortured by poverty and cold, Potosí [see chapter 1, Potosí] remains an open wound of the colonial system in America: a still audible 'J'accuse,'" Eduardo Galeano wrote in *Open Veins of Latin America*, a classic in the mold of dependency theory (see chapter 30.2, In Their Theories).[6] The precedent haunts those who write about oil in the twenty-first century. For example, Ricardo Soares de Oliveira concluded his discussion of African oil states in the Gulf of Guinea with a look across the Atlantic: "As happened to Andean countries haunted by their foundational silver booms centuries ago or Caribbean societies long shaped by the sugar needs of faraway consumers, the impact of the political economy of oil on the Gulf of Guinea will be acutely felt when oil profits are but a distant memory."[7]

It was a matter of institutions and power relations whether Saudi Arabia would follow down the path of Latin American countries. Charles David Kepner and Jay Henry Soothill outlined a potential scenario in their study of 1935, *The Banana Empire* (see chapter 10, United Fruit): "Politicians of small nations which are weak politically and undeveloped materially, anxious to stimulate agriculture and industry, fascinated by visions of interoceanic railways, and desirous of showing material results without much concern as to how these will be paid for in the future, have succumbed readily to the enticements offered by the fruit companies."[8] The commodity chains of the modern era were notorious for their unequal distribution of wealth among the various stakeholders (see chapter 2.1, Small Islands, Global Networks), and Ibn Saud was obviously at risk of falling for the lure of quick material gains. His empire was vast, fragile, and young—the Kingdom of Saudi Arabia was proclaimed as recently as 1932—and his meeting with Roosevelt revealed a penchant for tokens of appreciation. The presidential wheelchair ended up in Ibn Saud's private apartment for showing off.[9]

Just like the banana business, oil was firmly in the hands of large multinational corporations. John D. Rockefeller's Standard Oil Company monopolized the nascent oil industry in the late nineteenth century and became a symbol of capitalist excess in America's Gilded Age.[10] A conglomerate of multinational companies, commonly called Seven Sisters, controlled the global flow of oil and much else from the 1920s into the 1970s.[11] The oil business was about big money and big men, and it produced epic stories: Daniel Yergin wrote a nine-hundred-page best seller about the testosterone-fueled orgy of arm-twisting, plotting, and downright fraud that built the modern world of oil.[12] As quoted earlier, in his *Wealth of Nations* of 1776, Adam Smith remarked that conversations among merchants were bound to end in "a conspiracy against the publick."[13] Two centuries on, Calouste Gulbenkian, known as "Mister Five Percent" for his share in the Turkish Petroleum Company, offered an oil age addendum: "Oil friendships are very slippery."[14]

As commodities go, petroleum was also rather fluid in material terms. While guano (see chapter 8, Guano) was not good for anything beyond fertilization, the uses of oil changed tremendously. The nineteenth-century oil business was about kerosene for lamps, an important business before the invention of the incandescent lightbulb. Transport became the defining market in the twentieth century, though it did not start with unrestrained automobility (see chapter 34, Autobahn). When Winston Churchill fought for the acquisition of a majority stake in the Anglo-Persian Oil Company as First Lord of the Admiralty shortly before World War I, his guiding thought was about the speed of British battleships.[15] By the middle of the twentieth century, the burgeoning market for plastics (see chapter 40, Plastic Bags) claimed a growing share of petroleum production. Oil became part of the fabric of modern life in various forms, but distinctly masculine business practices remained part of the game into the twenty-first century. When East Timor became independent in 2002, Australia's prime minister John Howard came to the celebrations with a copy of the Timor Sea Treaty that gave Australia a stake in the maritime oil and gas resources of the nascent country. East Timor's prime minister signed the agreement and spent the following years in a ferocious battle with its big southern neighbor for a better deal.[16]

Against this background, it was anything but certain that Saudi Arabia's newly discovered underground riches would translate into the wealth of a nation. In order to make a killing with mineral resources, the right people had to sell the right commodity at the right time for the right price, and things had to work out on the first try: unlike organic resources, mineral deposits do not regrow. In other words, the deal between Roosevelt and Ibn Saud was only the beginning of a giant gamble, but neither side was afraid of a fight. Ibn Saud had personally led a daring raid when he captured Riyadh in 1902, and the petroleum industry had known about fierce opposition ever since the Tiflis Committee of the Russian Social Democratic Workers' Party had sent Ioseb Dzhugashvili (better known by his nom de guerre Stalin) to agitate among oil workers in Batumi on the Black Sea coast.[17] Oil was a trade for tough men, but in the long run, bravery in battle was arguably less important in the making of that rarest of beasts, the happy single-resource state, than a robust set of institutions.

2. THE RIGHT PRICE

Saudi Arabia had exactly two ministries when Ibn Saud met Roosevelt, and neither counted for much. It was the king's personal relations, his war chest, and his armed forces that held the young kingdom together.[18] The country was tantamount to a blank slate when it came to building a state, but some things played out to Saudi Arabia's advantage. First, oil production was a mature technology: almost a century had passed since the wild early days of the petroleum industry in western Pennsylvania and northeastern Ohio. Second, demand grew continuously in the postwar years, and petroleum was built into the fabric of modern societies in ways that assured stable demand. Third, Saudi Arabia's oil region along the Persian Gulf initially evolved in the absence of conflicts with neighbors. Emirates like Kuwait, Bahrain, Qatar, and Abu Dhabi were small and had plenty of oil themselves, and those that lacked oil, like Dubai, found other ways to tap into the wealth.[19] Saudi Arabia also enjoyed a more ambiguous advantage in that it never faced difficult choices between role models. The US government and the Arabian-American Oil Company (ARAMCO) provided the country with a sense of direction for better or worse.

Jointly owned by four American oil companies, ARAMCO held a monopoly on oil production in the Saudi kingdom, but it was far more than a resource company. It was an engine of development (see chapter 32.3, Planning Development): it launched programs as diverse as malaria abatement and agricultural improvement. The company's influence was at its greatest in the early years, when there was little in the way of a Saudi administration that might have competed with its efforts, and like so many development projects, results did not always meet expectations.[20] Scholars have also chastised ARAMCO for troubled labor relations and racial segregation in housing, but for all the legitimate criticism, a glance at the oil states in the Gulf of Guinea quickly puts these grievances into perspective.[21] In twenty-first-century Africa, oil companies are content with a skeletal sovereign state that provides little beyond islands of security (usually production regions and expat compounds) and a semblance of legitimacy.[22]

ARAMCO was not legally compelled to do better. "The terms set by the Saudi government and the supervision were light," Robert Fitzgerald has noted.[23] But it was the time of the Cold War, and with his Point Four Program of 1949, the US president Harry S. Truman promised technical assistance programs for developing countries in order to lure them into the Western camp. In this context, ARAMCO looked beyond the bare essentials of oil production. The company set up development programs to secure that "greatest prize," and ARAMCO "attempted to establish itself as what it called a vital partner in growth."[24] American oilmen wanted a stable country in their own image, or rather the closest approximation that an absolutist monarchy and a population with a 95 percent illiteracy rate would allow.

Some initiatives were remarkably successful. In 1952, American advisers convinced Ibn Saud to create the Saudi Arabian Monetary Authority (SAMA). Staffed by Lebanese accountants and run by a Pakistani for sixteen years—rigorous professional requirements ruled out domestic employees in the early years—the new agency took care of monetary and banking regulation, and "SAMA today is regarded as possibly the best central bank in the Middle East."[25] Americans were more skeptical of Abdullah Tariki, an ardent Saudi nationalist who took the helm of the newly created oil ministry in 1960 and helped create the Organization of the Petroleum Exporting Countries (OPEC) that same year. Sheikh Ahmed Zaki Yamani replaced Tariki in 1962 and held on to the post for twenty-four years, but the new oil minister provided little respite for America's oilmen. In the 1970s, Yamani engineered the gradual nationalization of ARAMCO (henceforth Saudi Aramco) and turned OPEC into a global powerhouse that shaped the price of oil.[26]

It spelled the end of the hegemony of the Seven Sisters, and it changed the world of oil. The large multinationals lost control over petroleum deposits in the wake of a wave of nationalizations, and today's private oil companies are mostly service providers rather than owners of resources.[27] For more than a decade, world leaders anxiously awaited the results of the next OPEC meeting, as higher oil prices rocked Western economies to the core, most memorably in the oil price shocks of 1973 and 1979. But in the 1980s, OPEC's hegemony collapsed along with the price of oil, and the future development of the oil market has been a popular topic among the world's pundits ever since. In 2018, the British weekly *The Economist* summarized its acquired wisdom after three decades of fluctuating prices: "Perhaps the most vexing thing for those watching the oil industry is not the whipsawing price of a barrel. It is the constant updating of theories to explain what lies behind it."[28]

By 1973 Saudi Arabia was the world's third-largest producer of petroleum, and unlike the world leaders, the United States and the Soviet Union, it did not have a large domestic market. That gave Saudi Arabia a powerful role in negotiations over oil prices and production quotas: as a swing producer, the country could increase and decrease oil production at minimal costs and nudge the market into the desired direction. But what was the right price for Saudi oil? The country had an obvious interest in maximizing returns, but its longtime security partner, the United States, had been wrestling with declining domestic production since 1970 and could ill afford a high price. And what if high prices prompted consumers to switch to other energy sources? As the country with the world's largest oil reserves, Saudi Arabia was inclined to take a long view, and Yamani knew that excessive prices would undermine future markets. During the 1973 oil crisis, Yamani summarized his concern in a remark that has entered the textbook literature: "The stone-age didn't end because we ran out of stones."[29]

Like every resource nation, Saudi Arabia was faced with conflicting rationales, but how it dealt with them remained anyone's guess. Decision making within the Saudi oil bureaucracy has always been notoriously opaque, and transnational contexts have not provided much clarity either: "OPEC never succeeded in agreeing on a 'long-term strategy' for prices," Giacomo Luciani observed.[30] We do not even know the precise extent of Saudi Arabia's petroleum reserves, which is perfectly normal in the world of oil. A Chatham House paper noted in 2004 that "probably half to three-quarters of the world's oil is in countries where the oil sector is a state monopoly and whose governments do not feel the need to explain the basis of their reserves estimates."[31] But Yamani provided a glimpse at the style of decision making when he delivered a speech on the 350th anniversary of Harvard University in 1986. Asked how Saudi Arabia mastered the conflicting goals and the multitude of stakeholders in decision making, the oil minister replied: "We play it by ear." The audience found it amusing, and it was probably honest. Perhaps too honest. Yamani was fired a month later.[32]

The reasons for his firing became subject to intense speculation, like so many things in the oil business. The one certainty about Saudi Arabia was that the regime's stability hinged on oil revenues. But what did that mean for the right price? A 2015 volume titled *Saudi Arabia in Transition* surveyed the full range of social, political, economic, and religious changes in the desert kingdom, and its editors offered a "rule of thumb" by way of conclusion: "As long as the price is high (roughly over $100 per barrel) and revenues are steady, the country will remain relatively stable."[33] As rules of thumb go, it was arguably one of the more convincing ones, but it suffered from one major flaw. The price per barrel fell below $100 before the book was out.

3. DUTCH DISEASE

Andrés Velasco was a professor at Harvard University's John F. Kennedy School of Government, but merely teaching international finance and development was not enough for him. He joined the government of Chile as finance minister in 2006, and his four years in office became a memorable experience. The price of copper, Chile's main export product, had quadrupled between 2003 and 2006, and the government faced enormous pressure to go on a spending spree.[34] But Velasco demurred and put most of the money into a rainy-day fund, which made him deeply unpopular until copper prices crashed in 2009. Chile had enough money for a generous stimulus package when the Great Recession hit, and Velasco became one of the country's most cherished politicians overnight.[35]

The economist Avinash Dixit proposed that the essence of Velasco's policy—"being a Keynesian means being one in both parts of the cycle"—should be "posted in huge letters on the walls of treasury departments in all countries."[36] A historian might suggest a less dogmatic reading. Resource history is full of stories about booming economies that eventually fell off the cliff. Peru's guano boom (see chapter 8.3, Running Empty) ended with a nation in default. On Nauru, the depletion of the phosphate reserves wrecked not only the national budget but also the land itself. The Soviet Union used windfall profits from petroleum sales to increase military spending eightfold during the tenure of Leonid Brezhnev, a remarkable growth rate in an otherwise sclerotic economy. Then Brezhnev died, and the oil price collapsed in the mid-1980s.[37] It would have been a challenging situation even in the absence of the Chernobyl nuclear disaster (see chapter 37, *Lucky Dragon No. 5*) and a budget-wrecking campaign against heavily taxed vodka sales (see Interlude, Opium).

Compared with these stories, the experience of the Netherlands was rather benign, and yet it produced the iconic term for the perils of resource-led development. "Dutch disease" was first diagnosed in an *Economist* article of 1977 that wondered about the correlation between natural gas discoveries and the decline of Dutch manufacturing, and economists subsequently used the term to discuss the paralyzing effect of natural resource exports on national economies.[38] But struggling non-Western societies were far more exposed to the risks of easy resource money than were postwar European welfare states, and Fernando Coronil suggested in his study of the Venezuelan oil state that "the Dutch disease should be renamed the third-world or neo-colonial disease."[39] In the case of Venezuela, the state acquired a magical quality, an imaginary of modernity underpinned by oil money and little else, and Venezuela's politicians "appear on the state's stage as powerful magicians who pull social reality, from public institutions to cosmogonies, out of a hat."[40] Coronil published his book *The Magical State* in 1997. One year later, Hugo Chávez was elected president of Venezuela and coined his own "Bolivarian" brand of Socialism. In Coronil's opinion, the Revolutionary Bolivarian Movement was not quite so revolutionary.[41]

Saudi Arabia used its oil money to feed a mushrooming bureaucracy. The number of civil employees grew to 344,000 in 1988, and the government also hired somewhere between 150,000 and 200,000 expatriates, but efficiency lagged behind the growth of bloated administrations: Steffen Hertog characterized the Saudi state as "a surprisingly fragmented, immobile behemoth."[42] Overlap was notorious. Saudi Arabia had six different units for economic planning as early as 1952, and cities saw veritable "battles of the bulldozers" when companies discovered that they had received building contracts for the same areas.[43] The regime also paid lavish sums for mediating brokers like Adnan Khashoggi and contractors, and it was not always about their work: when Mohammed bin Laden, best known to the world as the father of Osama, lost his life in a plane crash, the king made sure that his family received lucrative road contracts.[44] The 2004 *Lonely Planet* for Saudi Arabia described the net result as follows: "Negotiating the arcane workings of Saudi bureaucracy could test the patience of Mother Teresa."[45]

As befits the genre (see chapter 22.2, The Age of Handbooks), the guidebook authors tried their best to buff up the country's attractions, but they could not avoid the conclusion that "modern Saudi Arabia is a nation at odds with itself and with the outside world."[46] Few tourists enter the country anyway, though that makes Saudi Arabia "the last frontier of tourism" in the *Lonely Planet* world.[47] Besides coming to Saudi Arabia for hajj, most foreigners enter the country for business, and those who stay for a while live in carefully segregated communities: "Saudi Arabia is virtually the only country in the Middle East where most Western expatriates live on compounds which isolate them from the Saudi community at large."[48] It reflects a thoroughly immobilized society where change, to the extent that it exists, advances at a glacial pace. As the journalist Karen Elliott House remarked after decades of reporting on the Middle East, "Observing Saudi Arabia is like watching a gymnast dismount the balance beam in slow motion."[49] Oil fired up the economies of the industrialized world, but it had the opposite effect on Saudi society, and the same can be said about the cultural imagination in the rest of the world. Surveying the literary output of the age of oil, the Indian novelist Amitav Ghosh noted in 1992 that unlike the spice trade, "the Oil Encounter . . . has produced scarcely a single work of note."[50]

Of course, some people made good money in the oil business, and it was not just Saudi princes, investors, and managers. Hollywood stars of the postwar years put a lot of their money into oil drillings because a tax clause from the 1920s, the depletion allowance, permitted generous deductions on oil revenues.[51] A powerful government body, the Texas Railroad Commission, secured a niche for independent producers in the state, though the mythology of Texas oilmen, immortalized in the soap opera *Dallas*, did not exactly honor the significance of the institutional setting.[52] Whereas Ibn Saud slept in an improvised tent on the battleship's deck while en route to meet Roosevelt, today's princes travel in greater comfort. But being born into the Saudi royal family does not guarantee a life of luxury: there are simply too many offspring. Muslim polygyny produced a royal family unlike any other, with thousands of princes and an extended family circle that includes as many as thirty thousand people.[53]

Those who worked for Saudi Aramco were also paid handsomely, and they probably deserved it. National oil companies are renowned for inefficiency and corruption, but "Saudi Aramco stands out for its technical expertise and professionalism."[54] However, Saudi Aramco's employees are not a large group: in 2016, the company had 55,466 Saudis and 9,816 expatriates on its payroll.[55] This is characteristic of today's mining business: Morocco's OPC Group, which is to phosphates what Saudi Aramco is to oil, had 20,980 employees in 2016.[56] In his *Carbon Democracy*, Timothy Mitchell compared the slim roster of oil companies to the legions of workers in the heydays of coal and even argued that "an important goal of the conversion to oil was to permanently weaken the coal miners," but the laboring masses were a transitory phenomenon in the coalfields as well, as mass employment perished in the wake of mechanization.[57] But Saudi Arabia's population keeps growing and stands at about ten times what it was in 1945, and it shows in more than staggering youth unemployment rates. Lacking jobs and legitimate ways to vent their anger, young men have turned to joyriding, an illegal and often lethal automobilist subculture (see chapter 34.3, Dead Ends) that revolves around drifting cars.[58]

In Saudi Arabia, only one among five hundred citizens works for Saudi Aramco, but the company delivered 60 percent of the country's gross domestic product in 2008. Its contribution to the national budget is even greater: almost 90 percent of Saudi Arabia's government revenue came from oil, and other oil states like Nigeria, Venezuela, and Oman reported numbers in the same range.[59] Did it influence a country's governance when it relied overwhelmingly on resource wealth? More specifically, did it reduce democratic accountability and hence diminish performance when regimes did not depend on taxing their citizens? A lively academic debate arose over what came to be known as the "resource curse," and indications are strong that the answer must be affirmative. Surveying the literature between 2001 and 2013, the political scientist Michael Ross found "considerable evidence . . . that higher levels of petroleum income lead to more durable authoritarian rulers and regimes; that more petroleum income increases the likelihood of certain types of government corruption; and that moderately high levels of petroleum wealth, and possibly other types of resource wealth, tend to trigger or sustain conflict when they are found in regions dominated by marginalized ethnic groups."[60]

Resources can earn a lot of money, but numbers look different when placed in context. Saudi Arabia appeared fabulously wealthy at first glance, but it was to a significant extent for lack of people. It would have been a different story if two hundred million people had lived in the country, which is roughly the current size of Nigeria's population. Even Norway's legendary Statens Pensjonsfond, the world's largest sovereign wealth fund, holds less than €200,000 per citizen—barely enough for early retirement in a Western country, and certainly not enough in a country with Norway's cost of living expenses.

It all comes down to the fundamental paradox of the resource business. Modern societies have exploited mineral deposits in a frenzy that has no precedent in the history of planet earth, but the bonanza has left few people happy. As the political scientist Steve Yetiv observed, "A visitor from a faraway universe would be forgiven for concluding that oil was a disease, a scourge that needed to be eradicated and forsaken as soon as possible."[61] Even the big men of energy have sometimes taken refuge in a language of disease and addiction. Venezuela's Juan Pablo Pérez Alfonzo, a cofounder of OPEC, famously called oil "the devil's excrement" while Yamani rued in 1979: "All in all, I wish we had discovered water."[62] Even George W. Bush, who ran an oil company before his time in the White House, declared in his 2006 State of the Union Address: "America is addicted to oil."[63] It was the oil-age version of storing moral problems in the environment, a human coping mechanism that has been on display in countless natural disasters (see chapter 25.3, The Human Factor) or interpretations of the boll weevil (see chapter 12, Boll Weevil) in the US cotton South. A reference to forces of nature served to discourage painful questions about human responsibilities.

Cheap mineral resources are a pillar of modern life, but most people prefer to ignore the resource business as much as they can. They are even forgetful about men who mastered the flow for a time. When Andrés Velasco tried to draw on his success as finance minister and ran for Chile's presidency in 2013, he finished the primary with a paltry 13 percent.[64] It does not bode well for those who seek to confront the environmental and social repercussions of the resource business head-on. In a 2012 article for *Rolling Stone*, Bill McKibben proposed to keep 80 percent of the world's known fossil fuel reserves underground.[65] It might still happen, but maybe not by sheer force of will. Human agency is a very relative thing in the face of the combined momentum of masses and markets.

In his speech at Harvard University, Yamani noted that, as commodities go, oil was "special."[66] It is a statement that runs through the resource literature, and it is probably of interest mainly as an act of wishful thinking: it is akin to the rationale of the gambler who knows that the house will always win and nonetheless gives it another try. The price of resource extraction is well documented in history books, economies, and landscapes, but few players have shown an inclination to leave the casino. The latest entrant is Guyana, where oil was discovered offshore in 2015, and Guyana was projected to produce more oil per capita per day than Saudi Arabia within a decade.[67] Welcome to the next round.

PART IV

TECHNOLOGY TAKES COMMAND

PEOPLE ENJOYED COMMAND OVER MODERN ARTIFACTS.
BUT THEN THE ARTIFACTS TEAMED UP.

Whatever you think about modernity, two things are clear. First, technology has something to do with it, and second, defining that something is an enduring challenge. A plethora of new artifacts consumed a broad range of resources and produced an equally broad range of intended and unintended effect, and humans struggled to keep track and make sense of it. When a Dutch scholar compiled a "systematic overview of theories and opinions" about technological progress, the result ran to 1,400 pages.[1] Nineteenth-century imperialists drew a lot of their confidence about the superiority of the West from the command over advanced technology, but that confidence has given way to a more fundamental question: Do we have command over technology, or does technology command us?[2]

The use of technology is as old as human civilization, and every user of artifacts has learned sooner rather than later that things bite back.[3] But modern technology presented a new challenge: artifacts could work as intended and yet create problems through their interaction with other artifacts, humans, natural environments, or all of the above. Modern technology was interconnected on many levels and in many different ways, and that created tensions and synergies as well as puzzling links between benefits and problems. Sometimes connections were so strong that artifacts formed large technological systems with their own rules and challenges, but even innovations like synthetic nitrogen, where interaction with other artifacts went through material flows rather than systemic coupling, can create tremendous upstream and downstream problems. But whatever the nature of interconnections, it is characteristic that modern technology made it difficult to study artifacts in isolation. One of the founding fathers of the history of technology, Melvin Kranzberg, put it as follows: "Technology comes in packages, big and small."[4]

Modern technology presented societies with a type of power unlike anything that they were familiar with. The power of technology could not be inferred from or subsumed within existing power structures. Governments and socioeconomic elites played a significant role in the creation of new artifacts, but they never enjoyed full control over them, and certainly not for lack of trying. It did not suffice to inquire about benefits within the framework of classic economics, as the gains from modern technology turned into essentials that people described in fuzzy terms like "energy," a word that would have amazed a premodern individual in search of firewood. There was not even a clear causal relation between human needs and technological development: "Invention is the mother of necessity," was another one of Kranzberg's aphorisms that entered the collective memory of academic historiography as "Kranzberg's laws of technology."[5] Separating the technological sphere from the human sphere became difficult when new technologies intertwined with bodily routines: air-conditioning shaped how people felt comfortable, and using a water closet became a requirement for feeling clean. In short, modern technology was everywhere and nowhere, notably diffuse even in its physical presence and hard to grasp intellectually and politically. That did not bode well when it came to dealing with its environmental consequences.

The chapters in this section discuss various ways in which technology could create environmental problems. It could be about unintended by-products like the smoke particles that went along with the combustion of coal. It could be about intended effects in the wrong places: the growth effect of synthetic nitrogen was highly desirable on agricultural land and unwanted everywhere else. It could be about the transformation and relocation of stuff, a task that sewer systems and slaughterhouses performed in different ways. It could be about spatial segregation, like the microclimates that air-conditioning systems produced. It could be about energy and material resources, key factors in all the following chapters. And even if a large technological system operated efficiently and smoothly, it could trigger environmental concerns. Just ask any independent observer of slaughterhouse operations.

Many technologies have a multitude of environmental repercussions, and the definition of environmental problems mirrored people's priorities. London smog was not the only way in which coal combustion created problems. It was not even the only pollution problem that the use of fossil fuels brought about: dealing with sulfur dioxide emissions was about kicking the can down the road until the 1970s, and carbon dioxide came into focus even later. The shift from renewable to nonrenewable energy sources is one of the most consequential steps that modern societies took but also one that societies did not recognize until after the fact. However, people did notice that coal produced a new spatial order of energy. The relationship between coal-producing regions and the rest of urban industrial society was colonial in all but name and yet something that modern societies learned to live with.

Smoke was rarely welcome, but definitions of the problem shifted over time. They moved from a threat to cleanliness to matters of economics to a health issue, though the latter remains underrated: few people recognize that particulate emissions claim millions of lives every year. But perceptions were only the first step in a long path toward a solution. Urban pollution raised complicated issues about state authority and technological and administrative means, and it did not make things easier that polluters and victims were often the same people. Air pollution control was a deal that society made with itself, and yet people found it difficult to think about pollution in these terms. Moral outrage has been part of the pollution discourse throughout the modern era, and it is open to debate whether it has done more harm than good.

London smog came to an end after a devastating pollution episode in 1952 that claimed thousands of lives, and a popular narrative suggested a causal link between the disaster and the government's response. It said more about the predilections of Western societies than about the realities of pollution control. The narrative ignored decades of smoke abatement work that brought London to the point where a campaign could achieve results, and it gave scant attention to expert groups, technologies, and legal means, all of which are crucial for effective policies. Campaigns continue to define the public imagination of environmental policy nonetheless. Unlike technologies, modern societies suffer from a limited attention span.

Urban water systems were large technological systems with all their defining characteristics: size, interlocking components, and a momentum that dwarfed human agency. They also had a staff of experts and workers that kept the vast underground systems in operation, no small task in light of their unprecedented complexity and the huge investments that cities, national governments, or private corporations shouldered in the quest for the sanitary city. Water networks also extended into surrounding environments for freshwater and disposal of sewage, and approaches and infrastructures mirrored the hegemony of urban needs at the expense of all other concerns, including those of agriculturalists in search of nutrients.

The logic of the system was a concern for those who shouldered the investments and ran the network. Most urbanites had a much narrower perspective: they were focusing on appliances such as the water closet, a mass-produced essential since the late nineteenth century. It thrived thanks to the public health movement in combination with a new culture of cleanliness that made the private restroom the global standard when it comes to defecation. Differences in technology remain, but they pale in comparison with the enduring gap between those who have access to a private water closet and those who have not. In India, conversations about sanitation are also conversations about rape.

Water networks were large technological systems without individual system builders. The sanitary infrastructure of late nineteenth-century cities was built in bits and pieces and saw more than one managerial blunder. Chicago's slaughterhouses were different: they were designed and built by powerful corporate leaders. First invented in Cincinnati and then brought to perfection in the Windy City, American slaughterhouses thrived on economies of scale, speed, and scope, but they also relied on a favorable economic and institutional context. Modern slaughterhouses had repercussions up and down the commodity chain, and it was not a given that farmers, consumers, and other stakeholders would go along with its innovations.

This context was eminently American, and that made the Chicago slaughterhouse a US peculiarity for a number of decades. But the United States was changing, and so did slaughterhouse operation. Trust-busting ended the reign of the meat barons. Inspectorates improved hygiene and food safety. Workers organized and fought for a living wage. There were even improvements in the treatment of animals, though progress was perhaps least impressive on this front: the cruelty of the slaughterhouse was ultimately about the system and its mindset. But by the mid-twentieth century, the Chicago-style slaughterhouse could legitimately be seen as a model of how modern capitalism could be tamed.

Things fell apart in the postwar years. The industrial logic of the slaughterhouse became the dominant modus operandi along the entire commodity chain, and other countries rebuilt their food systems in light of the American way. Growing affluence showed in a growing demand for meat, and the carnivorous cravings of consumers were just as transnational as their mastery at selective perception: the slaughterhouse was never popular, but its products were. Just at the time when environmentalism grew into a global force, food systems all over the world fell for turbocharged, deregulated global capitalism, and it would never have happened without masses of consumers who were willing to consume industrialized meat. The global hegemony of Chicago's slaughterhouses is a cautionary tale for everyone who believes in ethical consumerism.

Synthetic nitrogen is probably the least well-known chapter topic in this section, but surely not for lack of significance. The Haber–Bosch process gave humans the power to turn atmospheric nitrogen into ammonia, a substance with a range of uses. It was a potent fertilizer and a key factor in the dramatic increase of yields per acre over the twentieth century. It was also important for the production of explosives, and the German army would have lost World War I much earlier if it had not been for Haber–Bosch. Beyond the intended effects, synthetic nitrogen revolutionized the global nitrogen cycle with dramatic repercussions for environments all over the planet. When it comes to effects and moral ambiguity, few chemicals can compete with synthetic nitrogen.

Inventing Haber–Bosch was a brilliant feat of engineering and chemistry, but the technology became an asset as well as a burden. It defined a new path for the German chemical industry with consequences beyond the economic sphere: the momentum of hydrogenation technology made the German chemical industry a willing ally of the Nazis. Haber–Bosch also changed the fertilizer market and, by extension, the view of agricultural soils, which increasingly looked like a mere temporary storage for nutrients on their way from the chemical factory to the plant. Positive feedback loops encouraged fertilizer use far beyond the real needs of agriculture.

Growing fertilizer use had its discontents. The quest for healthy soils was crucial for the making of organic agriculture, which did not turn into a comprehensive effort at superior environmental stewardship until late in the twentieth century. But the majority of farmers went down the route of technology-heavy high-input agriculture, and synthetic nitrogen became the leading cause of one of the world's most intricate pollution problems. Nitrates contaminated ground- and surface water with all sorts of consequences, but should environmental policy target it as a pollutant or rather as a part of a global nitrogen cycle that called for holistic management? The verdict is still out, for nitrates remain an unresolved problem. While scientific genius has brought us into the world of synthetic nitrogen, the only genius at work in dealing with eutrophication is about ducking the issue.

Just like the Haber–Bosch process, air-conditioning required a lot of energy. But whereas food was essential for life, cool air was a luxury. More precisely, it was a luxury that turned into a modern essential due to a transcultural conditioning of the modern body: once accustomed to the artificial chill, people found it difficult to go back. The path toward air-conditioning was usually a one-way street, as spheres that were sealed off and climatized usually remained so no matter what, and it was a memento for everyone who believed that modern technology increased our range of choices. Humans have found plenty of ways to cope with warm weather over the ages, but in modern societies, the range of options narrowed down to a single technology.

While the global spread of air-conditioning looks irresistible from a twenty-first-century viewpoint, scholars have pointed out that it was the cumulative result of numerous small steps, many of which were hotly contested. But is there really a difference between the two readings? Air-conditioning implied a tremendous change in infrastructures and corporal habits, but it was driven by a number of recurring factors: growing technological means, cheap energy, a global transformation of cultural norms that turned sweat, once an evolutionary advantage of the human race, into a taboo.[6] We can also see a growing expert community that built artifacts and cultural meanings in line with the industry of climate control and a dwindling of other types of climate expertise. It remains to be seen whether this will be the template for climate control on a global scale, but air-conditioning shows how the line between human mastery of technology and human entanglement with technology is perhaps just a figment of the imagination.

Air-conditioning was an American technology that went global, and the other artifacts in this section followed similar trajectories. They took shape in specific places in Germany, Great Britain, and the United States and then came into use all over and beyond the industrialized West, and they took on an air of irreversibility over time: as it stands, no country has managed to abandon fossil fuels, water toilets, industrialized meat, mineral fertilizer, or air-conditioning. But as the following chapters show, modern technology relied on certain requirements, down to the recurring episodes of rain that made sewers an attractive option in nineteenth-century cities, and problems escalated more rapidly if institutional, economic, or environmental conditions did not match the new artifacts. The gap between the industrialized West and the Global South is a recurring topic in the following pages; in fact, postcolonial scholars have argued that "the most obstinate line of demarcation between North and South is not income (criteria of wealth) but technology (criteria of skill)."[7] Modern technology is global, but its repercussions differ depending on contexts, and problems with particulates, agricultural chemicals, and air-conditioning units in the Global South seem to exceed those that we know from the industrialized world. Control of technology remained an unresolved issue in the twentieth century. But the issues of the twentieth century may ultimately pale in comparison with those of the twenty-first.

16

LONDON SMOG

IN THE AGE OF COAL

I. CITY PROBLEMS

Many painters were heading to the countryside during the nineteenth century in search of the best views that they could capture on canvas. But Claude Monet had other things in mind when he came to London in 1899. Visibility was not great in the center of the world's largest city, but getting a clear view was not all that important for an impressionist painter. Monet was more interested in a sense of mystery and the observer's subjective response, and in the case of London, a misty atmosphere, overcast skies, and occasional sunlight produced emotive sceneries that kept him busy for months. Lodged in a room with a view on the upper floors of the Savoy Hotel on the bank of the Thames, Monet produced thirty-four paintings of Charing Cross Bridge and forty-one paintings of Waterloo Bridge, many of which are now on display in the leading art collections of the world, and he was not alone in his infatuation with the city's atmosphere. London fog was the subject of countless cartoons and travelers' accounts; it supplied metaphoric raw material to the novels of Charles Dickens; it was the imagined backdrop to the crimes of Jack the Ripper and the cases of Conan Doyle's Sherlock Holmes; and it was the name of an American producer of raincoats. The dirty air of the world's largest city also inspired a long search for a name. Londoners talked about "pea-soupers," "London ivy," and "London particulars" for decades until they learned a new word that Henry Antoine Des Voeux coined in 1904: smog.[1]

The new word was a contraction of "smoke" and "fog," and that was a nice reflection of how urban pollution was the result of an interplay between man and nature. The British climate knows no dry season, as the present author knows from painful experience, and the valley of the Thames had a natural proclivity toward misty weather, but the smoke was entirely the result of human action. In scientific terms, smoke came from an imperfect combustion of coal, a fuel that London embraced earlier than other urban centers: the city was firmly committed to coal by 1600.[2] Compared with other fuels, coal had two crucial advantages. One was about energy density: it delivered more heat per pound than wood, its chief competitor on the early modern fuel market. The other was about supplies: while the amount of wood was constrained by the limits of an organic economy, coal was available in any quantity that mining and transport technology allowed. As it turned out over the course of the nineteenth century, technology and underground deposits allowed a massive growth of production figures, and coal emerged as the energy source that trumped all others: the extent of coal use was among the best indicators of industrial development by 1900.[3] But as coal became the energy equivalent of a killer application all over the world, smoke inevitably followed in its wake. Even Banaba, a tiny Pacific island that is part of today's Kiribati, recorded a smoke nuisance in the early 1900s when the local phosphate mine installed coal-powered rock crushers.[4]

Numerous scholars have suggested that the shift from renewable to nonrenewable energy was a watershed moment in the environmental history of the world. Energy supplies were henceforth a matter of geological assets, rather than the amount of biomass that farmers and foresters could produce, and modern societies were no longer constrained in their energy needs by a lack of space (see chapter 32.2, *Lebensraum*).[5] But for all its significance, this turning point was a retrospective construction. In the contemporary context, the move toward nonrenewable energy was about sleepwalking into a new age. Even the steam engine, hailed as a Promethean invention by old-school historians of

16.1 *Waterloo Bridge, Overcast Weather.* Painting by Claude Monet, on display at Hugh Lane Gallery in Dublin, Ireland. Image, Hugh Lane Gallery Dublin.

technology, spread across England and other countries at an eminently leisurely pace.[6] Conversations about coal were utterly pragmatic, with choices being made according to availability and costs rather than principles, and using the new fuel was certainly no matter of prestige. William Cavert has observed for seventeenth-century London that "while coal was generally *the* fuel of the poor, for the rich it complemented other varieties."[7] Wood and charcoal had their own advantages, one of which was smokeless combustion, and they were cherished among those who could afford the additional expense. The House of Commons stuck to charcoal until 1791.[8]

Fossil fuels changed the geographies of energy. It brought a spatial pattern to fuel that had been familiar for silver since the rise of Potosí (see chapter 1, Potosí). Coal's energy density facilitated long-distance trade, and declining transport costs (see chapter 3, Canal du Midi) in combination with the whims of geology brought a concentration of coal mining in specific realms. Coal could even create new regions such as the Donbass in Eastern Ukraine or Germany's Ruhr. Long supply lines allowed separate spheres for production and consumption, and coal regions became notorious for ruined landscapes and restless workers. It was a quasi-colonial situation, as coal regions carried a burden that would have been beyond the pale elsewhere, but it was a convenient arrangement for the modern metropolis. London certainly knew about the advantages of coal from afar. It had received "sea coal" via the North Sea from Newcastle since the Middle Ages, and London had a street called Sacoles Lane by 1228.[9] In the eighteenth century, the coal trade grew to such an extent that Adam Smith observed in *The Wealth of Nations* that it employed "more shipping than all the carrying trade of England."[10]

The City of London has an Old Seacoal Lane to the present day, and those who wish can read some symbolism into the fact that it is a short dead end. Fears of exhaustion ran through the age of coal from its inception, and they were completely rational: as E. A. (Tony) Wrigley has argued, historical experience in the organic economy suggested "that growth would grind to a halt."[11] When William Stanley Jevons published an essay on the coal question in 1865, his doubts about the sustainability of economic growth earned him an invitation to meet Prime Minister William Gladstone at 10 Downing Street, and Parliament set up a commission to investigate the future of coal.[12] At the other end of the world, the governor-general of Shanxi and Gansu Province, Tan Zhonglin, recorded his opposition to an expansion of coal mining in 1884. Faced with Chinese superiors who sought to boost coal production, Tan argued that by "using machines to mine, the benefits of a hundred years will be exhausted in ten years."[13] It made perfect sense in a place that had used coal for generations, but less so in a restless global economy that thrived on fossil fuels. In short, people did not use coal as if there was no tomorrow because they thought it was infinite. They used it because they could.

Against this background, the smog that wafted outside Monet's hotel room might seem a minor concern, but the people of London found it disagreeable enough. By the end of the nineteenth century, smoke was firmly established in the minds of Western people as an urban nightmare. Smoke was about dirt, and housewives struggled to maintain cleanliness and, by extension, bourgeois civility as best they could. And it was about money. Commercial goods lost value when they were soiled, and property owners harbored similar fears about real estate in smoky business districts.[14] Sanitary reformers had been trying to estimate the monetary costs of London smog since the mid-nineteenth century and had come up with sums in the range of £5 million per year.[15]

Smoke also claimed lives, and does so to the present day. The 2010 Global Burden of Disease Study put the death toll of ambient particulate matter pollution at more than 3.2 million people per year, which made it one of the top ten killers on a list of sixty-seven risk factors.[16] But in the nineteenth century, the health effects of coal smoke were contested terrain, and many people met it with incredulity. As Adam Rome remarked on contemporary opinions, "The medical argument against air pollution always was a hard sell."[17] Smoke was not a contagion, and medical authorities were inclined to see smoke as a disinfectant in the heydays of miasma theory.[18] And, in any case, the nineteenth-century city had plenty of other health problems.

Cities grew all over the Western world in the wake of industrialization, and so did the apprehensions of medical and other authorities. Urban centers brought many people from starkly different backgrounds into close contact, and they bred all sorts of environmental problems. The modern city was full of noise and trash (see chapter 5.2, The Waste Business), it allowed diseases to spread with unprecedented speed (see chapter 21, Cholera), and it challenged the olfactory sense on every occasion. Smoke fitted perfectly into a specter of urban gloom and doom, and it came to symbolize the darker side of the modern city. Smoky air was plain for everyone to see, it was an impediment to visibility and breathing, and it affected everyone regardless of wealth, ethnicity, or gender. When it came to smoke, all urbanites were in it together, or so it seemed.

But when it came to political solutions, it was one among many issues. Smoke was competing for attention with other city problems, and it typically did not fare well. Air pollution never received the level of attention that water did, and investments akin to those for freshwater supplies and sewer networks (see chapter 17, Water Closet) were out of the question. Few could afford to move to the countryside, and those who could faced a choice between fresh air and the amenities of urban society. City life was about a compromise between opportunities and dangers, and the smoke in people's noses and on their formerly white shirts was part of the price they had to pay for a front-row seat in the making of modernity.

But all the charms of the city did not prevent people from complaining. Britain recorded its first air pollution incident in 1257 when Eleanor of Provence, the wife of King Henry III of England, left Nottingham Castle because of the stench of coal smoke.[19] It was not the last time that a royal complained about smoke. In the years before the English Civil War, Charles I clashed with Westminster brewers in what William Cavert has called "the most sustained and serious attack on urban air pollution of the entire early modern period."[20] John Evelyn's *Fumifugium*, a pamphlet on "the inconveniencie of the Aer and Smoak of London" of 1661, was addressed "To His Sacred Majestie"—King Charles II—"and To the Parliament now Assembled."[21] By the late nineteenth century, discontent crystallized around formal organizations. Some of these crusaders did not lack energy or creativity: Henry Antoine Des Voeux was treasurer of the Coal Smoke Abatement Society when he coined his new word in a Christmas Day letter to the *Times* of London.[22] Other leagues had a flash-in-the-pan quality like Chicago's Society for the Prevention of Smoke, an initiative of business leaders who sought to clean up Chicago in preparation for the 1893 Columbian Exposition. The society hired five engineers and an attorney who took recalcitrant offenders to court, failed to gain convictions, and subsequently collapsed after only twenty months.[23] But for all their diversity, activists shared a common experience in that change in the urban atmosphere was always agonizingly slow. There was no easy solution to the problem of urban smoke. And there were far too many chimneys.

2. MATTERS OF STATE

Smoke was not an issue that Otto von Bismarck was passionate about. He fought three wars to unify Germany between 1864 and 1871, he spun the wheels of European diplomacy with passion and skill for decades, and he acquired Germany's colonial empire. But as the towering figure of German politics, he was also dealing with less prestigious issues, and the smoke nuisance reached his desk in the 1880s. Berlin's head of police called for an ordinance against smoke pollution, and Bismarck liked the idea. His minister of the interior, Robert von Puttkamer, agreed with Bismarck on principle, but he saw juridical trouble ahead: Prussia's new administrative courts would likely annul an antismoke ordinance. But Puttkamer knew a few things about Prussian statecraft, and he devised a complicated strategy in order to reverse the court's opinion. It might have worked, but it would have taken some time, and Bismarck was not a man who would patiently watch intricate bureaucratic maneuvering: he was, after all, the man who had unified Germany with blood and iron. When it came to smoke abatement, Bismarck sought quick results. He planned to send smoke offenders straight to criminal courts, effectively sidelining the new administrative courts, and thus get things under control.[24] Puttkamer was a loyal official, but that was too much for a Prussian bureaucrat. He

warned Bismarck that his approach was "poised to jeopardize the rule of law and the authority of the state."[25]

Smoke abatement was a Herculean task in technological terms. A city had thousands of chimneys with many different sizes and inclinations to pollute, and there was no way to put a police officer next to every smokestack. Furthermore, smoke abatement was to a great extent about attention and diligence. By the early 1900s, expert knowledge allowed for the elimination of a great amount of smoke from industrial establishments, but gains could quickly erode if firemen or their supervisors became negligent. However, the conversation between Bismarck and Puttkamer shows that the battle against smoke was about more than technology. Smoke abatement was a matter of prestige for the powers that be, and there were different ways to bolster that prestige. Did the authority of the state rest on an approach that was fair and judicially proper, as Puttkamer argued? Or was it ultimately about getting results, as Bismarck found? Smoke was just as taxing for the authority of the state as it was for the tissues of the lungs.

Smoke abatement called for a deal between different parties. It was about a compromise between those who used coal and those who suffered from pollution, and it did not make things easier that these were often the same people. The fight against the smoke nuisance was usually an elite endeavor, led and financed by people whose carbon footprint was far above the contemporary average, and questions about society's addiction to fossil fuels were beyond debate: smoke had to be fought within the age of coal. And then there were other stakeholders, the engineers, the jurists, and the medical experts, who would need to be part of the deal and yet had their own interests and concerns. It was one thing to hate smoke but quite another to bring all these parties together in a common strategy, as Bismarck learned when he chose to ignore Puttkamer's objections and ordered his subordinates to draft an ordinance. Engineers and jurists came up with widely divergent suggestions, and officials abandoned the project. Bismarck won wars against Denmark, Austria, and France, but his war against smoke ended in defeat.[26]

Many obstacles stood in the way of an effective deal, and one of the greatest was that it was hard to talk in these terms. Many people found it difficult to think about smoke abatement as a compromise between various stakeholders. Wasn't pollution really a moral issue, a matter of decency and proper behavior that called for determination and strength? "I have no more right to deluge my neighbor's premises with soot than I have to empty my garbage can over the fence line," an anti-smoke crusader declared at a meeting of the American Society of Mechanical Engineers.[27] Many laws were passed in this spirit, with enforcement a mere afterthought, and they typically fell by the wayside sooner rather than later. Moral fervor easily did more harm than good, and yet it was tempting to leave it at vociferous condemnations of the "smoke evil."[28]

Faced with a smoke nuisance that refused to go away, city dwellers tried to live with the problem as best they could. In some of the worst cities, businessmen used scarfs to keep their collars tidy on the street, or they brought a second shirt to work because the first one was too dirty for a respectable person beyond noon.[29] Those with money moved away or at least to quarters that were less prone to smoke. It left a mark on the development of cities. Many European cities had their less desirable parts in the east because the wind typically blew from the west on much of the continent. London's crusaders against smoke were typical in that they preferred to talk about pollution in general terms (see chapter 37.1, Global Pollution) though it was beyond debate that the East End had a bigger smoke problem than, say, Kensington and Chelsea. People were adjusting to smoke, and so was the insect world. Studies in England's industrial regions, one of which became known as the Black Country because of its smoke, found a growing population of black moths.[30]

The silver lining was that a reduction of smoke was not hard to achieve. Smoke abatement could even result in net savings for furnace owners if mistakes in design and maintenance were to blame, though the amount was usually too small to warrant much attention. When it came to industrial smoke, apathy was a greater problem than costs, and that opened the door for some progress in the first half of the twentieth century. The industrialists' quest for efficiency, sometimes helped by more or less haphazard enforcement of anti-smoke laws, brought down a significant part of urban pollution, and by the mid-1930s, air quality measurements showed clear improvements for a number of British cities including Cardiff, Glasgow, Newcastle, and London.[31] It was a significant achievement, and yet few people were jubilant. If pollution was evil, it was difficult to celebrate anything short of complete victory, and as improvements hinged on gains in efficiency, it was possible, if not highly probable, that the trend would peter out. After all, the industrial world remained firmly committed to coal.

And then there was the problem that activists liked to confront least: domestic smoke. By the middle of the twentieth century, there was no way to deny that households contributed in large measure to London smog, and yet policymakers were reluctant to intervene in people's private homes. In fact, it was a matter of discussion whether it was a good idea for governments to make smoke abatement a priority. Experience suggested that the outcome would likely be a partial victory at best, and the effort was bound to provoke grand hopes for government's ability to solve society's problems. Even a smog disaster in December 1952 that claimed thousands of lives did not trigger an immediate change of opinion. When the first anniversary of the "killer fog" was approaching, the British minister of health suggested keeping a low profile. He even advised against regular government meetings: "We would be under constant Press and House of Commons pressure to reveal what happened at our meetings and I fear that the statements that we would have to issue would be pretty meagre."[32] The minister of housing, Harold Macmillan, suggested more action on the political front, but both ministers were Tory politicians with an innate skepticism of grandiose government schemes: "Today everybody expects the Government to solve every problem. It is a symptom of the Welfare State," Macmillan groaned in a confidential cabinet-level memorandum. A committee had been investigating the air pollution problem since July 1953, but Macmillan felt that the state was facing limits in the face of smog: "We cannot do very much, but we can seem to be very busy—and that is half the battle nowadays."[33]

As it happened, the committee, named the Beaver Committee after its chairman, paved the way for the end of London smog. Based on the committee's recommendations, Parliament passed the Clean Air Act in 1956, which was the first British law that also addressed domestic smoke. Generous grants helped households switch to smokeless fuels, and the air of London and other British cities became noticeably cleaner within a matter of years.[34] It was a spectacular achievement after centuries of lamentations, and it spared the Swinging London of the 1960s the moody days of fog. Colors were now the order of the day, though they were not always deemed appropriate. A renovation of 10 Downing Street discovered in the 1950s that the familiar black exterior was the result of more than two centuries of London smog. The original color of the bricks was yellow, but that was too much for a building with a Right Honourable resident. In order to preserve the familiar smoke-stained look, the renovators painted the bricks black.[35]

The government put the death toll of the December 1952 smog at 4,000. It shocked the British public, though it likely underestimated the scale of the disaster; more recent estimates suggest that the real number was probably 13,500.[36] That made it tempting to frame subsequent reforms as a classic example of learning from disaster (see chapter 25, 1976 Tangshan Earthquake): faced with a dramatic threat, the state rose to the challenge and made sure that it could never happen again. It was a story that was almost too good to be true, and it did gloss over a few significant aspects. There was no place in this story for the decades of hard work by engineers and anti-smoke activists, nor for consideration of what would have happened without the Clean Air Act. Oil and gas were more convenient for domestic heating than coal, and one may surmise that people would have abandoned it sooner or later anyway.[37] The first natural gas field in British North Sea waters was discovered in 1965, and it is tempting to speculate about the place of the 1952 London smog disaster in collective memory if that had happened a decade earlier.[38]

These omissions matter not only for the sake of a complete account. Smoke abatement had raised inconvenient questions about social discrimination and state authority, but all that was dissipating along with the material reality of London smog by framing it as a story of learning from disaster. Smog was no longer looming as a marker of environmental inequality, and concerns about overtaxing the abilities of government became pointless: no longer would governments have to worry about looking weak in the face of smog. Even a Tory like Macmillan, who would move on to 10 Downing Street, could make his peace with a welfare state that brought such a persistent problem under control. And why would anyone care about the long tradition of anti-smoke efforts, except perhaps for a historian who gets overexcited about nuance?

As it happens, some historians can do without nuance, and those who naively search for stories of hope are particularly inclined to perpetuate the myth of the London smog campaign. Christof Mauch recently summarized it by noting that "awareness brought about action, and action resulted in change."[39] But awareness was the least of all worries when it came to smoke in London and elsewhere, and in the long run, success and failure came down to institutions. Everyday smoke-abatement work is neither spectacular nor heroic, but the London campaign would inevitably have failed without these long traditions; in fact, it is unlikely that it would have taken place at all. The 1956 Clean Air Act was about a combination of Chronos and Kairos, about painstaking work over decades and about seizing the moment, and yet it was convenient to frame the effort as a brief campaign. It smacked of determination and strength, and it served the interests of the government, anti-smoke activists, and the millions who switched to smokeless fuels. Whether it served the interests of those who wanted to emulate London's success remained to be seen.

3. AGENDA SETTING

The Beaver Committee discussed more than just smoke. It also looked into sulfur dioxide emissions from coal combustion at great length and recommended that "the most efficient practicable methods of removing sulphur from flue gases should be adopted at all new power stations in or near populated areas."[40] The idea did not make it into the Clean Air Act, and few people cared. No Western country had a serious policy against sulfur emissions from power plants at the time, and scrubbers were bound to be expensive. A generation later, sulfur dioxide was at the center of environmental policy as a leading cause of acid rain, and Britain did not look good among its peers in the industrialized world. The country became notorious as "the dirty man of Europe" in the 1980s, and its reluctance to clamp down on sulfur dioxide was part of the indictment.[41] Another generation on, climate change suggested yet another perspective, as the end of London smog changed nothing about modernity's fateful addiction to fossil fuels: the Western infatuation with London-type smog and acid rain looks suspiciously like an attempt to sustain narratives of progress by focusing on problems that have actually been solved. At the same time, particulate emissions linger as a severe problem in metropolitan regions, and the Western focus on global warming is at odds with the priorities in cities like New Delhi and Beijing. Even our preoccupation with ambient air, long a fixture in the air pollution discourse of modernity, looks dubious in a global context because the 2010 Global Burden of Disease Study ranked household air pollution from solid fuels among the three leading risk factors worldwide. In Southern Asia, household air pollution was risk factor number one.[42]

There are many ways to frame pollution problems, and agendas reflect more than the bare essentials of economics, technologies, and health. They also reflect historical experience. London smog continues to shape people's understanding of air pollution, and not just among those who read history books. Its resilience in popular culture is powerful enough. Christine Corton has shown how visual media gave London smog a potency that was at times larger than life. There was far less fog in Conan Doyle's writings than in the film versions of Sherlock Holmes, and the *Star Trek* science fiction series had a foggy episode with a Jack the Ripper theme regardless of the fact that the real murders occurred on smog-free days.[43] Wafting smog was a convenient way to evoke the urbanity of the Victorian age, a visible reminder of the bad old days that fortunately lay in the past. But while Westerners talk about London smog as a closed chapter of history, it lingers as an icon for those who are concerned about the air pollution problems of the Global South. Problems may differ in material terms, but the fight against London smog serves as a template for the kind of vigorous campaign that antipollution activists outside the West dream about.[44]

London smog is not a thing of the past, nor is the sense of moral clarity that has pervaded anti-smoke rhetoric throughout the ages. Pollution continues to loom as a form of evil, rather than a deal among stakeholders, and political rhetoric shows no sign of moderation, notwithstanding the fact that urbanites have been living with dirty air for centuries. When the World Health Organization published a report on outdoor air pollution in 2016, the Paris mayor Anne Hidalgo declared, "We cannot negotiate with Parisians' health."[45] But in the real world, these negotiations occur on the streets of Paris every day, though usually in actions rather than words, and so it is in the other metropoles of the world. People choose the opportunities of the modern city in exchange for noise, pollution, and other contestations of urban life, and yet they lack a language to talk about this deal. The preferred language of the health sciences, mortality statistics (see chapter 7.2, Numbers Games), certainly does not make the cut. The death toll of the 1952 London smog was arguably huge, but in the new millennium, more than 13,500 people die globally from particulate emissions every single day.

Whatever the causes, this speechlessness is certainly not due to a lack of research. Air pollution has attracted clever minds throughout the ages, and the amount of knowledge in earlier times is impressive. We can read about the urban heat island effect that traps pollutants in the city in Luke Howard's *The Climate of London* of 1818.[46] Yet centuries of research have not produced a consistent understanding of air pollution that people around the world would be willing to embrace collectively: agendas reflect differences in culture and socioeconomic status, and people are usually surprised when they learn about divergent understandings. Most people prefer to view pollution through a single window, just as Monet had during his stay at the Savoy Hotel. But unlike Monet, they lack a rationale for their constrained perspective.

17

THE WATER CLOSET

PRODUCING CLEANLINESS

I. PURITY

It was a busy day at the railroad station in Kanton, and the Chinese authorities sought to impose some order. A number of trains were about to depart, and security personnel defined an assembly point for each train where travelers were told to squat. They did as they were told and placed their buttocks on their heels, except for a German journalist who immediately toppled over. It left a deep impression on the foreigner who thought about the experience in terms of Chinese authoritarianism. It was a natural guess from a Western perspective, and one with background in the scholarly literature: Fernand Braudel, for one, has argued that traditional Chinese society knew "a sort of division between seated life and squatting life at ground level, the latter domestic, the former official."[1] However, his domestic friends suggested a more mundane interpretation. It was probably about defecating. In a country where toilets were sometimes just holes in the ground, squatting with your heels down was simply a necessity of life, and habits live on with people even when sanitation improves.[2] A survey about public toilet use in Taiwan found that almost half of respondents did not actually sit on sitting-type toilets, and a third expressed a preference for squatting on the rim.[3]

Using a restroom is an intimate affair, but in the twenty-first century, the setting is a matter of world politics. In 2013, the General Assembly of the United Nations passed a resolution on "sanitation for all" that, along with designating November 19 as World Toilet Day, urged member states "to end open defecation."[4] In light of some 2.5 billion people who lacked basic sanitation at the time, the issue was anything but trivial, and yet Western observers were blissfully unaware of World Toilet Day, not to speak of the consequences of squatting for lower body muscles. For those who live in an affluent society, few things are more normal than using a water closet, and if people have none at their disposal, they seek written instructions (see chapter 22.2, The Age of Handbooks). First published in 1989, the backpacking guidebook *How to Shit in the Woods* became a best seller that is in print to this day.[5]

The technology of flushing is probably as old as urban civilization. Lawrence Wright began his classic history of the water closet with a latrine in the Minoan Palace of Knossos in ancient Crete.[6] Archaeologists even found toilets in urban buildings of the Indus civilization, sometimes located in separate rooms, more than four thousand years ago and always connected to an elaborate water circulation system to take care of the waste.[7] The remains of the Roman Empire also hold plenty of toilets, somewhat to the embarrassment of nineteenth-century archaeologists who preferred to misread them as steam baths, prison cells, or machine chambers for hydraulic lifts.[8] Great Britain issued its first patent for a water closet in 1775, and the pace of tinkering and incremental innovation picked up over the following century.[9] Wright celebrates 1870 as "the *annus mirabilis* of the water-closet," and while condensing a prolonged development into a single year has a whiff of old-school history of technology, the remark suggests a fateful turning point in the global career of the water-flushed toilet.[10] Over the second half of the nineteenth century, the water closet morphed from a handcrafted plaything of the rich into a mass-produced essential.

Technology was changing, and so were the norms and values that guided its use. As the anthropologist Mary Douglas has argued in *Purity and Danger*, ideas and rituals about cleanliness are ripe with cultural significance.[11] Norbert Elias even cited the drive toward privacy in urinating and defecating as evidence of Europe's cultural exceptionalism in his *The Civilizing Process*, which looked at basic human needs in order to trace a long-term evolution of manners and a culture of self-restraint from the Middle Ages to the Victorian period. Commenting on toilet habits in the times of Erasmus, Elias observed a "behavior that can still be encountered throughout the Orient today."[12] *The Civilizing Process* became a classic of sociology, though it drew staunch criticism from the German anthropologist Hans Peter Duerr who argued for a more ambiguous concept of civilization and sought to dismantle the categorical dichotomy of primitive and modern cultures. His critique eventually grew to five volumes with 3,500 pages, and yet he never discussed the water closet.[13] It was probably too hard a case to argue.

Several trends made the nineteenth century a watershed in the production of cleanliness. Industrial technology reduced the price of pipes, valves, and porcelain. The growth of densely populated cities deprived urbanites of traditional ways to relieve themselves. Concerns about urban disease rates, spurred by waterborne epidemics like cholera (see chapter 21, Cholera), inspired a burgeoning public health movement. An ascendant bourgeoisie embraced cleanliness and privacy in intimate affairs. All over the Western industrialized world, the water closet became a matter of costs and space rather than principles. The American environmental historian Joel Tarr has estimated that by 1880, about a quarter of US urban households had access to a water closet.[14]

Building toilets was about building society. A German cultural critic, Julius Langbehn, argued in a late nineteenth-century best seller that Germany would have fewer Socialists if only it had more bathrooms.[15] Segregated toilets were standard practice in racially diverse societies from the American South to Germany's colonial empire.[16] When Bulgaria launched a Model Village program in 1937 in order to educate its peasant population, the benefits of running water, sewers, and modern hygiene were part of the agenda.[17] More recently, gender-neutral toilets became all the rage among Western liberals while other scholars sought to establish an enforceable constitutional right to safe drinking water for the First Nations people of Canada.[18] In the run-up to India's 2014 general election, Narenda Modi pledged to "build toilets first and temples later."[19] It probably helped him advance to the prime ministry, for toilets are about more than disease and status in India. Lack of sanitation forces women to relieve themselves outside at night, where they are vulnerable to rape.[20]

For all the privacy of a modern restroom, you were never really alone on a water closet. A web of traditions, customs, and distinctions connected the toilet user to the modern world, and the same held true for the artifact itself: it was shaped not only by the genius of inventors but also by governmental rules and regulations. Administrations and professional bodies devoted considerable time and energy to the development and enforcement of standards for everything from ventilation to the size of pipes, a peculiar challenge in light of the divergent goals in play: cleanliness, economy, reliability, consistency.[21] Water closets were complicated devices, where bad design or poor maintenance could easily cause sanitary and other calamities. And they were just one part of a technological network with its own rules and requirements.

2. CAN'T BEAT THE SYSTEM

Just like the water closet, sewers were around long before the dawn of modernity. Ancient Rome built the *cloaca maxima* around the fifth century BC, which kept the Roman Forum tidy and dry into imperial times.[22] Modern cities built great wastewater outlets as well, sometimes with a place name to match. Two giant storm sewers were built in New York City's Flushing Meadows, originally named after the Dutch harbor Vlissingen, when Parks Commissioner Robert Moses turned the tideland into a fairground in the 1930s.[23] But unlike their predecessors, the large sewers of modern cities were backbones for vast underground networks that extended into the private homes of millions of people. As of the nineteenth century, building the sanitary city was about building a system, or rather two systems: one for freshwater and one for wastewater.[24]

The sewers were the sturdier of the two systems, as the great underground canals had to cope with a number of unknowns. That was the inevitable price of building a system that reached into private quarters: sanitation departments had no control over the volume or quality of wastewater. Trouble was also lurking at the other end, as obstructed outlets were a classic source of problems. "Sometimes the backwash of the Tiber floods the sewers and makes its way along them upstream," Pliny the Elder's *Natural History* recorded on Rome's *cloaca maxima*.[25] A dry downstream river could cause other types of problems, as London learned in the hot summer of 1858. The British capital had embraced the water closet perhaps more enthusiastically than any other city in preceding years, but runoff accumulated in the Thames and produced an event that came to be known as the Great Stink. An awkward smell wafted through the Palace of Westminster where Members of Parliament displayed a sudden burst of interest in sanitation, but the path from interest to sustainable solutions was long.[26] Forty years after the incident, the Baedeker handbook (see chapter 22, Baedeker) found it "worthy of remark that this pollution of the most important river in Britain is at present made legal by an exceptional clause in the River Pollution Prevention Act."[27]

Climatic conditions in Western Europe and North America helped keep problems in check. In fact, it is tempting to speculate whether sanitation history would have taken a different course if the great cities of the nineteenth century had known months of dry weather or an extended monsoon season. Be that as it may, sewer technology faced limits beyond the Western world, and it was about more than malodorous water closets. "The lack of water is one of the principal obstacles to the extension of our domination," the *Inspecteur du Service des Ponts et Chaussées* lamented in 1860s French Senegal.[28] Saint-Louis, Senegal's capital at the time, had only four months per year with an abundance of water, but a shift to dry toilets, perhaps following the patented design of Henry Moule, the Vicar of Fordington in Dorset, was not the preferred option.[29] The governor ordered the construction of a reservoir (see chapter 29, Aswan Dam) with an intended capacity of 8.5 million cubic meters in 1866. However, a flood destroyed the barrage, and Saint-Louis was struck by a yellow fever epidemic and a cholera epidemic (see chapter 21, Cholera) before the end of the decade.[30] In Los Angeles, a desert city prone to flash floods, thousands of condoms washed up on the beach near the city's sewer outlet at Santa Monica in the late 1930s. The incident entered history books because it interrupted a conversation between Aldous Huxley and Thomas Mann about Shakespeare.[31]

The water system of Los Angeles did not look more attractive on the freshwater side. The water grab in Owens Valley, more than two hundred miles to the north of Los Angeles, even inspired a movie in the form of Roman Polanski's *Chinatown*.[32] The cinematic offshoot was exceptional, but bringing water in from far out in the countryside was perfectly normal for the growing cities of the world. Manchester built the Thirlmere Reservoir, now located in Lake District National Park (see chapter 26, Kruger National Park), in the face of staunch resistance from nature lovers; Munich diverted water from Lake Tegernsee on the northern fringe of the Alps; and New York City drew on the Catskill

Mountains.[33] Sydney secured its water through a system of dams and weirs to the city's south that was first tapped with a hastily built patchwork of pipes and timber flumes when the Australian city suffered from a severe drought in 1885.[34]

Sometimes conservation was an unintended beneficiary of the city's thirst for water. In Rio de Janeiro, nineteenth-century watershed protection eventually led to the creation of Tijuca National Park.[35] However, the more common result was a long and bitter conflict between the local population and a powerful urban water authority that typically ended with a resounding victory of metropolitan needs. Cities were disinclined to temper their thirst, and when they did, such as when New York City launched a comprehensive program to replace water-wasting toilets with low-flow models in the 1990s, costs were the defining issue.[36] Sewers and aqueducts were built affirmations of the city's hegemony in modern times.

In theory, it would have been a good idea to expand sewer and freshwater systems in sync. But in practice, both evolved more or less independently: coordinating the construction of supply and discharge networks was beyond the abilities of nineteenth-century authorities, and many of the early water closets emptied into cesspools rather than sewers.[37] Some cities did not even start both endeavors around the same time. Baltimore's city government took over a private water company as early as 1854 and embarked on a vigorous expansion of supply networks and reservoir capacity before the Civil War, but it postponed engagement with wastewater until after 1900, eventually becoming the last American city of its size to build a proper sewer system.[38] And even those who understood the systemic nature of the new technology, such as the English sanitary reformer Edwin Chadwick, did not necessarily embrace a management style to match. While running the General Board of Health from 1848 to 1854, Chadwick got bogged down in a bitter and ultimately counterproductive controversy over the merits of brick versus pipe sewers.[39]

Centralized urban planning was an uphill battle in most nineteenth-century cities, but sanitary systems had their own ways of imposing their will upon reformers. Gradients and the size of pipes and canals imposed limits on decision making, and the same held true for the vagaries of demand and outflow. In light of previously built infrastructures and the law of gravity, planners had to work with the flow and the circumstances. Construction became more orderly in the twentieth century, if only because the backbones of water systems,

17.1 Men standing in opening of unfinished sewer in the 1880s. Image, Edgar Sutton Dorr Photograph Collection, Boston Public Library.

long aqueducts and giant sewer mains, were in place at some point, and new suburbs typically received sewers before houses were built.[40] Aging networks and low-density suburbs presented new challenges, though they paled in comparison with the sanitary problems in the megacities of the Global South.[41] Water systems were expensive to build and maintain, and the level of investments reflected the wealth and poverty of nations.

Needless to say, the general trend left room for conflicts. Downstream cities battled with upstream cities over sewage treatment.[42] Engineers discussed the pros and cons of separate sewers for rainwater.[43] Thirsty cities competed with irrigation farmers who were even thirstier (see chapter 29, Aswan Dam). Public restrooms presented moral challenges.[44] Bathroom designers developed artifacts for every taste and level of affluence from squat toilets to Japanese high-tech solutions.[45] And then there were the divergences that national styles and geographic peculiarities inevitably produced. For example, Japan had only six sewer systems with treatment facilities by the end of World War II, and sewers served less than 5 percent of the total population.[46]

Designs and trajectories differed, and so did corporate structures. Mexico's president Porfirio Díaz directed the construction of the

Grand Canal that drained the Valley of Mexico, a project since Spanish colonial times that had grown more pressing with the growth of Mexico City. A British businessman, Weetman Pearson, completed the Grand Canal two years ahead of schedule, cementing a lifelong relationship with Díaz that brought Pearson lucrative contracts in harbor construction, railroads, and oil.[47] When Baku faced a water crisis, funds from Azerbaijani oilmen (and a widowed oilwoman, Nabat Ashurbeyli) allowed construction of a pipeline for Caucasian water.[48] In imperial Germany, municipalities owned 83 percent of the country's waterworks around 1900, as powerful city administrations sought to deal with sanitation issues on their own terms.[49] Saint Petersburg drew up forty-eight blueprints for a sewer network over forty-one years but failed to start construction before the October Revolution, and when the city took over a private water company in 1891, operation under municipal auspices focused on subsidizing the city's budget.[50] Istanbul had a French company for the European side of the city and a German company for the Asian side, and both focused primarily on affluent quarters.[51] However, both were deeply unpopular due to unclean water and lethargic management, and Turkey nationalized them in the 1930s.[52] And then, Istanbul at least had one single corporation per continent. London had eight companies until they were forced to merge into the Metropolitan Water Board in 1903.[53]

The divergence between private and municipal ownership was closely related to the supposed recipients of sanitary service. Were the water closet and other amenities intended as privileges of affluent people who could pay a private supplier handsomely? Or was sanitation a welfare issue, if not a right of every modern human, that was best delivered by a public body at low prices? In Berlin, 81 percent of households had access to a water closet by 1890 and 97 percent did a decade later, and while many of these toilets were used by more than one family, extensive coverage was quite an achievement in what was essentially a working-class city.[54] It was a litmus test for visions of society within government and the public at large. As Patrick Joyce has observed, "In the course of the nineteenth century modern versions of the 'social' in Britain emerged precisely around questions of the provision of infrastructure and public health."[55]

Water projects were difficult and expensive to build, and completion was duly celebrated. The inauguration of Mexico City's Grand Canal in March 1900 was an act of state with President Díaz, numerous ministers and diplomats, and extensive press coverage.[56] In 1873, Vienna celebrated the arrival of mountain springwater from the Alps by opening a large fountain that has paired with the Red Army's War Memorial since 1945.[57] Kaiser Franz Joseph was present to witness water shooting some sixty meters into the air, and water projects stimulated interest beyond opening days. Paris created a magnet for tourists of all stripes (see chapter 22, Baedeker) when it introduced guided tours through the city's sewers during the 1867 World's Fair, and it remains anyone's guess whether visitors came because of engineering features, Freudian motives, or Victor Hugo's *Les Misérables*.[58] Building the sanitary city was a show of force, and it left traces in collective memory. When the Beaver Committee reported on what to do about London smog (see chapter 16, London Smog), it argued that air pollution "needs to be combated with the same conviction and energy as were applied one hundred years ago in securing pure water."[59]

However, some projects were inaugurated on a quiet note. Chicago built a huge Sanitary and Ship Canal across the continental watershed in order to divert the city's sewage, which included the effluents of the stockyards (see chapter 18, Chicago's Slaughterhouses) on the city's South Side, away from Lake Michigan and toward the Mississippi. The earth-moving equipment was later used for the Panama Canal, and the American Society of Civil Engineers designated the project as one of the "Seven Wonders of American Engineering" in 1955, but the opening took place in undignified haste in January 1900.[60] Officials at the Sanitary District knew that downstream Saint Louis sought an injunction against the opening of the canal, and they wanted to create flowing facts before the case was formally filed with the US Supreme Court.[61] For all their technological sophistication, the sanitary systems of the nineteenth century were rather roughshod affairs, and it was particularly toward the end of the pipe that things were fizzling out.

3. THE DIRT OF CITIES

Sewer construction was not just about the needs of urban residents. It was also about the flow of nutrients between the city and the countryside. In the eyes of the German chemist Justus von Liebig, negligence about the use of human feces was beyond excuse at a time when agriculturalists paid hefty sums for guano (see chapter 8, Guano): "In the flesh and the produce of the field we have for centuries supplied to the large towns the constituent elements of guano, and have never brought this guano back again; and we now send vessels to Chili [sic], Peru, and Africa for this substance," Liebig wrote.[62] He pointed to China, where a flourishing trade in nightsoil showed that human feces were held in proper esteem.[63] Liebig even speculated about popular opinion in a country that he never saw: "Everybody knows the amount of excrements voided per man in a day, month, or year."[64] He was not alone in falling for Chinese fecal exoticism. "No Chinese peasant ... goes to the town without bringing back, at either end of his bamboo pole, two buckets filled with unmentionable matter," Victor Hugo wrote in *Les Misérables*, adding by way of explanation that it was "thanks to this human manure that the Chinese earth is as fruitful as in the days of Abraham."[65] And it was not just about how nutrients from excrements would boost yields per acre. Chadwick hoped that revenues from the sale of sewage would pay for expensive urban sewers.[66]

Water closets sent valuable plant nutrients down to an uncertain fate, and putting effluents to good use looked like a clever idea from a chemical perspective. But as sanitation turned from an idea into a built reality, the concept ran into a number of problems. Fields needed fertilization only during the growing season while sewage ran year-round. The composition of nutrients did not always match the needs of plants. Soil types and hydrology had to meet certain requirements. Sewage farms needed a lot of space on the urban periphery where land prices were typically rising. And as the twentieth century progressed, valuable feces in sewer outflow mixed with a growing array of chemicals that had more ambiguous qualities.[67]

Sometimes specific conditions played out to the advantage of sewage farming. In the United States, sewage farms thrived mostly in the American West, where arid conditions made sewage an alternative to irrigation.[68] Israel draws more than half of its irrigation water from effluents, which makes for a record-setting sewage recycling rate of 86 percent.[69] Sewage farming was a briefer episode in other parts of the world. In the late nineteenth century, Sydney maintained a sewage farm close to Botany Bay that struggled with unreliable lessees and complaints from neighbors for a few years until it succumbed to a disastrous swine fever outbreak in 1905.[70] And some cities eschewed sewage farming from the outset and went straight for sewage treatment, which ultimately changed rather than solved disposal problems.[71] When Baltimore finally came around to sewer construction in the early 1900s, it built a state-of-the-art treatment plant and dumped the sludge in the Chesapeake Bay.[72] The world was broadly moving in the direction of treatment, but that trend was not irreversible. As one of the few Pakistani cities with an operational wastewater treatment plant, Faisalabad faced court battles with farmers who won the right to use untreated wastewater.[73] And then, Pakistani farmers at least showed some interest in wastewater use, if only because it saved them the expenses for commercial fertilizers such as synthetic nitrogen (see chapter 19, Synthetic Nitrogen). Only about 10 percent of human excrement is reused in agriculture or aquaculture globally.[74]

And then there was the human element that typically received short shrift from chemical minds. Urban water systems were not just about bricks and pipes but also about many working hands that kept the system afloat. Those who worked in sewers faced all sorts of risks, and it was estimated that in Paris, where almost a thousand sewer workers labored underground around 1900, about one-third had died after ten years on the job.[75] Out in the countryside, studies have found that agricultural use of sewage played a role in a broad range of health problems.[76] In the twenty-first century, scientists are improving their analytical methods to detect dozens of hormones and pharmaceuticals in sewage sludge samples.[77] Just like scrap yards (see chapter 5, Shipbreaking in Chittagong), sewers reflect the material obsessions of urban societies with brutal honesty, and most people are glad if processing goes on as inconspicuously as possible.

From a distance, urban water systems have their own magic. Some five hundred years after the construction of the *cloaca maxima*, Titus Livius praised it profusely in his *History of Rome*: he described the work by saying that "even modern splendor can scarcely produce any thing equal."[78] But in everyday operation, it was more of a makeshift solution that only survived because of constant vigilance among maintenance crews and an apathetic general public. While underground systems are limping along, things are looking good from the seat of a properly operating toilet, and thoughts about the link between intimate needs and the wider world are eminently optional. Sometimes it did not even require a look at side effects to put the sense of pride into perspective. When a British engineer presented the results of water sampling in Tokyo to the members of the Asiatic Society of Japan in 1877, he came up with a striking discovery. Tokyo had a premodern water system with wooden pipes and a flourishing nightsoil trade in the years after the Meji Restoration, and yet a comparison with London, one of the hubs of the sanitary revolution, did not lead to a foregone conclusion. Tokyo's water was cleaner than London's.[79]

18

Chicago's Slaughterhouses

Animals on the Line

I. ANIMALS IN CHAINS

Sometimes it takes a lot of reading and reflection to make sense of things. And sometimes it takes just the right location and a contemplative mood. For Upton Sinclair, the gallery of a Chicago slaughterhouse was such a place for deep thoughts. Here visitors could observe how hogs were taken by their hind legs, hoisted upward, attached to an overhead rail, had their throats slit, and then traveled down a line of workers who dismantled them at breakneck speed. "It was all so very businesslike that one watched it fascinated. It was pork-making by machinery, pork-making by applied mathematics," Sinclair wrote in *The Jungle*, the book that shaped the public's view of Chicago's stockyards like no other. But then, this was also a place where animals were dying by the thousands each and every day, and that put everything in a different perspective. "One could not stand and watch very long without becoming philosophical."[1]

The Jungle was a novel, but the visitors' gallery was real enough. More than one million visitors watched the spectacle in 1893 alone, when the Columbian Exposition brought the world to Chicago.[2] After all, the stockyards were a marvel of industrial efficiency. Animals walked their final journey upward, as slaughtering took place on the top floor so that gravity could propel the carcass during dismantling. In a design that brought the division of labor to a new extreme, workers performed their single specified task within a matter of seconds. The assembly line, for which Henry Ford received most of the credit, actually made its debut as a disassembly line in Chicago.[3] Packinghouses typically had what Andrew Diamond has called "an ethnoracial pecking order" that ranged from Irish and Germans who formed a kind of "butcher aristocracy" to Black people for the dirtiest jobs; Eastern Europeans like Jurgis Rudkus, the Lithuanian-born protagonist of *The Jungle*, held places somewhere in between.[4] Overseers monitored workplace performance, far more so than meat quality, and fired workers who were too slow, botched their assignment, or otherwise created trouble. Replacements were hired and trained in no time, as the job on the line required little more than brawn and a tolerance for repetitive motion injuries.[5] Packinghouses also made sure that no part of the animal went to waste, and they took pains to find markets for everything, down to the production of nitrogen-rich fertilizer (see chapter 8, Guano, and chapter 19, Synthetic Nitrogen) from slaughterhouse waste.[6] The stockyards used "everything but the squeal," as the saying went, though that did not keep them from causing excessive water pollution (see chapter 17, Water Closet).[7]

Industrial efficiency was one part of the miracle, and economies of scale the other. Millions of animals exhausted their lives in the south of Chicago: four million hogs alone in 1877/1878, when the union stockyards were operating for a dozen years. Even Cincinnati, which had pioneered the mass production of pork to the west of Appalachia, was processing less than a fifth of this number by that time.[8] By the end of the century, some four hundred million animals had passed through Chicago's stockyards.[9] Entrepreneurs tried different business models until they found that focusing on packing brought the greatest profits.[10] Corporate concentration followed suit with a long series of expansions, acquisitions, and mergers to please the capitalist soul. By the 1890s, an oligopoly of Chicago-based companies was dominating the market, colloquially called Big Four, Big Five, or Big Six, depending on the state of corporate arm-twisting. On the eve of World War I, five companies controlled two-thirds of America's fresh-beef output: Swift, Armour, Morris, Cudahy, and Schwarzschild & Sulzberger.[11]

The slaughterhouse was at the heart of the trade, but the famous gate of Chicago's stockyards, a National Historic Landmark since 1981, did not designate the limits of corporate influence. Meatpacking was a corporate octopus long before United Fruit (see chapter 10, United Fruit) gained that moniker, with its tentacles stretching out to countless barns and fields. As William Cronon showed in his magistral *Nature's Metropolis*, the rise of Chicago's slaughterhouses hinged on the expansion of agriculture across the prairie lands that stretched all the way to the Rocky Mountains: the frontier, traditionally seen as a place of manly freedom, was coupled with an expanding urban market.[12] Telegraphs transmitted demand and prices, and railroads brought the animals to slaughter, with the return train supplying the farmers with plows, barbed wire, and everything else from the Sears catalog. Railroads (see chapter 3, Canal du Midi) were also crucial for distributing processed meat, with another world-class innovation, the refrigerated railroad car, coming out of the quest to get Chicago meat to East Coast consumers. Steamships extended the commodity chain overseas.[13]

In short, the visitors on the gallery had good reason to be amazed. And yet the industrial marvel hinged on a number of preconditions that commonly escaped their view: a vast agricultural hinterland full of market-oriented farmers, excess grain production that supplied cheap animal feed, railroad companies eager to find commodities, an immigrant population that kept wages low, a river that carried away the effluents, a lenient approach to air and water pollution control, and urban consumers willing to eat frozen meat. Perhaps most crucially, it took a city without a powerful network of artisanal butchers who cornered the meat market. In European cities, slaughtering remained firmly in their hand, with urban slaughterhouses, tightly controlled and municipally owned, supporting their trade. Even New York City had a guild of butchers until the city council liberalized meat sales in 1843. Butchering continued more or less unregulated until Chicago's meat, chilled and super-cheap, arrived in the 1880s. Not knowing what else to do, New York butchers abandoned slaughtering and focused on cutting up meat for retail.[14]

Chicago products traveled internationally, but for the time being, the Chicago slaughterhouse remained a distinctly American institution. An attempt to bring it south of the Rio Grande collapsed during the Mexican Revolution, and not just because the name of the company, DeKay, was probably the worst-ever in the meat business.[15] In tsarist Russia, Saint Petersburg and Moscow built large centralized slaughterhouses in the proximity of railway terminuses, but they were municipally owned, and reformers pushed for these projects with references to prototypes in the great European metropolises.[16] Looking at the industrial empires of Armour, Swift, and others, it is not difficult to identify the typical factors behind corporate growth in late nineteenth-century America: expanding urban markets, mass production, economies of scale, and a government averse to meddling with big business. But that was about to change.

2. CHAIN REACTIONS

The visitors' gallery was there for a reason. The meat barons knew that when it came to public opinion, there was room for improvement. Slaughtering animals had been a business fraught with rituals and taboos since biblical times, and slaughtering by the millions did not exactly diminish reservations. And who really knew what went into a sausage? There were also quite a few critical minds snooping around the slaughterhouses: public health officials targeting unsanitary conditions, politicians talking about reform, investigative journalists poking their noses into the mud. The gallery, in short, was an act of public relations avant la lettre.

Much has been made of Upton Sinclair's *The Jungle* and its dramatic description of what went on in the slaughterhouse: the brutal speed on the shop floor, the chemicals, the rats, the use of spoiled meat for sausages, the moonlight processing of deceased animals, and the man who fell into a rendering tank and ended up as pure leaf lard. But Sinclair's book was merely a catalyst for a long-standing campaign. *The Jungle* was the rallying cry that reformers of different stripes had sought, the kind of gut-wrenching critique that put the cause beyond reproach. Within eighteen months of the first publication of excerpts in *Appeal to Reason*, the leading Socialist paper of its day, and less than half a year after the first book edition, Congress had passed the Meat Inspection Act and the Pure Food and Drug Act.

The law was just to the taste of reform-minded citizens in a period that historians have called the Progressive Era: government control, an expansion of federal authority, and the use of scientific expertise merged in a quest for safer meat.[17] However, public health was only one of multiple fronts in the reformers' crusade. Another was trustbusting, an endeavor that found plenty of fodder in an oligopolistic market. After all, the reformers were not out to fight capitalism—quite the contrary, they sought to make it more efficient. Price fixing, market con-

18.1 Chicago's Union Stockyards in 1941. Image, John Vachon, Farm Security Administration—Office of War Information Photograph Collection, Library of Congress.

trol, joint companies—since the late 1880s, reformers and packers fought in public and in court, with allegations shifting but not the general thrust of charges. In the end, the Big Five submitted to a consent decree in 1920 that effectively ended the reign of the beef trust. Jimmy Skaggs has argued that by the 1920s, red meat was "the most regulated of American business endeavors."[18]

It took more time for working conditions to improve, much to the chagrin of Upton Sinclair, whose key concern had been Socialism rather than meat inspection. "I aimed at the public's heart, and by accident I hit it in the stomach," Sinclair noted, a phrase that people would subsequently quote ad nauseam.[19] The conflict included everything that the script book of US labor relations has to offer: coordinated strikes and spontaneous walkouts, union-busting and violence, tensions between immigrants and African Americans, strikebreaking and sabotage, and kind spirits trying to mediate (in this case, Mary McDowell and Jane Addams, in 1904). The Great Depression finally created the conditions for successful unionization, as common hardship facilitated organization across ethnic lines while federal legislation secured the workers' right to collective bargaining. Founded in 1943, the

United Packinghouse Workers of America became a pioneer of racially inclusive unionism and gave crucial support to Martin Luther King Jr. and the civil rights movement in the 1950s. By midcentury, the stockyards were unionized, the packers' jobs were secure, and they earned a living wage.[20]

Sinclair's book remained in print, and is so to the present day. But half a century after its original publication, it seemed as if the jungle had been tamed. Meat was still a capitalist business, but one with real competition, safeguards for human health, and respect for the working man. (Most working women remained underpaid, though, as if to remind the proletariat of its ongoing exploitation.) When it came to morals, the most glaring void was about the animals. The Illinois Humane Society and the American Humane Society kept an eye on transport and handling and encouraged a more circumspect use of the whip. Railroads eliminated the incentive to overcrowd by charging for weight instead of carloads. But laudable as these efforts were, they breathed the air of the nineteenth century: animal protection was about the elimination of abject violence. The problem was individual acts of cruelty and bad characters, not the system.[21]

In the slaughterhouse, the problem was much more essential. Here animals were simply units of production, optimized for economic performance and deprived of legitimate claims of their own. It was a new chapter in relations between humans and the animal world, a new kind of distance, and a new degree of disaffection and indifference. In fact, it was precisely this anonymity that prompted Sinclair's slaughterhouse epiphany: "Each one of these hogs was a separate creature. Some were white hogs, some were black; some were brown, some were spotted; and some were old, some were young; some were long and lean, some were monstrous. And each of them had an individuality of his own, a will of his own, a hope and a heart's desire; each was full of self-confidence, of self-importance, and a sense of dignity."[22] The Chicago slaughterhouse did not show respect for any of these things.

Of course, "respect" is an intricate word when it comes to human–animal relations. Animal husbandry before the dawn of modernity was cruel in its own way. Animal diseases were endemic, stalls were cold and damp if they existed at all, and feed reflected the exigencies of the moment. Hogs in particular were waste recyclers par excellence and specifically bred for fattening from human leftovers.[23] And yet there was also a sense of companionship, a readiness to watch and listen, and to do what one could for the animal's well-being, if only because farmers knew that they were in it together. Farmers gave names to their beasts, implying some kind of individual character, whereas the most that farm animals can hope for today is a digital code for tracking.[24] In short, the cruelty of premodern husbandry was largely a cruelty for lack of means. The new cruelty, that of the industrial age, was a cruelty of willful ignorance (see chapter 36.3, Liberation Movements).

The veterinary sciences already had a good understanding of animals' needs when Chicago's slaughterhouses went up, and the body of knowledge kept expanding as the twentieth century progressed. Epidemics, nutrition, excrements—the inputs and outputs of useful animals were meticulously studied, as were the processes in between, and scholars could identify good and bad conditions with growing precision. But at the end of the day, the insights that mattered most were those that made a difference in numbers (see chapter 7.2, Numbers Games), more specifically in numbers on the slaughterhouse scale and in the purse of producers and packers. The welfare of animals, their sense of curiosity and comfort, their instincts and their need for social interaction—it was hard to put these things into numbers, and harder still to measure their impact on the bottom line. The logic of industrialism, and its innate desire to subdue everything to the rationale of the market, was not dead. It was merely hibernating.

In 1944, Karl Polanyi published a book that, while not engaging with animals, was prophetic as to their future fate. Titled *The Great Transformation*, it argued that the new thing about the industrial age was not the logic of the market per se but its unrestrained hegemony. Respect, a sense of decency, and all other social conventions were yielding to the maximization of individual profit, and the new way of thinking put its stamp everywhere.[25] Farm animals would learn that this had very real consequences for them. The logic of industrialized meat was expanding massively in the postwar era, both along the commodity chain and beyond America's borders, ultimately making it difficult to escape its reign anywhere on the globe. In the decades after 1945, the American way of life beamed like never before or thereafter, and one of its messages was about meat. Statisticians would eventually find a stunning cross-cultural correlation between affluence and meat consumption.[26]

3. CONSUMERS IN CHAINS

There are a few places that witnessed key moments in the postwar transformation of meat. There is, for instance, the intersection of 14th and E Streets in San Bernardino, California, where Richard and Maurice McDonald opened their first restaurant in 1940, or the Chicago suburb of Des Plaines, where Ray Kroc opened a franchised restaurant in 1955, the man who would later buy out the McDonald brothers and drive the global expansion of McDonald's like a fast bus.[27] There is the slaughterhouse of Iowa Beef Packers (IBP) in Denison, Iowa, built close to a major cattle region. There is the Lewter Feed Yard near Lubbock, Texas, where a former county agent and a Dallas investor teamed up to raise 10,000 (later 34,000) cattle on 125 acres of land.[28] And there is Pushkin Square in Moscow, where McDonald's opened a branch in the dying days of the Soviet Union, causing a frenzy among the locals and consternation among left-wing intellectuals elsewhere.

Remarkably, none of this was actually new. McDonald did not invent the fast-food hamburger; that honor, if it is one, goes to White Castle in Wichita, Kansas.[29] Feedlots had spread in Illinois and Iowa in the wake of the Chicago slaughterhouse, but most of them were family farms that sought to overcome their sole reliance on wheat, as grain prices were more volatile than cattle prices.[30] While the feedlots of the nineteenth century grew out of a drive toward diversification, the new feedlots were bound to cattle for better or worse, and that was not the only difference. They bought grain on the market, depleted groundwater (as they were often in arid regions), focused on the final months in the cattle's life, produced year-round, and generally operated with the iron pulse of a disassembly line. Federal tax incentives encouraged investment from people who had no inner connection to agriculture or, like the Hollywood star John Wayne, earned their money acting as if they did.[31] The new feedlots were also much bigger than traditional ones, and they kept growing. Toward the end of the century, giant hog farms were spreading across the Great Plains, with the largest uniting more than a hundred thousand animals under a single roof. In his history of the Great Plains, Douglas Hurt describes the feedlot system as a colonial economy.[32]

The new feedlots also drew on the abundance of grain that stemmed from innovations such as hybrid corn (see chapter 28, Hybrid Corn) and nitrogen fertilizer (see chapter 19, Synthetic Nitrogen).[33] But then, feedlots did not rely on cheap grain alone. From drugs to the genetic pool, every factor was scrutinized for productivity gains, which made cattle akin to battery chicken (see chapter 36, Battery Chicken) without feathers. In the end, cattle even became carnivorous, thanks to the use of meat and bone meal for feeding. The public eventually took notice, but only because the practice was the suspected cause of bovine spongiform encephalopathy, better known as mad cow disease.[34] The meat business got out of step for a while.

Moving closer to livestock regions was not a new corporate strategy either. Armour had opened a dressed-beef plant next to its pork plant in Kansas City as early as 1884. But Kansas City was a railroad hub like Chicago, and transport volume made it prohibitive to build slaughterhouses elsewhere until construction of a new automotive infrastructure, with the Interstate System (see chapter 34, Autobahn) as its pinnacle, freed packers from their dependence on rail. Street transport was not the only innovation that made IBP's Denison plant a gamechanger in the business of meat. Whereas Chicago slaughterhouses had shipped carcasses for further dismantling by local butchers, IBP went all the way to household-sized portions, boxed and vacuum-sealed, ready for the new supermarkets. In doing so, IBP "took Gustavus Swift's original idea to its logical conclusion."[35]

As the new plants spread across the Midwest, they looked increasingly like the Chicago slaughterhouses at the turn of the century: they had an immigrant workforce, weak unions, and a strong aversion to government regulation (unless it helped to externalize social and environmental costs). It was as if Upton Sinclair's ghosts had risen from their graves, except that the workers were now from Latin America and Southeast Asia instead of Eastern Europe, and the new slaughterhouses were just as unpopular as Chicago's in the early 1900s: "IBP has come to symbolize, in the midwestern United States, the worst excesses of 1980s corporate arrogance," the anthropologist Deborah Fink wrote.[36] But it was immensely profitable, more so than the original, and that forced Chicago packers to cease operations in Chicago. The major plants had closed by 1960, and the Union Stock Yards shut down completely in 1971.[37] The companies of Swift and Armour were fading from public view and became prey for investors. A bewildering set of brands and faceless commodity chains took their place.

Enterprising businessmen were emulating the American slaughterhouse abroad. Karl Ludwig Schweisfurth, heir to a German meat company, saw Chicago's stockyards on a study trip in the mid-1950s, came back for a series of internships, and then used the acquired knowledge to revolutionize meat production in Germany.[38] Even Mexico, where Chicago's slaughterhouse had failed so spectacularly in the 1910s, gradually fell for US-style beef production and consumption habits, with the breakthrough coming with the wave of neoliberal deregulation in the 1980s.[39] All the while, the European municipally run slaughterhouse went into decline. The more scenic buildings could hope for an afterlife such as La Villette, the famous Paris slaughterhouse created at the instigation of Baron Haussmann in the 1860s, which closed its doors in 1974 and became a cultural center. At the same time, the factory farm (see chapter 36, Battery Chicken) emerged as the Siamese twin of the Chicago slaughterhouse. Both grew into global institutions and developed a kind of momentum that nourished a process of mutually assured expansion: slaughterhouses need a lot of animals for a decent load factor while factory farms need large buyers. One might suspect that the hunger for meat would expire at some point, but market saturation is an elusive concept in a globalized economy. When everything else fails, you can always dump excess production at bargain prices in the Global South.

Just as in the Progressive Era, authors jumped on the topic. The ethical critique of factory farming is as old as factory farming itself (see chapter 36.3, Liberation Movements), but it took the meteoric rise of environmental sentiments around 1970 to create a mass market for a pertinent book. Frances Moore Lappé hoped for a "small Berkeley publisher" when she started work on *Diet for a Small Planet* in 1969, but the manuscript landed with a paperback specialist, was published in 1971, and sold millions of copies.[40] Half a century later, readers can choose among many different levels of reflection and populist furor when it comes to unmasking the modern food business, slaughterhouse and all.[41] But despite the hopes of Progressives, grandiose aspirations for political reform have evaporated. Some slaughterhouses have hired animal behavior experts such as Temple Grandin, a professor of animal science at Colorado State University who specialized in "developing and implementing [animal] welfare auditing systems for major retailers and restaurants."[42] The decades-long struggle over battery cages (see chapter 36, Battery Chicken), a technology that was controversial from the start, demonstrates that to the extent that reform efforts exist, they have an air of incremental minimalism. Food scares and reports about dismal workplace conditions create a sensation for a few days, but they typically subside just as quickly. Reforms lack a powerful lobby, and they lack a focal point in the age of global supply chains. No longer can we pass a Meat Inspection Act and assume that all will be well.

In her foreword to Ruth Harrison's *Animal Machines*, a powerful rallying cry against factory farming of 1964, Rachel Carson (see chapter 38.2, A Matter of Humility) expressed her hope that the book would "spark a consumers' revolt."[43] Some people took to the task and sought to carve out alternatives. When Karl Ludwig Schweisfurth realized that none of his sons wished to be his successor, he took a more critical view of industrialized meat production, sold his company to Nestlé, and put his enterprising spirit to work in organic agriculture.[44] Others propose to forgo meat altogether and embrace a vegetarian or vegan diet (see chapter 23.3, Alternative Projects), an alternative that was already around in the Progressive Era: the vegetarian crusader John Harvey Kellogg had published a detailed, blood-drenched account of slaughterhouse work one year before Sinclair's *The Jungle*.[45] But as it stands, none of these ideas have carved out more than a niche, and it is anyone's guess whether that shows a penchant for cheap meat or merely the forgetfulness of the average consumer. For all the merits of alternative agriculture, informed consumerism has not become the magic weapon that Carson, Harrison, and countless other critics of the modern food business had hoped for. When it comes to the slaughterhouse, it would seem that humans are still struggling to understand that there is much more on the line than the corpses of animals.

More than a century ago, humans crossed a threshold in Chicago. It happened in the name of progress, the drivers included sophisticated technology, large corporations, and the rationalist quest for efficiency, and thanks to Sinclair and others, nobody could claim ignorance about what was going on. Maybe this was the kind of threshold that allows no way back.

19

Synthetic Nitrogen

Farming on Steroids

I. TECHNOLOGICAL MOMENTUM

On the evening of September 7, 1898, William Crookes took the podium in Bristol to deliver his presidential address to the British Association for the Advancement of Science. His topic was wheat, or the lack of it in the not-too-distant future. He sought to show "that England and all civilised nations stand in deadly peril of not having enough to eat."[1] The threat of starvation (see chapter 31, Holodomor) seemed remote in a country at the height of its imperial power, but as a chemist, Crookes saw things through a different lens. Population was growing while the supply of "virgin land" (see chapter 13, Little Grand Canyon) was dwindling, so everything depended on increasing yields per acre. Chemical fertilizer could boost productivity, but according to the law of the minimum commonly attributed to Justus von Liebig, which required a balanced dose of nutrients, a single deficiency limited growth even when all other factors were in abundance. Crookes was particularly concerned about nitrogen, which plants consumed in great quantities and hence had received particular attention since the days of

Liebig. It was freely available from the earth's atmosphere, but wheat could not take in atmospheric nitrogen directly. Guano (see chapter 8, Guano) offered available nitrogen, but its deposits were "so near exhaustion that they may be dismissed from consideration."[2] Chilean saltpeter deposits were already heavily mined, and other supplies such as urban sewage (see chapter 17.3, The Dirt of Cities) were unlikely to meet growing demand. For Crookes, the only viable solution was nitrogen fixation, "a gleam of light amid this darkness of despondency." But time was of the essence: if scientists did not find a way to turn atmospheric nitrogen into fertilizer soon, "the great Caucasian race will cease to be foremost in the world, and will be squeezed out of existence by races to whom wheaten bread is not the staff of life."[3] Crookes's conclusion was succinct: "It is the chemist who must come to the rescue."[4]

Fourteen years later, August Bernthsen, the director of research at the German chemical company BASF, took the podium at the Eighth International Congress of Applied Chemistry (see chapter 24.1, Nice to Meet) in New York City. Fritz Haber, a professor at the Technical University of Karlsruhe, had found a way to produce ammonia from the elements, and BASF's own Carl Bosch had designed equipment that allowed production on an industrial scale. All that the process required were nitrogen and hydrogen, two elements that the earth held in abundance, and a lot of energy. BASF was about to gear up commercial production for synthetic nitrogen, and Bernthsen had come to New York to make the announcement. On September 11, 1912, a packed audience heard Bernthsen shift into celebratory mode: "I am in the agreeable position of being able to inform you that the said problem has now been solved fully on a manufacturing scale, and that the walls of our first factory for synthetic ammonia are already rising above the ground at Oppau, near Ludwigshafen-on-Rhine."[5]

Ammonia synthesis was by all means an impressive invention. It required high temperatures and pressure in the range of 200 atmospheres. The search for the right catalyst alone required more than 6,500 experiments with 2,500 different substances. Engineers needed to find materials and designs that could withstand the extreme conditions. Industrial-scale ammonia synthesis was the result of a superb combination of chemical ingenuity and engineering skills, and it deservedly became known as the Haber–Bosch process after its main inventors. Fritz Haber received the Nobel Prize in Chemistry in 1918, to be followed by Carl Bosch in 1931.[6]

The accolades continue in the historical literature. Vaclav Smil even tried to calculate the number of lives that the Haber–Bosch process saved and came to the conclusion that 40 percent of the world population in 1996, or some 2.2 billion people, could not have been fed without synthetic nitrogen.[7] Counterfactual arguments are always open to debate, but there is no way to deny that the Haber–Bosch process marked a watershed in the history of agriculture. It removed the last remaining scarcity of a key nutrient and opened the door to a century of profligate fertilizer use. One of the leading global historians, Kenneth Pomeranz, has argued that chemical fertilizers, and synthetic nitrogen in particular, "may be the twentieth century's most important invention."[8]

But synthetic nitrogen was also a commercial product, and that implied a number of more worldly challenges. Patent issues needed to be resolved. In order to gain access to the fertilizer market, BASF sought an arrangement with other companies that produced ammonia as a coke oven by-product. Hydrogen production left major amounts of hydrogen sulfide as a waste product (see chapter 5, Shipbreaking in Chittagong), and boatmen complained about the foul-smelling substance when the company dumped it without treatment into the Rhine. In 1913, BASF opened the Limburgerhof agricultural experiment station to carry out extensive fertilizer trials, perhaps mindful of Liebig's fiasco seventy years earlier when he rushed his patent manure to market without testing (see chapter 8, Guano).[9]

In his New York speech, Bernthsen hoped for "a peaceful development of the various new industries for the combination of the nitrogen of the air side by side."[10] It turned out that there was not much peace in synthetic nitrogen's early years, though, as the Haber–Bosch process became a pillar of the German war economy during World War I. Ammonia can also be used for the production of explosives, and synthetic nitrogen essentially kept the German army firing after 1914. That provided the Haber–Bosch process with an air of moral ambiguity, though the link pales in comparison with Haber's other wartime work. Fritz Haber was the mastermind behind Germany's gas warfare and led the development of chemical weapons with a determination bordering on the maniacal. His Nobel Prize, the first to be awarded after the end of the war, drew international criticism.[11]

Four years of war made BASF a different company. Nitrogen production reached ninety thousand tons in the final year of the war, including production at a new plant in the central German brown coal region that BASF built from scratch. The Haber–Bosch process made BASF Germany's leading chemical manufacturer, with ammonia production generating no less than 59 percent of the company's turnover in 1919.[12] Even more, synthetic nitrogen "provided a common experience of great educational value for the engineers and scientists" on a par with Germany's rocket project in Peenemünde or the American Manhattan Project (see chapter 37, *Lucky Dragon No. 5*) during World War II.[13] Having mastered high-pressure, high-temperature catalytic hydrogenation within a few hectic years, the men of BASF (and yes, they were all men) sought to milk the new technology for all it was worth, an effort that led to products such as synthetic gasoline. This was no longer a company where research directors could take a podium and announce farsighted investments in new technologies. This was a company firmly in the grasp of what Thomas Hughes has called technological momentum.

Hughes's concept grew out of a sense of disaffection with traditional readings of technological progress that stressed the brilliance of inventors. In his study of the rise of electric power in England, Germany, and the United States, Hughes showed how supply networks developed their own systemic logic. Networks favored alternate over direct current as they grew, and the huge investment in cables and power lines paid off quicker when consumption increased, which drove utilities to advertise new electric appliances, particularly those appliances that consumed electricity at a time of day when demand was weak. Like a heavy object that is set in motion, the system acquired a momentum as it matured, though Hughes stayed clear of the shallows of technological determinism: "Technological momentum, like physical momentum, is not irresistible."[14] Instead of disputing that technologies are also shaped by political decisions and changes in economy and society, Hughes was concerned about how technological systems have a kind of inertia that imposed limits to the free will of experts, corporations, and consumers. In other words, Hughes showed that electric grids were networks of power in more than one sense.[15]

Before studying electric power systems, Hughes discussed technological momentum in the history of industrial chemistry. He showed how Haber–Bosch, the trailblazer for hydrogenation technology, changed BASF. Synthetic nitrogen left its mark on the company's equipment and the minds of its staff, and that set the company, which merged into the giant IG Farben conglomerate in 1925, on a fateful course: in Hughes's reading, technological momentum brought IG

Farben toward the synthetic gasoline project, which in turn drove it into the arms of the Nazis.[16] When Bernthsen was negotiating with Haber in 1908, he was still in a position to voice skepticism about the project.[17] A dozen years later, skepticism would have been pointless. The system was in the driver's seat.

In short, Crookes's hope for the genius of chemistry was only half the picture. The nitrogen question was not only about the white man's craving for wheat. It was also about technologies, experts, and corporations that developed a life of their own, and synthetic nitrogen captured a characteristic experience for people at the turn of the twentieth century. From electric power to sewage systems (see chapter 17, Water Closet), from Chicago's stockyards (see chapter 18, Chicago's Slaughterhouses) to automobility (see chapter 34, Autobahn)—large technological systems turned into behemoths with their own rationales as they matured, and one of the most important rationales was their inclination to grow. Farmers soon learned what this meant in practical terms.

2. A LOT HELPS A LOT

By the early 1900s, the fertilizer business was following the path that had taken shape during the age of guano (see chapter 8, Guano), but it operated on a different scale. Since about 1870, fertilizer use had ceased to be the privilege of a small entrepreneurial minority as farmers in Western and Central Europe embraced it as a normal part of agriculture.[18] In the United States, the number of fertilizer factories grew from 47 in 1859 to 478 in 1899, with the average capital per plant jumping from less than $10,000 to almost $150,000.[19] Nitrates from Chile, potash from German salt mines, phosphorus from the bones of slaughtered animals, and slag from steel mills—a growing range of fertilizers were clamoring for the farmers' attention. It was a tough business with a lot of infighting: sowing doubts about competing products was a routine part of the sales pitch. Liebig's law of the minimum stressed the need for a balance of nutrients, but many companies saw things differently. They treated the fertilizer business as a zero-sum market, where gains for one nutrient inevitably went at the expense of another. More than one manufacturer sold his product as the panacea for all plant nutrition woes, not to speak of the flying salesmen who roamed the countryside peddling miracle stuff. Bernthsen's remark about "peaceful development" revealed the naive hopes of a newcomer who was still learning about the rules of the game.

At first glance, ammonia was just one more product on the market, but it stood out in several respects. It came from a large company that worked with Nobel-acclaimed scientists. Thanks to the wartime expansion of Haber–Bosch, it was available in huge quantities, and BASF was keen to use capacity as best it could. Most crucially, it entered the market when German agriculture was in disarray. Soils were exhausted after the war, food was in short supply, workers were in revolt, and ammonia was the only ray of hope. The case for mineral fertilizer was stronger than ever, and all available authorities chimed in: scientists, manufacturers, salesmen, and the state. As a memorandum of the Prussian Ministry of Agriculture declared in 1920, "The necessary amounts of mineral fertilizer are available, and can be purchased and put into the soil. If that does not happen, people will starve" (see chapter 31, Holodomor).[20]

Few things capture the human imagination like a combination of fear and hope, and few things hurt more than when reality does not live up to expectations. Yields per acre stayed below prewar levels in Germany for years. Large doses of synthetic nitrogen even made some soils acidic and thus rendered them less fertile. The postwar chaos on many farms certainly did not help, and neither did the vast black market that made it difficult to obtain unspoiled products. But the debacle also revealed a flawed idea about soil fertility. Efforts focused narrowly on nutrients and ignored all the other factors that influenced plant growth. The propaganda for synthetic nitrogen betrayed a reductionism that had not been seen since the days of Liebig.[21]

The crisis gave a boost to what we now call organic agriculture. A series of lectures by Rudolf Steiner, the guiding spirit of anthroposophy, inspired his followers to develop a "biodynamic" farming system.[22] At the Institute of Plant Industry in Indore, India, Albert Howard compiled compost heaps and railed against what he called the "NPK mentality": "To-day the majority of farmers and market gardeners base their manurial programme on the cheapest forms of nitrogen (N), phosphorus (P), and potassium (K) on the market."[23] In Switzerland, Hans Müller founded a farmers' cooperative in 1946 that grew vegetables without nitrogen fertilizer and sold it through the Migros retail chain.[24] Whereas today's consumers associate organic food with healthier and tastier products (see chapter 38, DDT) and superior animal husbandry (see chapter 36, Battery Chicken), alternative farming originally grew out of concern for the fertile soil.

The fertilizer community was aghast and did not mince words, down to calls for a violent crackdown on German biodynamic farmers when the Nazis seized power.[25] But strong words did not convince everyone, particularly those who knew propaganda from the inside: two decades after his time at Stickstoff-Land GmbH, Heinrich Himmler was still fuming about how his superiors wanted him to write "doctored reports" about fertilizer.[26] Himmler developed an interest in biodynamic farming during World War II, and the experiments under his tutelage include an herb garden at the Dachau concentration camp on the outskirts of Munich.[27] His personal history notwithstanding, Himmler's stance was pragmatic rather than ideological: biodynamic farming had a productive potential that he sought to use—though that did not prevent later generations from suggesting more sinister motives.[28] Compared with Himmler's other agricultural interests, which include the infamous Generalplan Ost (see chapter 32.2, *Lebensraum*), synthetic nitrogen was a marginal issue.[29]

When it came to commercial fertilizers, the ideologues were clearly in the agrochemists' camp. They depicted organic farming as unscientific—in reality, it had firm roots in the academic network of the agricultural sciences—and ultimately pushed alternative agriculture into a secluded niche that it was unable to leave for decades.[30] Scathing attacks and strong convictions went along with new business practices. For one thing, agrochemists beefed up efforts at soil testing in order to give farmers a more precise idea about the needs of their fields. While scientists were previously loath to give detailed instructions, and some dismissed the demand for exact figures as "a silly request," they became more outspoken in the interwar years.[31] Agrochemistry confirmed Theodore Porter's observation about the use of numbers in modern societies (see chapter 7.2, *Numbers Games*): it went for quantification not when it was ascendant but when it was in trouble.[32]

The growing penchant for quantification was part of a broader realignment. Experts no longer sought to study soils closely and add what was needed. They became increasingly concerned with what one might call fool-proof fertilization: given the widespread sense of distrust in the farming community, it was crucial to avoid mistakes. One of the most popular solutions was a fertilizer mix with all the key nutrients that IG Farben had sold under the brand name Nitrophoska since 1927.[33] The fixed formula stood in obvious tension with the variable needs of individual plants, but at least none of the nutrients would be missing. Ready-mixed fertilizers are still popular. German gardeners can even buy a bio version with guano (see chapter 8,

Guano) as the main ingredient, courtesy of the Compo GmbH in Münster-Handorf.[34]

The popularity of ready-mixed fertilizers shed a revealing light on preferences in the farming community. A quick fix for all plant nutrition woes was just what many farmers were looking for. Soils were complex and capricious, close observation took time and expertise, and there were many other issues clamoring for attention on the farm. The talk about farmers as stewards of their soil was always stronger in expert circles than out in the fields, with the strongest voices coming from those experts that formed the cadre of the emerging soil conservation movement. Concerns about erosion (see chapter 13, Little Grand Canyon) allowed for the creation of distinct state agencies such as the US Soil Conservation Service, but their long-term perspective was always a difficult sell among farmers preoccupied with the upcoming growing season. In place of holistic soil expertise, modern agriculture developed two separate expert systems: short-term expertise that focused on commercial gains and foolproof solutions, and long-term expertise that embraced a more holistic view of soils and sought to conserve them to the best of its ability.

The fertilizer business exploded in the years after World War II. Global nitrogen consumption increased from less than 5 million tons in 1950 to 11.3 million in 1960 and 31.6 million in 1970.[35] "Like athletes pumping protein-building steroids into their bodies, industrial societies began to pump protein-building chemicals into their agricultural ecosystems, eventually allowing them to reach levels of productivity previously thought impossible," Hugh Gorman wrote in his history of nitrogen.[36] In the United States, fertilizer consumption grew so much that suppliers began building ammonia pipelines in the late 1960s, the first one running from Texas to the cornfields of Iowa (see chapter 28, Hybrid Corn).[37] For the farmers around the world, it came down to a stark choice between two different paths. They could refrain from the use of mineral fertilizer and join one of the schools of organic farming. Or they could go down the high-input path with friendly support from the fertilizer people.

Fertilizer use boosted yields per acre, but that is not the only yardstick for good farming. Short-term productivity per acre does not account for soil health or environmental side effects. It does not consider that according to the law of supply and demand, plentiful harvests tend to drive down prices. And focusing on per-acre yields was dangerously oblivious of the long-term dependencies that fertilizer use brought with it. Farmers ran the risk that corporations would seek to cream off profits, a notorious problem in markets with limited competition. In short, it was a completely open question whether rising yields per acre would really translate into higher income, and many farmers came to experience what economists call the "cost–price squeeze": growing costs eating away the gains of higher productivity. The case for mineral fertilizers was as much about growth effects as about the elimination of potential alternatives, as salespeople were disinclined to learn whether farmers might achieve competitive returns with more diversity and fewer inputs.

The ambiguous experiences of the early years disappeared from living memory with amazing speed. The collusion of commercial and scientific interests, concerns about erosion and acidic soils, the idea that soil fertility was really about more than nutrients—all this vanished in a remarkable act of collective amnesia. "Farmers of the 1940s and 1950s were far less concerned with ecological issues and with working together for total soil conservation and chemical-free farming than the farmers of a decade earlier," Randal Beeman and James Pritchard declared about American agriculture, and farmers in Europe looked equally forgetful.[38] Or maybe they just had other priorities? In his *Omnivore's Dilemma*, Michael Pollan met an Iowa farmer who described excessive fertilizer use on his cornfields as "a form of yield insurance": "You don't want to err on the side of too little," he replied when asked to explain why he used up to two hundred pounds of nitrogen per acre even though the recommended dose was only half that amount.[39] Just as technological systems had their own momentum, farmers had their own rationale for excessive fertilizer use: time was scarce, resources were cheap, and alternatives in the form of low-input agriculture were hard to come by. In a way, farmers preferred wasting resources to wasting ideas.

3. RUNNING CYCLES

Germany was not the only country that was concerned about nitrogen during World War I. The United States began construction of two facilities for nitrogen explosives when German submarines threatened the supply of Chilean nitrates in 1916. One plant sought to replicate the Haber–Bosch process but failed because BASF kept the catalyst secret. A second plant at Muscle Shoals, Alabama, used the older cyanamide process, which was more expensive and rendered it uncompetitive when the guns fell silent. Muscle Shoals languished over the following

19.1 Tennessee Valley Authority synthetic ammonia plant in Muscle Shoals, Alabama, 1942. Image, Alfred T. Palmer, Farm Security Administration—Office of War Information Photograph Collection, Library of Congress.

years, became part of the Tennessee Valley Authority (TVA) (see chapter 29, Aswan Dam) in the 1930s, and eventually shifted focus to phosphorus-based fertilizer. War preparedness then brought Muscle Shoals back to nitrogen, and a new Haber–Bosch facility produced synthetic nitrogen during World War II. It was one of ten facilities that stood ready after 1945 to supply American agriculture with synthetic fertilizer and a technological momentum propelling its use.[40]

The Haber–Bosch process became a global technology after 1945, the harbinger of high-input and high-energy farming. Thanks to its rich stores of hydrocarbons (see chapter 15, Saudi Arabia) and its pioneering role in the Green Revolution (see chapter 28, Hybrid Corn), it was Mexico that opened the first ammonia-producing plant in Latin America in 1951. Peru followed suit when it scrapped its guano industry (see chapter 8, Guano) in the 1960s.[41] By 1972, India had fourteen facilities with an annual capacity of 1.3 million tons of nitrogen, more than the United States had at the end of World War II; facilities for another 2.3 million tons were under construction.[42] At Aswan in the 1950s, Egypt built a plant that consumed almost all the country's hydroelectric power until the Aswan High Dam of the 1960s (see chapter 29, Aswan Dam) provided Egypt with a new energy cornucopia.[43] By the year 2000, the global nitrogen output stood at 85 million tons per year, more than 99 percent coming from Haber–Bosch facilities.[44] But Haber–Bosch was not the only thing that went global.

Nitrogen fertilizer stands out among plant nutrients for its stark inefficiency, as only a part of the input actually ends up in the crop. The precise amount depends on local conditions, particularly rainfall patterns, but scientists have put the typical loss rate between 50 percent and 70 percent.[45] Synthetic nitrogen dissolves in groundwater, and soil-based bacteria change it back into a gaseous state, and as fertilizer inputs increased, so did the amount of nitrogen that got lost in these ways. A global nitrogen cycle is at work on planet earth since geological times, but it changed tremendously in the modern era: the contribution of humans, which includes pollution from combustion processes and the cultivation of nitrogen-fixing plants such as legumes in addition to Haber–Bosch, is now on a par with the forces of nature. When Paul Crutzen argued that human intervention into planetary processes had grown to a scale that suggests the proclamation of a new geological epoch, the Anthropocene, nitrogen fertilizer was among the evidence.[46]

The Muscle Shoals facility received a new name in 1991: it became the TVA Environmental Research Center. It mirrored a growing awareness that excessive use of mineral fertilizers had serious environmental repercussions. An abundance of nitrogen jeopardizes ecosystems that depend on a meager supply of nutrients. A high level of nitrates renders groundwater unfit for human consumption. Michael Pollan's Iowa farmer had a reverse-osmosis water filtration system in his basement to make water from his well safe to drink.[47] Abundant nutrients in lakes and rivers lead to eutrophication and an explosive growth of plants and algae, and the effects extend far beyond agricultural regions: toward the end of the twentieth century, marine biologists were counting more than four hundred dead zones in coastal waters around the world, the number having doubled each decade since the 1960s.[48] Nonetheless, eutrophication is one of the lesser known pollution problems, probably due to its limited visibility. It rarely figures in the news unless green floating carpets make for some yucky pictures, though media types usually rank them lower than pictures of a decent oil spill (see chapter 39, *Torrey Canyon*). Sometimes the sheer size of these carpets generates excitement. In 2011, a bloom of algae on Lake Erie covered more than five thousand square kilometers.[49]

The European Union adopted a Nitrates Directive for the protection of water quality in 1991, but progress was agonizingly slow. Over the following ten years, the European Commission opened fifty-six legal cases against member states in order to speed up enforcement.[50] Did this reflect the intricate nature of the problem and the narrow profit margins in the food industry that constrained environmental initiatives in modern agriculture? Or was there an underlying conceptual problem in that environmental policy was reluctant to think in terms of cycles? Some scholars have argued that policies on nitrates should "be considered in the broad framework of resource management rather than in the narrower terms of pollution control."[51] However, perceptions and remedies typically remained wedded to the pollution framework, and the global nitrogen circle usually remained a background phenomenon. But was that really a mistake?

Environmentalists were typically fond of holistic thinking. The natural world was an interconnected one, after all, and human responses were supposed to reflect that complexity. The language of ecology thrived on interconnected thinking, and so did *The Limits to Growth*, the Club of Rome's 1972 report that showed a global audience what it meant to think in terms of cycles and feedback loops: the report's stark warnings drew on computer-based simulations of system dynamics that a team around Dennis Meadows conducted at the Massachusetts Institute of Technology.[52] But the discussion about *The Limits to Growth* got bogged down in disputes over figures and projections, and Meadows was not the first to have that experience. William Crookes suffered a similar fate after his call to arms on the wheat problem: when his speech went to press ten months after he took the podium in Bristol, the book included a defense of his figures that ran to almost double the length of his presidential address.[53] The effort did not exhaust the range of potential issues. For example, there is reason to doubt Crookes's contention that wheat bread was the only legitimate food for the Caucasian race.[54]

Crookes did not account for society's ability to adapt, and that was also the most popular argument against the doomsday predictions in *The Limits to Growth*: market pricing combined with scientific ingenuity would lead to new solutions that would forestall disaster. One economist, Julian Simon, even offered a famous wager to the American biologist Paul Ehrlich, and academics are debating the outcome to the present day.[55] Maybe innovation will allow the fertilizer business to shift from an extractive rationale to a cyclic rationale, where clever management of the flow of nutrients makes Nauru-style depletion (see chapter 8.3, Running Empty) appear to be an unfortunate excess in an unenlightened age. Or maybe innovation merely postpones the day of reckoning. Or maybe the question is false because neither markets nor innovators operate in a vacuum. Synthetic nitrogen was a product of war and technological momentum, both shaping the course of events in ways that Crookes and other heralds of scientific genius were never quite comfortable with.

Future outcomes are debatable, and so are the responsibilities of stakeholders. Crookes had a clear idea about his target audience: it was the chemist who was to come to the rescue. Agency was less clear with *The Limits to Growth*. The study was about exposing a problem rather than immediate solutions, and the target audience was humanity as such. It was a path that many environmental manifestos would follow, and they all suffered from the same lack of clarity as to who should do what. The nitrogen cycle was even more diffuse in terms of human agency: roles and responsibilities seemed to dissipate in a network of interconnections and feedback loops. Was the nitrogen cycle a problem, and if so, for whom? The people in charge of the nitrogen cycle were at the same time everybody and nobody, and that made the politics of thinking in cycles strangely opaque.

With that, policymakers faced a dilemma. They could either stick to the traditional pollution framework and treat eutrophication and groundwater contamination as if they were isolated problems, or they could embrace a much broader framework and work toward a comprehensive management of the global nitrogen cycle. In other words, policymakers could choose between a simplistic but time-honored approach with mixed results and a grand holistic vision where parties, authorities, and instruments were all up in the air. Pollution control (see chapter 16, London Smog) was an established field of policy where conflicts typically focused on speed and effectiveness rather than legitimacy. When it came to fights against pollution, the culprits were usually clear, and so were the spatial realms. Was it worth sacrificing both for the sake of a global approach that had the magic of holistic thinking but no real-world policy credentials? Modern societies have scant experience with the comprehensive management of material flows, but they have a clear idea about the politics of pollution and resource scarcity.

Nitrogen will not suffer from scarcity due to its abundance in the atmosphere, but the situation for other nutrients is less clear. Patrick Déry and Bart Anderson made global headlines in 2007 when they suggested that world society was probably past a "peak phosphorus" in production terms and thus in dire need of retooling its farming routines.[56] According to Liebig's law of the minimum, nothing can replace phosphorus if it is unavailable, and yet the stir of "peak phosphorus" faded away quickly. Perhaps that was due to other researchers who pointed out that the concept was misleading and that there are many ways toward more sustainable management of the global phosphorus cycle.[57] Or maybe that was due to an acquired skepticism about expert predictions. We have learned to live with the specter of resource exhaustion ever since William Stanley Jevons (see chapter 16.1, City Problems) penned his 1865 treatise "The Coal Question," and in the end, it was always not quite so dramatic.[58] And we have become familiar with overconfident scientists since the days of Liebig.

Our experience with a high-volume extractive lifestyle goes back only a few generations, but we do have experience with material scarcity in the fertilizer business. It is not very encouraging for those who believe in rational solutions. When the military's demand for explosives left the German farmers wanting for nitrogen during World War I, a popular solution was to increase the dose of potash. It did not make sense in light of the law of the minimum, but the German potash industry made a killing, and sales reached a historic peak in the spring of 1916—a perverse case of profiteering in a society of hunger and scarcity and yet precisely the kind of counterproductive behavior that a complex commodity chain tends to produce.[59] It was an object lesson about how the intricate web of farming practices, profit-seeking companies, and expertise that had grown around fertilizers since the days of guano (see chapter 8, Guano) could foster irrational solutions. The postwar critique was harsh. Franz Honcamp, the director of the Rostock agricultural experiment station, spoke of "angst-driven fertilizer use" that did more harm than good.[60] If the global commodity chain for fertilizers comes tumbling down one day, fear will likely be part of the picture.

20

Air-Conditioning

Engineering the Climate

I. Making Indoor Weather

The president was shot, but he was alive. The attack at a Washington, DC, railroad station left a nation in shock and James Garfield with a bullet in his body. He survived the following night despite expectations, and his bed at the White House became a place for freewheeling experimentation. Doctors poked into the president's body in search of the bullet. When they failed to locate it, Alexander Graham Bell brought a hastily invented metal detector to the White House, where the device malfunctioned due to an iron bed frame and uncooperative physicians. As the attack happened in July, Washington's oppressive summer heat became a matter of concern, and Navy engineers teamed up with the scientist Simon Newcomb to plug together an improvised air-conditioning unit. They combined a big fan with huge amounts of ice, fed the cool air through cotton screens to remove humidity and blew it into the president's bedroom to lower the temperature. America billed itself as a nation of inventors, but it failed to save the president's life. Garfield died on September 19, 1881, two and a half months after the attack.¹

It was a fitting overture to the age of air-conditioning. After all, the device at the White House captured some crucial features of indoor climate control. It was about technological innovation and expertise. It was about America and specifically the American South, where sweltering summer heat was more of an issue than in the industrial centers of nineteenth-century Europe. It was about tinkering: air-conditioning was about electric power rather than mountains of ice in subsequent generations, but improvisation and perennial adjustments remained part of the picture when it came to thermostats and unwarranted airflows. The laborious effort and the tremendous amount of resources foreshadowed an energy-intensive approach to climate control. Most crucially, air-conditioning was about the primacy of imminent needs and a fateful disregard for context ever since air-conditioning had come to the White House in 1881. Technology may have provided Garfield with some temporary relief, but it was ignorance about the unintended consequences of technology that claimed his life. He might have survived if it had not been for the infections that the president caught from his doctors, who were intruding into his body with fingers and instruments that lacked sterilization.²

Humans had sought ways to deal with high temperatures long before the machine age, and low-tech solutions included shade trees, naps during the hottest hours, and dips in cool waters. The transport revolution of the nineteenth century (see chapter 3, Canal du Midi) added mobility to the range of options, at least for those who could afford the trip to gentler climates. The hill stations at the foot of the Himalaya were a popular refuge for British colonialists during the summer months, and in 1864, one of these stations, Shimla, became the official summer capital of the British Raj in order to give officials a respite from Calcutta.³ Back in Europe, resorts thrived along the coasts, helped by royals who turned Brighton, Norderney, and other seaside towns into fashionable destinations.⁴ Resort towns were known for networking and a more leisurely pace. They were also the place for a dying president. After two months in agonizing pain at the White House, James Garfield spent his final days at the New Jersey seashore.⁵

Relocation implied long leaves of absence, which became more difficult to justify in an increasingly breathless modernity. But as the twentieth century began, air-conditioning offered another way to stay cool that worked regardless of geography. However, air-conditioning was not just about low temperatures but also about humidity. In fact, control of humidity was more important than temperature in the early years of air-conditioning, when many units were installed in factories with sensitive production processes where humidity levels made a difference.[6] Equipment also had to assure cleanliness and the proper distribution of air, and as rooms and buildings differed in size and shape, air-conditioning systems were typically tailor-made during the first decades of the twentieth century. Mass production of window air conditioners did not start until the industry set its eyes on the residential market during the Great Depression.[7]

Air-conditioning found a growing range of applications over the course of the American century, but regional and sectoral divergences were significant. As Raymond Arsenault wrote in a pioneering essay, "The so-called 'air conditioning revolution' . . . was actually an evolution—a long, slow, uneven process stretching over seven decades."[8] Cotton mills and cigar factories became climatized before World War I, and cinemas were air-conditioned in the 1920s, whereas most schools and university buildings had to wait into the postwar years (and some are waiting to this day, particularly where families are black and poor.) Railroads were air-conditioned before automobiles, and federal buildings before state and county ones, but by the 1970s, cool air was everywhere in America: in office buildings and private homes, hotels and hospitals, trucks and tractors. Even the Alamo Mission in San Antonio, remembered as a battle site in the Texan struggle for independence from Mexico, became climatized in the early 1960s.[9]

The southern United States led the march toward the cool indoors, and the reverse was no less true: air-conditioning was crucial for the rise of the Sunbelt after 1945. It is hard to imagine the mass migration from regions with more moderate climates without a technology that allowed people to cope with the long hot summer. The former cotton belt turned to manufacturing and service sector jobs, in the Arizona desert, suburban life was sprawling around Phoenix and Tucson, and while numerous factors played a role in the transformation, air-conditioning was a sine qua non.[10] Arsenault summarized his observations on the air conditioner and Southern culture as follows: "General Electric has proved a more devastating invader than General Sherman."[11]

The White House received a permanent cooling system during the Hoover administration in 1929, almost half a century after Garfield's

20.1 "No more hot air in Congress." This photograph of June 1938 shows the air-conditioning plant that supplied two million cubic feet of cool air per minute to the Capitol, the Senate and the new and old House Office Buildings in Washington, DC. Image, Harris & Ewing Photograph Collection, Library of Congress.

ordeal. Air-conditioning expanded dramatically in Washington during the New Deal, though Franklin D. Roosevelt, a patrician with an estate in the Hudson River valley, was less than enthusiastic about it.[12] Some seventy years later, American presidents enjoyed a pleasant microclimate even under open-air conditions. When George W. Bush visited Gorée Island off the coast of Senegal in 2003, a former hub of the transatlantic slave trade that Bush used as the backdrop for a passionate speech about the evil of American slavery, he evoked the "hot, narrow, sunless nightmare" of the middle passage while an air conditioner was going full blast at his feet. The inhabitants of Gorée Island were less fortunate, as security forces detained them on a soccer field for the duration of Bush's speech.[13] In the twenty-first century, Americans see air-conditioning as a natural birthright, and lack thereof as a sign of poverty, and they spare no effort for obtaining the artificial chill even under difficult conditions. A lot of things were dubious about Bush's war on terror, but the microclimate in tents and command posts was okay, thanks to the largesse of the US taxpayer. In 2011, a former Pentagon official revealed that the US military was spending more than $20 billion annually on air-conditioning in Afghanistan and Iraq.[14]

2. SEALED IN

For a while, travelers to the American South had a choice about air-conditioning. Many hotels offered climatized rooms for a $1 surcharge in the 1950s.[15] Managers typically dropped the surcharge when air-conditioning became common over the following decade, which nicely reflects how climatization advanced from luxury to normalcy to second nature. By 1957, the California Federal Savings and Loan Association of Los Angeles refused to offer mortgages for homes costing more than $20,000 if the design did not include provisions for an air-conditioning system.[16] More recently, homeowners associations have drawn up rules that prohibit shade trees and set up minimum requirements for air-conditioned floor space.[17] In twenty-first-century America, air-conditioning is an exemplary case of what Elizabeth Shove has called inconspicuous consumption.[18] It has moved beyond the point of consumer choices and cultural codes—it is a requirement, at times with the backing of the law. In *Mockingbird Song*, Jack Temple Kirby has called air-conditioning "the very foundation of contemporary southern (human) living."[19]

Air-conditioning changed the way houses were built. In fact, Gail Cooper has argued that it was architects and builders rather than homeowners who chose mass adoption in the residential market: the standardized suburban houses of the postwar years had indoor climate control built into their design.[20] Industry publicists had already rhapsodized in the 1930s about how air-conditioning would free architects from petty considerations regarding window light and ventilation. The natural environment would cease to be a matter of concern, and if they were so inclined, architects could "indulge in a passion for glass window-walls, provided clients were willing to pay the high costs of cooling."[21] Air-conditioning became part of a technological package with global appeal. According to the British architect Dean Hawkes, "It may be argued that the air-conditioned glass skyscraper is the most successful production of the construction industry in the twentieth century."[22]

One of the results was the seat of the United Nations in New York City. The skyscraper's glittering façade had its charm, but pervasive use of glass required a 50 percent increase in the thrust of air-conditioning, and manufacturing cool air consumed a lot of energy.[23] In the 1960s, declining rates for electric power drove the spread of residential air-conditioning.[24] But with perennial energy woes ever since the 1970s and the environmental critique of consumerism, the energy needs of air-conditioning became an enduring concern. It is about volume: in a 2010 book about our air-conditioned world, Stan Cox pointed out that the United States used as much electricity for air-conditioning as Africa used for everything.[25] And it is about peaks in demand. Air-conditioning units tend to power up in sync at certain hours of the day, and utilities are struggling to meet demand at these times.[26]

Air conditioners were also causing problems in other respects. Since the 1930s, manufacturers had used chlorofluorocarbons as safe and efficient refrigerants until researchers identified them as a prime cause of the ozone hole.[27] In 1976, the air-conditioning system at a hotel in Philadelphia became a disease vector when a heretofore unknown strain of bacteria bred in the water of a cooling tower. The outbreak killed twenty-nine veterans attending a convention of the American Legion, and the disease came to be known as Legionnaires' Disease.[28] Others became merely unwell from indoor air, and office workers learned a new word: sick building syndrome.[29]

None of this has changed America's infatuation with artificial coolness. Manufacturers have shifted to new refrigerants that do not damage the ozone layers, engineers have modified equipment to improve energy efficiency, and building managers have changed thermostats from freezing to merely cool in order to save some electric power, but the concept remains sacrosanct: air-conditioning has not relinquished a single part of America's public or private sphere from its grip. Climate control is deeply enmeshed in the design of modern buildings, and most architects take it for granted in light of a combination of technological path dependency, cultural norms, and a quest for control. "One of the most marked trends in architecture over the centuries has been that of replacing the functions of the building structure by engineering service systems," Hawkes wrote in 1996.[30] He felt that the trend was "not likely to be reversed . . . because the increased control which individual services afford over the built environment is now the norm expected by users of all building types."[31]

The American experience provided ample evidence that air-conditioning was an ambiguous one-way street, but that has not kept other countries from following the same path. It started as a luxury item and a way to mark social distinctions. During the construction of the Aswan High Dam (see chapter 29, Aswan Dam), Soviet engineers enjoyed air-conditioned housing while ordinary workers were sweating in the desert heat.[32] Then, in a remarkable coincidence with the globalization of environmentalism, the worldwide spread of air-conditioning gained momentum toward the end of the twentieth century, and some markets evolved within a matter of years. Climate control was a luxury feature in European automobiles into the 1990s, but by 2003, 70 percent of new cars sold within the European Union had air-conditioning.[33] In the United Arab Emirates, air-conditioning led residents to spend so much time indoors that it caused them to suffer from a deficiency of vitamin D, which humans usually produce in their skin from sunlight. A study in a Dubai hospital found that in one of the sunniest places on earth, only 2.1 percent of patients had an appropriate blood level of vitamin D.[34] On the plus side, powerful refrigeration units allow people to go skiing in a Dubai mall.[35]

Air-conditioning was a mature technology after a century of American engineering, but conditions in other countries presented new challenges. In Mediterranean towns, narrow roads with high buildings turn into heat canyons when numerous air conditioners operate close to each other.[36] In 2004, the state government in the Indian Punjab ordered a shutdown of air conditioners in order to free electric power for irrigation pumps.[37] It did not bring Indians to turn their backs on the American way of cool. Cox assumed that the energy consumption of India's air conditioners would increase tenfold from 2005 to 2020.[38]

In 2015, two American economists predicted "near-universal saturation of air-conditioning in all warm areas within just a few decades."[39] Global warming was poised to contribute to the trend, but the real driver was growth of income in the Global South. When it comes to air-conditioning, only money seems to stand in the way of the transition from luxury to normalcy. Even in Rio de Janeiro, where outdoor attractions are just as abundant as poverty, urban planners are wondering whether they should design outdoor space for a fully air-conditioned society.[40] The world may be beyond the point where decisions about cool air are a matter of the mind.

3. CORPORAL HABITS

American technology was equally about myths and engineering, and the producers of air-conditioning equipment reached out to people who could imbue their product with a cultural narrative.[41] One of their point men was Ellsworth Huntington of Yale University. As academics go, Huntington had a rather undistinguished career. He failed to obtain a doctorate from Harvard in 1907 and left Yale in 1915 after his promotion to a professorship was denied twice, and when he returned to Yale as a "research associate in geography," he paid far more for his personal secretary than he received by way of compensation from the university.[42] However, Huntington offered an enticing perspective on how climate underpinned the progress of the human race. He argued that "mankind as a whole has a definitive level of optimum temperature at which health and vigor are best."[43] Furthermore, Huntington felt that this optimum moved toward the cooler side when civilizations advanced and put a higher premium on innovation, and it just so happened that the climates of the West, and especially those of New England, were particularly conducive to intellectual work. Huntington spoke emphatically about "the coldward, stormward march of civilization."[44]

The collaboration between Huntington and the air-conditioning industry took various forms over the years, but it was a difficult relationship for both sides. Manufacturers had to supply Huntington with good evidence before he would offer an endorsement. Air-conditioning was also difficult to square with Huntington's concerns about human divorcement from nature and monotonous weather.[45] His office routines at Yale did not help: he opened the window twice a day regardless of the weather, making him a less than perfect role model for indoor climate control.[46] And in the end, Huntington's climatic determinism did his own discipline more harm than good. "Huntington gave geography a bad name," David Landes has argued, pointing to the abolition of geography departments at Harvard and elsewhere after World War II.[47] Global warming has produced an academic cottage industry that flirts with climatic determinism again, but it does not seem that the air-conditioning industry is soliciting endorsements this time around.[48]

Climates developed their own mythologies, and the same holds true for indoor climates. Gore Vidal famously dated "the end of the old republic and the birth of the empire to the invention, in the late thirties, of air conditioning."[49] In reality, the artificial cool came to Washington a few years earlier when American isolationism ran high.[50] In light of the global career of air-conditioning, it seems time to turn Huntington's question on its head: Why do different people in different cultural and socioeconomic settings in different parts of the world embrace the same technology? It may be about social status, or the enduring appeal of the American way of life, or a global visual culture that is uneasy about sweat. Or it may be about the environmental conditioning of the modern body.

For all the diversity of people around the globe, they have shown a common ability to adjust to air-conditioning. The experience of indoor coolness changes ideas about comfort zones, and it reduces the ability of the body to adjust to hot weather. In other words, air-conditioning is physically addictive, and the addiction may well be "the most pervasive and least noticed epidemic in modern America," as a Cambridge professor argued in 1992.[51] Air-conditioning has received its share of lamentations over the years, but its real history is about deeds rather than words.[52] That does not bode well when it comes to lessons from the experience.

After all, the history of climate control plays out not only indoors. Humans have sought to modify climates throughout the ages, and pertinent efforts include rainmakers as well as the Nobel laureate Irving Langmuir.[53] Endeavors were typically more powerful in people's dreams than in consequences, but that may change in the wake of global warming. Geoengineering has emerged as an expertocratic obsession in the twenty-first century, and if the history of air-conditioning is any guide, it will have unintended side effects.[54] But even if tinkering with the world's climate somehow works out according to plan, the outcome will inevitably separate people into those who benefit and those who do not. The indoor experience shows that climate control is rather effective in trapping and dividing humans. Whether it makes them happy remains open to debate.

INTERLUDE

Opium

THE INTRODUCTION ADVISED READERS TO PREPARE FOR A BUMPY RIDE in this book, and those who have made it to this point certainly know what I was talking about. The narrative seeks to tease out the common threads from the tremendous diversity of our planet, and that has required many daring jumps between countries and continents. But traveling the world inevitably takes its toll, and it may be good to interrupt the global journey in midpassage. Humans have an innate craving for relaxation, and human societies have developed plenty of pertinent routines from a quiet siesta in the middle of the day to a noisy evening at the pub. Breaks have a social dimension, but many routines involve the consumption of stimulants that bring people into the right mood. For centuries, alcohol was "the principal psychoactive substance of the Judeo-Christian world," but by the late eighteenth century, Europeans showed a growing interest in another stimulant, an extract from the poppy plant *Papaver somniferum*: opium.[1]

Stimulants are about small quantities with great effects. People do not consume them for their calories (see chapter 2, Sugar) or their enticing taste (see chapter 7, Breadfruit). Stimulants change the operating script of the body, usually by targeting the nervous system, and the effect can be anything from reduced anxiety to enhanced performance. Clever conveners know that a successful meeting needs more than good ideas and smart people. When West Germany's state governments sponsored a conclave of constitutional experts at the Herrenchiemsee Abbey in Bavaria in 1948, delegates enjoyed a generous daily allotment of twelve cigarettes or three cigars, half a bottle of wine, and one liter of beer per head.[2] The meeting produced a constitutional framework that has endured for more than seventy years.

The Herrenchiemsee convention was not the only working environment that thrived on the use of stimulants. The miners of Potosí (see chapter 1, Potosí) used coca leaves to make it through their workday.[3] Sex workers have used cocaine because it reduces self-control.[4] While running for president in 1960, John F. Kennedy received an amphetamine injection before he faced Richard Nixon for the first televised debates in the history of US elections.[5] Company-sponsored espresso machines have made it into a reference book entry (see chapter 22.2, The Age of Handbooks) on e-commerce.[6] Countless writers have used alcohol, nicotine, or mind-broadening drugs to meet their deadlines, and in books like Hunter S. Thompson's *Fear and Loathing in Las Vegas*, drugs stood center stage.[7] Economic historians have long recognized that it was probably more than a side show. Jan de Vries saw the growing use of coffee, tobacco, and other stimulants as part of what he called the "Industrious Revolution," which in turn was a corollary of the Industrial Revolution. In his words, "The industrious revolution floated like a cork on an expanding pool of alcohol."[8]

Stimulants are the spice of life, but just like spices, they call for judicious use. Humans had consumed opium for millennia—a commodity history begins with petrified poppy seeds from the late Stone Age—which gave them plenty of time to learn about effects and risks. People cherished opium as a painkiller, but they also knew that its sedative effect inhibited performance at work. They knew too that an overdose could kill even those who knew the drug well: the list of victims includes the Persian eleventh-century physician Avicenna, in his time "the Muslim world's unchallenged authority on opium."[9] Like most stimulants, opium created bodily dependencies and habits of an ambiguous nature. In short, opium was not dear to the heart of those who appreciated a clear look at material realities. Karl Marx's famous dictum about religion—"the opium of the people"—was not sympathetic toward opium either.[10]

Opium was less ambiguous on the side of the producers. It had a high value density, which made it a convenient commodity in the age of sail. Opium from India allowed the British to overcome their long-standing trade deficit with China and "to reverse the centuries-old flow of silver into the Middle Kingdom."[11] The repercussions culminated in two military conflicts, famously called Opium Wars, which ended with treaties that exposed China to further imports such as Indian-produced tea: Andrew Liu has argued that "the cultivation of Assam tea was the continuation of the opium wars by other means."[12] Later in the century, the German chemical company Bayer produced a synthetically modified opium alkaloid and sold it as a medicine for coughs and colds under the trade name heroin.[13] It happened to have side effects—a four-week trial had somehow failed to reveal that it was addictive—but moral purity was in short supply in the commodity business since the heydays of Potosí (see chapter 1, Potosí) and Caribbean sugar (see chapter 2, Sugar). The world's first cocaine syndicate was organized by German pharmaceutical companies, and it was perfectly legal.[14]

The syndicate was not the first attempt to corner the drug market. The British East India Company was running a tightly regulated, industrial-style production regime in Bihar and Benares by 1780, and other powers sought to cash in as well: most colonial regimes in Asia created an opium monopoly.[15] During World War II, Chinese collaborators collected taxes on opium to support the Japanese military.[16] Governments were fond of taxes on stimulants, which were less prone to conflict than taxes on essentials such as salt (see chapter 23, Gandhi's Salt). As Adam Smith wrote in *The Wealth of Nations*, "Sugar, rum and tobacco, are commodities which are no where necessaries of life, which are become objects of almost universal consumption, and which are therefore extremely proper subjects of taxation."[17] But just as stimulants had side effects, so did taxes on stimulants. Britain famously drove its North American colonies into revolt when it imposed a hefty tax on tea. Two centuries later, Mikhail Gorbachev launched an anti-alcohol crusade that caused a dramatic loss of government revenues from vodka sales at a time when the Soviet Union was struggling with a collapsing oil price (see chapter 15.3, Dutch Disease).[18]

But as the nineteenth century progressed, opium turned from a commodity among many into a moral scandal, and scholars have offered different readings for this change of affairs. Was it "the outcome of the class basis of Victorian society," as Virginia Berridge has argued with a view to working-class use in England?[19] Was it about the ascendant medical profession, which had increasingly embraced the concept of addiction since the 1860s?[20] Was it about cultural framing more broadly? In the words of Paul Gootenberg and Isaac Campos, opiates became "infused with Orientalistic, racist, degenerative, and gendered discourses."[21] Or was it about the misery of opium users in China that missionaries broadcast to a global public beginning in the 1830s?[22] Founded in 1874, the Anglo-Oriental Society for the Suppression of the Opium Trade argued that profiteering from opium was incompatible with the civilizing mission that should underpin the imperial project.[23] Whatever the reason, the campaign for the suppression of the opium trade gained momentum in the late nineteenth century, and the United States took the lead in drug diplomacy, starting with an International Opium Commission in Shanghai in 1909.[24]

Opium was henceforth illegal, or rather was meant to be, for producers and consumers were less than impressed. The world's leading producer of opium, Turkey, ratified the International Opium Convention in 1933, but enforcement was ineffective until the 1970s.[25] In 1906, China launched an eradication campaign that faltered after some early successes and played out to the advantage of warlords and others who cashed in on the inflated prices that illegalization brought.[26] In Afghanistan, opium production flourished in spite of eradication campaigns under the aegis of Western military powers, and it was not just about the money in an illicit business whose global revenues are on a par with tourism (see chapter 22, Baedeker) and the oil industry (see chapter 15, Saudi Arabia).[27] Afghanistan's farmers are wrestling with an increasingly dry climate and irrigation systems damaged by four decades of war (see chapter 29, Aswan Dam), and as it happens, "opium poppy is very drought resistant, requiring only one-fifth or one-sixth the water needed by traditional crops like wheat."[28]

The war against drugs is the longest war of the twentieth century, and it is probably the only war whose winners are not human beings. In his global history of narcotics, Richard Davenport-Hines described the American-led campaign "as requiring unconditional surrender from traffickers, dealers, addicts, and occasional recreational users," and his summary is succinct: "That surrender has not occurred."[29] The war was more successful on a different front. When Richard Nixon made the eradication of drugs a top priority of his presidency, the campaign targeted and tainted leftists and Black people, the two prime enemies in the Nixon White House.[30] It was not a coincidence. Nixon's adviser John Ehrlichman said in an interview in 1994, "Did we know we were lying about the drugs? Of course we did."[31]

While US-funded poppy eradicators performed their futile duty in Afghanistan, an opioid crisis spread back home. The victims were overwhelmingly white, and while the death toll was huge—the Centers for Disease Control and Prevention recorded more than 100,000 drug overdose deaths in 2021—arrests were not a method of choice.[32] Much of the blame focused on corporations and lax prescription policies on painkillers—legal proceedings against drug manufacturers are ongoing—while the responsibility of individual drug users remains opaque.[33] It mirrors a running theme of opium history since the late nineteenth century: there was moral vigor and iron determination from above, indifference and negligence in everyday use, and little that connects the two beyond the materiality of drugs. Everyone knows that drugs are dangerous, but on the ground, people have shown remarkable naiveté in the face of opium even in the absence of stimulants tampering with their nervous systems. The Afghanistan *Lonely Planet* (see chapter 22, Baedeker) cautioned against the latest version of "opium tourism" in 2007, which is about having your picture taken in a bright red field of poppies. "Most fields are guarded by armed men to protect the crop when it is growing and being harvested. It may be a lethal case of mistaken identity if you are confused as potential poppy eradication surveyor earmarking annual earnings for destruction."[34]

Opium is a sedative, but it has failed to make the world a more relaxed place. This also means that those who sought an actual break in this interlude will surely go away disappointed, not to speak of those who might have hoped for a more uplifting narrative. Stimulants provide perspectives on the big issues of history—an opportunity that a new drug history seeks to explore.[35] And the world of opium remains entangled with the rest of the world: materially, culturally, politically. It seems that the vortex has not left us with a place where people can actually get away from the world. That does not speak against taking a break. But it speaks volumes about life in the vortex.

PART V

RUPTURES

STUFF DOES ACTUALLY HAPPEN.

The Iraq War of 2003 was a disaster in so many ways, but it was rather successful on the aphorism front. One of the enduring phrases was born during a press briefing at the Pentagon on April 11, 2003, when the US secretary of defense Donald Rumsfeld remarked on looting in recently conquered Baghdad: "Stuff happens."[1] Like another memorable phrase, the "unknown unknowns," the remark had a specific context, in this case an attempt to excuse looting at Iraq's National Museum that American troops had failed to prevent. But the phrase stuck because there was more to it. It mirrored the blatant denial of responsibility for what was happening in Iraq, which ultimately emerged as a defining feature of the Iraq War. It also marked the beginning of the end of the neocon's political hegemony, and it was the contrast between the global project and the response on the ground that made the quote iconic. The Iraq War was about building a coalition, about gathering a powerful army in a desert environment and about leading it to military victory, and then it was an archaeological museum, of all places, that made the masters of the universe look clueless. As world politics goes, the looting was a minuscule event, but an event that made people realize the emperor did not have any clothes. Stuff does actually happen.

Events are the basic unit of the historian's craft, but most historians' work revolves around weaving these events into a plausible narrative. Sometimes the precise nature of an event becomes a matter of dispute (such as the extent of looting in wartime Baghdad), but usually it is all about context: political, social, economic, cultural, environmental—you name it. Sweeping syntheses are particularly prone to underestimating the contingency of events, as they legitimately focus on the great outlines, and yet sometimes events refuse to be subsumed in the grand scheme of things: they retain an erratic quality that defies even the best efforts to contextualize them. It is tempting to dismiss the erratic, or to contextualize against all odds, but such an exercise ultimately serves the scholarly craving for intellectual purity at the expense of real-world flavor. Stuff does happen, and it matters, and this section is an attempt to do justice to this. In other words, this section is about the environmental history of what was not meant to happen.

The vacuum of meaning is not just a scholarly concern. Contemporary observers noticed that some events simply did not make sense: they defied expectations, routines, and coping mechanisms, and yet there was no way to exorcise them from the record. However, there was one convenient exit. Events offered stories, and stories are one of the most important coping mechanisms of the human race. Narratives can bestow the erratic with a meaning or semblance thereof, and all the following chapters are also about the power of stories—and about the scholarly need to go against the grain. The final chapter on the Tangshan earthquake ends with a warning about the plethora of narratives that modern disasters produce, as narrative tsunamis drown big and important questions that so-called natural disasters raise, and the same warning applies to the other chapters as well: beware of sweet, simplistic narratives. In an electronic global village, where events can be recorded and publicized at lightning speed, narratives may well emerge as the new opium of the people.

The hunger for narratives is a transnational craving. The first chapter starts with the Nobel laureate Thomas Mann, who turned his own brush with cholera into a piece of world literature in his *Death in Venice*; the chapter ends with a more recent book by the Pakistani novelist Mohsin Hamid. As a disease that struck with extraordinary speed, cholera was particularly terrifying and in need of consolatory narratives. Or so it looked from the point of view of industrialized societies in the nineteenth century, as the experience of pandemic waves that might or might not strike was an eminently Western one. Cholera was endemic in India and other places and continues to kill in the twenty-first century, but that drama played out in a different disease world that few Westerners ever entered.

Exploring the mysteries of cholera was a defining challenge for nineteenth-century medical sciences, and the clash between contagionists and anticontagionists captured only a fraction of the known unknowns. It was about more than etiology: for one, medical perspectives were invariably tied to the pros and cons of quarantine measures. When Robert Koch returned from his successful journey to Calcutta, he was hailed in the style of a victorious general, and it was not just about militaristic rhetoric: he received a wartime decoration for the isolation of the cholera germ.[2] But knowledge about the cause of a disease was just one of many steps toward containment, which is the best that humans can achieve for most epidemic diseases. The number of diseases that medical progress has managed to eliminate globally stands at exactly two, smallpox and rinderpest.

An earlier generation of scholars gave cholera credit for boosting the urban sanitation movement. There can be no doubt that an epidemic heightened people's interest in public hygiene or the semblance thereof: when cholera hit Madrid in 1885, the Prussian ambassador was pleased to observe that it had triggered a much-needed street-cleaning effort.[3] But urban sanitation was also about costs and many other limiting factors, and the zealots of public hygiene never achieved the kind of expertocratic reign that they dreamed of. Many European cities did get cleaner in the second half of the nineteenth century, but that was the result of many small steps—though these efforts seemed much more heroic if framed in a narrative of learning from disaster. Progress was even slower when it came to transnational collaboration, and an international convention against cholera was not signed until 1903. A few years later, one of the signatories, Italy, kept a cholera epidemic under wraps, which violated the convention's letters and spirit. This was the cholera epidemic that Thomas Mann learned about during his vacation in Venice.

While nobody wanted cholera, people were seeking vacations if they could afford it, at times with religious fervor. It was a realm of personal freedom, or that is how it seemed: the Baedeker guides, along with their diligent use by a grateful traveling public, revealed the extent to which leisure time was prescribed even for those who did not book the package holidays of Thomas Cook. Originally conceived for travelers on the new steamship routes on the Rhine, Baedeker guides became the vade mecum for bourgeois readers who lacked time and abhorred the effort required to make their own explorations. The guidebook market diversified when growing affluence gave more people the chance to travel, and while the reality of traveling masses was not just the friendly meeting of cultures that the optimists had hoped for, the range of destinations and activities flourished and continues to expand to this day. Characteristically, more than two-thirds of the chapters in this book include a reference to a tourist destination.

Guidebooks held the power to make or break the careers of hotels, restaurants, and cultural institutions, and so did other types of handbooks. In fact, handbooks may be one of the hidden powers of modernity: people better have a good reason when choosing to go against the book. The genre offered crucial information, defined codes of conduct, generated a sense of identity, and generally conveyed a sense of cognitive certainty that many people were seeking amid the dynamism of the modern world. Of course, the handbook did not gain this status through the sheer power of print: handbooks were typically collaborative efforts, and authors were advised to temper their urge for creative expression and identify the conventional wisdom in their respective fields. In other words, handbooks were platforms for negotiations, gathering points for experts and stakeholders who sought to identify bodies of certified knowledge. Handbooks lacked power without the trust of their readers, and editors and authors pursued various ways to meet expectations. Karl Baedeker built his publishing empire on the pledge that he had seen everything in person.

The benefits of travel have always been disputed. For one, Adam Smith argued that travelers would learn more back home if the universities finally got their act together. Smith, who had worked as a tutor on a Grand Tour himself, found that a young man "commonly returns home more conceited, more unprincipled, more dissipated, and more incapable of any serious application either to study or to business, than he could well have become in so short a time, had he lived at home."[4] There was certainly reason to contest travel on environmental grounds. However, rather than reiterating pertinent failings—you can find an overview in every good guidebook today—the chapter focuses on the strange air of unreality that surrounds the environmental repercussions of modern tourism and the difficult choices that tourism presents for environmental policy in the twenty-first century. How much should we stress the real-world toll of environmental sojourns when the entire idea is to get away from the real world? And what is the point of criticizing tourism when that criticism is older than Baedeker? As environmental problems go, few are more conceptually blurry than the problems of tourism, and it is fitting that it currently contributes some 8 percent to global greenhouse emissions—a share large enough to be a matter of concern, but not quite large enough to be an obvious priority.

There is more than one way to read the title of the section's third chapter, "Gandhi's Salt," and there is more than one way to change the world. Gandhi's salt march to Dandi did not free India immediately, but it was a milestone on India's path to independence that resonates to this day.[5] There was even a "much-hyped reenactment" of the Dandi salt march on its seventy-fifth anniversary in 2005.[6] It was about the salt monopoly of the British Raj and about the mode of protest, as the salt march followed Gandhi's idea of *satyagraha*: a form of nonviolent protest, born out of a truth-seeking state of mind, that found the British Empire "trapped in the coils of their own ambivalence."[7] The idea continues to inspire discontents around the world, and the salt march was as close to the perfect *satyagraha* as Gandhi ever came.

The literature usually treats Gandhi as an exceptional individual, "a singular type of 'politician'" in the words of Judith Brown.[8] But much of what Gandhi was and achieved resulted from media coverage, and that is why this chapter analyzes him as a part of the modern cult of celebrities—not a popular word among environmental historians, and yet one that no comprehensive history of environmentalism can do without in the twenty-first century. In fact, the Mahatma, a real character if there ever was one, is a good starting point for a conversation about celebrity status that does not immediately conjure Hollywood clichés: modern societies like to condense causes in celebrities, and a dense flow of information allows ordinary people to develop a sense of familiarity with their stars. The Gandhi that we know was to a great extent a creation of journalists, with significant divergences between the man and the media myth and even more significant changes over time. As Sean Scalmer puts it, "For those familiar only with the time-honoured image of the saintly Mahatma, a glance at the contemporary press can concentrate attention with the force of a rude and perplexing shock."⁹ For imperial Britons, Gandhi was a dangerous agitator who stoked chaos and violence, and we should see this as just as much of a fabrication as the apostle of peace or the critic of technological hubris that environmentalists liked to read into the Mahatma. Celebrities were living myths, and myths can play out in different ways.

However, there was a man and a life beyond the myth, and the chapter seeks to capture something of the ambiguity that Gandhi had in more than one respect. In other words, the multitude of meanings in the title is by all means intentional: Gandhi made a difference by attacking the salt monopoly, by organizing the peaceful salt march, by sowing salt into the open woulds of imperial rule, and by showing his salt with a charismatic life of almost superhuman self-discipline. Scholars have tried to compress the experience of Gandhi into theoretical models, but that may say more about academic obsessions with modeling than about the real world: if anything, the Mahatma changed the world because he burst the realm of people's preconceptions. He was also responding to realities around him, far more than the latter-day myth tends to suggest. Changing the world is not a one-dimensional affair.

Gandhi has achieved a measure of popularity among environmentalists, but maybe for the wrong reason. He showed how individuals can chart their own path with vision and determination and thus make a difference—a powerful example for alternative projects around the world—but much of the interest focused on his ideas rather than his life course. More precisely, environmentalists focused on those ideas that they cherry-picked from the fifty thousand pages of Gandhi's writing, with particular attention to the *Hind Swaraj* of 1909. The approach is open to challenge on intellectual and political grounds: the concept of a great transformation deduced from a shining set of ideas is flawed even for a great soul like Gandhi. Reading ideas as mere blueprints for survival underestimates their role in the modern world. They are also icons, campaign tools, irritants, and other things, and while they stand along pathways, they are not paths themselves. Environmental challenges have provoked countless intellectual and real-world projects big and small, and these projects stand amid others in a modern world with many other concerns. When postimperial Britain honored Gandhi with a statue on Parliament Square in London, the Mahatma enjoyed the company of two prime ministers who fought him tooth and nail, Winston Churchill of Great Britain and Jan Smuts of South Africa.[10] It is more appropriate than one might initially think.

Gandhi will be world-famous for the foreseeable future. The same cannot be said of the Tokyo Resolution of March 12, 1970, but the resolution did not aim to reach the public at large. It was the result of an International Symposium on Environmental Disruption that reflects a quintessentially modern type of event. Numerous chapters in this book show the significance of international meetings and the agreements that they produced, and this chapter explores the peculiar intermingling of academic expertise and political virulence that has characterized international conventions since the mid-nineteenth century. The symposium at the Tokyo Prince Hotel created interlinkages between a number of ongoing developments over a few busy days: international academic exchange, a social science community in search of new issues and confirmation of professional relevance, preparations for the United Nations Conference on the Human Environment in Stockholm, Joseph Sax's pathbreaking public trust doctrine, and Japan's fight against pollution. It also linked all these things to generous funders and an event forty-one years later that might qualify as an ironic finale if it were not so tragic.

International meetings are inevitably ephemeral, and they are usually elite endeavors. Both features have made them suspect, and not just among those who had to stay back home: did the meeting really matter? Successful conferences did not operate in isolation: they made a difference because they fostered transnational webs of knowledge, power, and personal relations. International conventions do little that could not be achieved by other means, and yet they are amazingly effective as nodes in transnational networks. Since the 1970s, it has become difficult to imagine global environmental discourses without them, and they will be part of the future of environmentalism for better or worse. The Tokyo symposium was certainly not perfect—a recurrence today would likely include female speakers and more people from outside North America and Western Europe—but it mattered beyond the day, not least because it connected with events outside the room. However, results were never a given. Conference language can turn diffuse sentiments into words, or it can be empty rhetoric. There are plenty of examples for both scenarios.

The Tokyo symposium relied on the power of words, but other international conventions aimed for something more: binding agreements that would guide subsequent action. The number of environmental treaties has grown enormously since 1970, and few issues remain that do not have a corresponding international agreement, but achievements differ widely: some are amazingly effective, others are defunct, and many are somewhere in between. Global environmental governance has always been more alluring as a myth than as a reality, but its magic has not gone away. It is a matter of opinion whether that says more about people or about the state of the planet.

The final chapter returns to events with a natural cause, or rather those that were framed as such: the naturalness of the natural disaster was a narrative construct, one of several that helped to carry on in the wake of a seismic event. Even in a secular age, people embraced religious readings of diverse stripes: church-made, homespun, or party-sponsored. The latter was the official response in the aftermath of the Tangshan earthquake, for the Communist Party of China was supposed to believe the Maoist creed of human control over nature—no small challenge in the face of a death toll that likely exceeded half a million. It did not silence ruminations on what the Tangshan earthquake would augur for China after Mao. He died on September 9, 1976, six weeks after the disaster.

Responses to the Tangshan earthquake became a factor in the ensuing power struggle, which shows how natural disasters matter beyond the sphere of destruction. Disasters are character tests for individuals, nations, and regimes, and responses in crisis mode mattered beyond the day. Some leaders made a name for themselves through energetic action; others made a fortune, like Anastasio Somoza. Disasters suspended the common rules of politics and resistance from stakeholders, and crises opened a window for swift and dramatic change. They were less consequential when it came to patterns of social inequality. Earthquakes cut across common divisions in society, but they did not make everyone equal. Many natural disasters ended up reaffirming rather than reducing inequality.

Once the dust had settled, a mixture of scientific and administrative optimism took over: conventional wisdom held that one must learn from the disaster, just as cholera-stricken cities in the nineteenth-century West purportedly did. Responses were usually technological in nature and ranged from seismic upgrades and new urban designs to large systemic solutions, and more often than not, grandiose intention gave way to plenty of compromises. Rebuilt Tangshan was more similar to the previous built environment than intended, and whether it will do better in the next catastrophe is ultimately an open question. An article of 1995 suggested that a recurrence of the 1976 Tangshan earthquake would cause one-twentieth of the original fatalities, which may or may not rank as an achievement.[11] After all, it would mean a death toll in the range of 30,000.

Needless to say, these chapters cannot cover all types of events that have emerged as disruptive forces. However, they bring out the diversity of ruptures in history, and they show how the exploration of exceptional events leads back to the many different contexts that frame human history. Earthquakes and epidemics can kill, and bodies can relax during vacation or overcaffeinate while conferencing, but for all the environmental and bodily realities, meanings and consequences remain negotiable. Even time frames are matters of perspective: seismic shocks are over in seconds, but as Anthony Oliver-Smith has argued concerning a disaster in Peru, there can be such a thing as a "five-hundred-year earthquake."[12] But contextualization can turn into a secular exorcism if it brings us to deny the momentous, the disturbing, and the erratic that events insert into the Braudelian tides of history. Stuff does happen, and it may not make sense. And that is what makes it important.

21

CHOLERA

THE NATURE OF DISEASE

I. FEAR OF COLLAPSE

Gustav von Aschenbach was a famous man. The intellectual world knew him as an acclaimed novelist, and it was in shock when he passed away on a beach in Venice. The cause of his death was cholera, but that was only true in a strict medical sense. Aschenbach had heard rumors about the spread of the disease and read articles in newspapers. He also noticed the conspicuous smell of disinfectants in the streets of Venice and saw other guests at his grand hotel leaving in droves. When he inquired with an English travel agency on Piazza San Marco, the clerk told him in no uncertain terms to leave. But Aschenbach had reasons to stay. He had come to Venice in search of a cure for writer's block, arguably a more terrifying disease for an author than cholera, and he had fallen in love with a Polish teenager named Tadzio, an unfulfilled homoerotic romance that drove the aging man to make a fool of himself. In a way, his death on the beach was a convenient escape from a hopeless situation. Aschenbach had lost his dignity before he lost his life.[1]

Gustav von Aschenbach was fiction, the protagonist in Thomas Mann's *Death in Venice*, but the dangers of cholera were real enough. It was an infectious bacterial disease that spread via contaminated food or water, and unlike other diseases of the nineteenth century, cholera struck with lightning speed. People who felt perfectly healthy could be dead within a few hours, which left them with scant time for wills, prayers, or last words. The circumstances were equally horrific, as a bout of cholera showed in cramps, blue skin, and the uncontrolled emptying of the digestive system on both ends. It could happen in public, at work, or while visiting friends and family, and the disease could leave a reputation in tatters even when the body survived. Cholera did not just kill its victims. It degraded them.[2]

Cholera circled the nineteenth-century world in pandemic waves. Its native home was India, which has shaped European perceptions ever since the first epidemic hit Central Europe in 1831/1832. Johann Wolfgang von Goethe called cholera "that oriental monster" in his diary, and according to *Death in Venice*, the disease came from "the sultry morasses of the Ganges delta, . . . that primitive island jungle shunned by man, where tigers crouch in the bamboo thickets."[3] The spread of cholera followed some recurring routes such as the Volga in Russia, and seaports like Venice were natural pathways for a globe-trotting pathogen.[4] Experience also showed that cholera epidemics coincided with the movement of troops. Characteristically, the disease first made it to the heart of Europe during the Polish–Russian War of 1830/1831.[5] And then there were the sites of pilgrimage that doubled as disease hotspots: Mecca for Muslims, Karbala and Najaf in Iraq for Shiites, Haridwar on the Ganges for Hindus.[6] But nobody had certainty about the threat until calamity struck, and that made it a popular conversation topic: if you were interested in current affairs, cholera was a must. According to his diary, Goethe discussed it at least six times in 1831 alone.[7]

Sometimes cholera was a matter of communication in both a verbal and a medical sense. While enforcing a cordon sanitaire along the Polish Border in the spring of 1831, the commander of the Prussian army, Neidhardt von Gneisenau, wrote to his son-in-law Wilhelm von Scharnhorst that he found cholera "not very contagious or dangerous."[8] The disease killed him the following August, and his better-known chief of staff, Carl von Clausewitz, suffered the same fate in November.[9] Other prominent victims include the Bavarian Queen Consort Therese, wife of King Ludwig I (whose marriage occasioned the first Oktoberfest), the German philosopher Georg Wilhelm Friedrich Hegel, and the US president Zachary Taylor, who celebrated the Fourth of July of 1850 at the construction site of Washington Monument, helped himself to plenty of raw fruit and iced milk, and died five days later.[10] The nature of the disease invited storytelling, and it was difficult to separate fact from fiction. Was it actually true, as the German exile Heinrich Heine reported, that Paris ran out of coffins during the cholera epidemic of 1832 and that when cholera hit a masquerade ball, scores of victims were killed and buried with their costumes still on?[11] Even those with access to the best available information resorted to speculation and actually said so at times. In July 1855, a manager at the Illinois Central Railroad reported in his daily briefing for the railroad's president that in spite of "several sudden deaths from cholera within 24 hours," he had "a confidence, greater than I can give reason for, that we shall have very little cholera this year."[12]

The effects on countries and communities could differ, and so did perceptions of these effects. Dean Worcester, the Philippines' secretary of the interior during US colonial rule, opened his *History of Asiatic Cholera in the Philippine Islands* with ceremonial concerns: "Twice during the past year the presence of cholera in Manila has seriously interfered with important public events, necessitating the postponement of the Carnival and seriously interfering with the reception to the United States Battle-ship Fleet."[13] Others worried about the loss of precious labor. When more than 200,000 Brazilians died in a devastating cholera epidemic in 1855/1856, in which Black people suffered far more from the disease than white people, the country's landed elite bemoaned that Brazil had terminated the importation of slaves from Africa in 1850.[14] Cholera triggered riots in cities as different as Liverpool and Tashkent.[15] In Costa Rica, it cast a dark shadow over a great patriotic moment in the nation's history when an epidemic claimed about one in ten Costa Ricans after troops returned from the victorious battle against the US filibuster William Walker in 1856.[16] Cholera also doomed a world's fair when it hit Vienna in 1873.[17] Other places faced more existential issues than empty exhibition halls. Shelbyville, Illinois, had 1,600 residents when a crew of railroad workers brought cholera into town in June 1855. It killed 200, triggered a mass exodus, and left the town down to five families two months later.[18]

Cholera wrecked the worst havoc in big cities, which was unfortunate in a century of urban growth. It also claimed a disproportionate share of its victims among the urban poor, thus shedding an unflattering spotlight on the plight of the laboring classes in an age of industrialization. In fact, epidemics highlighted fault lines and social conditions so well that a medical historian, Charles E. Rosenberg, called cholera "a convenient and effective sampling device" for social and economic historians.[19] Epidemics were also unintended results of the transport revolution (see chapter 3, Canal du Midi), as steamships and railroads allowed cholera to spread in spite of its brief incubation period, and it left scientists and officials clueless. In other words, cholera revealed the dark underside of defining trends of the nineteenth century: urban life, industrial technology, social and geographic mobility, scientific and political authority. It was as if cholera were out to taint everything that Western modernity had achieved.

All this explains why cholera became such a terrifying disease. It was not the biggest killer in nineteenth-century Europe: measles, smallpox, and tuberculosis claimed far more lives.[20] Furthermore, a cholera epidemic usually killed less than 1 percent of the urban population.[21] But statistics (see chapter 7.2, Numbers Games) was only one way to determine the nature of a disease, and not always the most popular one. Cholera was not even the only epidemic that haunted the nineteenth-century world. Yellow fever triggered similar responses, and George Rosen has argued that it "was dreaded even more than cholera in the United States."[22] But unlike yellow fever, which spared much of Europe, cholera was truly global, a disease that only remote places could hope to escape, and that claimed an emotional toll around the world. "Much of cholera's story is a story of fear," Christopher Hamlin wrote in his "biography" of the disease.[23] From a Western perspective, cholera was a counternarrative to much of what the nineteenth century was proud of, a bane of civilization that threatened to bring out the worst in humans. In *Death in Venice*, Thomas Mann described a specter of moral collapse: rampant crime, open prostitution, and two proven cases of murder where "persons alleged to have died of the plague had in fact been poisoned by their own relatives."[24]

What could people do in the absence of dispensable relatives? They could try to flee, which many people did if they had the means. They could also make a note about purported remedies such as opium (see Interlude, Opium), a solution, or perhaps rather a palliative that researchers have found in the excerpts that Thomas Mann prepared while working on his novel.[25] They could adjust their baseline expectations and shrug off death by the dozens as the new normal. When cholera struck Barcelona in 1865, the Prussian emissary reported that the death toll had recently climbed to forty per day and cited this number as evidence that "the disease has not gained a malignant character as of yet."[26] Or they could resort to pseudo-scientific guesswork. When the local agent of the Illinois Central Railroad in Shelbyville reported on the effective collapse of the town, a manager offered his own cholera theory by way of consolation: "I regard those places the safest to live in for the next five years where they have had the cholera in the severest form within a couple of years—and those places the most unsafe where they have heretofore escaped the epidemic."[27] And if none of these options looked appealing, people could simply resign themselves to their fate like the Russians in Ivan Turgenev's *Fathers and Sons*, who enjoyed a beautiful day in June even though "there was once again a distant threat of cholera" because "the local inhabitants had already grown used to its visitations."[28] But in a self-declared age of progress, many people sought a more sophisticated solution.

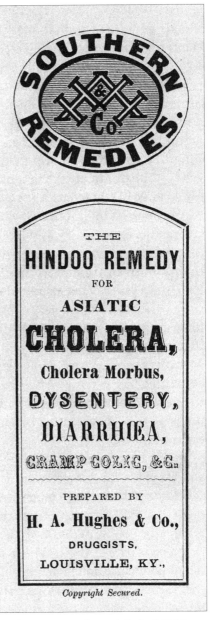

21.1 Patent medicine label of 1866. Image, Library of Congress.

2. CONTAINMENT EFFORTS

Medical science was not the only authority in play in the fight against cholera. Its professional credentials were still weak for much of the nineteenth century, and some people found its practices highly suspicions. The 1832 cholera riots in Liverpool targeted medical professionals because of rumors that they were killing cholera victims to claim their bodies for anatomical dissection.[29] Thoughts and prayers were a popular response in the United States, as cholera was obviously "a punishment . . . coming from God's hand," and matters of faith turned into a political football when President Jackson refused to endorse a "day of public fasting and humiliation" on constitutional grounds.[30] As in the face of a natural disaster (see chapter 25, 1976 Tangshan Earthquake), people turned to religion in a medical emergency, though not always in a pious mood. In 1892, cholera riots in tsarist Russia also targeted the Orthodox Church, including a mob attack on a priest in Voronezh "who tried to appease the masses with a cross in his hand" and was subsequently stoned to death.[31] And then there were the literary experts who stood ready to weave diseases into their narratives. Thomas Mann's *Death in Venice* won acclaim because it evoked a sense of decay, and literary critics have been debating ever since whether it was intellectualism, the bourgeoisie, or Western civilization that was going down the drain.[32] Storytellers can save lives in their own metaphoric ways, and their power was in particular demand when events lacked an obvious sense or purpose. Cholera did not have a meaning by itself, but writers have tried to provide it with one since the heydays of Romanticism. A cholera scholar, Olaf Briese, edited a full volume of "bad poetry" that Germans penned during the cholera epidemic of 1831/1832.[33]

But at a time when authority over disease was in flux, cholera was also a tremendous opportunity for aspiring medical professionals. A successful fight against a much-feared disease was the kind of achievement that built a reputation, and much of the ensuing debate revolved around the precise nature of cholera: Was it contagious, and if so, what exactly was the pathogen? The struggle between contagionists and anticontagionists came to an end in the late nineteenth century when the discoveries of Louis Pasteur and Robert Koch turned bacteriology into a cornerstone of the medical profession, but the path toward this conclusion was anything but straight. Bacteriology implied a sharp turn toward laboratory science in an academic field that had heretofore focused on sanitary conditions in all their diversity. The outcome also seemed unlikely around midcentury when the search for a contagium vivum looked like a dead-end street. As Erwin Ackerknecht wrote, "It was shortly before its disappearance that 'anticontagionism' reached its highest peak of elaboration, acceptance, and scientific respectability."[34]

The debate reached beyond medical circles. Those who sought a place at the forefront of scientific discovery threw themselves into the fray, and opinions reflected not just matters of pathogenesis. Justus von Liebig took the side of the anticontagionists because the miasma theory of disease, which focused on the emanations of decomposing matter, meshed well with his interest in fermentation.[35] Arguments revolved around a mixture of personal interests, evidence, and beliefs, and the latter could grow into quasi-religious proportions. "Like bornagain Christians, ardent germ theorists saw the world with new eyes, as places where air, water, and soil teemed with invisible life and their own skin and secretions swarmed with microbes," Nancy Tomes observed.[36] The debate came to a head between 1865 and 1895 when "Western medicine underwent a virtual civil war over the truth of the germ theory."[37] The war metaphor captures the bitterness of academic and personal rivalries, but the battle lines were far more muddled than in a military conflict, as observations, tropes, and models traveled rather freely between the two camps. Medicine was not a science where one paradigm determined mindsets and outcomes.[38] The contagion was merely an epistemic object, a concept that invited development and change. Thinking in terms of contagions was a path of inquiry rather than the end of all questions.[39]

As befits an epic quest, cholera research gained its own mythology. Perhaps the most enduring is the sharp distinction between winners and losers: miasmas do not actually exist, but miasma theory encouraged a rewarding look at sanitary conditions. Subsequent research showed the extent to which the spread of cholera bacteria hinges on favorable conditions, and germ theory "gradually . . . incorporated biologic, demographic, and environmental elements."[40] Michael Zeheter has argued that the real scholarly divide ran between complicators and simplifiers like Robert Koch.[41] A pub in London's Soho quarter commemorates John Snow, who allegedly stopped a cholera epidemic by disabling a nearby water pump. But Snow's epidemiological work looks far less spectacular in a contemporary context, and the story nourishes "the myth of the brilliant, solitary researcher" whereas actual research was collaborative in nature.[42] An Italian doctor, Filippo Pacini, observed and analyzed the cholera microbe three decades before Koch, but he did not receive credit for the discovery until after his death because his academic home, the University of Florence, was on the periphery of nineteenth-century science.[43] And then there was Max von Pettenkofer, whose lifetime achievements for public health faded into the background when he swallowed live cholera bacilli provided by his rival Robert Koch. He survived and claimed that the outcome disproved Koch's contagionism, and while he was fighting a lost cause, the act had a whiff of academic heroism. Admiring biographers took pleasure in quoting Pettenkofer's stern declaration that he was ready to die "in the service of science, like a soldier on the field of honor."[44] Didn't Francis Bacon, the patron saint of modern science, die from bronchitis or pneumonia after a fateful experiment in the cold that sought to explore the uses of snow for the preservation of chicken meat (see chapter 36, Battery Chicken)?[45] But in reality, Pettenkofer's self-experiment was an act of desperation. He stood with his back against the wall academically, and the result of his heroic endeavor was put into perspective when his assistant replicated the experiment ten days later and grew severely ill.[46]

The controversy over contagionism was about more than medical theories. If cholera was contagious, quarantines seemed a good way to stop the spread of the disease. But quarantines also strengthened the hand of authoritarian governments, particularly when enforcement lay in the hands of the military, and they inevitably brought a disastrous disruption of economic life. Consequently, anticontagionists were "not simply scientists," as Ackerknecht observed, "they were reformers, fighting for the freedom of the individual and commerce against the shackles of despotism and reaction."[47] It showed in reactions on the ground. Cities sought to delay announcements of cholera cases, just as Venice did in Thomas Mann's novel, while governments showed their muscles through cordons sanitaires, though enforcement was usually weaker than intentions. When Spain suffered from a cholera epidemic in 1885, local authorities were so overwhelmed that one place created a transhumance-style quarantine facility out of sheepfolds. The design turned tragic when one of the internees jumped the fence. He was killed when officials, already down to archaic means, tried to stop him by throwing stones.[48]

The commercial repercussions of quarantines prompted governments to keep an eye on the spread of cholera. When the Prussian consul in Santander remained silent in spite of rumors about an epidemic in the port on Spain's northern coast in 1865, he received a harsh missive that prompt reporting was crucial "on a matter that affects trade so profoundly."⁴⁹ A lack of medical knowledge left most emissaries ill prepared to separate facts from fiction, but diplomats knew how to look sincere when rumors were everywhere. Reports offered plenty of details when reliable information was not available, so the Prussian envoy in Florence informed his superiors in Berlin in July 1854 that trains departing from Livorno, an Italian seaside resort, were overcrowded because all foreigners were leaving.⁵⁰ In 1890, German diplomats in Constantinople sent the ruminations of a Greek tobacco merchant about cholera victims straight on to the desk of Bismarck. Nobody could check his assertion that the bodies of three Indian pilgrims had been thrown overboard on his steamship to Basra to avoid delays from quarantine measures. All that the diplomats knew was in a cable from Baghdad. But that was the best information they could offer.⁵¹

Medical professionals could have offered more than diplomatic hearsay, but governments did not like the idea of independent expertise on their own turf. When France convened the first international sanitary conference in Paris in 1851, countries were supposed to send *two* delegates, one physician and one diplomat.⁵² More than fifty years elapsed until the Paris Convention of 1903 (see chapter 24.3, Getting Serious), which "abolished the anarchic sanitary world where every state was free to impose health regulations of its own devising" and committed signatories to immediate and comprehensive information about the presence of cholera and plague, but the treaty's language was more impressive than institutional reality.⁵³ Enforcement fell to a small secretariat in Paris that Mark Mazower has called "a bureaucratic mouse."⁵⁴ Scientific research was transnational nonetheless, and not just in the form of grand international conferences. Robert Koch traveled to Egypt to study a cholera outbreak in 1883, and when the epidemic subsided before conclusive results, he went on to Calcutta, where he succeeded in isolating the cholera germ.⁵⁵ He was not the first medical expert to go the extra mile in a rather literal sense. On November 30, 1831, Goethe met a Bavarian physician in Weimar who had traveled all the way to northern Germany to study cholera and came back "with the comforting conviction . . . that it was not contagious."⁵⁶

Medical expertise found fertile ground in the mushrooming cities, and cholera figured prominently in urban debates over public health. Sanitary reformers in Australia even bemoaned the absence of cholera from their continent, as it would have provided a powerful catalyst for investments in urban sanitation (see chapter 17, Water Closet).[57] It seemed to work with amazing speed in places like Switzerland. Zurich suffered from the worst cholera epidemic in the country's history in 1867 and swiftly built a sewer system and a reliable freshwater supply over the following years. But the epidemic happened during a time of political turmoil, and the response drew more on democratic reform than on a mythical Swiss diligence.[58] Historians have grown weary of simple challenge-and-response models of cholera and sanitary reform, as the latter was about so much more: urban politics and administrative capacity, money and expertise, water resources and geographies.[59] Fear of cholera was genuine, but it was always just one of several factors in play.

The drive toward improvement was never self-evident, and it had its share of blunders. Hamburg was "a pioneer of sanitary reform on the continent," but negligence about filtration caused a devastating cholera epidemic in 1892.[60] Madras built a reservoir that lacked capacity, and outbreaks occurred regularly when the water table was low.[61] The 1892 Tashkent cholera riot occurred against the backdrop of a failed canal project whose chief design flaw, only discovered during construction, was that it required water to flow uphill.[62]

For all the money and energy that sanitary reform claimed in late nineteenth-century cities, improvements remained contested and incomplete, and the same held true for medical knowledge. "Many problems in connection with the epidemiology of cholera still remain unresolved," a review declared in the 1920s.[63] But cognitive uncertainty did not necessarily discourage professional ambition, and some experts argued for the sanitary city with a zeal that bordered on the religious. In a speech in his hometown of Illiers, the chair of hygiene at the Faculty of Medicine in Paris, Adrien Proust, declared that "hygiene should transform towns today, under pain of death, at the cost of aesthetic charm, and the beauty of streets and houses." But few were willing to embrace a worldview in which pathogens were all that mattered. Proust's son Marcel, for one, viewed the waterways of Illiers as more of a dreamscape with things like water lilies, which he immortalized in his *In Search of Lost Time*.[64] Public hygiene was ultimately about making deals with authorities and societies, and the deal that medical experts came up with in the late nineteenth century had a few requirements. As it turned out, some places could meet these requirements more easily than others.

3. DISEASE WORLDS

Thomas Mann drew on personal experience when he wrote *Death in Venice*. Cholera had come to Italy in 1910, a fact that the government concealed in blatant violation of the Paris Convention (see chapter 24.3, Getting Serious). A remarkably effective campaign censured the press and silenced medical professionals, and Mann was likely unaware of the threat when he arrived in Venice in May 1911.[65] But once in town, he heard rumors, just the way Aschenbach did in the novel. The German press discussed the fate of an Austrian man who died shortly after his return from Venice, another parallel to *Death in Venice*, where the travel agent reveals the case to Aschenbach. In fact, the conversation with the travel agent was itself probably authentic: Thomas Cook had a branch on Piazza San Marco at the time. Unlike Aschenbach, Thomas Mann left Venice for Munich on June 2, and the author could follow subsequent events from a safe distance. At the end of 1911, Venice recorded 247 cholera cases, 88 of them fatal.[66] As cholera epidemics go, it was a rather mild event, a small fraction of the estimated 18,000 Italians who died from cholera between 1910 and 1912.[67] It was also one of the last outbreaks of cholera on the European continent. Literary scholars will likely continue to debate whether *Death in Venice* mirrored the decline of the West, but it certainly did mirror the decline of cholera in the cities of the West.

Contagious diseases were still around in Western societies, but by the second half of the twentieth century, they no longer terrified people. The pathogens were known, vaccinations and treatments were at hand, and people were confident that medical authorities would tackle new challenges as they came along. Even the devastating influenza pandemic at the end of World War I was quickly forgotten, and most Westerners imagined the flu as a seasonal nuisance rather than a mortal threat.[68] Legionnaires' Disease was subject to frenzied media coverage in 1976 when a deadly strain of bacteria spread through the air-conditioning system (see chapter 20, Air-Conditioning) of a Philadelphia hotel, but when a book recounted the story five years, it closed on an upbeat note. The battle against contagious diseases would continue, but there could be "no doubt about the outcome. Man will win this confrontation, just as he has emerged victorious from so many other battles he has waged with that world."[69] Cholera was fading from collective memory, and if Western people encountered it at all, it was in books like Mann's *Death in Venice* or Gabriel García Márquez's *Love in the Time of Cholera*, where the disease is part of the setting for a lifelong romance.[70]

Medical knowledge has certainly advanced dramatically. When the World Health Organization published an authoritative monograph about cholera in 1959, the volume ran to more than 1,000 pages.[71] Medical experts have confidence about the causes of cholera, and they know how to treat those who fall ill, making death from cholera utterly preventable from a medical point of view. What they do not know is the actual number of deaths, as the vast majority of cholera cases are not officially recorded. However, we do know that about 1.4 billion people are at risk, that 51 countries are classified as cholera-endemic, and that virtually all these countries are in Africa and southern Asia.[72]

Modern medicine was a Western invention, and that left its mark on the imagination of disease. The idea of cholera as a devastating but brief scourge did not make sense in places where it was endemic. The disease never left places like Calcutta, which had more than 1,000 cases of cholera in every single year between 1841 and 1959.[73] The difference was about mindsets as well as numbers. Five cholera epidemics killed an estimated 130,000 people in Great Britain, and sanitary improvements reduced the toll in every epidemic since 1848.[74] British colonialism brought European policies to India, and the report of the health officer of Calcutta for 1893 recorded a "diminished prevalence of cholera" and praised "the sanitary improvements which have been effected of late years," but achievements were ultimately relative.[75] Between 1800 and 1925, the Indian death toll from cholera, with more than 25 million victims, exceeded the British by a factor of 200.[76]

India was not the only place where cholera was endemic. A memorandum for the League of Nations Health Organisation of 1925 identified "endemic centres . . . in Central Asia, in the Shat-el-Arab area, in certain parts of the Federated Malay States, in Java and the Dutch East Indies, in Indo-China, in the Philippines, and probably in Southern China."[77] The one piece of good news was that the disease somehow lost its inclination to travel the globe in the 1920s, but this proved to be a temporary phenomenon. A new pandemic, the seventh by most counts, began on Sulawesi in 1961 and quickly spread to other parts of Indonesia, southern China, and the Philippines. The seventh pandemic later reached the Middle East, Africa, and even a number of European port cities: Odessa, Barcelona, Lisbon, Naples. It has not ended to the present day.[78]

The new pandemic traveled erratically and unpredictably. Peru ex-

21.2 Cholera squad in the Philippines, ca. 1915–1920. Image, Bain News Service, George Grantham Bain Collection, Library of Congress.

perienced a major epidemic in 1991, the first one in South America since 1895.[79] After twenty years without a reported case, a major outbreak affected eight thousand people on Grande Comore in 1998/1999, or 3 percent of the population on the Indian Ocean island.[80] When UN peacekeepers from Nepal came to Haiti on a rescue mission after a devastating earthquake (see chapter 25, 1976 Tangshan Earthquake) in 2010, they brought cholera with them and accidentally killed thousands.[81] In October 2016, cholera hit Yemen on the heels of civil war.[82] Even airline food has been identified as the cause of a cholera outbreak.[83]

The World Health Organization opened its massive tome with a preface noting that "it was through cholera, and the fear to which its pandemic sweeps gave rise, that international solidarity in matters of health was born."[84] It is a rhetoric that international organizations (see chapter 24, 1970 Tokyo Resolution) like to produce when it comes to environmental challenges. Pathogens do not recognize national boundaries, and neither do pollutants (see chapter 16, London Smog) or endangered species (see chapter 11, Dodo), and we are all in this together. The reality was different. As the seventh pandemic embarked on its global journey, popular fears in Africa and Asia made for a striking contrast to the environmental discourse in Western societies, where fears of chemicals (see chapter 38, DDT) and nuclear radiation (see chapter 37, *Lucky Dragon No. 5*) were ascendant. While Western environmentalists were concerned about chronic poisoning, people in the Global South worried about a sudden death.

Will the seventh pandemic ever end? The cholera contagion can survive in water without human infection, and that makes it hard to exterminate.[85] A Global Task Force on Cholera Control, a global alliance hosted by the World Health Organization, published a "Declaration to Ending Cholera" in 2017, but its goal is actually more modest: it aims for "a 90 percent reduction in cholera deaths by 2030."[86] When it comes to cholera, doctors have an impressive array of tools at their disposal today. Even in Yemen, where only half the population had access to safe drinking water in 2014, health-care providers managed to limit the fatality rate to 0.5 percent.[87] But for all the advances of the medical sciences, total victory remains an elusive concept for cholera and many other pathogens. As it stands, humans have eliminated exactly two diseases, smallpox and rinderpest, from the face of the earth.[88]

In technical terms, fighting cholera is easy. Just using bottled water would eliminate much of the risk. But as in the nineteenth century, the simple demands of medical professionals face a more complicated world. For example, a cholera outbreak in South Africa has been linked to water privatization programs that deprived poor people of access to clean water.[89] And then, is bottled water perhaps just a shenanigan of unscrupulous businessmen? A Pakistani novelist, Mohsin Hamid, wrote a best seller, *How to Get Filthy Rich in Rising Asia*, that revolves around a man who makes a fortune as a bottled-water tycoon.[90] A world without cholera and unclean drinking water may be just as elusive as a world where novels no longer shape our reading of disease.

22

BAEDEKER

GUIDANCE FOR SEEKERS

I. MANUAL FOR A NEW AGE

You know that you have made it when you are immortalized in opera. The British version of *La Vie Parisienne*, loosely based on Jacques Offenbach's French classic, bestowed this honor on Karl Baedeker. It features British travelers visiting the Louvre with Baedekers in hand, and the libretto has them proclaiming that "Kings and Governments may err / But never Mr. Baedeker."[1] Almost a century had passed since Karl Baedeker published his first guidebook in Koblenz, a German town on the banks of the Rhine, and decades of prolific use had turned the name into a brand with appeal beyond Germany's borders. It was a synonym for the guidebooks that had become a cherished tool for travelers over the course of the nineteenth century. As Jules Verne advised in one of his novels, "When one does not know a place it is well to consult a guide-book."[2] The German translation replaced the word "guidebook" with a more specific one: "Baedeker."[3]

Guides in the vein of Baedeker were not the first books that were written for mobile readers. Eighteenth-century aristocrats on the

Grand Tour, the coming-of-age journey to round out their education, could choose from a broad range of titles, and some were written expressly "to inform the traveller on the spot."[4] But the transport revolution of the nineteenth century (see chapter 3, Canal du Midi) allowed travel at a new kind of speed, and that created a demand for a new kind of book. A new steamship line ran between Mainz and Cologne beginning in 1827, cutting travel time through the scenic valley of the Rhine to a single day downstream and two days upstream.[5] The following year, a history teacher from Koblenz published a book on the journey that specifically addressed "fast travelers" who "wish to view at a glance what is of interest in the region."[6] Scholars would subsequently grumble that guidebooks were "a debasement of an earlier and more sophisticated travel literature of the Enlightenment," but that said more about the critique of tourism than about flaws in the genre.[7] The new guidebooks were meant to be concise.

It was not that people on the Grand Tour invariably traveled at a leisurely pace. When Johann Wolfgang von Goethe came to Italy in 1786, he spent a mere three hours in Florence because he was eager to get to Rome by All Saints' Day.[8] But once in the eternal city, he stayed for some fourteen months, much of it spent in thrall with the built legacy of classical antiquity.[9] About 120 years later, the Baedeker *Berlin and Its Environs* noted with confidence that "a fair knowledge of Berlin may be gained in a single week": visitors should not amble north of the Spree River, south of Leipziger Straße, east of Alexanderplatz, or west of the lavish greenery of Tiergarten, except perhaps for a daytrip to Potsdam.[10] Such an itinerary called for a reliable digest on all the essentials of travel: transport, hotels, sights, local customs, and special events that deserved a change of schedule. If there had been a Baedeker in the time of Goethe, he might have learned in time that All Saint's Day was not a big thing in eighteenth-century Rome.[11]

Karl Baedeker had owned a bookstore in Koblenz since 1827 and observed what his customers were seeking. He acquired the rights to the Rhine journey guide for fast travelers and published a new edition in 1835.[12] He also learned from the *Red Book* guides published by John Murray for the British market and even copied their trademark red cover. Baedeker and Murray cooperated for a number of years, but while the British publisher moved on to other genres—John Murray survives as an imprint of the multinational Lagardère Group—Baedeker banked on guidebooks, and so did subsequent generations: family members led the publishing house until 1984.[13] Baedeker published the first French guide in 1846 and the first English one in 1861, and the books sold well in all three languages. The Rhine journey volume alone went through 32 editions in German, 18 editions in French, and 17 editions in English before World War I.[14] Baedeker also kept adding new countries to its portfolio: Belgium and the Netherlands in 1839, Switzerland in 1844, Italy in 1861, Sweden and Norway in 1879, Russia and Greece in 1883.[15]

Baedeker's growth mirrored an increasing band of travelers and a broadening range of destinations. Some of the new discoveries were about treasures of nature. Journeys to the seaside (see chapter 20, Air-Conditioning) and the Alps increased enormously during the nineteenth century, and national parks (see chapter 26, Kruger National Park) followed in due course. For those who sought an even more intimate experience of nature, Croatia became a mecca for nudists in the interwar years. The most famous practitioners were Britain's King Edward VIII and Wallis Simpson.[16] Entire regions appeared anew on the mental map of the West: Caribbean islands from Barbados to the Bahamas, heretofore known for sugar plantations (see chapter 2, Sugar) and devastating epidemics (see chapter 21, Cholera), "began to be transformed into playgrounds for itinerant Caucasians" by the early 1900s.[17] Just like its customers, Baedeker looked increasingly beyond traditional European destinations, and it published guidebooks on regions that Europeans were particularly interested in: Palestine and Syria in 1875, Egypt in 1877, the United States and Mexico in 1893, Constantinople and its environs in 1905, and India in 1914.[18]

New destinations raised new questions. What were the most important sights? Baedeker highlighted them with an asterisk. How could visitors use a bike in Berlin without violating the law? Guidebook users learned that "cyclists resident in Berlin must be provided with a permit (to be obtained from the district police-authorities), but for a short visit that is not necessary."[19] How do motorists use the autobahn (see chapter 34, Autobahn)? The 1936 Baedeker for Germany noted that "reversing and turning are forbidden."[20] How does one learn about the weather in Naples? Baedeker advised looking at the smoke from Mount Vesuvius, which doubled as "a gigantic barometer."[21] Should travelers carry a gun in Palestine? According to Baedeker, "Revolvers and other arms . . . add greatly to [the travelers'] importance in the estimation of the natives, but are not often brought into actual use."[22] And if you were not sure how to bribe someone in the Orient, Baedeker could help you do that, too.[23]

Baedeker was more reluctant to comment on political matters, a departure from the genre's traditions.[24] Travel writers typically offered their own assessments of peoples and governments, and scathing remarks ranked among the perks of many a narrative. Commenting on the mismanagement of the Papal States, a popular conversation topic among European protestants, Goethe wrote in his *Italian Journey* that the Papal States "seem to survive only because the earth does not want to swallow them."[25] More adventurous authors also commented on their own preconceptions. In *The Innocents Abroad*, Mark Twain confessed his bewilderment about the size of the Holy Land: "The word Palestine always brought to my mind a vague suggestion of a country as large as the United States."[26] He was also "a little surprised to find that the grand Sultan of Turkey was a man of only ordinary size."[27] Baedeker stayed clear of conflictual issues, and when they were impossible to avoid, guidebooks sought to treat them as tactfully as possible. The guidebook for the Paris Exhibition of 1889 noted that the event "has naturally been deprived of the official coöperation of the monarchical governments of Europe" because it was "intended mainly to celebrate the centenary of the Revolution of 1789," and went on to assure readers that it was nonetheless "one of the most imposing world's shows ever held."[28] The 1936 Baedeker for Berlin told readers about the main features of the Reichssportfeld but otherwise left them with their own thoughts about the Aryan Olympics.[29] Baedeker guidebooks did not tell readers what to think about the governments of the world. They just explained how to deal with them.

Some people ignored Baedeker because it did not match their ideas about travel. Phileas Fogg, the protagonist in Jules Verne's *Around the World in Eighty Days*, used a copy of Bradshaw's *Continental Railway Steam Transit and General Guide* because it "was to give him all the information needed for his journey."[30] But Baedeker's diligence, underscored by a pledge to visit everything in person, made his guidebooks trusted companions for the traveling classes. Even academics were swayed. Friedrich Ratzel, the German geographer who invented the *Lebensraum* concept (see chapter 32.2, *Lebensraum*), praised the quality of the maps, and a review in the *Journal of Hellenic Studies* declared that the 1888 Baedeker for Greece left "little, if anything, to be desired in thoroughness and in archeological accuracy."[31] Contra *La Vie Parisienne*, Baedeker guides did contain errors, and prefaces stressed that "infallibility cannot be attained."[32] A blunder even made it into the publisher's in-house history: an editor saw a French sign on the outskirts of Geneva that called for "silence" and "prudence," mistook it for a place name, and the next edition of the Swiss Baedeker mystified readers with an opaque reference to "Geneva's Silence-Prudence quarter."[33] However, the greater concern was that the information provided was far from exhaustive. When Lucy Honeychurch, the hero of E. M. Forster's *A Room with a View*, opened her guidebook to learn about the Basilica di Santa Croce in Florence, she was told that "we shall soon emancipate you from Baedeker. He does but touch the surface of things."[34]

It mirrored the fundamental paradox about the reception of the Baedeker guides. The manuals set out to make readers "as independent as possible of the services of couriers, guides, and commissionnaires," but many readers used their independence to stick to Baedeker's listings.[35] Guidebooks framed views and agendas and thus figured prominently in the making of what the sociologist John Urry has called "the tourist gaze": a highly visual and thoroughly commodified mode of perception that constrained the travelers' experiences.[36] Liberated by Baedeker, travelers spent much of their precious time verifying guidebook entries, notwithstanding mockery from more than one side: *La Vie Parisienne* has a British traveler starting his visit with page one of his Paris Baedeker and proclaiming with pride that he was "now at page 104, and so far we have missed nothing but the nudes."[37] In the end, the travelers' zeal terrified even the publisher. In the 1930s, a Baedeker guidebook warned readers that the "mass of detail" should not lead travelers "into a senseless hustle through all the 'sights' of the country."[38]

2. THE AGE OF HANDBOOKS

Baedeker guides held considerable power. Many anecdotes reflect how endorsements were coveted assets, and countless managers of hotels and restaurants saw omission from the book as a mortal threat. But the power of Baedeker was a peculiar type of power. Entries were inevitably subject to review on the ground, and inscrutable judgments jeopardized the readers' trust. Many travelers recorded their own observations and sent them to Baedeker, and as the number of editions grew, assessments changed from statements of authors to results of negotiations. Manuals like Baedeker's held power, but only because their producers were willing to share power with those who used them, and that gave handbooks a pivotal role in the discursive worlds of modernity. In its late nineteenth-century heydays, every decent traveler had an opinion on Baedeker's guidebooks, even those who never used them, and so it went with handbooks in many other fields.

Handbooks drew on the Enlightenment faith in the combined powers of public knowledge and the written word, and they had a number of features that defined them on the book market. They were meant to be authoritative, they circulated widely, and they were usually subject to selective reading in light of the occasion: few people read handbooks from cover to cover. The books focused on secure knowledge and avoided subjective or speculative opinions. Their prose was typically dry, though editors granted themselves an occasional escape such as the entry on the stone louse, originally an invention of the German comedian Loriot, in the *Pschyrembel* medical dictionary.[39] Handbooks defined rules for those who worked in a field, and they shaped the outward appearance of these fields. Some handbooks even created their own reality: the *Great Soviet Encyclopedia* declared in 1975 that "with the construction of a socialist society in the USSR, hunger and massive malnutrition have been completely liquidated (see chapter 31, Holodomor)."[40] Writing a handbook was typically a collaborative effort, and handbook editors sought legitimacy from a broad and esteemed circle of contributors even when sponsors claimed higher powers. The *Catechism of the Catholic Church*, commissioned by Pope John Paul II in 1985 and published six years later, noted by way of introduction that the text "was the object of extensive consultation among all Catholic Bishops, their Episcopal Conferences or Synods, and theological and catechetical institutes."[41]

Not every handbook commanded the assembled wisdom of a world religion. The first conservator of forests for the Gold Coast (see chapter 4, Sustainable Forestry) was appointed in 1909, and the department had a grand total of four employees in 1914, but it published a *Forest Officers' Handbook of the Gold Coast, Ashanti and the Northern Territories* in 1922. The officers conceded "that the information given is very small, and later research will doubtless modify many of the conclusions here set out, but the need of some reference work on the Gold Coast forests has been increasingly felt of late."[42] The remark neatly captured the way that handbooks served as milestones for the development of a field of knowledge. They sought to map and secure in writing what has been achieved, they were digests for everyday routines as well as starting points for future explorations, and they proclaimed to the world that something has come to stay. The *Cambridge Companion to*

Travel Writing opens by declaring that "travel has recently emerged as a key theme for the humanities and social sciences."[43]

Handbooks are not beyond competition. Baedeker's guidebooks were plagiarized with various degrees of shamelessness, and commercial survival hinged on the cultivation of a loyal constituency.[44] Some companies maintained a force of inspectors, who could be everything between the Harvard students who wrote the *Let's Go* budget travel guides and the mythical connoisseurs behind the *Guide Michelin*. Other manuals counted on their own buyers. England's *Good Food Guide*, first published in 1951, came with report cards that readers could fill out and return to the editors.[45] And then there was the confidence that came with an established brand. Consumers bought Chiquita bananas (see chapter 10, United Fruit) for the same reason that German historians (including globe-trotting ones) consulted their *Ploetz*: it was the safe thing to do.[46] Some manuals even preceded their own subjects. Hans von Wolzogen published his guide to Richard Wagner's *Ring of the Nibelung* four weeks before the world premiere. It became the vade mecum for Bayreuth's first festival in 1876, "a musical 'Baedeker' that no decent tourist here dares leave home without," and it sold well beyond the day. By 1910, 160,000 copies were in circulation.[47]

Handbooks were a cherished source of orientation, and they allowed people to stay calm even in exceptional situations. "I was not nervous," Mahatma Gandhi (see chapter 23, Gandhi's Salt) wrote in his autobiography about how he delivered his fourth child with the assistance of a medical book titled *Advice to a Mother*.[48] Handbooks even served as pillars of national identities. It is no coincidence that the all-encompassing encyclopedia, arguably the ultimate handbook, became subject to vigorous competition among the leading nations of Europe. Educated Germans were obliged to buy a *Brockhaus* encyclopedia, the English had their *Encyclopædia Britannica*, and Frenchmen pointed out that they were all sequels to Diderot's famous *Encyclopédie*. The Soviet Union later topped them all with the sixty-five-volume *Great Soviet Encyclopedia*. The leather-bound volumes looked good in strategically placed bookshelves, and they were the default source of information before the invention of the internet. When Thomas Mann sought to learn about cholera (see chapter 21, Cholera) for his *Death in Venice*, he penned a copy of the *Brockhaus* entry.[49]

Some handbooks served commercial purposes, such as the *Whole*

Earth Catalog that began defining the American counterculture in 1932.[50] Others occupied a twilight zone between education and commerce. Rupert Wheldon's book *No Animal Food* of 1910 argued for what we now call a vegan diet (see chapter 23.3, Alternative Projects), offered twenty pages of recipes, and then concluded with advertisements "from firms for whose integrity the author can vouch."[51] To highlight his independence, Karl Baedeker did not accept advertisements, but when his heirs had to rebuild the company after World War II, they accepted financial support from West German state governments and the industrialists of the Ruhr.[52] It was a commercial necessity as well as a balancing act, and Baedeker knew what powerful sponsors could wish for. In 1943, Baedeker published a guide on Nazi-occupied Poland under the aegis of Hans Frank, the head of the General Government, which justified the German occupation on racial and economic grounds.[53]

Baedeker did rebound after the war, but its tradition as a family business came to an end. Today's Baedeker is one of a dozen travel-related brands owned by MairDumont, which nicely reflects the ongoing diversification of readership.[54] Tourism had ceased to be a privilege of the bourgeoisie in the interwar years, and that had consequences for agendas and mindsets: no longer could guidebooks mention the Lorelei mountain on the Rhine, evoke "the well-known legend of the siren who had her dwelling on the rock," and trust that the educated reader would fill in the gaps.[55] The Dietz-Verlag in Berlin, which was closely affiliated with the Social Democratic Party of Germany, published the first guidebook for ordinary workers in 1932. Five years later, the Nazi's paramilitary Sturmabteilung published "a detailed guide to the sites of Nazi Party struggles against Communists and Jews in Berlin."[56] Different types of travelers called for different types of books, and today's handbook users can obtain written advice on virtually any issue that they may encounter en route. There is even a manual that devotes 134 pages to the many problems of defecating in the woods.[57] It is hard to conceive of a world without handbooks or their electronic equivalent in the twenty-first-century world. The same might be said for the world of travel.

3. TRAVELING MASSES

Baedeker guides were intended for peaceful travel, but they were open to other uses. T. E. Lawrence allegedly carried a copy of the *Palestine and Syria* edition with him during the Arab Revolt against the Ottoman

Empire. Nazi censors took a critical look at the celebrated Baedeker maps, and when Germany's air force bombed scenic English cities like Exeter, York, and Bath during World War II, the attacks were dubbed "Baedeker Raids."[58] But any damage from wartime abuses paled in comparison with the real-world effects of tourism since 1945. Growing affluence and vacation time turned holiday travel into a mass movement, and as the number of vacationeers increased, so did the environmental repercussions.

Some of the consequences were evident even to innocent observers who noted the changes in landscapes that mushrooming hotels and other tourist infrastructures brought with them. Others were hidden in resource flows, such as the water demand of shower-crazy tourists or the energy requirements of airplanes and air-conditioning units (see chapter 20, Air-Conditioning). Today many guidebooks make a point of touching on the environmental repercussions of tourism, but the consequences are so diverse that they are hard to acknowledge in full. A book on biological invasions in Australia (see chapter 14, Cane Toads) even worried about the dirt on the travelers' shoes: "Ecotourism, when you factor in all the spores, eggs and burrs falling from shoes and tyres, is not as eco-friendly as most people think."[59]

Most travelers are dimly aware of their environmental footprint in the twenty-first century, but the gap between intentions and actions is particularly wide in the travel business. Countless hotel guests have honored bathroom comments about the reuse of towels and found that the service staff changed them nonetheless. Vacations are the great other of the modern existence, the escape from the constraints of ordinary life back home, and that has imposed limits on the greening of tourism: the pleasures of a swimming pool were hard to reconcile with nagging thoughts about where all that water came from. Tourist organizations and managers produced a plethora of environmental pledges and proclamations—the output was "particularly prolific in the 1990s"—but the bottom line is open to debate. *The SAGE Handbook of Tourism Studies* found that "there is little evidence to point to a valuing of nature by either the industry or the consumer that moves beyond the extrinsic."[60]

Did tourism bring benefits that offset the environmental damage? John F. Kennedy was optimistic: "Travel has become one of the great forces for peace and understanding in our time," the US president declared shortly before his assassination during a trip to Dallas.[61] The sentiment still resonates in the travel literature, but today Kennedy's optimism is tempered by decades of experiences about the many things that can go wrong in cross-cultural communication. "Western feminist attitudes are simply irrelevant here," the Yemen *Lonely Planet* declared. "A woman's place in Yemeni society is, by and large, at home with the family and there is nothing you can do about this."[62] The Austrian town Zell am See, a popular summer destination for Arabs, published a pamphlet in 2014 that drew international ire due to insensitive remarks about purportedly Western achievements like garbage disposal (see chapter 40, Plastic Bags) and women not covering their faces.[63] Guidebooks also stress the risks of political conversations in authoritarian countries like Myanmar: "Talking politics can get not only you but also the locals you're speaking with into trouble."[64]

But for all its ambiguities, the idea of travel as a mind-broadening experience seems impossible to erase. It even colors assessments of postwar leaders: a 2002 study noted about Portugal's dictator, António Salazar, that his "narrowness of view was probably the result of his never traveling outside of Portugal further than Spain."[65] As it happened, restraint in long-distance travel did not benefit Salazar's health either. An incident during a domestic holiday—a toppling chair in Estoril—triggered the stroke that ended his rule.[66] Quality guidebooks today seek to enhance transcultural understanding, and unlike the Baedekers of bygone years, they no longer shun political issues. The 2007 *Lonely Planet* for Afghanistan pointed out after the American victory over the Taliban in 2001, "The country received less than a third of the aid per head ploughed into reconstruction efforts in Bosnia, East Timor or Rwanda."[67] But for all the good intentions, there are limits to what a guidebook can achieve in a few hundred pages. The same holds true for a few weeks abroad.

In 1889 the Baedeker for Greece declared emphatically that "even the shortest sojourn in the country itself will yield the richest rewards and contribute more than long years of study towards a thorough comprehension of [ancient Greek] civilisation."[68] And 130 years later, travelers are no longer sure whether their experience is more than a staged enactment. Cultural construction started long before the age of mass travel: scholars have analyzed the coevolution of tourism and local cultures in purportedly authentic places like Brittany and Scotland.[69] Hal Rothman has argued that tourism was "the most colonial of colonial economies" because it does more than physical damage. It invades the soul of peoples and places.[70]

As in many colonial settings, distant capital played a crucial role.

Travel Writing opens by declaring that "travel has recently emerged as a key theme for the humanities and social sciences."[43]

Handbooks are not beyond competition. Baedeker's guidebooks were plagiarized with various degrees of shamelessness, and commercial survival hinged on the cultivation of a loyal constituency.[44] Some companies maintained a force of inspectors, who could be everything between the Harvard students who wrote the *Let's Go* budget travel guides and the mythical connoisseurs behind the *Guide Michelin*. Other manuals counted on their own buyers. England's *Good Food Guide*, first published in 1951, came with report cards that readers could fill out and return to the editors.[45] And then there was the confidence that came with an established brand. Consumers bought Chiquita bananas (see chapter 10, United Fruit) for the same reason that German historians (including globe-trotting ones) consulted their *Ploetz*: it was the safe thing to do.[46] Some manuals even preceded their own subjects. Hans von Wolzogen published his guide to Richard Wagner's *Ring of the Nibelung* four weeks before the world premiere. It became the vade mecum for Bayreuth's first festival in 1876, "a musical 'Baedeker' that no decent tourist here dares leave home without," and it sold well beyond the day. By 1910, 160,000 copies were in circulation.[47]

Handbooks were a cherished source of orientation, and they allowed people to stay calm even in exceptional situations. "I was not nervous," Mahatma Gandhi (see chapter 23, Gandhi's Salt) wrote in his autobiography about how he delivered his fourth child with the assistance of a medical book titled *Advice to a Mother*.[48] Handbooks even served as pillars of national identities. It is no coincidence that the all-encompassing encyclopedia, arguably the ultimate handbook, became subject to vigorous competition among the leading nations of Europe. Educated Germans were obliged to buy a *Brockhaus* encyclopedia, the English had their *Encyclopædia Britannica*, and Frenchmen pointed out that they were all sequels to Diderot's famous *Encyclopédie*. The Soviet Union later topped them all with the sixty-five-volume *Great Soviet Encyclopedia*. The leather-bound volumes looked good in strategically placed bookshelves, and they were the default source of information before the invention of the internet. When Thomas Mann sought to learn about cholera (see chapter 21, Cholera) for his *Death in Venice*, he penned a copy of the *Brockhaus* entry.[49]

Some handbooks served commercial purposes, such as the *Whole*

Earth Catalog that began defining the American counterculture in 1968.[50] Others occupied a twilight zone between education and commerce. Rupert Wheldon's book *No Animal Food* of 1910 argued for what we now call a vegan diet (see chapter 23.3, Alternative Projects), offered twenty pages of recipes, and then concluded with advertisements "from firms for whose integrity the author can vouch."[51] To highlight his independence, Karl Baedeker did not accept advertisements, but when his heirs had to rebuild the company after World War II, they accepted financial support from West German state governments and the industrialists of the Ruhr.[52] It was a commercial necessity as well as a balancing act, and Baedeker knew what powerful sponsors could wish for. In 1943, Baedeker published a guide on Nazi-occupied Poland under the aegis of Hans Frank, the head of the General Government, which justified the German occupation on racial and economic grounds.[53]

Baedeker did rebound after the war, but its tradition as a family business came to an end. Today's Baedeker is one of a dozen travel-related brands owned by MairDumont, which nicely reflects the ongoing diversification of readership.[54] Tourism had ceased to be a privilege of the bourgeoisie in the interwar years, and that had consequences for agendas and mindsets: no longer could guidebooks mention the Lorelei mountain on the Rhine, evoke "the well-known legend of the siren who had her dwelling on the rock," and trust that the educated reader would fill in the gaps.[55] The Dietz-Verlag in Berlin, which was closely affiliated with the Social Democratic Party of Germany, published the first guidebook for ordinary workers in 1932. Five years later, the Nazi's paramilitary Sturmabteilung published "a detailed guide to the sites of Nazi Party struggles against Communists and Jews in Berlin."[56] Different types of travelers called for different types of books, and today's handbook users can obtain written advice on virtually any issue that they may encounter en route. There is even a manual that devotes 134 pages to the many problems of defecating in the woods.[57] It is hard to conceive of a world without handbooks or their electronic equivalent in the twenty-first-century world. The same might be said for the world of travel.

3. TRAVELING MASSES

Baedeker guides were intended for peaceful travel, but they were open to other uses. T. E. Lawrence allegedly carried a copy of the *Palestine and Syria* edition with him during the Arab Revolt against the Ottoman

On Jamaica, "tourism was sired by American banana traders" like United Fruit (see chapter 10, United Fruit).[71] Franco's Spain banked on beachgoers after 1945, ignored the dictator's personal reservations, and covered two-thirds of its trade deficit with tourism revenue in the 1960s.[72] Even North Korea sold itself as a travel destination, and it worked: Lonely Planet's Korea guide includes a rave description by Tony Wheeler, the founder of Lonely Planet, of thousands of schoolchildren acting in sync at the Pyongyang Mass Games.[73] In the twenty-first-century world, tourism is, in Rothman's words, "a panacea for the economic ills of places that have lost their way in the postindustrial world or for those that never found it."[74]

The imagination of tourism knows no limits, and so does its geographical spread. Intrepid travelers have risked their lives on freezing peaks and in sweltering deserts and sometimes lost them in due course, and the hunger for ever new destinations has produced ambiguous results in print as well. The magazine *Vanity Fair* even paid for an exploration of the tourist potential of Lebanon in 1984, when American visitors to Beirut had a good chance of receiving an undesired extension of their holiday chained to a radiator. The reporter returned with his limbs intact and produced a parody of tourism rhetoric. He recommended an Italian restaurant with "a spectacular view of military patrols and nighttime skirmishing along the beachfront" and advised "to tip the man who insists, at gunpoint, on guarding your car." The editors at *Vanity Fair* found the piece "much too weird to publish," but the author, P. J. O'Rourke, published the essay in *Holidays in Hell*.[75] The book became a classic of xenophobic travel writing.

It was meant to be funny, but it looked rather prophetic in light of Lonely Planet's Yemen guide a dozen years later. Published a mere eighteen months after the end of the civil war of 1994, the author, a native of Finland, rhapsodized about "the admirable optimistic Yemeni nature" in the preface: "We found a relaxed people busy trying to put their former divisions behind [them]."[76] Tourists were advised not to worry about lingering conflicts, as Yemenis were targeting each other rather than foreigners, and readers were left to reconcile these soothing remarks with his observations from a country with fifteen million people and sixty million firearms: "All kinds of armament from hand grenades to anti-aircraft Stingers can be freely purchased at special arms suqs around the country, so it is no wonder that whenever tribal disputes develop into full-blown conflicts the casualties are high."[77]

If armed conflict has failed to discourage travel, it remains doubtful whether anything can stop the flow. Tourists have their preferences—beaches, mountains, metropolises—but in the end, everything can turn into a destination. Even the site of a devastating earthquake can serve as an attraction today, as the debate over the monumentalization of the 1976 Tangshan Earthquake (see chapter 25, 1976 Tangshan Earthquake) serves to attest. Before the Russian invasion of Ukraine, Europeans could go on a trip to Chernobyl (see chapter 37, *Lucky Dragon No. 5*) from Brussels or Berlin for less than €500.[78] When it comes to travel experiences, the sky is the limit, though that metaphor has seen better days as well, courtesy of Richard Branson's Virgin Galactic and Elon Musk's SpaceX.

The critique of travel is probaby as old as travel itself. Goethe mocked the British as ardent seekers of "battlefields," "waterfalls," and "fallen walls" in his *Faust*.[79] The same can be said about guidebooks, which did not even please those who benefited from them. Richard Wagner was not amused about Wolzogen's signposting effort for his magnum opus.[80] But would we do better if we abandoned them? In Forster's *A Room with a View*, Lucy Honeychurch tries to find the Basilica di Santa Croce without a Baedeker and promptly gets lost, which turns out to be the start of a narrative journey that ends with a getaway to Italy.[81] One might object that this is a story rather than reality, but this line is arguably blurry in the tourist business. Is there really more to the experience of travel than the stories? That is, apart from a trillion-dollar industry that underpins these stories and produces 8 percent of global greenhouse gas emissions?[82]

The environmental history of tourism is the history of the unreal in more than one sense, and it is entirely appropriate that the new millennium has produced guidebooks on countries that do not exist. Jetlag Travel, an imprint of the Australian Hardie Grant media company, published three guidebooks on Molvanîa, Phaic Tăn, and San Sombrèro between 2003 and 2006. They offered stereotypical descriptions of Eastern Europe, Southeast Asia, and Latin America, and the narrative showed the extent to which environmentalism had become part of the tourist imagination. A glass of tap water in Molvanîa "contains 80% of your annual requirements of trace metals and e-coli," Phaic Tăn banned coral dynamiting by developers but allowed it "as a recreational pursuit," and San Sombrèro saw an environmental protest against a gas pipeline through a national park that was "successful in having the pipe painted green."[83] The publisher's intention was probably no grander than to make a killing with cheap books—Jetlag Travel sold them in the "pop culture" category—and yet it is tempting to read them as a fitting commentary on a wave of tourists that shows no sign of relenting in the twenty-first century. Faced with a reality of tourism that challenges common understandings of reality, and guidebooks that discourage stereotyping and invariably engage in it nonetheless, it might be the last resort to go over the top.

23

GANDHI'S SALT

TO CHANGE THE WORLD

I. CELEBRITY STATUS

On March 12, 1930, a group of men marched out of Ahmedabad, a major town in Gujarat in colonial India. They were heading for Dandi, a village some 150 miles away on India's west coast. When they arrived at the seashore on April 5, their march had turned into a mass movement. Thousands joined them for a few miles or all the way to Dandi, and once on the beach, people followed the group's example, tapped into the Indian Ocean, and drew salt from the water. The mood was celebratory, but in technical terms, the crowd was in violation of the salt monopoly of the British Raj, and tens of thousands were subsequently arrested by the police. But that had been the plan.[1]

The original group comprised seventy-nine volunteers, but only one of them actually mattered: Mohandas Karamchand Gandhi, known to his followers as *Mahatma* (Great Soul) or *Bapu* (Father). The salt march had been his idea from the beginning. He selected the cause, the date, and the route, and it was Gandhi who chose the members of the initial group from his own Sabarmati Ashram in Ahmed-

ably no grander than to make a killing with cheap books—Jetlag Travel sold them in the "pop culture" category—and yet it is tempting to read them as a fitting commentary on a wave of tourists that shows no sign of relenting in the twenty-first century. Faced with a reality of tourism that challenges common understandings of reality, and guidebooks that discourage stereotyping and invariably engage in it nonetheless, it might be the last resort to go over the top.

23

GANDHI'S SALT

TO CHANGE THE WORLD

I. CELEBRITY STATUS

On March 12, 1930, a group of men marched out of Ahmedabad, a major town in Gujarat in colonial India. They were heading for Dandi, a village some 150 miles away on India's west coast. When they arrived at the seashore on April 5, their march had turned into a mass movement. Thousands joined them for a few miles or all the way to Dandi, and once on the beach, people followed the group's example, tapped into the Indian Ocean, and drew salt from the water. The mood was celebratory, but in technical terms, the crowd was in violation of the salt monopoly of the British Raj, and tens of thousands were subsequently arrested by the police. But that had been the plan.[1]

The original group comprised seventy-nine volunteers, but only one of them actually mattered: Mohandas Karamchand Gandhi, known to his followers as *Mahatma* (Great Soul) or *Bapu* (Father). The salt march had been his idea from the beginning. He selected the cause, the date, and the route, and it was Gandhi who chose the members of the initial group from his own Sabarmati Ashram in Ahmedabad. He set the pace, allegedly outmarching some of the younger men despite his age of sixty-one, and he was among those who spent time in prison. The salt march became "one of the most dramatic and successful episodes in the history of the Indian freedom struggle," but its significance stood in inverse relation to the underlying mandate.[2] India's Congress Party had voted for independence and a campaign of civil disobedience at a meeting in Lahore a few weeks earlier, but it had left all the details to the Mahatma: there was never a vote on the salt march. It happened because Gandhi wanted it to happen.

Gandhi had held the rotating presidency of the Congress Party for a year in the 1920s, but formal positions said little about his standing in politics. Gandhi had launched a campaign against racial discrimination during his time in South Africa, and he joined the ongoing campaign against British rule upon return to his native India in 1915. He was arrested during the campaign of civil disobedience after World War I, but that made him one of many. In other words, Gandhi was barely alone in his cause or experiences, and yet he stood out as a person. He had a charisma that none of his political friends could match, and this put him in a league of his own among India's independence fighters. It is a familiar phenomenon to everyone who follows the news in the twenty-first century. In the terms of the new millennium, Gandhi was a celebrity.

Charismatic leaders have shaped social movements throughout the modern era, and the environmental movement was no exception. Joachim Radkau wrote an entire book on what he called "the age of ecology," which thrives on a broad tableau of charming environmental activists. More than one cause became tied to one pivotal figure: John Muir was Yosemite, Rachel Carson was DDT (see chapter 38, DDT), Jacques Cousteau was marine conservation, and Petra Kelly was the German Greens.[3] It was mostly a figment of the imagination, as all these causes were really much bigger, and yet modern societies revealed a strange inability to abandon their infatuation with charismatic individuals. If anything, the cult of celebrities increased toward the end of the twentieth century, and environmentalism was a major focal point for pertinent activities. The French film star Brigitte Bardot embraced animal protection as her pet project for the sunset phase of her career. Al Gore won the Nobel Peace Prize for his fight against climate change after losing the presidential election of 2000 because another environmentalist with celebrity status, the Green Party candidate Ralph Nader, deprived him of crucial votes. A few years on, electorates in different parts of the world fell for the cricket star Imran Khan and the reality television host Donald Trump.

The power of celebrities is a distinctly modern type of power, and yet one that scholars have been reluctant to discuss. Joachim Radkau does not even use the word "celebrity" in his *Age of Ecology*. It is arguably an intellectually awkward category, as celebrities challenge traditional understandings of political power. Their significance did not grow from family ties, elections, or monetary wealth, they did not make dramatic discoveries or invent new technologies, and if they had special skills, they often shared them with many others who lacked celebrity status. They did not even have to be consistently successful. In 1931, the British government invited Gandhi to the second stage of the Round Table Conference in London, an event that was a direct result of the electrifying salt march, and it ended in complete failure. In Maria Misra's words, Gandhi "returned from London empty handed and to a Raj well-prepared to crack down hard when Civil Disobedience was restarted in April 1932."[4] Gandhi even left the Congress Party in 1934. But he did not leave the scene.

Not everyone fell for the lure of charismatic individuals. The archimperialist Winston Churchill famously disparaged Gandhi as a "seditious fakir, striding half-naked up the steps of the Viceroy's palace, there to negotiate and to parley on equal terms with the representative of the King-Emperor."[5] But the mass media loved celebrities and seized on their various dealings with unabashed delight. It was a reciprocal relationship: constant media coverage built larger-than-life personalities, and celebrities delivered a constant stream of news or things that journalists could treat as such. Gandhi certainly understood the significance of good media relations and secured time for journalists even in the hectic days before the start of the salt march.[6] It paid off beyond his lifetime. After meeting the Mahatma in 1942, the American journalist Louis Fischer wrote an adoring biography that served as the basis for Richard Attenborough's Academy Award–winning movie *Gandhi*.[7] The Mahatma was a media myth, and never was coverage denser than during the campaign of 1930. "Years after his passing, the Mahatma's march to make salt at Dandi would be hailed as one of the founding events of global media history."[8]

From a twenty-first-century viewpoint, the salt march was custommade for the age of mass media. It was announced in time for journalists to make travel arrangements, it lasted for several weeks, it provided suspense and visual drama, and it drew on the time-honored plotline of David versus Goliath. It was more complicated in 1930. The British Raj controlled the Indian press, it sought to manipulate coverage abroad, and independent observers were scarce in the country: the salt march was the first event that prompted American newspapers to send their own reporters to India.[9] American journalists did not share the imperialist allegiances of their British colleagues, and they sensed that this was a big story. At the end of 1930, *Time Magazine* selected Gandhi as its "man of the year."[10]

As love affairs go, the relationship between celebrities and the press was always a difficult one, and Gandhi was no exception. Journalists continued to misunderstand his crusade, and it is open to debate whether this was due to geographic and cultural distance, the novelty of his approach, or the colonial gaze. They portrayed him as emotional, as an agitator, a blend of "the mystical Indian" and "the cunning Oriental," and as a man who was either blind to or willfully ignorant of the violence and chaos that followed in his path.[11] However, many of those who met Gandhi in person realized that there was something about the man. He stayed in London's East End during the Round Table Conference and befriended the local community, and his trip to Birmingham resonates in local memory to this day, down to the place where he ate a vegetarian meal.[12] He also charmed the press, not least because he was never shy about producing a good quote. When journalists inquired why he did not change his sparse Indian attire when he went to Buckingham Palace for tea with King George V, Gandhi replied that the king "had enough on for both of us."[13]

It was an experience to meet Gandhi, but most people knew him from a distance. It helped build a fame that transcended boundaries: after his death in 1948, eulogies came from people as diverse as Pope Pius XII and General Douglas MacArthur.[14] Gandhi was a more ambiguous figure for those who fought with him for independence. As friends in politics go, there were easier cases than an ascetic who made unpredictable decisions, and Gandhi was a political asset as well as an irritant. When he learned about the salt march plan, Motilal Nehru, then a leader of the Congress Party and the father of India's first prime minister, Jawaharlal Nehru, "was amused, even angered, by the apparent irrelevance of Gandhi's Plan."[15] At the Lahore meeting of the Congress Party that preceded his salt march, Gandhi introduced a motion deploring a recent bomb attack on the viceroy, a motion that generated a lot of bad blood and passed only narrowly.[16] Celebrity status was always about selective memory, about the choice of some charismatic scenes at the expense of many others, and it was inevitable even for someone like Gandhi, whose life had a stringency and a blend of action and belief that few people could match. It is perfectly fitting that Gandhi's best-known incarnation in the late twentieth century, the epic 1982 movie *Gandhi*, was built around carefully selected snapshots.

In short, Gandhi's salt march was political theater performed on a world media stage. The British certainly knew it, and not only because Gandhi had sent a long letter to the viceroy that laid out plans and motivations. The Raj was never concerned about the amount of illicit salt, but it was reluctant to look away when colonial subjects were openly challenging its authority. The initial response was muted, but tempers rose when the salt march struck a nerve, and yet Gandhi could stay a full month in Dandi until he was arrested. The British knew that it looked bad to put him in prison, they knew that it looked bad to let him go free, and they knew that Gandhi knew that they would not like either option.[17] When it came to celebrities, it was always hard to say what was right, what was important, and what did not matter, and that did not diminish their significance. In a way, that was the source of their power.

It is easy to dismiss the power of celebrities as a flimsy emanation of the age of mass media, and there is certainly no lack of evidence. It seems superficial that the movie director James Cameron, while living in Los Angeles, took a stance against Brazil's Belo Monte hydroelectric project (see chapter 29, Aswan Dam).[18] However, celebrity status may be more significant and more dangerous beyond the world of Hollywood. The rubber tapper Chico Mendes built a union under military rule in Brazil before he became a global celebrity that symbolized the fight for the Amazon rain forest, and he paid for it with his life: he was killed in 1988.[19] Gandhi was also murdered, but only after multiple fasts that he had threatened to continue until his death. After all, Gandhi fought for something he considered larger than himself.

2. THE POWER OF IDEAS

Gandhi was a modern saint, but he was not born as one. In his autobiography, Gandhi told "the story of my numerous experiments with truth" and noted that "my life consists of nothing but those experiments," and the narrative showed how Gandhi transformed past failings into searing memories that fueled his quest. He tried meat before he returned to the vegetarian diet of his childhood.[20] He opted for cel-

23.1 Exactly one hundred years after Gandhi's fateful night at the Pietermaritzburg railway station, Archbishop Desmond Tutu unveiled this monument in front of a colonial court building in downtown Pietermaritzburg. Image, Vishal Bhatia, Wikimedia Commons.

ibacy in 1906, but only after fathering four children and delivering the last one himself with the help of an advice book for mothers (see chapter 22.2, The Age of Handbooks).[21] His fight against racial discrimination began in 1893 when he was thrown out of a first-class railroad coach in South Africa and spent a winter night shivering in the Pietermaritzburg railway station.[22] His twists and turns in India's independence struggle were legion. Gandhi was on the move.

Gandhi did not invent civil disobedience. Embracing a sponge-like approach to ideas that knew no boundaries, he drew on Indian traditions and merged them with the ideas of Henry David Thoreau and Leo Tolstoy, and Christian ideas of suffering for a greater good.[23] But his charismatic example and ultimate success left a lasting impression on protest movements elsewhere on the globe. In his fight against racial discrimination in the United States, Martin Luther King drew explicitly on the Mahatma. "Again and again we must rise to the majestic heights of meeting physical force with soul force," King declared in his "I Have a Dream" speech in 1963.[24] In charting Gandhi's global legacy, David Hardiman cites the US civil rights movement along with the struggle against apartheid in South Africa and Petra Kelly's crusade for peace and ecology.[25]

Gandhi developed his own *satyagraha* brand of civil disobedience in South Africa, but it kept evolving in his mind and in political practice throughout his life. After all, *satyagraha* was more than a political tactic. It was a state of mind, an expression of willpower and discipline, and Gandhi proved perfectly able to cancel campaigns when he felt that his followers were lacking the right attitude. He called off civil disobedience in 1922 after twenty-two Indian police officers had burned to death in a police station in Chauri Chaura, and the campaign of 1930 was "carefully orchestrated so as to avoid any lapse into violence."[26] In fact, his choice of the salt tax was probably driven by a desire to keep emotions in check. As issues for freedom fighters go, it was a rather innocent one, at least for those who had never heard of *la gabelle*, the French salt tax that is widely regarded as a major grievance in prerevolutionary France.[27] As it turned out, most protesters stuck to the script even when the Raj resorted to violence. The Gandhi biographer Bal Ram Nanda remarked that "this campaign came closest to Gandhi's conception of a model satyagraha."[28]

But for all the significance of civil disobedience, Gandhi had a much broader range of concerns, and he raised them as he felt appropriate. The year before the salt march, Gandhi "returned to his long-running battle against public filth, public defecation and the total absence in India of a clean and ecologically productive system of sanitation" (see chapter 17, Water Closet).[29] However, Western environmentalists were more excited about the critique of modern technology that he voiced in his book *Hind Swaraj* of 1909. It is a reading that historians have been reluctant to share. "The question whether Gandhi was indeed an 'early environmentalist' is usually answered in the affirmative by his admirers, but rarely with supporting evidence," Ramachandra Guha and Juan Martinez-Alier wrote.[30] But intellectual honesty is usually a secondary concern for those who want to change the world, and Gandhi references served to support various endeavors. In his eco-best seller *Small Is Beautiful*, Ernst Friedrich Schumacher drew on Gandhi's remark that "Earth provides enough to satisfy every man's need, but not for every man's greed."[31] On the centenary of *Hind Swaraj*, an Indian Manifesto on Science and Technology argued for "knowledge democracy" and aimed "to liberate the West from a developmental mindset (see chapter 32.3, Planning Development) that alienates people and is deeply unsustainable."[32] And then there was iconic use. In a book on the environmentalism of the poor, Rob Nixon noted "the creative circuits of globalization from below," and argued that activists like Gandhi, Chico Mendes, and Ken Saro-Wiwa (see chapter 39.3, Creating Invisibility) "have assumed an allegorical potency for geographically distant struggles."[33]

Modern environmentalism hinges on ideas, perhaps more so than other causes, and the burgeoning movements of the last quarter of the twentieth century were bustling marketplaces of ideas, down to deep philosophical reflections on man's place in the natural world. An introduction to the environmental humanities even argued that ecocentric values were mandatory for every scholar in the field.[34] However, grand intellectual designs ran into trouble in the real world. In her book *Imagining Extinction*, Ursula Heise records how Bolivia's Law of the Rights of Mother Earth triggered a conflict with Indigenous groups over a road project that crossed a national park (see chapter 26, Kruger National Park).[35] Complications also worked in the opposite direction. When Bolivia's Indigenous president Eva Morales gave the inaugural speech at the World People's Conference on Climate Change and the Rights of Mother Nature in 2010 and proclaimed an alternative of "either Pachamama or death," the feminist Miriam Tola noted that "the normative gendering of Pachamama in the Bolivian context raises questions about the role of gender and sexuality in the power-laden relations that inform political ontology." She also found rights for Mother Nature "problematic when enacted within a project of state consolidation that heavily relies on the expansion of extractive industries" (see chapter 15, Saudi Arabia).[36]

Good ideas did not guarantee good results, and the risk of unexpected outcomes was particularly great when people embraced ideas without reflections about contexts and requirements. Gandhi's *satyagraha* would have failed in the absence of an enemy willing to enter into a competition for the moral high ground. Even a postcolonial critic of British rule like Shashi Tharoor has acknowledged as much: "It is ironically to the credit of the British Raj that it faced an opponent like Mahatma Gandhi and allowed him to succeed."[37] Nelson Mandela knew about *satyagraha* and discussed it with Gandhi's son Manilal, a member of the South African Indian Congress, but in the end, civil disobedience was "a practical necessity" for him. As he pointed out in *Long Walk to Freedom*, "The state was far more powerful than we, and any attempts at violence by us would be devastatingly crushed."[38] In authoritarian societies, civic groups are hopelessly outgunned, and contra the First Book of Samuel, David does not inevitably win against Goliath. Fighting against an overbearing enemy can be energizing, but the long-term effects are open to debate. Shortly before the 2000 presidential election, Ralph Nader justified his Green Party candidacy by arguing that "a bumbling Texas governor would galvanize the environmental community as never before."[39] As it happened, events took a different course.

Later in his life, Gandhi came to accept modern technology "in practical terms" and focused his ire on the fetishization of technology.[40] Nonetheless, subsequent generations read *Hind Swaraj* as a manifesto, and Hardiman has suggested that this was not so much about intellectual sloppiness as about the age of catastrophe: "If anything, the appeal of the tract increased over time, as the barbarities of world wars and fascism revealed a rottenness at the heart of Western civilisation."[41] It was the classic problem in the reception of Gandhi: interpretations diverged widely depending on whether he was read in the context of India's path to independence or as the transnational prophet of a different world. Gandhi's life had an otherworldly quality, but scholars have stressed that he was also a shrewd political player. "With all his singlemindedness and his rigid ethics of conviction Gandhi was not impervious to the world around him," Dieter Rothermund has argued.[42] He certainly knew about the ambiguities of human life. Even the acclaimed salt march had its ambiguities. Salt was an issue that transcended religious boundaries, thus bridging the divide between Hindus and Muslims in India, but *satyagraha* had deep roots in Hinduism and Jainism, and Misra has noted that the campaign for independence became more Hinduistic as a result. She also pointed out that salt was a minor issue for India's peasants, who were hit hard by the fall of commodity prices in the Great Depression.[43] The British Raj increased the economic plight when it refused to adjust tax rates and even increased the salt tax, which suggests, in Rothermund's judgment, "that the government was not much impressed with the Salt Satyagraha."[44] The salt campaign had charisma, and so did its leader, but it did not create a new world.

3. ALTERNATIVE PROJECTS

Economists are big-picture people. This book has cited a number of men who received the Nobel Prize in Economic Sciences, and they all worked on large topics. Robert Fogel studied railroads and American economic growth (see chapter 3.2, Connected), Amartya Sen developed the entitlement approach for the investigation of famines (see chapter 31.2, Hunger, Modern Style), and the neoliberalist Friedrich Hayek criticized the expansion of the welfare state in his *The Road to Serfdom* and inspired the chicken farmer Antony Fisher (see chapter 36, Battery Chicken). The Nobel laureate of 2008, Paul Krugman, even developed a theory on the economics of interstellar trade.[45] Elinor Ostrom stands out from the group, and not only because she was the first woman to receive the award, and the only one among more than eighty men until Esther Duflo became a Nobel laureate in 2019.[46] Her book *Governing the Commons* focused on groups that typically had between a few dozen and a few hundred members; the largest group comprised some fifteen thousand people.[47] They came from different parts of the world, and they had one thing in common. They were dealing with common-pool resources.

Management of the commons had heretofore provoked a large literature, and recommendations typically fell into one of two camps. Authors either cited Garrett Hardin's *Tragedy of the Commons* and argued that only comprehensive state management could prevent abuse, or they called for privatization.[48] Ostrom explored a different path. She took a close look at the institutional arrangements that allowed small groups to manage common-pool resources sustainably. Based on a set of empirical case studies, Ostrom traced the improvised deals that allowed these groups to succeed, and she showed how these solutions were anything but static. She took it as a cue to bemoan the economists' infatuation with models: "Policy analysts who would recommend a single prescription for commons problems have paid little attention to how diverse institutional arrangements operate in practice."⁴⁹ She sought to identify cross-cultural institutional features that underpinned long-term success and presented her findings "as a framework rather than as a model."⁵⁰ But in the male world of economics, you are really nothing without a proper theorem in your name, and so it came about that, on the twentieth anniversary of the publication of *Governing the Commons*, Lee Fennell proclaimed what she called "Ostrom's Law": "A resource arrangement that works in practice can work in theory."⁵¹

Gandhi was certainly averse to a primacy of ideas, and he was familiar with the benefits of small groups. His ideal India was a country of villages, and he used his spinning wheel as a powerful symbol of autonomous production as an alternative to India's dependence on Britain's textile manufacturers (see chapter 30.2, In Their Theories). Gandhi was also the founder of various autonomous communities, beginning with the Phoenix Settlement and Tolstoy Farm in South Africa.⁵² The quest for a different, "truer" life was in play in many movements with environmental concerns, and some initiatives even chose their names accordingly: the Monte Verità in Switzerland was a major hub of the life reform movement in the early 1900s.⁵³ Others launched alternative ventures with a commercial rationale from food stores to organic farms.⁵⁴ It could be the start of a global brand. Born in 1852, John Harvey Kellogg taught "biologic living" at the Battle Creek Sanitarium in Michigan, originally founded by Seventh Day Adventists, before he turned into a household name.⁵⁵ Fair trade bananas became such an attractive business case that Chiquita Brands International, successor to the infamous United Fruit Company (see chapter 10, United Fruit), considered buying Europe's leading importer in 2014.⁵⁶

Characteristically, food figured prominently in alternative projects. Few things influence people more intimately than their dietary habits, and Gandhi discussed it at length in his autobiography.⁵⁷ It also brought him one of his first social movement experiences when he organized a Vegetarian Club during his time in London.⁵⁸ Vegetarians had explored a broad range of new diets since the mid-nineteenth century, and life without meat was a part of the experience of modernity: the 1893 Columbian Exposition in Chicago featured not only trips to the stockyards (see chapter 18, Chicago's Slaughterhouses) but also a vegetarian showroom.⁵⁹ India figured prominently on the mental map of vegetarians long before the life of Gandhi. Gustav Struve, a leader of the German revolution of 1848 and later exile to the United States, argued for radical vegetarianism in a book of 1843 where he used a young Indian named Mandaras as his alter ego.[60]

Alternative projects were not invariably successful. Gandhi's spinning wheel is enshrined in India's flag to this day, but the country pursued a path that differed widely from his village utopia, and Misra notes that his experimental communities "proved a dismal failure, sowing discord not harmony and trying the temper of even the saintly Mahatma." Gandhi's communities relied on donations from business, "provoking the famous quip that it was costing a fortune to keep the Mahatma . . . in poverty."[61] But do we perhaps need to look beyond finances if we want to assess the achievements of alternative projects? Gandhi's ashram in Ahmedabad—not a village, by the way—was an excellent recruiting ground for the salt march, and other projects had benefits beyond the sphere of commerce as well. A study of cooperatives in Eastern European agriculture argued that they served to create a new rural elite beyond the old aristocracy, helped to build democracy, and effectively met a broad range of challenges.[62] It did not always work, and they did not succeed consistently, but maybe we should read the quest for consistency as an obsession of those who want to escape the sting of alternative projects. Gandhi certainly found that consistency was not what really mattered for those who sought to change the world. When asked whether he was contradicting himself, Gandhi replied, following Ralph Waldo Emerson: "Consistency is a hobgoblin."[63]

In introducing his reflections on interstellar trade, Krugman said that the article "is a serious analysis of a ridiculous subject, which is of course the opposite of what is usual in economics."[64] Interstellar commerce is arguably unlikely to become a reality anytime soon, but maybe we should be more careful with the word "ridiculous." Gandhi certainly qualified as ridiculous in more than one respect, not to speak of another celebrity who became president of the United States, but the word stops reflection at a point where it should actually begin. Alternatives were driven by people who envisioned a different world of various scales, who took the initiative, and who did not care too much about how it looked to innocent bystanders. "There are no supreme saviors," the lyrics of the Communist *Internationale* declared, "Neither God, nor Caesar nor Tribune." True, utopias have come a long way since the glory days of the sixty-five-volume *Great Soviet Encyclopedia* (see chapter 22.2, The Age of Handbooks), but should that discourage us from dreaming of a better world when even a great soul like Gandhi had ambiguities? "Let us save ourselves," the *Internationale* continued. Sing along![65]

24

THE 1970 TOKYO RESOLUTION

INTERNATIONAL CONVENTIONS

I. NICE TO MEET

Japan was an exciting place in 1970. The country had experienced a spectacular economic boom with annual growth rates around 10 percent over the two previous decades, Osaka hosted a world exposition, and Tokyo was the biggest city in the world.[1] The *Official Guide* of the Japan Travel Bureau (see chapter 22, Baedeker) described Tokyo as "a unique amalgamation of the tranquil past and the boisterous present" and noted that the city had "become steadily bigger, busier, and more crowded" since World War II.[2] But the academics who registered at the Tokyo Prince Hotel on March 8, 1970, had come for serious business. They did visit the Tokyo Tower across the road from their hotel, one of the city's landmarks and a symbol of Japan's postwar renaissance, but they went there for the Pollution Control Center. A field trip also brought them to a refuse reclamation center in the harbor, a public housing project, and the Ochiai Sewerage Treatment Station. The "International Symposium on Environmental Disruption in the Modern World" united forty-five people from Japan and the rest of the world, and after listening to twenty-three papers over the course of four days, the participants condensed their acquired wisdom in a document: the Tokyo Resolution.[3]

Intellectual exchange is probably as old as human civilization, but the International Symposium on Environmental Disruption was not just about informal conversation. It was not about public outreach either, unlike Earth Day in the United States a month later, which became a global awareness-raising event in the 1990s.[4] The International Symposium on Environmental Disruption included several receptions and meetings with Japanese officials, but it did not have a single public event. The symposium was aimed at a much smaller audience: a network of academics, policymakers, and functionaries that spanned the world.

The symposium was part of a string of events that showed this network in action. The Swedish ambassador to the United Nations had proposed to the United Nations Economic and Social Council in July 1968 that an international conference on the environment be held, and the General Assembly of the United Nations endorsed the idea in December.[5] As it happened, the executive committee of the International Social Science Council met in Paris a week later, and it seized the opportunity to brush up its professional expertise in the emerging field. The International Social Science Council created a Standing Committee on Environmental Disruption and appointed an American-trained Japanese economist, Shigeto Tsuru, as chairman pro tempore. Tsuru secured funding for an international meeting by July 1969 and then teamed up with Allen Kneese of a Washington-based nonprofit organization, Resources for the Future, to convene the symposium.[6] The results fed back into ongoing negotiations at the United Nations. The Tokyo Resolution proposed "the adoption in law of the principle that every person is entitled by right to the environment free of elements which infringe human health and well-being and the nature's endowment, including its beauty, which shall be the heritage of the present to the future generations."[7] Two years later, a United Nations summit on the environment in Stockholm adopted a declaration that proclaimed, in principle 1, that "man has the fundamental right to freedom, equality and adequate conditions of life, in an environment of a quality that permits a life of dignity and well-being, and he bears a solemn responsibility to protect and improve the environment for present and future generations."[8]

The Tokyo Resolution drew on the combined powers of academic expertise and the nation-state. Both were ascendant in the nineteenth century, and international meetings on environmental matters had become a feature of pertinent discussions since the middle of the century. Sometimes it mirrored a realization that problems transcended national boundaries. In response to recurring cholera epidemics (see chapter 21, Cholera), France convened the first international sanitary conference in 1851.[9] Sometimes migrating animals triggered conversation: birders have met every few years since the first International Ornithological Congress in Vienna in 1884.[10] And sometimes international conventions grew out of a mere realization that nations faced similar challenges. It was always nice to meet like-minded spirits, it was a chance to learn, and references to efforts in other countries were a way to stimulate government action. When nature conservationists from six European countries met in Paris for the first international conference on the protection of the countryside in 1909, the French president of the convention used the concluding banquet to call upon France to organize conservation based on the German model. At least that was what the German ambassador reported back home.[11]

During the interwar years, the League of Nations provided a platform for discussions on oil pollution, whaling, and the protection of primates, but these discussions were ultimately inconsequential and ephemeral, much like the league itself.[12] International exchange developed in more permanent form after World War II. It began as a Western project, and the United States was the undisputed leader. It showed in the list of speakers at the Tokyo symposium. Of the twenty-two participants from outside Japan, seven came from the United States, two from Canada, and nine from Western Europe. The rest of the world was represented by a grand total of four speakers from India, Yugoslavia, Hungary, and the Soviet Union. All the speakers were male, though the conference photograph showed thirty-five men and four women. The latter were the wives of Western professors.[13]

Presentations at the symposium differed widely in topics and scope. Some hands-on papers discussed air pollution control in Pittsburgh and water management in West Germany's Ruhr. Others were more interested in the big picture and discussed economic theory, ecological philosophy, and how the social sciences could "prove their usefulness and efficiency in a world which is not always taking them seriously."[14] Papers on economic incentives, city planning, and legal and legislative aspects explored the middle ground between the specialists and the generalists. The prevailing sentiment was grave concern about the state of the environment combined with cautious optimism. The Tokyo Resolution described environmental disruption as "the direct outcome . . . of the twin processes of industrialization and urbanization with the attendant progress in technology," but it was "not a necessary outcome of such processes: modern societies with command of science and technology have the means of countering, containing, and redressing the worsening trends of environmental disruption." The Tokyo Resolution called for more research and international collaboration, public education, and "appropriate legal, political and economic steps." It also called upon social scientists to study all these things.[15]

The Tokyo Resolution was adopted unanimously after the last papers had been presented, but it was not a collaborative effort. In its emphasis on individual rights, the text showed the handwriting of Joseph Sax, then a professor of administrative law at the University of Michigan.[16] Sax had just published a landmark article on "the public trust doctrine in natural resources law" that sought to establish an actionable right to a healthy environment.[17] Sax was also the man behind the Michigan Environmental Protection Act of 1970, the guiding thought of which was "that ordinary citizens should be able to play the role of attorney general."[18] Faced with administrations that were dragging their feet, Sax wanted to use the courts to push for more aggressive environmental policies. "It is time to recognize that the prospect of external scrutiny, such as the courts are prepared to employ, may itself be the most effective remedy for slothful administrators," Sax declared in his paper in Tokyo.[19] The Japanese in the audience knew what he was talking about.

2. BUZZWORDS

Japan enjoyed the fruits of a long economic boom in 1970, but it was also wrestling with the environmental consequences. Pollution was a particular matter of concern, and it was not only about the common problems of large urban centers, such as garbage and car exhausts. The situation was so bad in some places that new diseases were named after them. A petrochemical complex in Yokkaichi near Nagoya spewed sulfur dioxide and other toxic pollutants into the atmosphere, and local residents developed what came to be known as "Yokkaichi asthma." Minamata Bay on the west coast of Kyushu became infamous for "Minamata disease," a neural disease caused by a mercury compound coming from a local factory and accumulating in the aquatic environment. Cadmium poisoning caused a new disease that was named *itai itai*, or "it hurts, it hurts disease," in 1955. All these problems were known and under investigation by 1960, but determined responses were long in coming.[20]

Shigeto Tsuru was intimately familiar with these issues. He had founded the Research Committee on Pollution in 1963, a group of eight leftist academics who shared "a commitment to the plight of pollution victims and an unrelenting determination to expose the perpetrators."[21] The Research Committee was well represented among the Japanese participants at the international symposium in 1970, and the program was planned accordingly: participants went on a field trip to Yokkaichi and spoke with government officials and citizens.[22] Committee members also discussed presentations in light of their own experiences, the crucial skill of what Simon Avenell has called "rooted cosmopolitans": locally informed groups and people who can connect concerns to transnational and global debates.[23] After listening to the paper on air pollution control in Pittsburgh, Jun Ui, one of the more flamboyant members of the Research Committee, remarked that "the situation that prevailed in Pittsburgh in the 30's is quite similar to that prevailing in the present-day Japan."[24]

As it happened, the country's environmental governance improved notably over the following months. Japan's parliament passed or amended fourteen environmental laws in 1970, a legislative tour de force that entered history books as the "Pollution Diet." A new Environment Agency was created the following year in order to provide the new laws with teeth.[25] Japanese courts embraced the new mood and decided four big pollution cases in favor of the victims between 1971 and 1973.[26] Two decades later, Tsuru wrote that "the year 1970 has been aptly referred to in Japan as 'Kōgai Gannen,' or the year initiating the era of environmental challenge," and he gave some credit to the symposium of March 1970.[27] In his judgment, the symposium "had a strong impact on Japanese public opinion and on the bureaucracy," and the Tokyo Resolution "served as a springboard for a basic reorientation in the matters concerning the environmental rights of citizens."[28] Needless to say, the symposium was one of numerous factors that led to the change of tide, and it is difficult to specify its relative weight, but maybe that says more about how successful conferences work. They link up with ongoing trends. They put a spotlight on the host country. They raise awareness about efforts at home and abroad.

And if things go really well, they transform a heretofore diffuse sentiment into a new word.

In his opening remarks, Tsuru asked the participants of the symposium "to cooperate in deciding on the most appropriate term for the phenomena which tentatively had been referred to as 'environmental disruption' by the International Social Science Council." Would it be better to speak of "degradation," "desolation," "disorder," or simply "problems?"[29] The years around 1970 were watershed years for the global discourse on humans and nature, and a good part of the intellectual energies went into the search for new terms. As it turned out, "environmental disruption" became one of the less popular expressions, but "the environment" was a winner. It was the umbrella term that activists and policymakers were craving, a way to unite a diverse set of issues and reframe them as parts of something bigger. What had previously been fragmented debates over pollution (see chapter 16, London Smog), nature reserves (see chapter 26, Kruger National Park), or vanishing resources (see chapter 8.3, Running Empty) were now aspects of a comprehensive "environmental crisis" that commanded attention at the highest level.[30] Characteristically, the 1972 summit in Stockholm was the United Nations Conference on the Human Environment, a choice of words that "was indicative of the way in which humans and nature were beginning to be seen as sharing an inseparable destiny."[31]

As with every new term, "the environment" had advantages and disadvantages. For one, it was more abstract than previous words. For another, it lacked national and regional flavor. Tsuru cited the Japanese word for pollution, "kōgai" (literally "public damage"), in his opening remarks at the symposium, a word that was obviously dear to his heart after his long struggle against pollution but also one that had to yield to the new international language.[32] Environmental problems were also open to different readings. At the Stockholm summit, the Indian prime minister Indira Gandhi shocked Western environmentalists with a broader understanding of environmental grievances: "Are not poverty and need the greatest polluters?"[33] As seen from the Global South, environmental problems were intimately connected with social and economic issues.

Conflicts over buzzwords mirrored power struggles, and "the environment," Western-style, was no exception. It served to decontextualize environmental issues, which is how Western consumer societies, presumably classless and also devoid of other cleavages, liked to see ecological challenges (see chapter 37.1, Global Pollution). Segmented thinking also served the predilections of the bureaucracy, and setting up a new "ministry for the environment" became a transnational marker of getting serious. It reflects the cultural hegemony of the West that even critics of Western environmentalism could not do without the word. Joan Martinez-Alier outlined an "environmentalism of the poor" that thought and acted differently. But it was "environmentalism" nonetheless.[34]

Terminological discussions were not everyone's favorite, and there was a convenient exit. One could simply allow meanings to get blurry until they could mean almost anything. Sustainability was a rather clear concept in forestry (see chapter 4, Sustainable Forestry) until it became a global buzzword in the 1980s. The 1987 Brundtland Report defined sustainable development as "development that meets the needs of the present without compromising the ability of future generations to meet their own needs."[35] It was the result of extensive discussions among academics and politicians from around the globe, and it showed.[36] But another United Nations summit was around the corner, this time in Rio de Janeiro, and "sustainable development" became the shibboleth of the 1992 "Earth Summit." Many people used the word in earnest, and it arguably helped raise awareness, but it was ultimately a free-floating signifier for whatever countries, corporations, and other stakeholders had in mind. Academics are facing requests to fill it with meaning to the present day.[37]

The Rio Earth Summit shared many concerns with the Tokyo symposium, but it was different in a number of ways. It was much bigger: it had representatives from 175 nations including more than 100 heads of state and government, some 7,000 journalists, and a concluding document, Agenda 21, which was longer than the full set of papers and discussion minutes of the Tokyo symposium.[38] It was also more diverse: while the Tokyo symposium was the province of academics, the Rio summit was wide open for civil society. More than 1,400 nongovernmental organizations were registered at the conference, and thousands more took part in parallel events at a more informal Global Forum. The Earth Summit produced two landmark treaties that have framed global environmental diplomacy ever since, the United Nations Framework Convention on Climate Change and the Convention on Biological Diversity.[39] And just like the Tokyo symposium, the Earth Summit put a spotlight on the host country, specifically Brazil's engagement with the Amazon rain forest.[40]

The concept of sustainable development was born in long transnational debates that go back to the North–South Commission chaired by the former West German chancellor Willy Brandt.[41] However, it does not always take international commissions to coin a new environmental buzzword. The Anthropocene was a spontaneous idea of the atmospheric chemist Paul Crutzen at a meeting of the International Geosphere-Biosphere Program in Cuernavaca, Mexico, in 2000.[42] The word was soon seized upon by committees and publicists and became almost as opaque as sustainable development, but that was not the only problem with the burgeoning debate. While sustainable development was used by a broad range of people and arguably thrived on grassroots interest, the Anthropocene circulated mainly among globe-trotting elites. International conventions were never disembodied events: they were made of humans from vastly different backgrounds, and buzzwords gave them a joint cause or the semblance thereof. Faced with thousands of delegates and little time to meet, a shared vocabulary helped create some common ground, and this vocabulary could turn into second nature for those with a travel budget and a Western passport. Some families have been at home in transnational circles for generations, though not always for the same cause. The secretary general of the first International Ornithological Congress in Vienna was the grandfather of Friedrich August von Hayek.[43]

At the Tokyo symposium, Joseph Sax noted that his use of the courts amounted to "shock treatment to bureaucrats," but he did not depict bureaucrats as invariably lazy.[44] He did not even blame them for a lack of goodwill. It was just that "regulatory agencies have an interest and perspective of their own which is frequently at odds with that of significant segments of the public."[45] In a book that he published a year later, Sax described how administrators suffered from what he called an "insider perspective": focusing on staffs and budgets, laws and constituencies, momentary conflicts and priorities, bureaucracies could make decisions that baffled people beyond the office walls.[46] But officials and policymakers are not the only people who can succumb to tunnel vision. Intellectuals are just as prone to insider perspectives, and closed circles, a distinct transnational code language and a globe-trotting migration pattern go a long way in this direction. The Anthropocene may ultimately say less about the world than about those who invoke it prolifically.

3. GETTING SERIOUS

Sax eventually got a chance to view the world from an insider perspective when he became counselor to the secretary of the interior during the Clinton administration. But Sax spent most of his life in academia, and the public trust doctrine became his intellectual legacy. When he died in 2014, his obituary in the *New York Times* noted that the doctrine had been used in some three hundred federal and state decisions between 1997 and 2008. It also noted adoption in countries as different as Pakistan, Uganda, and Ecuador.[47] Using the courts for environmental reform was in the spirit of the Tokyo Resolution, except for one thing: decisions remained staunchly within the framework of national law.

International exchange became a defining feature of environmentalism in the 1960s, and not just in the form of meetings like the Tokyo symposium. The International Union for Conservation of Nature published the first Red Data books on endangered species in 1966 (see chapter 11.3, Conserving Diversity) and criteria for national parks (see chapter 26, Kruger National Park) in 1969, thus creating a body of knowledge that conservationists could draw upon all over the world.[48] The Stockholm summit helped conclude negotiations over the Convention on International Trade in Endangered Species (CITES), which was signed in 1973.[49] But CITES ultimately hinged on the authority of nation-states: it was a decision of a US court that doomed Antony Fisher's turtle farm in the Cayman Islands (see chapter 36, Battery Chicken).[50] Did global environmental problems perhaps call for responses that were not just national in scope? A hope for global environmental governance has run through international conversations since the 1970s.

International bodies were eager to respond, and not just out of idealism. UNESCO's reputation was in tatters when it launched its Man and the Biosphere Programme, and its array of activities, particularly the UNESCO biosphere reserves, helped to make it one of the more renowned branches of the United Nations. But for all the international exchange since the mid-nineteenth century, nation-states were traditionally reluctant to share power even in the face of concrete threats. It took more than half a century to move from the first cholera conference (see chapter 21, Cholera) in 1851 to a commitment to immediate communication and the creation of a small secretariat in Paris. The fragility of the treaty became clear just a few years later when Italy kept a cholera epidemic under wraps.⁵¹ During the Cold War, Aristotle Onassis and the Soviet Union defied international whaling agreements (see chapter 9, Whaling) in different ways. The International Whaling Commission later imposed a successful ban on commercial whaling, but only after being hijacked by environmental activists. Prodded by Greenpeace and other environmental groups, a number of countries joined the underlying convention with the goal of creating an anti-whaling majority. It was the government of the Seychelles that proposed the crucial moratorium of 1982. A member since 1979, the Seychelles did not have a tradition in whaling, but its delegation included an anti-whaling activist from South Africa.⁵²

The number of environmental treaties has grown enormously since the 1970s. For example, the European Union is a party to some thirty agreements from the Alpine Convention to the International Tropical Timber Agreement.⁵³ Some treaties are success stories. The Montreal Protocol of 1987 was an effective response to the depletion of the stratospheric ozone layer.⁵⁴ Others had ambiguities: the 2001 Stockholm Convention on Persistent Organic Pollutants ostracized a dozen dangerous chemicals, but made it difficult to use DDT (see chapter 38, DDT) in the fight against malaria.⁵⁵ And some treaties never went into force. The Hong Kong International Convention for the Safe and Environmentally Sound Recycling of Ships (see chapter 5, Shipbreaking in Chittagong) is a dead letter a dozen years after its adoption.⁵⁶

Global environmental governance was always a project in parts. It was also subject to the inevitable rivalries of the political world. The International Union for Conservation of Nature was lukewarm at best about the Stockholm summit of 1972.⁵⁷ And like all emanations of global diplomacy, it was a result of compromise, and that was a disappointment for environmentalists who hoped for more. On the last day of the Earth Summit, Japanese activists published a Japanese Citizens' Rio Declaration that declared the event a failure "because nations were not able to conclude treaties strong enough to protect the Earth from environmental destruction."⁵⁸ The treaty that saves the world will be part of the environmentalist imaginary for the foreseeable future, a dream that has been disappointed so many times and yet refuses to die. The Paris Agreement on climate change of 2015 was the most recent incarnation.

Many things have constrained global environmental governance, and one of the most persistent is the lack of a powerful lobby. The United States backed the environmental diplomacy of the early 1970s and was open to more powerful institutions like an international watchdog agency, but its leadership petered out during the Reagan administration.[59] The United States played a significant role in ozone diplomacy and the phaseout of single-hull tankers after the *Exxon Valdez* oil spill (see chapter 39.2, Drawing Up Lessons), but the default attitude was either indifference or open hostility.[60] Ever since the Stockholm summit, nations of the Global South have harbored a deep suspicion that global environmental agreements would deny them their right to development.[61] Global commitments also faced resistance from large multinational corporations (see chapter 10, United Fruit) in the age of globalization. The neoliberalism of Friedrich von Hayek has left a deeper impression on the twenty-first-century world than his grandfather's infatuation with birds.

Governing the blue planet is a romantic idea, but it faces a more complicated reality: divergent perspectives and definitions of problems, nation-states that carefully guard their autonomy, and transnational corporations that face stiff international competition. But just like administrators and globe-trotting intellectuals, corporations are not inherently evil. They can even provide crucial support for environmental initiatives. The Tokyo symposium received most of its funding from a local business organization chaired by Kazutaka Kikawada, then the president of the Tokyo Electric Power Company (TEPCO).[62] As the country's leading utility, TEPCO had the money for a conference that changed the world. It also had plenty of power plants, including a nuclear facility on the northeast coast of Japan. Fukushima Daiichi (see chapter 37.3, Up in the Air) changed the world in its own way when it was hit by a tsunami in 2011. Unit 1 exploded on March 12.

It was the fortieth anniversary of the Tokyo Resolution.

25

THE 1976 TANGSHAN EARTHQUAKE

SOMEWHAT NATURAL DISASTERS

I. SHAKEN

Tangshan was a city of one million people, but that did not count for much in the biggest country on earth. Two larger cities, Tianjin and Beijing, were just one or two hours away by train, and sights were rare in a town of coal and industry. Tangshan was home to one of China's first railroads, and a mining engineer named Herbert Hoover, later president of the United States, had worked there in 1899/1900, but few people knew about Tangshan until the early morning hours of July 28, 1976.[1] Geologists called it an exceptionally strong "continental intraplate earthquake" and ranked it at 7.5 on the Richter scale, which made it the third-strongest earthquake of 1976, and the force hit with devastating precision.[2] The epicenter was located just outside Tangshan, and the shock waves destroyed 78 percent of Tangshan's industrial plants and 95 percent of the residential buildings.[3] It was the deadliest seismic event of the twentieth century. The government recorded 242,000 victims, but that was only a fraction of the real number. In his book on the Tangshan earthquake, James Palmer put the death toll at "somewhere around 650,000."[4]

and was open to more powerful institutions like an international watchdog agency, but its leadership petered out during the Reagan administration.[59] The United States played a significant role in ozone diplomacy and the phaseout of single-hull tankers after the *Exxon Valdez* oil spill (see chapter 39.2, Drawing Up Lessons), but the default attitude was either indifference or open hostility.[60] Ever since the Stockholm summit, nations of the Global South have harbored a deep suspicion that global environmental agreements would deny them their right to development.[61] Global commitments also faced resistance from large multinational corporations (see chapter 10, United Fruit) in the age of globalization. The neoliberalism of Friedrich von Hayek has left a deeper impression on the twenty-first-century world than his grandfather's infatuation with birds.

Governing the blue planet is a romantic idea, but it faces a more complicated reality: divergent perspectives and definitions of problems, nation-states that carefully guard their autonomy, and transnational corporations that face stiff international competition. But just like administrators and globe-trotting intellectuals, corporations are not inherently evil. They can even provide crucial support for environmental initiatives. The Tokyo symposium received most of its funding from a local business organization chaired by Kazutaka Kikawada, then the president of the Tokyo Electric Power Company (TEPCO).[62] As the country's leading utility, TEPCO had the money for a conference that changed the world. It also had plenty of power plants, including a nuclear facility on the northeast coast of Japan. Fukushima Daiichi (see chapter 37.3, Up in the Air) changed the world in its own way when it was hit by a tsunami in 2011. Unit 1 exploded on March 12.

It was the fortieth anniversary of the Tokyo Resolution.

25

THE 1976 TANGSHAN EARTHQUAKE

SOMEWHAT NATURAL DISASTERS

I. SHAKEN

Tangshan was a city of one million people, but that did not count for much in the biggest country on earth. Two larger cities, Tianjin and Beijing, were just one or two hours away by train, and sights were rare in a town of coal and industry. Tangshan was home to one of China's first railroads, and a mining engineer named Herbert Hoover, later president of the United States, had worked there in 1899/1900, but few people knew about Tangshan until the early morning hours of July 28, 1976.[1] Geologists called it an exceptionally strong "continental intraplate earthquake" and ranked it at 7.5 on the Richter scale, which made it the third-strongest earthquake of 1976, and the force hit with devastating precision.[2] The epicenter was located just outside Tangshan, and the shock waves destroyed 78 percent of Tangshan's industrial plants and 95 percent of the residential buildings.[3] It was the deadliest seismic event of the twentieth century. The government recorded 242,000 victims, but that was only a fraction of the real number. In his book on the Tangshan earthquake, James Palmer put the death toll at "somewhere around 650,000."[4]

The numbers were horrifying, and so were the circumstances of death. Many residents were crushed in their sleep. Others perished slowly over hours or days while being caught in the rubble. Even those who survived with their limbs intact were not necessarily safe, as earthquakes could trigger veritable cascades of disaster. The Lisbon earthquake of 1755 wreaked havoc because the seismic shock was followed by a devastating tsunami.[5] When the Kanto earthquake struck Tokyo and Yokohama in 1923, toppled stoves caused fires that burned for two days.[6] Industrial technology extended the potential consequences of disasters into new dimensions. On March 11, 2011, a magnitude 9.1 earthquake off the Japanese coast caused a tsunami that destroyed the Fukushima Daiichi atomic power plant (see chapter 37.2, Nuclear Complications) and spread radioactive contamination.[7] The Tangshan earthquake damaged a nearby reservoir, and the city's ruins might have drowned in a deluge if it had not been for an army unit that opened a fifty-ton floodgate by hand.[8] Earthquakes put humans at the mercy of raw physical forces, and the same held true for other disasters that are commonly classified as natural: floods, wildfires, landslides, hurricanes, and volcanic eruptions.

Sometimes natural disasters had human beneficiaries. Two typhoons helped Japan to defeat Mongol invaders in 1274 and 1281, and the "divine wind," known in Japanese as *kamikaze*, became part of the nation's mythology.[9] But more often than not, people just perished: they were crushed, drowned, or burned to death with no discernible sense or purpose. The forces of nature killed in utter contempt for human sentiment, and that left survivors adrift as they searched for meaning in the midst of the rubble. Historians were just as clueless when they began to study natural disasters in history. In a pioneering essay on the Friuli earthquake of 1348, Arno Borst observed that dealing with earthquakes as recurring experiences in past and present "goes against the modern European ego."[10]

Humans are loath to accept a vacuum of meaning, and disaster-stricken people embraced emplotment long before Hayden White introduced the term into the philosophy of history.[11] Were disasters perhaps divine punishments for ethical failings? When a typhoon destroyed seawalls along the east coast of China in 1724, the emperor decreed that "the sufferers might have been in some measure responsible for their own misfortune, because of their moral defects."[12] Mindful of the ten plagues of Egypt recorded in the Book of Exodus, Christians viewed natural disasters as judgments from God and went through cycles of repentance.[13] Over in America, the Aztecs "had a pantheon of deities dedicated to disaster, ranging from Atlacoya, the goddess of drought, to Atlacamani, goddess of storms and hurricanes; Chantico, goddess of fire and volcanoes; and Tepeyollotl, 'heart of the mountains,' who was associated with earthquakes."[14] Enlightenment thinking and scientific research undercut these interpretations, but the resilience of parareligious readings remains subject to debate. Christian Pfister has argued that "magic-animistic tropes" down to a mythical revenge of nature continue to resonate in environmental circles.[15]

Natural disasters could also be a sign of change in the human sphere. According to the Gospel of Matthew, an earthquake shook Jerusalem after Jesus died on the cross.[16] When the Tiber River flooded Rome in December 1870, devout Catholics wondered whether this was God's commentary on the recent conquest of the Papal States and Rome's incorporation into the Kingdom of Italy.[17] And was the idea really absurd? Earlier that year, a thunderstorm had raged over Saint Peter's Basilica during the proclamation of papal infallibility.[18] China was well familiar with the concept. Tradition held "that changes in the physical world would mirror a new political era, just as it was commonly believed that earthquakes and floods signaled a dynasty's loss of the right to govern, or Mandate of Heaven."[19] The Cultural Revolution had not smoked out ideas about the mysterious intermingling of human and natural history, and that made the Tangshan earthquake a political event of the first order. Change was evidently under way in contemporary China: Zhou Enlai, the premier of the People's Republic since 1949, had died in January 1976, Mao Zedong was mortally ill, and a power struggle raged behind the scenes about who would succeed him.

The conflict ended with the victory of a group of reformers around Deng Xiaoping, but it did not look that way in the summer of 1976. The Gang of Four, which included Mao's last wife Jiang Qing, seemed set to assume power, and that framed the government's response to the Tangshan earthquake. Word went out from the Gang that survivors "should 'deepen and broaden' the criticism of Deng Xiaoping's revolutionary line."[20] Two days after the earthquake, *People's Daily* opened with the story of a senior cadre who had ignored the cries of his dying children and saved the local party chairman instead.[21] Control of nature was a running theme in Mao's teaching, and heroic tales of party service were a way to preserve the semblance of human supremacy in the face of carnage. Triumphalism defined official readings of Tangshan down to the memorial building that Hu Yaobang, general secretary of the Communist Party of China, dedicated to the fight against the disaster rather than the disaster itself. Opened in 1986, the monument depicts the earthquake as "an enemy to be defeated."[22]

With ideological purity preserved, the Gang of Four felt no need to make the short trip to the suffering city. The only high-level figure to come to Tangshan was Premier Hua Guofeng, whom Mao had appointed as Zhou Enlai's successor a few months earlier. He ditched party protocol and met with victims, and pictures of his visit were publicized widely.[23] Two months later, Hua led the arrest of the Gang of Four and opened the door for the men around Deng Xiaoping, who seized power over the following years.[24] There was a difference between order in the world of ideas and order on the ground, and both had their own challenges and opportunities.

2. CRISIS MODE

The Gang of Four members were not the first rulers to shun a place of disaster. When the Lisbon earthquake struck on All Saints' Day 1755, the Portuguese royal family fled to the countryside and camped out in tents for weeks, and King José I of Portugal battled claustrophobia for the rest of his life.[25] Some 130 years later, Alfonso XII of Spain clashed with his ministers over a trip to Murcia and Valencia. Cholera (see chapter 21, Cholera) was in the country, and royalists wondered whether a death-defying journey would bolster the king's reputation. Alfonso liked the idea, but the Prussian ambassador was aghast. He noted that the country's republicans were in favor of the trip, evidently hoping that the death of the young king would throw the monarchy into chaos, and, in any case, kings were not in the business of seeking praise from commoners. Alfonso should search for "completely different and more solid foundations" for his dynasty. At best, the trip would deliver "ephemeral popularity."[26] In the end, Alfonso made the trip nonetheless and died later that year from an unrelated cause, and his son became the last king of Spain before the Second Republic.

Disaster zones were dangerous, even for those who traveled with an entourage. A catastrophe was akin to a collective character test, and not everyone passed with flying colors. When a devastating earthquake hit Dubrovnik during the Holy Week of 1667, the moral fabric of the Republic of Ragusa fell apart in dramatic fashion. Rescuers asked for money before they dragged victims out of the rubble while a number of patricians robbed the treasury and set off by ship.[27] The Kanto earthquake triggered anti-Korean riots that killed between four hundred and six thousand people.[28] Tangshan lacked a sizable ethnic minority, but a trigger-happy militia established a brief reign of terror in the midst of the ruins. Places of disaster were places of fear, and years of indoctrination about traitors and saboteurs claimed an unknown toll.[29]

Restoring order was only the first of many challenges after a natural disaster. Survivors needed food, safe water, and protection from epidemics, and disaster managers faced many difficult questions. Where should rescue crews go first? When China mobilized the People's Liberation Army after Tangshan, most platoons marched to the city center, leaving outlying areas without help for days and weeks.[30] How should relief be distributed? When donations poured in after the Chicago fire of 1871, the Relief and Aid Society gave preferential treatment to the wealthy because it found that they had suffered more than the laboring classes.[31] And how did disaster relief relate to Indigenous coping mechanisms? In a study on the Philippines, Greg Bankoff argued that international aid marginalized local communities as "the hapless victims of an unruly nature and the needy recipients of foreign assistance."[32]

International aid began in haphazard fashion. When an earthquake destroyed Messina in southern Italy in 1908, the worst natural disaster in twentieth-century Europe, ships from the Imperial Russian Navy and the British Royal Navy happened to cruise nearby, and sailors joined the rescue effort.[33] International assistance became common after World War II, a routine part of the cultural script that kicks in whenever calamity strikes somewhere on the globe, but recipients did not always appreciate the humanitarian gesture. The experience of global solidarity can foster unrealistic expectations, and brief interventions such as food aid (see chapter 31.2, Hunger, Modern Style) can inflict long-term damage. Authoritarian regimes can also see international support as a threat to their legitimacy. China refused all humanitarian aid in 1976 and banned foreigners from visiting Tangshan for seven years, and Myanmar's military regime imposed a similar ban after cyclone Nargis in 2008 and revised it only to a limited extent after international pressure. A more self-confident China allowed international aid after the Sichuan earthquake in 2008, but it was more reluctant after the Yushu earthquake two years later. The disaster area was on the Tibetan plateau, and the Chinese rulers were wary of foreign humanitarians in an ethnic conflict zone.[34]

Premier Wen Jiabao went to Sichuan in the aftermath of the disaster, followed by the general secretary of the Communist Party of China, Hu Jintao.[35] Visits from politicians have become obligatory after a major calamity, a tribute to people's expectation of a caring leader, and some received a career boost as a result. The US secretary of commerce Herbert Hoover advanced to the White House after his energetic response to the Great Mississippi Flood of 1927, and the Hamburg flood of 1962 made Helmut Schmidt, later chancellor of the Federal Republic of Germany, a national figure.[36] Forty years on, the German chancellor Gerhard Schröder turned a national election around with a well-publicized intervention after a flood in the Elbe watershed.[37] However, Schröder lost the following election, and Hoover's presidency was not successful either. The White House was in the hands of the Democrats for the following two decades.

A visit to a disaster zone was a powerful symbol, but it was not always about genuine empathy. The Italian king Victor Emmanuel II took the Rome flood of 1870 as a good occasion to finally make the trip to the new capital, which his troops had conquered three months earlier.[38] When an earthquake hit Nicaragua's capital Managua in 1972, the dictator Anastasio Somoza took control of reconstruction and diverted much of the international aid into his family's pockets.[39] The practice was so common in modern Italy that people invented a new term, "earthquakism," for the illegal appropriation of reconstruction funds.[40] The usual rules no longer applied in times of crisis, and that opened the door for initiatives that would have fared poorly under normal conditions. Three months after hurricane Katrina, in the last op-ed of his life, ninety-three-year-old Milton Friedman called for radical privatization of New Orleans's school system, and private charter schools replaced most of the public schools over the following two years. Naomi Klein saw it as an exemplary case of what she called "disaster capitalism": "orchestrated raids on the public sphere in the wake of catastrophic events, combined with the treatment of disasters as exciting market opportunities."[41]

Sometimes disasters forged human bonds. Colonel Juan Perón met Eva Duarte (better known as Evita) during a charity event for victims of an earthquake in western Argentina in 1944.[42] Some scholars ventured that it might also work on a national scale. Pfister has argued for the Swiss case that natural disasters were the functional equivalent of war against a foreign enemy, an experience that Switzerland had lacked since the days of Napoleon.[43] But social status made for widely divergent experiences, and politicians were struggling to maintain the semblance of unity. After the Mississippi flood, Hoover made a deal with the head of the Tuskegee Institute, Robert Russa Moton, to keep the exceptional burden for African Americans under wraps.[44] Social inequality is a running theme in the history of natural disaster in America, as Ted Steinberg has shown in his *Acts of God*, and few things suggest that it was different in the rest of the world.[45] Disasters did not suspend the fault lines that ran through societies. They just made them less visible for a while.

When it came to rebuilding, modern planners put more faith in technology than solidarity. After Katrina, New Orleans received a flood-protection system worth $14.5 billion, in addition to a new school system.[46] The Netherlands built a system of storm surge barriers after the disastrous North Sea flood of 1953.[47] Other schemes never materialized: after the 1870 flood, the late Giuseppe Garibaldi advanced what a biographer called "a cyclopean project" to reroute the Tiber River around Rome.[48] Tangshan was rebuilt after the earthquake, but the blueprint changed notably due to a lack of funds, changes in economic policy, and other factors, and the new city was more similar to the old one than intended.[49] As with so many planned developments (see chapter 32.3, Planning Development), the confidence of masterplanners collided with a more complicated reality, and sometimes distrust of grand technological schemes was even on display in the design of new houses. The Lower Ninth Ward in New Orleans, the area worst hit by Katrina, features new buildings on stilts.

Politics in crisis mode comes to an end after a while, and disaster preparation is about building resilience as well as emergency services. But natural disasters defy human control, and their effects can go in many different directions, and that makes them an unnerving presence even for those who think that they have come to terms with them on an ideological level. And what becomes of the crisis mode if natural disasters occur in close sequence? A powerful narrative depicts the environmental crisis as "a spiraling race between human power and human-induced natural disasters."[50] If this vision becomes a reality, the crisis mode may turn into the default mode of politics.

3. THE HUMAN FACTOR

Many buildings turned into death traps during the Tangshan earthquake, but there were some significant differences. "Traditional village wooden houses were among the best places to be," Palmer observed, mostly because their building materials were less lethal than stone, steel, and cement, and people found it easier to dig out survivors.[51] Japanese engineers had made a similar experience in the aftermath of the Nobi earthquake near Nagoya in 1891: European-style brick buildings had fared particularly poorly while traditional Japanese temples and pagodas were still standing.[52] Humans made choices when they were building, and these choices had consequences for the chances of survival. Looked upon closely, the natural disaster was not quite so natural.

The relative weight of natural and human factors was usually a matter of perspective. A heat wave killed fifteen thousand people in France in 2003, but was that due to extreme weather, a lack of air-conditioning (see chapter 20, Air-Conditioning), or to poor retirees under Parisian tin roofs, social isolation, and an administration that spent August on vacation?[53] Even when a general deliberately blew up dikes, as Chiang Kai-shek did on the Yellow River in a desperate attempt to stop the advancing Japanese army in June 1938, the bottom line was not clear. It was "perhaps the most environmentally damaging act of warfare in world history," but it was also a disaster that was bound to happen.[54] Sediments made up a good part of the Yellow River water that traditionally settled in Henan's eastern plain—in fact, sediments gave the river its name— and keeping the Yellow River in check with ever higher dikes was not a permanent solution. In other words, the naturalness of natural disaster was a human myth, the terminological equivalent of a collective shrug. It helped deflect attention away from more inconvenient issues, and unlike other potential culprits, Mother Nature did not talk back. Or that is what people thought.

But naturalizing responsibility only got people to a certain point. Decisions were waiting to be made, and the most fundamental was about leaving or staying. It usually came down to the latter. The Tangshan recovery-planning task force considered relocation of the entire city but decided against it, not least with a view to the local coal deposits. Some places even depended on exactly the same forces that threatened them: volcanoes offered fertile soils at the expense of plenty of risks, and polders offered new land in exchange for life under the waterline. And then there was the resilience of the familiar. The Chinese planners sought to turn Tangshan's Lunan district into open space because of an active fault line, but there was also a desperate need for temporary housing, and the area housed an estimated 174,000 people in semipermanent structures four years after the disaster. The planners caved in three years later. The Lunan district was rebuilt as it was.[55]

When Hua Guofeng visited Tangshan in 1978, he pledged to make the new city the safest in the world.[56] But Hua, a lifelong Maoist, was known for rhetorical bluster, and like most builders in an earthquake zone, Tangshan's planners ended up juggling costs, risks, and probabilities. One of their solutions was geographic differentiation: safety standards for the city center were consistently higher than standards in surrounding areas.[57] Whether enhanced standards were carried out remains anyone's guess, for construction suffered from a lump-sum funding system that rewarded contractors for cutting corners.[58] And in the end, nobody really knows the size of the next seismic event. The Fukushima Daiichi nuclear complex (see chapter 37.3, Up in the Air) had a seawall, as did many places along the northeast coast of Honshu, and their design reflected the experiences with giant tsunamis in 1896, 1933, and 1960. But the tsunami of 2011 was bigger.[59]

While builders and seismologists have formulas to work with, the psychological consequences of disasters defy academic categorization. Based on interviews thirteen years after Tangshan, two Chinese psychologists came to the grand conclusion that "earthquakes do have long-term effects on people's minds," that these effects were "extensive and long-lasting," and that effects showed "not in the same way for different groups of people."[60] We know more about how survivors want to remember, for they spoke about this even in a totalitarian society like China. The Communist Party had set its preferred reading into stone with the Tangshan Earthquake Monument, but as it turned out, that did not settle the matter.

The monument had a central location and four pillars that reached thirty meters into the sky to symbolize human supremacy over nature, but it did not have a place where survivors could deal with their personal memories. They could not go to a tomb either: few earthquake victims have individual graves. People burned joss paper at various crossroads, a superstitious ritual usually performed in front of a tomb that the local government reluctantly tolerated. A private company saw a business opportunity and built a memorial wall where survivors

25.1 People mourn at the Tangshan Earthquake Memorial Park in July 2016, forty years after the disaster. Image, Imaginechina Limited / Alamy Stock Photo.

could immortalize the names of family members for a fee, but commercializing collective memory did not go down well. The city built its own memorial wall that features the names of the 240,000 officially recorded victims. It opened in 2008, and the private wall was torn down the following year.[61]

It is heartening to see respect for individuals in a country that long preferred to see them as mere constituent parts of faceless masses. The Tangshan monuments have also become featured tourist attractions (see chapter 22, Baedeker) in a city that lacks landmarks, and traveling to places of suffering and grief, commonly labeled dark tourism or thanatourism, has emerged as a popular pastime as well as a field of scholarly research with that quintessential ensign of academic coolness, an online journal project.[62] Publishers put out disaster books by the dozen that feature plenty of individual stories, but they are typically more reluctant to inquire about institutions and money, resilience and order, social status and social marginalization, master planners and master narratives, and all the other issues that disasters have historically raised. Disasters pose big questions, and they continue to do so in the new millennium. But you can always drown big questions in an endless chatter about individual stories.

PART VI

THE FINAL RESERVES

WHAT DO YOU DO WHEN YOU REACH THE ENDS OF THE EARTH?

That, in effect, was the question that Jean Brunhes, the rector of the University of Fribourg in Switzerland, was pondering in November 1909. The French geographer spoke at the opening of the academic year, typically an occasion for grandiose rhetoric and deep thoughts, and Brunhes used the opportunity for reflections on man's changing place on the planet. He identified "limits to our expansion everywhere."[1] Robert Peary had come close to the North Pole the previous spring, and Roald Amundsen and Robert Scott raced across Antarctica toward the South Pole two years later.[2] The expansion of colonial empires reached its final stage in the struggle over Morocco shortly before World War I, as there were few places left that had not been claimed yet. The Trans-Siberian Railway was nearing completion in tsarist Russia, making the epitome of remoteness, Siberia, accessible by coach. Telegraph and telephone networks sent information around the globe at unprecedented speed. The spheres of investigation and occupation, two developments that Brunhes saw as closely intertwined, faced natural limits. In short, Brunhes found that "everywhere we run up against the bars of our cage."[3]

It is a matter of debate whether the remark mirrored a contemporary awareness of planetary limits. Sabine Höhler has argued that Brunhes's comment was at least half a century premature.[4] Be that as it may, Brunhes's intellectual engagement with "the limits of our cage" did mirror a fundamental change in humanity's approach to the natural environment. Throughout human history, survival had often relied on what one might call hidden reserves: woodlands, fields, plants, and animals that people resorted to in an emergency. Some of these hidden reserves were tightly regulated commons (see chapter 23.3, Alternative Projects). Others were just difficult to access with contemporary means. Peripheral environments served as buffers against disaster, usually in the form of a vague idea rather than a contingency plan. After all, the point of hidden reserves was that no one knew when and how they would be used.

However, hidden reserves came under pressure over the course of modernity. Modes of communal management were increasingly seen as inefficient and wasteful, and the transport revolution in combination with scientific and technological progress greatly increased human options for resource allocation. Reserves were no longer just there. They called for careful planning with the power tools of modern societies: technology, expertise, and state power. It was not an idea without precedent, as the granaries of biblical times serve to attest, but it became more prevalent and potent than ever before. Premodern societies were about *having* reserves. Modern societies were about *managing* reserves. The chapters in this section aim to explore this shift.

As a geographer, Brunhes's thinking revolved around space, and the chapter on the Kruger National Park explores the transformation of spatiality through a favorite tool of the conservation community: nature reserves. With more than 200,000 protected areas around the world that together constitute more than 1,000 times the size of Belgium, it is easy to forget that nature reserves did not exist until the late nineteenth century. It was an outgrowth of what Charles Maier has called the age of territoriality: new state administrations extended control into all corners of their domain with equal force. Gone were the traditional imbalances of power between center and periphery, and natural treasures, located more often than not in peripheral regions, could now enjoy a level of protection that no government could have credibly enforced before the mid-nineteenth century. Natural environments differed enormously around the globe, and so did national and regional cultures that drew on treasures of nature, but they were all based on the authority of modern statehood.

Today nature reserves are criticized as an authoritarian gesture, but the underlying conflict is nothing new: control over contested space was at the heart of nature reserves from the beginning. The same holds true for the ecological rationale: reserves are about nature as well as culture, and both are social constructions that are open to debate. Visitors were a source of trouble and a necessity, both for cultural appreciation and to allocate revenues for a local population that typically faced constraints in access and use. Even Mother Nature created trouble because environments kept changing in utter contempt for government regulations. Designating national parks and other protected areas was fraught with ambiguities from the beginning, and yet there was no real alternative. In the world of global modernity, leaving natural environments alone was no longer an option. Remoteness was not protection anymore because, in ecological terms, remoteness had ceased to exist.

The one thing that kept conflicts within limits until the mid-twentieth century was size: nature reserves were vanishingly small when compared to total territory. But that changed in the postwar years: numbers and acreages increased dramatically, curiously at exactly the time when the powers of many nation-states were eroding. Was this a countervailing trend, or was it easier to designate reserves knowing that they were unlikely to become more than "paper parks?" In their hunger for space, nature reserves have long surpassed urban areas, though the comparison is misleading. A built-up area imposes its own rules through its physical properties. A nature reserve depends on rules that can be ignored, and often have been. In a world where doctors making life-and-death decisions found it difficult to enforce quarantines (see chapter 21, Cholera), conservationists faced dim prospects in trying to enforce access bans for their own purposes.

However, hidden reserves were not only about space but also about organisms, and pushing the limits of biology was a key trend of modern history. Organisms with superior features traveled around the world and transformed organic production regimes, and one of the beneficiaries was eucalyptus: unique to Australia when Joseph Banks collected samples in Botany Bay 250 years ago, now it delivers half of the world's wood fiber supply. Global modernity inaugurated a global competition among the species of the world, and eucalyptus stood out. Its thirst helped in swamp clearance, it was resilient to most insects and diseases beyond the Australian continent, but most of all, it grew fast—much faster than other trees that settlers, foresters, and businessmen might consider planting. It was not terribly good at other things and pretty bad in some respects—for example, eucalyptus trees are a notorious fire hazard—but one crucial feature was enough to boost its spread. Why care about undergrowth or groundwater tables when a dramatic growth rate promised quick results? It helped that growth rates offered the alluring simplicity of numbers while impoverished ecologies were more difficult to grasp. Forests were complicated, but the case for eucalyptus was not. The world needed wood fiber, and eucalyptus was good at producing wood fiber.

Biological potential was one part of the botanical exchange, and institutions the other. The travels of eucalyptus and many other species hinged on a network of gardens and people that revolved around exchanging and nurturing species. It included a remarkably diverse array of stakeholders—academics and plant lovers, farmers and estate owners, entrepreneurs and officials—and an equally diverse array of motives: profits were in play, but so was the quest for botanical curiosities and gardening prestige. All sorts of plants traveled in all sorts of ways in all sorts of directions, and that made the botanical exchange a messier affair than the neat garden plots tend to suggest. Growing expertise and new technologies were important, but so were improvisation and sheer luck, and in the end, it was probably the allocation of tremendous resources that lay at the heart of success. Before 1900, few investments in science exceeded those in botanical gardens, experimental plots, and academic institutions.

Eucalyptus was a winner of modern history, but it did not look that way for a long time. The spread of eucalyptus occurred in recurring cycles of boom and bust. Lack of experience played a role, but it was mostly about Indigenous opposition. Unlike native trees, eucalyptus did not offer much to those who lived in or close to forests, and that stimulated resistance. It was not until the postwar years that the global spread of eucalyptus turned into a sustained victory run, and the reasons deserve reflection. Maybe it was due to changes in demand, particularly the growth of a eucalyptus-based paper and pulp industry in an age of mass consumption. Or maybe eucalyptus prevailed because the system of political and economic powers was finally strong enough to marginalize local resistance. The presence of eucalyptus is not the worst biological marker of authoritarian societies.

However, pushing biological limits did not end with the global circulation of species. Humans have always sought to improve the performance of animals and plants, but as with the botanical exchange, modernity made a difference through institutions, knowledge, and the resources that it allocated to the task. Mendelian rules and genetics helped, but seed improvement was ultimately about experimentation on a grand scale. The topic of the third chapter, hybrid corn, was the result of countless trials and a major effort at genetic homogenization. After all, the quest for superior botanical potential was about performance as well as reliability, and the latter hinged on genetic uniformity.

Just like the botanical exchange, seed improvement brought a diverse array of people together, but powers were shifting toward the private sector. The most prolific seed improvers of all times, the farmers, lost out because hybrid corn did not allow reuse, and state agencies could not compete with the growing corporate might of seed companies. Genetic modification raised the stakes for corporate managers, and it opened a new front when patents on new seeds became a legal possibility. The field has been marked by a notable bifurcation of expertise ever since: companies need control of knowledge about the potential of seeds and about the patenting regime. All in all, the oligopoly of seed companies has been remarkably stable, rather unlike the claims of nation-states that seek to draw profits from genetic resources. Biopiracy is an issue of global conservation diplomacy, but unlike seed improvement for private corporations, it has not generated great profits for nation-states so far.

Knowledge shapes decisions through both content and format, and seed improvement was first and foremost about numbers. Yields per hectare increased dramatically, but these figures were only one of the parameters that farmers were interested in. What about the costs of seeds and fertilizers, the health of the soil, crop rotations, the machines, and the dependencies on experts and traders? These questions came to the fore in the controversy over the Green Revolution in the Global South, but they were on the minds of farmers everywhere: improved seeds only worked in an amenable context, as success was bound to a comprehensive project of chemical- and knowledge-intensive production. In short, whether hybrid corn boosted yields and farm incomes depended on a number of factors, but hybrid corn was hugely successful in sidelining other paths of organic development. With heavy investments in hybrids over many decades, exploring alternatives has faced huge obstacles, no matter whether the commodity in question is corn, chicken, or eucalyptus.

In other words, the final reserves are about more than materialities. They are also about the institutional framework and cultural customs that underpin development in a certain direction. Pushing the limits of biology requires a lot of scientific knowledge, but dependency goes both ways: a lot of scientific work would go to waste if corn, or chicken, or eucalyptus somehow fell out of favor. Seed improvement also needs obliging consumers, and fears of genetically modified organisms have put the public's acquiescence into doubt. The outcome is pending, and as it stands, concerns over genetically modified organisms are stronger in popular culture than in changing structures in the world of seeds.

Water is often a limiting factor when it comes to organic production, and while human manipulation of waterflows is probably as old as civilization, the high dams of modernity mark a new chapter. They gave people the chance to allocate water in unprecedented quantities, and the flow mirrored socioeconomic and political hierarchies. While hidden reserves were open to use from a broad range of people, often including the poor and the marginalized, dams subjected access to water to management decisions. It is crucial to recognize the social repercussions of decisions regarding water use, not least because high dams produced their own mythologies that drew attention away from patterns of inequality. Aswan Dam was built on a river that was ripe with memories ever since a desperate mother set her three-month-old baby (subsequently named Moses) afloat in a papyrus basket in the reeds of the Nile sometime around 1200 BC.[5] Reading the bible is easier than reading land registers or water distribution charts, but the latter decided who would profit from a dam and who would lose out.

The chapter starts hundreds of miles away from Aswan Dam, which is another way to confront the towering concrete majesty of high dams. Building a large barrier was no small engineering challenge, but it was only the most obvious element in a vast technological system whose consequences went in many different directions. Karl August Wittfogel has argued that control over water gave absolute power to a central bureaucracy, but that behemoth disintegrates upon closer inspection. Hydrological power was fragmented: it included political leaders and technocratic managers, hydraulic engineers and agricultural advisers, industrialists and fishermen, sanitarians, land owners, and other stakeholders, and bringing these concerns into a dialogue was a challenge. The history of high dam construction is also about men who managed to tie things together, but leadership was often more semblance than reality: water management was about living in a world of compromises. The story of Floyd Dominy, a major figure in the global boom of dam construction in the 1960s, is revealing. He was a power broker who initiated dam projects in the United States and beyond, but it is a matter of perspective whether this was charismatic leadership or a flight from the follow-up problems that inevitably materialized in hydroprojects. The magic that water projects had around 1900 has long faded, and we live in a twilight zone with plenty of experiences involving problems and side effects. But for those who live in darkness, perhaps even literally due to an ailing power grid, even a twilight can be a ray of hope.

Opposition to dam projects was a defining issue for environmental movements in many countries, but if environmentalism has made a difference, it was probably more in the way dams were built and water was used. Criticizing mega-projects on principle was usually more virtuous than effective. Aswan Dam was one of the world's most reviled water projects, but what mattered in and beyond Egypt was the judicious management of its many problems—from drainage to diseases. It is an ongoing challenge. Dams are Faustian bargains without a due date, gambles with environments that societies are taking everywhere on the globe, and the chances of a graceful exit are slim: dismantling dams has proved difficult even when they no longer serve a discernible purpose. With more than 55,000 large dams around the world, we are locked into a world of controlled water for better or worse.

A modern world that seeks to mobilize every remaining reserve leaves traces beyond institutions and materialities. It shows in the mental universe of communities, and the story of the rice-eating rubber tree sheds light on the mythology of interdependence. Circulating among smallholders on Borneo in the 1930s, the story gained currency against the backdrop of discrimination against Indigenous rubber production on world markets. Smallholders could not do much about global trade regimes, but they could limit their reliance on export revenues and bolster their subsistence base. That is what the story was all about.

The story of the rice-eating rubber tree thrived in a certain place and at a certain time, but it was not the only myth that people embraced to work through the experience of dependence from faraway forces. From an anthropological perspective, the story is remarkably similar to Latin America's dependency theory: both thrived in specific groups, they were about developments beyond people's control, they were about the economy and the feeling of deprivation, and they became the subject of frantic conversation for a number of years. It was just that one of these myths had more footnotes than the other.

When commodities and modes of production are similar around the globe, it should come as no surprise that cultural responses have similarities too. And then, myths are no privilege of the non-Western world. Managers on Southeast Asian rubber plantations were wrestling with their own mythology when they found out about the superior productivity of smallholders and came up with several face-saving explanations, all of which dissipated upon closer inspection. The one advantage that plantation owners held was access to the powers that be, which shaped the global trade regime. But sometimes trade regimes have unintended side effects. The chapter shows how the rise of Chico Mendes, an Amazon rubber tapper who received much acclaim as a transnational environmentalist icon, was a by-product of Brazilian autarky policy. An interconnected world does not allow for unambiguous mythologies.

In sum, it does not seem that humans were reaching limits in the early 1900s, and Brunhes did not suggest otherwise. It is only in retrospect that his speech turned into an intellectual precursor to the Club of Rome's *Limits to Growth*. Brunhes closed his lecture with reflections on the power of the mind to burst physical limits, and he was arguably onto something on this point.[6] In the following 114 years, humans proved remarkably successful in moving the bars of their cage, though rather less successful when it comes to dealing with the repercussions. In essence, mankind was not really living in an iron cage—it was more like life in one of those inflatable bouncy castles that funfairs offer to children. Humans can push limits, at times dramatically so, but in an interconnected world, it typically has unforeseen consequences elsewhere. And if we push the metaphor further, we can envision our modern predicament as akin to a bouncy castle that humans have stretched and modified as best they can, and if humans are less than proud about their achievement, it is because they know that they have transformed rather than obliterated limits. People know that pushing limits comes at a price that we need to pay in perpetuity: bouncy castles collapse in the absence of air being pumped into them. People also know that stretching cannot go on forever. For all the fun you can have in a bouncy castle, there is an underlying fear among the more mature users that things may simply pop. Nobody knows what will happen next, but it is likely that children will cry.

26

KRUGER NATIONAL PARK

RESERVED NATURE

I. BEST IDEAS

South Africa was a dispiriting place in the 1980s. Decades of apartheid rule had claimed a devastating toll on the country's society and its reputation abroad. South Africa was a living anachronism on a continent that had seen bastions of white rule collapsing in Angola, Mozambique, and Zimbabwe over the previous decade, and segregation had nurtured a brutalized society in which violence was endemic. The economy was in stagnation, social inequality was stark, tourists went elsewhere, and demographic projections foreshadowed a white minority of less than 10 percent in the new millennium. The future was bleak indeed, and the republic experienced net white emigration for the first time in 1977.[1] Even sports no longer offered the helpful distraction that they used to provide, as the country was banned from most international competitions. But there was always the Kruger National Park, the symbol of a different, more benign South Africa and a popular attraction in a country that needed one. By the mid-1980s, more than half a million people were visiting the Kruger National Park every year.[2]

They came for a wildlife experience of global fame. Visitors could look at elephants, lions, and other charismatic species from the safety of their cars, they could book safari tours and accommodation at lodges inside the park, and yet the Kruger National Park was about more than fierce creatures: the case for conservation was about culture as well as nature. A springbok replaced the king's head on South African stamps in 1926, precisely the year in which Kruger National Park was incorporated, and that nicely reflected how wild game was used in the construction of a national identity.[3] It was no coincidence that the drive toward national parks gathered momentum after the creation of the Union of South Africa in 1910, for unification turned the cultural gap between English-speaking and Afrikaans-speaking white South African people into a critical issue.[4] Historians have traced the ensuing mythology down to the park's name, which commemorates Paul Kruger, the president of Transvaal from 1883 to 1900. The Kruger National Park runs along the eastern border of Transvaal, but Paul Kruger never showed much interest in the conservation of wild game in that part of his country or anywhere else. As Jane Carruthers has argued, "The connection between the Kruger and national parks has been deliberately fomented to serve Afrikaner Nationalist political purposes."[5]

South Africa was not the only nation to recognize the cultural potency of nature. Nation-building had been in play in the creation of nature reserves ever since the Congress of the United States voted to protect Yosemite and Yellowstone in order to placate the cultural inferiority complex of white people in the New World. The two national parks, the world's first, featured trees, cliffs, waterfalls, and other natural treasures beyond anything that Europe had on offer.[6] Australia's Royal National Park, the first national park outside the United States, served the country's egalitarian sentiments, as it provided recreational space for nearby Sydney.[7] Mexico pursued similar ideas during the reform-minded presidency of Lárazo Cárdenas, and forty new national parks, most of them close to Mexico City, were open for visitors by the end of his six-year term in 1940.[8] The first national park in New Zealand, Tongariro National Park, goes back to a gift from the Maori people, symbolizing a merger of Indigenous and Western influences that resonates in the country's identity to this day.[9] East Timor held its declaration of independence in the coastal wetlands of Tasi Tolu in

2002, a site where Pope John Paul II had held mass during the Indonesian occupation of East Timor in 1989, and today the area doubles as a Peace Park (see chapter 35.3, Changes in the Land) and a bird reserve.[10] Ideas and landscapes differed, but the nexus worked all over the world. Nations built nature reserves, and nature reserves built nations.

National parks were not immune to nationalist excess. Slovenians celebrate Triglav National Park as an embodiment of the nation's spirit even though the park, designated for scientific purposes by the Kingdom of Yugoslavia in 1924 and elevated to national park status by the Socialist Federal Republic of Yugoslavia, went back to a Habsburg-era proposal that was modeled on a Bohemian wilderness reserve.[11] Maybe it was more than a fitting coincidence that the first Croatian victim of Yugoslavia's Civil War was a policeman who was killed over the issue of control of Plitvice Lakes National Park in 1991.[12] Be that as it may, national parks were, as nationalist projects go, one of the more open-minded endeavors. The discourse on national parks was transnational from the outset, and the United States held a special place in this conversation. In a speech before the American Civic Association, the British ambassador to the United States, James Bryce, praised the United States for having "led the world in the creation of National Parks" in 1912.[13] The compliment was later condensed in the notion that the national parks were "America's best idea," something that Bryce never actually said.[14] The speech nevertheless did mirror his admiration for American leadership in conservation along with an emerging notion that every self-respecting nation should commit itself to the preservation of nature. Setting aside some land for posterity became a statement of national character.

A century after Bryce's speech, the 2014 United Nations List of Protected Areas recorded more than 209,000 protected areas around the globe with a total size of more than 32 million square kilometers.[15] Some countries embraced the idea with particular vigor. Costa Rica and Belize placed about a quarter of their territory under protection and used their natural assets to attract visitors (see chapter 22, Baedeker).[16] Gabon created 13 large nature reserves in one fell swoop in 2002, surely helped by a contemporary upswing in oil prices that boosted the country's petroleum-dependent economy (see chapter 15, Saudi Arabia).[17] Some protected areas were so inhospitable to humans that their designation required little effort. The world's largest terrestrial reserve, Northeast Greenland National Park, adds almost a million square kilometers to the protected area of the world, and the second largest is in a part of the Saudi Arabian desert known as the Empty Quarter.[18] Other nature reserves are effectively inner-city parks such as Bukhansan National Park in Seoul or Tijuca National Park in Rio de Janeiro, which grew out of a nineteenth-century effort to secure the city's water supply (see chapter 17, Water Closet).[19] Some parks were acts of compensation for development projects such as Circeo National Park, whose 8,000 hectares were excluded from Fascist Italy's Pontine Marshes land reclamation project (see chapter 32, Pontine Marshes) upon Mussolini's command.[20] Other reserves were effectively accidents of nature such as the Rocky Mountain Arsenal near Denver, an abandoned US factory for chemical weapons that became a National Wildlife Refuge.[21] And some germs of nature did not gain official status, such as the demilitarized zone along the thirty-eighth parallel that South Korea proposed to UNESCO for designation as a biosphere reserve, an initiative that North Korea derailed in 2012 (though South Korea won the designation of a UNESCO biosphere reserve for land to the south of the demilitarized zone in 2019).[22] A nature reserve along a hot border might strike some people as odd, but many reserves feature relics of human civilization in their midst, and it can be anything from a log cabin to a Soviet-era missile silo, the latter being an attraction in Lithuania's Žemaitija National Park.[23] Kakadu National Park in Northern Australia is an Indigenous-owned reserve that contains an active uranium mine.[24]

Nature reserves were diverse, and so were national understandings of what defined a nature reserve. When Germany acquired its first national park by invading Poland in 1939, officials hurriedly downgraded it to a landscape reserve because it failed to meet the requirements of German law.[25] The International Union for Conservation of Nature (IUCN) now has six categories for protected areas, but that is a recent development: the IUCN, set up in 1948 as a platform for conservationists from all corners of the globe, did not come forward with a definition of a national park until 1969. Needless to say, a national park was supposed to possess particular "ecological, geomorphological or aesthetic features," but the IUCN also had two other criteria. It called for visitor access "for inspirational, educative, cultural and recreative purposes," and it required protection from "the highest competent authority of the country."[26] Both requirements came down to Faustian bargains.

2. IN THE AGE OF TERRITORIALITY

Nationalist sentiments typically flourish when something is under attack. The spread of reserves mirrored a transnational concern about an industrial civilization that was spreading its tentacles into every corner of the globe, and a feeling of an impending and irretrievable loss united conservation drives in different countries and landscapes. No longer could naturalists trust that a waterfall or a mountain was too remote to entice the interest of dam builders or quarry operators: traffic links were getting better, maps were getting more precise, and modern capitalism was insatiable anyway. In the case of the Kruger National Park, the defining issue was the growing toll that firearms were taking on the region's wildlife, and the park grew out of a game reserve that was set up in 1898. Game reserves owed their existence to fears of extinction (see chapter 11, Dodo), and yet it is rewarding to take a closer look. After all, fears of extinction were nothing new in Africa by the end of the nineteenth century.[27]

The global drive toward nature reserves did not come from a sudden outburst of environmental awareness. In the nineteenth century, affluent white people had various ways to demonstrate their appreciation of nature. People could scrutinize an almost limitless number of landscape paintings, they could read the literature of the Romantic era, and they could engage with the burgeoning natural sciences. Many zoological gardens and natural history museums opened their doors over the course of the nineteenth century, and curators took pride in rare specimens such as the Dodo (see chapter 11, Dodo), whose remains ended up in the Oxford University Museum of Natural History.[28] Natural history museums were a quintessential Western invention and typically grew under the auspices of an urban elite, but they were already turning into a global institution during the nineteenth century. The roots of Singapore's natural history museum, whose official opening year was 1878, go back to 1823.[29]

Against this background, nature reserves were just another way to appreciate and preserve nature. Sometimes older institutions helped in the creation of protected areas. For example, Westphalia had no fewer than fifty-six nature reserves by 1932, perhaps Germany's largest conservation network at the time, because Münster's natural history museum served as the hub for energetic conservation officials.[30] However, many designations did not need a scientist on duty. Spectacular scenery was a sufficient rationale for America's national parks until Everglades National Park was authorized in 1934.[31] In the case of the Kruger National Park, "scientists in South Africa were the only group which came out publicly against the park," mostly veterinarians who were concerned about livestock diseases.[32] Scientific backing was optional when the nature reserve was born, but something else was not. Protected areas were unthinkable without the state as it reinvented itself in the mid-nineteenth century.

Nation-building bolstered the authority of administrations all over the West, but other forces were also at play. Telegraph and railroad networks allowed information, officials, and soldiers to travel at unprecedented speed. Legislation and enforcement moved into new realms such as public health (see chapter 21, Cholera), and new technologies such as dams (see chapter 29, Aswan Dam) and sewer networks (see chapter 17, Water Closet) grew under the auspices of new government bodies. Recruitment for administrative posts, formerly a matter of tradition and personal allegiances, gave increasing weight to qualification and merit. While the power of the state had traditionally faded out at the periphery, the new nation-states sought a strong and uniform presence across their entire realm. Charles Maier spoke of an emerging age of territoriality, and nature reserves, often located in more remote areas, benefited directly from the new authority of the state. A protected area would have been an empty gesture in the days of the ancien régime unless the ruler showed some personal interest (usually out of a passion for hunting), but with the new ambitions and the new means of control, state jurisdiction over space could now be credibly enforced.[33]

The new authorities did not grow overnight, and some events in the early years of the conservation movement show the fragility of state power. When Congress discussed the creation of Yellowstone National Park, it had a rather faint idea about the precise location of its eminent features, which ultimately played out to the conservationists' advantage: lawmakers might have voted for a patchwork of individual sites rather than a large, rectangular space with better knowledge.[34] In South Africa and Swaziland, a number of game reserves were abolished after a few years for lack of enforcement, and even the game reserve that became the nucleus of the Kruger National Park had a turbulent early history. The first warden was killed in action one month after his appointment, the second refused to enter the game reserve on account of health, and it was only with the third successful candidate, James Stevenson-Hamilton, that the reserve achieved some stability. It helped that Stevenson-Hamilton knew a few things about controlling territory from his previous job. He was a career military officer who left the British army after the Boer War, became a highly regarded naturalist while enforcing the law of the land, and did not retire until 1946.[35]

The relationship between conservation and modern statehood was mutually reinforcing. New reserves called for administrative resources, and the imposition of spatial control bolstered the authority of officials, particularly when reserves were in regions with an underdeveloped presence. In the early 1900s, the Swedish state planted large national parks in Sami territory in order to consolidate national authority in the country's north.[36] To reassert its colonial authority after World War II, Great Britain pushed for new reserves in its African possessions, which resulted in the first national parks in Kenya in 1946, Tanganyika in 1948, Rhodesia in 1951, and Uganda in 1952.[37] Jeyamalar Kathirithamby-Wells has argued that Taman Negara, originally founded as King George V National Park, provided a showcase of accountable governance and the rule of law in postcolonial Malaysia.[38]

National parks usually had a human presence when they were designated, and curbing or eliminating land use was among the most important duties in new reserves. Residents did not necessarily lack social prestige. The National Trust for England and Wales gained many assets by acquiring estates from impoverished noblemen.[39] But more often than not, those on protected land were poor and not white, and many of the ensuing conflicts ran along class and ethnic lines. In the United States, Indian removal was part of national park history from the outset.[40] Finland created its first parks in 1938 in the face of vigorous opposition from the Peasant Party and the Communist Party, with all designations in the sparsely populated north and all supporters coming from the more urban south.[41] In the 1960s, Canada expropriated two hundred families to create Forillon National Park in Quebec and then selected "harmony between man, the land, and the sea" as the theme for its interpretative program.[42] At the Kruger National Park, evicting African residents was among the first actions of Stevenson-Hamilton, though the policy changed when it was realized that squatters on crown land also provided a welcome source of revenue for cash-strapped park administrators.[43] However, taxation did not win hearts and minds either. The Kruger National Park was an icon for white South Africans only.

Removal of Indigenous populations did not attract much attention for decades. Nations called for sacrifice in the name of a greater good, and that framed the discourse on nature reserves. They came across as land that humans had wisely set apart, a noble act of self-restraint that only the most heartless people would get excited about, and other readings triggered a furious response. Alfred Runte unleashed one of the first controversies of environmental history in the early 1980s when he argued that the national parks were actually worthless lands.[44] However, human use was not a static category. Runte built his argument on rhetoric at the time of designation, but some forty years after its creation, Yosemite National Park became a target for developers who sought and ultimately built a dam (see chapter 29, Aswan Dam) in the scenic Hetch Hetchy Valley.[45] Claims could appear out of nowhere, and they could dissipate just as quickly: more than one nature reserve grew out of military proving grounds that generals did not need anymore. Runte himself acknowledged some forms of use when he mentioned grazing sheep in Yosemite and alligator hunting in the Everglades, but that did not bring conservation up against powerful stakeholders.[46] In short, land in reserves was not literally worthless, but it was typically marginal. The eastern Transvaal Lowveld, where the Kruger National Park would eventually be located, certainly did not strike nineteenth-century settlers as terribly attractive. It was plagued by endemic malaria and horse sickness.[47]

Conservationists became more sensitive to the dislocation of Indigenous people in the new millennium, but the issue was part of conservation history from the start.[48] However, two factors kept the problem under wraps. One was context: Indian removal and the displacement of colonial subjects were a sad normality in the age of empire, and the conservationist's approach did not differ from those of dam builders (see chapter 29.2, Creations of Men) or creators of military reservations. The other was size: few regions had more than a few patches of protected land, and much of the land was state property anyway. A global inventory found only 9,214 protected sites by 1962, but that number tripled over the following two decades, and growth has been dramatic ever since.[49] Conservation has long surpassed urban growth in its hunger for fresh land, it makes extensive claims to private properties today, and it shows no sign of relenting. Target 11 of the Aichi Biodiversity Targets, adopted in 2010 by the Convention on Biological Diversity (see chapter 24.3, Getting Serious) and ratified by 196 countries, demanded protection on at least 17 percent of terrestrial and inland water areas by 2020.[50]

Several factors account for this remarkable change of affairs. Space emerged as a critical resource as road networks expanded down to the last village (see chapter 34, Autobahn), and nature reserves functioned as one of the few instruments that could keep an encroaching industrial civilization at bay. Scientists became the defining force in conservation over the course of the twentieth century and offered ever more precise information on the number of plants and animals, their needs, and their routes of migration. Tourism (see chapter 22, Baedeker) provided another boost, though few parks were actually turning a profit.[51] Even Kruger National Park, which has consistently attracted more than a million visitors per year in the new millennium, ultimately costs more money than it brings in.[52] And then there was charisma: sacrificing some space in the interest of a higher good touched a nerve among many people, particularly those who lived in urban areas, were poorly sensitized to land issues (see chapter 6, Land Title), and did not face the risk of constraints themselves. Nature reserves were among the most popular instruments of environmental policy, and environmental organizations were usually happy to support them. Interest in designations is often greater than interest in what they really mean, and the imagination knew no limits when it came to the area under protection. In January 1969, the energetic executive director of the Sierra Club, David Brower, placed a page-and-a-half advertisement in the *New York Times* that called for "a sort of Earth National Park."[53]

The one thing that did not keep pace was the power of the modern state. Everything seemed easy as long as conservation was a plaything for powerful men, who were running the show into the 1970s. The history of Auyuittuq National Park in the Canadian arctic began when Canada's minister of Indian affairs Jean Chrétien flew over Baffin Island, watched some spectacular fjords below, turned to his wife and promised to "make these a national park for you." Back in the office, he relates, "I asked for a map, and with a pen I circled off 5,100 square kilometres," and felt "very big" for the rest of the day.[54] Chrétien created ten national parks in this vein within four years. But formal negotiations eventually took over—even on Auyuittuq, designation eventually required a word with the Inuit.[55] Furthermore, expanding park administrations turned into bureaucracies with functional differentiation over time, though that created its own set of problems. The Kruger National Park was criticized for pursing sectoral policies for tourism, water, and elephant management with compartmentalized knowledge and poor integration.[56] In Kenya's Tsavo National Park, debates about the management of the elephant population ended in gridlock by the 1970s because different constituencies were unable to agree on a common understanding of the problem.⁵⁷ And then there were the dwindling resources, financial and otherwise, of government bodies in the final quarter of the twentieth century. The state was no longer what it used to be.

According to Charles Maier, the "technological, cultural, and sociopolitical scaffolding [of territoriality] began to corrode and fall apart in the late 1960s, initiating a process of profound transformation that continues today."⁵⁸ Many park administrators felt the fallout in that they were hamstrung by lack of means, and many of the new nature reserves were declarations of intentions that never materialized. "Paper parks" are a global concern: upon presenting a status report at the IUCN World Parks Congress in Sydney, Australia, in 2014, one of the lead authors conceded that "we know very little about the effectiveness of the worlds' protected areas."⁵⁹ And then, is strict enforcement really a good idea, or does it breed disaffection that ultimately turns nature reserves into a losing proposition? Conservationists have used the authority of the modern state to many good effects, and yet that authority had its charms as well, and conservationists cherished the opportunity to represent modern statehood even when circumstances suggested other priorities. When a conservation league in Southwest Germany set up a conservation watch in April 1940, complete with rosters, ID cards, and mandatory reports after each patrol, more than seven hundred members volunteered for service.⁶⁰

3. THE BUSINESS OF SEGREGATION

In 2016, South Africa had some 4,000 employees in 21 national parks, which meant, in statistical terms, that there was one staff member for every 1,000 hectares.⁶¹ With a surface area of 19,633 square kilometers, the Kruger National Park is larger than any park in Western Europe or the contiguous United States, and there is also no Western equivalent for some of the challenges that the park has faced over time. The eastern border of the Kruger National Park is also the border to Mozambique, and the South African military maintained a permanent presence inside the park during the 1980s in order to catch refugees from Mozambique's civil war. In addition, the military planted rows of spiky sisal plants along the border upon Israeli advice, provoking concerns among conservationists about the spread of an invasive species (see chapter 14, Cane Toads) and filling the stomachs of wild animals who ultimately devoured the green wall.⁶² After the end of apartheid, Mozambique set up its own national park across the border, and together with adjoining land in Zimbabwe, the protected areas add up to a nature reserve larger than Belgium.⁶³

Visitors were another source of headaches, and one that national parks around the world were well familiar with. As Patrick Kupper remarked in regard to Switzerland's national park of 1914, the park's sponsors were "calling all nature lovers," and those who came were tourists (see chapter 22, Baedeker).⁶⁴ The Kruger National Park was wrestling with overcrowding problems as early as 1953.⁶⁵ However, park administrators feared hordes of tourists as much as they coveted the stream of revenue that tourists brought, and they were not shy about building infrastructure to pave the way for visitors. America's National Park Service launched a comprehensive building program and teamed up with railroad companies that urged Americans to "See America First"—that is, to go to the national parks before that mandatory trip to Europe, which, in contemporary opinion, made one's education complete. Boosters even commissioned Swiss-style chalets and coaxed staff into pseudo-Alpine dirndls. When it came to America's national park, you did not have to be Indigenous to become a victim.⁶⁶

There was no lack of warnings and lamentations. The real goal of Bryce's seminal speech was to argue against allowing cars into Yosemite Valley (see chapter 34, Autobahn): "It is not merely that dust clouds would fill the air and coat the foliage, but the whole feeling of the spontaneity and freshness of primitive nature would be marred by this modern invention, with its din and whir and odious smell."⁶⁷ Today, a century later, World Wildlife Fund–sponsored tours in Indian tiger reserves provide a lightning rod for critics of conservation.⁶⁸ Visitors naturally expect the usual amenities of modern life down to air-conditioning (see chapter 20, Air-Conditioning) in remote lodges, and yet crowds were more diverse than they looked. A study of the Kruger National Park has shown that international tourists "were primarily interested in large predators and mega-herbivores," while local visitors had more of an eye for birds, plants, and less easily observable mammals like sable antelopes.⁶⁹

The natural world added another layer of complications. Droughts brought Kruger officials to launch a water provision program in the 1930s that had more than 300 boreholes at its peak.⁷⁰ Faced with a growing elephant population, park managers killed 16,201 animals in a controversial culling program from 1967 to 1994.⁷¹ Researchers took a critical look at ornamental plants in tourist villages and found that landscaping had led to the accidental introduction of invasive species (see chapter 14, Cane Toads).[72] After decades of intensive care, an academic publication of 2003 found that it was "generally accepted today that Kruger was overmanaged in the latter half of the twentieth century."[73] A poaching crisis put the rhino population at risk.[74] Kruger's rich biology called for constant attention, and other parks showed where negligence would lead. In the 1970s, an Italian environmentalist called Mussolini's Circeo (see chapter 32, Pontine Marshes) a nature reserve "born dead."[75]

With that, one big question was hanging over the expansion of nature reserves: Did it really work? Protected areas were not immune to pollutants (see chapter 16, London Smog), plastic bags (see chapter 40, Plastic Bags), and the many other emanations of industrial society, nor were plants and animals obliged to respect the borders that governments imposed. And what about private estates akin to nature reserves? Privately run game farms claim 13 percent of South Africa's territory whereas state and provincial reserves add up to just 6 percent.[76] All in all, conservation by government fiat looked increasingly suspicious, negotiations with local and Indigenous communities became more frequent, and community-based conservation became a discussion point in conservation circles in the 1980s.[77] As it turned out, community-based conservation tends to multiply the vagaries of nature by those of society, economics, and culture, and even where it succeeds, as in an ecotourism project in Waluma in Papua New Guinea, concerns remain about whether achievements could survive a surging number of visitors.[78]

When they run into trouble, modern political systems typically respond by invoking science.[79] Scientific knowledge became the dominant resource for nature protection policy in the second half of the twentieth century, but it was more than the neutralizing and objectifying force that decision makers had sought. Sometimes experts were simply wrong. A Japanese park planning team urged the Ethiopian government to create a protected area in the Omo Valley in 1978, failing to recognize that the landscape was the result of agricultural practices of the Indigenous Mursi people, whom they were trying to expunge.[80] In other cases, scientific buzzwords provided camouflage for vested interests. A study on Guinea revealed how "biodiversity" (see chapter 11, Dodo), a global shibboleth in conservation circles since the 1992 Convention on Biological Diversity (see chapter 24.3, Getting Serious), is really a loose assemblage of practices from computer-based modeling of ecosystem dynamics to the search for wild plants with economic value.[81]

In light of all these problems, maybe it was time to rethink the case for national parks. Derek Hanekom, a spokesman on agriculture for the African National Congress, proposed to abolish the Kruger National Park and put the land to more productive use in 1993, but the idea drowned in a storm of protest.[82] It is a matter of debate whether the endurance of Kruger National Park is due to its inherent charms or to the fact that annihilation would smack of surrender in the protection of planet earth, but the public response left park managers no choice but to sputter on. Rhino poachers were tracked and arrested, but traders of rhino products in East Asia escaped punishment.[83] Park managers reached out to surrounding communities.[84] Tourists were properly instructed rather than sent away. Visitors can still enjoy nature in the national parks of the world, but only after a primer on the code of conduct.

When Stevenson-Hamilton wrote his final report as game warden after forty-four years in office, he concluded by admonishing his successors to "keep it simple, keep it wild."[85] But wilderness was no less a human construction than were other perceptions of nature, and the same held true for the simplicity of nature reserves: it was a result of selective observation from the outset.[86] There is probably no way back to our former naiveté, and if there is, the path will likely be a painful one. Nature reserves are contested space, conservation is only one of many stakeholders, and clever managers strive to balance divergent claims as best they can. One of the more popular solutions is zoning, though it does not always achieve its goal. A development plan for the Pilanesberg National Park in Bophuthatswana failed in the 1980s because it proved "too complex to manage effectively."[87] Today, however, many park managers impose different rules for different places in order to concentrate the human footprint in some areas while leaving others undisturbed. Kruger National Park has zoned roughly half of its land as wilderness areas and has made a point of including the three million people living in its vicinity during planning, and yet zoning is an odd strategy in a post-apartheid South Africa.[88] Spatial segregation was a core element of apartheid policy, and while the rationale in nature reserves is ecological rather than racial, different zones for different people challenge humanity's natural craving for justice. And in the end, if experience has taught us anything about nature reserves, it would seem that borders will remain more permeable and more contested than their sponsors would like. Reserved nature will be a part of conservation policy for the foreseeable future, but maybe just because no one has come up with a better idea.

27

EUCALYPTUS

SUPERTREES

I. GROWTH RATES

The Arabs called it *shajarat al-Yahud*: the Jew's tree. When the first Zionists arrived in Palestine in the early twentieth century, eucalyptus trees were such a common feature of their settlements that they invited ethnic stereotyping.[1] Eucalyptus trees supported land reclamation by draining wetlands and decreased the incidence of malaria, and their fast growth promised a quick return. They had material benefits for the Jewish settlers, and they had symbolic power. The Jewish National Fund maintained an afforestation program since its foundation in 1901 because trees were both a valuable resource and "a source of spiritual renewal, a validating biological symbol of [the settlers'] hopes for a Jewish and Hebrew cultural renaissance."[2] It left its mark in the land as well as collective memory. When Israel's Ministry of Agriculture and Rural Development conducted a poll on the "most Israeli tree" in 2012, eucalyptus won, beating olive, cypress, and pine.[3]

It was quite an achievement for a tree from the other end of the world. The eucalyptus genus includes more than six hundred species, all of which come from Australia and adjacent islands, and they were virtually unknown beyond the region until the British natural scientist Joseph Banks, the future president of the Royal Society, went ashore in Botany Bay in 1770 and collected the first specimen during James Cook's first voyage to the southern Pacific.[4] Eucalyptus dominates the woods of Australia in singular fashion: according to the Australian environmental historian Tom Griffiths, "No other comparable area of land in the world is so completely characterised by a single genus of trees."[5] An extensive root system allows eucalyptus to draw in water, a critical advantage on the world's driest continent, and the tree grows well in poor and degraded soils. Only Australia has mammals that can digest eucalyptus leaves and insects that feed on the tree.

In short, eucalyptus would easily win the Australian equivalent to the Israeli contest, though only by default: as the historian Geoffrey Bolton noted, "Australians were finding that foreigners seemed to value the eucalypt more than they did."[6] Eucalyptus began its overseas career as a botanical curiosity, but people around the world eventually realized the tree's extraordinary potential. It had a reputation for swamp clearance and malaria prevention decades before the Zionists brought eucalyptus to Palestine. The 1851 Crystal Palace Exhibition in London showcased its value as lumber with two giant blocks of *Eucalyptus globulus* in the section on colonial produce.[7] The fast-growing tree promised raw material for the "tools of empire" such as telegraph poles and railroad ties and many other uses from mining to firewood.[8] Eucalyptus trees also served as windbreaks and helped stabilize land that was prone to erosion. Some species have produced eucalyptus oil for medical and other purposes since the mid-nineteenth century.[9] The tree's properties also matched an imperial desiccation discourse that depicted tree planting as a crucial part of the advancement of civilization.[10]

Eucalyptus trees had several uses, but they were equally valuable for the things that they *failed* to provide. In light of the perennial conflicts over customary forest use that characterized the rise of sustainable forestry (see chapter 4, Sustainable Forestry), it helped that eucalyptus trees decided conflicting claims through their very nature. Animals outside Australia could not eat its leaves, grass and underbrush did not grow well in its shadow, and illicit cutting was a challenge when the trunk grew to a diameter of more than a meter within a few years. In short, eucalyptus was the perfect tree for forest administrations that sought to produce for markets and eliminate Indigenous uses, and they served as powerful symbols that new times had arrived in the

27.1 Eucalyptus trunks near a sawmill in Khadera, Palestine, in the 1930s. Image, Matson Photograph Collection, Library of Congress.

land. The fast-growing trees were a statement that the woods were now first and foremost a capitalist resource.

With eucalyptus comprising more than seven hundred species, its global career was more of a group assault, and sometimes bad choices delayed the triumphant advance. The eucalyptus boom in India was postponed until after independence because foresters selected species that performed poorly on the subcontinent.[11] But by 1900, just a few decades after the first experimental plantings, eucalyptus had achieved a strong presence in a number of countries around the world. It was planted en masse in the French colony of Algeria, it drew water from Italy's Pontine Marshes (see chapter 32, Pontine Marshes), and plantations were standing in Portugal, Spain, Brazil, Argentina, and Hawaii.[12] Ethiopia introduced eucalyptus in the 1890s in order to deal with deforestation around Addis Ababa.[13] A frost-resistant variety even took root in the Otago region in New Zealand's south.[14] Eucalyptus also thrived in California, more or less fulfilling a prophecy of George Perkins Marsh who had learned about the tree's potential in Italy and predicted a boom in the American Southwest.[15] All over the world, people realized that eucalyptus could build biomass more rapidly than native tree species and they banked their fortunes on the neophyte (see chapter 14, Cane Toads). But eucalyptus owed its rise to more than its biological potential. The global expansion was the result of an equally global network.

2. BOTANICAL EXCHANGE

By the mid-nineteenth century, a complex web of institutions, societies, and individuals provided an infrastructure for the transfers of plants around the globe. The Dutch East India Company founded the Cape Botanical Garden in today's South Africa as early as 1694.[16] In the eighteenth century, France established a tightly organized network of gardens and *botanistes du roi* that was part of the French "scientifico-colonial machine."[17] The British network developed in bits and pieces but eventually grew to thirty-three stations from Fiji to Jamaica, and many of these stations were in turn hubs for regional networks. The botanical garden in Calcutta, founded with East India Company backing in 1787, drew support from a Horticultural Society and an Agricultural Society that were organized in Bengal in 1816 and 1820.[18] When Durban Botanical Gardens started distributing *Eucalyptus globulus* trees in Natal in the late nineteenth century, it was quickly joined by privately owned nursery gardens in nearby Pietermaritzburg.[19] Legions of gardeners and naturalists devoted their lives to the collection and propagation of new species, and some of their names are recorded in the language of biology to the present day. The Douglas fir is named after David Douglas, who found the tree during an expedition to the American Northwest in the 1820s and died on an excursion to Hawaii a few years later when he was unfortunate enough to fall into a pit trap where he was even more unfortunate to encounter a raging bull.[20]

Plant transfers were collaborative efforts by nature, and yet the botanical exchange offered plenty of opportunities for ambitious men. Eucalyptus made the career of Ferdinand von Müller, who was appointed government botanist for the Australian colony Victoria in 1853 and director of the Royal Botanical Gardens in Melbourne in 1857.[21] Robert Fortune helped build the Indian tea industry by smuggling plants, seeds, and trained tea workers out of China after the First Opium War (see Interlude, Opium), and several popular books about his time in China added to his fame.[22] France created a botanical garden on Mauritius at the urging of Pierre Poivre, an enterprising Jesuit turned administrator whose reading list ranged from Richard Cantillon's *Essai sur la nature du commerce en général* to Chinese natural history.[23] Sometimes only bold action could save the day. The transfer of rubber seeds from Amazonia to Southeast Asia succeeded only because the Royal Botanic Gardens at Kew devoted more than three hundred square feet of precious greenhouse space to the germination of more than sixty thousand *Hevea brasiliensis* seeds, a tour de force that paid off because just 4 percent of the seeds germinated.[24] But for all the skill and knowledge, there was always an element of chance that academic rigor was never able to exorcise. Even important tools of the trade were ultimately due to good luck. The Wardian case, a portable greenhouse that greatly improved the survival rate of plants on long sea voyages, was the result of an accidental discovery of the London physician Nathaniel Bagshaw Ward.[25]

The botanical exchange served national interests, but it was transnational in nature throughout the nineteenth century. Ferdinand von Müller was born in Germany, received a doctorate from Kiel University, left Europe for Australia in pursuit of a healthier climate, and never came back.[26] Nathaniel Wallich, who helped make tea an Indian commodity, was born to a Jewish merchant in Copenhagen and came to the Calcutta Botanic Garden via a Danish settlement at Serampore in Bengal.[27] Plants routinely crossed national borders, and botanists did not even hold back when species had obvious economic potential: Kew sent the precious rubber seeds to Ceylon and Singapore as well as the Dutch East Indies, German East Africa, and Portugal's Mozambique.[28] Not until the twentieth century did people begin to think about biological assets in terms of legal titles (see chapter 28.2, Legal Titles).

Underpinning the endeavor was a remarkably diverse set of ideas. Productive gains were an obvious motive: botany was "big science and big business," as Londa Schiebinger has noted.[29] However, the diligence in collection and classification work suggests that more was at stake than profit seeking. Wealth was probably more important as a general idea than as an incentive for individuals: most botanists settled for eternal fame in the form of a reference in the Linnaean taxonomy. Botanical curiosity was an important driving force, as was the quest for gardening prestige, with significant overlaps between both: botany was an academic field as well as a gentlemanly pursuit. There was also a desire to preserve and learn from nature. In fact, one environmental historian, Richard Grove, has pointed to these motives as proof that the origins of environmentalism lay in the tropics.[30] Botanic gardens were also about the display of power, as Kew's roots as a royal hobby serve to attest. Others came to the botanical gardens for relaxation and sociability, which did not necessarily interfere with academic ambitions: when William Hooker became director of Kew in 1841 and pushed for botanic professionalism, many of his allies "saw Kew's future . . . as a place for the amusement and edification of the nation," as Richard Drayton declared.[31] In fact, Drayton even traced the roots of the botanic garden to Christian myths of Eden.[32]

Regional problems added more worldly concerns to the mix of ideas. On Mauritius, anxieties about deforestation bolstered the case for a botanical garden.[33] Berlin's venerable Botanic Garden and Botanic Museum engaged in a hectic change of focus when Germany acquired colonies in 1884.[34] Sydney's botanical garden grew out of the unplanned cooperation of "British savants, Australian governors, commercial plant collectors and ambitious young botanists," a group held together by little more than the garden and its plants.[35] Concerned about its fuel supply, in 1903, the Brazilian Paulista Railroad Company appointed as director of forestry a young graduate of the University of Coimbra, Edmundo Navarro de Andrade, who spent the following thirty-eight years building an experiment station, numerous forest farms along the company's tracks, and a transnational reputation as a crusader for eucalyptus.[36] Navarro de Andrade's propaganda machine made a lasting impression on the Munich forestry professor Karl Leopold Escherich (see chapter 4.2, Specialist Trees, Specialist Minds), who visited the experiment station in 1926: "If I were to cultivate land in Brazil, and if their figures were just halfway accurate, I would not plant anything but eucalyptus."[37]

The diverse set of motives and agents turned the network of botanical exchange into a mix of hub and spoke with erratic threads that grew out of personal acquaintances, geographic proximities, individual hobbyhorses, or sheer chance. There were some command posts, but they never had full control of the threads. In fact, sometimes the threads seemed to control the commanders, as botanical networks had the power to turn self-confident directors into mere puppets of the system. Melbourne's Ferdinand von Müller was a fervent advocate of the acclimatization movement that sought to Europeanize Australia's nature, but international interest prompted him to devote much time and attention to plants moving in the opposite direction, and Müller ended up supporting the export of *Eucalyptus regnans* and other species that did not represent the perfection of nature in his personal judgment.[38] Some projects ended in complete failure. Britain never managed to get the Chinchona tree from Latin America to its Asian possessions.[39] Other projects were transferring pathogens along with the plants: the epidemic cane diseases that struck sugar plantations around the world in the mid-nineteenth century were an unintended by-product of the global exchange of new sugarcane varieties (see chapter 2, Sugar).[40]

In spite of the gardeners' best efforts, nature did not become irrelevant. For example, environmental conditions deprived Kew of any significant role in the global career of eucalyptus. The tree did not grow well in London's climate, and it is not a good idea to plant a giant tree in a greenhouse.[41] Sometimes environmental ignorance forced researchers to reinvent the wheel, as in the rubber transfer project, which Michael Dove argued "was far too uncoordinated and happenstance to merit the term *project*."[42] Transport was another source of trouble, as ships could sink or otherwise fail to arrive at their destination: the mutiny on HMS *Bounty* delayed the transfer of breadfruit trees from Tahiti to the West Indies (see chapter 7, Breadfruit).[43] And even when plants got to their destination intact, unexpected problems could arise in the final stretch. When the first eucalyptus trees arrived at a Rothschild-sponsored swamp draining project in Palestine in 1900, planting was such an awkward job that workers received extra compensation in the form of a daily bottle of cognac.[44]

Needless to say, the episode became the stuff of legends. Settler life is about overcoming hardships, frontiersmen can deal with an excess supply of alcohol (see Interlude, Opium), and both invite storytelling. However, the combination of careful planning and improvisation neatly captures how the botanical exchange worked: it was a global network whose operation relied on alert repairmen all along the way. More precisely, it was a global network operating at high speed: more than eight thousand plants were leaving Kew annually in all sorts of directions beginning around 1870, not counting the surplus bedding plants that went to London's poor.[45] While the core business was thriving at Kew, the satellites took on additional tasks on the periphery, nicely encapsulated in a string of new creations in the Caribbean. Kew had major hubs on Jamaica and Trinidad, and yet botanical gardens sprang up in Grenada and Barbados in 1886, in Saint Lucia and Dominica in 1889, and British Honduras in 1894. West Indian sugar planters called for subsidies when the world sugar price collapsed in the 1880s (see chapter 2, Sugar), and banking on science and the promise of higher yields offered a convenient way to placate their concerns.[46] The new institutions resembled agricultural experiment stations in practice, but botanical gardens were not a place for intellectual purists. More often than not, gardening was a messy job, and so was the botanical exchange.

In short, the transfer of plant species to new worlds was much more complicated than the first generation of environmental historians assumed. In his book *Ecological Imperialism*, Albert Crosby depicted the European flora and fauna as a staunch ally of the imperialists: in his reading, the global spread of European plants, animals, and diseases gave crucial support to their global hegemony.[47] But if that was the full story, an Australian upstart like eucalyptus would not have stood a chance.[48] Profits, national and personal prestige, and the joys of botanical exploration and experimentation came together in a collective effort full of surprises and unexpected side effects, and the tremendous change of ecosystems around the globe must not distract from the improvised character of the underlying network. The presumed master designers of the world's ecologies were really more akin to benevolent mudslingers who threw assets around as if in an improvement frenzy and then watched what stuck. As it happened, eucalyptus was one of the plants that stuck more often than not. But that was not always the end of history.

3. FIGHTING GIANTS

By the early twentieth century, eucalyptus was present around the globe, but knowledge about the tree was still far from sufficient. At least that is how Kew's *Bulletin of Miscellaneous Information* described the situation in 1903. Focusing on *Eucalyptus globulus*, one of the most popular eucalyptus species colloquially known as "blue gum," the bulletin bemoaned the "excessive trust" that ignorant people were putting into the neophyte. The botanists of Kew suggested a more critical perspective on the tree, for practical results had rarely lived up to expectations: "Few plants have been the cause of more disappointment than the Blue Gum."[49]

Hopes were typically high when eucalyptus entered a new country, and people were usually stunned when they discovered that there was more to the tree than growth rates. The tree's success was partly due to the absence of Australian pathogens in other parts of the world, but some domestic enemies caught up with the global spread after a while. In South Africa, plantations suffered serious defoliation when the eucalyptus snout beetle entered the country in 1916, and it took more than a decade until the introduction of a parasite from Australia brought the infestation under control.[50] But even when trees were perfectly healthy, their effects on groundwater and biodiversity and their posing a fire hazard gave reasons for concern. Ethiopia's Ministry of Agriculture issued a decree in 1913 ordering that two-thirds of the eucalyptus trees be uprooted and replaced with mulberries, and while it was barely enforced—eucalyptus was the most common tree on the Ethiopian highlands a century later—the proclamation mirrored the strength of concerns over the tree's thirst some twenty years after its arrival at the Horn of Africa.[51] In Palestine, 78 percent of the trees planted with Jewish National Fund support were eucalyptus before 1920, but the share declined dramatically during the 1920s, and hardly any were planted in the 1930s.[52] One contributing factor was that Jewish settlers often planted eucalyptus to drain swamps, but the tree was no longer needed when the water was gone. There was nevertheless more to the change of fortune. The boom-and-bust cycle was also about learning experiences.

Of course, eucalyptus was never popular among local populations who needed woodlands for their livelihoods. The tree was always the favorite of those who preferred to look at the natural world in monetary terms: characteristically, Navarro de Andrade had a degree in agronomy.[53] However, some politicians had second thoughts when they witnessed the reality on the ground. The South African statesman Jan Smuts opposed eucalyptus because it destroyed the unique native vegetation in the Cape Province.[54] In Lesotho, where the tree was planted extensively in and along gullies for soil conservation (see chapter 13, Little Grand Canyon), officials abandoned eucalyptus in the 1960s when they found that erosion continued even under the tree's cover.[55] Summarizing his observations on eucalyptus plantations in Brazil, Escherich declared that they were "no biological model, just like all monocultures." However, he was delighted to find a German expatriate who introduced *Dauerwald*-type forestry (see chapter 4, Sustainable Forestry) in Brazil. It was a sentimental moment for the widely traveled German professor when he observed mixed stands of eucalyptus and other trees, all the more so because they were growing on an estate that satisfied Escherich's penchant for German order and cleanliness: "We almost forgot that we were in Brazil."[56]

By the middle of the twentieth century, enough was known internationally about the tree's problems and side effects to warrant careful thinking and selective use. But in reality, the great boom was yet to come. Eucalyptus expanded globally to such an extent that it supplied

27.2 Members of Brazil's Landless Workers Movement occupy a eucalyptus plantation in Eunápolis in April 2011. Image, Joacy Souza / Alamy Stock Photo.

no less than 50 percent of the world's total wood fiber consumption in the new millennium.[57] It was due to new markets: pulp and paper, still listed under "miscellaneous uses" in a handbook of 1961, grew dramatically over the following decades as new production processes met with the inherent wastefulness of postwar consumer societies (see chapter 40, Plastic Bags).[58] And it was due to rampant indifference toward the demands of Indigenous populations, as eucalyptus was the favored tree of authoritarian governments and development agencies. In the tropics alone, forest plantations grew fivefold in size from 1950 to 1980. Of the twenty million hectares under plantation in 1980, eucalyptus claimed about a third.[59]

The expansion did not fail to provoke criticism. In India, politicians, scientists, and activists held what came to be known as the "Great Eucalyptus Debate" in the 1980s.[60] In Spain and Portugal, farmers embraced direct action when they destroyed eucalyptus seedlings after the collapse of Fascist rule.[61] Similar events happened in Thailand where farmers began a bitter struggle against the spread of commercial eucalyptus plantations in the late 1980s.[62] Discontent was huge, but most of it remained confined to the countries of the Global South. One of India's leading eucalyptus critics, Vandana Shiva, is better known in Western environmental circles for her opposition to genetically modified organisms (see chapter 28.3, Business Models).[63]

Protest movements were usually about more than eucalyptus. For all the passion that Thai farmers brought to ripping out saplings, their real concerns were about land rights and the appropriation of communal land.[64] Forests are easy targets if you want to attack authorities: it is usually less dangerous to storm a nursery than the halls of power. But forests are also resilient: when it comes to taking a stand, trees can easily outlast protesters. And then there is the fundamental asymmetry that runs through conflicts over modern forestry (see chapter 4, Sustainable Forestry): people need forests, but forests do not need people. As Vandana Shiva pointed out in her *Ecological Audit of Eucalyptus Cultivation*, the tree was perfect for absentee landlords with troubled labor relations: "In fact, the labour displacing potential of Eucalyptus was the first motivating force for large landowners to transfer from foodcrop cultivation to Eucalyptus farming."[65] The size and extent of eucalyptus plantations is not the worst indicator of the state of democracy in the countryside.

Some 250 years after Joseph Banks's discovery in Botany Bay, eucalyptus has established itself around the world, but it has also changed tremendously in the process: under the rules of global modernity, even a supertree is up for science-based improvement. In fact, human tampering has reached a new level in recent decades. In Brazil's pulp sector, the mean productivity of eucalyptus jumped from ten cubic meters per hectare per year in 1965 to thirty-eight cubic meters forty years later.[66] The use of mineral fertilizer (see chapter 19, Synthetic Nitrogen) is routine in Brazilian eucalyptus plantations, and so is genetic manipulation: half of the country's eucalyptus forests are clonal, and a tropical hybrid type (see chapter 28, Hybrid Corn) developed in the 1980s serves as "the world-class benchmark for clonal forest productivity."[67] Inspired by the growing interest in energy crops (see chapter 33, Chemurgy Movement), biotech companies are working on genetically modified eucalyptus trees that can be harvested after twelve to eighteen months.[68]

In *The World Without Us*, Alan Weisman describes eucalyptus as an invasive species (see chapter 14, Cane Toads). As "a ghost of the British Empire," eucalyptus will "bedevil the land long after we've departed."[69] However, after the transformations of recent decades, it is no foregone conclusion that eucalyptus will prevail: the supertree is now so dependent on humans that it may disappear as spectacularly as it has spread.

But with all the investments, monetary and otherwise, that people have made in the genus, eucalyptus is unlikely to lose human patronage anytime soon. Even England, the site of many failed experiments, is giving the tree another try, this time banking on demand for renewable energy (see chapter 33, Chemurgy Movement) and milder winters in the wake of global warming.[70] The magic of eucalyptus endures, for it is about one of modernity's obsessions. When modern societies catch sight of an impressive growth rate, they act as if nothing else matters.

28

Hybrid Corn

Breeding Ambitions

I. AGRICULTURE'S MANHATTAN PROJECT

Among the many mysteries of the modern world, few are as intricate as the path to the White House. Henry A. Wallace almost made it with a strong family, a media career, and a Russian connection. A native of Iowa, Wallace worked for many years as writer and editor for *Wallaces' Farmer*, a popular family-owned weekly in the rural Midwest. Roosevelt made him secretary of agriculture in 1933, a post previously held by his father from 1921 to 1924, and Wallace shaped the rural policies of the New Deal until Franklin D. Roosevelt selected him as his running mate in the presidential election of 1940. Along the way, a twisted religious journey brought him to befriend Nicholas Roerich, a Russian émigré and self-styled guru who fed Wallace's spiritual needs. Their correspondence became a scandal when it was made public in 1948, and Wallace's biographers have called the Russian connection "the most embarrassing chapter of his public life."[1] And at the start of his journey, Wallace went to Iowa State and conducted experiments with corn. He won a gold medal at the Iowa Corn Yield Test of 1924, set up the Hi-Bred Corn Company in 1926, and the company's product was all the rage among the region's farmers a dozen years later: hybrid corn.[2]

The quest for better seeds was probably as old as agriculture itself, and Wallace was well aware of this long history. As a teenager, he wrote an article for *Wallaces' Farmer*, "The Aztecs as Geneticists," which looked at corn improvement in pre-Columbian Mexico.[3] But as with so many things, a combination of new institutions and new insights transformed the endeavor over the course of the nineteenth century. A burgeoning network of academic institutions, Iowa State among them, brought unprecedented resources to seed improvement. The breakthrough of genetics around 1900, often framed as the rediscovery of Mendel's laws of inheritance, provided breeding with a conceptual backbone.[4] Biologists could henceforth work with a clearer understanding of inheritance, and one of the focal points of interests was hybrids, crosses between two dissimilar but related plants whose properties, including yields, could exceed the potential of either plant. The result was what Jack Kloppenburg has called "agriculture's Manhattan Project."[5]

As comparisons go, it was not a perfect one. While the Manhattan Project was about the basic research of nuclear physicists that bred practical consequences, seed improvement was about the opposite process: hybrids were a matter of breeding practices that were subsequently enhanced through a growing understanding of the genetic fundamentals. However, the comparison provides an idea of the scale of the breeding effort, the investments at stake, and the immense pressure that protagonists had to cope with. Mendelian laws helped in understanding how inheritance worked, but they did not specify which crosses would yield superb results. Finding a promising cross was a matter of experimentation on a grand scale, and when a winning combination was found after thousands of trials, it took another gargantuan effort to move from a small seed sample to a standardized mass product. Patience and discipline were crucial character traits for plant breeders, and so was a tolerance for frustration, as most crosses yielded disappointing results. The breeders also had to live in ignorance about the molecular basics of inheritance until the discovery of the double-helix structure of DNA. A German academic, Kurt von Rümker, called plant breeding "a step into the dark."[6]

But when a superior hybrid was finally found and brought to market, it could achieve hegemony within a matter of years. Hybrid

28.1 Demonstration plot of hybrid corn planted at the Iowa State Fair, Des Moines, September 1939. Image, Arthur Rothstein, Farm Security Administration—Office of War Information Photograph Collection, Library of Congress.

corn did not become a mass-produced commodity until the mid-1930s, "but by 1942 virtually all Iowa farmers were planting all their corn acres with it," and the increase in yields was so dramatic that many farmers were struggling to find adequate storage.[7] The contrast drew the attention of a graduate student at the University of Chicago named Zvi Griliches who used hybrid corn for a pioneering case study in the diffusion of technological innovation. It became a classic of econometrics.[8]

Hybrid corn differed from open-pollinated varieties in more than yield potential. Since the eighteenth-century experiments of the German botanist Joseph Gottlieb Kölreuter, hybrids between species were known as sterile.[9] This mattered for hybrid corn because sterility put an end to the standing practice among farmers to save a part of the harvest for the next growing season. Hybrid corn might grow again, but not with nearly the vigor of the original, as disbelieving farmers recognized when they replanted nonetheless. In other words, farmers who switched to hybrids ceased to be masters of their own seeds and had to buy seeds on the market, but what was a liability for some was a business opportunity for others. Commercial seed companies gained secure markets for hybrids, and they emerged as the defining authority on matters of seeds. As Deborah Fitzgerald has argued, hybrid corn was the first field of agricultural expertise where land-grant colleges and the United States Department of Agriculture surrendered to the private sector: "For the college, the success of hybrid corn signaled the end of an era."[10]

Hybrid corn was a modern marvel, and its fame reached beyond the Iron Curtain, where hybrids had acquired a different meaning under Lysenkoism.[11] In 1955, when Nikita Khrushchev befriended an Iowa seed-corn producer, Roswell Garst, Khrushchev urged his agriculturalists to abandon Lysenko, plant hybrid corn, and turn the Soviet Union into a major corn producer.[12] Other commodities followed the path of hybrid corn, and hybrids came to define organisms as different as eucalyptus (see chapter 27, Eucalyptus) and battery chicken (see chapter 36, Battery Chicken). It was a productivity revolution without precedent in the annals of human history. Jack Kloppenburg has noted that "since 1935, yields of all major crops in the United States have at least doubled, and at least half of these gains are attributable to genetic improvements."[13]

Initial breeding efforts focused mostly on the commodities of Western agriculture. Arthur Lewis has argued that "the only tropical crop to experience a scientific revolution before the First World War was sugar."[14] But non-Western crops caught up over the course of the twentieth century, at times with amazing results. In Malaysia's rubber industry, serious breeding did not begin until 1926, but half a century later, high-yielding trees produced up to six times as much latex as their original seedlings.[15] On the other hand, farmers in Malawi did not embrace hybrid corn until the early 1990s, somewhat later than farms in Zambia and Zimbabwe.[16]

Farming styles differed around the world, and so did uses of corn. Unlike the rest of the world that cherishes corn mostly as livestock feed, a major share of maize production in Southern and Eastern Africa goes into human consumption. Ethiopians even use the thick stalks of a hybrid variety as cooking fuel.[17] However, agricultural systems grew more similar in the 1940s when Cold War politics made maximizing production a global priority. Seed improvement in Mexico took off when the country's agriculture became part of the US war economy during World War II, and the Mexican Agricultural Program, which included a wheat program run by a forestry graduate from Iowa named

HYBRID CORN

Norman Borlaug, became the template for other countries with similar problems of "backwardness."[18] The US government and the Ford and Rockefeller Foundations offered similar support to rice in the Philippines and wheat in India and Pakistan, and the cumulative result won global fame as the Green Revolution.[19] In 1970, Norman Borlaug received the Nobel Peace Prize.

The award came almost 250 years after the first publication of *Gulliver's Travels*, in which Jonathan Swift wrote that "whoever could make two Ears of Corn or two Blades of Grass to grow upon a Spot of Ground where only one grew before, would deserve better of Mankind, and do more essential Service to his Country, than the whole Race of Politicians put together."[20] Agriculturalists would subsequently quote the remark ad nauseam, but the Green Revolution showed that improved seeds brought more than just higher yields.[21] The new seeds typically required a higher input of fertilizer and improved pest and weed control (see chapter 38, DDT), which called for additional outlays of capital from farmers. They also brought a dependence on outside experts that Western farmers had first experienced in the wake of the guano boom (see chapter 8, Guano). The new seeds came as part of a package that pushed farmers toward capital-intensive market-oriented monoculture, a leap of faith in financial and intellectual terms that many farmers were unwilling or unable to stomach. And then, farmers were not alone in struggling with competing expert systems. Nick Cullather has pointed out that the 1966–1967 Bihar famine, the pivotal event that defined the mythology of the Green Revolution, was ultimately about a conflict between US and Indian definitions of famine (see chapter 31, Holodomor).[22]

The Green Revolution has received criticism from many sides since the late 1960s. The range of contestations went from performance in the fields, where varieties bred for monoculture performed less well in intercropping, to the impact on diets, as malnutrition was about more than just calories (see chapter 7.2, Numbers Games).[23] And then there were the problems in a wider context. When agribusiness thrived in the wake of the Green Revolution in Mexico, it pushed peasants out of business and toward migration into city slums or across the US borders, and two Rockefeller grantees "were among the first to document the 'wetback problem' that would burst onto front pages in the 1950s."[24] But the Global South was not the only place that showed the ambiguities of higher-yielding seeds. When Henry A. Wallace became US secretary of agriculture, his overarching concern was about overproduction and the subsequent collapse of commodity prices. New Deal agricultural politics sought to stabilize farming, and the last thing Wallace needed in this struggle was an innovation that delivered higher yields per acre. As it happened, hybrid corn delivered exactly that.[25]

2. LEGAL TITLES

Government bodies transformed farm production in the nineteenth century, but they did not necessarily have agriculture in their name. The US Patent Office entered the seed business during the tenure of Henry Ellsworth in the 1830s. Collaborating with the Postal Service, the Patent Office collected and stored seeds, multiplied them in its own greenhouses, and sent them out to farmers free of charge. Like Wallace a hundred years later, Ellsworth dealt with agriculture simultaneously as a government official and as an investor. He owned land in the Midwest, and frontier conditions encouraged experiments with new varieties.[26] However, the seed distribution program remained in place after Ellsworth's ten years in office, and it became more international in scope after the annexation of northern Mexico: the new lands, so different from the eastern United States, called for new plant material, and the Patent Office drew on "diplomatic, missionary, military, and commercial agents to support the expansion and diversification of American plant resources."[27] When the newly created US Department of Agriculture took charge of the program in 1862, more than one million seed packages had left Washington.[28]

Free distribution of seeds was an odd task for an agency whose core business was the protection of technological innovation, but for Ellsworth, it was more of a complementary activity. The mandate of the Patent Office was about the promotion of science and the useful arts, seed propagation was good for farming, it met with the Jacksonian commitment to serve all white men, and, in any case, the Patent Office was the only federal agency in the knowledge business.[29] However, the patent system did not recognize seeds as an artifact worthy of protection, somewhat to the dismay of commercial breeders. The inventor of the double-cross method of hybrid seed production, Donald Jones, was particularly keen to gain legal protection and filed numerous patent applications beginning in the 1920s. The effort was finally abandoned in 1970, seven years after Jones's death.[30]

The nature of the seed business made it hard to enforce patents. Duplication took place on vast expanses of agricultural land that were difficult to police, and it required little effort until the rise of hybrid corn. Furthermore, different seeds could produce similar plants, which made it challenging to prove duplication before the invention of genetic sequencing. But beyond these practicalities, patent protection was fundamentally at odds with the nature of the seed business. Patents were invented to protect the brilliant technological genius, an archetype of the nineteenth-century history of technology, but plant breeders were not solitary lab workers. They were part of a network of collection, propagation, and exchange that transcended institutional and national boundaries. The quest for new seeds was collaborative in nature, and it brought together scientists, government bodies, and private companies with vastly different interests and resources. In an organic world that thrived on botanical exchange (see chapter 27.2, Botanical Exchange), species were notoriously mobile, and the same held true for the men who were marshaling them. In the eighteenth century, Carl Linnaeus traveled to Lapland, Holland, France, and England before he settled into a professorship in Uppsala in his native Sweden, and Kölreuter conducted his landmark experiments with hybrids in Saint Petersburg, Berlin, Leipzig, and the Württembergian town of Calw.[31]

The patenting issue received a new twist when another actor staked a claim for biological property rights: the nation-state. Plants grew naturally in certain regions that were within the sovereign territory of specific countries. So were these countries perhaps entitled to compensation when breeders were using seeds from their terrain? For example, should Japan receive some kind of reward for Norin 10, the semidwarf wheat variety that the United States acquired from Japan during the occupation and became a genetic cornerstone of the Green Revolution?[32] It was a conflict over money. In the 1990s, drugs derived from plants generated annual sales of $32 billion worldwide.[33] And it was a conflict over history, for the loss of genetic resources resonates in the collective memory of countries like Brazil to this day. As an Associated Press article declared in 2005, "Biopiracy haunts Brazilian history, beginning with Henry Wickham, an Englishman who smuggled rubber seeds out of the country in the 19th century and broke Brazil's global rubber monopoly."[34]

From a scholarly perspective, Wickham's theft is a classic case of retrospective construction. Wickham brought sixty thousand rubber seeds to the Royal Botanic Gardens at Kew in 1876, but it took decades of work to turn them into the foundation of a thriving rubber industry in Southeast Asia. The golden years of Brazilian rubber came *after* Wickham's feat, and the country was slow to recognize the competition: the rubber plantations of Southeast Asia were not mentioned in Brazil's Chamber of Deputies until 1906.[35] The argument also glosses over the ambiguities of a resource endowment (see chapter 15, Saudi Arabia), and the environmental historian Warren Dean has ventured that "Brazil might well be worse off" if it had retained its biological monopoly.[36] And in technical terms, Wickham did nothing illegal: Brazil did not impose an export ban on rubber seeds until 1918.[37] As Michael Dove has pointed out, "The only real thefts in the case of rubber have been in the opposite direction, not to but *from* the estate sector."[38] Smallholders liberally appropriated rubber seedlings from Southeast Asian plantations and built an Indigenous rubber industry that estate owners were unable to compete with (see chapter 30.3, Stuff of Legends).

However, historical accuracy yields to bigger things in the case of Brazilian rubber. Concerns about biopiracy thrived on notions of unfair treatment by distant forces, a popular sentiment in a world region that gave birth to dependency theory (see chapter 30.2, In Their Theories). It was also the mirror image to the British view of Wickham that entered postcolonial lore, a story from the good old days when real men could do big things unencumbered by petty laws. "There was always an air of the fantastic to Wickham's exploits, an extravagant blend of Edgar Rice Burroughs and Lord Dunsany, and even today, historians seem uncertain what to make of him," Joe Jackson wrote in his biography of Wickham.[39] And then, as resource endowments go, biological resources were arguably about cheap money (see chapter 15, Saudi Arabia). Unlike mining since the days of Potosí (see chapter 1, Potosí), they did not leave scars in the land. In order to preserve genetic resources, nation-states rarely had to do more than set up nature reserves (see chapter 26, Kruger National Park).

Perhaps most crucially, concerns about biopiracy grew from a glaring void in the moral scaffolding of modernity. Property is a Western concept if ever there was one (see chapter 6, Land Title), and yet biological resources are strangely exempt—and just as it happened, that void played out to the advantage of colonial and corporate powers. For centuries Western collectors could roam foreign lands in search of biological material without local obligations beyond the needs of their expedition, and as it stands, that has changed only marginally in recent years. A number of international treaties (see chapter

24.3, Getting Serious) including the Convention on Biological Diversity have sought to establish rules for the exchange of biological material, but they compete with another force of globalization: the large multinational corporation (see chapter 10, United Fruit).

3. BUSINESS MODELS

Henry A. Wallace served as vice president of the United States for four years, but there was never much love toward him among the Democrats' old guard. He lost the nomination to Harry S. Truman at the 1944 Democratic Convention, and Truman advanced to the presidency when Franklin D. Roosevelt died the following year.[40] Wallace ran for the presidency on the Progressive Party ticket in 1948, but revelations about his Russian connection and allegations of Communist sympathies derailed his campaign, and he won a paltry 2.4 percent of the popular vote.[41] But he always had his company, Hi-Bred Corn, renamed Pioneer Hi-Bred Corn in 1935. Wallace went back to genetics in 1949, and his test plots in South Salem, an hour away from downtown New York City, featured corn, gladiolus, and strawberries.[42]

Wallace was among the first to recognize the potential of hybrid corn, but his company struggled to survive and did not return a steady profit until 1933.[43] The boom of hybrid corn brought market consolidation as smaller seed producers closed or merged into larger companies, and Pioneer became a market leader.[44] Pioneer also moved into new fields and launched experiments with hybrid chicken (see chapter 36, Battery Chicken) as early as 1936. When DuPont bought the company's shares for $9.4 billion between 1997 and 1999, an agricultural economist from Iowa State suggested that the United States was "headed toward having only three or four companies, and maybe just two, control the sale of seed in this country."[45]

The merger of a giant seed company and a giant chemical company mirrored the interconnections between different branches of agricultural improvement. New seeds were not just about higher yields: they had to match developments in fertilizer use (see chapter 19, Synthetic Nitrogen), weed and pest control, and farm machinery. For example, mechanical corn pickers called for varieties that matured simultaneously and featured strong roots, straight stalks, and ears at a uniform height.[46] Few consumers cared about the outlook of a maize plant, but other innovations were not quite so innocent. The mechanical tomato harvester brought breeders to seek fruits that were firm and crack-resistant, held securely to vines, had limited foliage and constant quality, and ripened at the same time.⁴⁷ The innovation made history beyond the dinner table. When Jim Hightower published a scathing critique of how the American land-grant college complex had sold out to agribusiness in 1972, the hard tomato, custom-designed for mechanical picking and chemically ripened, symbolized the fall from grace.⁴⁸

The stakes grew higher still when molecular genetics acquired the tools to change DNA in the 1970s. Genetic engineering required even more capital and corporate might than hybrid seed production, and it led to a change in patent law when the Supreme Court of the United States heard a case about a genetically modified bacterium that General Electric sought to protect under the name of its Indian-born inventor, Ananda Chakrabarty. Decided in 1980, *Diamond v. Chakrabarty* opened the door to patents for living organisms, including seeds.⁴⁹ The result was an explosion of patents and conflicts over patents. In 2016, the US Patent and Trademark Office had more than one million patent applications pending, a quarter of them in biotechnology and organic chemistry.⁵⁰

Genetic engineering became agriculture's second Manhattan Project, but this time with an alert general public. While Hightower's critique of agricultural research had focused on the land-grant system that few really cared about, genetically modified crops brought critics to focus on the private sector, and the cause struck a nerve.⁵¹ Corporate action provided the movement with ample fodder. One of the giants of agricultural biotech, Monsanto, grew out of a chemical company that polluted neighborhoods with toxic PCBs (see chapter 38, DDT) and produced Agent Orange.⁵² Another US business, W. R. Grace & Co., acquired a patent for neem tree extracts, an Indian tree with many traditional uses that "to many Indians is fundamentally non-commodifiable."⁵³ A group of activists including the charismatic Vandana Shiva, formerly a campaigner against eucalyptus (see chapter 27, Eucalyptus), filed an appeal with the European Patent Office in Munich to challenge "the commodification of life."⁵⁴ In the twenty-first century, protecting high-yielding seeds involved courts beyond patent offices. In 2012, US marshals arrested a Chinese national after an FBI investigation with powers under the Foreign Intelligence Surveillance Act found that he had collected corn seeds in Iowa.⁵⁵

The controversy over genetic engineering drew on the Western experience with large technological projects in the postwar years. Industrial disasters like the *Torrey Canyon* oil spill (see chapter 39, *Torrey*

Canyon) taught affluent societies how small decisions could have devastating consequences, and the perils of nuclear power sensitized for invisible threats (see chapter 37.1, Global Pollution). The debate over genetic engineering resembled a rehash of the nuclear debate: while the 1962 comic version had Spiderman gaining superhuman powers after being bitten by a radioactive spider, it was a genetically engineered spider in the movie version of 2002.[56] It helped open a window for explorations of agricultural alternatives. Jack Kloppenburg supported biological open-source arrangements after the model of open-source software in order to undercut the patenting regime.[57] The appeals court at the European Patent Office ultimately revoked the neem patent, though not because of Shiva's eloquent vilification of "biopiracy" or the Sri Lankan farmer citing Sanskrit scriptures: the decisive testimony came from an Indian factory owner who had used a manufacturing process similar to W. R. Grace's since 1985.[58] Jim Hightower was elected Texas agriculture commissioner in 1982, and he became a champion of organic farming and environmental analysts targeting pesticides (see chapter 38, DDT). He was reelected in 1986 but lost in 1990 when a member of the Texas House of Representatives, Rick Perry, ran a smear campaign against him orchestrated by Karl Rove.[59] Perry moved on to three terms as governor of Texas, two runs for the presidency, and the secretary of energy under Donald Trump.

It was a difficult situation for those working in the world of seeds. When Norman Borlaug gave a lecture at the Norwegian Nobel Institute in Oslo thirty years after receiving the Peace Prize, he found himself sitting between two stools. He complained about "the current backlash against agricultural science and technology evident in some industrialized countries," but he was no less concerned about an oligopolistic market: "The high cost of biotechnology research is leading to a rapid consolidation in the ownership of agricultural life science companies. Is this desirable?"[60] The public's unease about genetically modified organisms did not go away, and neither did corporate concentration. In 2018, the German chemical company Bayer completed a takeover of Monsanto that left observers aghast, and it was about more than dizzying ten-digit dollar figures. Britain's *Telegraph* called it "the Frankenstein merger."[61]

Corporations were also under attack for what they failed to do. Agrobiodiversity (see chapter 11.3, Conserving Diversity) emerged as a matter of concern: while earlier generations of collectors could take the genetic diversity on farmers' fields for granted, the global reach of commercial seeds has put it under threat, inspiring a variety of responses. The Rockefeller Foundation and the Bill and Melinda Gates Foundation teamed up in 2006 to form the Alliance for a Green Revolution in Africa, whose funding scheme included startup grants for African seed companies that "deliver better seed to farmers via sustainable channels."[62] Others focused on preservation instead of use, when seed storage detached from seed improvement in the 1970s, and millions of specimens were dried, frozen, and locked away.[63] The Svalbard Global Seed Vault on Spitsbergen serves as the lender of last resort, and the banking metaphors probably speak about the realities of the twenty-first-century world: in the face of the ravages that global capitalism has wrought in the new organic (see chapter 2, Sugar), our best hope may be the spoils from the ravages of global capitalism. When the historian Jonathan Harwood wrote a book about an alternative to Green Revolution technologies that looked at peasant-friendly plant breeding at a state-run *Saatzuchtanstalt* in Bavaria, it had the air of an obituary for a world long gone.[64]

After a century of mergers and acquisitions, the corporate world of seeds looks unassailable, but its business will never be static. Pests and diseases will keep breeders busy, and then there are the unintended side effects that improvement can produce. Helped by a homogenized gene pool, an epidemic of Southern corn leaf blight swept America's cornfields in 1970, with losses of 50 percent and more in the Gulf region and a total shortfall of production of more than 700 million bushels.[65] Yet change will likely occur along a narrow corridor framed by corporate interests, modern technologies, and legal titles, supplemented by concerns of urbanites about genetic engineering. The human world of seeds lacks alternatives. The biological world may not.

Jack Kloppenburg has stressed that plant breeding suffered from an imbalance of funding since the heydays of hybrid corn. Open pollination was neglected in favor of hybrid breeding, and certainly not for lack of potential: "The tremendous 'success' evidenced by hybrid corn might have been achieved just as well through population improvement techniques in open-pollinated varieties."[66] But farmers would have reused these varieties year after year while the sterile hybrids provided commercial breeders with a secure market. From a biological perspective, the path toward open-pollinated corn is still open, but it would likely take many years and leave fields less uniform than they currently are. The reign of hybrids is unlikely to end anytime soon.

29

Aswan Dam

Damming and Developing

I. SHADES OF GRAY

In 1972, a field in Moshtohor north of Cairo commanded the attention of Egypt's Ministry of Irrigation, the Nile Delta Authority, and the Agricultural Projects Department of the World Bank in Washington, DC. A Dutch producer of agricultural machinery, A. H. Steenbergen, had run trenching equipment on a test plot in the Nile delta, and the results were important for ongoing improvement projects. The recently completed Aswan High Dam allowed more land to switch from seasonal to perennial irrigation, and drainage pipes were crucial to prevent waterlogging. Steenbergen claimed to have "experience in about any soiltype in the world where large subsoil tile drainage projects are being executed," but drainage in the Nile delta did not look like business as usual: "The soil conditions are about the most difficult and heavy known."[1] Machines would need to be "of particularly sturdy design and execution however be as uncomplicated as possible," they should not be too heavy lest they destroy the soil's structure, and as if that were not challenging enough, the results also suggested that the tender was up for revision.² As it stood, the tender asked for trenchers with a maximum depth capacity of five and a half feet, which was already one or two feet above common requirements, but Steenbergen's field trials found that equipment frequently crossed "small country roads, topping over the landsurface," where trenching would have to go down to a depth of six feet. Roads of this kind were everywhere in the delta, and while they failed to impress the innocent observer, they were a matter of concern to those who ran earthmoving equipment: "The negligence of this depth requirement will be extremely disturbing on operations and endanger correct gradient maintenance by contractors' personnel."³

Aswan High Dam was at the other end of Egypt, some five hundred miles away from the Moshtohor test field, and it was the pride of a nation. Conceived and built in the 1950s and 1960s, it was a monument to Gamel Abdel Nasser and the independence of Egypt, an embodiment of the fight against imperialism, and the purported engine of Egypt's economic and social transformation.⁴ It was a project of truly pharaonic dimensions, and rhetoric was framed accordingly: "Many Egyptian writers have stressed that the dam's construction utilized seventeen times the amount of material used to build the Great Pyramid at Giza."⁵ The dam rose 111 meters above the riverbed, with a grout curtain extending an additional 200 meters down to the granite bedrock, it was 980 meters wide at its base and almost 4 kilometers long, and the lake that it produced was so vast that it reached into neighboring Sudan.⁶ But for all its superlatives, the dam was only the most visible part of a large technological system the size of a nation. Other elements were less monumental or even buried in the ground, like drainage pipes, but that did not make them any less important. The post-Aswan Nile was, in the words of Richard White, an "organic machine," and like every machine, it operated smoothly only when the various components worked together likes cogs in a wheel.⁷ Six additional inches in depth for a drainage pipe looked like a trivial matter compared to the massive dam, but they could make a world of difference.

Dams are probably as old as human civilization, and they were certainly nothing new on the banks of the Nile. As Timothy Mitchell has written, "Long before the Aswan Dam, before all the irrigation work of the nineteenth century, the river was already as much a technical and social phenomenon as a natural one."⁸ Agriculture was traditionally the main beneficiary of water control along the Nile, but premodern dams could serve a variety of purposes. A dam at Saint-Ferréol supplied water to the Canal du Midi (see chapter 3, Canal du Midi), and a number of artificial lakes on the slopes of Potosí's Cerro Rico (see chapter 1, Potosí) helped keep the wheels of the silver mills turning.[9] The aqueducts of ancient Rome serve as a reminder that cities had drawn in water from beyond their perimeters long before the rise of the water closet (see chapter 17, Water Closet). Water was an essential of life, a word with hundreds of references in the Bible and the Quran, and seeking access was a natural urge.

However, dams grew in size in the late nineteenth century, and that made for a watershed in humanity's quest for hydraulic control. Mass-produced steel and reinforced concrete allowed dams to reach unprecedented dimensions, and large reservoirs were needed for a variety of purposes. They improved navigation on rivers and lakes and helped forestall natural disasters (see chapter 25, 1976 Tangshan Earthquake) by capturing floods. Urban consumers and expanding industries called for vast amounts of water that wells, brooks, and other traditional outlets could not deliver. With the rise of electric power grids, dams became prized producers of hydroelectric power. And then there was irrigation, a time-honored practice that gained new allure when global commodity markets showed an insatiable appetite for cash crops. The path to Aswan High Dam began in the 1820s when Muhammad Ali Pasha made cotton cultivation a cornerstone of his forced modernization of Egypt.[10]

The new dams were concrete manifestations of what David Nye has called the "technological sublime," and they inspired grandiose visions of human mastery of the natural world.[11] Winston Churchill fantasized about turning Uganda into a tropical commodities powerhouse with dams on the Upper Nile, blissfully unaware of the priorities of British colonial officials who wanted first to get a grip on the swamps in today's South Sudan.[12] Back in Europe, Herman Sörgel developed a hydraulic response to Oswald Spengler's lament about Europe's cultural decline with his plan to dam the Strait of Gibraltar and turn the Mediterranean into a managed reservoir.[13] Sörgel's Atlantropa project never moved beyond the paper stage, but it made an impression even on those with a more practical bent. The German architect Peter Behrens, a pioneer of the modernist movement, designed a glass-and-steel skyscraper taller than the Empire State Building for the locks at Gibraltar.[14] The great dams of modernity were great in capturing the minds of people, and they were designed and built intentionally for that purpose. When the US Bureau of Reclamation built one of its first dams on the Salt River near Phoenix, the bureau's founding director, Frederick Haynes Newell, insisted on a masonry gravity design with locally quarried sandstone. Other designs were cheaper but not quite as visually impressive, and looks were more important to him. In a letter he wrote toward the end of his tenure, Newell noted that his bureau favored solid dams "not only to have the works substantial but to have them appear so and [be] recognized by the public."[15]

But as dams and lakes grew into new dimensions, so did the side effects that water projects had always had. Dams changed plant communities, microclimates, and groundwater tables. Water from the bottom of a large reservoir was different in temperature and quality from river water. Silt settled in artificial lakes, and as Egypt learned after the inauguration of Aswan High Dam, clearer water allowed more sunlight to reach the bottom of irrigation canals, which in turn stimulated the growth of aquatic weeds.[16] Dams have long been implicated in the making of earthquakes, as masses of water weigh down on tectonic fault lines.[17] And dams could kill. A malaria outbreak struck Egypt in 1942, and it was due, in Mitchell's lucid analysis, to pools of stagnant water around Aswan, an invasive aquatic plant (see chapter 14, Cane Toads) that formed floating islands, and World War II.[18]

In short, large water projects were never really finished, and certainly not after ribbon-cutting ceremonies on high dams. The choice about maximum drainage depth in the Nile delta was only one of many decisions that were waiting to be made in the course of these projects, and benefits and side effects were the cumulative result of these decisions in combination with the mood swings of Mother Nature. But with so many decisions in play, water projects were also levers of power, and the flow of water reflected the priorities of rulers and societies. When British forces occupied Egypt in 1882, water policy was geared toward agriculture in order to boost cotton production whereas towns and villages suffered from a water shortage.[19] Hydrologists would later tout "multi-purpose dams," but that rhetoric left room for subtle hierarchies, and dam builders knew which buttons to press. After the devastating 1906 earthquake (see chapter 25, 1976 Tangshan Earthquake), San Francisco cited fire safety in its quest to build a reservoir in the scenic Hetch Hetchy valley of Yosemite National Park (see chapter 26, Kruger National Park), a project that preservationists fought tooth and nail, but the project was really about clean mountain water and hydroelectric power.[20] The Kariba Dam across the Zambesi

River in the Central African Federation supported the expansion of industry at the expense of the rural poor.[21]

The dam at Hetch Hetchy made national headlines, the Kariba region became a battlefield during Zimbabwe's war of independence, and Aswan became infamous all around the globe, but these dam conflicts were only the most visible in a veritable flood of controversies. Cities and utilities, landowners and industries, conservationists and fishermen, hydrologists and construction companies—dams united a vast array of stakeholders, and the precise issues and outcomes depended on the specifics of the cases at hand. They could even shape biological research: in the 1950s, Australian and American investors built Fogg Dam in Australia's Northern Territory for a rice cultivation project that never materialized, but the lake turned out to be heaven for snakes, which has made it the second home of the snake biologist Rick Shine since 1985, and Shine embarked on a second career as Australia's leading cane toads man (see chapter 14, Cane Toads) when the invasive species overran the area in the early 2000s.[22] The one commonality of the world's dams is that they have been popular enough to be built by the dozen all around the globe. As this book was going to press, the latest world register of the International Commission on Large Dams included 58,713 of them.[23]

With so many hydroprojects around the world, questions arose as to their political significance. Karl August Wittfogel famously stressed the despotic potential of hydraulic power in the Orient. According to his reading, control over water bestowed a central bureaucracy with absolute power.[24] Wittfogel's *Oriental Despotism* was part of a long European effort to unlock the mysteries of Asian governance, but Wittfogel has also resonated in readings of modern hydraulic regimes, and acolytes have rhapsodized over how water development was really about "achieving nothing less than total control, total management, total power."[25] They have missed the more exciting story. Centralized control over water met with other eminent authorities in modern history: stakeholders, many of them armed with land titles (see chapter 6, Land Title), the various political doctrines of the age of extremes, a growing body of experiences with environmental repercussions and technological options, a hydraulic profession with its own cognitive resources, liabilities, and sense of pride, and a medium that defied the grip of power through its natural properties. When it came to water, power relations were typically in a state of flux, and they were about much more than the flow of liquids.

2. CREATIONS OF MEN

In August 1969, the US commissioner of reclamation, Floyd Dominy, traveled to South Korea. It was his first visit to the Korean peninsula, but when the *Seoul Economic Daily Press* met him for an interview, it looked like he knew the place. Dominy praised the Soyang Dam, then under construction, and a prospective dam at Chungju as "the best dam sites in the Han River Basin from the viewpoint of physical condition and economic aspects." He also endorsed another six dam sites on the Han River that were earmarked "for future development," and he tossed in references to ongoing construction in Spain, Japan, Thailand, and Laos. Dominy was a widely traveled man, and when he entered new terrain, things looked familiar. "The Colorado River in the United States is similar in run off amount to the Han River, and the United States has built 15 dams on that river to attain economic development."[26]

The dams of modernity were also places for big men, and Nasser, whose name graces the lake behind Aswan Dam to this day, was not the only one who became immortalized in a water project. The Salt River Dam was named after Theodore Roosevelt in 1911 and ceremoniously opened by the former president himself, who headed out to Arizona just after returning from a safari in Africa.[27] The Portuguese dictator António Salazar even inaugurated his own personal dam while still in office.[28] But dams could also propel to prominence those who actually built them. Some were managing professionals like Newell, an MIT-trained engineer with a long career in government service.[29] Others were academics with stellar political connections such as Otto Hintze, the trailblazer of dam construction in Germany, who influenced key legislation, gave private lectures to Kaiser Wilhelm II, and gained a seat in the Prussian House of Lords.[30] Some builders even claimed higher powers such as Pierre-Paul Riquet, the father of the Canal du Midi (see chapter 3, Canal du Midi), who spoke of divine inspiration and a philosopher's stone.[31] He was not the last to mix hydrology and religion. In an interview in retirement, Dominy declared that he had "no apologies. I was a crusader for the development of water. I was the Messiah."[32] He was probably confusing Jesus with Moses, who struck a rock in the Sinai desert, found water pouring out, and thus saved the thirsty people of Israel.[33] Jesus just walked on water.[34]

However, rhetorical bombast went along with a life of compro-

29.1 Philae Island underwater in 1904. Image, Universal Photo Art Co., Library of Congress.

mises. Newell got his solid-looking dam built, but the Salt River project was delayed, exceeded cost estimates, and about half of the land was in holdings of more than 160 acres, something that was anathema to a reclamation program intended to support the small yeoman farmers of Jeffersonian fame.[35] It was fairly typical of the early projects of the Bureau of Reclamation, as troubles with cost overruns and irreverent farmers were legion, and Newell concluded toward the end of his tenure that "the problems of 'human nature' were far greater than the engineering problems of western reclamation."[36] Frontier knowledge on the American West was far inferior to that on the Egyptian Nile, and an engineer trained in colonial India, William Willcocks, devoted several years to a systematic investigation of the river's hydrology since 1889, but his proposals were subject to review by an international expert commission. War in Sudan delayed the start of construction, and when the first Aswan dam was finally opened in 1902, it was lower than intended in order to save a temple on Philae Island south of Aswan.[37] The archaeologists' relief was short-lived. Most of the island was submerged in high water, and the dam was raised twice, in 1912 and 1933.[38]

From 1917 to 1921, Willcocks became embroiled in a messy conflict with another engineer, ditching professional omerta in a public controversy that left all sides tarnished.[39] His writings met with more acclaim, for they skillfully blended British irrigation efforts into a landscape awash with biblical memories.[40] The gap between myth and reality was part of modern dam history from the beginning, and it did not shrink over time. A generation after Willcocks and Newell, the

Tennessee Valley Authority (TVA) built dams as a lever of regional development, armed with a flat budget and emphatic support from the New Deal president Franklin D. Roosevelt, but the TVA was soon at war with itself. The agency's three board members disagreed over priorities and management styles, and the most energetic, Arthur E. Morgan, worked in a way that has been described as "the antithesis of planning; he preferred to delve into a task without much forethought, to improvise solutions to problems discovered along the way."[41] The TVA remained a singular agency in the US federal system. A Missouri Valley Authority, proposed by Roosevelt in 1944, failed to materialize.[42]

After World War II, the TVA "was regarded with awe by the rest of the world for a considerable period," a curious turn of affairs after its tumultuous prewar history.[43] "There was growing interest in Europe for the TVA to serve postwar rehabilitation," David Ekbladh wrote, though plans for a Danube Valley Authority drew the ire of Friedrich Hayek.[44] Just like Nasser in Egypt, Kwame Nkrumah turned a preexisting British project into a comprehensive development program when he built a dam across the Volta River in Ghana.[45] Greece built the Kremasta Dam on the Achelous River, whose sediment load impressed Herodotus some 2,500 years ago: while "not as large as the Nile," the Achelous had "already turned half the Echinades islands into mainland."[46] Kremasta is still the country's largest artificial lake, but it is more famous for causing a 6.2 magnitude earthquake (see chapter 25, 1976 Tangshan Earthquake) half a year after the start of filling.[47] Spain built dams with vigor during the Franco years and became the country with the greatest number of large dams in Europe.[48] Over in Afghanistan, the Helmand Valley project sought "to immobilize the nomadic Pashtuns," the country's ruling ethnicity "whose migrations were a source of friction with Pakistan." The endeavor received praise from a visiting historian, Arnold Toynbee, in 1960.[49] In 1967, the head of the Aswan Regional Planning Project told a reporter from the *New York Times* that the goal of the high dam was "to make Aswan the Pittsburgh of Egypt."[50]

Contemporary opinion held that dams could deliver many things, including peace. In a speech at Johns Hopkins University in 1965, the American president Lyndon B. Johnson outlined a strategy for ending the Vietnam War with a "TVA on the Mekong."[51] After Israel's victory in the Six-Day War, a memorandum from the US Department of the Interior proposed to bring peace to the Middle East through "vigorous water statesmanship."[52] Others were just thinking about transportation problems. The Hudson Institute, a think tank led by the American futurist Herman Kahn, proposed a South American "Great Lakes" system for an inland waterway from the Orinoco to Buenos Aires.[53] Soviet engineers built one giant dam after another while the propaganda machine extolled the concrete virtues of Socialism.[54]

Dominy knew the reality behind these visions. He had traveled to Afghanistan in 1959 and found that "after only a few years of irrigation the land was white with salt."[55] In a letter to a US official in Ethiopia, he called the Helmand Valley project "an unfortunate example of project aid far beyond the capacity of the local government to finance and maintain after construction is completed."[56] But at the same time, Dominy knew that the project would be "politically significant because of its proximity to Soviet Russia and the adverse reaction that will prevail in official Afghan circles if an American-identified program of this magnitude continues its unsuccessful trend."[57] Cold War politics shaped the global boom of dam building after 1945, not least because the Aswan dam taught what might happen when a superpower said no. The United States and the World Bank withdrew their support in 1956, which led to the Suez crisis and Egypt's turn to the Soviet Union.[58] Memories of the debacle were still fresh when the World Bank was funding its drainage project in the Nile delta in the 1970s. The project had been on the bank's "Problem Projects list" since January 1972, and the problems included "major cost overruns, inadequate budgetary allocation of local funds, insufficient consultants' services, unsatisfactory contractual relationships, delays in equipment procurement and failure to make adequate use of existing equipment," but an internal World Bank memorandum urged taking it easy, on the grounds that it was "the first project since the severely strained relations of the 1950's."[59] And then there were the other conflicts beyond the Cold War framework. When the deadline for the partition of India and Pakistan approached, the British official who defined the new border frantically sought a line that squared demography with irrigation systems in the Indus River basin, only to conclude that there was no such line.[60]

After decades of dam construction, a lot was known about the many things that could go wrong in large water projects, and the same held true for the political expediencies that shaped project management. Dominy was intimately familiar with both because his career included a long stint in the Allocations and Repayment Branch of the Bureau of Reclamation, where dealing with past fiscal blunders was the daily bread.[61] Dominy knew that, beyond a certain point, there was not much one could do about a botched project, but at least one could learn from the experience, and he was religious about the formulas and procedures that his bureau used for new projects. He even said so in his boisterous interview in Korea. "A project should be planned patiently and carefully, considering the possible future outlook," Dominy told the *Seoul Economic Daily Press*. As an illustration of what he had in mind, he warned that it took "25 years on an average to plan and construct a major dam in the United States."[62] When the Department of the Interior, the parent agency of the Bureau of Reclamation, asked Dominy for his input on its watery Middle East peace plan, he stressed the bureau's expertise and warned that "the employment of either technology on a crash basis should not be considered as a substitute for more thorough study of alternatives."[63] He even dared to throw cold water on Johnson's "TVA on the Mekong." When the project was finally taking shape with a pioneering endeavor at Pa Mong, Dominy called on Cambodia, Laos, Thailand, and South Vietnam "to agree in advance on the elements necessary to fit operation of Pa Mong into a basin system."[64] It was an almost prohibitive requirement, but Dominy found treaties "essential . . . to avoid misunderstandings."[65]

Seen from the ground, the postwar dream of development (see chapter 32.3, Planning Development), the idea that state power, science and technology, economy and society would all pull together in a direction vaguely called "progress," was arguably dead on arrival. There were always multitudes of goals that claimed to bring "progress," and when it came to the allocation of limited amounts of water, there were inevitably winners and losers. Sometimes it was difficult to say which side people were on: many of the refugees from the 1960s Mangla Dam project in the Kashmir ended up as immigrants in the United Kingdom.[66] However, there was certainly no water authority "holding the desert and the river in its indefatigable grip," as a Wittfogelian reading would suggest: the power of water was far more muddled than that.[67] It was more a close entanglement of environments and technologies, residents and visitors, managers and water users that curtailed the range of options for all parties involved. The typical dam was far more a Faustian bargain than a tool of despotism.

But while Wittfogel's hydraulic power was a poor reflection of real-world dams, it did capture something of the charisma that dams had for those in the upper echelons of society. In an address to a symposium on "space age irrigation" in 1968, Dominy called on "the ambitious, the mentally alert and the skilled who are needed to become the leaders of tomorrow," and with retirement in sight, he was obviously thinking about his own reincarnation.[68] Nowhere was the man of energy more in his element (and no, there is no need for gender balance in this formulation) than in the planning stage, and Dominy banked on new projects knowing that they were facing the law of diminishing marginal utility. "The relatively simple water projects of the past decades have been built," Dominy told the convention of the Associated General Contractors of America in 1962.[69] Dominy built nonetheless, and he was a master at milking Congress for his projects. "In any one year of the 1960s, the bureau's construction budget exceeded all the expenditures of the Bureau of Reclamation from 1902 to 1933."[70]

Dominy ran the Bureau of Reclamation, in John McPhee's memorable phrase, "as if he were driving a fast bus," and it is perhaps time to recognize his tenure as an exemplification of what hydraulic messianism could lead to in spendthrift times.[71] It remains a matter of belief whether Dominy was really the Messiah, but he drove his agency to achieve the bureaucratic equivalent of running on water. Dominy's hydraulic regime thrived for the same reason that race boats fly above water, because of speed rather than buoyancy, but gravity eventually won the upper hand: the real-world entanglements of water projects caught up with hydraulic agencies sooner or later, and powerful bureaucracies turned into puppets of their own creations. The despotism of dams came back to haunt those who built them.

3. PATH DEPENDENCIES

As befits a water project, the construction of Aswan High Dam was an eventful affair. With US funding no longer available, Nasser nationalized the Suez Canal in 1956, a step that has been called "one of the most important African initiatives of the twentieth century."[72] The ill-fated military intervention of Israel, Great Britain, and France that sought to reverse nationalization is widely cited as a milestone in the demise of the British Empire.[73] Soviet and Egyptian engineers quarreled over what to make of the blueprints that the German construction company Hochtief had drafted before Suez.[74] A legendary UNESCO project saved some archaeological treasures from ancient Egypt while anthropologists rushed to study 120,000 Nubians, whose ancestral homeland was bound to drown in the lake.[75] On the con-

29.2 Gamal Abdel Nasser observing construction of the Aswan High Dam. Image, Bibliotheca Alexandrina, Wikimedia Commons.

struction front, Soviet specialists "declared a production crisis amid a desperate need for more skilled labour" in 1962, and the heat and the remote location created additional difficulties.[76] The casualty rate from accidents became subject to dark retrospective ruminations.[77]

Every dam project is a gamble that runs for an unspecified number of years, and megaprojects are the hydraulic equivalent of betting the farm. For Paraguay's dictator, Alfredo Stroessner, and his Colorado Party, the Itaipú Hydroelectric Dam, a joint project with Brazil that was the world's largest until Three Gorges Dam, was a tool to consolidate the regime's power, create and reward loyalists, and suppress opposition.[78] Completing Itaipú took almost two decades, but even smaller projects could drag on for a long time. The Bumbuna hydroelectric project in Sierra Leone was initiated in the early 1970s, site preparation began in 1982, civil war brought everything to a standstill in 1997, the World Bank approved a grant for completing the project in 2005, and when the power plant finally became operational in 2009, planning shifted to a second phase that will add more generation capacity.[79] Aswan advanced at a brisk pace by way of comparison, and the construction site made an impression even on those who knew what a dam under construction looked like. When a tour of the International Commission on Large Dams came to Aswan in 1963, the trip report of a US engineer declared, in rhetoric that only engineers can get away with, that from a distance, workers in a quarry "resemble a swarm of ants working on an ant hill."[80]

Work came to an end around 1970, which was not a good time to complete a hydroproject. Global environmentalism was approaching its first all-time high, and it was deeply skeptical of megaprojects with all sorts of negative repercussions. US environmentalists were particularly keen, as the American environmental movement had grown to a significant extent out of protests against water projects in the American West.[81] Aswan Dam was also an easy target because Egypt was not a US ally, and having been at war with Israel did not improve things. And then there was the combined backdrop of Pharaohs, Herodotus, and the Bible that made an intervention of brute force technology look like barbarism.[82] Echoes of contemporary stereotypes linger in the narratives of environmental historians. Joachim Radkau remarked on the Aswan project that "Soviet engineers . . . destroyed an irrigation culture that had ensured sustainable agriculture for five thousand years."[83]

The Aswan High Dam did cause a number of problems. An article in the *UNESCO Courier* called the weed invasion "perhaps the most serious side-effect" after twenty-five years.[84] Those who suffered from an infection with schistosomiasis were probably inclined to disagree, as the parasitic disease spread in stagnant water, and schistosomiasis became subject to international aid programs.[85] Lack of sediments increased the need for fertilizer use and caused severe coastal erosion. Soils became saltier and less fertile. Ancient Egyptian monuments suffered from an elevated groundwater level.[86] Brickmakers went out of business when the traditional use of silt for mud bricks was banned in 1984.[87] Perhaps most critically, even the resources of the mighty Nile were facing limits, not least through evaporation in one of the hottest places on earth, and studies have long realized "the threat of a water crisis in the near future."[88] And then there are the hypothetical risks. The Germans considered an attack on the old Aswan Dam during World War II, and the new dam will always be the Achilles' heel of Egypt.[89] It would not be the first wartime destruction of a high dam. During the Korean War, the US Air Force destroyed North Korean reservoirs in "a type of psychological and social warfare," and tens of thousands died in southern Ukraine from the hydraulic equivalent of friendly fire when the retreating Red Army blew up one of the world's largest dams across the Dnepr River on August 18, 1941.[90]

However, the lingering question is how to weigh these problems against the positive effects. The reservoir helped Egypt through several consecutive years of drought, something that the old dam could never have done. Aswan is a major source of electric power, some of which is used in the production of nitrogen fertilizer (see chapter 19, Synthetic Nitrogen).[91] The dam also improved navigation, allowing tourists (see chapter 22, Baedeker) to conclude their Nile cruise on time. Research will continue, and it will likely remain inconclusive. Ewald Blocher, who has studied measurements and modeling on the Nile, has argued that the sheer volume of accumulated data makes definitive results impossible.[92] And then there is the rift between different disciplines, as academic specialization has long taken its toll in hydraulic expertise. In 1965, officials from the World Bank and the Bureau of Reclamation discussed why their Indian partners were so relaxed about waterlogging and salinity problems in their Indus watershed projects, two serious issues downstream in Pakistan, and they came to the conclusion that they had talked mostly "with the people concerned with design and construction of engineering works, rather than those charged with obtaining the agricultural benefits from the completed system."[93]

And then there was the silt, an issue where everything depended on worldviews. When the International Commission on Large Dams came to Aswan in 1963, visitors were told that silt could accumulate "for a period of 500 years before the live storage of the reservoir will be encroached upon."[94] Twenty years after the dam's inauguration, the official estimate was 362 years, but uncertainties about two variables allowed life-span estimates between 299 and 535 years.[95] But should silt be a matter of concern? In *Cadillac Desert*, a scathing critique of water development in the American West, Marc Reisner argued that siltation shows how dams are "oddly vulnerable things": "Every reservoir eventually silts up—it is only a matter of when."[96] But for those who were living in the shadow of a dam, there were always other issues, such as the difference between 5½ and 6 feet maximum depth capacity for trenchers and whether it was a real or a cooked-up issue. Was Steenbergen flagging the issue out of pure self-interest insofar as revisions in the tender increased its chances of getting the contract? It certainly did not hurt Steenbergen to write the report, as the Dutch company, now incorporated as Steenbergen Hollanddrain Egypt, continues to hold a near-monopoly in the field. The company website claims that "90% of the trenchers operating in Egypt in the field of pipe laying and underground irrigation carry the Hollanddrain logo."[97]

For most people in Egypt, Aswan High Dam is not something that should be argued about on principle. It is a reality of life, a concrete barrier (with rocks and clay to boot) that channels water and profits, diseases and opportunities, and one had to work with it for better or worse. Ancient Egypt probably had a sustainable form of agriculture, but if we can trust Flavius Josephus's *Judean War*, it had some 7.5 million people in the times of the Roman Empire, and today's Egypt has about 100 million hungry citizens.[98] And besides, removing a dam is more difficult than one might think.

Modernity's love affair with dams has left plenty of barriers that no longer serve a purpose, but dismantling them is an ongoing struggle. New England has more than 14,000 dams, many of them leftovers from industrial uses in the nineteenth and early twentieth centuries that no longer exist. Removal would offer prospects for the revitalization of rivers, but in the twenty-first century, conservation projects require careful investigation. Many New England dams receive more scrutiny than ever before now that their lifespan is nearing its end, and the judgment of scientists may not be the final word. In 2008, after a five-year study involving 17 agencies and organizations, the town of Greenfield, Massachusetts, agreed to remove two dams for river restoration, one of which was labeled "high hazard." Six years later, the town reversed the decision in the face of community protests.[99] A mix of nostalgia, resistance to outside interference, and uncertainty about benefits has fueled opposition from New England locals, and by the mid-2010s, less than 1 percent of the region's dams had been slated for removal.[100]

New dams are facing resistance too, clearly a testament to collective learning in a globalizing world. We are beyond the point where megaprojects can be launched without critical questions being asked. Sometimes dams were built nonetheless, such as the Three Gorges Dam in China, a project that was shelved in the early 1990s after an eight-year-long environmental impact review, and realized ten years later.[101] Others were canceled after international protest, such as the Arun III project in Nepal that the World Bank killed in 1996.[102] Celebrities (see chapter 23.1, Celebrity Status) have rallied to the cause. James Cameron, the director of the blockbuster movie *Avatar*, traveled to the Amazon in 2010 and met with Indigenous leaders to protest against Brazil's Belo Monte hydroelectric project, though his stance might have been more convincing had it not been for his residence in a desert city named Los Angeles.[103] It is an existential issue for Indigenous people who live below an upcoming dam and a convenient one for Western activists, as dam projects have mostly petered out in the industrialized world. In the United States, it fell to the first commissioner of reclamation in the Clinton administration, Daniel P. Beard, to announce "that the grand construction phase of reclamation history had passed" and that it was time for new agendas.[104]

Belo Monte is now on the grid, and even Arun III is under construction, this time with Indian money to supply electric power to the subcontinent's air conditioners (see chapter 20, Air-Conditioning).[105] The rationale is more plausible from a non-Western perspective: the international campaign against Arun III had mostly ignored public opinion in Nepal, which had clearly favored the project in the 1990s.[106] Many countries in the Global South depend on hydroelectric projects to sustain their power grids, and they are willing to pay a price for it. When Burkina Faso built Kompienga Dam, the country's first hydroelectric project, it accepted a record-setting ratio of 1,426 hectares of inundated land for 1 megawatt of generating capacity; the ratio for the Three Gorges Dam was 317 hectares per megawatt, and it was 5 hectares per megawatt for the Grand Coulee Dam in the northwestern United States.[107] Other countries have accepted dependence on neighboring countries. Ghana has relied on imported electricity from Côte d'Ivoire when the water level behind Nkrumah's Volta River dam was too low, and Kenya had several years when it received a third of its electric power from a dam at Jinji in Uganda, which incidentally stands roughly where Churchill wanted a dam in 1908.[108] These dependencies have obvious risks, though the domestic politics of hydroelectricity can be perilous as well. Blackouts and rate increases in Kyrgyzstan, where 90 percent of domestic power production is hydroelectric, were major factors in the ouster of the Kyrgyz president Kurmanbek Bakiyev in 2010.[109] In January 1987, workers at Itaipú went on strike when Stroessner was about to arrive for the ceremonial initialization of several new turbines, and Paraguay's dictator was deposed in a coup two years later.[110] On the Brazilian side, Itaipú triggered rural protests that helped launch the Landless Workers Movement (see chapter 6.3, In Spite of All Doubts).[111]

Political conflicts run through the history of Aswan, and they were

ASWAN DAM

both domestic and international. An irrigation project at Gezira south of Khartoum became subject to extensive negotiations between colonial and postcolonial officials in Egypt and Sudan.[112] The newly independent countries signed a water-sharing agreement in 1959, one year after Egypt's contract with the Soviet Union, but that turned out to be a mere temporary settlement.[113] The most recent chapter is the Grand Ethiopian Renaissance Dam on the Blue Nile, which has been described as "a 'game-changer' that challenges Egypt's long-standing hegemony over the Nile Basin."[114] It is hardly the only standing conflict. In Central Asia, hydroelectric projects in Kyrgyzstan and Tajikistan generate tensions with downstream Uzbekistan, Kazakhstan, and Turkmenistan.[115] Burundi, Rwanda, and Tanzania signed contracts for initial construction work of the Regional Rusumo Falls Hydroelectric Project in 2016.[116] There are also ongoing negotiations about a collaborative project to replenish Lake Chad, with a 2,400-kilometer canal that would bring water from the Congo River watershed to the Central African Republic, where it would enter an existing river that flows into Chad. Twelve African countries are involved in the project, but only three, Cameroon, Nigeria, and Libya, are supposed to provide three-fourths of the funds, and the last two are high on any list of failing states.[117] Perhaps the best one can hope for in international water conflicts is an independent authority that investigates the issues at stake. The World Bank conducted an environmental impact study on dam projects in Mongolia that would affect Russia's Lake Baikal.[118] The knowledge acquired from over a century of high dam construction will never provide unambiguous conclusions, but as it stands, it has a better track record when it comes to sustainable solutions than does raw multilateral arm-twisting or civic activism from afar.

All the while, others are at work completing the projects of the past. The Helmand Valley Project crawled forward decades after Toynbee's visit in 1960 and reached a milestone in early 2001 when the Taliban linked the hydroelectric plant of Kajakai Dam to the city of Kandahar. American bombs destroyed the plant a few months later.[119] In 2008, the British military conducted its largest route clearance operation since World War II when it devoted several thousand soldiers to bring an 18.5-megawatt turbine to Kajakai Dam in Helmand.[120] The parts were still waiting to be installed three years later, as Chinese contractors had fled, 500 tons of cement were waiting to be delivered, and existing power lines could not handle the extra voltage.[121] The magic is long gone from the dreams of hydro-development, and so are the big men who purported to guide them to success, but their concrete legacies remain with us for better or worse. If dams are really Faustian bargains, it would seem that the devil is still out there collecting signatures.

30

THE RICE-EATING RUBBER TREE

SPEAKING OF DEPENDENCE

I. IN THEIR DREAMS

A dream was all the rage among the Indigenous people of Borneo in the 1930s. Nobody knew who spoke about it first, and it did not seem to matter who did, for the plot was galvanizing in its own right. It was about rice and rubber trees, two plants that most people on the Southeast Asian island were familiar with. The dream was about how rice mysteriously disappeared after being spread out in the sun to dry, only to be found again inside a hollow rubber tree. The idea of a rice-eating rubber tree touched a nerve among people who saw rice as their staple food, and Western ideas about the privacy of dreams were unknown to the tribal societies of Borneo. They felt that dreams were there to be shared and discussed in a common effort to decipher their meaning. The dream of the rice-eating rubber tree became the stuff of daily conversations, and some people felt that it took more than words to get to the bottom of the mystery. They cut down rubber trees to see whether they were really hiding some rice.[1]

Unlike rice, rubber trees were relatively new to Southeast Asia. They arrived in the late nineteenth century after a transfer from Brazil via Kew that entered the combined annals of botanical exchange (see chapter 27.2, Botanical Exchange) and biopiracy (see chapter 28.2, Legal Titles). Rubber trees of the *Hevea brasiliensis* variety gave off abundant amounts of high-quality latex from incisions in their bark, and latex turned from a biological curiosity into a priced raw material when Charles Goodyear discovered vulcanization and launched the modern rubber industry in the 1840s.[2] The director of Singapore Botanic Gardens, Henry Nicholas Ridley, devoted a good part of his time and reputation to the introduction of *Hevea brasiliensis* in Southeast Asia, and when the first plantation went up in Malaya in 1896 after almost two decades of botanical work, the global geography of rubber changed dramatically within a matter of years. Rubber plantations sprang up in Cameroon, German East Africa, and French Indochina, but Malaya and the Dutch East Indies were the ones that came to dominate the world market while Brazilian rubber fell off the cliff.[3] Rubber production had brought in a quarter of Brazil's export revenues between 1898 and 1910, but it relied on thousands of rubber tappers in the Amazon rain forest, and wild collection was no match for plantations in Southeast Asia.[4] An Amazon tapper produced about a quarter of the annual yield of an Asian worker, and the quality was inferior.[5]

But in the 1920s, plantations in Southeast Asia faced growing competition on their own turf. Smallholders in the region had long produced commodities for distant consumers, but rubber brought market-oriented production among smallholders to a new level.[6] Demand for rubber looked robust as automobilism was ascendant (see chapter 34, Autobahn), and cultivation required no special skills. Just like pepper and coffee in earlier days, rubber "could be woven directly into the fabric of swidden farming."[7] After clearing trees and undergrowth from a new plot, farmers planted *Hevea* seeds along with the first rice, thus giving rubber a head start over other trees. When the rice field was abandoned, rubber trees were big enough to keep undergrowth at bay.[8]

The combination of rice and rubber invited thinking in terms of tradition and modernity, and anthropologists have read the dream of the rice-eating rubber tree accordingly. Vinson Sutlive argued that it was about "the conflict between traditional rice farming and rubber gardening" with its lure of "large profits."[9] But from the smallholders' perspective, the dream was not so much about big overarching concepts as about the advantages of a dual economy. Rice and rubber combined the safety of subsistence production with the opportunities of a cash-based economy. The combination also provided a safety net if one of the crops ran into trouble, and biological contestations were a factor to be reckoned with in modern times (see chapter 12, Boll Weevil). The combination also made sense in ecological terms, as the two plants were largely complementary out in the field. Rubber trees were perfect for the labor regime of peasants, as tapping could be delayed when other tasks were more urgent. While rubber plantations were struggling to recruit and retain workers, smallholders relied on unpaid family labor, supplemented by hired hands if necessary.[10] Furthermore, rubber plantations held a lot of fixed capital, and they were slow to respond to price fluctuations because rubber trees took a few years to reach maturity. And as it happened, commodity markets of the time were notoriously unstable.[11]

However, the plantation lobby had the ear of colonial authorities, and stabilization of world rubber prices was the overarching goal of the International Rubber Regulation Agreement of 1934. Signed by the United Kingdom, India, the Netherlands, France, and Siam, it aimed to curtail production through taxes, sales quotas, and limits on planting. Like most agreements of its kind, the International Rubber Regulation Agreement triggered endless negotiations over details, but the bottom line was clear: "The benefits of restrictions were very unevenly divided between estates and smallholders, to the disadvantage of the latter."[12] The agreement sought to stabilize the estates, and lacking political pull, smallholders on Borneo had no choice but to live with discrimination on the rubber market. But they could adjust their behavior. That was what the rice-eating rubber dream was all about.

While the dream was agnostic about the benefits of rubber trees, it was keenly aware of the value of rice: it could only capture minds in a society that was terrified by the prospect of disappearing rice. Keeping a balance between subsistence needs and market production was traditionally at the heart of Indigenous agriculture on Borneo, and the dream was about the loss of balance in one direction only: rubber could eat rice, but rice was no threat to rubber. As Michael Dove has argued, "The dream was a mythologically condensed expression of an undesirable trajectory in agricultural development."[13] It told those who banked on rubber and neglected rice production that they were going down a dangerous path.

The rice-eating rubber tree was an Indigenous trope bound to a certain time: it never flourished again after its heydays in the 1930s.[14] It was more open in geographic terms, as it forged a link between choices in Bornean woodlands and world politics. "The dream of the rice-eating rubber illuminates Bornean tribesmen's consciousness of the threat posed by overcommitment to global commodity markets," Dove noted.[15] Indigenous people could not do much about international agreements, and it was obviously disturbing for smallholders to be at the mercy of shifting global markets, but at least they could work through the experience by debating a dream. It was not a unique situation in the global world of resources.

2. IN THEIR THEORIES

Dreams were not the only Indigenous response to the fall of rubber prices. In Sarawak in northwestern Borneo, where the British colonial government had encouraged planting, some Iban people saw their declining market prospects as an act of betrayal, refused to pay taxes, and launched a hapless rebellion.[16] Resistance runs through commodity histories from the slave uprisings on Caribbean sugar plantations (see chapter 2, Sugar) to Gandhi's salt march (see chapter 23, Gandhi's Salt), and for all the differences in tactics and outcomes, the tension between local conditions and global connections is a running theme in commodity history. Latin America had a long and tragic history in this regard, and it is more than a coincidence that Michael Dove, who solved the riddle of the rice-eating rubber tree, drew inspiration from Michael Taussig's *The Devil and Commodity Fetishism in South America*.[17] Taussig described how proletarianized peasants on sugarcane plantations in Western Colombia worked through their experiences with a trope about a deal with the devil.[18]

Conversations about commodities had a whiff of fatalism in Latin America, and a sense of victimization ran through societies. It inspired Eduardo Galeano to write *Open Veins of Latin America*, a classic of Latin American studies that chronicled, in the words of the subtitle, "five centuries of the pillage of a continent."[19] It was an indictment of global dependencies as well as national elites, and as a critical journalist, Galeano was forced into exile after a military coup in his native Uruguay in 1973, but the underlying sentiment was well familiar to those in power, and some politicians put it on paper in their own peculiar ways. One of the more dramatic expressions came from Getúlio Vargas, the erstwhile dictator of Brazil and the country's democratically elected president from 1951 to 1954. "I fought against the looting of Brazil,"

Vargas declared in a letter that he penned inside the presidential palace. Then he shot himself in the heart.[20]

In the years before his suicide, Vargas had embraced state-led industrialization and import substitution to turn Brazil into a developed nation. When these policies failed to achieve the desired result, dependency theory captured hearts and minds on the continent and beyond.[21] It was the Global Sixties at work.[22] The Cuban Revolution had opened a new sphere for visions of a different Latin America beyond a hegemonic United States. Universities were hiring, intellectuals took pride in engaging with real-world issues, and new institutions sprang up at the intersection of academia and politics. Located in Santiago, Chile, the UN Economic Commission for Latin America and the Caribbean became a hub for dependency theorists.[23] And in the 1960s, if you sought to make it in certain circles, you were nothing without a theory.

Dependency theory had a number of attractions. It provided an escape from the orthodoxies of Marxism, particularly its stage theory, which suggested that feudal societies had to go through a phase of thorough capitalist development before moving on to Socialism. Seen through the lens of dependency theory, agricultural production was not necessarily a relic of feudalism: it could be an integral part of a capitalist system. Dependency theory stressed that inequality was not only about social relations but also about geography, and the core–periphery concept became a cornerstone of Immanuel Wallerstein's world-systems theory.[24] It served as a caveat about Ricardo's theory of comparative advantages, as it naively assumed a level playing field that only existed in the mythology of free trade. In a wider context, dependency theory gave wings to political hopes, and Latin America lingers as a canvas for dreams about a better world to this day.[25] In the words of Andrés Velasco (see chapter 15.3, Dutch Disease), then a professor of international finance and development at Harvard University, dependency theory was "a religion that shaped the cosmology of a generation of Latin American leftists in the 1960s and 1970s and of leaders from Chilean President Salvador Allende to the Nicaraguan Sandinistas."[26]

Dependency theory has been criticized more recently for its disinterest in culture, though that may say more about scholarly predilections in the age of postcolonial theory.[27] Unlike orthodox Marxism, dependency theory looked beyond class relations and the collusion between capitalists and the state. As Louis Pérez has noted, "Dependency was seen to penetrate all levels of national institutions and assume a variety of forms."[28] Furthermore, dependency theory was part of a cultural self-assertion, a genuine Latin American contribution to economic thought and a critique of domestic elites who looked abroad for values, careers, and consumer goods. The trajectory of some defining books surely mirrored persistent cultural hierarchies. *Dependency and Development in Latin America* by Fernando Henrique Cardoso and Enzo Faletto was published in Portuguese in 1968 and in Spanish in 1969, but the English edition had to wait until 1979, "surely evidence of continuing North American ethnocentrism *vis-à-vis* Latin America."[29] But once beyond the language barrier, it met with a receptive audience. Velasco wrote that "U.S. college campuses embraced dependency with evangelical fervor."[30]

Like most intellectual trends, dependency theory was not a monolith. "There are almost as many currents of dependency analysis as there are major contributors to the debate," a professor of economics observed in a review article of 1980.[31] *Dependistas* looked at different countries, and they offered their diagnosis of Latin America's disease in different strengths. Andre Gunder Frank opened the preface of his *Capitalism and Underdevelopment in Latin America* by sternly declaring his belief that "it is capitalism, both world and national, which produced underdevelopment in the past and which still generates underdevelopment in the present."[32] In such a reading, Latin America could only overcome underdevelopment by breaking with the world capitalist system, and Frank was looking forward to "the successful pursuit of the revolutionary class struggle in Latin America."[33] Cardoso and Faletto offered a more open-ended view in *Dependency and Development in Latin America*, concluding that they did "not try to place theoretical limits on the probable course of future events."[34] In the introduction to the English edition, they even expressed doubts about the need for theory: "We refer to 'situations of dependency' rather than to the 'category' or the 'theory' of dependency."[35] Change was possible for Cardoso and Faletto, and they had an idea where to look. "In *Dependency and Development*, the national level is the site of potential agency; the global level is the source of structural constraint."[36]

As it happened, Cardoso pursued a political career on the heels of his academic one. He became a senator for São Paulo in 1983, federal minister of finance in 1993, and president of Brazil in 1995.[37] Economic policy during his presidency focused on a successful fight against inflation and a push for privatization, two issues that were more reminiscent of the neoliberal spirit of the 1990s than of writings from the 1960s, and when he left office in 2003, he became a voice in the critical discourse on globalization. Cardoso argued for "globalized social democracy," a combination of "openness to international markets, robust social policies that promote social justice, and a democratically accountable state."[38]

Along the way, Cardoso also served as president of the International Sociological Association. In his presidential address of 1986, Cardoso argued for an undogmatic sociology that was "unafraid to venture into fields where there may not be much scientific rigour."[39] Furthermore, he sketched the need for "a theory of change which does not assume that the destination—for developing countries, the safe haven already found by the developed countries—can be known in advance."[40] With knowledge of his two terms as president of Brazil, it reads like an anticipatory commentary on a political career that left observers dizzy. Did Cardoso abandon his intellectual roots for the presidency, and if so, what did that mean? Some felt betrayed. Cardoso's greetings to the 1997 meeting of the Latin American Sociological Association in São Paulo drew jeers from an audience more enamored of Ché Guevara, Nicaraguan Sandinistas, and Mexican Zapatistas.[41] Others were merely surprised. The *Penguin History of Latin America* described his anti-inflation program as "a spectacular volte face."[42] And some were gleeful. In an obituary for dependency theory published in *Foreign Policy*, Velasco recalled how Cardoso shocked a left-leaning crowd at Yale University in the early 1980s, first by wearing a neat blue suit and then by forsaking Socialism in favor of "perfecting capitalism."[43] Cardoso was not terribly good at playing to the instincts of academic audiences, and he was probably used to being misunderstood by that time. He published a scathing dissection of the US dependency-theory discourse as early as 1977.[44]

As an elder statesman, Cardoso was confident that he had remained true to his original commitment. Leftists and romantics were inclined to disagree, and those who fell into both categories disagreed with particular vigor, but in the end, it was all a matter of perspective. Cardoso was both an intellectual and a political entrepreneur, and the latter tends to teach lessons about the ambiguous merits of ideological purity. And what did it help to bask in a sense of victimhood when opportunities were out there in the economy?

Ideas were never pure (see chapter 23.2, The Power of Ideas), nor did they provide an exhaustive guide in everyday life, and dependency theory was as good an example as the dream of the rice-eating rubber tree. The latter was a powerful warning, but it was not as if everyone on Borneo was heeding it. Some regions were recording rice imports and scarcity of land during the rubber boom, sometimes to the point that rubber trees had to be felled.[45] The situation on Borneo was scarcely exceptional. In peasant societies, resilience and survival were goals rather than certainties, and whether they were actually achieved hinged on individual circumstances, environmental and commercial conditions, and luck. In any case, one may surmise that Cardoso would have found plenty of understanding in 1930s Borneo, where dreams were about balance rather than intellectual stringency. Maybe it was just that, in an interconnected world, myths were hard to retain, at least for those who refused to leave certain things out of the picture.

3. STUFF OF LEGENDS

Smallholders were not the only group on Borneo that was wrestling with a mental fixation in the 1930s. Officials and plantations embraced a mythology of their own, and like the rice-eating rubber tree, it was about the economy. Planters were aghast about the smallholders' ability to produce rubber at competitive costs. Surely something was wrong about their success. Were they overtapping rubber trees, or spreading disease, or destroying virgin forests? Scientists looked into the allegations and disproved them one by one, leaving planters with no good excuse for their inferior productivity. It was a disturbing insight for a group of people who saw plantations as the self-evident embodiment of modernity (see chapter 2, Sugar), and it was about more than economics: it went straight to the heart of the colonial project. Corey Ross has pointed out how the smallholders' success blurred racial hierarchies in a highly inconvenient way: "Any admission of being out-thought by 'natives' raised unsettling questions about why Europeans were there at all."[46]

It made for a nice symmetry of mythologies. As Michael Dove has noted, "This official myth of the diseased and threatening smallholdings was ... quite as structurally correct as the smallholder myth of the rice-eating rubber."[47] Time did not end the planters' humiliation, though a lot happened in the forests of Borneo after the 1930s. Independence for Indonesia and Malaysia changed power relations in the region, though not as much as smallholders might have hoped. Chinese rubber growers were violently evicted from West Kalimantan in Western Borneo, one of many conflicts in the Global South where a postcolonial legacy and postindependence power play intertwined: the Dutch colonizers had territorialized race and had denied ethnic Chinese access to land titles (see chapter 6, Land Title).[48] In the 1970s, the Indonesian government sought to bring rubber smallholders into its orbit with an extension program and nuclear government estates that catered to satellite smallholders.[49] Most recently, palm oil surpassed rubber as Malaysia's most important organic commodity.[50] But through it all, the smallholders' share of production continued to climb while plantations were losing out. By the twenty-first century, smallholders produced more than four-fifths of Indonesia's rubber.[51]

Unlike dependency theory, rubber production from smallholders received scant interest beyond the region, though it was not unique. In spite of the best efforts of the likes of Cadbury and the Lever Brothers, cocoa plantations invariably failed along Africa's Gold Coast, and the last European cocoa plantation collapsed in the early 1940s.[52] The region's cocoa boom was the work of Indigenous peasants. In Ecuador, a strict government policy gave independent farmers a foot in the banana business while other countries were in the clutches of United Fruit (see chapter 10, United Fruit).[53] None of these experiences gained much acclaim around the globe, helped by marketing departments that banked on consumers' ignorance: a German trading house sold its trademark chocolate drink Kaba as a *Plantagentrank* (plantation drink).[54] Maybe the Western imagination continues to bristle at the thought that peasants can compete with modern plantations. Or maybe they just stand in the shadow of other myths.

As iconic commodities go, rubber has shown potential. Several events entered the annals of modern history: the purported robbery of *Hevea brasiliensis* seeds (see chapter 28.2, Legal Titles), the genocidal reign of terror in King Leopold's Congo, the forced labor investigation on Firestone's rubber plantations in Liberia, and the wartime synthetic rubber projects in the United States and Nazi Germany. Others, like the Putumayo affair in the Western Amazon, at least made headlines for a few years. None of these events played out in Southeast Asia, which was quietly producing the lion's share of the world's natural rubber throughout the twentieth century. It might have gained some acclaim if Japan had pursued a scorched-earth policy at the end of its occupation during World War II, but Bornean rubber came out of the war largely unscathed. In collective memory, oil, not rubber, was the defining commodity of the Pacific War (see chapter 35, Pine Roots Campaign).[55]

Rubber was an iconic commodity from the mid-nineteenth century to 1945, down to *Blutgummi* (Blood Rubber), a commodity novel of 1938 that celebrated Germany's synthetic buna rubber as a "triumph of reason" because it freed the world from the barbarism of free trade.[56] But natural rubber production went out of focus in the postwar years, and when it reemerged, it happened on the other side of the world from Southeast Asia. The union president Francisco Alves Mendes Filho, better known to the world as Chico Mendes, became a charismatic figure of global fame in the mid-1980s when the Amazon rain forest emerged as an iconic region on the world's mental map.[57] Mendes was a clever leader of rural workers in Brazil's Acre Province who managed to forge links with environmentalists and human rights groups in other parts of Brazil, Europe, and North America. Under his aegis, the tapping business was rebranded as an act of environmental stewardship: "We demand to be recognized as [the] genuine defenders of the forest," declared a statement by the National Meeting of the Rubber Tappers of Amazonia at a conference in Brasília in 1985.[58]

Brazil's Indigenous people might have offered a different perspective, but they had no voice at the conference. Indigenous people had been involved in rubber tapping—the Putumayo affair was about abuse of native tappers—but much of the workforce, including Chico Mendes, had a migration background. However, Chico Mendes and his band of rubber tappers made an impression. Four years after the Brasília manifesto, Brazil's legislature amended the country's National Environmental Policy Act and allowed the creation of extractive reserves, a peculiar type of nature reserve (see chapter 26, Kruger National Park) where tappers could do their job in an otherwise protected rain forest.[59] Twenty years on, some fifty extractive reserves covered more than ten million hectares in Brazil.[60]

Chico Mendes did not live to see his struggle bear fruit. He was killed three days before Christmas 1988 after numerous threats from ranching interests, the prime cause of deforestation in Acre. The chief of Brazil's Federal Police oversaw the investigation into the murder, and the killers were caught, somewhat to their consternation. Mendes was the fifth union president murdered in Brazil in 1988 and one of more than a thousand people who were killed in land disputes (see chapter 6, Land Title) in rural Brazil since 1980, and according to estimates from Amnesty International, fewer than ten killers had gone to jail.[61] His death helped smother the tensions between tappers and environmentalists: a few weeks before his murder, Mendes yelled at a television when a report praised his crusade to save the "lungs of the world," declaring that he fought "because there are thousands of people living here who depend on the forest."[62] Chico Mendes simply had it all: he had worked as a tapper, built a union, befriended environmentalists and other activists at home and abroad, and he gave a face to the embattled Amazon rain forest. He had friends in high places, his picture in the *New York Times*, and, as of 1988, martyr status. And yet there was one question about Chico Mendes that remained largely unexplored: Why were there any tappers left in the Amazon?

The region had never recovered from the collapse of rubber extraction after 1910, and plantations came to naught in the Amazon; the most famous endeavor, Henry Ford's Fordlândia project, was literally eaten up by South American leaf blight, a lethal fungus (see chapter 12, Boll Weevil).[63] The last hurrah was during World War II when the Japanese advance in Southeast Asia put America's rubber supply at risk, but US interest was already receding during the war when synthetic rubber took off. However, Brazil's federal government kept subsidies for rubber tapping in place after 1945, which was perfectly in line with its import substitution policy.[64] It was these subsidies that made the difference. Without them, there would have been no tappers in the Amazon, no tappers' union, and no Chico Mendes.

One can find it an irony of history that a global icon emerged on the back of eminently anti-global policies. Or one can find it a case in point about how global interdependencies are neither one-way streets nor two-way streets. Commodities build multilayered networks that connect diverse places and people around the globe, and they produce outcomes in different forms, shapes, and degrees of significance. On an interdependent planet, it would be naive to assume that trends inevitably add up: they most likely do not. "The case of the rice-eating rubber dream suggests that, contra Wallerstein, incorporation—or, more accurately, an increase in the level of incorporation of local societies into the world system—does not initiate a sequence of predetermined change," Michael Dove wrote in his study on Borneo.[65] The same is true when incorporation produces not just resources but also legends.

PART VII

THE AGE OF CATASTROPHE

IT WAS AN AGE TO FORGET.
IT WAS ALSO AN AGE THAT WE MUST NOT FORGET.

Unlike previous parts of this book, the final two sections focus on specific time periods. The twentieth century is unusual in world history because it had clear turning points that left their mark everywhere on the globe. Two world wars and a Great Depression changed lives and livelihoods around the planet, and the same can be said for consumerism and the Cold War after 1945. Change was not only more universal but also more dramatic than anything that the modern era had previously seen: it was truly an "age of extremes," as highlighted in the title of Eric Hobsbawm's acclaimed history of the twentieth century. Hobsbawm identified three distinct periods for the seventy-seven years from 1914 to 1991: an "age of catastrophe" from 1914 to the aftermath of World War II, a "golden age" that lasted into the 1970s, and the 1970s and 1980s that Hobsbawm viewed as "the landslide." *The Age of Extremes* was first published in 1994, but a quarter-century later, we can probably extend the last period into our own time.[1] In fact, the landslide metaphor looks far more plausible than it was at the time of writing. Hobsbawm, who died in 2012 and published a book under the cheery title *How to Change the World* in the year before his death, would probably be appalled about all the mud that has moved in recent years.[2] Enthusiasm for freedom, democracy, and market forces defined the 1990s, but seen from 2020, that decade looks more like a temporary slowdown in a global avalanche.

The difference between the golden age and the landslide is open to debate in a vortexian history, as the environmental problems of the late twentieth century were intrinsically linked to mass consumption. In fact, the age of the landslide was exactly the time when environmental debates and environmental policies changed in ways that resonate to this day. The 1970s and 1980s were watershed years for the global environmentalism that has transformed virtually every issue in this book, and all this was part of an even bigger transformation discussed in the final part of this book as "the great entrenchment." Environmental historians have fewer problems with the age of catastrophe, though scholars need to move a bit beyond Hobsbawm's narrative. He does acknowledge "pollution and ecological deterioration" as byproducts of the golden years, but by and large, Hobsbawm's age of catastrophe is staunchly anthropocentric in nature.[3]

The age of catastrophe is defined by the beginning of World War I and the aftermath of World War II, and Hobsbawm's discussion starts with a chapter on total war. This section concludes with a discussion of the pine roots campaign that explores the environmental dimension of total war, but that was only one of numerous manifestations of war between 1914 and 1945. The first two chapters focus on events that were not strictly military in nature but defined by its spirit: the "war against the kulaks" that plunged Ukraine and other parts of the Soviet Union into a devastating famine and the "battle for land" that Fascists fought in Italy's Pontine Marshes and continued elsewhere. The interventionist legacy of war economies was the backdrop for a campaign for the use of agricultural raw materials launched by the chemurgy movement in New Deal America. War also shaped the mythology of contemporary projects, and some of these myths have kept historians busy ever since. It is perfectly clear that Hitler's divided highways were not built for military purposes, but that may not become popular knowledge anytime soon.

The exigencies of warfare bolstered the authority of nation-states, one of the most consequential developments for the environmental history of the twentieth century. In 1914, it was still undecided how state authorities would fare on environmental matters compared to the municipal authorities that built the sanitary city (see chapter 17, Water Closet), supranational bodies like the Brussels Sugar Convention (see chapter 2, Sugar), and private endeavors such as United Fruit (see chapter 10, United Fruit) and Chicago's slaughterhouses (see chapter 18, Chicago's Slaughterhouses). By 1945, the matter was settled: nation-states had more money and more power than ever before, and they would be defining political agents until their powers shrank toward the end of the twentieth century. Goals and strategies could differ, as states could be everything from engines of war to nascent welfare states, but there was no longer a real competition between the different levels of government.

State authorities served an unprecedented range of ideologies. The age of catastrophe saw the first Socialist and the first Fascist regimes of world history, and this section covers the full range of political systems. Ukraine's Holodomor was tied to Soviet collectivization, the defining policy of Socialism in the countryside. The section also looks at the three Axis powers at different stages of their historical trajectory: at peace with Nazi Germany's autobahn project, at war with the Japanese pine roots campaign, and somewhere in between with Fascist Italy's Pontine Marshes. The chemurgy movement in the United States mirrored the divergent impulses in open societies during and beyond the interwar years: democracy, market forces, scientific prowess, and the burgeoning state interventionism of the New Deal years. It is a mix that Western democracies are wrestling with to this day.

It makes for a striking contrast to the postwar years. All the chapters in the final section are defined by material trends: the craving for meat (battery chicken), the rise of nuclear power (*Lucky Dragon No. 5*), dangerous chemicals (DDT), the oil glut and the inherent risks of large technological systems (*Torrey Canyon*) and an ephemeral but popular consumer product (the plastic bag). But in this section, each event hinged on a political decision or, in the case of the chemurgy movement, the hope for one. Even more, decisions hinged to a significant extent on individual leaders: there might have been no Holodomor without Stalin, no land reclamation project in the Pontine Marshes without Mussolini, no autobahn project without Hitler, and no manic quest for pine roots without Japan's generals. The following chapters show that their decisions were shaped by contemporary trends and tropes, and yet structures and contexts only get us to a certain point. The age of catastrophe was also an age of big men.

This section is also the most Eurocentric in this book. For all the global repercussions between 1914 and 1945, they usually went back to events and decisions in Europe and North America. But these events were not distinctly European, and the Great Famine of 1932-1933, more recently reframed in Ukrainian nationalist mythology as the Holodomor, is a case in point. It mirrored the new nature of hunger crises in the modern age: they were no longer about material scarcity—the sheer absence of food—but about access to food. Governments and organizations have learned a lot about the management of food supplies, and the notable shift from emergency aid to nutrition regimes that has taken place in recent decades is stronger in policies than in public awareness, but authorities have pursued other goals at times. War and famine were closely linked throughout human history—the Book of Revelation depicted them as fellow horsemen—but in the twentieth century, war became the primary cause of mass starvation. It did not have to be a military conflict between nations. With collectivization and dekulakization in full swing, the Soviet Union of the early 1930s was arguably in a state of war.

The collective farm was a defining feature of the Communist world. None of the world's Socialist governments could afford to dispense with an attempt to dispossess the peasantry. But the Soviet kolkhoz was scarcely a role model: born out of the chaos of the Great Famine, it was an unstable compromise between state and peasant interests. It tried to square central planning with individual initiative, and the result was no more satisfactory than similar efforts in other realms of the Socialist economy. Accountability was a notorious problem of Socialist rule right to the end: the East German dissident Rudolf Bahro described the Socialist economy as "a bureaucratic mechanism with the tendency to kill off or privatize any subjective initiative."[4] It constrained the ability of Communist governments to confront environmental problems, and it has framed the global memory of collective farming. Collectives fell apart along with the world of Communism, and where governments failed to reorganize corporate structures or redistribute the land (see chapter 6, Land Title), agribusiness seized the initiative. Ukraine's oligarchs may be disinclined to acknowledge it, but they owe a good part of their assets to Stalin's obsession with collectivization.

The Holodomor became a pillar of Ukraine's post-Soviet identity. It has carried the imprint of transnational commemoration of the age of extremes—in a nutshell, Ukraine's pro-Western governments sought to draw legitimacy from the Holodomor in the same way that Israel drew legitimacy from the Holocaust—but famines have raised questions about rulers wherever they have occurred in modern times: starving people heralded the ultimate failure of governance. Amartya Sen famously wrote in the 1990s that famines have never occurred in a functioning democracy, but some caveats are in order twenty years later, and not just because Sen's list of resilient democracies included Zimbabwe.[5] Only the most ruthless governments allow their people to starve, but they can leave them malnourished or obese and get away with it.[6] And then there are the enduring hierarchies in the conscience of the world. Ukraine's Holodomor received attention because Western audiences could relate to peasants in Europe. Contemporary Kazakhstan, where the victims were nomads and Asian, was a more difficult sell.

Ukraine and Kazakhstan were not the only places where food was a particular concern in the interwar years. Agricultural products were among the most globalized commodities before 1914, but war and the collapse of world trade in the wake of the Great Depression threw many nations back on their domestic resources. One of the consequences was a quest for new farmland, a particularly significant quest in countries like Italy and Germany that were struggling to feed their populations from their own soil. Land reclamation in the Pontine Marshes sought to boost domestic grain production, and it served as a template for Italy's African colonies as well as Nazi Germany's conquests in Eastern Europe. *Lebensraum* was an ideology, an economic doctrine in the pursuit of autarky, and a concept for a mushrooming array of scientists and planners who sought to forge the new space and the new men who were supposed to work the land. *Lebensraum* was also a growing anachronism in a world that was inching toward higher yields with new seeds (see chapter 28, Hybrid Corn), fertilizers (see chapter 19, Synthetic Nitrogen), and other tools of biological improvement (see chapter 27, Eucalyptus), but agricultural progress failed to stop the quest for space. Military defeat did.

However, the battle for food was only one of several battles that the Fascists were fighting in the former wetland south of Rome. They also sought ruralization, a higher birth rate, an improved Italian race, a solution for Italy's landless people that did not require stepping on the toes of landowners, and a setting for a propaganda show within driving distance from the capital. Mussolini took a personal interest in the Pontine Marshes project, and it had a distinct Fascist flavor—it helped to be a Fascist if you wanted to believe in an Italian race—and yet the campaign also mirrored a new style of government action: bold, comprehensively planned under centralized leadership with mass mobilization and carefully crafted media work. Italy's *battaglia del grano* was plagiarized as the *campanha do trigo* in Portugal and as the *Erzeugungsschlacht* in Nazi Germany, but it was not just Fascists who developed a taste for politics in battle mode. Born in the age of catastrophe, state-led campaigns became such a defining feature of modern politics that most people forgot that "campaign" was originally a military term.

The Pontine Marshes was an emblematic development project: funded by the state, shaped by scientific expertise, realized with an energetic push, and rationalized with a utopian goal (or multitude thereof) whose legitimacy was beyond debate. While the quest for *Lebensraum* became moot after 1945, decolonization and the Cold War made development projects a defining feature of postwar politics, particularly in the Global South. And just as in the Pontine Marshes, development met with resistance on the ground, including from those who were supposed to benefit from projects. Development typically brought about change, but rarely in the exact way that the blueprints envisioned, and the experience cast a shadow over development projects while the tropes that underpinned them melted into air. The only clear winners were the experts and their institutions. The age of catastrophe saw the rise of new academic professions and growing powers for existing ones, and more often than not, ambitions were greater than knowledge and skills. It mattered for the outcome of development projects, but it mattered far less for the fate of professions. Apart from ideological purebreds like Lysenko, few experts were pushed off the stage in the postwar years.

However, the place of academic expertise in politics and society was still open to debate. Professions always seek to claim jurisdiction for a field of expertise, but the crisis years of the age of catastrophe probably asked for more: a new polity where everything would yield to the supreme authority of the experts. The chemurgy movement tried to gain that kind of authority in 1930s America, and one should not see failure as preordained: nobody could know whether capitalism would survive the Great Depression. Chemurgists sought to turn agricultural commodities into industrial raw materials and thus make fears of looming resource exhaustion pointless, but they offered more than scientific visions and technological skills. They wanted a new economy where commodities were traded according to chemical values (as opposed to old-fashioned monetary ones). Some also thought about moving farm production to new expert-run agricenters. The chemurgy movement certainly pursued an ambitious agenda, but the times did not reward the reluctant expert.

It did not work as planned, but the outcome was more akin to a stalemate than a defeat. Professions occupy an uncertain place in political systems of various stripes to this day: they are more than service providers, but they failed to secure chemurgical omnipotence. Chemurgy had a more definitive outcome on a different front. Its technologies, and particularly its flagship fuel alcohol project, challenged the barrier between food and nonfood resources with a chemical vision of universal transformability. Chemists had undermined that barrier for generations, but the chemical industry of the twentieth century could manipulate compounds on an unprecedented scale, and the depression years were not a good time to raise sentimental questions about ethical implications. The consequences resonate in today's debate about biofuels, where the moral case against burning food is both omnipresent and strangely inconsequential.

The chemurgy movement developed the technology for fuel alcohol production on an industrial scale, but it fared less well in contemporary politics. The New Dealers refused to support the project, something that is remarkable in itself: a generation earlier, the fate of new resources would have been up to market forces rather than politics. The reluctance of the New Deal had nothing to do with ethical qualms—its agricultural policy began with an intensively publicized mass slaughter of six million pigs—and everything to do with the interests of the large oil companies, which had a competing product in the form of tetraethyl lead. Politicians stood center stage in the age of catastrophe, but one should not overlook the power of corporate octopuses like Big Oil (see chapter 10, United Fruit).

But some tasks were inexorably tilting toward the nation-state in the interwar years, and road construction was one of the most consequential. Private money built Fascist Italy's *autostradas*, and local and regional authorities were traditionally planning and paying for new roads, but Hitler's autobahn project relied on a national government in a spending mood. It was a disaster in almost every respect during the Nazi era (though a buzzing propaganda machine suggested otherwise), but it pointed to a future with state-funded and ever-expanding street networks: a good part of the road to the welfare state was paved with asphalt. Maintaining the crumbling leftovers of the building spree is one of the great challenges for the twenty-first century.

The new streets were purpose-built for automobiles, and it takes a historical perspective to recognize the extraordinary nature of this decision. Streets were traditionally spheres of many uses, and transforming them into mere traffic arteries for wheeled transportation took comprehensive rebuilding and several generations of social disciplining. Anti-automobile sentiments decreased over time, or at least lost much of their political and cultural sting, and today the drivers of the world get away with an annual death toll in the seven digits. It is one of numerous problems of pervasive car use, but the critiques, environmental and others, seem no match for the enigma of automotive mobility. Even the Soviet Union abandoned its flirtation with collective car use and surrendered to automobile individualism. The freedom of driving is perhaps the world's most universally acclaimed type of freedom, though the reality is more akin to a fading afterglow of bygone utopias.

The final chapter in the section returns to warfare and focuses on what might be the most dramatic among the many environmental consequences of military conflicts in the age of catastrophe: resource use. Total war required the total mobilization of all available resources, and so it came to pass that legions of Japanese schoolchildren were digging for pine roots in the final months of World War II. The campaign has often been portrayed as a giant folly—not least because it sought to produce aircraft fuel that was never used in a plane—but it may be more adequate to view it as the brutal final stage of a type of warfare that relied on ever more resources. The problem was material as well as cultural. Modern nations have narratives that account for defeat in the face of overwhelming force or lack of heroism, but they do not have narratives for defeat due to a lack of stuff.

The pine roots were cooked in purpose-made kettles, which produced a type of crude oil that served as raw material for aircraft fuel. With that, the pine roots campaign relied on the mobilization of chemical and technological knowledge, another defining feature of total war. Oil production never worked as intended, but it was an intriguing idea, and Heinrich Himmler set up a short-lived research group within his mushrooming SS empire to explore the idea in the final days of the Nazi empire. The mobilization of science had consequences that went beyond the day when the guns fell silent. Total war was a catalyst in the making of the cheap, abundant, and faceless resource of modernity.

The new world of resources was a rather inconspicuous legacy of warfare in the age of catastrophe. The more obvious legacy was about the land: the traces of war showed in battlefields and bomb craters as well as in toxic substances in the soil. Sometimes war produced landscapes that are as close to wilderness as things get in global modernity. And sometimes it has left landscapes that veterans and park managers want to preserve as much as possible. The effort runs against the inherent dynamism of nature and can only produce partial results, even under the best of circumstances, and maybe that should not be an issue of concern. Biological change on former battlefields provides a fitting mirror for the legacy of the age of catastrophes.[7]

The case for a periodization can be structural as well as moral, and the latter has particular relevance for the age of catastrophe. The experiences acquired between 1914 and 1945 are the closest thing the world has to a collective memory. It is a rather Eurocentric reading, but it defines historical imaginations across national borders. War and genocide, hunger and expulsion, Fascism and antisemitism trigger a multinational response that comes down to a simple point: never again. The age of catastrophe left us with inspiring and powerful narratives, but whether they provide guidance for the political challenges of the day is open to debate. Charles Maier has argued that they may serve as "a prick to conscience" in the face of ethnic cleansing, but that was about it. "As a narrative of annihilation, the Holocaust is hard to apply to political challenges that seem to fall short of its horror."[8]

Commemoration of the age of catastrophe is obviously selective. Japan forgot about the pine roots campaign in an act of postwar amnesia, and the iconic program of Fascism is still around in the Pontine Marshes as if Mussolini were still in power, down to a beachfront hotel whose name commemorates a genocidal war. But perhaps the greater concern should be the yawning gaps that the age has left in our collective imagination. The age of catastrophe sensitized us to the perils of hunger, but malnutrition is perhaps a greater worry in the twenty-first-century world. It produced ideas of science-based development that became overgrown with sobering experiences and yet proved impossible to exorcise. It kept the ambitious scientist of chemurgic fame away from the levers of power, but it gave plenty of room to burgeoning professions. It also marked the breakthrough of an unrestrained automobility that can get away with everything, including mass murder. The legacy of the age of catastrophe will be with us for the foreseeable future. But it is not the legacy that we usually think of when we remember the years between 1914 and 1945.

31

Holodomor

The Politics of Hunger

I. AGRICULTURE, SOVIET STYLE

The harvest was great in 1932. Or that is what Stalin told the joint plenum of the Central Committee and the Central Control Commission in January 1933. If food supplies were less plentiful than expected, that was entirely due to hoarding and subversive behavior. But Stalin had already taken appropriate steps, and Ukraine, whose rich dark soils made it a perfect place for grain, was a region of particular interest. Several requisitioning drives culminated in a decree of December 29, 1932, which threatened farms that failed to meet quotas with cancellation of credits, bans on tractor use, and seizure of seed corn. A few weeks later, the secret police sealed off the region, effectively trapping the peasants of Ukraine with few things to eat.[1] It failed to allocate the desired amounts of grain, but it set the scene for one of the great humanitarian catastrophes of the twentieth century. Millions died in what came to be known as the Great Famine of 1932–1933. In 1997, the *Black Book of Communism* put the death toll at four million for Ukraine's peasantry and six million for the entire Soviet Union.[2]

The Great Famine of 1932–1933 was not the first devastating famine that happened on the Soviets' watch. An estimated five million people died from starvation in the Volga provinces and the southern Urals in 1921–1922.[3] It was the direct result of War Communism, the authoritarian style of ressource allocation during the Russian Civil War. The Bolshevik government ordered the seizure of surplus grain from peasants, who responded by curtailing production beyond subsistence needs or stowing away produce for sale on rampant black markets.[4] Faced with an existential crisis for his regime, Lenin changed course and somewhat relaxed the iron grip of the state, and the New Economic Policy provided agriculturalists with incentives to restore production. It helped feed the Soviets through the rest of the 1920s, but the improvised New Economic Policy was ultimately an ideological oddity, an ill-defined temporary phase on the path to a Communist utopia where individual entrepreneurship would no longer have a place.[5] The New Economic Policy came to an inglorious conclusion at the end of the decade when Stalin seized power and promoted the collectivization of agriculture and rapid industrialization. In an article in *Pravda* on the twelfth anniversary of the October Revolution, Stalin called it "the Great Break."[6]

Collectivization was a grand experiment in state-led development (see chapter 32.3, Planning Development), and it took the form of the quintessential political action mode of the age of catastrophe, the campaign. It spelled the end of family farms and peasant village communities, whose assets were seized and placed under purportedly collective ownership in large kolkhoz farms. The Bolsheviks also sought to weed out kulaks, a conveniently vague term for peasants who were richer than others, held power outside party structures, or otherwise created trouble. Those who were deemed kulaks were swiftly dispossessed and deported or killed while the rest of the peasantry was coaxed to join the new collective farms. Violence was a pillar of collectivization from the outset, but it looked different from the commanding heights of the economy. Large units allowed for the use of modern technology and economies of scale. It also linked up with scientific progress: the geneticist Nikolai Vavilov received generous funds in the wake of Stalin's Great Break and had 185 Soviet institutions engaged in plant breeding by 1932–1933.[7] Collectives were certainly easier to steer for central planners, they gave the government control of grain reserves that it could use for exports, and the revenue was earmarked for the purchase of machinery that the Soviet Union needed for rapid industrialization. At least that was the plan.

In reality, collectivization threw the countryside into turmoil. Few peasants viewed the kolkhozes as their own endeavors. Many fled to the cities or slaughtered their farm animals before joining a collective, and some launched desperate revolts. The state replied with repression and draconian interventions, propelled by a perennially suspicious Stalin at the top. The cumulative result was a downward spiral: crop rotations shrank or disappeared, shoddy practices in plowing, sowing, and harvesting reduced yields, draught power declined as fodder perished and disgruntled peasants did away with their animals, the new tractors failed to materialize or stalled due to inexperienced operators, and, moreover, the weather was not good in 1931 and 1932.[8] The country ran out of food, and the death of millions was only part of the disaster. According to Stephen Kotkin, "Upward of 50 million Soviet inhabitants, perhaps as many as 70 million, were caught in regions with little or no food."[9]

In his history of Communism, Gerd Koenen wrote that collectivization was "a disaster with far-reaching consequences in light of every yardstick that one can possibly conceive."[10] Central planning did not reward judicious use of human and natural resources on the ground, and output remained notoriously below expectations. The collective farm was not even Socialist. Farm workers remained tied to their collectives in ways that reminded Koenen of serfdom or sharecropping (see chapter 12, Boll Weevil), and the Soviet government grudgingly accepted some entrepreneurial elements to keep the regime afloat.[11] While production of staple crops limped along on collective farms, eventually making the Soviet Union dependent on grain imports from the capitalist world in the 1970s, small private gardens thrived, and kolkhoz farmers fed the country by selling their produce on local markets.[12] There was nothing illicit about these markets—kolkhozniks were taxed for their private ventures—but devout Communists never warmed up to the concept.[13] As late as 1990, delegates at the twenty-eighth party congress "voted by a large majority, to Gorbachev's annoyance, to omit the word 'market' from the name of the economic reform commission which it appointed."[14] It served to illustrate Robert Service's argument about deviant behavior in Communist societies: "These phenomena were not the grit in the machinery of totalitarianism but the oil."[15]

It was not that planning was invariably doomed to fail. The Soviets even managed to introduce new seeds on a mass basis in the midst of the chaos of collectivization.[16] But in the long run, central planning and good farming proved impossible to reconcile: fulfilling the plan was fundamentally at odds with the entrepreneurial spirit that brought high yields. Like much of Socialist reality, the collective farm was an unstable hybrid of planning and market forces, of Soviet sclerosis paired with perennial improvisation, and reforms could backfire in spectacular fashion. When Gorbachev launched a campaign against "unearned incomes," local officials took it as a cue to crack down on kolkhoz markets and peasant gardens.[17]

In short, collectivization already had a dismal track record when it went global after 1945. In fact, the kolkhoz went global in the absence of a substantial literature on how it actually worked.[18] That left some room for adjustments, but while the collectives of the Communist world were not just carbon copies of the Soviet original, the enthusiasm of their creators was typically their most impressive feature. The Communist governments of Eastern Europe dispossessed their peasants as rapidly as possible and, in Robert Service's words, "competed to out-Stalinise each other"; Bulgaria had collectivized 56 percent of its agricultural land by 1953, beating Czechoslovakia with 54 percent.[19] They ended up with sclerotic farming sectors that became the target of endless reforms, none of which managed to resolve the fundamental paradox.

Problems did not end with a dismal return on investments. When East Germany's government, a latecomer by Eastern European standards, voted for comprehensive collectivization in 1960, it triggered a mass exodus that did not stop until the construction of Berlin Wall the following year.[20] In Ethiopia, part of the Communist sphere since a coup in 1974, the collectivization of agriculture was one of the causes of the great famine of 1983–1985, though the horrors of this well-publicized famine paled in comparison with Mao's Great Leap Forward.[21] The revisionist historian Frank Dikötter, surely one of the world's best-named scholars, has argued that "at least 45 million people died unnecessarily between 1958 and 1962."[22] In 1988, the Romanian dictator Nicolae Ceaușescu announced a megalomanic plan to erase some seven thousand villages and move eleven million people to new "agro-industrial complexes" under the banner of "systematization." However, Ceaușescu died in front of a firing squad on Christmas Day the following year, and the plan died with him.[23]

But for all the ideological zeal, some Socialist leaders showed remarkable flexibility when it came to revolution in the countryside. Poland dragged its feet and ultimately abandoned collectivization, in part because Władysław Gomułka, Poland's leader since 1956, had seen Ukraine in the early 1930s.[24] In 1932, Stalin torpedoed a collectivization drive in Mongolia that zealots in the Mongolian People's Party had launched upon encouragement by Comintern advisers: in light of the turmoil in the Soviet Union, Stalin wanted stability in a buffer state to war-torn China.[25] It was perhaps more than the pragmatism of the man who would sign a pact with Hitler seven years later. Collectivization carried the hallmarks of Socialism—central planning, distrust of individual initiative, and a supersized faith in technology and human control over nature—but it was not a pillar of Socialist orthodoxy. It was more about filling an ideological void: Karl Marx, who wrote scornfully about the "idiocy of rural life" in his *Communist Manifesto*, was far more interested in industrial production, and the kolkhoz was never a star in the pantheon of the Reds.[26] Post-Socialist nostalgia seized on many things after the world of Communism collapsed around 1990, but collective farming was always a minor concern.

However, a lot of land and equipment remained while the idea of the collective melted into air, and the tragedy of Soviet-style agriculture gave way to what Marx might have called a post-Socialist farce. The large agricultural production unit with industrial technology was not a Red exclusive—William Hale had already written about "agricenters" in his blueprint for the chemurgy movement (see chapter 33, Chemurgy Movement) when Stalin had yet to purge the politburo—and many collectives lived on with a new operating script in capitalist times. In his archaeology of the Soviet century, Karl Schlögel wrote about "large estates in excess of the nobility's holdings in the times of the Tsar, yet now in the form of joint-stock companies, joint ventures, and agro-holdings."[27] Oligarchs control large swaths of land in Ukraine, in part because post-Socialist governments were reluctant to create a national cadastre for the registration of land titles (see chapter 6, Land Title).[28] Ukraine's government was more proactive in the commemoration of the Great Famine of 1932–1933, and it embraced a new term that heralded a new reading and a new urgency of remembrance: "Holodomor."[29]

2. HUNGER, MODERN STYLE

Amartya Sen grew up in a famine region. Born on a university campus in India in 1933, he spent his childhood years in Dhaka, today the capital of Bangladesh, a region where between 3.5 million and 3.8 million people died in the Great Bengal Famine of 1943.[30] Some four decades later, after a stellar academic career that brought him to Calcutta, Cambridge, Delhi, London, and Oxford, the Bengal famine was a case study in his landmark book *Poverty and Famines*. Sen criticized the delayed and ineffective response of the British colonial government that saw distribution as a merely technical issue and instead spent much time trying to quantify supplies. But the region had enough food in spite of the Japanese invasion of neighboring Burma—in fact, supplies were 13 percent higher than two years earlier—and Sen likened the official attempts to define the shortage to "a search in a dark room for a black cat which wasn't there."[31] Sen used the Bengal Famine to develop what he called "the entitlement approach": in his reading, the crucial issue was not the sheer amount of foodstuff but the ability to access it through legal means. "The entitlement approach concentrates on each person's entitlements to commodity bundles including food, and views starvation as resulting from a failure to be entitled to a bundle with enough food," Sen wrote in *Poverty and Famines*, which became a classic of development economics.[32] Sen received the 1998 Nobel Memorial Prize in Economic Sciences, and while his entitlement approach remains under debate, as academic theories should be, it is credited for shifting "the analytical focus away from a fixation on food supplies—the Malthusian logic of 'too many people, too little food'—and onto the inability of particular groups of people to acquire food."[33]

It shattered received wisdoms that associated famines with failing harvests and depleting reserves (see part VI, Final Reserves). In the old biological regime, starvation was a sad part of normal life. "Famine recurred so insistently for centuries on end that it became incorporated into man's biological regime and built into his daily life," Fernand Braudel noted in the first volume of *Civilization and Capitalism*.[34] Environments and the biological potential of plants and animals imposed rigid limits on economies and populations. People might ride out one bad harvest, but two consecutive failures spelled disaster, and there was not much that premodern societies could do about it. In 1696–1697, a famine in Finland claimed between a quarter and a third of the population.[35] The hinterland failed Florence every fourth year on average, and the affluent Renaissance city had to buy grain from places as distant as England, Flanders, and Poland in the sixteenth century.[36] The cold years during the early modern Little Ice Age increased the risk of bad harvests and contributed to spikes in grain prices, and merchants stood ready to take advantage. During the Dutch Golden Age, "Dutch merchants could earn lucrative profits by selling the Baltic grain they had stockpiled in Amsterdam's great warehouses."[37] It was against this backdrop that Thomas Malthus penned his gloomy vision of populations outgrowing their food base.

The one option in premodern times was migration. Bad harvests were often regional phenomena, and hungry people have moved to places with more abundant resources since ancient times. According to the Book of Genesis, Jacob sent his sons to Egypt when famine befell the land of Canaan, which happened to reunite them with their brother Joseph.[38] The Irish Potato Famine of the 1840s triggered mass emigration, as did the devastating famines in Finland two decades later.[39] However, the Finnish famines of the 1860s became the last peacetime famines in Europe outside Russia.[40] The growth of agricultural production (see chapter 2, Sugar) provided the world with sufficient calories (see chapter 7.2, Numbers Games) even in an age of population growth, and the transport revolution of the nineteenth century (see chapter 3, Canal du Midi) allowed bulk shipments of foodstuff to starving regions. The food supply of modern societies was about human decisions rather than the vagaries of Mother Nature, though some regions moved more quickly into the orbit of transnational food systems than others. When the German ambassador to China reported on a famine in the inland province of Sichuan in 1897, he knew little of substance and relied on hearsay from the press, but he was well informed about the financial shenanigans behind the donations of Empress Dowager Cixi.[41]

The report mirrored a crucial change in the politics of hunger. It was no longer about having the necessary foodstuff but about knowing where to deliver it. Information was the crucial challenge even in the absence of a paranoid Stalin ruminating about hoarding and subversive behavior, and it was about far more than available supplies. Authorities had to keep an eye on the health problems that typically went hand in hand with a lack of food. "Famine or scarcity, with the resulting lowered vitality in the population, produces 'suitable conditions' for epidemic outbursts of cholera" (see chapter 21, Cholera), a memorandum for the League of Nations Health Organization noted in the 1920s.[42] Shipments also put local food markets in disarray, which could jeopardize incomes of local peasants and discourage production. Deficiencies in vitamins and micronutrients could create problems in the most unexpected ways. Women and children in Pondoland, a coastal region in the Union of South Africa, suffered from malnutrition when expanding herds of cattle and sheep devoured a variety of wild spinach that supplied people with crucial vitamins.[43]

Governments could even use food aid for sinister purposes. When international organizations swept into Ethiopia in the 1980s, it dawned on the humanitarians that the famine relief camps they were catering to were part of a forced resettlement program that the government had launched in its battle against rebels in the country's north.[44] Food aid looked like a simple matter from the donors' perspective, but it faced moral dilemmas on the ground, and there was no way to postpone decisions. The fight against starvation was always a race against time.

As with so many challenges in modern times, famine prevention grew more complicated with the accumulation of experience, down to concerns about the project itself. Toward the end of the century, an expanding number of international organizations competed for aid in a shrinking number of famine states, raising doubts about what Cormac Ó Gráda has called "the fire-brigade methods of international aid."[45] Be that as it may, we are well past the naiveté of late nineteenth-century observers, who felt that it was just a matter of goodwill and individual initiative. When Russia was wrestling with a devastating famine in 1891–1892, the German ambasssador Hans Lothar von Schweinitz, a general of the infantry, lambasted the botched responses of the government's "pashas" and called for action from "men in the fullest sense of the word."[46] He would have been surprised to learn about the challenges that food aid and manhood were bound to face in the twentieth century.

Sen was not dogmatic about the entitlement approach, and his writings do not show the kind of mathematical overkill that befell so much economic writing in the late twentieth century. *Poverty and Famines* acknowledged that the economic forces leading to starvation could be "most diverse," and Sen described his approach as "a general framework for analysing famines rather than one particular hypothesis about their causation."[47] But for all the diversity of factors, he was adamant about the pivotal role of government policies. "It is not surprising that no famine has ever taken place in the history of the world in a functioning democracy," Sen wrote in the 1990s: accountable rulers could not afford the loss of legitimacy that came with mass starvation.[48] Ó Gráda added a further qualification when he observed that "Mars in his various guises accounted for more famines than Malthus": in the twentieth century, times of hunger typically went along with warfare or civil strife.[49] Unfortunately, 1930s Ukraine failed on both counts. The Soviet Union was not a democracy, and in Stalin, it had a leader who—in the words of Alec Nove—"knew he was at war."[50] Internal enemies count.

Some relief efforts were success stories. A Commission for Relief run by Herbert Hoover, later president of the United States, supplied German-occupied Belgium with food during World War I.[51] When the famine of 1921–1922 hit the Soviet Union, the Norwegian Fridtjof Nansen, formerly a famed explorer and scientist, ran a relief campaign under the umbrella of the League of Nations, and together with the American Relief Administration under Hoover, the campaign fed eleven million Russians at its peak.[52] No such effort came to the aid of Ukraine's peasants in 1932–1933. Soviet propaganda broadcast the dramatic progress of forced industrialization and collectivization to an international audience and could not afford the embarassment of a famine. The regime also needed the revenues from grain exports. Quantities fell dramatically as the disaster unfolded, but the Soviet Union was still sending foodstuff abroad in 1933 while its citizens were starving.[53]

The Soviet Union suffered from another famine in 1946–1947, this time with epicenters in Ukraine, Moldova, and the middle and lower Volga.[54] But in the long run, Eastern Europe joined the global trend from acute crises to chronic food problems. Toward the end of the twentieth century, the conventional wisdom was that "while chronic hunger doesn't make the evening news, it takes more lives than famine," and then there were the problems of plenty.[55] More than half of the population has been diagnosed as overweight or obese in Ukraine, and vitamin A deficiency and anemia among children were also above international standards.[56] But while Ukraine's society has faced difficult questions regarding its current food system, it has preferred to keep things simple when it came to the food systems of the past.

3. BLAME GAMES

Ukraine's Holodomor was a state secret in the Soviet Union. The *Great Soviet Encyclopedia* (see chapter 22.2, The Age of Handbooks) declared in the 1970s that "socialism eliminates the causes of hunger and creates social and productive conditions for fully overcoming food shortages" and cited the USSR as proof.[57] But like so many things in the Communist sphere, this reading crumbled in the 1980s, and the Ukrainian Communist Party leader Volodymyr Shcherbytsky made the first official reference to the Great Famine in a speech of December 1987.[58] It was a hapless attempt at damage control, and not just because the government lacked credibility in Ukraine after the botched cover-up of the Chernobyl nuclear disaster (see chapter 37, *Lucky Dragon No. 5*) the previous year.[59] In 1986, the British historian Robert Conquest published a monograph, *The Harvest of Sorrow*, that drew international attention to what the author called "the terror-famine," and the United States Congress convened a Commission on the Ukraine Famine that delivered its report in 1988.[60] Discussions in the Soviet Union advanced from tepid explorations to broad conversations within a few years, and when Ukraine became independent with the dissolution of the Soviet Union toward the end of 1991, "the famine quickly came to symbolize the period of Soviet rule at its cruelest."[61] The Great Famine, now rebaptized as the Holodomor, became a focal point of collective memory: "For independent Ukraine, no event has greater significance in the history of the developing nation state," David Marples remarked.[62] When a German foundation conducted a study fifteen years after independence, it found 342 monuments in Ukraine, including some in Western regions that had been part of Poland until 1939.[63]

As collective memories go, famines are difficult topics. They offer little in the way of heroism and plenty of harrowing stories. Traditions and customs did not count for much when people were struggling to feed themselves, and many survivors shared a feeling of guilt. Even the proud bourgeoisie of Amsterdam dispensed with moral reservations when the city was cut off from supplies during the last winter of World War II: Geert Mak wrote that "in order to survive, even the most respectable citizens had to do things that they would rather not remember."[64] But as governments played a pivotal role in the response to modern famines, blaming the authorities became a convenient way to externalize guilt. It did not always work. Finnish authorities tried and

31.1 "Holodomor—Genocide Against the Ukrainian People." The National Bank of Ukraine issued this commemorative coin on the seventy-fifth anniversary of the Great Famine in 2007. Image, Wikimedia Commons.

failed to respond effectively to the famines of the 1860s, an experience that nationalists preferred to forget when the tsar's Russification policy threatened Finland's autonomy later in the century, and Finnish nation-building framed the famine "as a learning process or a hardship that stimulated and increased the stamina and resourcefulness of the people."[65] When Lebanon suffered from a devastating famine during World War I, locals blamed locusts because it absolved them of the need to choose between two potential villains, the Ottoman rulers and the Entente Powers who cut the region off from supplies through a naval blockade.[66] However, blame games flourished when they could seize on running themes in national mythologies. Ireland's Great Famine entered collective memory as a particularly dark chapter in a long history of quasi-colonial rule from London, and Ukraine's Holodomor became a political football in a country divided between East and West.[67]

All presidents of independent Ukraine have paid tribute to the famine's victims, but emphasis has differed depending on the direction that they envisioned for their country. Commemoration was a particular priority during the presidency of Victor Yushchenko, who came to power in the wake of the Orange Revolution of 2004. He built a Holodomor museum in Kyiv, had Ukraine's parliament pass a declaration in 2006 that recognized the famine "as a deliberate act of genocide against the Ukrainian people," and subsequently lobbied the United Nations, the Council of Europe, and other nations to follow suit.[68] It came together in a significant change in the framing of the narrative. Hunger in Ukraine was no longer part of a pan-Soviet experience. The Holodomor was now a distinctly Ukrainian affair, most evidently in the efforts to get it recognized as a genocide. But did Stalin really seek to eliminate the Ukrainians by starving them to death?

It showed the extent to which the age of extremes has left us with a transnational frame for collective memories of individual nations. As seen from the West, genocides were far more powerful experiences in the twentieth century than failing states, and one genocide in particular captured the public imagination. Nazi references figured prominently in the Holodomor discourse from its inception: Conquest opened his landmark book with a sentence that compared famine-stricken Ukraine to the Nazi concentration camp Bergen-Belsen.[69] The term "Holodomor" is not etymologically related to the Holocaust—it is a blend of the Ukrainian words *holod* (famine) and *moryty* (to kill)—but the connection runs through the literature, down to attempts to inflate the number of Ukrainian victims to six million or more.[70] The metaphor was also in use for famines outside of Europe. Mike Davis famously described the El Niño famines of the late nineteenth century as "late Victorian holocausts."[71]

The differences between the Holocaust and the Holodomor remain hard to expunge, but they become a matter of degrees if narratives seize on the excessive cruelty on display. Like all major famines, Ukraine's Holodomor had all sorts of gut-wrenching events, and some authors were happy to share them with the world. Timothy Snyder's *Bloodlands*, which opens with the Soviet famines of 1932–1933, records that "at least 2,505 people were sentenced for cannibalism" in the years of the Holodomor, and as a diligent chronist of every horror in stock, Snyder goes on to note that "the actual number of cases was certainly much greater."[72] Its impact on the world's conscience remains subject to negotiations—for one, Canada's prime minister Justin Trudeau spoke of genocide in his Holodomor Memorial Day statement in 2017—but the international scholarly literature has remained unimpressed.[73] "No hard evidence has so far come to light of the regime's intention to kill millions through famine, let alone of a genocide campaign against the Ukrainians," Orlando Figes wrote.[74] Other scholars took issue with the "sacralization of the Holodomor" and pointed out that scholarship was rooted in a Soviet legacy, the traditional fatherland approach that discouraged transnational perspectives.[75] However, authors from Ukraine and elsewhere continued to press the Holodomor into an interpretative frame that did not quite fit, and it was open to debate whether this was due to the white heat of nation-building or the ghosts of the twentieth century.

The Holodomor was a political trump card for those who sought to distance Ukraine from Russia, and the Kremlin played its anticipated role and viewed commemoration as detrimental to its international prestige.[76] The trope surely helped mobilize Ukrainian resistance when Russia invaded the country on February 24, 2022, but up to that date, the Holodomor was remarkably inert as a national myth. It deprived Ukrainians of the chance to view the famine as a window into the nature of authoritarian regimes, something that Ukraine, not exactly a model democracy, could well need. It also did not help the country to reform its agricultural policy, not to speak of insights that would have inspired Ukrainians to improve their diets. And like many national myths, the Holodomor encouraged introspection and made citizens insensitive to similar experiences elsewhere.

The Great Famine of 1932–1933 claimed the majority of its victims in Ukraine, but it hit people across the Soviet Union. In fact, another Soviet Republic, Kazakhstan, suffered even more if one considers the number of victims as a share of the total population: a quarter of Kazakhstan's citizens perished in the wake of collectivization, or about 1.5 million people.[77] But Kazakhstan is much farther away from Europe than Ukraine, which shares a border with four EU member states, and that has allowed Western publics to categorize Kazakhstan's Great Famine as just another chapter in the epic tragedy of hunger in Asia. And while Ukraine's victims were peasants, most of those who starved in Kazakhstan were nomads. In the Kazakh Soviet Socialist Republic, collectivization was also about sedentarization, which made it part of a pet project of global modernity: there was a widely held opinion that the lifestyle of nomads did not have a place in the modern world. Even the kingdom of Saudi Arabia (see chapter 15, Saudi Arabia) sought to crack down on the autonomy of Bedouin nomads in the process of state-building.[78] Ukrainian scholars have counted more than twenty thousand publications on the Holodomor, and articles zealously noted that "it is crucial to distinguish the All-Union famine of 1932–33 . . . from the Holodomor."[79] Meanwhile, over in Kazakhstan, the Great Famine has yet to obtain an iconic name.[80]

32

THE PONTINE MARSHES

FIGHTING FOR SPACE

I. BATTLE MODE

Fascism was a product of war. Originally a band of veterans from World War I, Italy's National Fascist Party adopted a program that emphasized its "bellicose character" in 1921.[1] The movement grew in the midst of the ruins of a disintegrating political system, and once in power, the Fascists were not in the mood for a different approach. "I consider the Italian nation to be in a permanent state of war," Benito Mussolini declared in parliament in 1925.[2] The regime had just launched a "battle for grain" to make the country independent of food imports, and it was about to start the "battle for the lira" that was followed by a "battle for births."[3] But even in permanent battle mode, Fascists had their priorities. War was good, but some wars were better than others, and so it happened that, in 1932, Mussolini went to the inauguration of a town named Littoria and declared, "This is the war that we prefer."[4]

Mussolini had come to Littoria to celebrate a flagship project of Italian fascism. Littoria was at the center of the Pontine Marshes, a malaria-infested wasteland on the Tyrrhenian coast south of Rome that the Fascists sought to claim for agriculture. Following a comprehensive survey of the land by the Army Geographical Institute, thousands of workers moved 5 million cubic meters of soil to drain the marshes with more than 700 kilometers of canals. They also built a road network, homesteads for peasants, and a number of towns while medical professionals contained the threat of malaria. Settlement fell into the hands of the national association of war veterans (Opera Nazionale Combattenti, or ONC), which took charge of 41,600 hectares of land in 1931.[5] The first 100 families, all from the Veneto region in northern Italy, arrived in October 1932.[6] Two months later, Mussolini went to Littoria and proclaimed, "We have conquered a new province."[7]

Land for peasants was a transnational concern in interwar Europe, and it was not inherently reactionary. Agrarians all over Europe drew on the work of the Swiss agronomist Ernst Laur, secretary of the Union of Swiss Peasants and professor at ETH Zurich, who had studied the accounts of thousands of Swiss farms and concluded that peasant farming was economically viable.[8] Conflicts over landownership and tenancy agreements (see chapter 6, Land Title) escalated in Italy after World War I, and some veterans took matters into their own hands with land occupations on the *latifondos* in the country's south. Italy's parliament discussed a comprehensive land reform bill in the months before the March on Rome, and the Fascists passed a land reclamation law in 1924 that allowed for the confiscation of unreclaimed land, but they were reluctant to use it: the Fascists were firmly on the side of landowners on matters of land policy. Against this background, reclamation was a technological fix for a social problem.[9]

The Pontine Marshes project was about new land as well as new people. It was part of the Fascist quest for ruralization, the attempt to curb the growth of cities and boost the agricultural population. It was largely unsuccessful, but it was pursued in earnest: "We are witnessing a return to Mother Earth," a book on land reclamation in Italy declared in 1936.[10] Italy's Fascists were also concerned about a declining fertility rate, and land reclamation was one of several pronatalist measures that included a ban on contraceptives and a bachelor tax first levied in 1927. Settler families were selected for their inclination to procreate, and Fascist authorities were not shy about reminding them of their duty. A Fascist land reclamation project on Sardinia grew around a new town named Fertilia.[11] It all came together in a comprehensive endeavor to make a new Italy. In the words of Mussolini, the goal was "to reclaim the land, and with the land the men, and with the men the race."[12]

Saving a nation seemed a tall order for a land reclamation project, even if it had the size of the Agro Pontino. Some thirty thousand people moved to the Pontine Marshes in the 1930s, a marginal number in a country of forty million. As Carl Ipsen has noted, "More Italians left the country in search of work in a single year than were settled in the entire Agro Pontino project."[13] Italy's Fascists struggled to achieve goals, but it did not really matter: "For us Fascists, the battle has even more importance than victory," Mussolini declared upon the inauguration of Littoria Province in 1934.[14] The Pontine Marshes project never counted for much in material terms, but the regime's propaganda suggested otherwise. Countless reports extolled the project's virtues, Mussolini's visits were recorded in word and in film, and an exiled Italian historian, Gaetano Salvemini, quipped that "to-day any foreigner who is not a complete nonentity cannot remain in Rome three days without having some high-placed personage of the regime invite him to go in an automobile, or even an aeroplane, to visit the reclamation works in the Pontine Marshes."[15] Dignitaries could admire canals and pumping stations, plenty of new buildings, and an iconic program that was less than subtle. The biggest canal in the Pontine Marshes was named after Mussolini; village names recorded the battle sites of World War I. Littoria was named after the lictors of ancient Rome, officers who carried the fasces, the bundles of wooden rods that gave the Fascist movement its name.[16]

The effort made an impression beyond Italy's borders. The *Geographical Review* of the American Geographical Society published an article in 1934 that called the transformation "little short of miraculous. Everywhere one is struck by the beautiful order and the Utopian quality of the work."[17] Five years later, Britain's Royal Geographical Society devoted an evening meeting to Italy's agricultural colonization program where Sir John Russell, director of the Rothamsted Experimental Station, rhapsodized about "the unbounded faith and optimism of the Italians."[18] Turning a malaria-infested wetland into an agricultural province within a matter of years served as living proof to the world about what Fascism could achieve, and it looked even more glorious with a sense of history. After all, the Fascists were not the first to launch a land reclamation project in the Pontine Marshes.

In his 1932 speech in Littoria, Mussolini boasted that "what was attempted in vain in the past twenty-five centuries is now being translated by us into a living reality."[19] The time frame is debatable, for evidence suggests that the Pontine Marshes "had a flourishing agricultural economy in the fifth and early fourth centuries BC."[20] But by the Late Roman Republic, the region figured as an inhospitable place that defied reclamation by Nero, Julius Caesar, and others.[21] A reference to ancient Rome was always welcome in Fascist Italy, particularly when it came to conquering space. As every Italian pupil knew, war and conquest were not just metaphors in the days of the Roman Empire.

2. *LEBENSRAUM*

Mussolini's "war that we prefer" became "one of the most popular maxims of the *Duce*."[22] It struck a nerve immediately. The people of Littoria broke into spontaneous applause when they heard the words and stopped the dictator in mid-sentence. Mussolini decided to skip the rest of his phrase, which was less peaceful: it called for employing the same methods and the same spirit "in other fields."[23] There was no mystery as to what he meant to say. Just as reclamation of the Pontine Marshes was shifting into high gear, years of genocidal warfare came to a close in Libya when Italian forces conquered the Kufra oasis in the country's south. Mussolini targeted Libya for settler colonization, and the same held true for a second African country, Ethiopia, which Mussolini invaded in 1935. Both countries were much bigger than the Pontine Marshes, and they allowed for migration on a different scale. Alessandro Lessona, Italy's minister of colonies, considered admitting about 500,000 settlers to Libya alone.[24]

The colonialist dimension of *bonifica* (the Italian word for reclamation) was obvious even in the titles of periodicals: Italy had a journal titled *Bonifica e Colonizzazione* beginning in 1937. Those who did not follow the literature could learn about the link through the movie *Scipio the African*, winner of the Mussolini Cup at the 1937 Biennale di Venezia, which was about ancient Rome's victory over Hannibal. A celebration of the Roman conquest of Africa produced during the Italo-Ethiopian War, it was filmed in the Pontine Marshes, a fact diligently recorded in contemporary newsreels.[25] Russell's presentation to the Royal Geographical Society covered the Pontine Marshes and Libya without much comment on the different settings.[26]

Colonization and conquest were nothing new in the interwar years, but the Italian efforts had a new quality. Jürgen Osterhammel spoke of "an intensification of ideology and state intervention in the opening up of new farming settlements" after 1918: "The idea that strong nations needed living space to escape the danger of resource shortage that came with overpopulation, and that they had a right and duty to take inadequately 'cultivated' land from less efficient or even racially inferior peoples, can be found among numerous far-right movements and opinion makers in the early twentieth century."[27] The idea became state policy in Italy's African colonies, in Japanese-occupied Manchuria and Nazi Germany's war in Eastern Europe. All these efforts were about resources as well as new men and new types of societies, with the fate of local populations as a marginal concern even to the extent of genocide. "The settlers of fascist imperial dreams—whether in Africa, Manchuria, or on the Volga—were guinea pigs for a state-directed *Volkstumspolitik*."[28]

Contemporary views were different. Russell found that "one can only admire the courage of the Italian nation in boldly applying new methods to this old problem of colonization." He had some doubts about the Libyan project, but they were first and foremost about soil erosion (see chapter 13, Little Grand Canyon) and the water supply (see chapter 29, Aswan Dam). Displacement of natives was an issue for Russell, but the Italians had created watering holes for nomads "at places farther inland," and that would probably do the job. Other things excited him more. "The scientific services are admirable," Russell observed, and there was "splendid human material in the colonists." For the British agriculturalist, the settlement of Libya was an endeavor that "will be watched with the deepest interest by all concerned with colonization, and certainly by many administrators in the British Empire."[29]

In the twentieth century, claiming new land had become more difficult than in previous centuries, when the great land rush (see chapter 13.1, The Virgin Land) transformed environments and agricultural commodity chains around the world. Most of the easy land had already been claimed for the world of modern agriculture, and what remained took a dedicated effort on the part of the state. The Pontine Marshes was not the only environment that was transformed in a state-led campaign in the interwar years. The Netherlands claimed the Wieringermeer in the 1930s, the inaugural project in the transformation of an entire North Sea bay, the Zuiderzee.[30] Nazi Germany drained moorland and created an Adolf Hitler Polder and a Hermann Göring Polder on the North Sea, but it was all small change: the Nazi's 100-year plan foresaw 43 new polders for 10,000 people.[31] The real space, the new *Lebensraum* for the hungry Aryan race, lay in the east.

Lebensraum had an academic pedigree. Drawing on Social Darwinism and a sense of claustrophobia, the German geographer Friedrich Ratzel, a zoologist by training, had coined the word in the late nineteenth century. A generation on, Karl Haushofer, a professor at Munich University, saw the solution for Germany's *Lebensraum* problem in hegemony over continental Europe.[32] Historians found it easy to connect Haushofer's ideas to the Nazis: one of Haushofer's students in Munich was Rudolf Hess, an early member of Hitler's inner circle and later Deputy Führer.[33] The fateful connection between people, food, and land was starting to look dubious at the time, as new seeds (see chapter 28, Hybrid Corn) and mineral fertilizer (see chapter 19, Synthetic Nitrogen) promised to boost yields per acre, but Hitler vigorously dismissed this alternative in his writing. "Hitler understood that agricultural science posed a specific threat to the logic of his system," Timothy Snyder has argued. "If humans could intervene in nature to create more food without taking more land, his whole system collapsed."[34]

Academic expertise was also in play on a more practical level. The mastermind behind Italy's land reclamation program was the agronomist Arrigo Serpieri, who twice served as undersecretary of agriculture but always thought of himself more as an academic than a politician.[35] The Nazi's drive to the east inspired a comprehensive multidisciplinary research effort that culminated in the ghastly *Generalplan Ost*, a blueprint for colonizing Eastern Europe.[36] Earlier generations of historians depicted Hitler's quest for *Lebensraum* as "irrational," a figment of the imagination that served as camouflage for his Machiavellian quest for power.[37] But recent research has shown that it was about something concrete: it was about land in need of development. However, Nazi planners had scant experience in designing settlements from scratch, and as they looked abroad in search of ideas, they found Italy's settler colonialism an inspiring model. They were particularly impressed by the emphasis on community life in newly built towns like Littoria, an aspect that had heretofore escaped the attention of German planners. They even copied the characteristic bell tower, much coveted by Italian Fascists, into their own designs.[38]

The quest for *Lebensraum* imagined space as a blank canvas. Existing populations were essentially obstacles waiting to be enslaved or removed, and people could experience much worse than the previous landowners in the Pontine Marshes, who were swiftly expropriated in spite of Fascism's antipathy to land reform.[39] But for all the determination, there was something strange about a social-engineering project in the pursuit of ruralization: the combined powers of science, technology, and the state set out to create something that was supposed to grow organically. The colonialist blueprint of the *Generalplan Ost* was worlds away from Friedrich Ratzel's vision, which conceptualized colonization as a process of "natural growth" with many small steps.[40] The design of *Lebensraum* was a contradiction in terms. Maybe that was why people believed in it so feverishly.

The quest for *Lebensraum* did not rule out pragmatism on the ground. In spite of the Fascists' emphasis on wheat production, planners in the Pontine Marshes allowed the cultivation of sugar beets (see chapter 2, Sugar) and built a sugar mill in 1935.[41] The Fascists also left some land in the Pontine Marshes unreclaimed to show "how the area was supposed to appear in the 'mythical' times of the Roman Empire," though the result, Circeo National Park (see chapter 26, Kruger National Park), might have looked more authentic if the Fascists had refrained from reforestation with eucalyptus (see chapter 27, Eucalyptus).[42] Blueprints meant even less in German-occupied Eastern Europe, where endless turf wars, administrative fiefdoms, racial disdain for local populations, and the demands of total war came together in a genocidal mess.[43] And in the end, it all hinged on a military victory that never materialized. Nazi plans for a new Eastern Europe dissipated with the German defeat in World War II while Italy lost its colonies, its duce, and a good part of what it had accomplished in reclamation.

The suppression of malaria was one of the great achievements of the Pontine Marshes project. But the war took its toll, and medical authorities noticed a resurgence of malaria in 1942. In the fall of 1943, with Allied forces advancing northward on the Italian mainland, the German army stopped the drainage pumps in the hydraulic equivalent of scorched-earth policy, and much of the region was underwater again by the next spring. Military action came to the region with the Allied landing in Anzio in January 1944, and many homesteads were destroyed in subsequent battles because they could provide cover for snipers.[44] As Frank Snowden remarked, "The Pontine Marshes had been restored to the conditions before reclamation, except that now, instead of being uninhabited, they were thickly populated with homeless refugees shivering with fever."[45] According to official statistics, which likely underestimate the real number, Littoria province had more than 100,000 cases of malaria in 1944 and 1945.[46] As the Prussian general Hellmuth von Moltke famously said, "No plan survives contact with the enemy."[47]

3. PLANNING DEVELOPMENT

The Pontine Marshes project was rather successful on the fertility front. For a number of years, Littoria was the province with the highest birth rate in Italy.[48] Achievements were less impressive in other respects. Administration was a notorious source of trouble. The ONC ran the project while the Ministry of Agriculture was in charge of finances, and the Ministry of the Interior, the Ministry of Public Works, and the Commissariat of Internal Migration had their fingers in the pie as well.[49] Farmers were not happy either, and the new men of Fascism were an unruly lot. Some forty colonists went on strike in 1934, followed by one hundred more two years later. The Fascists expelled the troublemakers, but that failed to restore calm, and it took a personal intervention from the secretary of agriculture to avert a mass strike in 1938. Harvests were less plentiful than anticipated, and many peasants who had arrived hoping for their own land ended up heavily in debt.[50]

State-led development projects like the Pontine Marshes were a key feature of the age of extremes, and they were notorious for discontent on the ground. In fact, James Scott argued that marginalizing local opinion was a defining feature of what he called "high modernism." He argued in *Seeing Like a State* that the top-down approach of state authorities was more important than specific political ideologies: "The troubling features of high modernism derive, for the most part, from its claim to speak about the improvement of the human condition with the authority of scientific knowledge and its tendency to disallow other sources of judgment."[51] The ONC responded to criticism in characteristic style and complained about laziness, cheating, and a lack of peasant mentality among the colonists. But the limits of the ONC's expertise became visible as early as 1933 when seeds from Northern Italy failed in the warmer climate of the Pontine Marshes. When Mussolini handed out cash payments to families later that year, it was more apology than paternalistic benevolence.[52]

To err is human, but this mistake was clearly about the gap between skills and ambitions. High modernists were not the kind of leaders who were deterred by a fragile knowledge base: Hugh Hammond Bennett promised to save soils and civilization while simultaneously building a profession and a federal agency from scratch (see chapter 13.2, Saviors of the Soil). When Portugal's Fascists launched their own

32.1 The new man of Fascism and his family. Statue on a building in Latina. Image, Frank Uekötter.

battle for wheat and claimed unused land in the southern part of the country, cultivation caused severe erosion and declining yields, and the regime's attempt to extinguish Communism through irrigation (see chapter 29, Aswan Dam) did not work either.[53] For all the rhetorical bluster of high modernism, knowledge about the environment was always incomplete, and sometimes knowledge about institutional hierarchies was not much better. Italy's land reclamation campaign was not the only development program in which political responsibilities remained opaque. When the US secretary of the interior Stewart Udall visited water projects in the Soviet Union in 1962, his delegation "found it impossible to learn where in the Russian Government and political system basic policy decisions are made."[54]

Scott's "high modernism" is one of several readings of state-led development. Many scholars have stressed the political context of the Cold War. Against the background of mass poverty in what was then called the Third World, development "was invariably tied to the U.S.-led campaign to counteract communist influence."[55] Others looked at development through a Foucauldian lens and stressed the suffocating effect of the bureaucratic discourse. In his analysis of an agricultural project in Lesotho, James Ferguson called development "an 'anti-politics machine,' depoliticizing everything it touches, everywhere whisking political realities out of sight, all the while performing, almost unnoticed, its own pre-eminently political operation of expanding bureaucratic state power."[56] More recent studies place postwar projects in a wider chronological context and view development as an amalgam of expertise, political power, and modern technology that came together in projects like the Aswan Dam (see chapter 29, Aswan Dam).[57]

A healthy skepticism toward achievements and ambitions runs through the literature on the history of development, and yet outright condemnation was surprisingly difficult. Even notorious projects had their ambiguities. The Stalin Plan for the Transformation of Nature is typically considered the epitome of Socialism's environmental hubris, but Stephen Brain has argued that it was "a basically conservative project designed to restore the Russian landscape to its prehistoric ideal [that] was twisted into a promethean endeavor dominated by Trofim Denisovich Lysenko."[58] Failure was usually a matter of perspective. The effects of large state-led projects went in so many different directions that even the most hopeless of projects could claim some positive outcomes. Britain's infamous Groundnut Scheme in colonial Tanganyika ultimately consumed more peanuts than it produced, but it did inflate wages in the region and stimulated the local economy.[59] Land reclamation in the Pontine Marshes failed to boost ruralization, and the program in the rest of Italy was so shoddy that opinions diverged into the postwar years over how many hectares had actually been reclaimed, but the new towns played an important role in the development of urban planning in Italy.[60]

The Pontine Marshes also showed how different time frames made a world of difference. In 1945, with much of the land underwater and malaria back with a vengeance, the project's outcome looked disastrous indeed. But some twenty years later, malaria was gone (see chapter 38.3, Banner Slogans), and the Pontine Marshes were an agricultural region again. And then there was the small matter of costs. Serpieri saw the Pontine Marshes project as a model, but it took investments of about 1 billion lire between 1922 and 1940, which made it the exception among the 1,150 reclamation projects that Italy had in 1933.[61] For Salvemini, who spent his years in exile teaching the history of Italian civilization at Harvard, it was a waste of money whose sole purpose was to stage a show for innocent dignitaries from abroad:

"Nobody tells the bamboozled foreigner that in order to put up that show at the gates of Rome the Government starves the land reclamation works elsewhere in Italy."[62]

The conquest of *Lebensraum* was a thoroughly discredited concept after World War II. It had led the Axis powers into a moral abyss, and it did not even make sense in material terms anymore: most postwar societies in the industrialized West became more concerned about agricultural overproduction than about food scarcity. The fate of development was different. The concept looked dubious in so many ways, and yet it showed remarkable resilience. For all the disappointments and complications, heads of state continue to fall for large development projects. In the 1990s, the Indonesian president Suharto launched the Mega Rice Project on Borneo and sought to turn a peat swamp forest roughly a third the size of Belgium into the country's rice bowl, but all that he created was "a smouldering heap of ash": more than 2.7 million hectares of peat land burned in 1997 alone.[63] Developmentalism also survives in transnational discourses and the institutional scaffolding of the global economy. "The central tenets of the development discourse continue to persist and permeate the minds of policy makers and analysts, seemingly impervious to criticism and meaningful reform," Joseph Hodge has noted.[64] Gone are the days when planning was a charismatic endeavor, and yet we still live in a world of plans and planners, including occasional calls for five-year plans in some places.[65] State-led development and comprehensive planning have left a tarnished legacy that is at the same time alive, with millions of people living amid the ruins, and those who revisit developmentalist tropes in the twenty-first century have experiences akin to those who visit the Pontine Marshes.

Much of the land is still in agricultural use, but the Pontine Marshes are not the food-producing colony that Mussolini envisioned. For one thing, the towns are much bigger than intended: the Fascists conceived them as mere service centers for the rural population. The region also has a decommissioned nuclear reactor (see chapter 37, *Lucky Dragon No. 5*), a project that did not serve ruralization either. A relic from the pre-reclamation days, the water buffalo, has rebounded as the producer of milk for prized *mozzarella di bufala*.[66] Local farming includes vegetables, grain, and grapes as well as Demeter-certified organic agriculture. Visitors continue to go to the Pontine Marshes, but those who do not head for the beach come to look backward rather than forward nowadays. The region is attracting tourists (see chapter

32.2 A mosaic on the façade of the Catholic church in Sabaudia shows Mussolini helping with the harvest. Image, Frank Uekötter.

22, Baedeker) in search of something that they have not seen yet: the built legacy of Fascism.[67]

The leftovers from Mussolini's rule are visible throughout the Pontine Marshes. Littoria was renamed Latina after the war, as was the province that bore the town's name, but a Mussolini quote adorns the bell tower to the present day. Much of the monumentalist architecture still stands, ready to impress or intimidate depending on viewers' perspectives, and the restaurant *Impero* (Empire) in the center of Latina has continued to welcome guests "since 1934."[68] Sometimes the Fascist past becomes a matter of political debate: Latina made headlines in 1996 when it sought to rededicate an urban park to the memory of Benito Mussolini's brother Arnaldo.[69] But more often than not, it is just there, in an undefined twilight zone of memory that befits collective memory of planned development. Tourists can stay in a fancy hotel on the beach in Sabaudia, and yet the peacefulness of the seaside is strangely at odds with a name that commemorates a genocidal war for *Lebensraum*. The hotel is the Oasi di Kufra.[70]

33

The Chemurgy Movement

The Business of Biofuels

I. EXPERTS ON THE MARCH

The 1920s were crisis years for American agriculture. After a commodity boom during World War I, prices for farm products slumped in 1920–1921. Growing yields per acre met with saturated markets at home, and foreign markets that had previously consumed America's excess production did not recover throughout the decade.[1] With numerous farmers succumbing to bankruptcy and many more tightening their belts, a national debate arose as to how to improve their lot. One of the more unusual suggestions came from an article in the *Dearborn Independent*, a widely circulated periodical under the tutelage of Henry Ford that is now infamous for its antisemitic content. The author, William Jay Hale, was a trained chemist, and that shaped his view of the farm: he saw it "simply as an organic chemical manufacturer." But, Hale continued, this offered a whole range of new opportunities. Since industrial chemistry could extract many different products from raw materials, it was poised to open new markets for agriculture. American farmers should sell their underpriced commodities to industry, and that would bring a "new undreamed-of era of prosperity for the farmer." And Hale did not stop at this point. He saw industrializing agriculture as a matter of markets as well as production methods, and he sketched a future with huge "agricenters" where farmers would work under the direction of farsighted experts. He had little hope for the family farm of yore: "The single farm is so small a unit that between its intake and output no leeway is permitted for any employment of specialized talent."[2]

It was a bold, almost utopian vision; but then, Hale was not a man of modesty. Born in 1876, he had degrees from Harvard and spent some time at German universities before the University of Michigan hired him to teach chemistry. One of his students was Helen Dow, the daughter of the founder of Dow Chemical. They married in 1917, and Hale built an organic research laboratory at his father-in-law's company.[3] Preaching ran in the family for this son of a Presbyterian minister, and as Hale embraced the gospel of chemistry, he eagerly deduced solutions for all kinds of human woes from its scriptures. Hale also liked to juggle with words, and so it was by all means characteristic that he coined a new term for the passion of his life, the work with and for chemicals. He combined the Greek words *chemeia* (chemistry) and *ergon* (work) and thus created a literary compound that would stick: chemurgy.[4]

"In chemurgy lies the hope of the world," Hale declared in a speech in 1949, when the word had been in circulation for a decade and a half. He actually meant it: "No longer need we fear famine and pestilence or strife and turmoil when once we have adjusted our economy to the chemurgic way under Divine law."[5] Hale's claim to authority extended far beyond his professional sphere, and in that he was characteristic of a generation: during the 1920s and 1930s, experts of various stripes felt that it was time for them to take charge. Engineers promised to solve the problems of societies with technological means, culminating in dreams of technocratic rule.[6] Dam builders celebrated their concrete manifestations as the ultimate panacea for development woes (see chapter 29, Aswan Dam). And then there were the experts of more infamous memory, the eugenicists, who felt that societies' gene pools were up for science-based improvement.[7] Scientific overreach was common among the experts of the interwar years, and sooner or later they were all in for a hard landing.

All over the West, scientists and engineers asked for more money, more authority, and a new place in society: they were, in a nutshell, the self-proclaimed priests of a new epoch. And they were not shy about offering evidence. In his 1934 book, *The Farm Chemurgic*, Hale gave an enthusiastic account of the rise of chemistry, culminating in the Haber–Bosch process for synthetic ammonia (see chapter 19, Synthetic Nitrogen)—"the final blow that threw the world into chemical cataclysm," as he put it.[8] From Hale's point of view, the farm crisis was essentially an unexpected result of chemistry's progress. As naturally growing commodities such as indigo and silk were replaced by synthetic alternatives, "the farmers were left to drift": they lost their accustomed markets without a prospect of new ones.[9] But chemurgy would come to the rescue and develop new industrial uses for main products, by-products, and wastes. After all, chemurgists argued, the human stomach was limited, but industry's hunger for raw material was not.[10]

Hale's firebrand style of preaching chemurgy was peculiar, but his idea was not. The US chemical industry expanded dramatically after World War I, and with that came a growing number of pertinent projects. The movement also had a powerful patron in Henry Ford, who nurtured a penchant for a new agricultural crop, the soybean, and had created a chemurgy lab as early as 1929.[11] The various strands came together in 1935 when the American Farm Bureau Federation, the National Grange, the National Agriculture Convention, and the Chemical Foundation sponsored a conference in Dearborn, Michigan. With more than three hundred people in attendance, the meeting witnessed the creation of a National Farm Chemurgic Council that would henceforth serve as the movement's hub.[12] Delegates also attended a melodramatic ceremony in Henry Ford's replica of Independence Hall with the signing of a "Declaration of Dependence Upon the Soil and the Right of Self-Maintenance."[13] Chemurgy was an unabashedly political movement, and not just in that it had political demands. It aimed for a new kind of polity, with experts in the driving seat and everyone else yielding to their supreme authority. In the presidential election of 1944, the president of the National Farm Chemurgic Council, Wheeler McMillen, launched a bid for the Republican nomination.[14]

Chemurgy was a remarkably flexible concept. In 1937, the Farm Chemurgic Council entertained committees on cellulose, insecticides, fertilizers, plastics, soybeans, and tung oil and discussed themes ranging from the uses of Jerusalem artichokes to the merits of woven cotton mesh in bituminous roads.[15] Furthermore, the movement was not shy about claiming credit for preexisting work such as George

Washington Carver's research on the many uses of peanuts.[16] Nonetheless, a common thread ran through these activities. Chemurgy sought to break down traditional patterns of resource use and specifically worked to annihilate the barrier between food and nonfood uses. Of course, chemists had been gnawing away at that barrier at least since the times of Liebig, but the chemurgy movement acted as if it had ceased to exist: commodities were plentiful, and there were no moral differences that chemurgists would need to care about. The traditional world of commodities was revving up, and whatever order existed in patterns of use was now subject to the miraculous powers of scientific innovation.

But for all the diversity of the chemurgy movement, one issue stood out above the others: the use of alcohol in motor fuel.[17] The immediate aim was to establish alcohol as a fuel additive that improved gasoline combustion and reduced engine knock, but the movement's dreams were much bigger than that. It envisioned "the construction of nearly a thousand fermentation plants of eight to ten thousand gallons daily output," offering work to a million men in these plants and on the farms and securing another two million jobs "in allied industries."[18] Chemurgists also stressed the need to look beyond the exigencies of the day: "Someday, when petroleum resources are depleted, power alcohol may become a replacement fuel of superior quality."[19] And then, alcohol was a versatile molecule, and its use as a fuel additive was "only the first step in inauguration of the greatest industry man is destined to know for a century."[20] Who could possibly object to a cause that would open new markets, make resource use more efficient, secure energy independence, and even "eliminate all unemployment?"[21] It all came down to whether society would heed the wisdom of the experts: "Common sense alone should dictate the course to be pursued by chemically advancing nations."[22]

For those who looked beyond expert dreams, things were slightly more complicated. Fuel alcohol held the promise of a large and expanding market, but it was also a field with a range of stakeholders. A successful commodity chain for fuel alcohol had to unite farmers, industrial-size fermenting plants, and a network of gas stations for retail. The outlook for fuel alcohol was best when all these groups could expect good profits from the new line of business, and history did not bode well for that. Modern societies have a lot of experience mastering long supply chains since the heydays of Potosí (see chapter 1, Potosí) and the Caribbean sugar plantation (see chapter 2, Sugar).

But as these and other examples suggest, modernity's record is less impressive when it comes to distributing the gains equitably along the commodity chain.

2. MAKING MARKETS

By the time that the Farm Chemurgic Council was formed, a different agricultural policy was already in place. The farm crisis was high on the agenda when Franklin D. Roosevelt moved into the White House in 1933, and his administration drafted radical new farm legislation during his first one hundred days in office. The federal government bought large amounts of commodities to take them off the market, paid farmers to retire land, and later offered rewards for erosion prevention (see chapter 13, Little Grand Canyon). Whereas chemurgists sought to find new uses for overabundant products, the New Dealers aimed to curb overproduction and boost farm income through higher commodity prices.[23]

As a matter of principle, the two approaches were somewhat at odds; but then, the New Deal was not known for conceptual dogmatism. Officials from the US Department of Agriculture reached out to chemurgists behind the scenes when they developed their first policies in the spring of 1933. The collaboration produced draft legislation that offered a favorable tax rate for gasoline blended with alcohol and a blueprint for "a power alcohol industry consisting of 200–300 plants spread throughout the grain belt."[24] However, the New Dealers changed their minds when the proposal encountered fierce opposition. The petroleum industry set out to fight fuel alcohol tooth and nail, for the simplest of reasons: it had a competing product.[25]

Engine knock had emerged as a hot topic in the car and petroleum industries in previous years. After a frantic search for solutions, three corporate giants, DuPont, General Motors, and Standard Oil, focused on a fuel additive named tetraethyl lead, and the latter two formed the Ethyl Gasoline Corporation for the additive's production and sale in 1924. Tetraethyl lead was not a harmless substance—the United States War Department had explored its potential as a nerve gas—but the Ethyl Corporation was powerful enough to dominate not only the market for tetraethyl lead but also research on its health hazards, and the level of public concern remained below a critical threshold.[26] In any case, the public controversy over tetraethyl lead, fueled by a number of gruesome workplace deaths, was mostly over by the early 1930s, and that made fuel alcohol a challenge to a flourishing business. Not known for pulling punches, the petroleum industry was determined to fight all competition to tetraethyl lead, and that rendered preferential legislation for fuel alcohol a dead issue on Capitol Hill.[27] It also poisoned the relationship between the chemurgy movement and the New Deal, though the US Department of Agriculture eventually came around to setting up centers for chemurgic research in Peoria, New Orleans, Philadelphia, and San Francisco.[28] From the federal point of view, research on the uses of agricultural commodities was welcome, but it should keep its hands off the delicate fuel issue.

With that, the chemurgy movement was facing two formidable enemies, but it did not look terribly frightened. The president of the Farm Chemurgic Council, Francis Garvan, "cheerfully fanned the fires of conflict, all the happier because the foe was rich and powerful," and that probably betrayed more than a belligerent character.[29] For the chemurgy movement, the fight for fuel alcohol was not just about a business proposition: it was a crusade for a scientific categorical imperative. From this point of view, opposition could only result from backwardness and ignorance, and rhetoric mirrored that classic dream of warriors, the upcoming decisive battle, the quicker the better. Chemurgists in battle mode found that everything should bow to the experts' judgment, including market prices. "Never again should prices be allowed to transcend chemical values," Hale wrote in an article of 1933.[30]

When federal politicians remained unimpressed, the Farm Chemurgic Council decided to proceed on its own account. It built an experimental plant on the Missouri River at Atchison, Kansas, and it became a disaster. The plant ran into a host of technological problems, and when it was finally running smoothly, it produced at prohibitive costs. As if to add insult to injury, the priests of a new epoch also clashed with another supreme power, the tax authorities. In spite of the repeal of prohibition in 1933, alcohol production remained a tightly controlled business in New Deal America, and the tax collectors were so suspicious of the Atchison plant that they dragged operators into long and costly negotiations. When the plant closed its doors, the financial loss was somewhere between $300,000 and $600,000.[31]

The Atchison project also revealed the precarious power base of chemurgy. The movement had a few members from the farming community, most prominently Wheeler McMillen, a journalist with a lifelong passion for farming.[32] But most chemurgists came from backgrounds in science and industry, and many farmers viewed the movement with suspicion.³³ Hale's idea of shuffling farmers into new "agricenters," which was insensitive at best from an agricultural viewpoint, revealed a profound ignorance of rural realities: he never cared that autonomy and possession of land (see chapter 6, Land Title) were fundamental to the farmers' identity. Neither did chemurgists realize that their business case was deeply flawed from a farming perspective. They banked on inexpensive commodities because that made their products cheaper, but these depressed prices were the cause of the farm crisis. When local farmers refused to sell raw material to the Atchison plant because they were speculating for better markets, the director was so disturbed that he suggested buying land for future projects.³⁴

For a movement that had banked its hopes on fuel alcohol, the collapse of the Atchison project was an unsettling experience, and chemurgists tried their best to change the outcome through creative accounting and wishful thinking. In fact, Hale got so carried away by his own enthusiasm that he envisioned alcohol production for free: in a 1941 prospectus for investors, he wrote that there was "every possibility" that ethyl alcohol production "may record a cost approaching nothing."³⁵ All the while, McMillen buried the issue for the Farm Chemurgic Council: "We realize keenly that no effort can succeed unless it is entirely sound in its economic basis," he declared at the council's annual dinner in 1941, adding that "the chemurgic program has been hampered by the unfortunate controversy . . . over power alcohol."³⁶

Fuel alcohol failed on the market; but then, that market was not simply a given. Gasoline additives are a textbook example of the social construction of markets: supply and demand were shaped by political decisions, styles of regulation, corporate interests, and the resilience of technological paths (see chapter 29.3, Path Dependencies). Neoliberal mythology suggests that these things are unpleasant intrusions into the world of free enterprise, but the reverse is more convincing: rules and institutions create markets in the first place.³⁷ Chemurgy did act accordingly, as its push for preferential taxation was a call to bend some rules in its favor. But in the end, that initiative fell prey to hardball politics except for an unimportant Nebraska law that offered a state gasoline tax refund for alcohol blends.³⁸ When push came to shove, all chemurgy could offer was its professional creed, and preachers are nothing without a grateful audience.

Markets are subject to unexpected change. Chemurgy learned this in a somewhat ironic way during World War II when agriculture supplied more alcohol to the synthetic rubber project than the petroleum industry.[39] It was a remarkable achievement for a business model that the federal government had previously shunned, but the boom was not to last. World trade resumed after the war (to the chagrin of Hale and others, who felt that "international trade in agriculture must cease in a modern world").[40] One of the results was a cheap and abundant supply of oil from Saudi Arabia (see chapter 15, Saudi Arabia) and other countries that pushed all dreams about renewable alternatives into oblivion. Hale continued to warn of "our dwindling petroleum supplies," but his clarion calls fell on deaf ears.[41] Shortly before his death, Hale put his last hopes in "a revitalized scientific government." After the 1956 presidential election, a "revolutionary leadership" would nationalize the petroleum industry and "spend a billion dollars or more on chemurgic research."[42]

It was a bitter and slightly otherworldly vision, but the imagination was the only recourse when powerful stakeholders kept an iron grasp on real markets. It was probably a mistake for the chemurgy movement to put so much hope into the notoriously tough fuel business. But resource markets have historically shown dramatic fluctuations in volume and price, and business plans inevitably carry the air of a speculative gamble. In a capitalist economy, you never really know what will pay. When Henry Ford II set out to kill the pathetic soybean project after his grandfather's death in 1947, he discovered that large swaths of undeveloped land close to a major city were actually a pretty good investment. The former soybean fields found willing buyers among Detroit's suburbanites, and Ford made a killing on the old man's folly. It was an ironic postscript to a movement beyond its prime. At last chemurgy could claim a terrific commercial success, if only in suburban real estate.[43]

3. THE ETHICS OF BURNING FOOD

In his *Farm Chemurgic*, William Hale discussed how much alcohol America would need for a 50 percent blend in the country's gasoline. His calculation found that it would take 10 billion gallons of alcohol per year, which in turn would require 4 billion bushels of corn, or double the contemporary American harvest. Given an average yield of 25 bushels per acre, alcohol production would claim 160 million acres, which was close to half of America's total farmland, and basically all the American land where corn cultivation was feasible. None of these figures seemed to shock him, and he went on to calculate the labor requirements.⁴⁴ Scarcity was clearly not an issue for Hale, and not only in his own time: "We need entertain no fear for future generations that must feed, clothe, and house themselves comfortably."⁴⁵

Hale was writing in a land of plenty, and he harbored no doubts about future yield growths.⁴⁶ Other countries were less fortunate. In his famous *An Essay on the Principle of Population*, Thomas Robert Malthus had raised the specter of populations outgrowing agricultural production, and that scenario began to haunt people again when the global population exploded in the postwar years. Once more, the global experience was a divided one. Population growth was particularly strong in the Global South while birth rates declined in the industrialized world, and that made it delicate to speak about what Paul Ehrlich controversially called a "population bomb."⁴⁷ Maybe concerns over population growth were a way to deflect blame away from Western consumerism, arguably a bigger cause of environmental disruption. The ensuing debate was a bitter and complex one, but it brought one fundamental insight: when we talk about food and population growth, we should reflect not only on statistics but also on morals.⁴⁸ This changed the rules for discussions about what would soon be called biofuels. If agricultural supplies were getting scarce, was it ethical to burn food?

In retrospect, the chemurgy movement was fortunate in that it never had to grapple with these moral issues. It focused on scientific methods of problem solving, and chemurgists would have looked rather clueless in ethical debates—not good for self-proclaimed priests of a new epoch. The movement's perennial opponents, the New Dealers, did not raise the issue either, and for good reason. In a move to reduce the meat surplus, the federal government had bought six million hogs in August 1933 and turned them into fertilizer and lard. The purchase program became "a lightning rod for supporters and critics" of New Deal agricultural policies, and the controversy offers some striking parallels to today's debates over biofuels.⁴⁹ Should farm animals die prematurely for the sake of higher market prices? Opinions diverged widely, particularly in light of mass starvation during the Great Depression. It was a clash between two different normative worlds, between the logic of commerce and moral reasoning, and the debate never produced a clear result during the years of the New Deal. Then came the war, and people faced other moral issues.

Chemurgy and mass slaughter during the New Deal revealed a moral void, questions were left hanging in the air, and it seems that the ongoing debate over biofuels may head toward an equally inconclusive result. Many people harbor doubts, but as it stands, no country has banned the production of biofuels. In fact, many governments have implemented policies that support biofuels through subsidies and tax breaks. They seek to promote renewable energies in a quest to reduce their dependence on imported fossil fuels. And sooner or later, they have come to realize that the issue is more complex than in the heydays of chemurgy, and not only because of moral concerns. For one, the geographic scope has changed. While the chemurgists were staunch economic nationalists at a time when world trade had mostly collapsed, today's discussions take place in an age of globalization.

Global warming has added another dimension to ongoing debates, as opinions differed sharply on whether biofuels really help in the fight against greenhouse gases. Farm production hinges on energy-intensive chemicals such as synthetic nitrogen (see chapter 19, Synthetic Nitrogen) that have their own carbon footprint. Farming for biofuels also changes carbon storage in soils, particularly where land is cleared for agricultural use. The overall balance depends strongly on local conditions, and it can even vary dramatically in one place: for example, a study of palm oil production on a single plantation in Colombia showed that the greenhouse gas balance can be positive or negative depending on previous land use and the choice of fertilizer.[50] With that, the debate over biofuels has gained several new layers, and nobody knows how to resolve the ensuing tensions.

It does not help that the time span is potentially infinite. Advocates of biofuels look far beyond the immediate needs of their time. Similarly to those in the chemurgy movement, they point to the day when oil will run out or otherwise become untenable. But energy needs have increased enormously since the 1930s, and this means that model calculations about future demand lead to astronomic figures that are even more appalling than Hale's. Once more, the vast time scale and the unknowns invite optimistic speculation. One study suggested that by the year 2050, the maximum potential of energy crops may be up to 1,272 exajoules per year, which is around twice the current global energy consumption.[51] Others, such as the environmental scientist and energy historian Vaclav Smil, are more skeptical: "I feel strongly that the recent proposals of massive biomass energy schemes are among the most regrettable examples of wishful thinking and ignorance of ecosystemic realities and necessities."[52]

The debate entered a new phase when global food prices were skyrocketing in 2007. In the preceding years, a notable increase of energy prices had stimulated interest in renewables, and that nourished speculations about a connection. "The role of biofuels in the 2007–08 food price episode is probably one of the most controversial issues in any discussion on both the causes of the crisis and the appropriate policy responses," a report of the UN Food and Agriculture Organization declared.[53] It was yet another reminder about the capriciousness of resources prices: they reflect world politics, corporate power, path dependencies, technological choices, and speculation, in addition to the customary rules of supply and demand, and the exact mixture is anyone's guess.

Economists will surely continue to dissect the causes of the price hike with advanced computer models for some time. But there is also a real-life example for the possible future of biofuels. Brazil began to support alcohol production from sugar cane (see chapter 2, Sugar) in the wake of the 1973 oil price shock, and the industry survived the end of the country's military dictatorship, the collapse of oil prices in the 1980s, and the shift of economic policy from import substitution to neoliberalism and globalization. Some twenty years after the launch of the National Alcohol Program, the industry employed about one million people, produced thirteen billion liters of ethanol per year, and had saved the country $28.7 billion in foreign exchange for oil imports.[54]

However, the program had its ambiguities. It was vulnerable to environmental shocks, as shown in the 1980s when drought conditions coincided with the collapse of global oil prices.[55] Brazil is the last remaining country with a vast land reserve that is suitable for agricultural use.[56] Sugar cane cultivation is tough on agricultural workers, evoking awkward reminiscences of slavery and sugar plantations: Brazil alone accounted for 41 percent of the transatlantic slave trade.[57] Is it more than a coincidence that biofuels flourished in the world's most unequal society? Eric Hobsbawm famously called Brazil "a monument to social neglect," and disinterest in side effects was arguably crucial for the program's resilience.[58] As Jennifer Eaglin has noted, "National pride in the alternative fuel program papered over the realities of the program's social and environmental costs."[59]

The sugar cane region in Northeastern Brazil became a hotbed of Brazil's Landless Workers Movement (see chapter 6.3, In Spite of All Doubts) in the 1990s when cheap oil and free trade governments put the biofuels business on the defensive.[60] It was not an unexpected revolt. When the use of ethanol-powered cars was spreading around

1980, a Brazilian engineer had suggested that rural workers would pay the price: "Now the wheels of the entire world's cars will turn on the hunger of the Northeast."[61] There are even doubts about the effect on climate change, as sugar cane requires lavish doses of synthetic nitrogen (see chapter 19, Synthetic Nitrogen). With the efficiency of nitrogen use in the range of 30 percent, runoff may contribute to a rapidly changing nitrogen cycle in the tropics, which may lead to an increase in N_2O and tropospheric O_3 production that will probably cancel out any carbon dioxide savings.[62]

Brazil made a conscious choice to enter the biofuels business. It was a more muddled decision in other parts of the world. Jimmy Carter launched the US ethanol program "as part of an effort to woo Iowa voters during the 1980 presidential election." It did not work as planned—Reagan carried the state by a margin of more than 12 percent—but the Iowa caucus, the first electoral test for presidential hopefuls, has stabilized political support for fuel alcohol for decades.[63] The European Community came into the biofuels business as a by-product of an attempt to reform its infamous Common Agricultural Policy. Trying to come to grips with chronic overproduction, it launched a scheme that paid farmers to retire some of their arable land and gave them the option to use it for nonfood crops. By 2006, energy crops claimed more than 95 percent of this officially retired land.[64] Many biofuels have environmental balance sheets that look ambiguous at best, though there are some exceptions. In 2007, an OECD report came to the conclusion that one of the few biofuels whose overall environmental performance was better than gasoline was biodiesel from used cooking oil.[65] And in light of modernity's enduring love affair with meat (see chapter 18, Chicago's Slaughterhouses), used cooking oil arguably qualifies as an inexhaustible resource.

Biofuels have an array of potential futures, and they will probably differ between countries and continents. But whatever the outcome, the door between food and nonfood uses will likely remain open. Imposing moral limits is never easy, but it becomes exceedingly difficult when they seize on something that people have been doing without much ado for a long time. Chemurgy pushed that door wide open about a century ago, and it did so with the backing of the abundant cultural capital of modern science.[66] It would take a mighty countervailing force to close that door again and ban all uses of food besides the needs of the human belly. As it stands, our collective embarrassment about the world's hunger is probably not enough.

34

Autobahn

The *Endsieg* of Automobilism

I. CLAIMING THE STREET

Berlin was an exciting city in the 1920s. One of its attractions was a new racetrack on the western outskirts. Construction of the new *Automobil-Verkehrs- und Übungsstraße*, which everyone referred to as AVUS, began before World War I, but wartime and postwar woes delayed completion until 1921. The masses came for a grand opening in September 1921, and events such as the first German Grand Prix in 1926 drew up to 300,000 spectators. Daredevils drove cars with an average speed of more than 200 kilometers per hour on the straight course, and when a steep embankment was added to the northern hairpin turn in 1937, top speeds came close to 400 kilometers per hour. On ordinary days, common drivers could push their own vehicles to the limit if they paid the entrance fee.[1]

The German elite of the time was well schooled in Greek and Latin and thus knew that *avus* was also the Latin word for grandfather. It was a prodigious choice: the AVUS racetrack was the granddaddy of divided highways all over the world, the first road with separate lanes and no petty intersections that was built exclusively for cars.[2] It is equally fitting that racing gave birth to the iconic road of modernity, for speed was what the early automobile was all about. Cars were what Gijs Mom called "the adventure machine," a playtoy for those with enough money to make a scene, go for a joyride, and sooth frail urban nerves.[3] Like many Western countries, Germany had its share of rich enthusiasts, including the Kaiser himself, who had some twenty-five cars, fifteen drivers, and a brother in competitive racing.[4]

However, rich people with cars faced a hostile reception among the other 99 percent. It was about the state of the street. Cars of the early 1900s were notorious for the dust clouds they left behind on country roads, and the street networks of European towns, often dating back to medieval times, presented all sorts of obstacles to the new vehicles. And it was about the meaning of the street, as most people felt that transportation was only one of its many functions. Streets were where children were playing, where adults were trading and socializing, and where political movements were making their case. A good part of nineteenth-century political life played out on the street, from the barricades of revolutionary times to the marches of Socialists for a better tomorrow. Before the automobile, claiming the street was about far more than a technological choice.

Thanks to growing numbers of horses, streetcars, and bicycles, most cities were already familiar with traffic woes when automobiles entered the fray around 1900. It was more of a shock in the countryside, where streets were calmer, horses more irritable, and cars faster. Newspapers were full of reports about clashes between drivers and the rural population, and quite a few turned violent. Countless children threw stones at speeding cars, and drivers faced a serious risk of being roughed up if they caused an accident. In fact, violence was so common that German lawmakers legalized hit and run: since 1909, drivers were allowed to flee the scene of an accident if they turned themselves in to the police the following day.[5] And it was about more than scratches and bruises: cars could kill, and they did in growing numbers. In 1913, Great Britain recorded 2,099 fatal and 42,544 nonfatal accidents, a stunning figure in light of the fact that the country only had some 200,000 motor vehicles at the time.[6] In 1906, Woodrow Wilson, future president of the United States, surmised in a widely quoted speech that "nothing has spread Socialistic feeling in this country more than the use of automobiles."[7]

It seemed an open question whether automobilism had come to

stay. Maybe the need for speed would just peter out. "Mankind ruins itself to go fast," Jean Brunhes argued in his lecture on "the limits of our cage."[8] He pointed to the fact that a high-speed steamer consumed six thousand tons of coal to get from one side of the Atlantic to the other, which added "economic limitations" to his specter of ubiquitous limits: "Just as the genius of man pushes the limits of the possible, so the necessary expenses paralyze progress and reduce it to exceptional events."[9] But energy became cheap in the twentieth century, and it was not much of a constraint on people's quest for automotive mobility or the joys of high velocity. Having a car was a good thing, and speed was good, too—at least when it came to covering distances rather than the speed of crossing the road. In one country after another, automobilism thrived with a vigor that made it akin to a force of nature.

Authorities resorted to a comprehensive project of social disciplining that involved fists, fines, and a plethora of educational campaigns. In order to forestall punitive legislation, automobile associations preached the virtues of careful driving.[10] Policemen struggled to maintain respect for the authorities in spite of endemic transgressions of the law, and the few incidents that were actually recorded were clogging up the courts. In early 1930s England, more than 40 percent of criminal proceedings revolved around offenses against the Highway Acts.[11] A Swiss canton, Graubünden, even imposed a comprehensive ban on automobiles in 1900. The ban was finally repealed after much debate in 1925, a decision that mirrored the declining vigor of anti-automobile sentiments in the 1920s.[12] People were learning to live with the automobile for better or worse.

The car claimed the street all over the industrial world, and yet it was a contested hegemony that people accepted only grudgingly. A German intellectual, Werner Sombart, even fantasized about ending the automotive age in 1928. When a literary magazine asked what he would do if he came to power, Sombart said he would outlaw cars and motorcycles, except perhaps for large cities and other places where there was "nothing left to be spoiled."[13] Others took a more pragmatic view. With the car's ascendance beyond debate, designating roads for its exclusive use was simply the next step. France was the first European country to reserve a road for motorized transport when the German army attacked Verdun in 1916. Supplies hinged on a single road, and the military swiftly cleared it of all pedestrians, cyclists, and horses and put up improvised dividers, the crucial tool for separating traffic that serves as the hallmark of a proper highway today. Every day 13,600 trucks used the makeshift highway, or one every six seconds, and the road entered military history as La Voie Sacrée.[14]

Costs were not much of an issue in a military emergency, but civil governments took a more sober view after World War I. Why spend public money on roads that only a few privileged drivers could use? It was private funds from one of Germany's richest businessmen, Hugo Stinnes, that secured the completion of Berlin's AVUS.[15] Even in Fascist Italy, where Mussolini boasted about long-distance roads that would replicate those of ancient Rome, most of the money came from private investors, and a government agency took over only when these concessions faced financial collapse for lack of traffic.[16] But the role of the state changed when Hitler came to power in Germany and made autobahn construction his pet project.

After World War II, remorseful Germans pointed to the autobahn as the good side of the Nazis, and historians have gone to great lengths in pointing out that few things were really new about the project. Nazi Germany's autobahns followed Italy's autostrada projects, and in the United States, the first country to achieve mass motorization, new parkways "rose to prominence in the 1920s and '30s as an international model for the harmonious integration of engineering and landscape architecture."[17] Hitler also drew on German blueprints from the Weimar years, when the Nazi Party had criticized highway projects as a waste of money for the privileged few, and the AVUS racetrack was unceremoniously connected to the autobahn network in 1940.[18] The harmonious integration into the landscape, a favorite theme of Nazi propaganda and much subsequent commentary, was patchy at best.[19] Hitler was not even the first to use autobahn construction to combat unemployment. That distinction goes to the Cologne mayor Konrad Adenauer, later chancellor of the Federal Republic, who oversaw construction of a limited-access highway connecting Cologne and Bonn between 1929 and 1932.[20]

The one thing that was actually new was the Nazis' determination to build: all other contemporary projects paled in comparison with the scale, scope, and speed of the autobahn. Construction began with a ground-breaking ceremony near Frankfurt in September 1933, the first section was inaugurated in May 1935, 1,000 kilometers were complete by the fall of 1936, and at the end of 1938, the car-owning public could drive on some 3,000 kilometers of autobahn.[21] Hitler's right-hand man was Fritz Todt, a civil engineer who rose from obscurity in

34.1 A stretch of the Autobahn with a service station and without traffic in the 1930s. Image, Wolf Strache, Library of Congress.

1933 to become the head of German war production in 1940. Hitler was so pleased with Todt's breakneck speed that he endearingly called him "a fanatic" in 1937.[22] However, the network fell short of the 7,000 kilometers that the regime had announced as its goal in 1933, and the target length grew to 20,000 kilometers during the war.[23] Documents captured after the German defeat reveal plans for major highways radiating from Berlin to Finland along the Baltic coast and to the Persian Gulf via Baghdad.[24]

In light of these plans, the Nazis' autobahn project was a disappointment, but it also had more mundane failures. It was never the presumed panacea for unemployment, though Nazi propaganda did its best to create a different impression. At its peak in 1936, 124,483 people worked for the Reich's autobahn corporation, a small fraction of the 6 million people who were without a job when the Nazis took over.[25] Finances were a mess, and costs were skyrocketing: work continued only because the regime feared losing face in a bankruptcy.[26] Since most of Todt's engineers came from a railroad background and had never built a roadway, they made embarrassing mistakes, as a red-faced Todt discovered during a bumpy nighttime ride over a new stretch of autobahn in the company of the British minister of transport.[27] In the quest for scenic views, Nazi planners chose routes with steep inclines, and some of the notorious trouble spots in today's traffic updates, such as the Irschenberg ascent on the autobahn between Salzburg and Munich, go back to fateful choices of the 1930s.[28] The Nazis did not even pursue a coherent motorization policy: together with tax incentives and the promise of a cheap Volkswagen motorcar for the masses, the autobahn was meant to encourage automobile use, but fuel prices remained exceedingly high due to the government's support of synthetic fuel production in the quest for rearmament.[29] It showed in the dismal results of traffic counts in the late 1930s: Hitler's pet project was the first white elephant of the automobile age. In fact, it was a white elephant in quite a literal sense: the bright gray lines were clearly visible from the sky, and they led to the major cities, to the dismay of air force officials who feared that they would facilitate navigation for enemy pilots.[30] And as it turned out, Germany's airspace would soon have its share of enemy warplanes.

2. A MATTER OF FREEDOM

When autobahn construction ground to a halt toward the end of 1941, the autobahn was a disjointed network with a total length of 3,819.7 kilometers and no discernible use. Three decades later, many of the gaps were closed, the bridges that the retreating Germany army had blown up in the final stages of the war were rebuilt, and the autobahn was the backbone of German road transport. It even entered music history when Kraftwerk published its enigmatic *Autobahn* album in 1974.[31] Germans today have more than 13,000 kilometers of autobahn at their disposal, and that is just a fraction of the vast road network that has grown since 1945. Mushrooming suburbs called for good traffic links while even the remotest villages asked for an asphalt path as if it were a natural birthright.[32] Unlike the Nazis, the automobile had won.

It was the German version of a trend that united all Western societies. Car ownership became the norm, and it was about more than getting from here to there. It was about freedom. Mass motorization allowed people to escape the grime of cities (see chapter 16, London Smog) to the new suburbs, it brought all sorts of tourist destinations into reach (see chapter 22, Baedeker), and it liberated the people from the improprieties of mass transportation such as schedules and fellow travelers: the world looked different from behind the steering wheel. It was a peculiar type of consumerist freedom, as the car of choice revealed a lot about an owner's income, family size, driving habits, and ego problems. It was also a distinctly masculine type of freedom, as it was typically men who occupied the drivers' seat. For women, automotive freedom was a more complicated matter.[33]

Not only people experienced a new type of freedom. The new road networks also carried goods, effectively breaking the monopoly that the railroads had enjoyed in many places. The fate of the Chicago stockyards (see chapter 18, Chicago's Slaughterhouses), which closed in 1971 because of new slaughterhouses closer to the livestock regions, showed how roads redrew the map of commerce and jeopardized even long-standing champions. It was another blunder of Hitler's autobahn, as the Nazis had focused on passenger cars and enacted a restrictive policy for truck traffic.[34] Even after building thousands of kilometers and conceiving many more, it did not occur to Nazi planners that their roads might open the door to a new age of logistics. When blueprints were drawn for a new autobahn from Breslau to Vienna after the conquest of Czechoslovakia, the plans included a rest stop in Moravia that was completely self-sufficient, doubling as a biodynamic farm with vegetable gardens, pigs and chickens, a dairy, and a sewage treatment plant.[35]

Mass motorization changed cities, economies, and lifestyles akin to a force of nature, and like many trends that people find overwhelming, it gained its own set of mythologies. Was it all maybe a capitalist conspiracy? In 1944, a consortium that included General Motors, Standard Oil of California, and Firestone bought the Los Angeles Railway, and the new owners swiftly replaced streetcars with buses. The deal became the subject of rumors, but the trolley system was already long past its prime and deeply unpopular among locals, and the phase-out decision dates back to 1940.[36] On the East Coast, New York's infrastructure tsar Robert Moses was rumored to have built low bridges on Long Island parkways so as to prevent poor people, who were using buses, from reaching state parks.[37] But buses were generally not allowed on American parkways, and bus service existed on parallel roads.[38] As for the autobahn, speculations about military uses have proved impossible to eradicate, and even books that should know better declare that "thanks to Todt's motorways, 300,000 men could cross the Reich from east to west in just two days."[39] In reality, the German army was firmly committed to railroads for long-distance transport, and when the first major stretches of autobahn opened, it turned out that many cars were overheating when they drove long distances at high speed.[40] With equipment of the 1930s, driving entire divisions across Germany was tantamount to demilitarization.

To be sure, security interests left their mark in the history of automobility. Dwight D. Eisenhower talked about quick evacuation "in case of an atomic attack on our key cities" when he urged Congress to adopt a national highway program in 1955, though this was just one of four arguments that he put forward.[41] Cold War politics brought the United States to export highway expertise to countries like the Philippines, Turkey, Jordan, and Yemen.[42] In the 1950s, a Greek architect planned a new quarter in Baghdad that featured wide avenues and low-density housing, hoping this would scatter residents and allow security forces to squash unrest swiftly. It did not quite work out as planned, as the settlement, later renamed Saddam City and Sadr City, became a center of resistance after the American-led invasion of Iraq in 2003.[43] In Saudi Arabia, streets were used in burgeoning Riyadh to split and isolate potentially troublesome neighborhoods.[44] It was not the only case in which traffic links were conceived as barriers. When West Germany built the Elbe Lateral Canal along the inner-German border from 1968 to 1976, the military asked for a design that would have served as an obstacle to Soviet tanks. As it turned out, tanks did end up in the canal, though they were Western models and they sank on purpose. A dam broke five weeks after the canal's opening, and in a desperate effort to stop the flow, rescue crews built an improvised barrier with Bundeswehr tanks.[45]

The new mobility did not make everyone a winner. Highways claimed a lot of space, and negotiations with landowners (see chapter 6, Land Title) were not always as easy as in the case of the AVUS, where the Prussian Ministry of Agriculture was happy to make the land available when the Kaiser endorsed the project.[46] Those who lived close to busy streets suffered from noise, pollution, and declining property values. Interstate Highways cut through many neighborhoods in the United States, much to the surprise of President Eisenhower, the father of the Interstate System, who did not realize how much Interstates intervened in urban areas until his motorcade got stuck in a construction site on the outskirts of Washington in 1959.[47] The West German chancellor Konrad Adenauer, who rode in a black Mercedes on his daily commute to Bonn, mused in private in the mid-1950s that if he were not already chairman of the Federal Republic's largest party, he "would found a party against automobilism, for that party would be even stronger."[48] Twenty years later, Portugal had a brief flirt with Communism after the fall of the Fascist Estado Novo and eagerly discussed whether the automobile was an illegitimate bourgeois privilege.[49] The Portuguese were probably unaware of the real-world attempt to establish collective car ownership in the Soviet Union. At the Twentieth Party Congress in 1956, Nikita Khrushchev denounced Stalin's crimes as well as wasteful use of passenger cars by state officials, but a subsequent experiment with rental cars collapsed when reckless individuals destroyed a good part of the fleet.[50] Since the 1960s, cars had been problems of supplies rather than principles in the socialist sphere, as waiting periods grew to legendary lengths.[51]

Like all types of freedom, the automobile variety had some requirements that tacitly underpinned the mobility of people and goods. For one, the bonanza relied on abundant and cheap oil from countries like Saudi Arabia (see chapter 15, Saudi Arabia), particularly before the 1973 oil price shock. For another, it built on national governments that could shoulder the enormous costs of the new roads. Investments in transportation were traditionally local and regional matters, but that policy was facing limits in the early 1900s when authorities sought to pave dusty overland roads.[52] National governments took over, helped by schemes that committed automobile taxes to road construction. Great Britain struck what James Flink called a "gentlemen's agreement between Parliament and British motorists" in 1909; the breakthrough in the United States was the Federal Aid Road Act of 1916.[53] Stephen Goddard has argued that Eisenhower's Interstate System was really akin to "interstate socialism," as financing through the Highway Trust Fund was "the political version of a perpetual motion machine."[54] Against the backdrop of billions of dollars, the rhetoric of freedom came in handy, as it fostered a sense of entitlement rather than concerns about costs. Streets were political statements, though slogans were now set in stone (or asphalt, for that matter) rather than written on banners.

The leading role of national governments produced distinct national styles. The German autobahn did not get a speed limit and remains toll-free for passenger cars to this day while other countries were less generous. Belgium's Ministry of Public Works decided in 1969 to put up lighting along its motorways and all roads with a traffic density of more than six thousand vehicles per day.[55] The United Arab Emirates, formerly a British protectorate, felt that eight-lane freeways go along nicely with four-lane roundabouts. And then there were other rules, legal and customary, that made for multiple mobilities (see

chapter 3.3, Unsettled). But for all the diversity of constructions and rules, users of divided highways found out sooner or later that there were limits to automotive freedom. Congestion is a truly global experience in the twenty-first century, with widely diverging reactions depending on individual tempers, national cultures, and the state of air-conditioning (see chapter 20, Air-Conditioning). And then, traffic jams at least dissolved sooner or later. Other repercussions of mass mobility were more terminal.

3. DEAD ENDS

Mercedes clinched the 1926 German Grand Prix on the AVUS, but the race also claimed the lives of four assistants, including two students on a timekeeping assignment.[56] It was not a great shock in contemporary terms. Casualties were simply part of car racing, and certainly not enough reason to stop the fun. When a Mercedes flew into the grandstands at Le Mans in 1955 and killed 81, the worst racing accident to this day, the cars kept logging laps for another 21 hours, except for the remaining Mercedes crews who withdrew from the race after several hours of frantic phone calls with company headquarters in Stuttgart, and spectators were standing at the site of the carnage the following morning amid the stench of blood.[57] The toll of everyday automobilism was no less extreme. In West Germany alone, 383,951 people were injured in car accidents in the year of the Le Mans disaster, and 12,791 of them died.[58] It was against this background that Adenauer joked about his new anti-automobilist party.

Lobbyists were quick to point out that the new roads were less prone to accidents, and this was one of many steps toward safer roads. Authorities beefed up driving schools. Educational campaigns evoked the horrors of reckless driving. Pedestrians were taught to meet vehicles with caution and respect.[59] Carmakers embraced new designs with crumple zones and rollover protection, introduced new features like seat belts and airbags, and gradually phased out unsafe cars like the Volkswagen Beetle, which the Nazis had conceived for the masses. The results were dramatic, particularly in light of steadily growing mileage. While almost 200 people died on every 1 billion kilometers of German roads in 1955, that number is now below 5.[60] But for all these efforts, it is equally remarkable that some options stayed beyond debate. High-speed driving on the autobahn remained a national pastime in Germany in spite of overwhelming evidence that it cost lives. When the United States imposed a national speed limit of 55 miles per hour in the wake of the 1973 oil price shock, the fatality rate fell from 4.28 deaths per 100 million miles of vehicle travel in 1972 to 3.33 in 1974.[61]

People kept dying in and under cars, and the same held true for other effects of automobility. Cars remained a noise problem in spite of quieter engines and ubiquitous noise barriers. Filters and new fuels reduced lead emissions and urban smog, but other pollution problems persist, including some that rarely make the news. Soil samples from the banks of the AVUS show that heavy metal concentrations are up to thirty times higher than regional background levels.[62] Cars continue to devour fossil fuels, and they stand still en masse during rush hour, as new roads often created additional traffic rather than relief. Road networks continue to grow while neighborhoods and landscapes get under the wheel, and only some of the most egregious project were actually canceled: it took years of bitter fighting to kill an elevated "Vieux Carré Expressway" between the French Quarter in New Orleans and the Mississippi riverfront.[63] Even the idea that taxation entitles people to new roads is still around, even though Winston Churchill demolished it almost a century ago. As chancellor of the exchequer, he revoked the 1909 deal with motorists, calling the idea that motor taxes should be reserved for construction "nonsense," "absurd," and "an outrage upon the sovereignty of Parliament and upon common sense."[64]

As seen from the West, solutions to the problems of automobility are eminently halfhearted. And then, industrialized nations have a relatively benign perspective, having seen most of the parameters go down over the years. The real drama is playing out in the Global South, where Western-style cars roam without a century of road construction and social disciplining. Europe and North America account for less than 40 percent of global car registrations today, having rescinded the majority of the market for personal automobiles to the rest of the world in the depression of 2009, and it shows in appalling death rates.[65] The World Health Organization estimates that 1.25 million people are killed on the globe's roads each year, and most of them die outside the Western world. A car in Benin, Guinea, or the Democratic Republic of the Congo is over a thousand times more likely to kill someone than a car in Sweden, Switzerland, or the United Kingdom.[66]

Indictments run through the history of the automobile, and inspired far more than the lunatic fringe. In the early 1970s, the car's reputation was so bad that small-car producers like Fiat and Honda invoked the litany in advertisements.[67] It was about the many problems of freewheeling automobilism, and it was about a symbol. "No panorama of urban degradation is complete without a representation of dying automobiles piled in a deserted scrap yard," Emma Rothschild wrote in *Paradise Lost: The Decline of the Auto-Industrial Age*.[68] Paul Virilio depicted Paris as under a "permanent 'state of siege' of the urgent stream of automobiles" due to the *périphérique*, the inner city ring road around the twenty arrondissements.[69] But for all its sins, the car has shown a remarkable ability to sputter on, and so has the road network that underpinned its rule. Few roads have been dismantled or abandoned since the dawn of the automobile age, and if they were, it was usually because something better was recently completed nearby.

Road networks kept expanding in the Western world, but the rate of growth declined notably. Once builders started digging, they found themselves wrestling with a wide array of stakeholders, including archaeologists who quickly realized the opportunities of large earthmoving projects. Even the Nazis were willing to temper their longing for ever more autobahn mileage if it clashed with their Germanophile instincts. When autobahn construction between Hamburg and Berlin hit on prehistoric burial sites in 1938, authorities allowed for a yearlong excavation project.[70] Resistance gathered strength in the 1970s, as protests sprung up even in the unlikeliest of places. In Belfast, an urban motorway managed to create a common cause for Catholics and Protestants, quite an achievement in light of the sectarian violence in Northern Ireland during the 1970s.[71] Managers devoted more and more time and resources to dealing with conflicting issues, and the heroic builders of early years turned into perennial negotiators. It took its toll on the pace of construction, and projects that began as freeways to modernity produced trouble for decades and meager results. The German city of Bielefeld dreamed up inner-city motorways with a direct link to the next autobahn in the 1950s, faced countless delays and cutbacks over the years, and finally completed the project in a state of exasperation in 2012.[72]

The glamour of roads has long been fading, and so has the expertise that grew with them. Traffic planning became a science of its own, and yet dreams of technocratic control have long evaporated. As Enrique Peñalosa, the mayor of Colombia's capital Bogotá, declared, "It is not a problem for traffic engineers to solve transportation problems, it is a political decision."[73] In fact, the mythology of builders was already crumbling in the original autobahn project once it hit the ground: in overheating cars on empty roads, in escalating costs, in Germanic fantasies about prehistoric heaps, and in an Irschenberg summit that Hitler selected for Todt's mausoleum and today has a McDonald's rest stop.[74] No longer can we entrust road projects to the likes of Fritz Todt and Robert Moses and believe that all will be well, and with the wisdom of hindsight, it was not such a good idea in the first place.

Divided highways invariably evoke thoughts about cost overruns and delays nowadays, and time will tell whether the growth of road networks in the West finally grinds to a halt or goes into reverse. Demobilization may seem an unlikely prospect on a globe that has mostly moved in the opposite direction throughout modernity, but it may not be a matter of choice. It may be a matter of money. The booming, financially promiscuous nation-states of the postwar years no longer exist, and the governments of the industrialized world are stuck with the escalating costs of aging infrastructures. Maintaining and rebuilding the existing transport network may just prove too expensive, though Western welfare states have shown reluctance to renege on commitments, both on infrastructures and otherwise.

Streets are still political, but it is all too convenient to merely take them as a given. The French anthropologist Marc Augé listed them among what he called "non-places" of "supermodernity," anonymous space that he depicts as culturally inert.[75] But streets can turn into political hotbeds in a flash, and not just in the style of the nineteenth century. In the 1990s, Riyadh experienced what Pascal Menoret has called "a car insurgency of sorts."[76] Joyriders were let loose on the highways of Saudi Arabia's capital, adrift both in their cars and in society, as most drivers were without jobs or prospects due to the geriatric structures of the Saudi oil state (see chapter 15, Saudi Arabia). It had most of the ingredients of twentieth-century automotive mobility: cheap oil, new roads purpose-built for cars, a good dose of testosterone, a police force struggling to maintain order, and the lure of infinite jest. The one thing missing was a destination, whether real or utopian. The mobility of Saudi Arabian joyriders was thin camouflage for the fact that they were stuck.

35

THE PINE ROOTS CAMPAIGN

THE TOTALITY OF WAR

I. NATIONAL MOBILIZATION

The outlook was dim for Japan's military leaders in the spring of 1945. US forces had seized the Philippines over the previous months and conducted devastating bombing raids on Japanese cities. Propaganda prepared the population for an upcoming invasion, but supplies were running short on the mainland, and it was by all means uncertain what defenders could offer beyond fierce determination.[1] Oil was a particular matter of concern. Japanese forces had occupied the oil fields of Southeast Asia during the first months of the Pacific War, but American control of the sea and air had caused a steady decline of oil shipments since 1943. The Japanese abandoned their main oil ports in Balikpapan and Surabaya in December 1944 and ended all attempts to supply the mainland in March 1945, and fuel shortages emerged as the ruling constraint on military operations. The air force cut flight training to the bone and embraced new tactics such as Kamikaze suicide missions. After all, Kamikaze planes did not need fuel for a return trip.[2]

The US military did not quite grasp the extent of the shortage, and many of the refineries and oil storage areas that bombers were targeting had actually run dry.[3] But Japan's military was under no illusion that it was facing demobilization for lack of oil, and it sought to boost domestic substitutes. Military leaders looked at agriculture and tried soybeans, peanuts, and alcohol from various sources. When the battleship *Yamato* embarked on its final one-way mission to Okinawa, "edible refined soya bean oil was used as bunker fuel."[4] The military also looked at the forests, specifically at the roots of pine trees. If cooked in purpose-made kettles for twelve hours, pine roots produced a crude oil that could serve as raw material for aircraft fuel. Thirty-four thousand of these kettles were distributed all over the country, and thus began the big dig, wartime forestry version. The propaganda machine jumped into action: "Two hundred pine roots will keep a plane in the air for an hour," a slogan ran.[5] As labor was scarce in Japan's wartime society, scavenging fell to old people and schoolchildren.[6] The government launched the program in October 1944, and after a few months, it was in an upbeat mood. In March 1945, the Cabinet of Japan decided to increase the production goal by 150 percent.[7]

The target was twelve thousand barrels of crude per day, but even that figure would not have changed the stark imbalance of resource endowments in the Pacific War.[8] The American airbase on Guam had command over ten times the amount in aviation fuel alone.[9] As a result, scholars have offered harsh assessments of the pine roots campaign, effectively treating it as the resource equivalent of the death-defying resistance that Japanese soldiers showed in battle. *The Cambridge History of the Second World War* speaks of "desperate, hugely expensive measures," Francis Pike of "the economics of pure desperation," and Daniel Yergin, in his commanding synthesis of the history of oil, called the pine roots campaign "fantastic."[10] However, Japan was not the first wartime society to seize on the woodlands in the quest for oil. During World War I, the German government had launched a call to collect spruce cones for that purpose, followed by a call to collect pine resin.[11] In fact, when news of the pine roots campaign came to Germany in December 1944, the leader of the SS, Heinrich Himmler, initiated a study by one of his underlings, SS-*Obersturmführer* Dr. Lipinsky. The assignment was canceled after a month "in light of the present circumstances," and the project did nothing to avert the German defeat, but it earned Dr. Lipinsky a deferment from front duty and a promotion to *Hauptsturmführer*.[12]

35.1 "Dig for Pine Roots!" Japanese propaganda poster of 1944. Image, Shōwakan, ed., *Sejō o Utsusu Shōwa no Posutā: Posutā ni Miru Senchū Sengo no Nihon* [Posters as a Mirror of the Showa Era: Japan of the War and Postwar Years in Posters] (Tokyo: Shōwakan, 2011), 23.

Mobilizing resources for war is probably as old as war itself, but three trends made the wars of modernity a chapter of their own. The first was geographic scope. The modern era was the first capable of launching world wars that were truly global, extending and complicating supply lines in unprecedented fashion. The eighteenth-century competition between the Dutch and the British East India Company over the Bihar saltpeter trade was a harbinger of things to come: never before had powers competed so fiercely for a critical military resource in such a distant land.[13] Second, the volume of material grew to new dimensions with the use of industrial technology, a trend that came to be associated with trench warfare during World War I. The technology of war became akin to a force of nature, an experience infamously captured by Ernst Jünger in his evocation of the "storms of steel."[14] Third, critical resources grew not only in volume but also in number. Advancing technologies relied on an ever-widening range of materials with specific properties, and military planners realized that insufficient stocks of only one critical commodity could have devastating results. When Japan cut off America's rubber supplies from Southeast Asia in 1942, the United States became the first country in world history to be haunted by the fear that it might a lose a war for lack of rubber.[15]

Resource flows shaped the outcome of wars, and they shaped the path toward them. Concerns about oil supplies were among the factors that put Japan on the road to Pearl Harbor.[16] Wartime decisions also shaped resource use long after the guns fell silent. For example, the stellar rise of the aluminum industry during the twentieth century "cannot be understood without considering the vital importance of aluminum to fighting and winning modern wars."[17] Germany did not have a single aluminum smelter in 1914, and global production more than doubled from 84,000 tons to 180,000 tons during World War I, instilling frantic postwar efforts to find new uses.[18] Synthetic nitrogen followed the same trajectory in even more dramatic fashion, as the decisive breakthrough happened on the eve of the Great War (see chapter 19, Synthetic Nitrogen). Ammonium from the Haber–Bosch process kept the German army firing, and when military demand collapsed after Armistice Day, the fertilizer market was swamped with copious amounts of a product that was potent and problematic in equal measure. Oil consumption increased by 50 percent over the four years of war, inspiring the first spate of depletion warnings after 1918.[19] Industrialists were disinclined to write off the new production capacities, and consumers were not in the mood for restraint after wartime exigencies. Total war was a catalyst in the making of the cheap, abundant, and faceless resources that linger as a hallmark of modernity.

Against this background, the pine roots campaign was no singular excess of sylvan fanaticism. It was the result of an escalating resource crisis that had occupied the Japanese people to an ever-growing degree for years, and resource crises were perfectly normal in total war. The onslaught on Japan's forests certainly did not start with the quest for pine roots. Woodlands were already overused before the Pacific War, and some fourteen thousand square miles, or 15 percent of Japan's forests, were logged from 1941 to 1945, and more than two-thirds underwent clear-cutting.[20] As a schoolgirl, Tsutsui Ayako spent much of her fourth form digging for pine roots in the mountain forests around her hometown and spoke about it decades later for an oral history project, and it does not sound like the work came as much of a shock to her. She had spent the previous year in factories producing silk and on farms catching grasshoppers, "not only to protect the crops from their ravages, but also because grasshoppers, boiled with soy-sauce and sugar, make a strongly flavoured and protein-filled sweetmeat."[21] Japan was short on material resources, but it had plenty of narratives about coping with scarcities. It also had narratives about heroism in battle, which helped cope with the foreboding of defeat. But like most militaries of the industrial world, it did not have narratives about losing a war for lack of stuff.

The pine roots campaign nonetheless looked dubious even in contemporary Japan. In a diary that was published posthumously after the war, Kiyoshi Kiyosawa figured that the labor requirements did not make sense. Kiyosawa wrote on March 19, 1945, "It takes three hundred people to obtain one ton of it. To obtain 100,000 tons requires thirty million people."[22] Kiyosawa was a well-informed journalist, and his diary "is considered the most thoughtful, perceptive, and courageous account kept by a Japanese liberal during the war," and yet his critique is just as remarkable for what it left unsaid.[23] He did not worry about the future of Japan's forests, the conscription of society's weakest for labor service, or the country's long tradition of sustainable forestry.[24] He was worried whether the campaign would achieve its goal.

As it turned out, the pine roots campaign fell short of its promise. Production reached seventy thousand barrels per month in June 1945, which was roughly 2 percent of supplies on Guam. But that was the yield in crude pine root oil, and refining was more difficult than expected. By the end of the war, Japan had produced a paltry three thousand barrels of aviation-grade gasoline. We do not have a record of how it performed under actual flight conditions. However, we do know what happened when the US Army made a trial run with pine root oil in some of its jeeps. The fuel gummed the engines beyond repair.[25]

2. MOBILIZING SCIENCE

The failure of the pine root project was certainly not due to a lack of enthusiasm among the collectors. Occupying US forces reported that "monumental piles of roots and stumps lined many of the roadways," but claiming possession of a resource was only half the effort in a twentieth-century war.[26] It took the tools of science and technology to turn raw materials into military assets, and research and development became a cornerstone of modern warfare. The work of scientists shaped the course and outcome of military conflicts, and their results lingered in more peaceful times along with their institutions and mindsets.

Mobilizing science for war had its own set of challenges. Sometimes matters of conscience made themselves heard, such as when scientists watched the first nuclear explosion in the New Mexico desert in 1945 (see chapter 37, *Lucky Dragon No. 5*). However, the more common obstacles were bad organization and lack of expertise, and both played a role in the pine root oil fiasco. The split between the navy and the army, a familiar topic for students of Japan's military in World War II, left its mark on the pine roots campaign as well, as both branches set up their own distillation plants.[27] A lack of qualified personnel and high-grade equipment had already hampered previous efforts to produce synthetic fuels. The endeavor began with a Synthetic Oil Industry Law in 1937, but production lagged far behind projections. Japan's German ally sent some equipment and a few engineers who stayed until the end of the war, and yet annual production was just 8 percent of the target in 1943. "The synthetic fuel industry in Japan, in terms of its absorption of materials and manpower and its meager product, was more of a liability than an asset during the war," Jerome Cohen wrote in his influential assessment of the Japanese war economy.[28] It is barely surprising that the refining of pine root oil did not go according to plan.

The pine roots campaign was unusual in scale, but the quest for substitutes was a popular field for inventors of all stripes between 1914 and 1945. At the end of World War I, German ingenuity had produced more than eleven thousand ersatz products alone.[29] One of the creative spirits was Cologne's mayor Konrad Adenauer, who worried about the food supply during the war and invented a bread whose mediocre taste was good enough to fill stomachs but not so good as to encourage overconsumption. It earned him a patent in 1915, but the Imperial Patent Office balked when Adenauer followed up with a soy-based sausage with "peace flavor."[30] Nobody was supposed to eat either product in normal times. The bread is still on sale in a Cologne bakery, but only because Adenauer is a local icon. His bread-making skills pale in comparison with his fourteen years as chancellor of West Germany.[31]

Emergencies stimulate research and development, but only in light of the needs of the day. The long-term fate of wartime inventions was usually beyond the horizon, and that has framed scholarly opinion. Ulrich Wengenroth has argued that the quest for substitutes derailed the German innovation system as a whole and contributed to the demise of what had arguably been the world's leading research network before World War I. According to Wengenroth, autarky production claimed tremendous intellectual and financial resources from 1914 to 1945, and Germany gained the ability to turn inferior resources into second-rate products.[32] However, ersatz products had a range of careers.[33] Sometimes commodities returned after a while. Corn and rapeseed, two crops that the Nazis pushed aggressively in their quest for autarky, disappeared from German fields after 1945, but they reemerged two decades later as pillars of Germany's industrial agriculture.[34] Sometimes wartime conditions allowed new products to thrive. The American military used a natural insecticide, pyrethrum, for delousing troops, but as more than 90 percent of the raw material came from Japan, it shifted to a little-known inorganic compound named DDT during World War II (see chapter 38, DDT).[35] And sometimes autarky was genocidal. During World War II, the quest for rubber led German scientists to conduct field experiments with kok-saghyz plants in occupied Eastern Europe that relied on forced labor.[36]

The imprint of war was not limited to the products themselves. Wartime experiences also left their mark on visions and mindsets. Pesticides were closely intertwined with military experiences from the outset, and not only because the pilots of Huff Daland Dusters, who spread calcium arsenate on cotton fields in the 1920s in order to kill boll weevils (see chapter 12, Boll Weevil), were veterans of World War I. Ideas of overwhelming force and total victory framed pest control as a war between humans and insects until it dawned on exterminators that a favorable truce might also be an option.[37] Cold War science gave researchers tools and concepts that helped them to understand anthropogenic global warming.[38] When the United States stopped listening for enemy submarines after the end of the Cold War, the underwater microphone network found a new use when naval scientists employed them to monitor blue whales (see chapter 9, Whaling).[39]

The role of science received yet another dimension when the environmental impact of war became a research topic in its own right. It probably began with the boredom that befell soldiers even in total war: a 1916 article in the journal of the Bavarian Botanical Society discussed "the death of spruce trees caused by artillery shells" with all the earnestness that a German academic can muster.[40] Foresters conducted more comprehensive studies of the French woodlands later on and estimated that World War I had cost France some 2.5 billion board feet of lumber.[41] When environmentalism became a global force toward the end of the twentieth century, military agencies launched research projects of their own, with motives ranging from environmental stewardship and risk assessment to greenwashing. The British Ministry of Defence has published a conservation magazine, *Sanctuary*, since 1976.[42]

The relationship between science and the military has many aspects, but one trend runs throughout the endeavor: it turned matters of war into specialist subjects. Understanding articles in *Sanctuary* takes more background knowledge than, say, Erich Maria Remarque's *All Quiet on the Western Front*, and this inevitably constrained the range of participants. In other words, science was a precursor to the social segregation of military affairs that turned war from a national experience to an occupation for trained professionals. Many countries have abolished conscription in recent decades, but segregating the environmental repercussions of war will only work to a point. Conversations on these matters have turned into the business of experts, but the material consequences will be with us for the foreseeable future. The age of total war may be over, as other forms of warfare have taken hold, but the legacy of total war has left its mark in the land.

3. CHANGES IN THE LAND

The Japanese forests did not look good after extensive logging and the pine roots campaign, and things did not improve after Japan's capitulation in August 1945. A pine bark beetle infestation reached its climax in the postwar years, and the demands of reconstruction, including timber for housing that had been destroyed or damaged during the war, meant that the rate of deforestation actually increased in 1946 and 1947.[43] "Throughout Japan a totally cleared forest plot is a familiar sight," an American observer noted, and Japanese lumbermen were frighteningly thorough: "Everything disappears from a plot which is being cut, including all underbrush and slashings, even where it is on a slope as steep as 60 degrees."[44] It showed in floods and an increase of erosion (see chapter 13, Little Grand Canyon), and contemporary accounts took note that the pine roots campaign carried some of the blame. As a Japanese forester wrote in a 1947 article in the *Nippon Times*, "The digging out of pine roots, which went on promiscuously and frantically on every hill and mountain during the war, in order to extract oil therefrom, further ravaged forest-lands to add to the frequency of landslides."[45]

Exhaustion and disorientation were common among humans and environments after World War II, particularly when nations had to stomach the humiliation of defeat, and yet devastation was not the full story. Those who looked closely found some puzzling ambiguities. Carpet bombing had left many German cities in ruins, but fresh vegetation grew on the rubble, and biologists identified distinct new plant communities in different parts of the country.[46] A British academic spent the summer of 1945 studying water-filled bomb craters from the Battle of Britain that had become home to a range of plants and snails.[47] He sought to identify "some factors governing colonization in ponds of known age," but his findings were open to readings beyond the sober classifications of biological taxonomy.[48] It was as if Mother Nature were out to write a comforting epilogue to the horrors of war.

Total war brought changes in the land, but even ravaged grounds showed signs of life after some time, and that struck a special chord among humans. The Demilitarized Zone separating North and South Korea, where wildlife flourishes largely undisturbed, became a tourist attraction that guidebooks (see chapter 22, Baedeker) tout as "an environmental haven."[49] In the Andean highlands, the Revolutionary Armed Forces of Colombia (FARC) have claimed green credentials for keeping agriculture out of pristine forests, though their interest in camouflage probably ranked higher than their purported concern for the beauty of the woods.[50] After the end of the Cold War, conservationists all over Europe worked to turn the former wasteland along the Iron Curtain into a "European Green Belt."[51] Elsewhere on the globe, diplomats and conservationists have established nature reserves in formerly war-torn regions, named them "Peace Parks," and tried to turn them into engines of reconciliation.[52] In places that have seen the worst of humans, it seemed opportune to step back and let nature take its course.

Needless to say, the fate of peace parks, like that of all nature reserves (see chapter 26, Kruger National Park), depends on fair and informed management on the ground. However, the penchant for peace parks stands in marked contrast to the other popular approach to landscapes of war. Restoration rules supreme on many former battlefields and other places that rank high in collective memory: the place is supposed to look exactly as it was on its great day in history. After World War I, reforestation on the battlefields around Verdun met with protest from veterans.[53] Landscape management at Gettysburg has sought to restore battlefield conditions at the time of the US Civil War ever since the National Park Service took control of the site in 1933.[54] Full restoration is obviously elusive, not least because locals have mixed feelings about the constraints that come with living in a museum landscape, and the quest for authenticity does not mesh easily with the demands of mass tourism (see chapter 22, Baedeker), but the guiding idea is remarkable enough. Former battlefields are viewed as monuments for eternity, and any changes in the land, even those at the hand of nature, are tantamount to heresy.

Biologists occupy an uncomfortable place somewhere between these extremes. Researchers have long established that many military training grounds are assets for conservation, and not only because they keep civilians at bay. The use of military equipment can jeopardize ecologies, but it can also open up niches for new ones, and conservationists have argued for the continuation of destructive activities in abandoned military training areas in order to save disturbance-dependent species.[55] As an academic discipline, ecology does not hand out value judgments, and yet it deserves reflection that the real-world fates of plants and animals on war-torn land can play out in all sorts of ways. It shows that both visions of militarized landscapes, the ecological and the heritage views, draw on an exceedingly simple understanding of the environmental legacy of war.[56]

The visuality of landscapes obscures a legacy that is far more complex and far more disturbing. The environmental legacy of war lingers in residues that are radioactive or toxic, such as the dioxin that stays in Vietnamese soils decades after the infamous Operation Ranch Hand made Agent Orange a household word.[57] It lingers in conditions on military bases that Richard Cheney, secretary of defense under George H. W. Bush, found so appalling that he launched a "Defense and the Environment Initiative" in 1990.[58] It lingers in the cattle ticks that were first identified in French New Caledonia in 1944, likely after coming from Australia on horseback, and in the brown tree snake plague on Guam (see chapter 14, Cane Toads).[59] It lingers in bomb crater ponds that replace lost habitats for amphibians in Laos, Hungary, and elsewhere.[60] It lingers in disturbed soils in the Mojave Desert, where tank tracks may not disappear until another ice age.[61] It may also linger in the forests of Japan, though a graceful act of postwar amnesia seems to have kept the pine roots campaign and its environmental legacy beyond academic scrutiny.[62]

The legacies of total war do not just go away, and then, these wars at least come to an end at some point. Total wars have an innate tendency to burn out after a while whereas low-level warfare can continue for decades, and never-ending wars have another set of environmental repercussions that are pernicious in their own ways. A powerful narrative depicts war as the dark underside of human civilization, which has made modern societies reluctant to engage with its environmental legacy in full. Involvement of modern science has scarcely made warfare more ennobling, and experts have looked back at their lifetime projects with a sense of remorse. In congressional testimony a few years before his death, Hyman Rickover, who led nuclear development in the US Navy for decades and built the pioneering Shippingport Atomic Power Station (see chapter 37.2, Nuclear Complications), confessed that he was "not proud" of his work. He felt that nuclear-powered ships were "a necessary evil." If it had not been for national security, "I would sink them all."[63]

Many environmental legacies of war defy a quick fix, but the rhetoric of militarized landscapes can cope with all sorts of disfigurations. Take, for instance, the "Pool of Peace" in Spanbroekmolen, close to the border between Belgium and France.[64] It stems from one of the nineteen underground mines that the British Army ignited in the early morning hours of June 7, 1917, an explosion that killed an estimated ten thousand German soldiers in an instant and marked the beginning of the Battle of Messines. But no soothing name commemorates the mine that failed to explode on that day and continues to lodge in the ground to this day. Its precise location is not known, and its presence was "exciting periodic local nervousness," as Martin Gilbert found during personal visits in 1970 and 1971, and that makes it a fitting *lieu de mémoire* for the environmental legacy of war.[65] It is a diffuse threat, unsettling and puzzling, and it can go in all sorts of directions, including none at all. But if it strikes, the effects may be beyond recall.

PART VIII

THE GREAT ENTRENCHMENT

THE TIMES WERE CHANGING, AND THAT
MADE PEOPLE LESS AMENABLE TO CHANGE.

"Most of our people have never had it so good," the British prime minister Harold Macmillan declared at a Tory rally in Bedford in 1957. It struck a nerve: Macmillan made the phrase the running theme of his next election campaign and he won.[1] It was the British version of a common trend in Western postwar democracies. Decades of strong and unrelenting economic growth transformed the societies of Western Europe and North America as large swaths of the population came to enjoy the fruits of mass consumption. It was a boom without precedent, and coming against the backdrop of two world wars, it left a lasting impression on collective memory. West Germans spoke of an economic miracle (*Wirtschaftswunder*) while the French called it *les trente glorieuses*, and Eric Hobsbawm retrospectively categorized the years from the end of World War II to the early 1970s as "a brief Golden Age."[2] The West had never been so good.

Environmental historians have suggested less enthusiastic readings of the postwar years. A quarter century ago, Christian Pfister coined the term "1950s syndrome" to highlight the environmental toll of consumerism.[3] More recently, John McNeill and Peter Engelke argued that 1945 was the start of a "great acceleration" that catapulted mankind into a new age, the Anthropocene, where the impact of humans had reached such an extent that it stood on a par with the forces of nature.[4] Their book drew on findings from scientists, in which the discussion on the "great acceleration" goes back to a synthesis project of the International Geosphere-Biosphere Programme from 1999 to 2003.[5] Pfister, McNeill, and Engelke built their arguments on a wealth of statistics, and scientists have engaged in similar efforts to make the geological case for the Anthropocene. As a result, we have a clear idea of the unprecedented scale, the growing speed, and the many dimensions of the transformation of earth systems since the middle of the twentieth century. However, these numbers spoke less clearly when it came to specifying agents and underlying causes: the "great acceleration" looked strangely faceless, as if it was a force of nature rather than a historical process. Furthermore, it did not need the wisdom of hindsight to discover that the new affluence came at a price. People in the industrialized world noticed the toll of consumerism, and complaints about pollution, changing landscapes, and other environmental problems grew notably during the postwar years. New organizations took up the fight, scientists looked into problems in growing numbers, and governments created new departments for environmental affairs. Few issues were actually new, but they certainly felt that way, as the visibility and urgency of environmental problems reached unprecedented heights. Critical minds stress that the response was woefully inadequate in light of the overall problem, but there can be no doubt that environmentalism has changed the way humans see the world. Joachim Radkau has argued that the impact on hearts and minds has been so profound that it "may be conceived as a New Enlightenment."[6]

Both consumerism and environmentalism came from the industrialized world, but they resonated all over the globe. Resource flows circled the planet in new quantities, and the same held true for pollutants and wastes. Activists and ideas crossed borders, books and media reports found audiences far beyond their points of origin, and many an environmental campaign focused on distant countries or the seas in between. There was no escape from consumerism and environmentalism anywhere on the globe by the late twentieth century, but this globalization was not a one-way street, as people outside the industrialized world were remarkably creative in shaping their own responses. In fact, environmentalism in the Global South was so different that Joan Martinez-Alier has identified it as a distinct brand, the "environmentalism of the poor," where socioeconomic and cultural issues merged with environmental concerns.[7] Consumerism and environmentalism meant different things to different people, and while both evolved according to their own rationales, the two developments became entangled on many different levels.

A mutual entanglement does not require equal significance. It is quite clear that consumerism has captured hearts and minds among the people of the world to a greater extent than environmentalism. And, needless to say, consumerism and environmentalism are both hugely diverse. There are significant differences in levels of affluence and styles of consumption around the globe, and the same holds true for the definition of environmental problems and the vigor of environmental sentiments—an obvious point in a way, but it deserves attention in light of Anthropocene narratives that conflate the diversity of opinions and impacts around the globe into "an abstract humanity uniformly involved."[8] But for all the ambiguities and the diversity of perspectives around the globe, it is striking how consumerism and environmentalism have become locked in an interdependent relationship since 1945. Both trends grew in size and scope to such an extent that it became increasingly difficult to think about one without the other.

Entanglements were a more general phenomenon of the postwar year, and this has left its mark on the trajectory of the following chapters. The transnational exchange gained a new dynamism, and it changed the way trends and events were experienced. Most previous chapters were about local events and processes that only traveled after a time lag. However, the *Torrey Canyon* and the *Lucky Dragon No. 5* disasters became world news instantaneously, and the hope for the peaceful atom spread with amazing speed (technology transfer took a bit longer). The world was watching, and that makes the following stories less material and more cultural: few people had firsthand experience with uranium, and yet they had an opinion about it. It was a result of new infrastructures that facilitated the circulation of goods, people, and information. And it was the result of the Cold War, which brought the two superpowers to monitor developments all over the world with keen interest. It might just have been the opponent's next fateful move.

Consumerism and environmentalism emanated from an urban world, but it is rewarding to start this discussion in the countryside. The transformation of food production was one of the most consequential trends of the postwar years, and it was in many ways the cumulative result of several developments that have been discussed in previous chapters: the industrial logic of the Chicago slaughterhouse, the systematic, science-based breeding that led to hybrid seeds, the farmers' new chemical helpers such as synthetic nitrogen, and state administrations that learned through events like Ukraine's Holodomor that scarce food was a first-rate threat to their legitimacy. But battery chicken brought all this to a new level, for it was an unprecedented, quasi-totalitarian system that grew around unprecedented masses of animals in captivity. This was not about how agricultural production changed in the wake of new technologies. This was about how technology *defined* agricultural production. Battery chickens were at the mercy of technology at virtually every stage of their lives, plus for some time before and thereafter.

As it turned out, control over all stages in the life of a chicken brought results that pushed other modes of production to the margins. When it came to eggs and chicken meat, factory farming was in a league of its own. The output of factory farming has shaped the world market for agricultural commodities for decades, and production is going global, too: in the new millennium, chicken breeders were busy making the animal fit for tropical climates.[9] It had tremendous consequences for the animals, for agricultural labor, and for the people formerly known as farmers: unlike traditional ways of agricultural production, factory farming did not require land, let alone a sentimental attachment to the rural world, but it did require capital, sophisticated technology, and a distinct type of masculine brutality. It also took consumers who cared more about costs than about fellow creatures. To some extent, the postwar boom was a result of declining food prices and the ensuing increase in disposable income, and everyday meat consumption became a feature of the new affluent societies with remarkable cross-cultural appeal.

The battery chicken was the pioneer species of factory farming, the first animal to undergo comprehensive industrialization. It was about building and fine-tuning new technologies: artificial light and ventilation, automatic feeding and removal of feces, pharmaceutics and disease control. And it was about building new arguments, for battery cages raised concerns about animal welfare from the outset. Critics pointed to the cramped conditions, the separation from nature, and the painfully deformed bodies that breeders produced in the quest for top performance in the most prized body parts. Chicken farmers replied that they had a genuine interest in the animal's health, that they drew on the latest insights of researchers, and that animals were generally better off in the custody of humans. As the debate progressed, it showed that animal welfare was a more ambiguous concept than both sides liked to suggest, but it may have been deeds rather than words that shaped the ultimate result. Finding arguments against battery chicken was easy, but building a different production regime was a different matter. The chicken complex was remarkably successful in constraining the room for alternatives commercially and biologically: other production regimes had higher unit costs, and alternative producers struggled to find the right biological material, as commercial breeders were slow to come up with new chicken varieties for the more complex demands of organic farming. Other animals followed the path of the industrialized chicken, but the approach did not work for every species, and it faced particular difficulties with animals that Westerners deemed exotic. The British chicken pioneer Antony Fisher learned this at great financial loss when he sought to replicate his achievement with sea turtles on the Cayman Islands.

While battery chicken was a commercial success story, nuclear power became an economic folly. It did not look that way initially because hopes for the "peaceful atom" were truly global in the 1950s. Nuclear power seemed to represent the future, an embodiment of science-based progress and the genius of humans, but enthusiasm was fading when reactors became a technological reality. Escalating costs and spectacular disasters made nuclear power a gamble, economic and otherwise, and once these risks sank in among the utilities, it took lavish government support to lure them into building yet another reactor. Nuclear power would have long collapsed if it had not been for two factors. One is institutional momentum: reactors typically run for decades, and they sustain a community of experts and government officials whose careers depend on the technology's survival. The other is the quest for the ultimate weapon, as the line between civil and military uses has always been stronger in the imagination than in technological reality.

The fate of the *Lucky Dragon No. 5*, a Japanese fishing vessel that got caught downwind from a US nuclear test, is a good place to explore the intermingling of military and civilian dimensions. The contamination of the Japanese crew made world news, and when the Americans responded in a less than graceful way, the *Lucky Dragon* became the trigger for a transnational movement against fallout from nuclear weapons testing. But fears of radiation did not keep Japan from ordering nuclear reactors whose design mirrored the priorities of the US Navy. Light-water reactors had a head start thanks to the USS *Nautilus*, the world's first nuclear-powered submarine, and other reactor types never managed to catch up, notwithstanding the fact that some features of light-water reactors were commercially dubious and operationally dangerous.[10] One of these features, a reliance on emergency cooling systems, was the underlying cause of the 2011 Fukushima disaster.

Fallout from nuclear tests circled the entire globe, and that shaped a new perspective on pollution problems. Whereas traditional problems such as London smog were local and by and large a property issue rather than a health problem, the new pollutants were invisible, carcinogenic, and present all over the world. It was about new substances—DDT was another example—and it was about affluent societies that were loath to think about pollution as a measure of social inequality. The new pollution problems seemed to expose everyone on the globe, though the fate of the *Lucky Dragon* crew suggested otherwise.

Just like nuclear power, DDT was among the scientific tools that won World War II, as it helped keep American soldiers healthy in disease-prone environments. But innovations could have military significance beyond the battlefield. The global campaign against malaria was not just a humanitarian gesture but also a strategic effort to woo the poor countries of the world at a time of decolonization. Moreover, DDT helped boost agricultural yields. The use of DDT in the fight against malaria was less controversial than its use in agriculture, which in turn was less controversial than its use against invasive species such as the fire ant (see chapter 14, Cane Toads). As it turned out, indiscriminate spraying on private land led to a permanent shift in the lines of discourse.

DDT was a panacea initially, or at least used that way, as its remarkable effectiveness seemed to spell the end of all insect woes. Farmers soon learned about resistance problems and other side effects of lavish use, but discussions took a different turn when Rachel Carson published *Silent Spring* in 1962. DDT figured prominently in Carson's best seller, and the book's success changed the lines of discourse. Discussions concerning DDT were henceforth about problems as well as a symbol, and when environmentalism grew into a global force in the 1970s, banning DDT became a symbolic act par excellence, the crucial test for Western governments as to whether they meant to be serious about the ecological crisis. The material properties of DDT and its symbolic value have coexisted uneasily ever since, and the ban stands mostly because the symbolic costs of a repeal far outweigh potential gains.

In light of the book that started it all, it was a bit of a misunderstanding, as Carson's ambitions lay beyond the specific case of DDT. Her real concern was the intricate web of life and the folly of indiscriminate brute force interventions, and DDT was merely a particularly glaring case in point. *Silent Spring* was one of numerous blueprints that were drafted or rediscovered during the postwar years for a new environmental philosophy, at times helped by a good dose of pot (see Interlude, Opium), and these blueprints inspired countless passionate discussions, particularly during the movement's early years. But when it came to pollution control and government policies, the guiding philosophies all over the globe were clearly instrumentalism, incrementalism, and symbolism. Technological solutions were typically the favored answers to the environmental crisis, particularly when they relieved society of the need for major change, and it was understood that these new technologies were best phased in gradually. We cannot understand the symbolic appeal of the ban on DDT and other persistent organic chemicals if we fail to account for the fact that environmental politics as usual smacked of perennial compromise.

Some corporations reacted viciously to the publication of *Silent Spring*, but the business case for DDT was more ambiguous than headlines would suggest. DDT was cheap and hence commercially unattractive compared to more expensive alternatives, and use within the United States had actually begun to decline before Rachel Carson. It is important to stress these ambiguities, not least with a view to our current infatuation with the wealth of elites, for entrepreneurial energies of the postwar years were arguably about more than getting rich. Many promoters of nuclear technology truly believed in its Protean promise, and the fathers of factory farming were also driven by the Schumpeterian urge to demolish the old and create the new: Antony Fisher was not only a chicken farmer but also a passionate apostle of neoliberalism. Some readers may cringe when I suggest that today Fisher would put his entrepreneurial energies to work in the organic camp, but it is important to recall that creative destruction is not an inherently antienvironmental concept. The future of environmentalism depends to a significant extent on whether clever businessmen can build an economic case around it.

Humanity's dependence on technology surely increased in the postwar years, which made the prospect of failure ever more terrifying. A disaster like the 1967 *Torrey Canyon* oil spill was inconceivable before the rise of supertankers and it provoked a sense of shock, but learning from technological failure was easier said than done. For one, priorities were open to debate. Characteristically, liability issues were at the top of the agenda while changes in tanker design were postponed until the *Exxon Valdez* disaster some two decades later. For another, even the best precautions could not eliminate the possibility of disaster, and in an age with almost limitless confidence in scientific progress, neatly captured by Vannevar Bush's 1945 report *Science: The Endless Frontier*, people found it hard to define limits to technological advancement.[11] And to the extent that learning did take place, it was usually within a small group of faceless insiders with tenuous accountability.

For the wider public, the *Torrey Canyon* disaster was not about liability law or hull design. It was about images. Never had visual communication been more important than after 1945, and never had its ambiguities become clearer. Policymakers soon learned about the power of images: a 1964 memorandum from the British Ministry of Agriculture warned that cruelty to animals "could really cause a stir especially bearing in mind the fact that it lends itself to pictorial treatment."[12] The *Torrey Canyon* oil spill defined a cultural script that subsequent disasters have followed, down to the frantic efforts to save oil-stained seabirds against all odds as if it were an act of repentance for our collective obsessions with carbon fuels. Images can move people, but visual media rarely tell the full story, or even half of it. And then, some parts of the world are more visible than others.

Authoritarian regimes and vested interests seek to control visual media, and that is not the only difference between disasters in Western democracies and those in the Global South. While supertanker oil spills are temporally and spatially limited events, oil pollution is a chronic problem in countries like Nigeria, and frail or nonexistent state administrations deprive disaster victims of resources that many Westerners take as a given. It was relatively easy to bring modern technologies—from automobiles to plastic bags—into the non-Western world, but fixing the ensuing problems was a different matter.

PART VIII

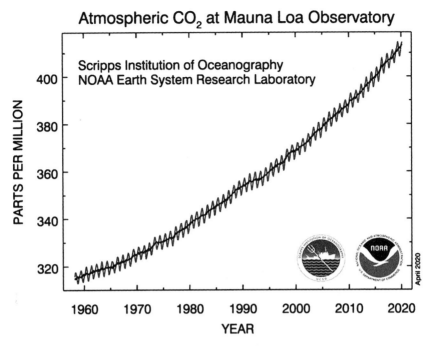

VIII.1 The Keeling Curve records the global atmospheric carbon dioxide concentration since 1958. Image, Scripps Institution of Oceanography.

Western approaches had requirements such as a functioning state administration, and they framed problems in ways that did not match the conditions of other countries. Plastic bags look different when they serve as breeding ground for mosquitoes and when municipal waste collection is unreliable at best.

Plastic bags surely rank among the less glamorous of postwar innovations, but they provide a showcase for the different dimensions of consumerism. Should the environmental assessment focus on the resource base, or energy, or recycling? And what about the human dimension, which included suffocating toddlers as well as Indian cotton farmers? It says a lot about Western environmentalism that plastic bags generated so much excitement, and it says a lot about modernity's infatuation with expertise that scientists figured so prominently in the quest for answers even though value judgments and human habits played a major role. It also says a lot about environmentalism's shortcomings in the face of mass consumption that government policies were slow in coming.

As befits a history of modernity, the final chapter moves into outer space, though not in the heroic mood of bygone times. A mysterious plastic bag adrift may not rank high in the grand scheme of things, but it shows how dissipation may be the most enduring legacy of the project of modernity. It is a legacy that may one day emerge as the supreme irony of the postwar years. Mass consumption was firmly committed to the here and now, and yet it created a material legacy from plastics to plutonium that will last for a long time.

All this makes consumerism a strangely ambiguous force of the postwar years. On the one hand, mass consumption spread akin to a force of nature: the famous Keeling curve, which measures ambient carbon dioxide concentrations, has shown an increase year by year since 1958. But on the other hand, consumer satisfaction was fluctuating widely, and that was just one of many forces that had unexpected consequences in the real world. Consumers found that they could live with industrialized chicken, but they abandoned turtle soup. Nuclear power fueled utopian hopes and cancer fears in equal measure and then fell behind its promise due to the trivialities of economics, and plastic bags triggered a crisis for the mighty chemical industry when they got into the hands of small children. And when it comes to neoliberalism, the favorite ideology of economists and policymakers since the 1970s, its ascendance stands in marked contrast to the persistence of a maze of rules, and not just because of overeager governments. Antony Fisher was a man of freedom, but his chicken business relied on numerous agreements with scientists, regulators, and supermarket chains.

The same holds true for environmentalism. It grew into a global force during the postwar years, and yet it was fraught with ambiguities. Environmentalists criticized battery chicken from the start, but animal welfare was a difficult topic conceptually and politically. They depicted pollution as threats without borders in spite of overwhelming evidence that some groups suffered more than others. They celebrated Rachel Carson and distorted her stance, and they got excited about supertanker oil spills despite the fact that they were only a small fraction of the problem. And they abhor plastic bags and go for cotton, although it only makes a difference when they are used more than one hundred times.

So what should we call the postwar years? It is obviously a matter of value judgments whether one speaks glowingly of a golden age or remorsefully of the great acceleration, but the dilemma is not just about morals. The problem is that it is impossible to identify a prime mover, for the real story of the postwar years was about how consumerism and environmentalism became locked in a mutual dependence that shows no sign of relenting in the twenty-first century. The consumerist inclinations of the world's people persist, and so does awareness that consumption has huge environmental consequences. Faced with the diverging imperatives of consumerism and environmentalism, no country has come to the conclusion that it should ditch one and focus on the other. The outcome was always a notoriously unstable hybrid.

The postwar years were years of growth, both in material impacts and in awareness of problems. Even more, societies have shown themselves unable to drop some of the accumulated baggage: while concerns about *Lebensraum*, so painfully virulent in the age of catastrophe, have lost much of their thrust, we have not managed to abandon any significant part of the post-1945 legacy. Even nuclear energy, a botched technology if ever there was one, is still alive. Environmentalists honor the legacy of Rachel Carson, and they do not find it odd that an American book of 1962 continues to shape global policy in the new millennium. Environmental institutions and agreements have established themselves, and while their real-world achievements are a mixed bag, they have been tremendously successful in defending their own turf. Even climate diplomacy has proved remarkably resilient in spite of scant achievements over the thirty years that have passed since the Rio Summit of 1992 agreed on the UN Framework Convention on Climate Change. Consumerism and environmentalism created an entangled legacy that was—and is—hard to supersede materially, institutionally, and intellectually. Both trends show entrenched lines of debate, conflict, and action, and as it stands, all attempts to open a new chapter have failed. Whether that is a good thing or a bad thing is a matter of political inclinations. But it seems that we are stuck with this legacy for better or worse.

36

BATTERY CHICKEN

THE INDUSTRIALIZED ANIMAL

I. BUSINESS VENTURES

Antony Fisher was a man of freedom. A graduate of Eton and Cambridge and a fighter pilot in the Battle of Britain, he read a condensed version of Friedrich Hayek's *The Road to Serfdom* in the April 1945 issue of *Reader's Digest*. After meeting Hayek at the London School of Economics later that year, he became a lifelong campaigner against government intrusions.[1] He published a pamphlet of his own in 1949, *The Case for Freedom*, in more than one respect a slim book that was never published again.[2] However, there were ways to boost freedom that did not require a command of the written word. Six years later, he founded the Institute of Economic Affairs, which Daniel Stedman Jones has called "the most important of all the think tanks for the development of transatlantic neoliberal politics to be set up after World War II, certainly in Britain."[3] After keeping the institute financially afloat during its first years, he traveled the world and pushed for the creation of similar think tanks abroad. Having witnessed his crusade bearing fruit with Thatcher in Britain and Reagan in the United States, he died shortly after receiving a knighthood in 1988. But freedom was not for everyone under his watch. Fisher made a fortune with mass incarceration, and the inmates were among the most helpless of creatures: chickens.[4]

Fisher's company, Buxted Chicken, was the British version of a general trend in postwar consumer societies. Growing affluence typically led to growing demand for meat, and chicken was the big winner. A luxury food a century ago, only consumed on a regular basis by minorities such as the Jews of New York City, chicken became the most popular meat in many Western countries, and nowadays some sixty billion chickens are slaughtered globally every year.[5] Most of the supply came from new companies such as Buxted, and their industrial production methods led to a dramatic fall in the price per unit, which in turn stimulated demand. The decline of food prices and the corresponding increase in disposable income were among the hidden factors behind the postwar miracle years.

Chickens were gratifying objects for industrial-style rationalization. Raising chickens was traditionally more about eggs than meat, and many Western farmers saw it as a mere side business that they gladly left to their wives. Flocks were small and sometimes fed with leftovers.[6] As late as 1957, 40 percent of British chickens were still in flocks of less than two hundred birds.[7] A short lifespan and a high reproduction rate made commercial breeding easy, and researchers had extensively studied the needs and diseases of chickens long before the boom.[8] There was also little in the way of sentimental attachments to the animal, and neither did tampering with chicken raise the ire of a powerful profession. The "chicken farmer" is a standing joke in the literature on Antony Fisher and the rise of neoliberalism.[9]

As so often when it comes to mass production and agricultural technology, American farmers were the pioneers, and their model inspired others to follow. Fisher entered the chicken business after seeing fifteen thousand birds in a single building at Cornell University.[10] The path of individual countries invariably differed on some finer points. British poultry producers were already banking on frozen food in the 1950s when American consumers still preferred their chicken chilled.[11] Italian boosters invented a new cooling method in order to give their countrymen the dry chicken meat that they preferred.[12] The German Democratic Republic invented its own gold-laced brand of fried chicken, the Goldbroiler, in order to add some glitter to bland Socialist consumerism.[13] McDonald's created a transnational dimension when it glued reconstituted chicken meat into a handy breaded piece, put it on the menu as Chicken McNuggets in 1983, and became America's second-largest purchaser of chicken after KFC within a single month.[14] Diversity increased even further when McDonald's introduced an "all white meat" chicken nugget in 2003, prompting consumers to wonder what they had been eating all along.[15]

The boom created new geographies of food production (see chapter 2, Sugar). Unlike other types of agriculture, chicken farming did not depend on the availability of sufficient land. Feed could come from Peruvian fisheries (see chapter 8.3, Running Empty) or any other place on the globe, and the expanding road networks of the postwar years (see chapter 34, Autobahn) allowed commodity chains beyond rail networks. Chicken farming clustered in specific areas as a result, and they were typically regions with a tradition of poverty. The hub of West German chicken production was the Oldenburg Münsterland in the northwestern lowlands, a region renowned for its meager soils.[16] Chicken replaced cotton farming in northeastern Georgia during the interwar years, a transition made all the easier as the system of contract farming, which gave individual farmers chicks, feed, medicine, and rigid instructions on what to do with them, bore a striking resemblance to the sharecropping system in the post–Civil War South (see chapter 12.2, Size Matters).[17] With no land constraint holding expansion in check, overproduction became a notorious problem and exports the much-needed safety valve. By 1962, one of Georgia's leading corporations, Gold Kist, was already selling frozen chicken to Switzerland, Austria, Greece, Hong Kong, Puerto Rico, Bermuda, and Kuwait.[18] As overproduction was a transatlantic problem, tariffs on chicken meat led to the first trade conflict between the United States and the European Community in the early 1960s, a conflict that entered history books as the "chicken war."[19]

In such a setting, the air grew thin for Antony Fisher. Individual entrepreneurs could not easily muster the capital for large, vertically integrated companies (see chapter 10, United Fruit), and they struggled to meet the demands of expanding supermarket chains, which asked for high volume and standardized quality.[20] After a few years of dynamic growth, the chicken industry developed increasingly rigid structures that defied personal initiative. Writing on the American broiler industry, William Boyd and Michael Watts have argued that chicken meat "became one of the most tightly coordinated and institutionally dense commodity systems in US agriculture."[21] Fisher also suffered from recurring bouts of depression and became estranged from his wife who was, horror of horrors, a run-of-the-mill Tory who did not share his enthusiasm for the free market. He sold his company and filed for divorce in 1968.[22]

The inspiration for his next business venture came from an unlikely source: *National Geographic*. The June 1967 issue had an article on the imperiled Caribbean green turtle that caught the eye of Antony Fisher's son.[23] It was not quite chicken, whose domestication occurred in the fog of prehistory, but green turtles had long served human consumption. Indigenous turtle hunting among the Miskito people of Nicaragua and Honduras goes back to pre-Columbian times, and seafarers in the age of sail were fond of a type of meat that could be stored alive on deck for weeks. When Englishmen grew sick on Jamaica, they often went to the Cayman Islands and restored their health on a turtle diet.[24] Turtle soup was a well-known delicacy, particularly in the English-speaking world, and when demand increased with postwar consumerism, Nicaragua's Somoza government geared up export capacities in the late 1960s.[25] Green turtles offered "an expansible food resource for the future," as *National Geographic* declared, but uncontrolled hunting was driving them toward extinction (see chapter 11, Dodo). A more sustainable mode of production would not suffer from a lack of demand: "There is a ready market for frozen turtle meat, a growing demand for clear green turtle soup, and a rising commerce in turtle hides for leather."[26]

It looked like the perfect business opportunity for a man who had made a fortune with factory farming. Fisher had a lot of free money and partners who shared his enthusiasm. Furthermore, he had the endorsement of conservation biology: the author of the *National Geographic* article, Archie Carr, was the world's foremost authority on sea turtles.[27] The investment gave Fisher, ever the networker, access to the Brotherhood of the Green Turtle, "a loosely knit group of influential people bound together only by the obligation to think how to save the green turtle, and thereby to insure to Caribbean *costeños* more protein in their diet, and to Winston Churchill his nightly cup of green turtle soup."[28] Fisher was also fond of snorkeling, his son was a trained diver, and one of the animal's habitats, the Cayman Islands, was famous for its coral reefs.[29] And when it comes to places for doing business, the Cayman Islands are not a bad choice if you are afraid of overbearing governments.

Mariculture Limited was incorporated in the fall of 1968 and built a facility on the west coast of Grand Cayman. It faced numerous open questions. Nobody knew the right food for baby turtles or how to supply the herbivorous animal with sufficient amounts of their coveted sea grass. Water temperature and cleanliness were additional unknowns, and it was anyone's guess how the migratory turtle would respond to mass captivity.[30] But business ventures were about getting a difficult job done, and within a decade Mariculture had gained control over all the steps in the life of green turtles from mating to maturity, effectively making the farm independent of supplies from the wild.[31] It did not fail to leave an impression on the investors, who soon fantasized about expansion throughout the Caribbean and an annual production in the range of 100 million turtles.[32]

Fisher left daily operations in other hands and focused on trade and customs issues from his estate in England. It was not the greatest task for a champion of free markets, but someone had to make sure that exports would run smoothly once the breeding program met with success. It initially looked like the job was merely about turning a gentlemen's agreement into the letters of the law. He had the world's leading turtle man on his side, and the International Union for Conservation of Nature had unanimously endorsed the general idea at its 1963 Assembly in Nairobi, Kenya.[33] But negotiations turned into a protracted affair. There was a difference between the speed of business and the speed of conservation diplomacy. There was also a difference between chickens and green turtles.

2. THE PRICE OF FREEDOM

On March 17, 1962, a German veterinary official paid a visit to a chicken farm in Spreda, a small town in the heart of the Oldenburg Münsterland. Upon his arrival, he saw two young men folding protective clothing. When they caught sight of the official, they jumped into their Volkswagen and drove away. The official discovered that the mysterious men had forgotten a leaflet with instructions for an inoculation that only approved veterinarians were allowed to use. When he confronted the farmer and his family with these observations, they fell silent. Münsterländers have a well-deserved reputation as tight-lipped people, but this was a suspicious type of silence. Under pressure, the farmer's wife cracked, but she claimed to know the men only by their first names: Heinz and Wilfried. The official made some inquiries and found that the men had done the same job on many other chicken farms. He also revealed the true identities of Heinz and Wilfried, their employer, and the license plate of their car. He filed a report with the police.[34]

The new chicken farms were great places for economies of scale, but they were equally great for breeding disease. Infections could spread rapidly when thousands of animals were crammed into a tight space. The traditional responses to diseased animals—isolation, special care, and premature slaughtering—were not suited to the realities of large production units. With thousands of animals in their care, farmers struggled to identify individual animals and easily failed to notice disease problems until it was too late. And then there were the all-important costs: special care allowed animals to survive, but it took time and money. Against this background, farmers could either rethink the path toward ever greater units and economies of scale. Or they could react in the way that German forestry reacted to insect damage (see chapter 4.2, Specialist Trees, Specialist Minds) and call in specialized expertise in order to tackle the symptoms. For factory farming, pharmaceutics was the discipline of choice.

Large pharmaceutical companies were glad to respond. Merck entered the animal business in the late 1940s when it came up with a treatment for chicken coccidiosis, a parasitic disease. Merck followed up with antiparasitics for turkeys and livestock.[35] In Great Britain, expenses for veterinary medicines, which stood at £329,000 in 1930, increased dramatically after the war and reached £225 million in 1992.[36] Factory farming relied on the advances of pharmacology, and yet the boom was about more than scientific progress and a new set of tools. It was also about fear: the chickens that underwent inoculation in Oldenburg were perfectly healthy, and many other animals being treated with drugs did not show symptoms either. Just as with nitrogen fertilizer (see chapter 19, Synthetic Nitrogen), farmers would rather be safe than sorry, and that invited lavish use: after all, they might pay a hefty price for negligence. Antony Fisher certainly knew what was at stake when it came to animal diseases. He came into the chicken business after a devastating bout of foot-and-mouth disease had decimated his dairy herd.[37]

The case for pharmaceutics moved beyond prevention and cures. Researchers found shortly after World War II that low doses of antibiotics had a significant effect on the growth of animals. The discovery invited pervasive use, and farm animals consumed a quarter of antibiotics production in the 1950s, notwithstanding vigorous protest from physicians who feared microbial resistance problems.[38] It did not even take a conscious decision on the part of the farmer to use antibiotics for better performance: many feed companies offered ready-made mixtures. In fact, some products were hard to obtain *without* antibiotics.[39] The trend continued with growth hormones that found enthusiastic buyers among farmers. When the US Food and Drug Administration approved stilbestrol as a growth hormone for cattle in 1954, 20 percent of Iowa cattle farmers were using it within the first forty days of sales.[40]

It came down to a new rationale for the use of pharmaceuticals. Farmers no longer sought healthy animals but animals that brought top performance. Eliminating a disease was optional, but the gains in productivity were not. For example, the dairy cow disease mastitis remained prevalent in Britain in spite of intensive scientific research, but the effort succeeded in keeping a potential obstacle to increased milk output at bay, and this was ultimately all that mattered.[41] Competitive results without pharmaceutical support became a thing for romantics or farmers without business sense. In short, the inoculation in Oldenburg, while technically illegal, was simply a part of the new normal. It was surely not something that people should be questioned about in a court of law. At least that is how the German police saw things in 1962. They never brought charges against Heinz and Wilfried.[42]

The use of pharmaceutics was part of a comprehensive redesign of chicken rearing. Heretofore an animal that roamed freely on many farms, chickens were now living in an environment where everything was under human control: temperature, ventilation, water, feed, light, and the removal of feces. Of course, one could legitimately inquire whether battery cages, where space per animal was usually less than the size of a standard letter paper, gave chickens enough room to spread their wings, or reflect on whether animals should be more than a production unit. But if you did that, you were out of business.

Chicken farming was not a place for sentimental souls. Brutality was part of the business, though that was ultimately a matter of perspective. It took a lot of skill and knowledge to turn a chicken farm into a well-oiled machine that churned out eggs and meat at discount prices. A manual from the 1960s warned that compared with traditional egg production, cages called for more attention and diligence: farmers would pay a high price for the use of unskilled labor.[43] This was no longer a job that housewives could do in their spare time, and those who successfully ran a chicken farm could see that as a source of professional pride. One of the British pioneers, Geoffrey Sykes, relished the prospect of an upcoming "Golden Age of International Agribusiness" in 1963.[44] As the number of chickens in confinement multiplied, their fate hinged on a small group of self-confident professionals.

Factory farming relied on hidden resources, and one of the least recognized was energy. Agriculture had traditionally been a net producer of energy: it transformed solar radiation into plant growth and products for human consumption. However, the use of fuel-powered machines and energy-intensive inputs such as nitrogen fertilizer (see chapter 19, Synthetic Nitrogen) pushed the energy balance into the red, and the postwar oil glut (see chapter 15, Saudi Arabia) allowed it to go further. The factory farm was an engineering marvel where everything relied on cheap and abundant energy, and dependency did not end at the farm gate: in the United States, the energy requirements of processing and retailing are almost twice as high as those of agriculture alone.[45] Energy was not a problem in the industrialized countries that pioneered factory farming, but things looked different beyond the Western world: a 1971 poultry husbandry textbook published under the aegis of the United Nations Food and Agriculture Organization warned that battery systems required "an absolutely dependable electric power supply," for the engineering marvel would turn into a lethal trap if the lights went out.[46] When air strikes during the 1991 Iraq War caused blackouts in the country's 8,400 poultry houses, 106 million birds suffered a gruesome death.[47]

Change was equally dramatic when it came to the gene pool. Of course, humans had left their mark on the outlook of animals for many generations through selective use, and some interventions were dramatic long before the age of industry. Western agriculturalists introduced a new waste-fed pig breed from China in the early 1700s because European varieties, shaped through centuries of pannage, could not compete with its feed conversion rate.[48] But science-led breeding offered far more radical options. Once more, hybridity was a winning formula: as with hybrid corn (see chapter 28, Hybrid Corn), hybrid chicken breeds were in a league of their own.[49] It was no coincidence that the chicken was the first animal whose genome was sequenced in full.[50]

Just like pharmaceutics, breeding produced a skewed perspective on the animals in factory farming. Efforts focused on egg production with females and breast meat with males while other body parts were more or less surplus to requirements. Breeding pushed the limits of biology, and heretofore unknown problems arose: giant breasts caused roosters to topple over and chickens died from cardiac arrest because hearts could no longer cope with supersized bodies.[51] And then there was the fateful question of what to do about chicks that were born with the wrong gender. The biological profile of hens and roosters diverged in the 1950s, which meant that laying breeds performed poorly in meat production and vice versa.[52] There was no place for males in industrial-style egg production.[53] Under the factory farming regime, every other chick did not survive its first day.[54]

The industrialized chicken defined a path that other animals would follow: cows, pigs, turkeys, and fish in aquaculture. As soon as the modern agribusiness seized on a specific animal, the same rationale came into play: a systematic improvement of everything that could boost productivity, coupled with systemic disinterest in every other aspect of animal life. It all hinged on a powerful agreement between expertise and entrepreneurship: scientists stood ready to obliterate biological constraints, and farmers stood ready to adopt improvements swiftly, no sentimental questions asked. It was a relationship of trust, based on a shared style of thought and a shared language: both parties could say everything that mattered about animals in numbers (see chapter 7.2, Numbers Games). It was the foundation of a powerful network, and yet it was a consensus that worked only for certain animals. As it turned out, it did not work for green turtles.

Archie Carr had endorsed Mariculture, but it was not an emphatic commitment. He talked about Caribbean turtles as "a resource of major importance" in his 1963 book *The Reptiles* and declared that his nongovernmental organization, the Caribbean Conservation Corporation, was "planning pilot projects" in "turtle farming," with fenced areas in shallow waters serving as "natural pastures" where "green turtles could be kept like aquatic cattle."[55] But in the same breath, he confessed his "mixed feelings" about the idea: "No price can be set for the things that have to be preserved."[56] His reservations grew when turtle farming turned from an idea into corporate reality. Carr did not enjoy the time he spent lobbying on Mariculture's behalf, nor did he appreciate businessmen who talked passionately about markets and profits. Wasn't conservation really a moral issue (see chapter 26, Kruger National Park)?[57] There was also no guarantee that turtle farming would really help the cause of conservation. Maybe a thriving turtle business would stimulate demand and thus encourage poaching? After all, it was not difficult to hunt turtles. You just had to wait on a beach at night during nesting season and watch out for turtles coming ashore.

Industrial rationalization could achieve many things in animal husbandry, but competing with fishing of wild stocks was beyond its means. In an article of 1974, David Ehrenfeld, a pioneering figure in conservation biology, argued that domesticated turtle would always be an expensive product: "No amount of production efficiency will eliminate the extra costs imposed by the biological and ecological peculiarities of green turtles; they are not economically homologous with chicken."[58] So did Mariculture perhaps need a new business strategy? In an article for *Audubon* magazine in 1972, Carr argued that when it came to turtle farming, "the only effort to be encouraged should be a nonprofit, government-sponsored campaign."[59] It was not an idea that went down well with the investors.

Carr eventually grew tired of the debate and withdrew from the frontlines of politics. Views among other conservationists differed, and the debate took a nasty personal turn when research at the Grand Cayman turtle farm became a discussion point.[60] And then there was the wider context of a burgeoning environmental movement that shaped discussions during the 1970s: a profit-seeking turtle farm was a difficult sell in this community, particularly among American environmentalists who routinely chastised businessmen for pollution, trash, and their general indifference toward the natural world. All the while, operating costs were eating up Mariculture's resources, and the company went into receivership in 1975.[61] A German couple, Judith and Heinz Mittag, bought the remaining assets the following year and continued the turtle farm project on a nonprofit base.[62] Just like Fisher, the Mittags had made a fortune during the postwar years, though their product lacked the moral ambiguity of chicken farming. Judith Esser Mittag had earned her money and eternal gratitude from the world's women by inventing the o.b. tampon.[63]

Decisions on turtle farming had long moved beyond Caribbean beaches at the time. The Convention on International Trade in Endangered Species (CITES) was finally signed in 1973 after a decade of diplomatic wrangling (see chapter 24.3, Getting Serious), but its meaning for turtle farming was undecided until two US agencies, the National Marine Fisheries Services and the Fish and Wildlife Service, imposed a complete ban on selling and trading turtle products within the United States in 1978. A lawsuit to reverse the decision failed in 1980.[64] The age of turtle soup was over, though more as a matter of practice than principle. In 1981, the Conference of the Parties to CITES passed guidelines for ranching operations with endangered species, and while the requirements were difficult to meet, the door remained open. In the 1980s, CITES even gave tentative approval to a green turtle farm in Suriname that never materialized.[65] When a social scientist interviewed thirty-eight marine turtle conservation experts in 1995, she discovered that a minority still found farming an option worth exploring. But at the same time, all these experts voiced reservations on moral and practical grounds.[66] As it stands, turtle farming is a conservation strategy that remains untested rather than discredited, and doubts linger about whether things would develop a life of their own once the commercial spirit was out of the bottle. The experience of battery chickens is not poised to dispel these doubts.

3. LIBERATION MOVEMENTS

Great Britain was the pioneer in battery cage technology. It was due to the weather. Unlike egg producers in California, British farmers could not keep their chickens outside year-round, and key innovations such as multiple-story cages and automatic feeding systems grew out of a need to optimize the use of indoor space, a need that did not exist in milder climates.[67] Conditions in other European countries were similar, and in the early 1950s, the first continental chicken farmers were buying British equipment. One of the first West German facilities stood on the outskirts of Düsseldorf, and officials came for a visit in the fall of 1952. They found much to admire. The chickens were cackling happily, they rarely grew sick, and they were prolific layers of eggs. It looked like a promising endeavor in a society where memories of the hunger years after World War II were still fresh, and yet the officials felt that battery cages also raised an issue of an entirely different nature. Was this a case of cruelty to animals?[68]

Battery chickens were controversial from the start, and they invited the worst of comparisons. A columnist for the British magazine *Spectator* found broiler houses "powerfully reminiscent of concentration camps" in 1963: "There is inevitably something squalid and degrading about the poultry business because it isn't possible ultimately to handle large numbers of birds efficiently without falling into the sin against the Holy Ghost of treating living creatures as things."[69] The following year, a British Quaker, Ruth Harrison, published *Animal Machine*, a book that was quickly translated into Danish, German, and Dutch. Meticulously researched and arguing in a calm yet determined manner, it showed how the new farming methods confronted Western societies with issues beyond traditional notions of cruelty to animals.

And just like Rachel Carson (see chapter 38.2, A Matter of Humility), who wrote a foreword for the book, Harrison was better at raising questions than at providing answers: she was wondering, "How far have we the right to take our domination of the animal world?"[70] *Animal Machines* prompted the British government to set up a committee, and the product of its deliberations, named the Brambell Report after its chairman, became a landmark document "that launched animal welfare as a formal scientific discipline."[71] Soon after bringing battery cages into this world, the motherland of animal protection was exporting its critique.

However, the path from sentiments to policies was fraught with complications. As political concepts go, animal welfare was amenable to a particularly broad range of interpretations. The Brambell Report declared that animals "should at least have sufficient freedom of movement to be able without difficulty, to turn around, groom itself, get up, lie down and stretch its limbs," but there were other dimensions of animal welfare.[72] Shouldn't animals be healthy in the first place? Producers of battery cages argued along these lines from the beginning, insisting for instance that "the health of the birds is improved because of the controlled feeding" and that automatic watering at regular intervals "obviates boredom and monotony."[73] Was traditional chicken husbandry really superior? When the British government created the Brambell Committee, the British Chicken Association sheepishly suggested that it "should also investigate cruelty arising out of keeping animals extensively with particular reference to the effects of bad weather, and attacks by foxes etc."[74] And what about animals that hurt each other? Pecking orders are not just a matter of metaphors in the chicken world. Or was the entire debate pointless, as it revolved around "an interpretation of animals' feelings which can only be a matter of individual opinion," as a British official argued when the Ministry of Agriculture began to reflect on its response to Harrison's book?[75] When an animal welfare activist suggested in 1970 that the behavior of battery chickens suggested a high stress level, Siegfried Scholtyssek, a German professor of agricultural science and coauthor of *The Chicken and Poultry Bible*, dismissed the observations on methodological grounds: they lacked "exact quantification" (see chapter 7.2, Numbers Games).[76]

The chicken farmers certainly found that there was no need for discussion. From their point of view, Harrison's book was a threat to their business model, if not to their masculinity.[77] Chicken was a business for real men, and an American chicken magnate, Frank Perdue, even came up with a testosterone-fueled advertising slogan: "It takes a tough man to make a tender chicken."[78] In the view of Britain's National Farmers Union, *Animal Machines* called for a return to the dark ages of premodern agriculture: "It is unrealistic to expect the sophisticated food needs of a rapidly increasing and affluent 20th century population to be provided for by 16th century methods."[79] As to Antony Fisher, he deflected criticism by pointing "to the care his company had taken in keeping his chickens warm, fed, watered and submissive as a perfect example of the welfare state."[80]

The debate lacked a moral consensus, and it was about a lot of money. The Brambell Report jeopardized the investments of British producers, and it raised inconvenient questions about national regulation in a globalized market. The National Farmers Union felt that the import question was the crux of the matter: "Unless the Government can find some way of imposing comparable conditions on our overseas competitors, it has no right whatsoever to bring in legislation to shackle our own producers."[81] The government's response was to create a second committee, this time with staff members, that quietly watered down Brambell's suggestions. Whereas Brambell had put the suffering of animals at the core of its argument, the second committee focused on what farmers were already doing: it codified—as an official noted in an internal letter—"present-day practice followed by those poultry keepers whom we regard as meeting a satisfactory standard."[82] Other governments followed a similar line of incremental minimalism: long delays, calls for ever more research, and much understanding for the needs of a highly competitive business became transnational characteristics of animal welfare policy.[83] A DDT-style ban on battery cages (see chapter 38.3, Banner Slogans), a long-standing demand of animal welfare organizations, remained beyond the scope of government policies until mad cow disease put the agricultural establishment on the defensive.[84] In 1999, a council directive of the European Union prohibited chicken rearing in battery cages by 2012.[85]

Factory farming was never popular, but its products found willing buyers, and concerns about public opinion were tempered by the experience that most consumers cherished inexpensive meat no matter what. Food scares were recurring events, but people usually returned to their carnivorous routines after a while.[86] But consumer apathy (see chapter 18.3, Consumers in Chains) was only half the truth: when alternative agriculture became a significant branch of agriculture in the late twentieth century, it turned out that factory farming was also remarkably effective in burning the bridges for other production systems. It was about the biological material: organic chicken farmers struggled with severe herd management problems because breeders had neglected animal behavior for decades—social life just was not a big issue in battery cages.[87] And it was about the large corporations that controlled breeding: they looked at the small size of the organic market and the development costs for an organic chicken breed and found that the investment would not pay.[88] And then there was the question whether consumers, accustomed to cheap food for decades, were actually willing to pay more for animal welfare. In short, organic farming suffered from a chicken-and-egg problem: farmers were reluctant to embrace other production methods because they were not sure about the consumer, and the consumer could not convince them otherwise for lack of alternatives.

What Antony Fisher's take would be on these recent developments remains anyone's guess. He left agribusiness after the turtle farm fiasco, restored his fortune by marrying a rich widow he met at a regional meeting of the Mont Pèlerin Society, and spent the rest of his life as "the Johnny Appleseed of the right-of-center think tank movement internationally."[89] The Atlas Economic Research Foundation that he established in 1981 "claimed to have had a hand in the formation of more than 275 free-market think tanks in 70 countries" some 30 years later.[90] Fisher would probably be surprised about the plethora of rules that surround the chicken business nowadays. Like other pioneers of chicken farming, Fisher told war stories about how the genetic stock of Britain's poultry industry grew from smuggling. He broke the law when he personally carried 24 fertilized eggs (draped as Easter eggs) from the United States to Britain in 1953.[91] But there are grounds to suspect that he would gravitate toward the organic business today. It would not be about sentimental attachments or even a later-in-life sense of remorse à la Karl Ludwig Schweisfurth (see chapter 18.3, Consumers in Chains). It would be about the spirit of enterprise. After all, there is very little freedom in the conventional chicken business these days: producing for supermarkets is mostly about the precise execution of predefined rules. But building a different kind of chicken business, less ruthless and more sustainable, calls for creativity and venturesome businessmen.

Of course, the organic chicken market is not a neoliberal's dream. It depends on rules for production and commodity chains that governments need to enforce. But maybe Fisher would prefer these policies over the lavish subsidies for which agricultural policy is notorious (see chapter 2.2, Power Games). It does not take a membership in the Mont Pèlerin Society to think that these subsidies support dubious projects. The government of the Cayman Islands continues to pour millions of Caymanian dollars into a turtle farm every year, and it is precisely the turtle farm that Fisher and his associates set up half a century ago.[92] As a British overseas territory, the Cayman Islands have a reputation for a hands-off approach to economic affairs, but governments can create subsidies for all sorts of reasons. In the case of the turtle farm, it was due to the Falklands War.

The British government set up a fund for veterans after the war, and Caymanians made the largest overseas donation. That earned the island a royal visit in February 1983. At this time, the turtle farm was scheduled for liquidation on May 1, 1983, but as the island's only tourist attraction above water, the government of the Cayman Islands made it a part of the royal itinerary. His Royal Highness Prince Philip, president of WWF International at the time, visited the facility while Queen Elizabeth preferred a nap on board the Royal Yacht *Britannia*, and the government bought the farm a few weeks later.[93] The project remains open to paying visitors, and the farm's website boasts about "world renowned research and conservation activities" (though some features—a petting pool, a water slide, and a crocodile trained to jump—raise doubts about the sincerity of academic ambitions).[94] Cayman Turtle Farm is one of the island's tourist traps—*Lonely Planet* (see chapter 22, Baedeker) has called it "the closest thing Cayman has to Disneyland"—and amenities include a dining option with a seating capacity of 140. The website does not say whether turtle stew, a local favorite, is on the menu. However, it does mention chicken.[95]

37

Lucky Dragon No. 5

Atoms without Limits

I. GLOBAL POLLUTION

Snow is not what you expect on a fishing boat in tropical waters. But March 1, 1954, was not a normal day on board the *Lucky Dragon No. 5*. It all began at a quarter to seven when Shinzo Suzuku disturbed his breakfasting crew with a startling observation: he had seen the sun rising in the west. Two hours later, the crew witnessed a slight drizzle that could have been snow if it had not been for the temperature. They went on with their work, only to develop a set of mysterious symptoms: itching, headaches, nausea, and diarrhea. It was all a bit unnerving, and the day's catch was not good either. They decided to return to Japan, where doctors were quick to come up with a diagnosis: radiation sickness.[1]

The mysterious sun was courtesy of the United States Atomic Energy Commission. America had detonated a hydrogen bomb, code-named Castle Bravo, on the Bikini Atoll, now part of the Marshall Islands, and the blast was stronger than expected. Fallout traveled beyond the predefined 50,000-square-mile exclusion zone and fell on the *Lucky Dragon*, some 160 kilometers to the east of Bikini. Within days of returning, all 23 crewmembers were hospitalized, one of them died a few months later, and the Japanese public faced an inconvenient truth. Nine years after the horrors of Hiroshima and Nagasaki, Japanese citizens were again victims of an American nuclear bomb.[2]

Japanese newspapers wrote about "ashes of death," letters from disturbed readers poured in, and the sale of tuna plummeted.[3] It did not help that the Atomic Energy Commission was secretive about the circumstances and that it reacted to press reports with suggestions that the *Lucky Dragon* was probably a Soviet spying outfit. The debate even made movie history when Godzilla hit the screen later that year as the reptilian equivalent to the mushroom cloud.[4] The debate spread beyond Japanese borders, and protest sprung up around the globe. Albert Schweitzer and Pope Pius XII expressed deep concern. The British Labour Party called for a summit with the Soviet Union to suspend nuclear testing. Jawaharlal Nehru asked for an immediate test ban in a formal address to the Indian Parliament.[5] His ambassador to the United States, Gaganvihari Lallubhai Mehta, told dinner party guests that "because the atom bomb was dropped on Asians, and the H-bomb tested in Asian waters," there was a feeling in Asia that Americans "did not value colored people's lives" in quite the same way as "white people's."[6]

The dangers of nuclear radiation had come into view soon after the German physicist Wilhelm Conrad Röntgen accidentally discovered the mysterious "x-rays" while working in his lab at Würzburg University on a Friday evening in 1895. Researchers found that exposure to these "x-rays" led to burned skin, loss of hair, and worse, and Clarence Dally, an assistant to Thomas Alva Edison, became the first person to die from ionizing radiation. A researcher at a General Electric lab, Elihu Thompson, even decided which part of his body he would miss the least, exposed the little finger of his left hand to radiation, and subsequently told everyone not to duplicate the experiment.[7] But these events ultimately fed heroic tales about science: the eminent researcher giving his life, at times literally, to the cause of progress. Marie Curie's four-year ordeal to produce radium made her a Nobel laureate, the heroine of more than one hundred children's books, and a victim of radiation-induced anemia.[8] In 1936, the German Roentgen Society unveiled a monument in Hamburg that carried the names of 169 men and women from 14 different countries who had died as martyrs for the advancement of nuclear science.[9]

37.1 The X-Ray Martyrs' Memorial in Hamburg, Germany. The inscription celebrates the men and women as "heroic trailblazers." Image, Frank Uekötter.

Things became more complicated during World War II when the United States launched the Manhattan Project. Wartime secrecy and fears about a German bomb project discouraged critical reflections during four hectic years, but when the scientists witnessed the world's first nuclear explosion in the New Mexico desert on July 16, 1945, they had second thoughts. Under the impression of the fireball and the shockwaves, Robert Oppenheimer, the wartime director of the Los Alamos laboratory, thought of a verse from the *Bhagavad Gita*: "I am become Death, the shatterer of worlds."[10] Another Los Alamos physicist, less familiar with Hindu holy scriptures, summarized his emotions more bluntly: "Now we are all sons of bitches."[11]

After Hiroshima and Nagasaki, concern was a transnational phenomenon, but it was difficult to isolate the feeling in 1945: the bomb was also the harbinger of victory or defeat. It was easier to keep things separate nine years later. More than thirty million people signed a petition opposing nuclear tests in Japan alone.[12] Concern focused on the terrifying prospect of thermonuclear war as well as on nuclear fallout, a problem that was virtually nonexistent until July 16, 1945. The unlucky fishermen showed the dangers of nuclear fallout, but contamination did not stop in the vicinity of mushroom clouds. Nuclear explosions sent fallout into the upper layers of the atmosphere, and measurements showed that it was circling the globe. The health effects were anyone's guess, but some estimates were frightening. Linus Pauling, a Nobel laureate in chemistry, suggested in 1957 that the tests had already caused ten thousand cases of leukemia and would lead to millions of birth defects in generations to come.[13]

Fallout was more than a new pollutant. It was also a pollution problem unlike anything people knew. So far pollution had been a strictly local problem. It was an issue for industrial neighborhoods or, at worst, entire cities. London smog (see chapter 16, London Smog) was bad, but one could keep it at bay by moving to the suburbs—which is what postwar citizens did in growing numbers, helped by cars and expanding road networks (see chapter 34, Autobahn). Ever since the rise of the water closet (see chapter 17, Water Closet) in the nineteenth century, containing and removing waste was standing practice in the industrialized world. There was no relief, spatial or other, from nuclear fallout, and that gave the issue a new quality: it was now possible to imagine pollution as a problem that affected virtually everyone on the planet. It is difficult to understand the concern over the global spread of DDT (see chapter 38, DDT) in the wake of Rachel Carson's *Silent Spring* without the preceding debate over nuclear fallout.[14] And DDT was not the last pollutant that matched the template. It worked equally well for other persistent organic chemicals, for the chlorofluorocarbons that were identified as the root cause of the ozone hole, or for the greenhouse gases behind global warming.

37.2 The *Lucky Dragon No. 5* on display in its museum in Tokyo. Image, Frank Uekötter.

The notion of global contamination was a perfect match for the waning class instincts of Western societies. Faced with rising affluence and a Cold War enemy that touted socialism, people lost interest in the fault lines that ran through society. A problem that transcended boundaries of class, nation, and ethnicity was exactly what these societies needed: fallout allowed Western societies to imagine themselves as part of one human race where everyone was invariably exposed. The notion of unlimited contamination struck not only an environmental nerve but also a social one. And as the crew of the *Lucky Dragon* knew firsthand, that notion was always more appealing as an idea than as a description of reality.

As a pollution problem, fallout was in a class of its own, and this provoked a remarkably swift reaction. For a start, the Atomic Energy Commission exerted more caution with future explosions, and Castle Bravo remains the largest bomb that the United States ever detonated in the atmosphere. In 1958, President Dwight D. Eisenhower announced a moratorium on nuclear tests, and the United States, Great Britain, and the Soviet Union agreed on a permanent ban on atmospheric testing in 1963. It was the typical halfheartedness of Cold War agreements (see chapter 24.3, Getting Serious). The United States engaged in a veritable orgy of testing before the deadline and set off more explosions in 1958 than the Soviets had since 1949.[15] Furthermore, the treaty placed no limits on underground testing, and the global count for nuclear explosions stood at more than two thousand half a century later.[16] However, the radioactive contamination of the global atmosphere declined, and fallout was fading from public view.[17] Even Godzilla turned around. Movies of the 1960s portrayed him as a superhero fighting against a diverse set of enemies that included a monster spider, a giant crab, aliens, and King Kong.[18]

Commemoration eventually advanced to more formal structures. After serving ten years as a training vessel for Tokyo University of Fisheries, the *Lucky Dragon* ended up in its own museum in Tokyo.[19] UNESCO inscribed the Bikini Atoll nuclear test site on its World Heritage List in 2010. Mass tourism (see chapter 22.3, Traveling Masses) is taking its toll: a travel website calls the Tokyo museum "a must-see for serious Godzilla fans," and the Marshall Islands opened the Bikini Atoll for scuba diving in 1996 since the place allowed "some of the most historical and fantastic wreck diving in the world."[20] But the enduring legacy of the *Lucky Dragon* stands in sharp contrast to the brevity of the diplomatic follow-up. The United States and Japan concluded negotiations over a settlement agreement within ten months of the incident, and the US government paid $2 million dollars in exchange for legal immunity. Memories of the *Lucky Dragon* stood in the way of another project.[21]

2. NUCLEAR COMPLICATIONS

In early 1956, the Hiroshima Peace Memorial Museum was not a place of peace. The museum had opened its doors the previous August as part of a comprehensive plan to commemorate the first nuclear attack, and residents were surprised that after just a few months, more than two thousand artifacts were removed from display. The museum sought to make room for a temporary exhibit, and it was a rather peculiar exhibit that was coming to town. The Hiroshima Peace Memorial Museum was about to host the "Atoms for Peace" exhibit, a joint initiative of the Japanese government and the United States Information Agency whose stated goal was to tell the Japanese people about the promise of nuclear power.

Was the Hiroshima Peace Memorial Museum the right place for such an exhibit? Faced with public outrage, the city authorities were on the defensive, arguing that they lacked space elsewhere in the city and that the removal was only temporary. The sponsors held a public symposium in March 1956 that allowed critics to let off steam. Tempers calmed down, and the museum management got its way. Atoms for Peace opened its doors in the heart of Hiroshima on May 27, 1956.[22]

The exhibition became a rousing success. It was the event of the year in Hiroshima, and visitors poured in by the thousands. Local media extolled the virtues of the "peaceful atom." It even swooned over the *hibakusha*, the survivors of the 1945 atomic blast. When they formed an organization a few months later, it enthusiastically embraced nuclear energy. The city decided to keep material about atomic energy in the museum after Atoms for Peace closed its doors. It was exactly the response that the sponsors of the exhibit had hoped for. Japan ordered its first nuclear reactor.

Atoms for Peace was an embodiment of American leadership in the 1950s. Announced in a speech by President Eisenhower before the General Assembly of the United Nations in December 1953, it offered support for the development of nuclear energy under the auspices of a benevolent US government. It drew attention away from military uses, showcased America as the world's leader in science, and strengthened international relations. It provided camouflage for resource dependencies, as America relied on imports of nuclear raw material. The networks that grew from the initiative were also helpful for intelligence gathering. As John Krige has argued, "International scientific exchange deftly reconciled the universalistic appeal to the pursuits of truth with the particularist needs of national security."[23]

Japan was one of many countries that fell for the lure of the peaceful atom. By 1961, the US Atomic Energy Commission had signed 39 bilateral agreements.[24] The Soviet Union set up collaboration programs of its own with the fraternal countries of Eastern Europe and sent a large delegation to the International Conference on Peaceful Uses of Atomic Energy (see chapter 24.1, Nice to Meet) that took place at the Palace of Nations in Geneva, Switzerland, in 1955.[25] The world was watching: more than 900 journalists came to observe 1,400 delegates from 73 countries, and 3,000 papers were published in the proceedings.[26] It all came down to a new transnational consensus. Every self-respecting advanced nation would need to engage with nuclear power.

But all the enthusiasm could not distract from one crucial point: in the 1950s, the blessings of the peaceful atom were more imagined than real. Even more, it gradually dawned on the nuclear community that practical experiences would be long in coming: Japan broke ground for its first reactor in 1956, but the facility did not come on line until 1965.[27] The United States had run an atomic power station in Shippingport, Pennsylvania, since 1957, a spinoff from an abandoned project to build a nuclear aircraft carrier. But with an electrical power capacity of 60 megawatts, Shippingport was a far cry from later facilities in the 1,000 megawatt range, and the station never produced electricity at competitive rates.[28] The economic case for nuclear power was completely open, and the famous promise of Lewis Strauss, chairman of the Atomic Energy Commission, that nuclear electricity would be "too cheap to meter" was essentially a shot in the dark.[29] All the while, the more enthusiastic proponents of the peaceful atom fostered dreams about nuclear locomotives and nuclear airplanes.

The nuclear community learned a lot as projects moved from drawing boards to reality, and some insights were frightening. Safety was a field for particularly inconvenient discoveries. Nuclear power hinged on chain reactions, and chain reactions could get out of control. When Enrico Fermi assembled a pile of uranium pellets and graphite blocks under the viewing stand of a University of Chicago football stadium and unleashed the world's first controlled chain reaction on December 2, 1942, he placed three staff members on a platform above the pile. They had buckets with a neutron-capturing cadmium solution at their feet that would have served as the last line of defense, which turned these men into all-time exemplars for the heroism of research assistants.[30] Later reactors had automatic control rods and emergency cooling systems. But would that rule out disasters?

From a technological point of view, emergency cooling systems are an engineering nightmare. They sit idle most of the time, but they need to work perfectly in an emergency: a light water reactor is otherwise bound for a meltdown. The nuclear community did not recognize the problem until the mid-1960s, and legions of experts agreed over subsequent decades that scientific wisdom, strict supervision, and redundancy would make a failure of emergency cooling a merely theoretical prospect.[31] The discussion came to a close on March 12, 2011, when the world watched unit 1 of the Fukushima Daiichi nuclear complex on the northeastern coast of Japan going up in smoke. A tsunami had triggered a collapse of the plant's power supply, and frantic efforts of operating crews, who even looted the parking lot for car batteries, could not stop the unfolding disaster.[32]

In theory, the late 1960s would have been a good time to pause for a moment and explore other reactors with superior safety features.[33] In reality, the late 1960s saw the worldwide breakthrough of light water technology. It was what the utility executive Philip Sporn called "the great bandwagon market." American power companies ordered twenty reactors in 1966 and thirty-one more in 1967.[34] It was a gamble, driven by wishful thinking and generous subsidies, with managers and politicians in the background who had grown increasingly nervous about previous investments. All the while, scientists closed the books on other visions for the use of the peaceful atom. Henceforth nuclear power would be nothing more than a somewhat complicated way to boil water.

Many utilities ended up repeating BASF's experience with synthetic nitrogen and the Haber–Bosch process (see chapter 19, Synthetic Nitrogen): if you try to get a large technological system running, chances are that the system will ultimately run you. Nuclear reactors became notorious for delays, cost overruns, and hair-raising accidents: a worker set the reactor control cabling at the TVA's Browns Ferry Nuclear Power Plant ablaze in 1975 when he searched for a ventilation leak with a candle.[35] Westinghouse and General Electric suffered heavy losses when they offered to build facilities at a fixed price.[36] Utility managers broke into a sweat when they considered that they would need to write off hundreds of millions of dollars if an operator turned the wrong switches. And since the 1970s, many countries had antinuclear protest movements that broadcast the latest failings of the nuclear industry far and wide.

The precise sequence of events varied from country to country. The pioneering US market was also the first to dry up, and some utilities converted nearly finished nuclear plants to fossil fuels.[37] France nurtured a distinct gas-graphite reactor system until the French electric power giant EDF switched to light water technology after Charles de Gaulle resigned from the presidency in April 1969.[38] Japan's quest for energy independence brought the country to invest more than $10 billion into the Monju breeder reactor project that produced costly accidents and little else.[39] Austria built a complete reactor at Zwentendorf near Vienna, held a plebiscite on nuclear power in 1978, and then mothballed the entire facility after the project's defeat at the polls.[40] Brazil signed an agreement with West Germany for the construction of eight 1,300 megawatt units in 1975, broke ground for two, finished one in 2000, and the second is still under construction.[41] Individual trajectories differed, but the outcome did not: a thriving nuclear community was nowhere to be found. As Leo Tolstoy might have said, every nuclear nation became unhappy in its own individual way.

Even the line between military and civil uses turned out to be more imagined than real, as nuclear programs could serve both purposes. The International Atomic Energy Agency set up a complex monitoring system to provide some clarity, but at the end of the day, military abstinence was a matter of good manners rather than technological capabilities. West Germany was a role model: it acquired all the essentials for a bomb including a stack of plutonium but never bothered to assemble one (until an antinuclear government engaged in a bunker-clearing effort).[42] Other countries such as Iran and North Korea maintained nuclear programs that have kept diplomats around the world on edge. The peaceful atom could all too easily turn into the sum of all fears.

By the end of the twentieth century, the former energy of the future looked more like the energy of the past. Any sense of utopianism had long evaporated, and the reality of nuclear power, apart from nuclear tests in India, Pakistan, and North Korea, was a fleet of aging reactors. Nuclear power was basically a solution in search of a problem, and proponents seized on global warming and energy independence as best they could. Some Western countries gave in and started new reactor projects—Olkiluoto in Finland, Flamanville in France, Hinkley Point in England, Vogtle and Summer in the United States—and they all showed the familiar symptoms: exploding costs, huge delays, and, in the case of Summer, cancellation when a third of the project was complete.[43] Even the most fervent proponents did not promise electricity too cheap to meter.

And then there was the issue clouding nuclear power development from the start: What to do with nuclear waste? In his book of 1963, *Change, Hope, and the Bomb*, the first chairman of the Atomic Energy Commission, David Lilienthal, had argued that it would be an irresponsible use of public and private funds to build nuclear power plants "until a safe method to meet this problem of waste disposal has been demonstrated."[44] More than half a century later, none of the world's nuclear nations has a permanent storage facility for spent nuclear fuel, and certainly not for lack of trying. Pursuant to the Nuclear Waste Policy Act of 1982, the US government launched a comprehensive search for a deep geological repository that focused on the Yucca Mountain in Nevada since 1987. The endeavor grew into an industry of its own that even inspired a new academic discipline, nuclear semiotics, as warning signs on a repository would need to reach many generations to come. Nuclear waste is a mortal danger for thousands of years, which inspired discussions on how to tell future civilizations to keep off.[45]

After spending nearly $15 billion on the project, and faced with prospective additional costs for Yucca Mountain between $41 billion and $67 billion, the Obama administration pulled the plug in 2009. It also eliminated funding for the Office of Civilian Radioactive Waste Management, resulting in a significant loss of expertise and institutional memory.[46] The decision was rife with symbolism, and it certainly matched a nuclear history beset by an endless cataclysm of complications. Some sixty years after Eisenhower set out to win the "hearts and minds" of the civilized world, the only chance for the nuclear community was to forget.

3. UP IN THE AIR

Eight decades have passed since Enrico Fermi removed the control rods from his improvised Chicago reactor, and the business case for nuclear power is clear: no commercial reactor has ever been built without lavish government support. The picture is more complicated with a view to the human toll. The British medical journal *Lancet* published an assessment of the health effects of nuclear weapons testing on the seventieth anniversary of the bombings on Hiroshima and Nagasaki, but it was fraught with ambiguities. The article was based on models with a number of unknowns including the baseline cancer rates for radiogenic malignant diseases. Its 95 percent confidence interval for thyroid cancer was so wide that it allowed for a range of more than 200,000 potential victims. It focused only on US residents who were living at the time of testing even though fallout never cared about America's borders. And then, what does it mean that nuclear fallout probably caused 49,000 extra cases of thyroid cancer and 11,000 deaths from non-thyroid cancers, with maybe a tenth of the latter figure claimed by leukemia?[47]

For cancer researchers accustomed to viewing mass death through numbers (see chapter 7.2, Numbers Games), these figures were little more than a statistical blip. In light of more than 400,000 cases of thyroid cancer in the absence of fallout, the effect was indeed "small."[48] But this perspective was open to challenge, and not only if you happened to be a member of that small group. There is no moral consensus on nuclear power, and plenty of experience with the ensuing problems. For one thing, nuclear power defies conventional ideas about individual responsibilities: no one was held accountable for the Fukushima disaster.[49] For another, issues like nuclear fallout defy the routines of democratic decision making. Most of the victims never had a chance to vote on a government that detonated nuclear bombs in the atmosphere. Even if you lived in the United States, Great Britain, or France, you were not necessarily in a position to hold your rulers accountable: babies received by far the highest doses of radiation during the 1950s, and nuclear testing had moved underground long before they could cast a vote. Nuclear power was always a challenge for democratic accountability, as the consequences of decisions extend far beyond election cycles. It was tempting to achieve short-term gains at the expense of long-term liabilities: the Cold War that inspired Eisenhower's Atoms for Peace initiative is history, but its radiating leftovers are not. We may never know the precise toll of nuclear power in terms of cancer rates and cancer deaths. But we do know that it takes its toll on democracy's soul.

Nuclear science has always been unabashedly elitist, down to fantasies about the need for a "nuclear priesthood."[50] Proponents banked on the myth of scientific and technological progress, and they usually stayed aloft of costs and other petty details: the allure of nuclear power was always at its best on the level of principles. Historians have taken a lot of air out of claims of scientific rationalism and showed how decisions grew out of negotiations between experts, politicians, and stakeholders.[51] The producers of *The Simpsons* took it from there, and the vicious Mr. Burns and his Springfield Nuclear Power Plant became so notorious that the US Department of Energy has a web page with "7 Things The Simpsons Got Wrong about Nuclear."[52] There is probably no other industry that has done so much to undermine popular beliefs in the inherent link between science and civilization—the supreme irony of nuclear history.[53] But these scholarly efforts have not succeeded in exorcising the nimbus of scientific rationality, and advocates of nuclear power continue to invoke notions of "irrationality" and "emotionality" in the face of opposition, arguments that are frequently laced with anti-feminine stereotyping and whiffs of technological determinism. Jacob Darwin Hamblin has argued that "nuclear fear" was "the most prominent and apparently immortal motif of nuclear history," and the trope's resilience to historical experience is remarkable indeed.[54] It provides camouflage for a peculiar kind of speechlessness.

In the aftermath of Japan's 2011 nuclear disaster, much attention focused on the "Fukushima 50": the handful of employees who stayed amid the smoking ruins in the days after the explosions. Their heroism earned them worldwide praise, and a US documentary set out to tell their "true story" a year later.[55] However, the Fukushima 50 were only a small part of the post-disaster workforce. About 30,000 people were engaged in cleanup operations during the first year alone, and almost 90 percent of them were subcontract workers. The Japanese nuclear industry had a standing practice of hiring workers on short-term contracts who were typically exposed to higher levels of radiation than ordinary employees, and the ranks of jobseekers had swelled in the wake of the 2008 financial crisis. Accountability dissipated in a maze of contracting and subcontracting—one study found 618 companies involved in Fukushima, including some engaged in fourth-level subcontracting—and the labor safety board cannot provide independent oversight. Access to Fukushima is tightly controlled, and the agency needs to announce upcoming inspections to the utility.[56]

A thick tapestry of narratives surrounds nuclear power nowadays, but none of them matched the fate of these workers. They were no scientists, engineers, or otherwise notable people. Quite the contrary, it was the complete absence of social status that landed them in their jobs. They worked on the site of a traumatic disaster. They did not contribute to Japan's military power, the undisclosed ambition behind the country's nuclear policy: in light of nuclear bombs in China and North Korea and tensions in the South China Sea, a Japanese bomb is possible, notwithstanding vigorous protests from the dwindling cadre of *hibakushas*. And when it comes to radioactive contamination, Western societies preferred to stick to the cliché of pollution without limits from the fallout days. It had happened before: whereas Japan mobilized its reserve army of labor, the Soviet Union mobilized a real army in the aftermath of the 1986 Chernobyl disaster, and the ranks of the so-called liquidators grew to at least half a million. They received gratitude in the style of the late Soviet Union: the government commissioned a special commemorative medal, produced a grand total of 5,400, and then defined quotas for their distribution.[57] Western environmentalists did not show much interest in the liquidators either and focused on recreational stays for children.[58] When the German sociologist Ulrich Beck published his 1986 book *Risk Society*, which stressed "the universality and supra-nationality of the circulation of pollutants," the response was enthusiastic, and the term entered sociological theory as well as popular parlance.[59]

Modern societies are remarkably reluctant to accept environmental discrimination as a dimension of social inequality. They can even ignore victims when the world's media are on full alert. When the United States finished its 1954 nuclear test series and the Japanese Ministry of Welfare concluded its inspections of fishing vessels, it turned out that it was misleading to focus all attention on the fate of a single crew. The *Lucky Dragon* was by no means the only ship that had experienced direct nuclear contamination.

It was one of ninety-seven ships.[60]

38

DDT

Learning from a Book

I. PANACEAS

Few substances can claim a more dramatic entrance on the stage of history than DDT. Paul Müller, a Swiss scientist in the employ of the Geigy chemical company, discovered the substance's power to kill insects in 1939. His superiors had instructed Müller to search for a new pesticide, and dichloro-diphenyl-trichloroethane—a chemical that surely called for a shorthand—was one of some 350 substances that he explored. Filed for patent in 1940 and subjected to field trials, Geigy started selling DDT commercially in 1942. The peasants of Switzerland became prolific buyers, but DDT's fame quickly reached beyond farming circles.[1] As enthusiastic users of science-based tools (see chapter 35.2, Mobilizing Science), the Allied powers became profligate users and employed DDT to fight malaria and yellow fever in the military theaters of World War II. In 1944, DDT brought a potentially devastating outbreak of typhus under control in newly occupied Naples, the first time that a typhus epidemic had been stopped in midwinter.[2] Millions of people gained firsthand experience with the substance in delousing campaigns, and its effectiveness made it a showcase for the miraculous power of modern science. Within a decade of his seminal discovery, Paul Müller was in possession of a Nobel Prize.

Pesticides were anything but new in the 1940s, but DDT stood out on a number of points. First of all, it was effective. Even minuscule quantities killed insects by wrecking their nervous system.[3] Furthermore, it was persistent and kept its toxic qualities over long periods of time. In fact, it was this characteristic that had alerted Müller to DDT's potential, as he kept finding dead flies in a container that he had treated just once.[4] DDT was also cheap, particularly when compared with competing products that were often based on precious metals such as copper and arsenic. The DDT molecule comprised only carbon, hydrogen, and chlorine, which were among the most popular ingredients of modern chemistry. Getting the raw material posed no problem, and synthesizing the compound was not particularly difficult either. American production reached a million pounds per month by 1944, and in 1945, annual production of DDT totaled 36 million pounds.[5]

Output continued to increase through the postwar years, as DDT found a growing range of applications. Immediately after World War II, it was used systematically in the fight against malaria-bearing mosquitoes in places as far-flung as Greece, Venezuela, Ceylon, and Sierra Leone. The effort became global when the World Health Organization (WHO) launched a campaign to eradicate malaria in 1955, a program that became "the largest undertaking to date of the WHO and the first-ever attempt to eradicate a disease" (see chapter 21, Cholera).[6] DDT also became popular in agriculture as farmers embraced the wonders of chemistry. Wherever we look on a postwar farm, we find agriculturalists exploring new technologies with unprecedented enthusiasm. Gas-guzzling tractors and combines revolutionized fieldwork, synthetic nitrogen (see chapter 19, Synthetic Nitrogen) rendered labor-intensive fertilizing techniques obsolete, battery chicken (see chapter 36, Battery Chicken) produced meat by using factory methods, and new herbicides did to weeds what DDT did to insects. Industrial chemistry played a role in the making of new technologies, and its solutions were cheap. A new range of chemicals became the farmers' little helpers, and as with a popular drug, the meaning of help changed with prolific use.

DDT provides a good illustration of how low costs transformed the general approach to pertinent problems. As long as insecticides were expensive, farmers and public health officials were inclined to focus on cases of imminent danger. But with a cheap substance at hand, there was no longer any need to wait until trouble arose. In fact, waiting and observing looked increasingly unwise. While the costs for pesticides plummeted, outlays for machines, seeds (see chapter 28, Hybrid Corn), and other investments grew. Agriculture became a capital-intensive business, and the financial risks grew accordingly. So why should farmers take a chance and risk crop losses when prolific spraying absolved them of all worries? It all came down to a new rationale for the use of chemicals on the farm: lavish use of pesticides such as DDT became a no-nonsense insurance that no biological agent would jeopardize the investment.[7]

Compared with arsenic- and copper-based solutions, DDT looked relatively benign. In fact, it had stellar credentials as a savior of human lives since the well-publicized Naples campaign: *Life* published a picture of a public health officer spraying a half-naked toddler for delousing.[8] Yet warning signs were not long in coming. During World War II, researchers at the Food and Drug Administration and the National Institutes of Health conducted feeding experiments and painted the bellies of rabbits with DDT solutions: the results gave reason to be concerned.[9] Experiments studying the impact on wildlife were no more encouraging.[10] But then, long-term effects were understandably a minor concern for marines who stormed beaches that DDT had previously cleared of malaria-bearing mosquitoes. In fact, DDT was a perfect match for the mindset of military men: a chemical that promised comprehensive annihilation of deadly agents looked like the ideal companion in a fight for total victory.[11]

Elimination was a military concept, and those familiar with the inner workings of nature were hesitant to embrace it. "It is fully realized that such a powerful insecticide may be a double-edged sword, and that its unintelligent use might eliminate certain valuable insects essential to agriculture and horticulture," an otherwise enthusiastic article of January 1945 proclaimed, adding that DDT "might conceivably disturb vital balances in the animal and plant kingdoms."[12] Pervasive spraying could even render the pesticide ineffective, as it ran a huge risk of fostering new generations of resistant insects. In Switzerland, farmers observed the first resistant flies as early as 1945, and sales of DDT declined notably over the following years.[13] In the 1950s, officials imposed restrictions on feeding silage when crops had been sprayed, as DDT, readily soluble in fat, showed up in beef and dairy products.[14] It turned out that DDT was no panacea for insect problems, no more than guano (see chapter 8, Guano) or synthetic nitrogen (see chapter 19, Synthetic Nitrogen) had spelled the end of plant nutrition woes.[15]

However, the magic of DDT died slowly. Eradication remained a powerful goal, and it is a matter of debate whether this was due to the enigma of science, the militarization of American society, or simply to the fact that when you have a hammer, everything looks like a nail. When the WHO launched its malaria eradication campaign in 1955, the guiding sentiment was nervousness rather than optimism: the decision was driven by concerns about DDT's dwindling efficiency, and fears that the interest of governments might wane without a major initiative.[16] Two years later, the United States Department of Agriculture started a campaign to eradicate the red imported fire ant, *Solenopsis invicta*, an invasive species (see chapter 14, Cane Toads) that had entered the United States through the port of Mobile, Alabama, in the 1930s. Officials spent more than $200 million and sprayed more than 57 million acres before they gave up. By the end of the century, fire ants occupied some 300 million acres.[17]

In retrospect, it seems that Paul Müller's discovery was probably the easy part. The real challenge was to use DDT judiciously and encourage farmers to adopt best practices. However, getting to that point commonly took a long process of trial and error, of learning about intended and unintended effects, of trying different technologies and alternative chemicals, and of listening to the views of all stakeholders. With a bit of luck, people would learn over time how to maximize benefits and minimize side effects and then look back on the wild early days of pesticide use with a mixture of bewilderment and bemusement. The history of modern agricultural technology is essentially about learning out in the fields, about fine-tuning machines and work routines in a collective effort to gain experience and abandon unrealistic dreams.[18] These processes can span several generations, and the government's penchant for massive spraying campaigns suggests that when it came to pesticides, humans had barely begun their learning curve by the 1960s. But in the case of DDT, things took a different turn.

2. A MATTER OF HUMILITY

Pervasive spraying sounded like a good idea in government rooms, but things looked different on the ground. Unlike pesticide use in agriculture, eradication campaigns not only targeted farmland. They covered all potentially infested greenery irrespective of ownership, including private gardens. One such garden belonged to Marjorie Spock and Mary Richards of Brookville, Long Island. They had bought the garden in the early 1950s to practice organic agriculture and were dismayed when they observed spraying planes dumping DDT mixed in fuel oil on their garden plots in the summer of 1957—allegedly fourteen times in a single day. They filed a suit and did not relent until the US Supreme Court dismissed their case on a technicality in 1960. They had lost, but their trial caught the attention of a well-known author: Rachel Carson.[19]

Carson had previously heard that there were problems with DDT. She had even proposed to write an article for *Reader's Digest* as early as 1945 when she learned about the first experiments on DDT and wildlife. However, *Reader's Digest* was not interested, and Carson focused on other topics. A trained marine biologist with a degree in zoology from Johns Hopkins University, she wrote passionately about the sea. Much of her early writing was public education material for the US Bureau of Fisheries, but she also managed to publish two books, *Under the Sea-Wind* in 1941 and *The Sea Around Us* in 1951. The latter book became a best seller, which allowed Carson to quit her job at the federal government and devote herself fully to writing.[20]

Given her career, DDT was not an obvious topic for Carson. It brought her away from the seashore and into the line of fire of powerful corporations, and she had little experience in either setting: the American tradition of investigative, "muck-raking" journalism was foreign literary terrain to her. But once she was on the topic, she showed a number of important skills. She was familiar with academic publications and knew how to weigh evidence. She had learned to combine beautiful writing with scientific rigor throughout her career. She knew how to capture the attention of readers. And she stayed on the job. Five years after spraying planes flew over that garden on Long Island, her book, enigmatically titled *Silent Spring*, was available in print.

It became an immediate sensation. There was none of the awkward silence that can plague so many authors after publication. The *New Yorker* published a condensed version over three issues in June 1962, and the topic touched a nerve among the American public. Scientists, lobbyists, and policymakers started trading opinions. As early as July 22, 1962, the *New York Times* noted in one of its trademark plays on words that *Silent Spring* was making for a noisy summer. In August, president John F. Kennedy answered a press question about the administration's take on Rachel Carson. And the book was not even out at that point.[21]

In a correspondence while she was writing, Carson confessed that she "shall rant a little, too."[22] But for all the passion that she brought to the cause, *Silent Spring* was an eminently rational book: the balance of nature that she invoked found a match in clear, nuanced prose. The opening fable, in which Carson imagined a springtime village in eerie silence due to the absence of birds, set the tone. When she played with emotions, concern and compassion were more important than fear or hate. After almost two decades of DDT use, she could also draw on a wide range of insights from other researchers. There was no need for further investigations: she made her case based on the numerous reservations and caveats in the literature. As James Whorton wrote in *Before Silent Spring*, Carson "raised the previously simmering discontent with DDT to a boil."[23]

When it came to emotions, things looked different on the side of her opponents, who resorted to frantic attacks. The Velsicol Chemical Corporation even tried to pressure Houghton Mifflin into not publishing the book. Mocking her evidence became common sport in trade journals and parts of the press, and the more vicious critiques targeted her personally: an unmarried woman outside the institutional structures of science and corporate America was a threat to established hierarchies.[24] But the massive response could easily be taken as a sign of a bad conscience, and it surely fed a time-honored narrative. When it came to Rachel Carson and the chemical industry, the plot was familiar ever since a Jewish sheepherder named David volunteered to fight against Goliath.

By the 1960s, expert committees had taken the place of slingshots when it came to settling an argument, and Carson's book gained a particularly prominent one. John F. Kennedy asked his President's Science Advisory Committee to look into the matter, and the ensuing report, "The Uses of Pesticides," shared many of Carson's concerns. A sympathetic CBS broadcast, an appearance on *The Today Show*, and two Senate hearings put the finishing touches on Rachel Carson's public image. Against the background of hyperventilating lobbyists ranting about bogus science and hysteria, her calm presentations made a lasting impression: she sounded reasonable, and her concerns clearly deserved deeper investigation. In December 1963, Rachel Carson received prestigious awards from the National Audubon Society and the American Geographical Society and was inducted into the American

Academy of Arts and Letters. But she could not enjoy her vindication for long. She died of cancer on April 14, 1964, less than a year after the Senate hearings.[25]

Her untimely death sealed Carson's transfiguration into the patron saint of the nascent environmental movement.[26] She simply had it all: meticulous research, beautiful writing, personal modesty, and a proven readiness to speak truth to power. Her death also absolved her from the perils of drafting policy proposals. Carson had been notably vague in these respects. She offered some ideas in the final chapter of *Silent Spring*, but it read more like a laundry list of ongoing research: her conclusion was that there was really "a truly extraordinary variety of alternatives to the chemical control of insects."[27] She never called for a ban on DDT, and not just because she knew that the vested interests were only waiting for such a remark. After all, Carson was not just concerned about DDT. She was concerned about nature, about the web of life, and about the folly of intervening in it with brute force.

That vision continues to inspire people around the world. It also continues to escape those who criticize her. Ted Nordhaus and Michael Shellenberger called *Silent Spring* a "polemic against chemical pesticides in general and DDT in particular"; others labeled it "a diatribe against chemicals."[28] Carson might have smiled knowingly: her book was, if anything, about why people *fall for* diatribes. In her view, diatribes were tantamount to intellectual capitulation, as nature undercut intellectual obsessions through its sheer complexity. She retained a sense of wonder for the natural world throughout her life, and while she carefully screened the insights of scientific research for her book, she felt that science only went so far in understanding the inner workings of nature: "The fabric of life," Carson wrote upon concluding *Silent Spring*, was "on the one hand delicate and destructible, on the other miraculously tough and resilient, and capable of striking back in unexpected ways." In the end, it was all a matter of mindsets. What she was missing among those who ordered pervasive spraying was "humility before the vast forces with which they tamper."[29]

In essence, *Silent Spring* was not a book about politics. It was about the intellectual foundations on which wise environmental decisions would build. Her Senate testimonials showed that she held no contempt for the world of politics, but it was all part of something bigger. DDT, and pesticides in general, were merely a showcase to demonstrate that it was futile and dangerous to seek "the control of nature"—which was the draft title of her book. But then, an example can develop a life of its own.

3. BANNER SLOGANS

For all the excitement swirling around Carson and her work, it had no immediate impact on pesticide use. Several studies and counterstudies were in the news over the following years, either exonerating pesticides or calling for tougher controls, but decisions were long in coming.[30] The sale and use of DDT remained legal, and fantasies of pest elimination emerged from the brawl as if nothing had happened. From July 1963 to April 1964, the US Department of Agriculture sprayed more than 1.2 million acres in yet another eradication campaign, this time banking on a pesticide named mirex as the weapon of choice.[31] In the United States, total use of synthetic pesticides increased from 320 million pounds in 1964 to 880 million pounds in 1982, when markets finally showed signs of saturation.[32]

It took a change of the general context to instill action. In the United States and elsewhere, environmentalism emerged as a public force during the sixties, and DDT was one of its chief concerns. The United Nations decided to hold a large environmental conference in Stockholm in 1972 (see chapter 24, 1970 Tokyo Resolution), and that prompted Western governments to brush up their environmental credentials. Sweden and Norway enacted a complete ban on DDT in 1970. Under pressure from a campaign by the Environmental Defense Fund and other groups, the United States outlawed its use in 1972; the founding head of the US Environmental Protection Agency, William Ruckelshaus, announced the decision at the Stockholm summit. By that time, 1.35 billion pounds of DDT had been used in the United States alone.[33]

The rationale for the ban was political as much as scientific. Pollution was a defining issue of environmentalism during those years, and those deemed responsible, from factory owners to car manufacturers, were facing vigorous attacks.[34] But at the end of the day, it usually came down to a compromise: emission standards were set, filters were installed, and oversight was tightened. Negotiations were long and complicated, and the vested interests were flexing their muscles on every occasion. The automobile industry, the prime culprit for photochemical smog (see chapter 16, London Smog), became particularly notorious for its hardball approach.[35] It was all very frustrating for a movement that felt pollution was evil. A comprehensive ban, if only for a single chemical, was just right to its taste.

The clash was ostensibly between the manufacturers of pesticides and the environmental community, but it also betrayed a gap between agriculture and an increasingly urbanized society. Rachel Carson had not written her chapter on alternatives to chemicals for nothing: she understood the need for pest control. But fewer and fewer people were working in agriculture, and Western societies grew increasingly oblivious of these issues: from the viewpoint of Western urbanites, pesticide use appeared as irresponsible tampering with toxics. Malaria and yellow fever did not scare them either, as these diseases were mostly gone from the West (see chapter 21.3, Disease Worlds). With that, attacking DDT looked like a simple case, a matter of common sense. In 2006, the senior news editor of *Nature Medicine* wrote that DDT was "possibly the most reviled chemical on the planet."[36]

What was usually forgotten in all these discussions was that DDT was already past its prime when Rachel Carson published *Silent Spring*. In the United States, the year of maximum use was 1959, as increasing competition and decreasing effectiveness made for a shrinking market share.[37] Other pesticides, less persistent than organochlorines but often more acutely toxic, began to claim the market, and manufacturers did not mind: the new formulas were usually more expensive. "DDT was a relatively easy target," the political scientist Christopher Bosso has argued.[38] The ban on DDT suggested forceful action on the pesticides front while keeping the more intricate issues in the background. In a world of monoculture, control of pests (see chapter 12, Boll Weevil) was beyond debate.

Things looked equally ambiguous outside agriculture. The campaign against malaria managed to reduce the extent of the disease, but like most human crusades against disease (see chapter 21, Cholera), it fell short of the stated goal of eradication. The campaign unraveled in the 1960s, not least due to a growing resistance of *Anopheles* mosquitoes to DDT and other insecticides.[39] The magic of the panacea was already gone before it was outlawed, and in a way, the ban provided DDT with a graceful exit. It did not leave the stage because experience had shown its limits. It left because environmentalists, citing Rachel Carson, wanted it to go. Time would tell that it made a difference whether one learned from experience or learned from a book.

For the environmentalists, the ban on DDT and other pesticides was meant to be the end of history. It merely became the opening of a new chapter. In 1978, frustration over the Environmental Protection Agency's ban on mirex prompted Tom DeLay, a pest exterminator from Houston, to launch a successful bid for election to the Texas legislature. Seventeen years later, he became the Republican majority leader in the US House of Representatives and pursued his conservative agenda with a zeal befitting his original occupation.[40] Free-market pundits pointed to the comeback of malaria in the Global South and suggested that by banning DDT, environmentalists were depriving people of their best defense. It did not take long for the debate to turn ugly. In his novel *State of Fear*, the late Michael Crichton had one of his protagonists declare: "Banning DDT killed more people than Hitler."[41]

The issue at stake was obviously more than pesticides. It was about cultural hegemony. The ban on DDT was one of the first great victories of the environmental movement and a resounding defeat for corporate America: the image of the chemical industry was never quite the same. A repeal of the ban would tarnish an environmental icon, reverse a landmark decision, and it held the promise of a return to the good old days when better living through chemistry was a real-life utopia. Whether it would help in the fight against malaria was completely open.

The new infatuation with DDT replicated the mythologies of the early years. DDT helped free the Pontine Marshes (see chapter 32, Pontine Marshes) of malaria after World War II, following up on the Fascist reclamation effort that had reduced the disease without quite vanquishing it, and the campaign was sold as evidence of American superiority.[42] However, success was due to more than DDT. Authorities conducted a broad campaign that included food aid (see chapter 31.2, Hunger, Modern Style), economic reconstruction, medical services, and political stability. "DDT made a vital contribution, but it was not the magic bullet that many observers thought they perceived," Frank Snowden has noted.[43] The effort also "took full advantage of the institutional and cultural accomplishments of the national campaign that had begun in 1900 and had continued almost uninterrupted for half a century," another context that official readings were loath to mention: after all, it meant giving due credit to the Fascists.[44] Malaria only disappeared from the Pontine Marshes because authorities relied on more than DDT: "Latina demonstrates instead the importance of interlocking multiple initiatives."[45]

The campaign is of more than historical interest, as malaria is an enduring problem in the Global South. In 1998, the World Health Organization, the United Nations Development Programme, the United Nations Children's Fund (UNICEF), and the World Bank launched the Roll Back Malaria Partnership to resume the global campaign that had collapsed in the 1960s. However, using DDT was not in the cards initially. The organizations had little choice, as the development agencies of Western governments refused to fund malaria control initiatives that employed DDT.[46] The WHO reversed its stance in 2006 and endorsed indoor spraying of DDT for malaria control, but it went to great lengths to minimize environmental exposure: DDT was only applied in small amounts to inside walls and ceilings.[47] It was a tightrope walk, both in malaria-infested environments and in the collective memory of the world. Nobody wanted to sound as if they liked DDT or failed to appreciate the legacy of Rachel Carson.

Using DDT in such a limited fashion did not calm the heralds of free enterprise, nor did it inspire them to support the notoriously underfunded fight against malaria, and neither should come as a surprise. The bigger question is how environmentalists will react to the return of DDT. It is, after all, a textbook case on regulatory strategy. Should environmentalists strive for a complete ban, or rather content themselves with controls that minimize impact? Answers usually hinge on resources, technological options, and political clout, and yet DDT highlights the symbolic overtones of this choice. A complete ban sounds determined and has a whiff of eternity; a management plan reeks of compromise and remains subject to the vagaries of future enforcement. A ban looks heroic. Management rarely does.

Characteristically, the issue did not become moot in US environmental policy when DDT use was banned in 1972. Five days before leaving office, President Carter signed an executive order that prohibited DDT exports. The Reagan administration quickly reversed that decision in order to liberate enterprise, but the corporate response was weak: in 1985, America had only two manufacturers who produced some three hundred tons of DDT, all of it for export, and both abandoned production a few years later.[48] DDT was ostracized again through the 2001 Stockholm Convention on Persistent Organic Pollutants (see chapter 24.3, Getting Serious), which aims to eliminate dangerous chemicals that may accumulate in the food chain. The convention specifically targets DDT, mirex, PCBs, and nine other substances; the list became known as the "dirty dozen."[49] The scientific case against persistent chemicals is sound, and yet one cannot help feeling that, from a political standpoint, a well-publicized ban on organochlorine pesticides is a nice way to distract attention from the far greater quandaries entailed by the regulation of other pesticides.[50] Even more, the treaty brought all those who want to talk about potential benefits into an awkward situation.

The Stockholm Convention allows DDT use for disease control but at the same time stipulates that elimination is the ultimate goal. It is a clause that fittingly illustrates the moral dilemma of bans: once you have defined something as evil, it is difficult to negotiate about appropriate use. The clause leaves those who want to fight malaria mostly in the dark as they try to come to terms with environmental sensibilities. Can we still define ways of responsible use for a substance that environmentalists have fought to ban multiple times, the latest ban taking the form of a landmark global treaty? Or is using DDT with a bad conscience the most that environmentalism will allow? Answers may differ, though it is certainly a testimony to global power relations that the global responsibility to ban DDT rests on much firmer legal footing than the global responsibility to fight malaria.

Those who seek a world without DDT are fond of citing Rachel Carson and *Silent Spring*; but then, what her take would be about ongoing discussions is anyone's guess. Her persistence as an environmental icon makes her stand in a strangely disconnected way from a world that has changed enormously. That goes for the issue of DDT: it is a long way from protests against airborne eradication campaigns to excitement about a few grams per square meter inside a house. And that goes for environmental rhetoric: it is difficult to imagine Carson embracing testosterone-fueled talk about a "dirty dozen." One of her biographers, Mark Hamilton Lytle, ventured to suggest that "Carson would have been among the first to support limited applications of DDT in order to save lives."[51] Others will surely disagree, not least because Carson was not a person who rushed to judgment. But through her literary work and through her life, she made clear that she would rather get wise than get loud.

39

TORREY CANYON

COPING WITH TECHNOLOGICAL FAILURE

I. THE WORLD WAS WATCHING

The weather was calm, visibility was good, and the captain had 40 years' experience at sea. The safe channel between the coast and the reef was 12 miles wide, and the area was dotted with lighthouses and radio beacons. Ships could check their position 40 miles away from land, and the *Torrey Canyon*, a tanker with a capacity of 120,000 tons, had all the equipment one would expect on a ship that had received the highest rating of seaworthiness from Lloyd's Register. Navigating between the Scilly Isles and the British mainland was a routine job, one of many on the long journey from the Persian Gulf to Milford Haven, a deepwater port in the southwest of Wales. But the *Torrey Canyon* never made it to Milford Haven. Its journey ended on March 18, 1967, around 8:50 in the morning on the Seven Stones Reef, an obstacle clearly marked on every navigational map.

Shipwrecks were nothing new on the British coast, and authorities reacted swiftly. A Royal Navy helicopter hovered over the scene within two hours, several tugboats arrived during the day, and salvage crews got to work on board. It proved to be a frustrating job. The *Torrey Canyon* lost about a quarter of its oil within the first few hours. An explosion killed the captain of a Dutch salvage team. The weather changed. The ship broke into three sections. After ten days, the situation had grown so desperate that the government called in the Royal Air Force. Dozens of warplanes flew bombing missions over three days to set the wreck ablaze and burn any oil that was still on board while officials were trying to figure out what to do about the oil slick that was washing up on the shores of Cornwall and Brittany. The world learned about a new type of industrial accident: the supertanker oil spill.[1]

Of course, industrial accidents were a familiar experience. When steam power emerged as the prime mover of industrialization, boilers exploded with unnerving frequency. Countless trains have derailed, clashed with obstacles, or killed innocent bystanders ever since, during the opening ceremonies for the world's first intercity railroad between Liverpool and Manchester, a locomotive ran over William Huskisson, a Member of Parliament for Liverpool.[2] The hazards increased with the growth and spread of large technological systems in the twentieth century. A few years after the invention of the Haber–Bosch process for synthetic nitrogen (see chapter 19, Synthetic Nitrogen), an exploding ammonium silo took more than 500 lives at a BASF plant in Ludwigshafen, Germany.[3] Two exploding ships destroyed one-third of the houses in Texas City, Texas, in 1947 and left 3,500 people injured, 405 identified and 63 unidentified dead, and 113 missing without a trace.[4] And these were peacetime events. When Japan was about to conquer the oil fields of Borneo in the British East Indies in January 1942, Royal Dutch/Shell set its refinery complex at Balikpapan ablaze.[5]

Disasters were disturbing, but they also held a certain fascination, and voyeurism was already a problem before the age of mass media. On September 11, 1881, curious locals flocked to the Swiss village of Elm where a quarry was about to collapse, and when the avalanche was bigger than expected many of the spectators were among the 114 victims.[6] A Texas railroad company even staged a train collision in 1896. Forty thousand people traveled on reduced tickets to a place near Hillsboro and watched two locomotives smash head-on, a spectacle that was slightly marred when unexpected boiler explosions killed two; Scott Joplin immortalized the event in a ragtime song.[7] The *Torrey Canyon* eventually won its own song as well, courtesy of Serge Gainsbourg, but only after making world news for weeks. The oil spill had all the ingredients of a great story. It was about the largest shipwreck to date, it offered suspense and drama, it unfolded over several days, and it helped fill the news gap over the Easter holidays. And it happened within a few hours' drive from one of the world's media centers.

Most crucially, the *Torrey Canyon* offered dramatic pictures: a huge ship breaking apart, military airplanes in action, polluted beaches, and dying seabirds. It became a new chapter in modernity's infatuation with disaster images. Whereas Voltaire and Rousseau learned about the 1755 Lisbon earthquake from written sources, modern observers called for photographic evidence.[8] As Susan Sontag wrote, "Being a spectator of calamities taking place in another country is a quintessential modern experience."[9] The *Torrey Canyon* was one of the first industrial disasters to gain extensive television coverage, and the images created a community of eyewitnesses that spanned the globe. Even more, the event produced the first draft of a cultural script that observers would follow for similar events in the future.

The most enduring legacy is the environmental icon of the devastating oil spill: the soiled seabird. Whenever a major oil spill occurred, photographers and television crews sought pictures of birds in oil, usually paired with shots of rescue workers trying to save them against all odds.[10] When the *Deepwater Horizon* disaster produced oiled birds with some delay, the media got visibly nervous.[11] Industrial disasters have a peculiar potential for iconic images, and in a global village, someone was usually there to take that definitive picture: chloracne on the face of an Italian girl after the Seveso disaster in 1976; the face of a dead baby in the mud in Bhopal; the Red Rhine after the 1986 Sandoz fire; and the exploding nuclear reactor in Fukushima (see chapter 37.3, Up in the Air). These images needed no further comment. But did they say it all?

Pictures are inevitably snapshots, devoid of sounds and smells, subject to the vagaries of light and perspective and framed by cultural codes. Pictures can lie and often have, as they are notoriously poor on context. They also offer cheap morals: viewers can engage in a kind of visual long-distance empathy, knowing that they may never meet someone involved. Even when pictures capture an authentic experience, their meaning can shift through repetitive use. Susan Sontag, in her critical reflections on the ambiguities of images as documents and moral icons, put it as follows: "While an event known through photographs certainly becomes more real than it would have been had one never seen the photographs, after repeated exposure it also becomes

39.1 The *Torrey Canyon* on Seven Stones Reef, March 1967. Image, PA Images / Alamy Stock Photo.

less real."[12] Modern societies adore pictures in disaster communication, but maybe just for lack of something better.

The moral ambiguities of images are particularly stark when they clash with firsthand experience. Dramatic visuals circle the globe within seconds, but they usually have a hard time locally: residents want other pictures or, better yet, none at all. Seen from the ground, global attention is an ambiguous thing, particularly when it brandishes your home as a disaster zone. The *Torrey Canyon* was a case in point. When the French navy cleared the last spots of oil in early June and journalists lacked fodder for more headlines, a notable silence spread along the beaches of Cornwall. "The summer brought few complaints either from holiday-makers or fishermen," a British government committee declared just a few months after the greatest oil spill to date.[13] "There is hardly any memory of the public anxiety which prevailed during the period of crisis in the latter part of March and early part of April."[14] It was as if the disaster had never happened.

There were good reasons for forgetfulness. Hotel managers wanted to attract guests (see chapter 22, Baedeker). Fishermen wanted to sell their catch. Tourists did not want to hear tales of human suffering during the most precious weeks of the year. For the people on the ground, disaster memory was rarely the primary concern: they were more interested in getting their old lives back.[15] And it was not just about stubborn localism. People sensed that short-term solutions could turn into long-term liabilities. In the case of the *Torrey Canyon*, British authorities dispersed lavish amounts of detergents to dissolve the oil, which ultimately increased the damage to marine life.[16]

So what was the televised disaster really good for? It was easy to engage in visual long-distance empathy, but it also produced a distinct type of speechlessness. Disaster images stir emotions, but where do these emotions go? This was not a war or a crime with a clear culprit on which to focus one's anger. Nobody wanted the *Torrey Canyon* to end up on a reef, and the same held true for the fateful explosion in Fukushima and the deadly gas leaks in Seveso and Bhopal: the modern industrial disaster was completely and utterly senseless. For all the visual drama, reports on faraway disasters left most observers somewhat clueless. The last resort was usually a faint hope that someone, somewhere, might learn something from the incident.

2. DRAWING UP LESSONS

Four days after the crash, the British government convened an emergency committee of scientists.[17] Five days later, when the magnitude of the spill was becoming clear along the Cornish coast, the Plymouth Laboratory of Britain's Marine Biological Association decided to devote its full resources to a comprehensive investigation of the effects on marine life, and other expert circles joined them over the following weeks.[18] A flurry of reports and statements were published in the aftermath of the disaster, with issues and quality depending on the stakeholders involved. A Board of Investigation convened by the government of Liberia, where the ship was registered, focused on the immediate causes, questioned a grand total of six witnesses in closed session, and put all blame on the captain.[19] The British government studied the salvage effort in order to do better next time.[20] The Inter-Governmental Maritime Consultative Organization, precursor of today's International Maritime Organization (IMO), began looking at liability and compensation.[21] The scale of the event called for a thorough inquiry, and a desire to learn typically mixed with remorse about earlier failings. A former tanker captain offered his own reflections on the *Torrey Canyon* five years after the disaster and confessed up front that he himself felt "guilty of imprudences which no experienced master should commit."[22]

It says a lot about the priorities of twentieth-century societies that the most determined efforts focused on financial issues. Holding someone accountable for the damage turned out to be difficult. The *Torrey Canyon* was flying the Liberian flag, was owned by the Bermuda-based Barracuda Tanker Company, a subsidiary of the Union Oil Company of Los Angeles, the crew was Italian, and the ship operated on a single-voyage charter for BP.[23] When the British government tried to send the bill for the cleanup effort, the endeavor turned into a global cat-and-mouse game with a whiff of James Bond, culminating in a British lawyer's boarding a Barracuda-owned tanker in the port of Singapore and affixing a writ to the mast.[24] Seafarers liked their traditions, but that was clearly no way to deal with multimillion-dollar claims. To forestall government action, oil interests hastily created a Tanker Owners Voluntary Agreement concerning Liability for Oil Pollution.[25] However, the maritime nations of the world did not feel like relying on the goodwill of big oil, and two IMO conventions of 1969 and 1971 (see chapter 24.3, Getting Serious) created a framework for future oil spills. Amended in 1992 and 2000, they remain in force to this day and are widely acclaimed as a model for transnational law beyond oil. The *Torrey Canyon* disaster wrote the book on maritime liability.[26]

Progress was slower on the technological side. A 1973 International Convention for the Prevention of Pollution from Ships ignored hull design and focused on tank cleaning, a chronic source of water pollution. It took another world-class event, the *Exxon Valdez* oil spill in Alaska's Prince William Sound, to pave the way for double-hull tankers, and the US Congress took the lead with the Oil Pollution Act of 1990.[27] The IMO established similar standards in 1992, and the phaseout of single-hull tankers accelerated after two more tanker accidents, the *Erika* in 1999 and the *Prestige* in 2002.[28] However, the case for large supertankers remained off-limits in global negotiations. Supertankers were an outgrowth of the post-1945 oil boom, as they allowed for economies of scale (see chapter 10, United Fruit) on the long journey from Middle Eastern oil states (see chapter 15, Saudi Arabia) to Western consumers that traditional ships could not beat. The supersized tanker will likely remain a part of the maritime world until the demise of the age of oil.

But why did the *Torrey Canyon* end up on a reef in good weather? There was obviously no external cause like the iceberg of *Titanic* fame, so investigators focused on the chain of events on the bridge. There were tensions between the captain and his officers. A BP agent pressed the captain to catch the high tide at Milford Haven that evening. The captain had an undiagnosed tubercular condition. The crew failed to take a strong eastward current into account. Fishing vessels constrained navigation in the hour before impact. The third officer took bad bearings and failed to fix the ship's position for twenty minutes. An automatic steering system thwarted a last-minute change of course.[29] A number of factors contributed to the crash, and yet none of them doomed the ship: it was the interaction that turned a routine trip into a disaster. It was an unsettling conclusion for those who sought clear lessons, and an altogether characteristic one for complex technological systems. And then, as large artifacts go, there were more complicated ones than tankers.

The events on the *Torrey Canyon* bear a striking resemblance to those in the control room of the Three Mile Island nuclear reactor (see chapter 37, *Lucky Dragon No. 5*) in Harrisburg, Pennsylvania, during the early morning hours of March 28, 1979. A leaky seal in the water-cleaning equipment triggered an automatic shutdown of the main turbine, and the crew's subsequent decisions, all rational and harmless by themselves, led to a partial meltdown. It was the worst accident of a nuclear reactor to date. For the Yale sociologist Charles Perrow, it was a classic example of what he called "normal accidents": a complex technological system overwhelming human control. In Perrow's reading, accidents in nuclear power plants were bound to happen, and so it was with industrial chemistry and genetic engineering (see chapter 28.3, Business Models). For those reflecting on high-risk technologies, Perrow's *Normal Accidents* became mandatory reading.[30]

The great advantage of *Normal Accidents* was that it moved the debate beyond simplistic notions such as human error. It was the inherent complexity of modern technology, the combination of the close coupling of components and their nonlinear interaction, that bred disaster. The great disadvantage was that Perrow's model did not account for history: his narrative suggests that there is no escape from complexity. Perrow gave more room to learning experiences in subsequent books, for instance, noting that the number of serious incidents in US nuclear reactors fell from 0.32 per reactor year in 1988 to 0.04 in 1997.[31] Accidents often happen during the early years of a technology when operators are still early in their learning curve. The meltdown at Three Mile Island occurred after numerous shutdowns during the reactor's first few months, and the *Torrey Canyon*, built in 1959, was supersized to almost double capacity two years before it ended up on the Seven Stones Reef.[32] However, aging technological systems have their own challenges as well, and the US nuclear incident rate jumped up to 0.213 in 2001.[33] Technological risks know no place of grace.

People can learn from disasters. But then, it is always a small group of insiders who experience learning: industry experts, government regulators, and operating personnel. Outsiders usually find it hard to gain sufficient knowledge, and even harder to get heard in a community of insiders who know each other, and this makes for a gap that runs through the collective imagination of technological disasters. Insiders fear bad standards, improper procedures, and a loss of control; outsiders merely fear that the insiders will screw up. Insiders can do something; most outsiders remain passive, if only for lack of a choice. And in many fields, the circle of insiders grows smaller with the advancement of technology and a growing level of sophistication. When it comes to technological risks, most people are stuck with an uneasy sense of technological anxiety. Should they really trust a group of people whom they may never get to know? Those who have sat in an airplane during a stormy landing, or installed a mysterious security upgrade on their laptops, know the feeling.

3. CREATING INVISIBILITY

When a chemical plant spewed toxic fumes over the Italian town of Seveso in 1976, the disaster was front-page news across Europe. Six years later, the European Community passed a Seveso directive on the major-accident hazards of certain industrial activities, supplemented by a Seveso II directive in 1996 and Seveso III in 2012. No directive was ever named after Bhopal, the town in India where a Union Carbide plant blanketed adjoining slums with lethal methyl isocyanate. While the survivors of Seveso eventually gained a new hometown and monitoring by physicians, we do not even know the number of Bhopal victims. The Indian government put the death toll at 1,754, but various estimates suggest that the real number was up to and beyond 10,000, and somewhere between 200,000 and 300,000 people sustained injuries.[34] The quest for paper proof of victimization turned into a shoddy business that featured, among others, political activists, American liability lawyers, and corrupt middlemen.[35]

The age of electronic media created visibility within a matter of seconds. When an engine exploded on Southwest Airlines flight 1380 in April 2018, passengers pulled out their credit cards and bought in-flight internet access so that friends and relatives could follow their fate in real time.[36] But visibility in the global village was rather selective. The Santa Barbara oil spill of 1969 gained extensive television coverage, and so did the toxic waste scandal in Love Canal in upstate New York. Attention was already scarcer in other parts of the United States. It took a burgeoning environmental justice movement to draw attention to the disproportionate burden for African American communities in the American South.[37] In the Global South, cameras rarely found their way into the vicinity of dangerous plants, and that was about more than the number of journalists per square mile. Nondemocratic countries such as China do not appreciate their environmental toll being exposed, and they are not shy about enforcing invisibility.

Authoritarian policies are only one way to move technological hazards out of sight. In the 3 years before the *Torrey Canyon* disaster, 91 tankers had become stranded all over the world and 238 had collided with other vessels, resulting in 9 fires, 16 total losses, and 39 oil spills.[38] In other words, the *Torrey Canyon* was by no means the first disaster of its kind but merely the first one that passed a certain media threshold. And tanker accidents are only a small part of the overall problem. According to the *Oil Spill Science and Technology* handbook, they contribute less than 5 percent to the general problem of oil pollution. In the United States alone, there were "about 25 spills per day into navigable waters and an estimated 75 spills on land" in 2010, of which most spread more than 1,000 gallons.[39] And that, to be sure, is the result of several decades of incremental improvements along the entire commodity chain. The everyday oil spill will be with us for the foreseeable future.

And then the United States at least has reliable figures. Nobody knows the number of oil spills in the Niger delta, the center of Nigeria's onshore oil production. The plight of people in the oil region has become world news ever since the execution of Ken Saro-Wiwa and eight other leaders of the Ogoni people in 1995, which provoked worldwide protests against Nigeria's military rulers and Royal Dutch Shell, the most important multinational in the country. Western ideas about oil spills became fuzzy in the region. The concept of innocent bystanders was a dubious one in the Niger delta: gangs tampered with oil equipment in order to extort money from Shell, and the precise number of these incidents was hugely contested. Under the shadow of the Ogoni struggle, oil companies invested heavily in corporate social responsibility, though it was a matter of debate how much support actually reached the victims. State authority is notoriously weak in the oil region, and many researchers and nongovernmental organizations are equally loath to accept the risks of a field trip, particularly when they would need to ask questions about gang behavior. The two certainties are that pollution is chronic and that Shell has no plans to pack up. Corporate executives will not let go of Nigeria's oil, any more than shipping companies want to scuttle supertankers.[40]

The situation holds relevance beyond the Niger delta. Oil spills routinely provoke questions about the state as a guardian of public safety. One of the attractions that the *Torrey Canyon* held for journalists was the obvious failure of authorities: the conservative *Spectator*, never shy of criticizing the sitting Labour government, scoffed at "a nation unprepared."[41] But no such state exists in Nigeria. In fact, the situation is exactly the opposite. While technological failure cast doubt on the British state, the Nigerian case was about the state of public order (or lack thereof) challenging a large technological system. The quest for safety is turning into a two-front conflict: for better technology and more expertise—and for the material and immaterial resources that underpin the experts' work.

The Western imagination conceives disasters such as the *Torrey Canyon* as temporally and thematically limited events. "The *Torrey Canyon* disaster is, to all intents and purposes, a thing of the past," the committee of the British government declared in late 1967.[42] That notion is a dubious one in the Global South: disasters tend to endure, and they interconnect with all sorts of other issues. The Bhopal disaster was not just about industrial safety. It was about poverty, activism, and opportunism. It was about the Green Revolution (see chapter 28, Hybrid Corn) because the Union Carbide factory was in the pesticides business (see chapter 38, DDT). It was about economic policy: the contemporary Indian government was turning away from state planning and sought to build an open, neoliberal economy when the fateful cloud spread. And it was about doing business in an age of globalization. The legal repercussions of the disaster, which included prison sentences for company executives after a quarter century, have made investors wary of putting money into India.[43]

Industrial accidents know many losers, and they make few people look good. In the aftermath of the *Torrey Canyon*, the most heroic work was arguably that of people who took care of oil-stained birds. Thousands of ordinary citizens lent a hand, including many children who sacrificed their school holidays.[44] However, their work was tarnished by an abysmal success rate. Of 7,849 birds brought to a cleaning station, only 450 were still alive by mid-April, and the casualties continued to mount.⁴⁵ By autumn, the number of avian survivors was less than 100.⁴⁶ The effort was probably more important for humans than for animals.

Just like natural disasters (see chapter 25, 1976 Tangshan Earthquake), industrial accidents produced an overabundance of individual stories, and quite a few revolved around the rescue efforts. Aerial bombing remained the exception in the fight against oil spills, but the use of military resources became standard practice, and it was about more than manpower and equipment. Military discipline and military heroism were among the favored routines and tropes that people resorted to in the face of a disaster, down to notions of sacrificial death, though much depended on cultural context. For example, *Opfertod* fantasies held particular sway in Germany, as "an infatuation with hopeless situations has long been one of the characteristic features of at least one strand of German thought."⁴⁷ Those who had watched too many Wagner operas were particularly at risk. Hitler's suicide in a Berlin bunker was the farcical reenactment of a quintessentially German script.

But heroism in the face of disaster was always more myth than reality, a comforting narrative for survivors rather than an authentic driver of action, and that included those who made the ultimate sacrifice. Generations of German schoolchildren learned Theodor Fontane's poem *John Maynard* about a heroic helmsman who died a cruel death while steering a burning Lake Erie steamer to shore. The poem was based on a true story, but that concept was already a dubious one in the nineteenth century. In Fontane's account, all the passengers were saved except for John Maynard, who subsequently received a tearful funeral. In Lake Erie reality, a fire consumed a steamer departing from Buffalo on August 9, 1841, and the ship sank 4 miles off shore. Some 250 of the 300 passengers perished, and the first mate, whose real name was Luther Fuller, did not get a hero's funeral. He was among the survivors.⁴⁸

40

PLASTIC BAGS

EPHEMERALIA

I. GOING INTO PLASTICS

Atlantis had a problem. The crew aboard the space shuttle *Atlantis* was in the midst of preparations for its return to earth in September 2006 when it discovered a mysterious object floating nearby. Three years earlier, a chance collision with debris during takeoff had caused the disintegration of the space shuttle *Columbia* during reentry, and staff at NASA was edgy about everything that it found in the vicinity of a space shuttle in orbit. The crew conducted another examination of the shuttle's exterior and took pictures of the mysterious object, and when experts studied these pictures, they came to the conclusion that it was a plastic bag. NASA was wondering where the bag came from, both because it feared for the safety of *Atlantis* and because it was generally concerned about debris in space. After all, a plastic bag has enormous destructive power if it travels at a speed of more than 17,500 miles per hour.[1]

The bag was small, but NASA knew that size was a relative thing ever since Neil Armstrong had urged mankind to reflect on the size of his lunar footstep. NASA also knew a few things about synthetic compounds and their inherent risks. A malfunctioning rubber seal had caused the explosion of the space shuttle *Challenger* in 1986, and the *Columbia* disaster was due to insulating foam from the shuttle's main tank that hit the leading edge of its left wing.[2] The 2006 *Atlantis* incident added another chapter to a shuttle program that ended up providing insights into the ambiguities of plastics. But then, earthlings probably did not need manned spaceflight to learn about these ambiguities.

It is fitting that plastics figured so prominently among the 2.5 million parts that the shuttle system had on liftoff. After all, plastics, like the space shuttle, were a towering achievement of human ingenuity. Most types of plastics did not exist on planet earth until scientists found a way to build long polymers from small organic molecules. Bakelite, which ruled the market for a quarter century after its invention in 1907, is commonly cited as the breakthrough, as it showcased plastics' "protean versatility, its unique ability to become whatever one wanted."[3] The pace of innovation picked up markedly in the 1930s, and plastics entered the mass market in the postwar boom years as the quintessential marker of modernity and the American way of life. It was everywhere in affluent societies, in items from radios to pink flamingos, and it was understood that all this was just the beginning. As Dustin Hoffman learned in *The Graduate*, there was a great future in plastics.

There were many types of plastic, and they could be machine molded into all sorts of products. Earl Silas Tupper designed a brand of kitchenware in the 1940s, and his sales representative in Detroit, Brownie Wise, invented a new sales strategy, the Tupperware party, which became a fixture in suburban American life and made Wise the first woman to appear on the cover of *Business Week*.[4] Other synthetics revolutionized the market for textiles, and not just in the form of nylon stockings for women.[5] Plastics changed the way cars were built, and its share of vehicle weight in American cars grew from 0.6 percent in 1960 to 7.5 percent in 2000.[6] Plastics even revolutionized the face of money thanks to the credit card, an item so momentous that Neil MacGregor selected it for his *History of the World in 100 Objects*.[7] In 1979, American plastics surpassed steel in volume of production terms.[8]

The United States led the world into the plastic age, and other countries joined the bonanza. The new material also found friends behind the iron curtain. The GDR invested tremendous resources into the development of a plastics industry and made *Plaste* a defining part of Socialist consumer culture.[9] Like every postwar mystery, plastic called for some deep thoughts from a French intellectual, which Roland Barthes supplied in a chapter of his 1957 book *Mythologies*. He argued that plastic was "in essence the stuff of alchemy," for a common injection-molding machine easily achieved "the magical operation par excellence: the transmutation of matter."[10] Barthes depicted plastic as an unpoetic, "disgraced material, lost between the effusiveness of rubber and the flat hardness of metal," and yet it had abolished "the hierarchy of substances" and was out to rule supreme: "The whole world *can* be plasticized."[11] There was a sense of ambiguity in Barthes's remarks, but it was easy to miss if you were not into deep French thoughts, and this arguably held true for the majority of postwar consumers. Plastics were cheap, they developed no patina over time, and if they looked worn or broke, they went into the trash without much ado. Plastics were the perfect material for an age that sought to live in the here and now.

Among the many uses of plastics, packaging was one of the less prestigious ones. It was also among the uses that developed with some delay. Most types of plastic were expensive initially because companies sought to recoup the costs of innovation, and companies preferred packaging material that was cheap. In other words, the plastic bag was the ugly duckling of the new age of plastics, or rather the ugly duckling with scant hope for a transfiguration along the lines of Andersen's fairy tale. It was slightly unfair in technical terms, as it was no mean feat to develop a synthetic material that could serve the purpose. The new material had to be less rigid than Bakelite and less crisp than cellophane, tough enough for heavy loads and yet flexible enough for different cargoes. In the end, polyethylene became the material of choice, an invention of the British chemical giant ICI.[12]

Polyethylene was discovered by chance in 1933, and it took ICI six years to master the challenges of production on a commercial scale. Output increased dramatically during World War II, and companies looked for new ways to use capacities after 1945.[13] The fate of polyethylene was characteristic of synthetic polymers in general, as plastics were one of the winners of the war. Annual production nearly tripled in the United States from 1940 to 1945.[14] But military concerns already shaped the path of plastics in the years before World War II. The Nazis' quest for autarky pushed the German shoe industry into experiments with synthetic leather, which happened to provide them with a head start on world markets after 1945. The Nazis also provided shoe manufacturers with new ways of product testing. They built a test track at the Sachsenhausen Concentration Camp in 1940, where prisoners were forced to march on a meticulously maintained course with seven different surfaces from concrete to mud. With daily ordeals of forty kilometers and more, always under brutal supervision and with mandatory singing of German songs, few inmates survived for more than a few weeks.[15]

Plastics were a booming material, and yet consumer satisfaction was never a given. It took consumers time to warm up to new materials, fashions came and went, and then there was the symbolism. Plastics stood for a new civilization of consuming masses, and it served as a popular lightning rod for those who took issue with superficial materiality. In his best-selling book of 1960, *The Waste Makers*, Vance Packard argued that producers were replacing metals with plastics not just because they were cheaper "but also because their built-in colors help[ed] promote selling on the basis of style and impulse," which, in Packard's opinion, showed the manipulative powers of corporate America.[16] A few years later, Norman Mailer went on record suggesting that there could be "a malign force loose in the universe that is the social equivalent of cancer, and it's plastic."[17] From Mailer's point of view, plastic was more sinister than the FBI or the CIA: "Plastic tends to deaden people. It deadens their nerve ends. And when the nerve ends are dead, the mind is much more susceptible to manipulation."[18]

However, concerns about plastics also grew from more worldly concerns. In fact, it was the humble plastic bag that caused the first existential crisis for the burgeoning plastics industry. It was about the transparent polyethylene bags that DuPont had been selling to laundries across the United States since the mid-1950s. Introduced as a protection for clean clothes, unsuspecting parents allowed these bags to get into the hands of small children, who knew no better than to explore them with their mouths. The American Medical Association issued a warning in April 1959 after four children died from suffocation while playing with polyethylene bags, and the press was reporting further incidents. In the end, the death toll stood at eighty. Newspapers were calling for a ban, and legislators began to look into the matter. A polyethylene bag looked rather insignificant compared to the life of a toddler, but the industry sold a billion of them in 1958, and they brought in some $20 million.[19]

The Society of the Plastics Industry hurriedly put out advertisements warning of the hazard, but that did not settle the matter. The society went on to hire a Washington lawyer, Jerome H. Heckman, who traveled more than forty thousand miles to represent the industry's stance at hearings all over the country until he could report back that legislation had either been dropped or confined to warning labels.[20] Heckman continued to defend plastics against legal challenges and was eventually inducted into both the Plastics Hall of Fame and the Packaging and Processing Hall of Fame, and the accolade on the former's websites notes without a hint of irony that Heckman "has certainly been as creative as any inventor in his work in building and maintaining a business environment that has permitted the tremendous growth achieved by the plastics industry."[21] His work characterized the emerging default approach of US corporations to environmental regulation. American companies could marshal tremendous scientific and legal resources in the defense of their interests, and if that failed to make an impression, there was always money. When the Seattle City Council passed legislation for a $0.20 fee on paper and plastic bags in 2008, the American Chemistry Council put $1.5 million into a referendum that defeated the law, outspending environmentalists during the campaign by 15 to 1.[22]

2. CONSUMER CHOICES

Cotton bags were in use before the age of plastics, and they were about more than transportation. They also served as raw material for sewing done by women, particularly on farms, where fertilizer (see chapter 19, Synthetic Nitrogen), seed (see chapter 28, Hybrid Corn), animal feed, and other products came in sturdy large bags. Domestic sewing machines made bag reuse a standing practice in the mid-nineteenth century, and magazines for farm women offered helpful advice on cuts and paint removal. The National Cotton Council put out instructive booklets during the Depression in order to encourage reuse, and some millers and sugar producers made it part of their sales pitch. State fairs featured bag-sewing contests into the 1950s.[23]

Plastic bags were less amenable to the fashions of the day, but they had other advantages. They were light and clean. People could use them after shopping to collect their trash or to pick up dog feces. They could carry all sorts of commercial or political messages. And best of all, they were cheap, thanks to the advances of industrial chemistry and the postwar oil glut (see chapter 15, Saudi Arabia). They also met with the predilections of affluent societies where consumers no longer worried about discarding things after a few minutes of use.

Reuse of bags was always the province of those in need, but it mirrored a common appreciation of material resources. As long as bags were made of cotton or other textiles, people did not throw them away lightly. Bags could even turn into matters of grave concern in times of crisis. When German fertilizer producers ran short on bags during World War I, authorities set up a special agency for the procurement of sacks that they named, with all the earnestness that German authorities can muster, the *Reichssackzentrale*.[24] But this tradition of reuse and conservation came to an unglamorous end with the spread of plastic bags, which were so cheap that many retailers gave them away for free. As supermarkets revolutionized retailing, customers learned to expect complementary bags as a natural birthright, and they thought no more about the costs than they thought about lighting and air-conditioning (see chapter 20, Air-Conditioning) in the new cathedrals of mass consumption.

Mass production of plastic shopping bags began in the mid-1960s, and use continued in spite of the growing vigor of environmental sentiments and the increased costs of raw material in the wake of the 1973 oil price shock.[25] Environmentalists found it easy to criticize the boom, as plastics were firmly established in the public mind as a material of wastefulness, but it dawned on them that shopping bags called for more than the familiar critique of mass consumption. In 1975, a Swiss organization, Erklärung von Bern (renamed Public Eye in 2016), found a women's cooperative in Bangladesh that was stitching jute bags, set up a supply chain, and made "Jute statt Plastik" a standing phrase in the German language. The jute bag became an icon of the counterculture in Switzerland and Germany.[26] Using textiles instead of plastics promised moral clarity. But did it fulfill the promise?

The jute bag found both enthusiastic buyers and hostile reactions from vested interests, including the false allegation that it was tainted with DDT (see chapter 38, DDT).[27] The more pertinent questions were about the relative merits of different resources. Reusable bags are heavier than plastic ones for single use, and this is not the only difference that makes shopping bags a surprisingly complex issue. Researchers need to compare bags with different size and different materials, which in turn have different production methods, pollutants, and recycling potential. A 2011 study by the Environment Agency for England and Wales used nine different aspects, from resource depletion to smog formation, and assessing their relative importance was a difficult task. In line with the priorities of Western environmentalism in the new millennium, the Environment Agency put particular emphasis on global warming.[28]

The result was clear when it came to single-use bags. The common polyethylene bag easily beat the paper carrier and the biodegradable starch-polyester bag. The findings were more ambiguous when it came to the comparison with reusable bags, not least because the study focused on the material and ignored the human dimensions of the issue, and ambiguities increased even more beyond national contexts. In the age of globalization, Egyptian waste collectors followed the fluctuations of plastic prices in Shanghai while discarded plastic bags served as breeding grounds for malaria-bearing mosquitoes in Mali, and South Africa's minister of environmental affairs made bitter jokes about drifting plastic bags as the country's new national flower.[29] The global plastic bag is about livelihoods as well as energy balances and material flows, and the distance between places and lifestyles helps consumers ignore things like the environmental footprint of cotton. According to the report of the Environment Agency, shoppers need to use a cotton bag 131 times to get below the global warming potential of single-use polyethylene bags. A sturdy reusable polypropylene bag needs only eleven trips to the supermarket to break even, but unlike cotton and paper, petroleum-based products evoke a sense of unease among Western consumers. The study did not look into jute, which is more environmentally benign than cotton but suffers from a coarseness that has confined jute bags to the ecological shopping scene.[30]

For a long time, Western governments were reluctant to impose rules regarding plastic bags. They have been equally reluctant about the single-use concept more generally, and Western consumers can now buy one-way cameras and one-way cell phones.[31] Things look different in countries of the Global South, where one issue stands out: in places without an efficient waste management system, littering is the cardinal problem. Rwanda imposed a complete ban on plastic bags that the government enforced with an iron fist, and other African countries have followed suit.[32] It may be due to particular challenges. When Mauritania banned the use of plastic bags, they made up a quarter of waste in the country's capital, Nouakchott, and eating a plastic bag was the cause of death for more than 70 percent of the cattle and sheep that died in the city prematurely.[33] Or maybe African countries found it easier to enact policies because they did not have domestic producers of plastic bags. When Ireland became the first European country to introduce a €0.15 fee for plastic bags, the consequences for Irish industrialists were marginal, as the country was importing four out of five shopping bags from abroad.[34]

Researchers will likely continue to develop complex calculations of material flows and energy balances for the different types of bags, but the case for the Irish law was much simpler than that. It was about the landscape: Ireland is a windy country with a lot of hedgerows, in which discarded plastic bags were accumulating in great numbers. An energetic minister for environment and local government made the fee a reality after consultation with all stakeholders, and the result provided insights into the thoughts behind the plastic bag's popularity, or rather the lack thereof: use of plastic bags fell by more than 90 percent.[35] It would seem that most people were previously using plastic bags not because they needed them but because they were free.

3. DISSIPATIONS

In the end, the space shuttle *Atlantis* landed safely, but the drifting plastic bag remained a mystery. For all its diligence and brainpower, NASA never managed to get a clear idea of where the piece came from or where it was going.[36] It was an outcome that resonated with experiences back on planet earth. Plastic bags go many different ways, and their path can make a world of difference. It is an aesthetic problem when the bag ends up in a tree or on a beach. It is a flood hazard when the bag clogs a storm sewer. It threatens wildlife if the bag ends up in an animal's stomach. Yet few bags generate as much excitement as the ominous one outside the space shuttle *Atlantis*, and it is a matter of debate whether this is due to numbers, the decline of morals, or the fact that most people do not have resources on a par with NASA.

Drifting plastic bags are poised to evoke passionate comments about consumer carelessness, but they are about more than human will. They are about the laws of nature. According to the second law of thermodynamics, every isolated system is bound for a growing state of disorder. Dissipation is the natural tendency of everything from plastic bags to plutonium, and the best that humans can do is stem the flow. As Jens Soentgen has noted, "A pullover produces fluff, but fluff does not produce a pullover."[37] As it happens, fluff is the technical term among American car dismantlers (see chapter 5, Shipbreaking in Chittagong) for the shredder residue that includes plastic as well as rubber, glass, paint, oil, and dirt. Fluff is what is left of cars when the metals are removed, and it is considered a hazardous substance that goes either to the landfill or the incinerator. The United States produces some five million tons of fluff every year.[38]

Dissipation occurs easily in moving water. The Norwegian adventurer-explorer Thor Heyerdahl noted plenty of floating refuse, notably plastic, when he crossed the Atlantic in papyrus rafts in 1969 and 1970.[39] Marine debris became a running concern in environmental circles after the US National Marine Fisheries Service held a conference in Honolulu in November 1984 (see chapter 24.1, Nice to Meet).[40] Plastics figured prominently in this discourse after an early study found that 86 percent of marine debris in the North Pacific Ocean was plastic.[41] The region became notorious for what came to be known as the Great Pacific Garbage Patch after the sailboat captain Charles Moore noticed the enormous amount of plastics on his return trip from a yacht race and made the fight against marine debris the cause of his life.[42] It is safe to assume that a good part of the junk comes from plastic bags, as polyethylene is one of the few types of plastics light enough to float, and it is equally safe to assume that marine debris causes a range of problems for marine wildlife, though the extent of these problems remains a mystery. As the leftovers of our industrial civilization dissipate, so does our knowledge about the ensuing problems.[43]

A good part of environmental policy is about slowing dissipation. After all, chemicals can create unexpected problems when they drift to new realms, as the world learned when chlorofluorocarbons, an inert gas long used for spray cans and air conditioners (see chapter 20, Air-Conditioning), traveled to the upper atmosphere and damaged the ozone layer.[44] Landfills (see chapter 5, Shipbreaking in Chittagong) have emerged as a popular way to immobilize stuff, but they merely slow down dissipation, and their quality depends on constant monitoring and the efficiency of water collection and cleaning. Plastic bags will surely remain a fixture in the landfills of the world, for they disintegrate slowly in the absence of sunlight, and they are joined by many other types of plastics that may emerge as one of the more lasting remnants of our industrial civilization.[45] Archaeologists, who are famous for their love of waste heaps, may go into plastics in their own peculiar way a few centuries from now.

The plastic bag outside the space shuttle *Atlantis* may no longer be with us at that time. Objects in low orbit gradually lose speed and enter the atmosphere, where a light plastic item is bound to burn without a trace. If it does not collide with a spacecraft, the bag is unlikely to have an impact on human history, and yet it deserves consideration in light of the Western master narrative about technology and human supremacy. "Despite many setbacks, human victories over nature have outnumbered their defeats," Daniel Headrick proclaimed in his global environmental history.[46] But like so many things discussed in this book, that is a matter of perspective. It took humans thousands of years of technological progress to put a plastic bag into space. But once spaceflight was a reality, humans were unable to control the bag's movement, or even to conceive it with the cognitive certainty that NASA experts typically seek, and that arguably calls for humility when it comes to human power over the natural environment. If an institution with the scientific and financial prowess of NASA is unable to account for its plastic bags, maybe the narrative of humanity's triumph over nature was flawed from the outset.

CODA

The Pandemic

While this book was under review, a virus embarked on a global journey. First identified in the Chinese city of Wuhan, COVID-19 was an exceptionally lethal infectious disease, and on March 11, 2020, the World Health Organization declared the outbreak a pandemic. Governments around the world enacted drastic measures: borders were closed, public life came to a halt, and people were coaxed to follow social-distancing rules. There was no precedent in living memory, at least for those who lived in the Western world, and yet things looked eerily familiar to me. Unfolding events were the high-speed version of the type of processes that I had studied in previous years.

The following comments are not intended as a first draft of coronavirus history. I have sought to keep current affairs at bay throughout the writing process—distance is one of the more important intellectual assets of academic life—and I have no intention of changing into presentist mode toward the end. But with pandemic events fresh on everybody's mind, and with a pervasive speechlessness that ran through all the pandemic chatter, it might be opportune to highlight some of the basic narrative threads that any future history of COVID-19 will need to engage with. Most of all, this chapter seeks to debunk the myth that we have gone through exceptional times unlike any other. The events of 2020/2021 were what typically happens to environmental challenges in an interconnected modern world. It just happened faster than usual in this case.

Like every disease, COVID-19 had some peculiarities. It was more contagious and more insidious in its effects than other respiratory diseases, and global travel networks being what they are, the virus was soon present in many corners of the world. But none of these features determined the course of events. They merely provided the material context in which the drama of human history unfolded. Pathogens spread in utter disregard for human sentiments, but they interact with societies in all sorts of ways. As a result, the story of COVID-19 developed a multitude of narrative threads within weeks, and these threads developed lives of their own while entangling in several ways. In other words, COVID-19 produced another one of those webs of materialities, institutions, and meanings that evolve inside the vortex.

The story of COVID-19 began in a manner that is entirely characteristic for advanced modern societies: it fell to the experts to speak up first. In the twenty-first century, every challenge has a designated respondent in the system of professions, and the medical profession developed a decent first assessment with breathtaking speed. Within a matter of weeks, medical experts identified the virus, analyzed its genetic code, agreed on a set of symptoms, specified modes of transmission and effective prevention measures, and developed a test kit. In the nineteenth century, it took decades of medical research to understand the nature of cholera (see chapter 21, Cholera), and researchers were mystified for months when they discovered Legionnaires' Disease in 1976, but the mysteries of COVID-19 vanished so quickly that the room for debate narrowed down tremendously. Epidemiologists disagreed over details, and physicians tried different treatments, but there was nothing like the erstwhile debate over cholera, where different scientific opinions on miasmas and quarantines were politically charged. The expertise of the medical profession was beyond serious competition, and its assessments provided politicians with a sobering outlook: normal life would cause death rates to grow exponentially. The only remaining question was whether leaders were willing to face up to that reality or develop their own bespoke expertise. The latter approach inevitably ended with intellectual embarrassment.

CODA

Concerns over COVID-19 did not just grow from medical advice. A second narrative thread was about the media. As had been the case with the *Torrey Canyon* disaster of 1967 (see chapter 39, *Torrey Canyon*), it was the pictures that defined coverage, and photographs and videos provided the disease with a visual urgency that made it hard to ignore. Pictures of Italian military trucks with coffins went global, followed by images of mass graves in New York City, and they resonated against the backdrop of another set of images from overtaxed hospitals. It was the global village at work, including the hierarchies of attention that are sadly familiar: death by the thousands is not unusual in today's world, but it really should not happen to white people in a Western metropolis. Once iconic images were in circulation, governments had a mandate for drastic measures.

In the Western world, it fell to affluent welfare states to implement restrictions, and this became a third narrative thread. Western democracies were supposed to enact schemes to keep families and businesses afloat. Many voters viewed disaster relief as something akin to a birthright, even if that required rescue packages of staggering proportions. People naturally expected their governments to intervene in the event of a hurricane or an earthquake (see chapter 25, 1976 Tangshan Earthquake), they knew about the massive interventions on behalf of banks in the global financial crisis of 2007–2008, and that left elected leaders no choice but to allocate billions. Historians will surely dissect these policies in the future and analyze who suffered and who profited unduly—the first drafts of that history hit bookstores when this manuscript went to press—but a legacy of state-run disaster relief called for responses from governments that were virtually unthinkable a century ago.

Political context made for a fourth narrative, though one with plenty of individual threads. COVID-19 became enmeshed with different regimes and political systems, and the one communality was that the pandemic spread at a time when diplomacy and international cooperation were in crisis. Responses became entangled with the powerplay in totalitarian China, a polarizing president and an upcoming election in the United States, long-standing tensions between Italy and the European Union, the woes of Brexit Britain, and so forth. Conspiracy theorists had a field day, as did the snake oil salesmen of our day, and the battle against rumors and false information kept authorities on edge—unless they were spreading nonsense themselves.

The virus was first identified in China—the precise origins may always remain mysterious, as do those of the first boll weevil (see chapter 12, Boll Weevil) on US soil—but the history of COVID-19 quickly turned into a sequence of events in which the West was calling the shots. It had many of the leading experts, it had the money for welfare systems and the service economy that made things like working from home feasible, and this framed responses all around the world. In fact, some governments in the Global South seem to have followed Western templates not because they made sense but because they looked like the default approach. Once vaccines became available, distribution became a testament to global hierarchies in that rich people were offered shots in arms first, and readers of this book will not be surprised. It would have been nice if the world had come together in the fight against the pandemic and acted in a way that offered the same measure of respect for every human life. But it would have been a global first.

There was another level of narratives: the individual stories. People confronted the new pandemic world in many different ways. Some people went into lockdown, others did essential work; the latter included medical staff in overtaxed hospitals and the everyday heroes who delivered parcels and mail. Thanks to the electronic media, we learned about many of these experiences, and this made for a veritable tsunami of personal stories about suffering, loss, and survival. It was overwhelming, even for the most compassionate media consumer. Years earlier, I had written in the introduction to part V that individual narratives may emerge as the new opium of the people. It could be read as prescient in the days of COVID-19.

But for all these narrative threads, they did not add up to some kind of master narrative. In the early days, some political leaders resorted to the rhetoric of war, but it merely underscored the narrative void that the virus revealed: if war rhetoric made sense at all, this was the messiest guerrilla war of all time. People lacked words and narrative templates to grasp what was going on, and it was not about lack of information or imagination: it was about how linear narratives fare poorly in an interconnected world. Things did not come together in a way that made intuitive sense because there was no obvious hierarchy in the medical, visual, socioeconomic, and political narratives that came into play. Different narratives and imperatives stood next to each other without an obvious idea of how they might fit together, and in everyday life, the different rationales interlocked in a rather brutal way: the path of legitimate action was exceedingly narrow, even taking into account that we live in what the conclusion calls the Great Narrowing. The pandemic is certainly not open to a traditional chronological narrative with a beginning, an ending, and a more or less clear path between the two. Once more, people felt the peculiar dizziness that comes with life in a vortex.

When the writing of corona history begins in earnest, researchers will likely argue over additional rationales and the narrative threads they produced. For example, there is room for debate about whether the mutations of the virus really were as important as some took them in 2021, or they were mere surface disturbances, "crests of foam" in the vortex—to use Braudelian terminology once more. But for all the additional layers, historians may come down to a limited number of narrative threads that they untangle retrospectively. Vortexian history is not about an unbridled complexity where everything is a matter of perspective. Quite the contrary, it is about identifying the few threads that mattered most in global perspective, knowing that there are a few other, finer threads that matter less in the grand scheme of things.

As so often in modern history, efforts started from a simple premise. Few goals are as self-evident as saving lives from a deadly virus. But then things became complicated fairly quickly, and some experts had a field day. Historians were not among the authorities in the spotlight, notwithstanding some notable efforts from environmental historians.[1] By and large, the historical profession was mostly speechless throughout the pandemic, and it was not just because people were busy with the essentials of life. COVID-19 did not seem to fit into established modes of historical narration, and thus scholars were searching for words and concepts that allowed them to make sense of it all.

As the pandemic recedes, it is a popular wish that we be better prepared next time. Maybe vortexian history can help us come to terms—metaphorically and literally—when the next global crisis unfolds.

THE MESS WE'RE IN

An Inconclusive Conclusion

It was tempting to end this book with COVID-19. The pandemic was not over when the manuscript went to press, and the same held true for most of the other topics. This volume leaves a lot of things hanging in the air: as we arrive in our time on one issue after another, we see standing conflicts, loose threads, plenty of moral dilemmas, and nothing resembling a blueprint of how things might work out. There is no real ending in any of the chapters, and that makes it challenging to end with a conclusion in the true sense of the word. The best one-sentence summary of this volume would be that it is all a big mess, and we are stuck in the middle of it.

I have made the case for nonlinear narration in the introduction, and nowhere is the subversive nature of nonlinear narration more evident than toward the end, when things do not come together in a clear and morally unambiguous way. While linear narratives have a conclusion, nonlinear narratives have outcomes. They lack consistency, not to mention narrative grandeur, and more often than not, the only certainty is that we will keep muddling through for lack of a choice. And just to be sure, this is not the old adage that the future is uncertain and not for mortals to know. In fact, this book comes down to the opposite view. We actually know a lot about the future, and perhaps a bit more when we take the history of our engagement with the nonhuman world into account, and we certainly know that we are bound to swim in a vortex of truly monstrous proportions. What we do not know is whether we will keep our heads above water.

Apocalyptic scenarios permeate the modern imaginary of environmental challenges, but maybe it is time to take apocalypticism in the spirit of everything else in this book: as a legacy that should be analyzed and put into perspective. We may not expunge the ghosts of the past, but we are unlikely to see any big bangs outside of Hollywood: the more likely scenario is death by a thousand cuts.[1] The following comments seek to identify some overarching patterns that define the modern way of stuff, but these patterns differ from the scenarios and narratives that typically populate the conclusions of history books. They are mere aggregates of individual trends, attempts to identify some defining currents in troubled waters. Vortices have a driving force, ways to channel the flow, and a general inclination to increase or decrease in velocity. I will review these components in this order, but not without a word on the literature. While this volume is mostly about the real world and not about the secluded realms of the academic ivory tower, there is also a scholarly message at the heart of this book.

I convey this message with a deep sense of gratitude. Every synthesis draws on preexisting scholarship, but few books of this kind draw to such an extent on a literature that is only one or two decades old. The number of empirically rich and methodologically sophisticated publications has grown tremendously in the new millennium, and I say this against the backdrop of a long track record of chastising the methodological naivetés of my peers.[2] We still have weak and mediocre books—every academic field churns them out by the dozens—but we should not worry about them too much: we have enough to show that environmental history has finally grown up. But as parents know well, young adults have a natural urge to act out their independence, and maybe that is the part we need to work on in the years ahead.

BEYOND PROFESSIONAL TRIBALISM

There is only one earth and only one history, but you would not know that from much of what historians write today. Specialization is a defining feature of expertise in the modern world, and the historical profession is no exception. The result is a community of historians that has long ceased to be an actual community. It is more akin to a set of tribes that identify as political historians, business historians, medical historians, and so forth. Job descriptions typically specify the field in demand, scholarly journals and professional organizations cater to the needs of subdisciplines, and peers and colleagues usually hail from the same academic field. Professional silos are comfortable places nowadays, if not the natural place to be for an academic historian, and reaching out to other subdisciplines, or even thinking about the patchwork of subdisciplines as a whole, is entirely optional. Scholars have probably heard about total history in graduate school, but it is one of those things from France that you do not really need to care about.

Specialization is a typical response to the growth of an academic field, and the size of the historical profession today is staggering indeed. For those who seek a reading list of manageable length, staying within the confines of a subdiscipline is the reasonable thing to do. But historians have always been eager to stress the need for context, far more so than other academic pursuits, and that raises questions about the fragmentation of historical research. Is a role played by the fact that linear narration is much more plausible if historians stick to a specific subdiscipline? Most scholarly fields have a defining paradigm: profit for business historians, equality for gender historians, health for medical historians, and so on. This provides helpful guidance when it comes to assessing events, individuals, and other objects of historical research, a moral clarity that allows for more or less straightforward stories with neat, unambiguous conclusions. Other subdisciplines may heed a different set of coordinates, but this is not much of a headache if those scholars are unlikely to review one of your drafts or grant applications.

The paradigms of environmental history have always been more diffuse than those of other subdisciplines, but this is probably due to the inherent diversity of the natural world. Humans interacted with the natural environment in many different ways, and that made for an exceptionally broad range of issues and approaches. However, environmental historians were certainly not immune to groupthink. For the first few decades, a defining interest was about the downside of industrial modernity, the "price of progress," a desire to offer counternarratives to the stories that circulated among agricultural historians, forest historians, mining historians, and historians of technology. It came down to a convenient division of labor among subdisciplines. While other historians talked about resource allocation, environmental historians talked about side effects.

The founding generation of environmental history consisted mostly of self-declared environmentalists, and activism, spiritual and other, left its mark in the field's intellectual code. Environmental historians might have grown skeptical of linear narratives much earlier if it had not been for their deep roots in the collective experience of environmentalists in the 1970s and 1980s. Contemporary conflicts fostered linear thinking in terms of problem identification, mobilization, and outcomes—at least for those who were lucky enough to live in a Western country with a flourishing environmental movement.[3] Linear narratives were far less plausible in the Global South, where environmental conflicts were typically perennial struggles rather than short campaigns, and recent experience in Western countries suggests that the rapid advances of Western environmentalism in the 1970s and 1980s were probably exceptional. Linear narratives are a tough sell if environmental challenges from climate change to the extinction of species are intrinsically linked to modern life in its full complexity.

Over the past two decades, environmental historians have grown skeptical of a narrow focus on side effects and counternarratives, and a significant part of the recent literature is located in a scholarly no-man's-land at the intersection of environmental history, agricultural history, forest history, mining history, business history, and the history of science and technology. With that in mind, it may be time for the next step, a rearrangement of scholarly boundaries that reflects the realities of an interconnected world. In the twenty-first century, it seems anachronistic to draw a line between a core process of resource allocation and the repercussions for humans and natural environments, or another line between organic and mineral resources. In fact, some of the chapters in this book are about how these lines have eroded over the course of modern history. It is time to conceive of dealing with material resources as a field of its own, with everything from prospecting to pollution as mere facets of a comprehensive history that revolves around supplying the world with stuff.

Such a history would also help close a hole in the mental universe of the historical profession. Allocating resources was never easy, but there was a point in time when it looked that way. If you lived in a Western country in the 1960s, it was tempting to think of material resources as essentially hassle free: food was cheap and abundant, and so were energy and minerals, which made it tempting to view resource allocation as a mere technical pursuit that required some money, modern technology, and little else. As it happened, the 1960s and early 1970s were the time when history departments expanded like never before, and it seems that a disdain for material resources was hardwired into our historical imagination at the time. In hindsight, it is plain that this disdain never made much sense in rural regions, or the Global South, or really anyplace beyond the metropolitan regions of the Western world where many universities were located. In the new millennium, the dream of carefree stuff is over even in affluent countries. The resource and energy woes of recent decades have alerted Western consumers that the resource base of our modern existence is fragile and contested, and this has a price. Historians ignore it at their own peril.

Agricultural history and mining history have flourished in recent years, though the legacy of their erstwhile niche existence is still evident. This book has drawn on the insights this boom has produced, but it also seeks to place them in a wider context. The narrative makes numerous forays into scholarly terrain that is commonly the province of business history, mining history, agricultural history, and other subdisciplines of history, all in an effort to obliterate the boundaries that have constrained our understanding of what the nonhuman world really meant in the modern era. Side effects and unintended outcomes are important, but they are parts of a much bigger story about the material basis of our modern existence: the quest to make a living on a planet whose natural environment was often less than inviting for human beings, and certainly was not meant to sustain eight billion of them.

This book bills itself as an environmental history, though I am less than sure that this is an adequate description. But what would be a better term? One might speak of a "materialist" or "new materialist" history, but that would probably raise more questions than it answers.[4] The same holds true for the Anthropocene, which is the academic equivalent of a fashion victim today. A cacophonous academic debate gave the Anthropocene a notorious mushiness as the shibboleth for any number of intellectual pursuits—and besides, an elite term is a bad choice for an endeavor that is historically anything but elite. I was tempted to suggest "vortexian history" for the title, but it felt like a bridge too far. For a time during the lengthy review stage, the subtitle promised "a history of human survival in the modern world," but it smacked of the apocalyptic rhetoric that I seek to put into perspective. In the end, I stuck with "environmental history" in the hope that readers would go away pondering whether their received understanding of "environment" is still adequate.

In any case, this is a book about stuff—and the people and the environments that provide it. Many commodity histories begin with a stern declaration that their respective protagonist was "special," but the time has come for a wider lens. Every material resource is special in its own peculiar way, and yet it is also part of a web of chains and entanglements that this book seeks to understand, and any distinctive features, significant as they may be, pale in view of the fundamental novelty of our modern ways to satisfy material needs. If we step back and look at the big picture that emerges from modernity's engagement with the natural world, it seems that five big trends have framed the ways in which modern societies have secured the material basis of their existence. I propose to call them the Great Need, the Great Externalization, the Great Reckoning, the Great Regulation, and the Great Narrowing.

THE GREAT NEED

The narrative began with a statement in New Testament style: in the beginning was the stuff. It has not lost its pivotal significance ever since. Global resource flows have increased dramatically in size and complexity throughout the nineteenth and twentieth centuries, which makes for the most fundamental among the five trends. If it were not for all the material that humans put into motion, our environmental challenges would be completely different and probably a smaller issue in global debates. Legions of scientists have modeled planetary processes in recent decades, and their verdict is clear. We have overshot planetary boundaries, and we will pay a price for lifestyles that are inherently unsustainable. This also means that the growth in the stream of material resources is bound to end rather soon, one way or another.

As concepts go, "need" is a rather vague one, and at the risk of sounding paradoxical, this made it attractive for the purposes of this book: explaining human needs is probably beyond the analytical capabilities of an environmental history. The previous narratives are often deliberately diffuse on the precise nature of needs, as much of the discussion focuses on the *consequences* of the human craving for ever more stuff. There was a growing demand for metals and energy, a growing hunger for meat that seemed to defy cultural differences, and a growing range of commodities in demand, but the underlying reasons remain underexplored: deciphering human needs remains the province of other intellectual pursuits, assuming for the moment that decoding human needs is in fact a feasible endeavor.[5] Going through the topics of this volume seems to reveal an intellectual and moral vacuum at the heart of the modern hunger for ever more stuff. Certainly for more than one product, the defining rationale for consuming it was that we could.

Be that as it may, the discussion of the Great Need has focused mostly on exploring the variety, the size, and the chronology of human demands. Urban needs differed from those in the countryside, rich countries had predilections different from those of poor ones, and it mattered whether resource flows were under the control of state authorities, large vertically integrated corporations, or no one at all. The discussion also traced the evolution in the human understanding of essentials. Humans have spent thousands of years on planet earth without air-conditioning and package holidays in Ibiza or Cancun, but affluent societies treat both as something that consumers are entitled to by nature.

In short, the Great Need has been first and foremost a driving force in the previous pages—a giant sucking sound of sorts, and one that the narrative has by and large treated as mere background noise. For the purposes of this inquiry, it is sufficient to acknowledge that the Great Need has fueled the accelerating speed of the vortex, though one cannot help but marvel about what needs, even the most essential ones, have done to humans. We live in a world where more people are obese than underweight, likely a first in human history, and it happened against the backdrop of extensive global debates about the right diet. Human appetites are mysterious and limitless in equal measure, and it is hard to conceive a decent future inside the vortex that does not include some kind of moderation. But as it stands, moderation does not seem to be one of the things that modern humans really need.

THE GREAT EXTERNALIZATION

Modern technology created a few environmental problems that were in fact new. Humans never had a problem with plutonium before the advent of nuclear power, and it took industrial chemistry to produce the persistent organic pollutants that the 2001 Stockholm Convention sought to ban. But many challenges have probably been with us since the dawn of civilization. Soils have been eroding as a result of agricultural use since the Neolithic Revolution, mines and cities have always spread pollutants of different sorts, and plants and animals have traveled along with migrating humans since prehistoric times. But modernity made a difference, and it was scale. Humans have interacted with the natural environment in many different ways throughout history, but never has humanity's footprint reached the intensity of the modern age.

The natural world has plenty of feedback loops, and humans have encountered the results of their own behavior in many different ways. Exhausted fields, dwindling woodlands, disappearing animal populations—people have always faced limits in their interaction with the natural world. Given the dramatic growth of the human footprint, one might assume that feedback mechanisms grew in scale in roughly similar measure, but by and large, this does not seem to have happened: modern societies did not feel the ecological consequences of their style of resource allocation, at least not in similar measure or in ways that triggered adequate responses. So how did modern societies get away with the repercussions of an ever-growing impact on the natural world? A big part of the answer is that modern societies were extremely good at externalization. Rather than dealing with problems as they came along, people found many different ways to push them into realms that did not call for a direct response. As the previous pages have shown, this Great Externalization has taken a multitude of forms: spatial, social, cultural, temporal, and expertocratic.

The modern world was global from its inception, and this provided modern societies with ample opportunities for spatial externalization. Mining produced environmental devastation at the sites of excavation and downstream along rivers, but many of those who profited from the bounty lived elsewhere. Food production intensified in places far away from consumers since the heydays of Caribbean sugar, and in the twenty-first century, few people have firsthand knowledge of factory farms and slaughterhouses. In short, resource allocation caused plenty of problems on the ground, but these problems were in some other place.

Spatial externalization worked on different levels. Sometimes it was a matter of a few feet. In the world of air-conditioning, externalization occurred on the level of individual rooms and buildings, as everyone who has walked through a heat canyon in a Mediterranean village will know. Cities developed distinct spatial orders as they grew, with the better quarters typically in the direction of prevailing winds, and sewer networks transformed city problems into problems at sewage farms, treatment plants, or wherever urban effluents left the belowground world. Nations were inclined to view industrial regions as special areas with special rules: what was normal in Pittsburgh or the North of England would have been beyond the pale elsewhere. And then there was the stratification of what Immanuel Wallerstein has called the world-system with its excessive burden for the Global South.

Spatial segregation went hand in hand with social segregation. Those who worked in mines, monocultures, or slaughterhouses were not a representative cross-section of the population, and their disproportionate burden continued after the end of the working day. Poor city dwellers typically suffered from more noise and more pollution than affluent people, who could also afford better food than those who were stuck with things like breadfruit. Automobiles were different threats for those seated inside cars and those who got hit. Hybrid corn contributed to the cornucopia on modern food markets, but it looked more ambiguous when you were a smallholder. Environmental problems were often a marker of social inequality—though the affluent societies of the postwar years were reluctant to acknowledge this.

Just like other modes of externalization, cultural segregation took many different forms, and it did not end in the age of ecology. In fact, some environmentalist tropes provided camouflage for social inequality. The global contamination trope, so evident in the response to the *Lucky Dragon* incident, relieved people of awkward questions about whether some people faced a greater burden from pollution than others—as was the case on ninety-seven Japanese fishing boats in 1954, of which ninety-six never received any coverage in the world's media. Conversations about nature reserves prioritized visuals and acreage under protection over the fate of those who happened to live in one of these areas. The environmental justice movement and activists all over the Global South have challenged these received wisdoms of Western environmentalism, but from a global perspective, it seems that their work has barely begun. Externalization, including cultural externalization, is too deeply ingrained in modernity's approach to man and the natural world as to be overcome within a matter of years.

Temporal externalization is arguably as old as the waste heap—a fixture of human civilization, as archaeologists know—but modern people have brought it to a new level. Excessive use of synthetic nitrogen brought rich harvests, but it eventually produced groundwater hazards. Introducing cane toads to Australia served the imminent needs of Queensland sugar farmers (or so they thought until they saw bugs flying over the big mouths of the ground-dwelling animals) while wrestling with the toll of an invasive species became a job for generations. Dodos filled the stomachs of sailors for a while and were then gone forever. States like Saudi Arabia banked on the exploitation of finite mineral reserves and faced an uncertain future. Landfills turned into veritable mountain ranges with toxic effluents from things like decomposing plastic bags. And I will not even mention nuclear waste.

However, modernity's most popular mode of externalization was probably about the creation of expert groups. As one looks through the previous pages, it almost seems to be the default mode of environmental management in the modern era. Dam builders, sanitary engineers, pollution control officers, erosion specialists, foresters, entomologists, rangers in national parks—whenever a problem arose, new professions evolved, offered special cognitive and practical skills, claimed jurisdiction for a certain realm, and pledged to take care of things.[6] They had scientific knowledge, academic prestige, the ear of the state and large corporations, and a growing body of experience. Many expert groups clustered around agencies or organizations that were never shy about telling everyone what was best. Furthermore, they never went away. This book has covered plenty of expert groups and traced how they evolved, but it has not discussed a single group of experts that disappeared.

These modes of externalization jeopardized the feedback loops that previously shaped the interactions between humans and the natural world. More precisely, they short-circuited these feedback loops: negative effects went in directions that deprived them of their sting. Externalization played a crucial role in the making of what Jason W. Moore has called the Four Cheaps: labor, food, energy, and raw materials. As Moore has argued, the resilience of industrial capitalism hinged on Cheap Natures, "a rising stream of low-cost food, labor-power, energy, and raw materials to the factory gates," and externalization of costs was crucial for making them inexpensive in contemporary markets.[7] After all, the Four Cheaps were not literally cheap. They just did not cost all that much money.

It has been seventy years since Karl William Kapp published his seminal book *The Social Costs of Private Enterprise*, and internalizing external costs has long been a rallying cry of modern environmentalism.[8] However, the Great Externalization is a much more complex and multidimensional process than Kapp had thought. In retrospect, his approach suffered from the classic problem of linear thinking in the modern world: calculating social costs hinges on clear parameters and frames of reference, and in the twenty-first century, we can no longer define either in the self-evident manner of earlier activists and researchers. Every assessment of social costs is open to contestation on multiple fronts: geography, politics, economics, society, and culture. This does not speak against pertinent calculations, but it does show that sheer numbers cannot capture the Great Externalization in an adequate manner, let alone in monetary figures. Some environmental challenges will always defy monetarization, though economists will surely try to suggest otherwise. There will never be an indisputable price tag for the extinction of species or the hegemony of certain professions, but both have real costs.

Although externalization frequently came across as a natural process, it was anything but that: it was made and renegotiated on a daily basis. Spatial externalization relied on long commodity chains that required a robust institutional framework across countries and continents. Social externalization needed property laws, comprehensive land use planning, and ways to maintain social order. Expertocratic externalization hinged on resources for burgeoning professions. And for all the pertinent efforts, externalization rarely worked perfectly. Problems were noticed nonetheless, and they stimulated responses. That is what the next two trends in the environmental history of the modern world were all about.

THE GREAT RECKONING

Many historians have stressed that researchers recorded the repercussions of human actions long before the dawn of modern environmentalism. But as a rule, it did not take expert knowledge to recognize environmental challenges. The divided highway taught the hierarchies of transportation through its sheer physical properties: for everyone with eyesight and a sense of smell, highways served as concrete evidence that cars rule the world. Humans got addicted to air-conditioning irrespective of culture and skin color, nitrogen fertilizer boosted plant growth visibly all around the world, and wherever hybrid breeds prohibited reuse of seeds in the next season, farmers realized a fateful dependency on companies and scientific authorities. Sheer materialities could trigger reckonings, and more than once, these nonverbal responses dwarfed the impact of scientific studies—though the Great Reckoning included plenty of those as well.

At first glance, experiences were tied to a specific context, but this book has sought to tease out the global similarities. The boll weevil was framed as particular to the US South, but many agriculturalists around the world can identify with some basics of the plot—not because they have ever seen a boll weevil but because they experienced something similar. Like many other monocultures in peril, cotton production in the US South became more capital-intensive, smaller producers were driven out of business, social hierarchies came under threat, and there was an enduring sense of nervousness. Every environmental challenge produced recurring patterns in human responses, and this book has reviewed them extensively. Every famine of the modern era has challenged the legitimacy of the powers that be, and wherever you have installed water closets, you must find a reliable supply of water and a way to dispose of effluents. Smoke was a danger to lungs wherever people lived, and nuclear radiation did not discriminate between cultures in its damage to human genes. Invasive species multiplied rapidly in the absence of natural enemies, whether it was an insect, a mammal, or the charming cane toad.

In the modern era, many of these experiences were recorded, shared, and analyzed, and this led to a growing body of transnational knowledge. However, the accumulation of experiences hinged on a number of requirements. For one, experiences did not translate into learning curves unless they were properly measured and stored. For another, transfer processes were framed by the state of transport, communication networks, and the power relations in play. It was also important that problems did not suddenly disappear, as enduring challenges served to refresh the memory of the next generation—one of the cardinal reasons why this volume has focused on the years since 1500. Gathering experience also took time, though some learning curves advanced at amazing speeds. For example, DDT, initially touted as the panacea for pest problems, was already in crisis more than a decade before Rachel Carson penned *Silent Spring*.

As this book has stressed throughout, learning was no faceless process that involved all of humanity. It took place in specific groups that shared certain experiences. As a result, numerous fault lines run through the body of acquired wisdoms, and one of the most persistent involved the gap between the industrialized world and the Global South.[9] It was not necessarily about different material burdens. Chronology could also make a world of difference. Smog provides a case in point: particulate pollution was a growing menace in the Global South around the time when Western environmentalists felt that the problem had been solved and they had moved on to global warming. While car safety has improved in Western countries since the 1970s, appalling casualty rates in the Global South meant that the global death toll of automobilism was actually skyrocketing. Opinions diverge on other fronts as well, and many learning curves are ultimately a matter of perspective. If you like Big Macs, you can read the chapter on Chicago's slaughterhouses with a sense of pride, as inexpensive fast food hinges on the comprehensive industrialization of the commodity chain for meat. But then there are other creatures on this planet, both human and other, who have a different point of view. There is no such thing as a free lunch in an interconnected world, and certainly not at McDonald's.

As a result, this book has been reluctant to use the word "problem"—far more so than other environmental history books. Problems do not exist without some kind of norm or value, and normative assumptions are always open to challenge. The word should never be used without qualification: What was the material essence, what were the norms and values at stake, for whom was this a problem, and in which way? In other words, there is a difference between a problem and a "challenge," a word that is used prolifically throughout the book. A challenge is a material condition that has effects on humans and environments, and there is usually more than one way to frame a challenge as a problem. Synthetic nitrogen can be a crucial plant nutrient, a pollutant, an agent of ecological change, or just another thing that circulates in the world.

All this suggests that we should be cautious in speaking of collective learning curves. Like Paul Sutter, I am wary about "reintroduc[ing] whiggishness into narratives of environmental management."[10] Speaking of learning curves evokes indisputable insights, lessons that humanity has learned once and for all, reckonings that only the know-nothings of our time will wish to challenge. Learning curves are always tied to specific groups and their respective viewpoints, and this makes them ambiguous in so many ways. Yet some experiences have left a permanent mark in people's minds, and maybe we can speak of a long process of decreasing intellectual innocence that runs through modern history. Once upon a time, we could dream about dams as the ultimate development tools, or monocultures as clever specialization à la David Ricardo, or divided highways as the panacea for traffic woes, or tourism as a driver of global understanding. But as dreams turned into concrete realities, we learned that dams have side effects, that monocultures breed biological problems and fateful dependencies, that divided highways are not immune to congestion, and that there are different types of tourists. This does not mean we no longer build divided highways or dams, or we cease to maintain those that already exist, but it does mean that the magic is gone for good. Even a well-tuned propaganda machine is nowadays struggling to bring it back. The Chinese government learned as much when it embarked on the construction of the Three Gorges Dam on the Yangtze River.

The Great Reckoning is an ongoing process. We do not know all the species that are currently threatened by extinction, and we are just starting to experience what it means to live on a warming planet. But many problems were recognized fairly quickly, and usually long before scientific research started in earnest. It did not require academic expertise to learn that wastewater accumulated in cities, that mineral deposits ran out after some time, or that mass tourism changed destinations. But knowing the rough outlines of a challenge is only the first step on the long way to actual responses, and that process was at the core of the fourth defining trend in the environmental history of the modern world.

THE GREAT REGULATION

The vortex was more than a raw force of nature that drew on the sheer force of materiality. It was shaped and modified throughout modern history, and the vortex was always tightly regulated—though not always with a view to minimizing impacts on the natural world. The flow of resources, pollutants, and leftovers evolved along with institutional, cultural, technological, and material frameworks that served a broad range of purposes. Modern societies reacted to challenges in many different ways, and that variety has been under discussion throughout this book. Few of these reactions spelled the definite end of problems, but they were remarkably effective in creating some kind of order.

Some reactions were targeted responses, such as legislation against pollution or for the protection of species. But more often than not, responses to environmental challenges were part of a bigger story. The chemurgy movement was about the place of science in modern society, exuberant entrepreneurial visions, and the politics of the New Deal; paving the way for biofuels was an almost incidental by-product. The botanical exchange allowed targeted endeavors like the breadfruit transfer, but it also drew on a sense of curiosity, a spirit of freewheeling experimentation, and the assets on the ships that happened to drop anchor in the proximity of botanical gardens. Hybrid seeds were about money for seed companies, and the Green Revolution that hybrids underpinned in the postwar years was fueled and shaped by the Cold War. And then there were the responses that, if anything, exacerbated environmental problems. Nobody thought that the pine roots campaign was good for Japan's woodlands, but it served a higher purpose.

Some initiatives sought to eliminate challenges completely, such as the ban on DDT and other persistent organic pollutants. However, the more common response involved modifications and constraints. Meat inspection reduced the risks that consumers were facing from Chicago's slaughterhouses, but it did not eliminate the consequences of a meat-heavy diet for human health. Smoke abatement curtailed emissions, but it changed nothing about humanity's dependence on fossil fuels. The boll weevil became the target of a multipronged approach that included crop dusting, adjustments in cultivation methods, and constant monitoring, but all this needs to continue as long as farmers plant cotton in the American South. And then there were responses that targeted minds rather than matters. The dream of the rice-eating rubber tree was a way to work through the experience of dependency that many people were wrestling with at the end of faceless commodity chains.

Responses mirrored the shifting paradigms of the powers that be. Sustainable forestry and the Canal du Midi grew out of the spirit of mercantilism, but land titles and large integrated corporations served capitalist interests. Fascist regimes pursued autarky projects such as the draining of the Pontine Marshes while postwar consumer societies embraced battery chicken, plastic bags, and dreams of an upcoming atomic age. However, responses developed lives of their own. States built forestry while forestry built the state, but sustainable forestry moved on to turn forests into coniferous plantations, something that Hans Carl von Carlowitz never dreamed of doing. As reactors were built and used, nuclear power shrank from a utopian technology to an expensive way to heat water. Synthetic nitrogen was intended as a product for farmers, then served as raw material for explosives and prolonged World War I for years, and ended up changing the global nitrogen cycle.

By 1970, the modern world had plenty of ideas, rules, and technologies in circulation that sought to curtail or modify the environmental excesses of the vortex. However, many environmentalists were not particularly interested in what was there before them. Environmentalism resembled a cause without history and certainly viewed itself as a new movement that seized upon heretofore neglected issues, and all that seemed to matter was scientific knowledge and a readiness to speak truth to power. It fueled a spectacular boom, one of the most dramatic among the social movements in the history of the West, and yet the great green awakening was an act of self-deception: environmentalism was not made in a vacuum. Largely unacknowledged by activists and policymakers, the prevailing ideas, approaches, and policies reflected a long legacy that this book has tried to dissect. Clean air and water, energy and resources, the protection of species and landscapes, whaling and water management—there was no blank canvas for any of the challenges taken on by environmentalists, and preexisting scribblings were about more than laws and agencies.

The environmentalists' amnesia was closely related to the tremendous change that the burgeoning movement achieved in the 1970s and 1980s. Within just a few decades, environmentalism left its mark in virtually every field of humanity's engagement with the natural world, and at the risk of stating the obvious, it was usually for the better. Garbage collection and pollution control became tightly regulated endeavors. Species were saved from extinction, sometimes with 24/7 monitoring of nests and habitats. Anonymous commodity chains came under scrutiny. Organic farming and fair trade gave ethical consumers a choice. Clean water and access to toilets became a reality beyond the Western world. None of this worked perfectly, and change rarely occurred without a fight, but environmentalism helped define new paths toward new solutions. Things looked as if they were being sorted out once more, but that impression was short-lived.

Some issues evaded satisfactory solutions. Western shipyards began to fall under environmental regulation, but it did not matter for shipbreaking because dismantling moved to Taiwan and South Korea and on to India, Pakistan, and Bangladesh. Many mining landscapes were abandoned rather than restored, and when authorities did step in, it frequently had an air of improvisation, like the ice rink technology that seeks to freeze mind-boggling amounts of arsenic trioxide in its place at Canada's Yellowknife gold mine. Concerns about plastic bags led to the spread of cotton shopping bags, whose environmental footprint is beyond the radar of the common Western consumer because Western environmentalism sensitized them for petroleum-based products but not for the environmental footprint of cotton production. Dams delivered water and electricity to cities and farmers and helped control floods, but they have all sorts of side effects even when they operate smoothly on a technical level, which is no certainty even in a flagship project like Aswan Dam.

This volume has identified many moral and practical dilemmas, and perhaps the nastiest ones involved legitimate and effective responses that allowed other, bigger issues to linger. Divided highways were safer than other types of roads, but they underpinned a global automobilism whose annual death toll runs to more than one million in the twenty-first century. Incremental improvements made slaughterhouses and chicken batteries better places for animals, but they distract from more fundamental questions about the inherent cruelty of the factory farming system. The growing intensity of land use reduced the pressure to use other areas with conservation value, but plantations have plenty of pernicious problems. International conferences helped advance mutual understanding, but these meetings also nourished utopian and sometimes counterproductive dreams of global environmental policy.

Like the other trends, the Great Regulation is a process rather than a single event, and it certainly did not follow an overarching blueprint. Humans have influenced the flow of modern history in an eminently haphazard way, and they have produced a patchwork that we can only understand as a product of history. Regulation evokes ideas about laws and treaties, but the Great Regulation was not just about authorities. The Great Regulation was about the full range of formal and informal institutions that guided the use of materialities: from laws and treaties to company policies to the everyday decisions of consumers that kept slaughterhouses and air-conditioning units in business. A lot of stakeholders played a role in the making of these rules, and they did so in many different places, in many different ways, and over many generations. At its core, the Great Regulation is a web of constant negotiations between governments, corporations, experts, activists, and consumers.

Sometimes the Great Regulation led to a clear conclusion. For example, the separation of food and nonfood resources is gone since the heydays of the chemurgy movement. Air-conditioning has shaped construction styles, power grids, and bodily habits. Like other invasive species, cane toads will occupy a prominent place in Australian environments for the foreseeable future. Plastic bags will circulate in the ecosystems of the world and beyond for many years. The Chicago slaughterhouse became the dominant mode of satisfying the carnivorous cravings of affluent urban societies, and farmers around the world are stuck with synthetic nitrogen and hybrid seeds and the dependencies that they brought. In the absence of some dramatic event, these innovations, and the routines they begot, will continue to shape the modern world.

But many negotiations over formal and informal rules led to less definitive outcomes. Nature reserves looked like the perfect response to an encroaching industrial urbanity a hundred years ago, but it showed over time that they produced all sorts of problems for human and nonhuman stakeholders. Battery chickens supply consumers with meat and eggs as well as a sense of unease about animals in confinement. London's smog is a thing of the past, but particulate emissions still claim an obscene death toll in Britain's capital and many other cities of the world. In spite of several generations of soil conservationists, erosion is still a global challenge of the first order, international conventions are both more powerful and more problematic half a century after the Tokyo Resolution of 1970, and like other countries with a generous resource endowment, Saudi Arabia has found oil a blessing and a curse. The Great Regulation did not inevitably lead to solutions. More than once, learning curves went from euphoria to aporia.

What all this comes down to is that the Great Regulation has left us both more capable and more constrained in our response to environmental challenges. On the one hand, our body of knowledge, our technological and political means, and our experiences with intended and unintended effects has expanded enormously, leaving us wiser than ever as to what we are doing. But on the other hand, we have acquired an institutional, cultural, and material legacy that imposes limits on our range of responses. Cultural tropes and formal and informal institutions have left us with a dwindling array of options. It remains to be seen whether the Great Regulation will pave the way to a sustainable future, but it was remarkably effective in defining things that we cannot do. The environmental crisis is also about humans who have painted themselves into a corner: materially, technologically, politically, economically, and culturally.

We have seen the shrinkage of humanity's maneuvering room throughout this book. We cannot ban biofuels because, some ninety years ago, the chemurgy movement eliminated the barrier between food and nonfood resources under the flag of scientific advancement. We will continue to eat factory-farmed chicken because we have banned other animals from similar treatment and because it would take many years of development to bring other creatures to the same level of industrial efficiency. We can discourage the use of plastic bags among consumers, but we struggle to enlighten them about cotton bags. We have banned commercial whaling for the foreseeable future, but we cannot move on to a comparable response to commercial fishing. Every city dweller makes a daily bargain between the pleasures of urban life and exposure to urban pollution, but we cannot talk about this bargain because a hegemonic environmental discourse depicts pollutants as evil. We will be stubborn about DDT as long as Rachel Carson's *Silent Spring* is in living memory. We can moan about these things, and if we are Michael Crichton, we can write a bestselling novel to vent our anger, but otherwise, it seems that we are stuck.

There does not seem to be a good metaphor that captures the impact of the Great Regulation on the vortex. In terms of managing the flow, one might talk about weirs, sieves, siphons, and similar devices, but those terms might overtax the reader's tolerance for hydraulic rhetoric. For those in search of a conceptual metaphor, it might be helpful to think in terms of guardrails in car traffic. Drivers know that guardrails leave room for different vehicles, speeds, and styles of driving, but there are limits. And like so many guidance systems along the streets of the world, these guardrails are only seemingly permanent. They change over time as people have new experiences, they may even disappear in the wake of a dramatic event, and sometimes a gaping hole reminds us of the unfortunate people who fell into the abyss. The accumulation of experiences does not preclude the repetition of mistakes, but it does increase awareness of the price of ignorance. After all, experiences, like guardrails, differ enormously in strength. Some collapse on impact while others are rock solid, and you do not know until you bump into them.

Guardrails do not dictate the number or size or speed of cars, but they constrain the room for driving, and many discussions in this book focus on the *limits* of the contemporary range of opinions. No world history of reasonable size can explore the full diversity of views around the world, but it can trace how materialities and technologies, formal and informal institutions, and cultural tropes reduced maneuvering room. In that sense, this book is about what one might call the great paradox of modern environmentalism. While the extent and depth of our knowledge and the range of technological and political solutions has grown over time, it has also left us increasingly constrained in what we do.

THE GREAT NARROWING

Looking across the range of topics and experiences, it is hard to avoid the impression that things are moving ever faster. The vortex revolves at mind-boggling speed, space gets tight, and we are drifting toward the core. Scholars have started to speak of a Great Acceleration, and while this book emphatically agrees with the general diagnosis, it has used the term reluctantly.[11] As it stands, the Great Acceleration is closely tied to the Anthropocene thesis and a threshold around 1950, which seems less than useful for the purposes of understanding a global network of flows running amok. There is a threshold around 1950, and another one around 1970; we have several thresholds that mark the boom of globalization around 1900, and we have a peculiar path in China, which may be the point where thinking in threshold terms ceases to make sense. The vortex has accelerated with variable speed and with significant regional differences, but the bigger point is that the vortex never really stopped. Humans did not so much cross a single threshold as drift into a new age, and if they were in the right place, it was kind of fun until people discovered that they were at the mercy of elements.

The Great Acceleration also conveys an exceedingly Eurocentric view of world history. It focuses on resource flows, particularly those taking off in the wake of mass consumption in Western societies, but this inadvertently suggests a bygone world of leisurely change that is a tough sell beyond the Western world. The miners of Potosí and the slaves on Caribbean sugar plantations would have been surprised to learn that they lived in times of peaceful resource flows. Conversely, pressures did not necessarily increase in the Global South after 1950. For the countries under the thumb of United Fruit, the 1960s were a time of relative deceleration, as the crisis of the company and of interventionist US policies in Latin America made *el pulpo* a somewhat less aggressive creature. Cold War innovations such as the international coffee agreements brought a measure of stability. The Great Acceleration also pays scant attention to the growth of environmental activism, government policies, consumer awareness, and international agreements since the 1970s.

None of this is meant to suggest that the growing speed of material trends that the Great Acceleration seeks to flag were not real or not significant. They merely serve as reminders of what I have called the Anthropocene delusion: we should not let awareness of material trends overwhelm our discussion of socioeconomic, cultural, and political developments. If we think of the growth of resource flows in vortexian terms, we might speak of greater velocity and an increasing lack of space as we move closer to the core. In other words, all the tensions, struggles, and dilemmas that this book has chronicled are bound to increase in severity—there simply is not much room left to keep different perspectives, different interests, and different needs apart. We have seen a Great Narrowing in recent decades, and whatever you say about the outlook, it does seem that the range of options is shrinking. In fact, if you are deep down into a vortex, you may not have much of an outlook at all.

LIVING WITH AMBIGUITIES

In many chapters of this book, there was a moment, typically early on, when things looked amazingly simple. Practical means and moral or political expediencies seemed to converge and point in a specific direction, but whatever these certainties were about, they did not last. People usually wrestle with more than one challenge, environmental and otherwise, and these challenges tend to overlap and intertwine. As a result, issues that looked simple initially became increasingly complex, ambiguous, and multidimensional. When guano was first introduced to agriculturalists in Europe and North America in the 1840s, it seemed to be a straightforward affair: it was about enriching the soil in the quest for higher yields. But as guano use became common, and with it the use of artificial fertilizers, farmers realized that the new stuff came with some baggage. A long commodity chain made for fluctuations in supply and quality. Paying for guano made farming more capital-intensive. Agriculturalists became dependent on the expertise of people beyond the farm, as they could no longer make decisions on fertilizer quality and fertilizer needs without recourse to scientific expertise. Greater soil fertility rewarded farmers for improvements in related farming activities such as plowing, pest control, and seed improvement. Some plants used guano more efficiently than others, prompting reflections on whether production should focus on specific crops, and if a farmer went down the path of specialization, the problems of monoculture were usually not long in coming.

It was about the growth of experiences and the expanding range of perspectives. The guano commodity chain united different stakeholders, and then there were people such as urban consumers who had their own view of agricultural affairs, or perhaps no view at all. City dwellers shaped the world of agriculture through consumer choices and political interests or lack thereof—the latter being particularly glaring when land reform failed to excite citizens in urban societies. And then there were the other stakeholders in rural society who held views that diverged from those of farm managers on account of profession, status, or gender. Battery chicken was a great business model if you were in command of the essentials—capital, know-how, and a good dose of testosterone—but things looked different to the farm women who had previously drawn a share of their household income from selling eggs. The views of the animals remain anyone's guess.

Environmental challenges could play out in different ways, and some were utterly unexpected. Even a pest was not necessarily a bad thing. While American cotton planters feared the boll weevil's arrival in their fields, sharecroppers could use the infestation to gain leverage in their negotiations with landlords. Responses to environmental problems reflected the divisions that run through societies, and they often ended up reinforcing them. London smog did not affect all residents in the same way, as those in the western parts of town were breathing cleaner air than the East End poor. Sustainable management of forests bolstered the authority of state administrations and caused countless conflicts with other forest users from Central Europe to the African Sahel. Opinions on land titles typically diverged between those who owned land and those who did not.

In highlighting the diversity of perspectives, this book does not mean to suggest that there was some kind of equivalence between views. In fact, the discussion has stressed throughout that different groups faced different burdens and they had different chances to make themselves heard, compete with other stakeholders, and influence authorities. Environmental experiences reflect and reinforce the fault lines that run through modern societies: class, race, gender, ethnicity, wealth, and power. Different views had different levels of authority, and sometimes none at all, and this should make readers skeptical of every endeavor that looks exclusively at the stories within one of these groups. Rather than retelling these stories, historians should ask why they were told.

Conversations on environmental issues typically focus on things like policies, international agreements, and consumer information while the power of narratives tends to get short shrift. However, narratives are among the most important human coping mechanisms. The boll weevil found its way into blues songs, the gloom of London smog was a fixture in nineteenth-century English culture, cholera played a major role in Thomas Mann's *Death in Venice*, and the dodo became immortalized in Lewis Carroll's *Alice in Wonderland*. Narratives were also tools of power. Louis XIV had the Canal du Midi incorporated into the decoration of the Hall of Mirrors in Versailles, land reclamation in the Pontine Marshes figured prominently in the propaganda of Fascist Italy, and the autobahn was one of the most enduring myths of the Nazis. Narratives could also be the weapons of the weak. The story of the rice-eating rubber tree circulated among Indigenous rubber tappers on Borneo when the International Rubber Regulation Agreement sought to protect the rubber plantations of Asia against competition from smallholders, and from a global perspective, it was not the only dream people drew on to work through the experience of dependency.

Ascendant expert groups were particularly keen to employ the power of linear narratives. Professions needed public acclaim and plenty of money, and powerful stories were important parts of their sales pitch. They often seized on specific places, artifacts, and events. The Little Grand Canyon symbolized soil erosion, DDT showcased better living through chemistry, and Aswan Dam stood for the promise of development through large technological systems. The cultural hegemony of experts inevitably lost some of its shine when these icons fell short of their promise, as DDT and Aswan Dam did all too soon, but the wisdom of hindsight is not always needed to cast doubts on these narratives. Sometimes the divergences between lives and messages spoke loudly enough. Hugh Hammond Bennett preached soil conservation in the style of an Old Testament prophet, but his institutional creation, the US Soil Conservation Service, worked through subsidies and technical assistance rather than moral suasion. The forestry professor Karl Escherich, the father of applied entomology in Germany, dedicated his life to finding ways to kill insects, but he really dreamed about mixed forests that would make pest control obsolete. Ferdinand von Müller played a major role in the global spread of eucalyptus trees even though he had come to Melbourne to Europeanize Australia's nature. In short, this book is also about big men who knew better, who were powerful advocates for a certain cause while their own lives told a different story. And yes, in case you are concerned about gendered language, they were all men. That was part of the problem.

Environmentalism has its own set of narratives, and they are no less deserving of critical perspectives. Generations of environmentalists have celebrated Carson's wake-up call, but looking at her book in context brings in crucial nuance: *Silent Spring* also disrupted an ongoing process of learning on the ground. It gave a new symbolic dimension to ongoing processes of environmental learning about pesticides, which is a phenomenon that deserves more attention beyond the scholarly community: the natural environment was also a place for the negotiation of experiences that were eminently human. In fact, it was a helpful medium in which to deal with awkward experiences because it provided camouflage for human complicity. Environmental challenges could be placed in the "acts of god" category, and invoking them served to play down the role of humans. For example, it was convenient to frame the transformation of the Cotton South after 1890 as the result of the advancing boll weevil. For people in the American South, and specifically those who were white, blaming a tiny insect was easier than confronting bad credit, working conditions, or endemic racism.

For a long time, it seemed as if every self-respecting historian was obliged to take the air out of dubious stories. But in the new millennium, it may be equally important to understand their resilience. The narratives of expert communities are part of a global legacy that we have acquired over the years as we have wrestled with environmental challenges of many different kinds. It is not something inherently good or bad, nor is it neutral: it is just where we have come to at the end of a long way. There is plenty of history in our ongoing engagement with the nonhuman world, and it matters. When it comes to environmental challenges, the past is not dead. It is not even past.

AND WHAT DO *YOU* THINK?

Writing a synthesis is similar to mapmaking. You can make choices about scale, the boundaries of your map, and the features you wish to highlight, but beyond that point, the cardinal duty is to show what is there. For every topic under debate, I have sought to map the full range of perspectives and interests and their change over time, and to focus the discussion on contexts and rationales rather than sympathy points. But the mapping of debates inevitably raises the matter of where the author stands on these issues, and that question is particularly pressing when it comes to environmental challenges. There can be no doubt that the capitalist world-system has produced the most destructive system of resource allocation the world has ever seen, and this makes reflections on a shift toward a different, more benign system akin to a natural urge. It is also becoming clear that engaging with these challenges will not be something we can choose to do or not. There is plenty of evidence—from climate change to COVID-19—that the nonhuman world will be among the defining troublespots of the twenty-first century. So what is to be done?

Statements typically fall into one of two categories. Authors either go personal and implore their readers to recycle paper, fly less, and save string, or they come up with ambitious blueprints for a green future. Since the 1970s, scores of environmentalists have made the case for comprehensive management of the global commons, and the approach has not disappeared in the new millennium.[12] We even have technocratic versions in the form of geoengineering visions to curb global warming.[13] I have reservations about both approaches. I do think that the earth would be a better place if the people of the industrialized world ate less meat and thought more about their consumer choices, but I doubt that they were waiting for a historian to tell them so. And as to the great green masterplan to save the earth, authors and sponsors should never lose sight of the law of unintended consequences. It is a good idea to think big in the face of huge challenges, but I know what happened to grandiose visions in the Pontine Marshes, at Aswan Dam, and with nuclear power.

However, historians can take a third approach: they can help build a culture of self-observation. It is sorely needed. Looking in the mirror might seem a trivial thing, but history shows otherwise. For all the diversity of the challenges and responses in this book, acts of self-reflection are exceedingly rare throughout history, and even rarer were acts of self-reflection that had consequences. Karl Ludwig Schweisfurth reversed his stance on factory farming in dramatic fashion, but most stuck to their guns, "like that Roman soldier whose bones were found in front of a door in Pompeii, who, during the eruption of Vesuvius, died at his post because they forgot to relieve him." Oswald Spengler celebrated the memory of this soldier because he showed how we are destined "to hold on to the lost position, without hope, without rescue," but I disagree.[14] Sticking to an untenable position is an embarrassment for *Homo sapiens*.

Historians can contribute greatly to a culture of self-observation. Many environmentalists have looked to the natural sciences for advice, but understanding the inner workings of nature is only one part of the challenge. Effective and equitable solutions require an understanding of the institutional, socioeconomic, cultural, and technological factors that frame our engagement with environmental challenges, and a good part of that legacy is semiconscious at best. History resonates in many details: if you want to draft a policy on DDT or plastic bags, you will not get very far if you ignore that we have come a long way on these issues. And history provides at least one big lesson for environmental debates: we need to readjust the balance between ideas and agents of change.

Since the 1970s, ideas have been abundant in the environmental discourse—to such an extent that these ideas are now a major part of our environmental legacy. As Theodor Adorno suggested in the opening sentence of *Negative Dialectics*, "Philosophy, which once seemed obsolete, lives on because the moment to realize it was missed."[15] Environmentalists have been more reluctant to talk about agents of change, particularly when they failed to meet the criteria of either heroes or villains. The history of environmentalism speaks loud and clear about idealism at work, but I doubt that altruism alone will save the planet. Humans are selfish creatures: they want to make money, pursue a career, and find personal satisfaction, and our fate in the twenty-first century hinges on whether we can align these motivations with the quest for a sustainable future.

When this book went to press, a popular hope held that COVID-19 would be a game changer in the fight against global warming. You can read my assessment in the chapter on the 1976 Tangshan earthquake: change is easier as long as societies are in crisis mode, but for the most part, visions for a better future, about "learning from disaster," are a mental coping strategy that serves to fill the narrative void of an event without a discernible sense and purpose. It is a classic example of an idea that sounds enticing but lacks clarity on who would wish to push for change in the wake of a disaster. Many people are not in the mood for grandiose plans when calamity has struck. They are busy trying to get back their former lives.

Careers and profits are not motivations that sound good in historical narratives, but there is a Gandhi quote for everyone who would like things to be more straightforward: "Consistency is a hobgoblin."[16] When it comes to environmental challenges, life will be about ambiguities, and we should not pretend otherwise. We have multiple rationales and plenty of perspectives in play for all the issues before us, and the only question is how we deal with them. In fact, we can find many ambiguities in the lives and works of the men and women who have shaped paths in the modern world for better or worse. We can learn more from their stories if we let the ambiguities in.

It is in this spirit that I have looked at the heroes and the villains of environmentalism: the women and men who populate so many of the stories that activists, publicists, and policymakers have been telling each other. I am fully aware that some readers will question my morals when they note the ambiguities that my discussion inserts into the careers of Antony Fisher, Karl Escherich, and Rachel Carson. After all, we are talking about a godfather of neoliberalism, a Nazi, and a writer who faced vicious attacks and is widely hailed as a patron saint of environmentalism. But heroism has always been about two different archetypes, people whose righteousness was beyond human doubt and people who searched for the righteous path when none of the options looked good. We can learn from both types of heroes, but usually about different things. The paths of righteous people tell us about what is good and bad, and the paths of those who search tell us about how to live a righteous life. Maybe it is the latter type of heroism, the ethics of the quest, that the twenty-first century needs more.

In a roundtable for the *Journal of American History*, Paul Sutter talked about his "search for an ethics of environmental entanglement" that he hoped would "someday inform all history writing."[17] To the best of my knowledge, this search is ongoing, and maybe that is precisely the point. The vortex does not allow for solutions in the classic sense of the word, but it does allow for clever, carefully calibrated moves, and a culture of self-observation is poised to increase the chance of clever moves. In other words, maybe the key thing about Sutter's search was to embark on it, to avoid the siren calls of simple moral tales, and to monitor the paths taken. We no longer have that one lodestar, but we have plenty of experience navigating with more than a paradigm. We even have a name for that treasure trove of experiences. We call it modern history.

APPENDIX

MAKING CHOICES

IN THE CONVERSATIONS THAT THIS PROJECT INSPIRED OVER THE YEARS, one question came up time and again: How did you choose your topics? It is a perfectly legitimate question, but also one that defies a comprehensive answer. There are few things I have thought about more during writing than the list of topics, and I am keenly aware of the fact that every choice is open to debate in more than one respect. Reflections on the table of contents continued almost to the end of the project, and they happened in places as different as air-conditioned conference rooms and a swimming pool in Munich. The list coevolved with the book, and revisions took the form of occasional tweaking as well as comprehensive revisions. New topics emerged as others fell out of favor, and the number of topics that were under serious consideration at one point or another runs into the three digits. I have a list of April 2014, about half a year after writing had started in earnest, and it includes sixteen chapter titles that did not make it into this book.

As a start, it might be helpful to remind readers of the exemplary principle at the heart of this volume. In other words, topics do not stand for themselves: they serve as examples for environmental challenges that we can identify all over the world. For example, the chapter on cane toads represents invasive species in general, United Fruit stands for large, vertically integrated companies, and the boll weevil exemplifies the biological challenges to monocultures. Some chapters cover two issues: London smog is about pollution as well as fossil fuels, Baedeker is about tourism and the handbook genre, and eucalyptus is about plants with superior properties and about the botanical exchange, as the globalization of eucalyptus and other species hinged on both genetic potential and an institutional network that underpinned its spread around the world. This means that choices about topics raised questions about whether they provide a window onto general challenges. Does whaling adequately mirror the perils of overfishing, and does the dodo provide a good springboard for a discussion of the extinction of species?

Good topics provide a multitude of perspectives. The *Lucky Dragon* incident is a great example because it sheds light on several threads of nuclear history: nuclear testing and the military roots of nuclear power, the fear of radiation, and Japan's choice to build a huge fleet of reactors. The incident also served as the template for a new type of global pollution where contaminants knew no limits. The Kruger National Park stands for iconic nature, but it also provides insights into the repercussions that the concept of reserved nature brought about. A national park in South Africa also evokes the experience of apartheid, and as the chapter argues, national parks are about a structurally similar and equally conflictual mode of social segregation. And some choices are equally attractive for *not* being typical: the dodo is world famous, but many species that became extinct are not. The clumsy bird from Mauritius drives home how we are utterly selective, if not willfully ignorant, of the real extent of extinction.

But for all their individual charms, decisions about topics were never made in isolation. All chapters serve as nodes where several narratives meet, and good examples need to offer significant insights for each individual thread. As a result, topics were particularly attractive when they encapsulated a multitude of narrative paths. Synthetic nitrogen is a great topic because it is a potent fertilizer, it highlights the energy hidden in products, the power of industrial chemistry, and the impact of war economies, it is one of the lesser-recognized pollutants, and Ludwigshafen, the place where BASF pioneered the production of synthetic nitrogen, was the site of a catastrophic explosion that entered the annals of industrial disasters. Choices about topics need to take these connections into account, and replacing one topic with another typically involved a good deal of reweaving.

Nonlinear narration was a challenge as well as a boon, as it provided an elegant way to deal with doubts about some topics. A chapter on the Royal Botanic Gardens at Kew came under debate when I grew disaffected with Kew-centric narratives of botanical transfers, and the solution was a subchapter on the botanical exchange in the eucalyptus chapter. The opera house in Manaus was a personal favorite (being human, I had those), as it provided a gold-plated exemplification of a resource boomtown. But in the end, it was better to discuss it as one aspect of the broader phenomenon of resource cities, and that is why the brief spectacle of opera in the Amazon rain forest comes up in the Potosí chapter. After long reflections over how to discuss the shift to fossil fuels, I decided to make this a subchapter in London smog. Energy historians might argue that the topic calls for a grander place, but putting fossil fuels in a somewhat inconspicuous place was part of the rationale. As I argue in the chapter, the fossil fuel revolution is a retrospective construction. In the contemporary context, it was about sleepwalking into a new age.

As the project advanced, a third aspect came to frame the range of choices. It was about chronology: none of my choices are before 1500 or after 1970. Experiments with older examples ran into a number of problems. The aqueducts of ancient Rome were spectacular technological achievements, but they provide a poor template for the sewer revolution of the nineteenth century (though sanitary engineers liked to treat them as such because it bestowed their projects with a sense of cultural grandeur). China built the Grand Canal more than a millennium before Louis XIV commissioned the Canal du Midi, but it fell into disuse in the nineteenth century, just as the global transport revolution was gathering steam. And then there was the large shadow of Jan Assmann, whose seminal work on cultural memory can also be read as a suggestion that commemorating ancient civilizations calls for peculiar methods.[1] I remain skeptical about Assmann's distinction between social and cultural memory, but elaborating would require a different kind of book. I am also a bit weak when it comes to reading hieroglyphs and Ugaritic script.

The chronological barrier around 1970 might seem counterintuitive, as this book makes a point of following the material, institutional, and cultural legacies of events up to our own time. Virtually all chapters show how environmental debates over the past half century have influenced the meaning of previous events, created new interpretations, and questioned preexisting ones. As Michael Bess observed in his environmental history of France, "Practically every facet of French society eventually came to acquire an environmentalist tint."[2] I would revise the metaphor to reflect that environmentalism often provided more than one shade of green, as conflicts over environmental agendas, priorities, and definitions of problems run through this book, but this must not distract from environmentalism's fundamental importance. As it emerges in this volume, environmentalism appears as an irreversible threshold that left no sphere of human existence unchanged. When it comes to the environment, we have lived in a different world since 1970, for better or worse.

In short, if I did not select events that occurred after 1970, this does not reflect a disdain for the past fifty years. It was about avoiding the shallows of short-termism. Legacies need some time to unfold, and assessing legacies needs some additional time. When I began this project, the fiasco of the 2009 Copenhagen climate summit looked like the defining event of global environmental policy in our time.[3] Six years later, the Paris Agreement eclipsed the memory of Copenhagen. But the United States withdrew from the agreement in 2017, the 2019 meeting in Madrid failed, the 2020 meeting was postponed due to COVID-19, and global climate policy looked more dubious than ever. Or maybe it will rebound, now that the United States rejoined the Paris Agreement under President Biden. Historians are ill advised to compete with the pundits, and they should not burden their books with observations that may have a short shelf life. Syntheses age soon enough anyway.

I have chosen the year 1970 as a cutoff date because the following years mark a watershed in world history. As numerous scholars have argued, global interconnections of all kinds have grown markedly in speed and vigor since the 1970s, the intellectual and economic hegemony of the West has been fading, and people all around the planet have been living in societies defined by globalization ever since. The consequences show in the chapters on DDT and the *Torrey Canyon*, which were respectively the third and seventh chapters I wrote: Rachel Carson's *Silent Spring* and the oil spill off the coast of Cornwall were global media events from the beginning. While most events in this book have a legacy that is a mix of materialities, institutions, and cultural icons, the legacies of *Silent Spring* and *Torrey Canyon* are overwhelmingly cultural. This does not mean they were less important, but maybe the age of globalization asks for a different approach than the one I have pursued here.

Needless to say, this time frame does not suggest equal attention to every year between 1500 and 1970. It is clear that my selection tilts toward the late nineteenth and twentieth centuries, and that is perfectly in line with the idea of this book: it is during these years that the vortex was speeding up and spreading into every corner of the planet. It is equally clear that numerous things happened during these years that were not part of the process of modernization. Global modernity was always a work in progress, and it was never the full show. This book is an inquiry into the making of the modern world, and topics were chosen with a view to how they might illuminate this process.

But even within these general parameters, numerous choices were waiting to be made, and five aspects came to guide my decisions: conciseness, significance, geographic balance, diversity, and specificity. I do not need to elaborate about the first point: in light of the scope of this volume, avoiding overlaps and repetitions was a matter of common sense. I deal with some issues in a single chapter: maritime resources (whaling), erosion (Little Grand Canyon), total war (pine roots campaign), and hydraulic engineering (Aswan Dam). Others are discussed in twin chapters: transport (Canal du Midi and Autobahn), fertilizers (guano and synthetic nitrogen), factory farming (Chicago's slaughterhouses and battery chicken), waste (shipbreaking in Chittagong and plastic bags), and botanical exchange (breadfruit and eucalyptus). Several chapters dissolved when key arguments found a home in other chapters. My thoughts about Witwatersrand and the gold rush are now in the chapters on Potosí and Saudi Arabia, and Los Angeles smog became subsumed under London smog, if only for the purposes of this book.

Significance is a more difficult category, and arguably the one that is the most prone to subjective judgment. In a global history project, significance can be about two different things: it can be about the *representation* of the entire planet and about the *effects* on the planet. But from either perspective, the globality of environmental challenges was always a matter of degrees. As the *Lucky Dragon* chapter argues, even radioactive fallout, the poster child of global pollution, did not affect all humans on the planet in equal measure. For the purposes of this volume, global environmental issues are challenges that people encountered in similar situations all over the world, and these situations had to be present in different types of countries. For example, a chapter on Nowa Huta or Magnitogorsk fell by the wayside after much deliberation because Socialist forced industrialization was not truly global—it was about a specific political sphere. The cowboy would have provided plenty of material for an exciting chapter about masculinity and the world as seen from horseback, but I came to the conclusion that there were not enough charros and gauchos around the world to justify their inclusion. It also smacked too much of nostalgia for a bygone world.

The cowboy might have made the book if I had realized my plan for a full section on nostalgic memories. Frontier life was an important yet fading part of the experience of modernity, and it was not difficult to find other topics in the "significant but temporary" category: the range of potential candidates runs from El Dorado to the beautification movements that sprang up wherever scenic charm and a sizable bourgeoisie linked up in the late nineteenth century. And how about nomadism, a widespread way of life whose status changed from threatening to threatened over the course of modernity? But in the end, two points ruled against nostalgic chapters. Nostalgia produces a legacy that is exceedingly cultural, which means that material and institutional legacies would have received short shrift, and nostalgia rarely draws on global cultural currents—it thrives much better in regional, national, and social status settings. And if you would like a third reason that is probably less quotable, I will say that there is already an overdose of nostalgia in the Western world of our time.

Novelty is an obvious criterion for significance, and global "firsts" stood a good chance of inclusion if they subsequently shaped institutional and material developments around the world. Guano, Chicago's slaughterhouses, hybrid corn, Germany's autobahn, and battery chicken were selected in this vein: all these innovations were trailblazers that became transnational icons for a reason. Yet sometimes it was rewarding to choose examples because they were *not* new. The Kruger National Park was not the first national park or the first on the African continent, but that was precisely the point: as the chapter argues, we should not treat the world's national parks as carbon copies of Yosemite and Yellowstone. Saudi Arabia was not the first state to be shaped by a single resource, but it is the recent development of the Saudi resource state and the broad and costly array of welfare policies that it took on board that make it a particularly glaring case in point.

Many of my examples are famous, but occasionally it was good to opt against transnational fame. I have abandoned a Dust Bowl chapter in favor of the Little Grand Canyon, knowing from Paul Sutter's book that "it is one of the lesser-visited parks in Georgia, and attendance is in decline."[4] The Little Grand Canyon is closer to key themes in the history of soil erosion such as long-term use, the institutionalization of expertise, and the irreversibility of damage. It is also about water erosion, which is a bigger issue globally than wind erosion of the Dust Bowl type. In fact, the Dust Bowl is burdened with so much cultural baggage that it forces the environmental historian to cut through multiple layers of collective memory: it was about so much more than erosion.[5]

In other words, one might say that some examples fell through because they were *too* famous. As sinking ships go, the *Torrey Canyon* will never be as famous as the *Titanic*, but can you really engage with the *Titanic* today without Celine Dion's *My Heart Will Go On* in your ear? Only connoisseurs of French chansons know about Serge Gainsbourg's "Torrey Canyon," and in the absence of a Hollywood blockbuster and an endless series of *Titanic* exhibitions, the discussion can focus on underlying causes, the iconography of disaster communication, and learning from disaster. The 1970 Tokyo Resolution provides a much better case study for a targeted and nuanced discussion of the ambiguities of conferencing than mega-events such as the Rio Earth Summit of 1992. Similar reasons ruled against the iconic pictures of planet earth taken by the *Apollo* astronauts. Space agencies and science fiction writers have invested heavily in the cultural significance of human space travel, and in light of this propaganda machine, I found that it was more appropriate to let the space age drift in—and if you have read the chapter on plastic bags, you know why I use the word "drift."

Significance is also ambiguous because it is at odds with the quest for geographic balance. The "firsts" of modern history typically involved countries like Great Britain, Germany, France, Spain, and the United States. It would have been easy to fill the entire list with places in the domain of these five countries, but that would be unthinkable against the backdrop of a global history community whose shibboleth is lashing out against Eurocentrism. Giving a voice to concerns and experiences beyond the Western world has been a concern throughout this book, and it is plain that this should find reflection in the choice of examples. The rice-eating rubber tree, representing myths that help people work through the experience of dependence on distant forces, was one of the topics that was never in serious doubt.

Readers will have to decide whether this book reflects a Eurocentric worldview. In fact, I am emphatically Eurocentric when it comes to accountability: no critique of Eurocentrism must distract from the pivotal role played by the industrial world in the transformation of the planet over the past five hundred years. But concerns over Eurocentrism are usually about cultural bias, and it obviously matters for my scholarly perspectives that I am a native of Germany who lives and works in England. I embarked on this intellectual journey from a strong base in German history and a weaker one in US history, and those who search for bias in light of this background will easily find evidence. If you wish to go further, you can read my six references to Winston Churchill, none of them called for, as an attempt to please readers in my home country. I also added six quotations from Adam Smith's *The Wealth of Nations* in order to placate our Scottish cousins—so yes, being a migrant in Brexit Britain has taken its toll. Most of the literature I have consulted was in English and German, and private matters limited my archival work to Germany, England, and the United States (though materials consulted also spoke about Argentina, Belgium, Bulgaria, China, Egypt, Ethiopia, Italy, Mexico, Poland, Russia, Spain, and South Korea). I have tried to work against these biases as best I could, and I have sought to obtain firsthand knowledge of places under discussion, but at the risk of stating the obvious, this book was not written from some kind of omega point.

Some of the topics lack a specific location: sugar, the land title, cholera, the water closet, air-conditioning. Others bring countries in different parts of the world together. Guano was a resource from Peru that sold mostly in the United States and Western Europe, and the *Torrey Canyon* flew the Liberian flag, had an Italian crew, was owned by a Bermuda-based subsidiary of a US oil company, and polluted beaches in England and France. However, the chapters on the Canal du Midi, the Holodomor, and the Pontine Marshes show that themes with a distinct national flavor have their charms, too. I took care that my list included at least two topics from every continent except Antarctica, and I made sure that China, India, Japan, the Soviet Union, and the empires of Britain and Spain all have "their" entries. It would have been desirable to include additional countries, particularly Indonesia, Brazil, and Pakistan—though, incidentally, Chittagong, today a city in Bangladesh, belonged to Pakistan when the *M.D. Alpine* ran aground in 1960. But with 193 member states of the United Nations and 40 chapters plus 1 interlude, there is only so much one can do. A quest for balance is inherent in every synthesis, but, quite frankly, geographic gaps were one of the issues I worried about the least: they were impossible to avoid anyway. I did not worry too much that Africa's most populous country, Nigeria, lacks a chapter of its own, but it was important to map the experience of the African oil state.

Geography was one of three dimensions of diversity that I sought to respect. There was also thematic diversity, as every environmental challenge from diseases to dams was meant to receive its due—not a surprising goal in a world history, but one that provoked endless ruminations over how to cover what with how much intensity. And then there was the diversity in the types of examples. My list includes events, places, plants, animals, artifacts, types of work, campaigns, corporations, commodities, intellectual concepts, and a title that is deliberately open to more than one reading—Gandhi's Salt. Other types of examples fell prey to my fifth and final criterion: specificity.

None of my chapters is dedicated to an individual. It is not because I am distrustful of the agency of single persons. I paid close attention to Karl Escherich, William Jay Hale, and Antony Fisher in the chapters on sustainable forestry, the chemurgy movement, and battery chicken, and it is arguably impossible to discuss DDT without Rachel Carson. But reducing individuals to one book or one project does violence to the complexity of human lives, and my discussions of outstanding men and women provide biographical snippets rather than comprehensive pictures. I was also hesitant to discuss commodities that lack a clear historical profile. There can be no doubt that coffee, oil, and concrete are materials of modernity, but their careers have so many facets that chapters would need to be either brutally selective or terribly diffuse.[6] And then there was Fernand Braudel's warning, written with a view to coffee but probably a general challenge to commodity histories, that these stories "may lead us astray" because "the anecdotal, the picturesque and the unreliable play an enormous part in it."[7] There were some borderline cases: aluminum, asbestos, and TNT. They might have made it if there had not already been six commodities on my list.

Needless to say, the other four criteria also brought up many difficult decisions. Should I devote a chapter to hunting or to food additives like monosodium glutamate, or are these challenges ultimately less important than other ones? What about simple inventions with complex repercussions such as the lightbulb? How about a chapter on McDonald's, or the supermarket, or the ill-fated groundnut scheme?[8] I had reasons for my decision in all these cases and many others, but I would never rank them as beyond debate.

It bears recognition that all these decisions had to be made within different contexts. Choices had to fit the general frame of this book as well as the more specific frames of individual paths, which effectively turned the selection game into a kind of three-dimensional chess. To give just one example, diversity management assured that the "path of pollution" includes pollutants (London smog), products that turned into pollutants (DDT, plastic bags), contagions (cholera), events (*Lucky Dragon No. 5*), technologies (water closet), activities (shipbreaking in Chittagong), and places (Potosí). But these paths are no fixed frames of reference, as evident in the fact that there are both two twinned chapters on mining (Potosí and Saudi Arabia) and a "mining path" with four chapters. And then there is the one issue that bursts every subcategory in this book: anthropogenic climate change. I found it impossible to deal with this challenge in the form of twin or triplet chapters or a "climate path," and so references to the many aspects of global warming are interspersed throughout the book without an overarching plan. I dearly hope that readers who miss climate change in the table of contents will make it to this paragraph, as I think that this is a result that one cannot stress enough. It shows how climate change defies the geographic, topical, and political boundaries as well as the imaginary that we have acquired throughout the modern era.

Choices did not end with a general list of topics. The precise wording was another matter that deserved careful reflection, and sometimes adjustments were made even while writing was under way. It makes a difference whether a chapter is about air-conditioning or air conditioners: the latter is an artifact, the former a service of a technological device that meshed with bodily routines. I changed battery cages to battery chicken because it brings the animal (and both hens and roosters) into focus. It also creates a nicer complement to the technology-heavy chapter on Chicago's slaughterhouses. Barbed wire has an important place in modern history, but putting an artifact into the title would have drawn attention away from the conceptual issues that the chapter on the land title dissects. Naming the DDT chapter after Rachel Carson's *Silent Spring* would have constrained the discussion by nourishing the old myth that the demise of DDT was all about a book. Potosí brings up the effects of mining on an entire region while "silver from Potosí" or "the Potosí mine" would have encouraged narrower perspectives.

And then there is the prevalence of death and human misery in my choice of topics. It would have been easy to select a less devastating earthquake than Tangshan or an invasive species with better looks than cane toads. It would also have been possible to speak more about the charms of wild nature in the Kruger National Park chapter, to heap praise on the intelligence of whales, or to elaborate on the heroism of antinuclear campaigning in the chapter on *Lucky Dragon No. 5*. Sugarcoating is always possible, but it rarely advances understanding. And besides, if you have read this book, sugarcoating should make you think of Caribbean plantations.

NOTES

INTRODUCTION
A NEW HISTORY FOR A YOUNG CENTURY

1. John R. McNeill, "Future Research Needs in Environmental History: Regions, Eras, and Themes," in Kimberly Coulter and Christof Mauch, eds., *The Future of Environmental History: Needs and Opportunities*, 13–15 (Munich: Rachel Carson Center, 2011), 13.

2. John R. McNeill, *Something New Under the Sun: An Environmental History of the Twentieth Century* (London: Penguin, 2000); Joachim Radkau, *Nature and Power: A Global History of the Environment* (New York: Cambridge University Press, 2008). The original German edition was *Natur und Macht: Eine Weltgeschichte der Umwelt* (Munich: Beck, 2000).

3. For perspectives on different types of entanglements, see Roland Wenzlhuemer, *Connecting the Nineteenth-Century World: The Telegraph and Globalization* (Cambridge: Cambridge University Press, 2013); Gunnar Folke Schuppert, *Verflochtene Staatlichkeit: Globalisierung als Governance-Geschichte* (Frankfurt: Campus, 2014); and Robert Brier, ed., *Entangled Protest: Transnational Approaches to the History of Dissent in Eastern Europe and the Soviet Union* (Osnabrück: fibre, 2013).

4. Paul Warde, Libby Robin, and Sverker Sörlin, *The Environment: A History of the Idea* (Baltimore: Johns Hopkins University Press, 2018). For a discussion with a wider focus on environmentalism avant la lettre, see Etienne S. Benson, *Surroundings: A History of Environments and Environmentalisms* (Chicago: University of Chicago Press, 2020).

5. See Peter Burke, *The French Historical Revolution: The Annales School, 1929–2014*, 2nd ed. (Stanford, CA: Stanford University Press, 2015), 36–48.

6. Fernand Braudel, *The Mediterranean and the Mediterranean World in the Age of Philip II*, vol. 1 (London: Fontana/Collins, 1981 [1949]), 20n.

7. Braudel, *Mediterranean*, 267.

8. Braudel, *Mediterranean*, 267.

9. Braudel, *Mediterranean*, 329, 333 (quotation).

10. See p. 16.

11. William H. McNeill, *The Rise of the West: A History of the Human Community* (Chicago: University of Chicago Press, 1991 [1963]).

12. Barry Commoner, *The Closing Circle: Nature, Man and Technology* (New York: Bantam Books, 1972), 29.

13. For more on global warming, see my comments on p. 648.

14. See p. 366.

15. Julia Adeney Thomas, Mark Williams, and Jan Zalasiewicz, *The Anthropocene: A Multidisciplinary Approach* (Cambridge: Polity, 2020), xii. See also Erle C. Ellis, *Anthropocene: A Very Short Introduction* (Oxford: Oxford University Press, 2018); Christophe Bonneuil and Jean-Baptiste Fressoz, *The Shock of the Anthropocene: The Earth, History and Us* (London: Verso, 2016); Clive Hamilton, *Defiant Earth: The Fate of Humans in the Anthropocene* (Cambridge: Polity Press, 2017); Jeremy Davies, *The Birth of the Anthropocene* (Oakland:

University of California Press, 2016); and Dipesh Chakrabarty, "Anthropocene Time," *History and Theory* 57 (2018): 5–32.

16. Jan Zalasiewicz et al., "The Working Group on the Anthropocene: Summary of Evidence and Interim Recommendations," *Anthropocene* 19 (2017): 55–60; here, 56.

17. For more on this issue see p. 630.

18. Zalasiewicz et al., "Working Group," 59.

19. See Ronald E. Hester and Roy M. Harrison, eds., *Geoengineering of the Climate System* (Cambridge: Royal Society of Chemistry, 2014).

20. See p. 489.

1: POTOSÍ

1. Bartolomé Arzáns de Orsúa y Vela, *Historia de la Villa Imperial de Potosí*, ed. Lewis Hanke and Gunnar Mendoza, vol. 1 (Providence, RI: Brown University Press, 1965), 3.

2. Lewis Hanke, *The Imperial City of Potosí: An Unwritten Chapter in the History of Spanish America* (The Hague: Martinus Nijhoff, 1956), 30.

3. Adam Smith, *An Inquiry into the Nature and Causes of the Wealth of Nations*, ed. R. H. Campbell and A. S. Skinner, vol. 1 (Oxford: Clarendon Press, 1976), 220.

4. Karl Marx, *Capital*, vol. 1, in Karl Marx and Frederick Engels, *Collected Works*, vol. 35. London: Lawrence and Wishart, 1996), 651.

5. See Christoph Bartels and Rainer Slotta eds., *Der alteuropäische Bergbau: Von den Anfängen bis zur Mitte des 18. Jahrhunderts* (Münster: Aschendorff, 2012).

6. Philipp Strobl, *Das Leben mit dem Silber: Die Bergbauregion Schwaz in der Frühen Neuzeit* (Munich: Meidenbauer, 2011), 151n.

7. Christoph Bartels and Lothar Klappauf, "Das Mittelalter: Der Aufschwung des Bergbaus unter den karolingischen und ottonischen Herrschern, die mittelalterliche Blüte und der Abschwung bis zur Mitte des 14. Jahrhunderts," in Bartels and Slotta, *Der alteuropäische Bergbau*, vol. 1, 111–248; here, 175; and John Temple, *Mining: An International History* (London: Ernest Benn, 1972), 52.

8. Robin Hermann, *Böhmischer Erzbergbau: Der Altbergbau im böhmischen Erzgebirge* (Chemnitz: Hermann, 2013), 28; and Cedric E. Gregory, *A Concise History of Mining* (New York: Pergamon Press, 1980), 214.

9. Georgius Agricola, *De Re Metallica*, trans. Herbert Clark Hoover and Lou Henry Hoover from the First Latin Edition of 1556 (New York: Dover Publications, 1950).

10. Lewis Mumford, *Technics and Civilization* (London: Routledge, 1934), 69.

11. Hubert Kiesewetter, *Das einzigartige Europa: Zufällige und notwendige Faktoren der Industrialisierung* (Göttingen: Vandenhoeck & Ruprecht, 1996), 162.

12. D. A. Brading and Harry E. Cross, "Colonial Silver Mining: Mexico and Peru," *Hispanic American Historical Review* 52 (1972): 545–79; here, 571.

13. Ulrich Pfister, "Silber," in *Enzyklopädie der Neuzeit*, vol. 12 (Stuttgart: Metzler, 2010), cols. 1–8; here, col. 1n.

14. Alistair Hennessy, *The Frontier in Latin American History* (London: Edward Arnold, 1978), p. 75. See also Peter Bakewell, "Mining in Colonial Spanish America," in Leslie Bethell, ed., *The Cambridge History of Latin America*, vol. 2: *Colonial Latin America*, 105–51 (Cambridge: Cambridge University Press, 1984).

15. Dennis O. Flynn and Arturo Giráldez, "Born with a 'Silver Spoon': The Origin of World Trade in 1571," *Journal of World History* 6 (1995): 201–21; here, 201. See also Dennis O. Flynn and Arturo Giráldez, "Born Again: Globalization's Sixteenth Century Origins (Asian/Global versus European Dynamics)," *Pacific Economic Review* 13 (2008): 359–87.

16. Erich W. Zimmermann, *World Resources and Industries: A Functional Appraisal of the*

Availability of Agricultural and Industrial Materials (New York: Harper, 1951), 15; emphasis in the original.

17. Renate Pieper, "Amerikanische Edelmetalle in Europa (1492–1621): Ihr Einfluß auf die Verwendung von Gold und Silber," *Jahrbuch für Geschichte Lateinamerikas* 32 (1995): 163–92; here, 190n; Carlos Marichal, "The Spanish-American Silver-Peso: Export Commodity and Global Money of the Ancien Regime, 1550–1800," in Steven Topik, Carlos Marichal, and Zephyr Frank, eds., *From Silver to Cocaine: Latin American Commodity Chains and the Building of the World Economy, 1500–2000*, 25–52 (Durham, NC: Duke University Press, 2006), 35, 42n.

18. William S. Atwell, "International Bullion Flows and the Chinese Economy circa 1530–1650," *Past and Present* 95 (1982): 68–90; here, 77, 83.

19. Flynn and Giráldez, "Born with a 'Silver Spoon,'" 209.

20. Brading and Cross, *Colonial Silver Mining*, 566.

21. Immanuel Wallerstein, *The Modern World-System I: Capitalist Agriculture and the Origins of the European World-Economy in the Sixteenth Century* (Berkeley: University of California Press, 2011), 100.

22. Kendall W. Brown, *A History of Mining in Latin America: From the Colonial Era to the Present* (Albuquerque: University of New Mexico Press, 2012), 24.

23. Bakewell, "Mining," 115, 120.

24. Daniel R. Headrick, *Humans versus Nature: A Global Environmental History* (New York: Oxford University Press, 2020), 181.

25. Jürgen Osterhammel, *The Transformation of the World: A Global History of the Nineteenth Century* (Princeton, NJ: Princeton University Press, 2014), 187.

26. Brown, *History of Mining*, 30.

27. Brading and Cross, *Colonial Silver Mining*, 546.

28. See Peter Bakewell, *Miners of the Red Mountain: Indian Labor in Potosí, 1545–1650* (Albuquerque: University of New Mexico Press, 1984); and Jeffrey A. Cole, *The Potosí Mita, 1573–1700: Compulsory Indian Labor in the Andes* (Stanford, CA: Stanford University Press, 1985).

29. Eduardo Galeano, *Open Veins of Latin America: Five Centuries of the Pillage of a Continent* (New York: Monthly Review Press, 1997 [1971]), 40.

30. Lewis Hanke and Gunnar Mendoza, "Bartolomé Arzáns de Orsúa y Vela: Su Vida y Su Obra," in Arzáns de Orsúa y Vela, *Historia de la Villa Imperial de Potosí*, vol. 1, xxvii–clxxxii; here, xxxiv.

31. Malcolm J. Rohrbough, *Aspen: The History of a Silver-Mining Town, 1879–1893* (New York: Oxford University Press, 1986), 30, 115, 168, 173.

32. Barbara Weinstein, *The Amazon Rubber Boom, 1850–1920* (Stanford, CA: Stanford University Press 1983), 193; and Joe Jackson, *The Thief at the End of the World: Rubber, Empire and the Obsessions of Henry Wickham* (London: Duckworth Overlook, 2009), 253–55.

33. Hennessy, *Frontier*, 154.

34. Robert V. Hine and John Mack Faragher, *Frontiers: A Short History of the America West* (New Haven, CT: Yale University Press, 2007), 96; Anne Kelk Mager, *Beer, Sociability and Masculinity in South Africa* (Bloomington: Indiana University Press, 2010), 55n. In a revealing bow to postcolonial male sentiments, Castle Lager advertisements of the 1980s ignored the actual brewer, Lisa Glass, and focused on her husband Charles.

35. Hine and Faragher, *Frontiers*, 96n.

36. Brown, *History of Mining*, 116.

37. Lewis Hanke, *Bartolomé Arzáns de Orsúa y Vela's History of Potosí* (Providence, RI: Brown University Press, 1965), 28.

38. Ad Knotter, "Mining," in Karin Hofmeester and Marcel van der Linden, eds., *Handbook: The Global History of Work*, 237–58 (Berlin: De Gruyter Oldenbourg, 2018), 237.

39. Nell Irving Painter, *Standing at Armageddon: The United States, 1877–1919* (New York: Norton, 1987), 15, 180; Thomas Nipperdey, *Deutsche Geschichte 1866–1918*, vol. 1: *Arbeitswelt und Bürgergeist* (Munich: Beck, 1994), 332; Peter Clarke, *Hope and Glory: Britain 1900–1990* (London: Penguin, 1997), 139n; and Leonard Thompson, *A History of South Africa*, 3rd ed. (New Haven, CT: Yale University Press, 2001), 239.

40. Thomas G. Andrews, *Killing for Coal: America's Deadliest Labor War* (Cambridge, MA: Harvard University Press, 2008), 6.

41. Consider, for instance, Timothy Mitchell's debatable argument that "workers were gradually connected together not so much by the weak ties of a class culture, collective ideology or political organisation, but by the increasing and highly concentrated quantities of carbon energy they mined, loaded, carried, stoked, and put to work." (Timothy Mitchell, *Carbon Democracy: Political Power in the Age of Oil* [London: Verso, 2013], 27.)

42. Bakewell, *Miners*, 180n; and Cole, *Potosí Mita*, 123.

43. Ricardo A. Godoy, "Technical and Economic Efficiency of Peasant Miners in Bolivia," *Economic Development and Cultural Change* 34 (1985): 103–20; here, 107n.

44. Galeano, *Open Veins*, 39.

45. Michael T. Taussig, *The Devil and Commodity Fetishism in South America* (Chapel Hill: University of North Carolina Press, 1980), 202n.

46. John L. Phelan, "Review Article: The History of Potosí of Bartolomé Arzáns de Orsúa y Vela," *Hispanic American Historical Review* 47 (1967): 532–36; here, 535.

47. Richard L. Garner, "Long-Term Silver Mining Trends in Spanish America: A Comparative Analysis of Peru and Mexico," *American Historical Review* 93 (1988): 898–935; here, 909.

48. John Lynch, *Simón Bolívar: A Life* (New Haven, CT: Yale University Press, 2007), 200, 206.

49. Paul Gootenberg, *Between Silver and Guano: Commercial Policy and the State in Postindependence Peru* (Princeton, NJ: Princeton University Press, 1989), 143.

50. Gootenberg, *Between Silver and Guano*, 143.

51. Timothy J. LeCain, *Mass Destruction: The Men and Giant Mines that Wired America and Scarred the Planet* (New Brunswick, NJ: Rutgers University Press, 2009), 126.

52. Zimmermann, *World Resources*, 15.

53. James H. Bamberg, *The History of the British Petroleum Company*, vol. 2: *The Anglo-Iranian Years, 1928–1954* (Cambridge: Cambridge University Press, 1994), 522.

54. See Gabrielle Hecht, *Being Nuclear: Africans and the Global Uranium Trade* (Cambridge, MA: MIT Press, 2012).

55. LeCain, *Mass Destruction*, 204.

56. Alan Weisman, *The World Without Us* (New York: St. Martin's Press, 2007), 221.

57. Kevin O'Reilly, "Liability, Legacy, and Perpetual Care: Government Ownership and Management of the Giant Mine, 1999–2015," in Arn Keeling and John Sandlos, eds., *Mining and Communities in Northern Canada: History, Politics, and Memory*, 341–76 (Calgary: University of Calgary Press, 2015), 343, 346. See also John Sandlos and Arn Keeling, "The Giant Mine's Long Shadow: Arsenic Pollution and Native People in Yellowknife, Northwest Territories," in John R. McNeill and George Vrtis, eds., *Mining North America: An Environmental History since 1522*, 280–312 (Oakland: University of California Press, 2017).

58. Marco Meniketti, "Sugar Mills, Technology, and Environmental Change: A Case Study of Colonial Agro-Industrial Development in the Caribbean," *IA. The Journal of the Society for Industrial Archeology* 32 (2006): 53–80; here, 79.

59. Kevin H. Henke, "Preface," in Henke, ed., *Arsenic: Environmental Chemistry, Health Threats and Waste Treatment*, xvii–sviii (Chichester: Wiley, 2009), xvii.

60. Henke, "Preface," xvii.

61. Jerome O. Nriagu, "Mercury Pollution from the Past Mining of Gold and Silver in the Americas," *Science of the Total Environment* 149 (1994): 167–81; here, 177.

62. Nicholas A. Robins, *Mercury, Mining, and Empire: The Human and Ecological Cost of Colonial Silver Mining in the Andes* (Bloomington: Indiana University Press, 2011), ix.

63. Hennessy, *Frontier*, 75.

64. Brown, *History of Mining*, 177.

65. See http://whc.unesco.org/en/list/420, accessed November 3, 2021.

66. "Bolivia's Cerro Rico, the 'Mountain that Eats Men,' Could Sink Whole City," accessed November 3, 2021, https://www.theguardian.com/world/2014/jan/10/bolivia-cerro-rico-mountain-sink-city-potosi.

2: SUGAR

1. Laurent Dubois, *A Colony of Citizens: Revolution and Slave Emancipation in the French Caribbean, 1787–1804* (Chapel Hill: University of North Carolina Press, 2004), 35n.

2. Robert Fogel has argued that in the seventeenth and eighteenth century, "sugar was the single most important of the internationally traded commodities, dwarfing in value the trade in grain, meat, fish, tobacco, cattle, spices, cloth, or metals." (Robert William Fogel, *Without Consent or Contract: The Rise and Fall of American Slavery* [New York: Norton, 1991], 21.)

3. See Thorstein Veblen, *The Theory of the Leisure Class* (Boston: Houghton Mifflin, 1973 [1899]), 60–80.

4. Sidney W. Mintz, *Sweetness and Power: The Place of Sugar in Modern History* (New York: Penguin, 1986), 6.

5. Sucheta Mazumdar, *Sugar and Society in China: Peasants, Technology, and the World Market* (Cambridge, MA: Harvard University Press, 1998); J. H. Galloway, "Sugar," in Kenneth F. Kiple and Kriemhild Coneè Ornelas, eds., *The Cambridge World History of Food*, vol. 1, 437–49 (Cambridge: Cambridge University Press, 2000), 442.

6. See Andrew M. Watson, "The Arab Agricultural Revolution and Its Diffusion, 700–1100," *Journal of Economic History* 34 (1974): 8–35. For two recent assessments, see Michael Decker, "Plants and Progress: Rethinking the Islamic Agricultural Revolution," *Journal of World History* 20 (2009): 187–206; and Paolo Squatriti, "Of Seeds, Seasons, and Seas: Andrew Watson's Medieval Agrarian Revolution Forty Years Later," *Journal of Economic History* 74 (2014): 1205–20.

7. B. W. Higman, "The Sugar Revolution," *Economic History Review* 53 (2000): 213–36; here, 221; Philip D. Curtin, *The Rise and Fall of the Plantation Complex: Essays in Atlantic History*, 2nd ed. (New York: Cambridge University Press, 1998), 23; and Stuart B. Schwartz, *Sugar Plantations in the Formation of Brazilian Society: Bahia, 1550–1835* (New York: Cambridge University Press, 1985).

8. Wallerstein conceives of the Caribbean islands as part of an "extended Caribbean" that includes Brazilian gold and tobacco from Virginia and Maryland. (Immanuel Wallerstein, *The Modern World-System II: Mercantilism and the Consolidation of the European World-Economy, 1600–1750* [Berkeley: University of California Press, 2011], 167.)

9. B. W. Higman, *A Concise History of the Caribbean* (New York: Cambridge University Press, 2011), 103. The concept of a "sugar revolution" has received some criticism. See, for instance, Russell R. Menard, *Sweet Negotiations: Sugar, Slavery, and Plantation Agriculture in Early Barbados* (Charlottesville: University of Virginia Press, 2006). However, some of the criticism, such as Ira Berlin's concept of a "plantation revolution," concurs with the main

thrust of this chapter. (Ira Berlin, *Many Thousands Gone: The First Two Centuries of Slavery in Northern America* [Cambridge, MA: Belknap Press of Harvard University Press, 1998], 97.) For a passionate defense of the term, see Higman, *Sugar Revolution*.

10. Aubrey C. Land, "Economic Base and Social Structure: The Northern Chesapeake in the Eighteenth Century," *Journal of Economic History* 25 (1965): 639–54; here, 647. Nonetheless, or maybe because of that, Barbadian planters liked their English traditions. (See Larry Gragg, *Englishmen Transplanted: The English Colonization of Barbados 1627–1660* [Oxford: Oxford University Press, 2003].)

11. Immanuel Wallerstein, *World-Systems Analysis: An Introduction* (Durham, NC: Duke University Press, 2004), 11.

12. Giorgio Riello, *Cotton: The Fabric That Made the Modern World* (Cambridge: Cambridge University Press, 2013), 7.

13. Curtin, *Rise and Fall*, 78, 83 (quotation).

14. Michael Craton, *Testing the Chains: Resistance to Slavery in the British West Indies* (Ithaca, NY: Cornell University Press, 2009), 89n.

15. See Laurent Dubois, *Avengers of the New World: The Story of the Haitian Revolution* (Cambridge, MA: Belknap Press of Harvard University Press, 2004); Jeremy D. Popkins, *You Are All Free: The Haitian Revolution and the Abolition of Slavery* (New York: Cambridge University Press, 2010); and Nick Nesbitt, *Universal Emancipation: The Haitian Revolution and the Radical Enlightenment* (Charlottesville: University of Virginia Press, 2008).

16. Seymour Drescher, *Abolition: A History of Slavery and Antislavery* (New York: Cambridge University Press, 2009), 347. For ranking lists of Caribbean sugar producers, see David Watts, *The West Indies: Patterns of Development, Culture and Environmental Change since 1492* (Cambridge: Cambridge University Press, 1987), 500.

17. Peter Blanchard, *Slavery and Abolition in Early Republican Peru* (Wilmington, DE: Scholarly Resources, 1992), 186, 203. For an overview of sugar production in Peru, see Peter F. Klaren, "The Sugar Industry in Peru," *Revista de Indias* 65, no. 233 (April 1, 2005): 33–48.

18. Ulbe Bosma, *The Sugar Plantation in India and Indonesia: Industrial Production, 1770–2010* (Cambridge: Cambridge University Press, 2013), 89–95, 99–100.

19. Peter Griggs, "Sugar Plantations in Queensland, 1864–1912: Origins, Characteristics, Distribution, and Decline," *Agricultural History* 74 (2000): 609–47.

20. Bruce Knapman, "The Rise and Fall of the White Sugar Planter in Fiji, 1880–1925," *Pacific Studies* 9, no. 1 (November 1985): 53–82.

21. César J. Ayala, *American Sugar Kingdom: The Plantation Economy of the Spanish Caribbean, 1898–1934* (Chapel Hill: University of North Carolina Press, 1999).

22. Justin Roberts, "The 'Better Sort' and the 'Poorer Sort': Wealth Inequalities, Family Formation and the Economy of Energy on British Caribbean Sugar Plantations, 1750–1800," *Slavery & Abolition* 35 (2014): 458–73; and Richard B. Sheridan, "The Formation of Caribbean Plantation Society, 1689–1748," in Peter James Marshall, Alaine M. Low, and William Roger Louis, eds., *The Oxford History of the British Empire*, vol. 2: *The Eighteenth Century*, 394–414 (Oxford: Oxford University Press, 1998), 405.

23. Paul J. Dosal, *Doing Business with the Dictators: A Political History of United Fruit in Guatemala 1899–1944* (Wilmington, DE: Scholarly Resources, 1993), 9n.

24. William Kelleher Storey, "Small-Scale Sugar Cane Farmers and Biotechnology in Mauritius: The 'Uba' Riots of 1937," *Agricultural History* 69 (1995): 163–76.

25. John R. McNeill, *Mosquito Empires: Ecology and War in the Greater Caribbean, 1620–1914* (New York: Cambridge University Press, 2010), 23.

26. Stuart McCook, *States of Nature: Science, Agriculture, and Environment in the Spanish Caribbean, 1760–1940* (Austin: University of Texas Press, 2002), 22, 83.

27. Richard B. Sheridan, *Sugar and Slavery: An Economic History of the British West Indies*

1623-1775 (Baltimore: Johns Hopkins University Press, 1973), 141; and Sheridan, "Formation," 401, 407.

28. Sheridan, *Sugar and Slavery*, 160.

29. Reinaldo Funes Monzote, *From Rainforest to Cane Field in Cuba: An Environmental History since 1492* (Chapel Hill: University of North Carolina Press, 2008), 148n. J. H. Galloway reports that "the adoption of bagasse as the fuel [replacing firewood] may have been complete throughout the Lesser Antilles by 1719." (Galloway, *The Sugar Cane Industry: An Historical Geography from Its Origins to 1914* [Cambridge: Cambridge University Press, 1989], 98.)

30. Mark Overton, *Agricultural Revolution in England: The Transformation of the Agrarian Economy 1500-1850* (Cambridge: Cambridge University Press, 1996), 117-21.

31. Frank Uekötter, *Die Wahrheit ist auf dem Feld: Eine Wissensgeschichte der deutschen Landwirtschaft* (Göttingen: Vandenhoeck & Ruprecht, 2010), 383n.

32. David Ricardo, *On the Principles of Political Economy and Taxation* (London: John Murray, 1817).

33. Christoph Maria Merki, *Zucker gegen Saccharin: Zur Geschichte der künstlichen Süßstoffe* (Frankfurt: Campus, 1993), 13.

34. Merki, *Zucker*, 52, 56, 83-86, 200.

35. Max-Ferdinand Krawinkel, *Die Rübenzuckerwirtschaft im 19. Jahrhundert in Deutschland: Analyse und Bewertung der betriebswirtschaftlichen und volkswirtschaftlichen Entwicklung* (Cologne: Botermann & Botermann, 1994), 7.

36. Dirk Schaal, *Rübenzuckerindustrie und regionale Industrialisierung: Der Industrialisierungsprozeß im mitteldeutschen Raum 1799-1930* (Münster: LIT-Verlag, 2005), 29.

37. Hans-Heinrich Müller, "Landwirtschaft und industrielle Revolution: am Beispiel der Magdeburger Börde," in Toni Pierenkemper, ed., *Landwirtschaft und industrielle Entwicklung: Zur ökonomischen Bedeutung von Bauernbefreiung, Agrarreform und Agrarrevolution*, 45-57 (Stuttgart: Steiner, 1989), 54.

38. Schaal, *Rübenzuckerindustrie*, 23, 43, 221.

39. Merki, *Zucker*, 108.

40. Ricardo, *Principles*, 348.

41. Hauptstaatsarchiv Dresden 10736 Ministerium des Innern no. 16517, "Bericht über die argentinische Zuckerproduktion," Buenos Aires, February 1, 1897, pp. 51, 53, 82, 84 (quotation). See also no. 16519, "Die Entwickelung der argentinischen Zuckerindustrie seit 1897," Buenos Aires, April 29, 1902, p. 3.

42. Hauptstaatsarchiv Dresden 10736 Ministerium des Innern no. 16491, "Die bulgarische Zuckerindustrie," Bucarest, May 20, 1902, p. 1.

43. Hauptstaatsarchiv Dresden 10736 Ministerium des Innern no. 16516, "Die Zuckererzeugung in Mexico," Mexico, May 15, 1900, p. 81.

44. Hauptstaatsarchiv Dresden 10736 Ministerium des Innern no. 16500, "Die Entwicklung der russischen Zucker-Produktion, ihre heutige Lage und Zukunft," undated (ca. 1899), 57.

45. Hauptstaatsarchiv Dresden 10736 Ministerium des Innern no. 16491, "Die bulgarische Zuckerindustrie," Bucarest, May 20, 1902, p. 1.

46. See Ulbe Bosma and Roger Knight, "Global Factory and Local Field: Convergence and Divergence in the International Cane-Sugar Industry, 1850-1940," *International Review of Social History* 49 (2004): 1-25.

47. Bartow J. Elmore, *Citizen Coke: The Making of Coca-Cola Capitalism* (New York: Norton, 2015), 77, 98n.

48. Michael Fakhri, *Sugar and the Making of International Trade Law* (Cambridge: Cambridge University Press, 2014), 25.

49. Merki, *Zucker*, 149.

50. Hauptstaatsarchiv Dresden 10736 Ministerium des Innern no. 15500, "Aufruf an die Deutsche Landwirtschaft, Zucker-Industrie und -Handel zur Gründung einer Vereinigung zur Hebung des Zuckerkonsums," undated (ca. 1908), 1.

51. Ben Richardson, *Sugar* (Cambridge: Polity, 2015), 4.

52. Diana Paton, "The Abolition of Slavery in the Non-Hispanic Caribbean," in Stephan Palmié and Francisco A. Scarano, eds., *The Caribbean: A History of the Region and Its Peoples*, 289–301 (Chicago: University of Chicago Press, 2011), 290.

53. Francisco Arturo Rosales, *Chicano! The History of the Mexican American Civil Rights Movement* (Houston, TX: Arte Público Press, 1996), 130–51.

54. See p. 115.

55. Jonathan T. Thomas, "And the Rest is History: A Conversation with Sidney Mintz," *American Anthropologist* 116, no. 3 (September 2014): 1–14; here, 6.

56. David Grigg, *The Agricultural Revolution in South Lincolnshire* (Cambridge: Cambridge University Press, 1966).

57. Kimiko de Freytas-Tamura, "As More Immigrants Arrive, Some Britons Want to Show Them and E.U. the Door," *New York Times*, March 14, 2016.

58. "Lincolnshire Records UK's Highest Brexit Vote," accessed November 3, 2021, http://www.bbc.co.uk/news/uk-politics-eu-referendum-36616740.

59. Peter T. Marsh, *Joseph Chamberlain: Entrepreneur in Politics* (New Haven, CT: Yale University Press, 1994), 324.

60. Justin Roberts, "Surrendering Surinam: The Barbadian Diaspora and the Expansion of the English Sugar Frontier, 1650–75," *William and Mary Quarterly* 73 (2016): 225–56.

61. For comprehensive accounts of the sisal project, see Nick Smart, *Neville Chamberlain* (London: Routledge, 2010), 9–25; Robert Self, *Neville Chamberlain: A Biography* (Aldershot: Ashgate, 2006), 21–26; and David Dilks, *Neville Chamberlain*, vol. 1: *Pioneering and Reform, 1869–1929* (Cambridge: Cambridge University Press, 1984), 37–75.

62. Gregg Mitman and Paul Erickson, "Latex and Blood: Science, Markets, and American Empire," *Radical History Review* 107 (2010): 45–73; here, 59.

63. Francesco Monastra and Elisabetta Raparelli, *Inventory of Almond Research, Germplasm and References* (Rome: Food and Agriculture Organization, 1997).

64. See John Asafu-Adjaye et al., "An Ecomodernist Manifesto," April 2015, p. 18, accessed November 3, 2021, http://www.ecomodernism.org/manifesto-english/.

65. NCD Risk Factor Collaboration, "Trends in Adult Body-Mass Index in 200 Countries from 1975 to 2014: A Pooled Analysis of 1698 Population-Based Measurement Studies with 19.2 Million Participants," *Lancet* 387 (2016): 1377–96; here, 1383n.

66. John Yudkin, *Pure, White and Deadly: The Problem of Sugar* (London: Davis-Poynter, 1972).

67. Ed Pilkington, "Caribbean Nations Prepare Demand for Slavery Reparations," *The Guardian*, March 9, 2014, accessed November 3, 2021, https://www.theguardian.com/world/2014/mar/09/caribbean-nations-demand-slavery-reparations.

68. Zachary J. Foster, "Why Are Modern Famines so Deadly? The First World War in Syria and Palestine," in Richard P. Tucker et al., eds., *Environmental Histories of the First World War*, 191–207 (New York: Cambridge University Press, 2018), 198.

69. Philip T. Hoffman, *Growth in a Traditional Society: The French Countryside, 1450–1815* (Princeton, NJ: Princeton University Press, 1996), 197.

70. Suzanne Moon, "Empirical Knowledge, Scientific Authority, and Native Development: The Conflict over Sugar/Rice Ecology in the Netherlands East Indies, 1905–1914," *Environment and History* 10 (2004): 59–81.

71. Beth Fowkes Tobin, *Colonizing Nature: The Tropics in British Arts and Letters, 1760–*

1820 (Philadelphia: University of Pennsylvania Press, 2005), 59–62; and Judith A. Carney and Richard Nicholas Rosomoff, *In the Shadow of Slavery: Africa's Botanical Legacy in the Atlantic World* (Berkeley: University of California Press, 2009), 123 (quotation).

72. The phrase "ends of the earth" is borrowed from Robert D. Kaplan, *The Ends of the Earth: A Journey at the Dawn of the 21st Century* (New York: Random House, 1996).

73. Norman Davies, *Heart of Europe: The Past in Poland's Present* (Oxford: Oxford University Press, 2001), 49.

3: THE CANAL DU MIDI

1. Philippe Delvit, "Un Canal au Midi," in Conseil d'Architecture, d'Urbanisme et de l'Environnement de la Haute-Garonne, ed., *Canal Royal de Languedoc: Le Partage des Eaux*, 204–24 (Portet-sur-Garonne: Loubatières, 1992), 206.

2. As befits a defining project of La Grande Nation, there is a broad range of publications on the Canal du Midi. The most comprehensive treatise is Jean-Denis Bergasse, ed., *Le Canal du Midi*, 4 vols. (Cessenon: Jean-Denis Bergasse, 1982–1985). Other works of note are André Maistre, *Le Canal des Deux Mers: Canal Royal du Languedoc, 1666–1810* (Toulouse: Édouard Privat, 1968); Conseil d'Architecture, d'Urbanisme et de l'Environnement de la Haute-Garonne, *Canal Royal de Languedoc*; and L. T. C. Rolt, *From Sea to Sea: The Canal du Midi* (London: Allen Lane, 1973). Chandra Mukerji, *Impossible Engineering: Technology and Territoriality on the Canal du Midi* (Princeton, NJ: Princeton University Press, 2009), is instructive on the canal's symbolism and Riquet's marshaling of local expertise. Books directed toward a popular audience include Michel Cotte, *Le Canal du Midi: "Merveille de L'Europe"* (Paris: Belin-Herscher, 2003); and René Gast and Jacques Debru, *Le Canal du Midi: Histoire d'un chef-d'œuvre* (Rennes: Editions Ouest-France, 2006).

3. David Parker, *Class and State in* Ancien Régime *France: The Road to Modernity* (London: Routledge, 1996), 35.

4. Mukerji, *Impossible Engineering*, 16.

5. Cotte, *Le Canal du Midi*, 13, 19; Gast and Debru, *Le Canal du Midi*, 2; Uwe Burghardt, "Zu Wasser über Land: Der Canal du Midi—ein merkantilistischer Verkehrsweg," *Kultur und Technik* 17, no. 2 (1993): 23–29; here, 23.

6. Mukerji, *Impossible Engineering*, 28.

7. Andrée Corvol, "Autres Regards, Autres Perspectives," in Corvol, ed., *La Forêt Malade: Débats Anciens et Phénomènes Nouveaux XVIIe–XXe Siècle*, 35–50 (Paris: L'Harmattan, 1994), 36.

8. Mukerji, *Impossible Engineering*, 26. See Sharon Kettering, *Patrons, Brokers, and Clients in Seventeenth-Century France* (New York: Oxford University Press, 1986), 234.

9. Mukerji, *Impossible Engineering*, 7.

10. Thomas Babington Macaulay, *The Complete Works of Lord Macaulay: History of England*, vol. 1 (London: Longmans, Green, and Co., 1898), 389.

11. Jacques Colin, "The Premises of Logistics: The Organisation of Warships in France in the 17th and 18th Centuries," in Alberto Ochoa-Zezzatti et al., eds., *Handbook of Research on Military, Aeronautical, and Maritime Logistics and Operations*, 1–12 (Hershey, PA: Business Science Reference, 2016), 4.

12. Jonathan I. Israel, *Conflicts of Empires: Spain, the Low Countries and the Struggle for World Supremacy 1585–1713* (London: Hambledon Press, 1997), 45–62.

13. Maxwell G. Lay, *Ways of the World: A History of the World's Roads and of the Vehicles That Used Them* (New Brunswick, NJ: Rutgers University Press, 1992), 96.

14. Dennis E. Showalter, "Railroads, the Prussian Army, and the German Way of War in the Nineteenth Century," in T. G. Otte and Keith Neilson, eds., *Railways and International Politics: Paths of Empire, 1848–1945*, 21–44 (London: Routledge, 2006), 22–24.

15. See Michael Salewski, "Die militärische Bedeutung des Nord-Ostsee-Kanals," in Rainer Lagoni, Hellmuth St. Seidenfus, and Hans-Jürgen Teuteberg, eds., *Nord-Ostsee-Kanal 1895-1995: Festschrift*, 341-64 (Neumünster: Wachholtz, 1995).

16. The 1888 Convention of Constantinople gave free passage to warships of all nations, but Britain effectively blocked the canal during World War I. (Douglas Anthony Farnie, *East and West of Suez: The Suez Canal in History 1854-1956* [Oxford: Clarendon Press, 1969], 337, 548.)

17. See pp. 511-12.

18. David Gugerli and Daniel Speich, *Topografien der Nation: Politik, Kartografische Ordnung und Landschaft im 19. Jahrhundert* (Zurich: Chronos, 2002), 11, 19, 52.

19. See Todd A. Shallat, *Structures in the Stream: Water, Science, and the Rise of the U.S. Army Corps of Engineers* (Austin: University of Texas Press, 1994).

20. Daniel R. Headrick, *The Tools of Empire: Technology and European Imperialism in the Nineteenth Century* (New York: Oxford University Press, 1981), 12.

21. For Riquet's biography, see Monique Dollin du Fresnel, *Pierre-Paul Riquet (1609-1680): L'incroyable aventure du Canal des Deux-Mers* (Bordeaux: Éditions Sud Ouest, 2012).

22. Michael S. Mahoney, "Organizing Expertise: Engineering and Public Works under Jean-Baptiste Colbert, 1662-83," *Osiris* 25 (2010): 149-70; here, 151, 157n.

23. Chandra Mukerji, "The New Rome: Infrastructure and National Identity on the Canal du Midi," *Osiris* 24 (2009): pp. 15-32; here, 18, 23.

24. Jean-Denis Bergasse, "Le 'Culte' de Riquet en Languedoc au XIXe Siècle," in Bergasse, *Canal du Midi*, vol. 1, 217-51; here, 220. As a literary monument, see Les Descendans de Pierre-Paul Riquet de Bonrepos, *Histoire du Canal de Languedoc, Rédigée sur les Pièces authentiques conservées à la Bibliothèque Impériale et aux Achives du Canal* (Paris: L'Imprimerie de Carpelet, 1805).

25. Lydia Beauvais, "An Allegorised History of the First Eighteen Years of Louis XIV's Reign," in Jeanne Faton and Louis Faton, eds., *The Hall of Mirrors: History and Restoration*, 214-87 (Dijon: Faton, 2007), 286.

26. Michel Adgé, "Chronologie des Principaux Èvénements de la Construction du Canal (1662-1694)," Bergasse, *Le Canal du Midi*, vol. 4, 171-92; here, 190-92; Janis Langins, *Conserving the Enlightenment: French Military Engineering from Vauban to the Revolution* (Cambridge, MA: MIT Press, 2004), 68-70; and Bernard Pujo, *Vauban* (Paris: Albin Michel, 1991), 130n.

27. See David Edgerton, *The Shock of the Old: Technology and Global History Since 1900* (London: Profile Books, 2008), 75-102; and Andrew L. Russell and Lee Vinsel, "After Innovation, Turn to Maintenance," *Technology and Culture* 59 (2018): 1-25.

28. Thomas Wakefield Goodspeed, *A History of the University of Chicago: The First Quarter-Century* (Chicago: University of Chicago Press, 1972 [1916]), 9-11; George W. Corner, *A History of the Rockefeller Institute 1901-1953: Origins and Growth* (New York: Rockefeller Institute Press, 1964), 67-69; and Linnea A. Nelson, Elma S. Herradura, and Eliza U. Griño, *Scientia et Fides: The Story of Central Philippine University* (Iloilo City: The University, 1981), 1.

29. See Dollin du Fresnel, *Pierre-Paul Riquet*, 331-72.

30. Jean Servières, "Du Canal du Midi à la Compagnie des Chemins de Fer du Midi: Concurrence ou complémentarité?" in Bergasse, *Le Canal du Midi*, vol. 4, 387-410; here, 406.

31. Emmanuel Le Roy Ladurie, *The Ancien Régime: A History of France, 1610-1774* (Cambridge, MA: Blackwell, 1996), 150n.

32. Emmanuel Le Roy Ladurie, *The Peasants of Languedoc* (Urbana: University of Illinois Press, 1974), 224.

33. Richard White, *Railroaded: The Transcontinentals and the Making of Modern America* (New York: Norton, 2011), 517.

34. Jamie Monson, *Africa's Freedom Railway: How a Chinese Development Project Changed Lives and Livelihoods in Tanzania* (Bloomington: Indiana University Press, 2009), 22, 123; Richard Hall and Hugh Peyman, *The Great Uhuru Railway: China's Showpiece in Africa* (London: Victor Gollancz, 1976), 57, 188.

35. Steven G. Marks, *Road to Power: The Trans-Siberian Railroad and the Colonization of Asian Russia 1850–1917* (Ithaca, NY: Cornell University Press, 1991), 218.

36. "History of the Bridge to Nowhere," accessed November 3, 2021, https://www.doc.govt.nz/parks-and-recreation/places-to-go/manawatu-whanganui/places/whanganui-national-park/bridge-to-nowhere.

37. Philip T. Hoffman, *Growth in a Traditional Society: The French Countryside, 1450–1815* (Princeton, NJ: Princeton University Press, 1996), 183.

38. Robert William Fogel, *Railroads and American Economic Growth: Essays in Econometric History* (Baltimore: Johns Hopkins University Press, 1964).

39. Fogel's approach is arguably more helpful when it comes to the economic effects of construction itself, as the railroads' demand for iron was lower than commonly assumed. However, the railroad's impact was bigger in other countries. (See Rainer Fremdling, "Railroads and German Economic Growth: A Leading Sector Analysis with a Comparison to the United States and Great Britain," *Journal of Economic History* 37 [1977]: 583–604.)

40. Fogel takes the extra costs for monthlong storage into account, but it seems that he did not find this satisfactory. He also reports on ideas to keep canals free of ice "by the application or [sic] artificially generated heat." (Fogel, *Railroads*, 224.)

41. Charles Hadfield, *British Canals: An Illustrated History*, 6th ed. (Newton Abbot: David & Charles, 1979), 261, 291.

42. Roger Pilkington, "Pierre-Paul Riquet and the Canal du Midi," *History Today* 23, no. 3 (March 1, 1973): 170–76; here, 175.

43. Quoted in Klaus Augustin, *Die Umgehungsstraße: Telgter Zankapfel seit 177 Jahren* (Telgte: N.p., 1987), 23.

44. Frederick Allen, *Atlanta Rising: The Invention of an International City 1946–1996* (Marietta, GA: Longstreet Press, 1996), 24.

45. Per Högselius, Arne Kaijser, and Erik van der Vleuten, *Europe's Infrastructure Transition: Economy, War, Nature* (Basingstoke: Palgrave Macmillan, 2015), 8.

46. See Ashley Carse, *Beyond the Big Ditch: Politics, Ecology, and Infrastructure at the Panama Canal* (Cambridge, MA: MIT Press, 2014), 37–68.

47. Mukerji, *Impossible Engineering*, 226.

48. Robert B. Marks, *China: Its Environment and History* (Lanham, MD: Rowman and Littlefield, 2012), 119, 240.

49. Cotte, *Canal du Midi*, 139–42.

50. Francois Bordry, "A Canal in Southern France," *UNESCO Courier* 50, no. 9 (September 1997): 37.

51. For context, see Richard P. Mitchell, *The Society of the Muslim Brothers* (New York: Oxford University Press, 1993), 7; Brynjar Lia, *The Society of the Muslim Brothers in Egypt* (Reading, UK: Ithaca Press, 1998), 30; and Gudrun Krämer, *Der Architekt des Islamismus: Hasan al-Banna und die Muslimbrüder: Eine Biographie* (Munich: Beck, 2022). Ironically, the Suez Canal Company was among the earlier supporters of the Muslim Brothers, donating a quarter of the money for a new mosque. (Lia, *Society*, 41.)

52. James L. Gelvin, *The Modern Middle East: A History*, 2nd ed. (New York: Oxford University Press, 2008), 300.

53. Valeska Huber, *Channelling Mobilities: Migration and Globalisation in the Suez Canal Region and Beyond, 1869–1914* (Cambridge: Cambridge University Press, 2013), 153.

54. Alan R. Longhurst, *Ecological Geography of the Sea*, 2nd ed. (Amsterdam: Elsevier, 2007), 182. Migration in the opposite direction is less frequent because conditions in the Red Sea provide its fauna with an evolutionary advantage.

55. Huber, *Channelling Mobilities*, 28.

56. David McCullough, *The Path between the Seas: The Creation of the Panama Canal 1870–1914* (New York: Simon and Schuster, 1977), 172, 203.

57. Walter LaFeber, *The Panama Canal: The Crisis in Historical Perspective* (New York: Oxford University Press, 1978), 32, 38; and McCullough, *Path*, 334, 388.

58. Bertrand Gabolde, "Les Ouvriers du Chantier," in Bergasse, *Le Canal du Midi*, vol. 3, 233–39.

59. Huber, *Channelling Mobilities*, 252.

60. Philippe Oriol, *L'histoire de l'affaire Dreyfus. I: L'affaire du Captaine Dreyfus 1894–1897* (Paris: Éditions Stock, 2008), 250.

61. For an inspiring discussion of these issues, see Valeska Huber, "Multiple Mobilities: Über den Umgang mit verschiedenen Mobilitätsformen um 1900," *Geschichte und Gesellschaft* 36 (2010): 317–41.

62. Le Roy Ladurie, *Ancien Régime*, 151.

63. Delphine de Mallevoüe, "Le sacrifice des plantanes du canal du Midi," accessed November 3, 2021, http://www.lefigaro.fr/actualite-france/2011/07/26/01016-20110726ARTFIG00584-le-sacrifice-des-platanes-du-canal-du-midi.php. See also https://www.replantonslecanaldumidi.fr/notre-programme-de-replantation/, accessed November 3, 2021.

4: SUSTAINABLE FORESTRY

1. Hans Carl von Carlowitz, *Sylvicultura oeconomica oder Haußwirthliche Nachricht und Naturmäßige Anweisung zur Wilden Baum-Zucht*, edited by Joachim Hamberger (Munich: oekom, 2013), 189, 190, 216. Page 105 is the page number in the original edition of 1713.

2. World Commission on Environment and Development, *Our Common Future: Development and International Economic Co-operation; Environment* (N.p.: United Nations, 1987). See also Iris Borowy, "Sustainable Development and the United Nations," in Jeremy L. Caradonna, ed., *Routledge Handbook of the History of Sustainability*, 151–63 (London: Routledge, 2018), 156.

3. Paul Warde, *The Invention of Sustainability: Nature and Destiny, c. 1500–1870* (Cambridge: Cambridge University Press, 2018), 168.

4. Carlowitz, *Sylvicultura oeconomica*, 189.

5. Sam White, "From Globalized Pig Breeds to Capitalist Pigs: A Study in Animal Cultures and Evolutionary History," *Environmental History* 16 (2011): 94–120.

6. See Elisabeth Weinberger, *Waldnutzung und Waldgewerbe in Altbayern im 18. und beginnenden 19. Jahrhundert* (Stuttgart: Steiner, 2001), 85.

7. Ulrich Gruber, *Die Entdeckung der Nachhaltigkeit: Kulturgeschichte eines Begriffs* (Munich: Kunstmann, 2010), 111.

8. This sentence is a play on Charles Tilly, *Coercion, Capital, and European States, AD 990–1992* (Cambridge, MA: Blackwell, 1992), 67–95.

9. Cornel Zwierlein, *Der gezähmte Prometheus: Feuer und Sicherheit zwischen Früher Neuzeit und Moderne* (Göttingen: Vandenhoeck & Ruprecht, 2011).

10. See Bernward Selter, *Waldnutzung und ländliche Gesellschaft: Landwirtschaftlicher "Nährwald" und neue Holzökonomie im Sauerland des 18. und 19. Jahrhunderts* (Paderborn: Schöningh, 1995); Bernd-Stefan Grewe, *Der versperrte Wald. Ressourcenmangel in der baye-

rischen Pfalz (1814–1870) (Cologne: Böhlau, 2004); and Richard Hölzl, *Umkämpfte Wälder: Die Geschichte einer ökologischen Reform in Deutschland 1760–1860* (Frankfurt: Campus, 2010).

11. Joachim Hamberger, "Einführung in die Edition," in Carlowitz, *Sylvicultura oeconomica*, 18–87; here, 23.

12. Paul Warde, "Fear of Wood Shortage and the Reality of the Woodland in Europe, ca. 1450–1850," *History Workshop Journal* 62 (2006): 28–57; here, 31; and Karen Brown, "The Conservation and Utilisation of the Natural World: Silviculture in the Cape Colony, c. 1902–1910," *Environment and History* 7 (2001): 427–47.

13. Ramachandra Guha, *The Unquiet Woods: Ecological Change and Peasant Resistance in the Himalaya* (Delhi: Oxford University Press, 1989), 105; and Richard P. Tucker, *A Forest History of India* (New Delhi: Sage, 2012), 46n, 58.

14. William Beinart and Lotte Hughes, *Environment and Empire* (Oxford: Oxford University Press, 2007), 128.

15. Robert Pogue Harrison, *Forests: The Shadow of Civilization* (Chicago: University of Chicago Press, 1993), 201.

16. Harold K. Steen, *The U.S. Forest Service: A History* (Seattle: University of Washington Press, 1991), 24, 48; S. Ravi Rajan, *Modernizing Nature: Forestry and Imperial Eco-Development 1800–1950* (Hyderabad: Orient Longman, 2006), 46n; and Tamara L. Whited, *Forests and Peasant Politics in Modern France* (New Haven, CT: Yale University Press, 2000), 33.

17. Rajan, *Modernizing Nature*, 82–90; here, 89.

18. Shirley Ye, "Business, Water, and the Global City: Germany, Europe, and China, 1820–1950" (PhD diss., Harvard University, 2013), 275, 281.

19. Karl Hasel, Ekkehard Schwartz, *Forstgeschichte: Ein Grundriß für Studium und Praxis* (Remagen: Kessel, 2002), 333–50.

20. For the origins of the revisionist reading of the timber famine trope in Germany, see Joachim Radkau, "Holzverknappung und Krisenbewußtsein im 18. Jahrhundert," *Geschichte und Gesellschaft* 9 (1983): 513–43; and Joachim Radkau, "Zur angeblichen Energiekrise des 18. Jahrhunderts: Revisionistische Betrachtungen über die 'Holznot,'" *Vierteljahresschrift für Wirtschafts- und Sozialgeschichte* 73 (1986): 1–37. English-language summaries of subsequent reflections include Warde, "Fear of Wood Shortage"; Richard Hölzl, "Historicizing Sustainability: German Scientific Forestry in the Eighteenth and Nineteenth Centuries," *Science as Culture* 19 (2010): 431–60; and Bernd-Stefan Grewe, "Power, Politics, and Protecting the Forest: Scares about Wood Shortages and Deforestation in Early Modern German States," in Frank Uekötter, ed., *Exploring Apocalyptica: Coming to Terms with Environmental Alarmism*, 12–35 (Pittsburgh, PA: University of Pittsburgh Press, 2018).

21. Reginald Cline-Cole, "Scientific/Empire Forestry and Late-Colonial Northern Nigeria: Exploring Transnational (Inter-)Connections" (unpublished manuscript [2017]).

22. Peter Linebaugh, "Karl Marx, the Theft of Wood, and Working-Class Composition: A Contribution to the Current Debate," *Social Justice* 40 (2014): 137–61; here, 140n.

23. Thomas Sikor, Johannes Stahl, and Stefan Dorondel, "Negotiating Post-Socialist Property and State: Struggles over Forests in Albania and Romania," *Development and Change* 40 (2009): 171–93.

24. David Clay Large, *Hitlers München: Aufstieg und Fall der Hauptstadt der Bewegung* (Munich: dtv, 2003), 200. For insights into Escherich's state of mind in 1923, see Karl Escherich, *Der deutsche Wald und die feindlichen Mächte* (Hamburg: Hanseatische Verlagsanstalt, ca. 1923).

25. Ute Deichmann, *Biologists under Hitler* (Cambridge, MA: Harvard University Press, 1996), 45.

26. As an example of this type of whitewashing, see Erwin Schimitschek, "Escherich, Karl Leopold," in Historische Kommission bei der Bayerischen Akademie der Wissenschaften, ed., *Neue Deutsche Biographie*, vol. 4 (Berlin: Duncker & Humblot, 1959), 649.

27. Archive of Ludwig-Maximilians-University Munich Y-XVI-14, Staatswissenschaftliche Fakultät der Ludwig-Maximilians-Universität to the Senat, February 20, 1914, p. 2.

28. Karl Escherich, *Leben und Forschen: Kampf um eine Wissenschaft*, 2nd ed. (Stuttgart: Wissenschaftliche Verlagsgesellschaft, 1949), 20, 59, 88, 197.

29. Karl Escherich, *Die Forstinsekten Mitteleuropas: Ein Lehr- und Handbuch*, vols. 1–3, 5) (Berlin: Parey, 1914/1923/1931/1942).

30. Karl Gayer, *Der gemischte Wald, seine Begründung und Pflege, insbesondere durch Horst- und Gruppenwirtschaft* (Berlin: Paul Parey, 1886), 7.

31. Karl Hasel, *Studien über Wilhelm Pfeil* (Hanover: Schaper, 1982), 160, 176 (quotation). It is important to recognize this diversity of opinions as some accounts of German forestry depict it as a monolithic campaign for monoculture. (See James C. Scott, *Seeing Like a State: How Certain Schemes to Improve the Human Condition Have Failed* [New Haven, CT: Yale University Press, 1998], 11–22.) Christian Lotz has rightly pointed out that research on the state's claim to the woodlands is much more extensive than research on nineteenth-century forest management. (Christian Lotz, "Expanding the Space for Future Resource Management: Explorations of the Timber Frontier in Northern Europe and the Rescaling of Sustainability during the Nineteenth Century," *Environment and History* 21 [2015]: 257–79; here, 259.)

32. Joachim Radkau, *Holz: Wie ein Naturstoff Geschichte schreibt* (Munich: oekom, 2007), 166.

33. Karl Escherich, *Biologisches Gleichgewicht: Eine zweite Münchener Rektoratsrede über die Erziehung zum politischen Menschen* (Munich: Albert Langen und Georg Müller, 1935), 16.

34. Archive of Ludwig-Maximilians-University Munich Y-XVI-14, Karl Escherich, Stellungnahme zur Besetzung forstzoologischer Professuren, January 2, 1939, and Karl Escherich to Dekan der Staatswirtschaftlichen Fakultät, January 3, 1939.

35. Escherich, *Leben und Forschen*, 314; Archive of Ludwig-Maximilians-University Munich Y-XVI-14, Dekan der staatswirtschaftlichen Fakultät der Universität München to Staatsministerium für Unterricht und Kultus, November 21, 1938, p. 2.

36. See the pertinent correspondence in Archive of Ludwig-Maximilians-University Munich M-IX-13.

37. Karl Escherich, *Die angewandte Entomologie in den Vereinigten Staaten: Eine Einführung in die biologische Bekämpfungsmethode. Zugleich mit Vorschlägen zu einer Reform der Entomologie in Deutschland* (Berlin: Parey, 1913), esp. 24, 167, 172.

38. Escherich, *Leben und Forschen*, 146.

39. Karl Escherich, *Die Erforschung der Waldverderber: Drei Jahrzehnte im Kampf gegen Forstschädlinge. Rückblick und Ausblick* (Berlin: Parey, 1936), 18.

40. Escherich, *Leben und Forschen*, 306.

41. Paul Warde, "The Invention of Sustainability," *Modern Intellectual History* 8 (2011): 153–70; here, 162.

42. Henry E. Lowood, "The Calculating Forester: Quantification, Cameral Science, and the Emergence of Scientific Forestry Management in Germany," in Tore Frängsmyr, John L. Heilbron and Robin E. Rider, eds., *The Quantifying Spirit in the Eighteenth Century*, 315–42 (Berkeley: University of California Press, 1990), 337.

43. Wilhelm Pfeil, *Die Forstgeschichte Preussens bis zum Jahre 1806* (Leipzig: Baumgärtners Buchhandlung, 1839), 274.

44. T. C. Smout, "Land and Sea: The Environment," in T. M. Devine and Jenny

Wormald, eds., *The Oxford Handbook of Modern Scottish History*, 19–38 (Oxford: Oxford University Press, 2014), 35.

45. Scottish Executive, *The Scottish Forestry Strategy 2006* (Edinburgh: Forestry Commission Scotland, 2006), 10, 15, 16 (quotations).

46. Oliver Rackham, *Woodlands* (London: William Collins, 2015), 365.

47. Frank Uekoetter, *The Green and the Brown: A History of Conservation in Nazi Germany* (New York: Cambridge University Press, 2006), 69–71. Some scholars mistook rhetoric for reality and falsely presumed a dedicated policy of forest reform. For one example, see Simon Schama, *Landscape and Memory* (London: HarperCollins, 1995), 119.

48. Aldo Leopold, "Deer and Dauerwald in Germany, I. History; II. Ecology and Policy," *Journal of Forestry* 34 (1936): 366–75, 460–66. See also Susan L. Flader, *Thinking Like a Mountain: Aldo Leopold and the Evolution of an Ecological Attitude toward Deer, Wolves and Forests* (Madison: University of Wisconsin Press, 1994), 139–44.

49. Wilhelm Bode and Martin von Hohnhorst, *Waldwende: Vom Försterwald zum Naturwald* (Munich: Beck, 2000), 74.

50. Lotz, "Expanding the Space," 263n. See also Christian Lotz, *Nachhaltigkeit neu skalieren. Internationale forstwissenschaftliche Kongresse und Debatten um die Ressourcenversorgung der Zukunft im Nord- und Ostseeraum (1870–1914)* (Cologne: Böhlau, 2018).

51. Uwe Eduard Schmidt, "Steinkohlenbergbau und Forstwirtschaft: eine Schicksalsgemeinschaft im Saarkohlenrevier," in Karlheinz Pohmer, ed., *Der Saarländische Steinkohlenbergbau. Dokumentation seiner historischen Bedeutung und seines kulturellen Erbes*, 370–413 (Dillingen: Krüger, 2012).

52. Cyril E. Hart, *Royal Forest: A History of Dean's Woods as Producers of Timber* (Oxford: Clarendon Press, 1966), 207n, 223n.

53. Rackham, *Woodlands*, 360.

54. As befits the metaphor, some people previously saw the forest without the help of satellite pictures: see R. A. Cline-Cole, H. A. C. Main, and J. E. Nichol, "On Fuelwood Consumption, Population Dynamics and Deforestation in Africa," *World Development* 18 (1990): 513–27.

55. See the divergent assessments in Martin Brandt et al., "Ground- and Satellite-based Evidence of the Biophysical Mechanisms Behind the Greening Sahel," *Global Change Biology* 21 (2015): 1610–1620; and Cecile Dardel et al., "Re-greening Sahel: 30 Years of Remote Sensing Data and Field Observations (Mali, Niger)," *Remote Sensing of Environment* 140 (2014): 350–64.

56. Reginald Cline-Cole, "Knowledge Claims and Landscapes: Alternative Views of the Fuelwood-Degradation Nexus in Northern Nigeria," *Environment and Planning D: Society and Space* 16 (1998): 311–46; here, 317.

57. Abasse Tougiani, Chaibou Guero, and Tony Rinaudo, "Community Mobilisation for Improved Livelihoods through Tree Crop Management in Niger," *GeoJournal* 74 (2009): 377–89; here, 378.

58. See Michael Mortimore, *Roots in the African Dust: Sustaining the Sub-Saharan Drylands* (Cambridge: Cambridge University Press, 1998), 170n.

59. Michael J. Mortimore and William M. Adams, "Farmer Adaptation, Change and 'Crisis' in the Sahel," *Global Environmental Change* 11 (2001): 49–57; here, 55.

60. Joachim N. Binam et al., "Effects of Farmer Managed Natural Regeneration on Livelihoods in Semi-Arid West Africa," *Environmental Economics and Policy Studies* 17 (2015): 543–75; here, 544.

61. Tougiani, Guero, and Rinaudo, "Community Mobilisation," 381; Eric Haglund et al., "Dry Land Tree Management for Improved Household Livelihoods: Farmer Managed

Natural Regeneration in Niger," *Journal of Environmental Management* 92 (2011): 1696–705; here, 1704.

62. Peter Weston et al., "Farmer-Managed Natural Regeneration Enhances Rural Livelihoods in Dryland West Africa," *Environmental Management* 55 (2015): 1402–17; here, 1406. See also Gérard Buttoud, "Politiques et Pratiques Forestières en Afrique Sèche," *Économie Rurale* 191 (1989): 3–7.

63. Stefanie M. Herrmann and G. Gray Tappan, "Vegetation Impoverishment Despite Greening: A Case Study from Central Senegal," *Journal of Arid Environments* 90 (2013): 55–66; here, 64.

64. Reginald Cline-Cole, "Woodfuel Discourses and the Re-framing of Wood Energy," *Forum for Development Studies* 34 (2007): 121–53; here, 145.

65. Cline-Cole, "Knowledge Claims," 334.

66. Kieko Matteson, *Forests in Revolutionary France: Conservation, Community, and Conflict, 1669–1848* (New York: Cambridge University Press, 2015), 263.

67. Matteson, *Forests*, 318.

68. See Eike Luedeling and Henry Neufeldt, "Carbon Sequestration Potential of Parkland Agroforestry in the Sahel," *Climatic Change* 115 (2012): 443–61.

69. Ralf Straußberger and Nicola Uhde, *Deutschlands Forstwirtschaft auf dem Holzweg: BUND-Schwarzbuch Wald* (Berlin: Bund für Umwelt und Naturschutz Deutschland, 2009), 3.

5: SHIPBREAKING IN CHITTAGONG

1. Deborah Breen, "Constellations of Mobility and the Politics of Environment: Preliminary Considerations of the Shipbreaking Industry in Bangladesh," *Transfers* 1, no. 3 (Winter 2011): 24–43; here, 24.

2. Md. M. Maruf Hossain, Mohammad Mahmudul Islam, *Ship Breaking Activities and Its Impact on the Coastal Zone of Chittagong, Bangladesh: Towards Sustainable Management* (Chittagong: Young Power in Social Action, 2006), 12. Bangladesh was part of Pakistan until it became an independent country in 1971. For an overview of the nation's history, see Willem van Schendel, *A History of Bangladesh* (Cambridge: Cambridge University Press, 2009).

3. "Hard to Break Up: Ship Breaking in Bangladesh," *The Economist* 405, no. 8808 (October 27, 2012): 44.

4. Renate Pieper, "Amerikanische Edelmetalle in Europa (1492–1621): Ihr Einfluß auf die Verwendung von Gold und Silber," *Jahrbuch für Geschichte Lateinamerikas* 32 (1995): 163–92; here, 166.

5. Kendall W. Brown, *A History of Mining in Latin America: From the Colonial Era to the Present* (Albuquerque: University of New Mexico Press, 2012), 36.

6. S. M. Mizanur Rahman and Audrey L. Mayer, "How Social Ties Influence Metal Resource Flows in the Bangladesh Ship Recycling Industry," *Resources, Conservation and Recycling* 104 (2015): 254–64.

7. Chad Denton, "'Récupérez!' The German Origins of French Wartime Salvage Drives, 1939–1945," *Contemporary European History* 22 (2013): 399–430; here, 419n; Susanne Köstering, "'Pioniere der Rohstoffbeschaffung': Lumpensammler im Nationalsozialismus, 1934–1939," *Werkstatt Geschichte* 17 (1997): 45–65; here, 47.

8. Carl A. Zimring, *Cash for your Trash: Scrap Recycling in America* (New Brunswick, NJ: Rutgers University Press, 2005), 58.

9. R. Scott Frey, "Breaking Ships in the World-System: An Analysis of Two Ship Breaking Capitals, Alang-Sosiya, India and Chittagong, Bangladesh," *Journal of World-Systems Research* 21 (2015): 25–49; here, 31. Shipbreaking has never been the dominant industry in

Chittagong. (Md Aslam Mia et al., "Chittagong, Bangladesh," *Cities* 48 [2015]: 31–41; here, 33.)

10. Hossain and Islam, *Ship Breaking Activities*, 33.

11. Zimring, *Cash*, 118.

12. Muhammod Shaheen Chowdhury, "Compliance with Core International Labor Standards in National Jurisdiction: Evidence from Bangladesh," *Labor Law Journal* 68, no. 1 (Spring 2017): 78–93; here, 87.

13. Ruth Oldenziel and Heike Weber, "Introduction: Reconsidering Recycling," *Contemporary European History* 22 (2013): 347–70; here, 349.

14. Quoted in Mark Pendergrast, *For God, Country, and Coca-Cola: The Unauthorized History of the World's Most Popular Soft Drink* (London: Phoenix, 1994), 227.

15. See Saujanya Sinha, "Ship Scrapping and the Environment: the Buck Should Stop," *Maritime Policy and Management* 25 (1998): 397–403; here, 397n; Mohammad Sujauddin et al., "Characterization of Ship Breaking Industry in Bangladesh," *Journal of Material Cycles and Waste Management* 17 (2015): 72–83; here, 73; and Peter Rousmaniere and Nikhil Raj, "Shipbreaking in the Developing World: Problems and Prospects," *International Journal of Occupational and Environmental Health* 13 (2007): 359–68; here, 359n.

16. Sinha, "Ship Scrapping," 401.

17. Susanne Köstering, "'Millionen im Müll?' Altmaterialverwertung nach dem Vierjahresplan," in Köstering and Renate Rüb, eds., *Müll von gestern? Eine umweltgeschichtliche Erkundung in Berlin und Brandenburg*, 138–49 (Münster: Waxmann, 2003), 140.

18. Peter Thorsheim, *Waste into Weapons: Recycling in Britain during the Second World War* (New York: Cambridge University Press, 2015), 2.

19. Christopher M. Davidson, *Dubai: The Vulnerability of Success* (New York: Columbia University Press, 2008), 35.

20. Ruth Oldenziel and Milena Veenis, "The Glass Recycling Container in the Netherlands: Symbol in Times of Scarcity and Abundance, 1939–1978," *Contemporary European History* 22 (2013): 453–76.

21. Denton, "'Récupérez!'" 428.

22. Dieter Schott, *Europäische Urbanisierung (1000–2000): Eine umwelthistorische Einführung* (Cologne: Böhlau, 2014), 118.

23. Frank Trentmann, *Empire of Things: How We Became a World of Consumers, from the Fifteenth Century to the Twenty-First* (London: Allen Lane, 2016), 633.

24. William B. Cohen, *Urban Government and the Rise of the French City: Five Municipalities in the Nineteenth Century* (Basingstoke: Macmillan, 1998), 161.

25. Benjamin Miller, *Fat of the Land: Garbage of New York—The Last Two Hundred Years* (New York: Four Walls Eight Windows, 2000), 42n.

26. Martin V. Melosi, *Garbage in the Cities: Refuse, Reform, and the Environment* (Pittsburgh, PA: University of Pittsburgh Press, 2005), 52–54.

27. Melosi, *Garbage*, 42, 65.

28. Cohen, *Urban Government*, 253n.

29. Ted Steinberg, *Gotham Unbound: The Ecological History of Greater New York* (New York: Simon and Schuster, 2014), 241.

30. Melosi, *Garbage*, 153n.

31. Anne Berg, "The Nazi Rag-Pickers and Their Wine: The Politics of Waste and Recycling in Nazi Germany," *Social History* 40 (2015): 446–72; here, 464.

32. Joel A. Tarr, "The Search for the Ultimate Sink: Urban Air, Land, and Water Pollution in Historical Perspective," in Tarr, ed., *The Search for the Ultimate Sink: Urban Pollution in Historical Perspective*, 7–35 (Akron, OH: University of Akron Press, 1996).

33. Peter Payer, "Die Säuberung der Stadt: Straßenreinigung und Müllabfuhr," in Karl

Brunner and Petra Schneider, eds., *Umwelt Stadt: Geschichte des Natur- und Lebensraumes Wien*, 274–78 (Vienna: Böhlau, 2005), 277n.

34. Steinberg, *Gotham Unbound*, 241.

35. Tarr, "Search," 25.

36. Roman Köster, *Hausmüll: Abfall und Gesellschaft in Westdeutschland 1945–1990* (Göttingen: Vandenhoeck & Ruprecht, 2017), 256.

37. Geoffrey Blight, "Landfills: Yesterday, Today and Tomorrow," in Trevor M. Letcher and Daniel A. Vallero, eds., *Waste: A Handbook for Management*, 469–85 (Amsterdam: Elsevier, 2011), 484.

38. Sabine Barles, *L'Invention des Déchets urbains: France 1790–1970* (Seyssel: Champ Vallon, 2005), 167–69.

39. Trentmann, *Empire of Things*, 622.

40. Thorstein Veblen, *The Theory of the Leisure Class* (Boston: Houghton Mifflin, 1973 [1899]), 118.

41. Susan Strasser, *Waste and Want: A Social History of Trash* (New York: Henry Holt, 1999), 278.

42. Melosi, *Garbage*, 215.

43. Vandana Shiva, "International Institutions Practicing Environmental Double Standards," in Kevin Danaher, ed., *50 Years Is Enough: The Case Against the World Bank and the International Monetary Fund*, 102–6 (Boston: South End Press, 1994), 102; Robert D. Bullard, "Anatomy of Environmental Racism and the Environmental Justice Movement," in Bullard, ed., *Confronting Environmental Racism: Voices from the Grassroots*, 15–39 (Boston: South End Press, 1993), 20. On Summers's memo, see Simone M. Müller, "Rettet die Erde vor den Ökonomen? Lawrence Summers' Memo und der Kampf um die Deutungshoheit über den internationalen Giftmüllhandel," *Archiv für Sozialgeschichte* 56 (2016): 353–71.

44. See Robert D. Bullard, *Dumping in Dixie: Race, Class, and Environmental Quality* (Boulder, CO: Westview, 1994).

45. Shawkat Alam and Abdullah Faruque, "Legal Regulation of the Shipbreaking Industry in Bangladesh: The International Regulatory Framework and Domestic Implementation Challenges," *Marine Policy* 47 (2014): 46–56; here, 47n.

46. International Maritime Organization, "Status of Conventions," accessed November 3, 2021, https://www.imo.org/en/About/Conventions/Pages/StatusOfConventions.aspx.

47. See Alam and Faruque, "Legal Regulation," 50–52. See also Tony George Puthucherril, *From Shipbreaking to Sustainable Ship Recycling: Evolution of a Legal Regime* (Leiden: Martinus Nijhoff, 2010).

48. Frank Uekötter, *The Greenest Nation? A New History of German Environmentalism* (Cambridge, MA: MIT Press, 2014), 125.

49. Frank Uekötter, "Ökologische Verflechtungen: Umrisse einer grünen Zeitgeschichte," in Frank Bösch, ed., *Geteilte Geschichte: Ost- und Westdeutschland 1970–2000*, 117–52 (Göttingen: Vandenhoeck & Ruprecht, 2015), 121n.

50. Mike Crang et al., "Death, the Phoenix and Pandora: Transforming Things and Values in Bangladesh," in Catherine Alexander and Joshua Reno, eds., *Economies of Recycling: The Global Transformation of Materials, Values and Social Relations*, 59–75 (London: Zed Books, 2012), 63.

51. "1998 Pulitzer Prizes," accessed November 3, 2021, http://www.pulitzer.org/prize-winners-by-year/1998.

52. Regulation (EU) No. 1257/2013 of November 20, 2013, and https://ec.europa.eu/environment/waste/ships/list.htm, accessed November 3, 2021 .

53. Alam and Faruque, "Legal Regulation," 53n. See also S.M. Mizanur Rahman and

Audrey L. Mayer, "Policy Compliance Recommendations for International Shipbreaking Treaties for Bangladesh," *Marine Policy* 73 (2016): 122–29.

54. Breen, "Constellations of Mobility," 38.

55. Breen, "Constellations of Mobility," 38.

56. Puthucherril, *From Shipbreaking*, 102.

57. Mike Crang, "The Death of Great Ships: Photography, Politics, and Waste in the Global Imaginary," *Environment and Planning A* 42 (2010): 1084–1102; here, 1098n. See also George Cairns, "Postcard from Chittagong: Wish You Were Here?" *Critical Perspectives on International Business* 3 (2007): 266–79.

58. Hossain and Islam, *Ship Breaking Activities*, 14, 18.

PART II
APPROPRIATIONS

1. Niall Ferguson, *Civilization: The West and the Rest* (London: Allen Lane, 2011), 12.

6: THE LAND TITLE

1. World Bank Archives, box 117339 B, Agricultural Development Project, Ethiopia, August 1967, annex 5, p. 1.

2. World Bank Archives, box 117339 B, Stanford Research Institute, A Proposed Program of Research and Planning Studies for the Agro-Industrial Survey in Ethiopia. Phase I Report, June 1967, p. 37.

3. C. B. Macpherson, "The Meaning of Property," in Macpherson, ed., *Property: Mainstream and Critical Positions*, 1–13 (Oxford: Basil Blackwell, 1978), 8.

4. H. R. Loyn, "The Beyond of Domesday Book," in J. C. Holt, ed., *Domesday Studies: Papers presented at the Novocentenary Conference of the Royal Historical Society and the Institute of British Geographers, Winchester, 1986*, 1–13 (Woodbridge, UK: Boydell Press, 1987), 4n; and Frank Barlow, "Domesday Book: An Introduction," in Christopher Holdsworth, ed., *Domesday Essays*, 17–39 (Exeter: University of Exeter, 1986).

5. See Dieter Schwab, "Eigentum," in Otto Brunner, Werner Conze, and Reinhart Koselleck, eds., *Geschichtliche Grundbegriffe: Historisches Lexikon zur politisch-sozialen Sprache in Deutschland*, vol. 2: E–G, 65–115 (Stuttgart: Klett, 1975), 74.

6. Jeremy Bentham, *The Theory of Legislation* (London: Routledge and Kegan Paul, 1931), 113.

7. Johnson v. McIntosh, 21 U.S. (8 Wheat.) 543 (1823).

8. Saliha Belmessous, "Introduction: The Problem of Indigenous Claim Making in Colonial History," in Belmessous, ed., *Native Claims: Indigenous Law against Empire, 1500–1920*, 3–18 (Oxford: Oxford University Press, 2012), 8n.

9. Patricia Nelson Limerick, *The Legacy of Conquest: The Unbroken Past of the American West* (New York: Norton, 1988), 59n.

10. Mark P. Thompson, *Modern Land Law*, 5th ed. (Oxford: Oxford University Press, 2012), 119.

11. Rudolf Isay, "Bergrecht," in Franz Schlegelberger, ed., *Rechtsvergleichendes Handwörterbuch für das Zivil- und Handelsrecht des In- und Auslandes*, vol. 2, 437–87 (Berlin: Franz Vahlen, 1929), 441.

12. Tom Stephenson, *Forbidden Land: The Strugle for Access to Mountain and Moorland* (Manchester: Manchester University Press, 1989).

13. Reviel Netz, *Barbed Wire: An Ecology of Modernity* (Middletown, CT: Wesleyan University Press, 2004), 24.

14. Jürgen Osterhammel, *The Transformation of the World: A Global History of the Nineteenth Century* (Princeton, NJ: Princeton University Press, 2014), 107.

15. Theodore Steinberg, *Slide Mountain or The Folly of Owning Nature* (Berkeley: University of California Press, 1995), 135–65. See also Seymour I. Toll, *Zoned American* (New York: Grossman, 1969).

16. David Cannadine, *The Decline and Fall of the British Aristocracy* (New York: Vintage Books, 1999), 55.

17. See Daniel J. Boorstin, *The Lost World of Thomas Jefferson* (Chicago: University of Chicago Press, 1993 [1948]).

18. Daniel W. Bromley, "Private Property and the Public Interest: Land in the American Idea," in William G. Robbins and James C. Foster, eds., *Land in the American West: Private Claims and the Common Good*, 23–36 (Seattle: University of Washington Press, 2000), 25. See also Schwab, "Eigentum," p. 79

19. Theodor Bernhardi, *Versuch einer Kritik der Gründe, die für grosses und kleines Grundeigenthum angeführt werden* (St. Petersburg: Buckdruckerei der Kaiserlichen Akademie der Wissenschaften, 1849), 654; v. Bernhardi, "Bernhardi, Felix Theodor von," in *Allgemeine Deutsche Biographie*, vol. 46, 424–30 (Leipzig: Duncker & Humblot, 1902), 426.

20. John Quentin Colborne Mackrell, *The Attack on "Feudalism" in Eighteenth-Century France* (London: Routledge and Kegan Paul, 1973), 173.

21. Peeter Maandi, "Land Reforms and Territorial Integration in Post-Tsarist Estonia, 1918–1940," *Journal of Historical Geography* 36 (2010): 441–52; here, 446.

22. Klaus Richter, *Fragmentation in East Central Europe: Poland and the Baltics, 1915–29* (Oxford: Oxford University Press, 2020), 252–54; 253 (quotation).

23. Wojciech Roszkowski, *Land Reforms in East Central Europe after World War One* (Warsaw: Instytut Studiów Politycznych PAN, 1995), 120, 226; and Hans Jörgensen, "The Inter-War Land Reforms in Estonia, Finland and Bulgaria: A Comparative Study," *Scandinavian Economic History Review* 54, no. 1 (2006): 64–97.

24. Robert Ryal Miller, *Mexico: A History* (Norman: University of Oklahoma Press, 1985), 319.

25. Ian Talbot, *Pakistan: A Modern History* (London: Hurst, 2005), 165n.

26. R. W. Davies, *Soviet Economic Development from Lenin to Khrushchev* (Cambridge: Cambridge University Press, 1998), 19, 47n; Alec Nove, *An Economic History of the USSR 1917–1991*, 3rd ed. (London: Penguin, 1992), 41, 101, 160; and Paul R. Gregory, *The Political Economy of Stalinism: Evidence from the Soviet Secret Archives* (Cambridge: Cambridge University Press, 2004), 23, 27, 40n.

27. Marius B. Jansen, *The Making of Modern Japan* (Cambridge, MA: Belknap Press of Harvard University Press, 2000), 682n.

28. Richard Gott, *Cuba: A New History* (New Haven, CT: Yale Nota Bene, 2005), 180.

29. Julie A. Charlip, "Central America in Upheaval," in Thomas H. Holloway, ed., *A Companion to Latin American History*, 406–23 (Malden, MA: Wiley-Blackwell, 2011), 410.

30. Ghislaine Alleaume, "An Industrial Revolution in Agriculture? Some Observations on the Evolution of Rural Egypt in the Nineteenth Century," in Alan K. Bowman and Eugene Rogan, eds., *Agriculture in Egypt: From Pharaonic to Modern Times*, 331–45 (Oxford: Oxford University Press, 1999), 338.

31. Cannadine, *Decline and Fall*, 70.

32. F. M. L. Thompson, "Epilogue: The Strange Death of the English Land Question," in Matthew Cragoe and Paul Readman, eds., *The Land Question in Britain, 1750–1950*, 257–70 (Basingstoke: Palgrave Macmillan, 2010), 259.

33. See, for instance, Hans Rosenberg, "Die Pseudodemokratisierung der Rittergutsbesitzerklasse," in Rosenberg, ed., *Machteliten und Wirtschaftskonjunkturen: Studien zur neueren deutschen Sozial- und Wirtschaftsgeschichte*, 83–191 (Göttingen: Vandenhoeck & Ruprecht,

1978); Arno Mayer, *The Persistence of the Old Regime: Europe to the Great War* (New York: Pantheon Books, 1981); and Cannadine, *Decline and Fall*.

34. Robert C. McMath Jr., *American Populism: A Social History 1877–1898* (New York: Noonday Press, 1993); and Lutz Raphael, *Imperiale Gewalt und mobilisierte Nation: Europa 1914–1945* (Munich: Beck, 2011), 102. See also Eduard Kubů et al., eds., *Agrarismus und Agrareliten in Ostmitteleuropa* (Prague: Dokořán, 2013).

35. Heinz Haushofer, "Die internationale Organisation der Bauernparteien," in Heinz Gollwitzer, ed., *Europäische Bauernparteien im 20. Jahrhundert*, 668–90 (Stuttgart: G. Fischer, 1977), 674.

36. Arthur Young, *Travels during the Years 1787, 1788 and 1789* (London: W. Richardson, 1794), 88.

37. Russell King, *Land Reform: A World Survey* (London: G. Bell and Sons, 1977), 24.

38. Harold C. Marcus, *A History of Ethiopia* (Berkeley: University of California Press, 2002), 192.

39. Andro Linklater, *Owning the Earth: The Transformative History of Land Ownership* (London: Bloomsbury, 2014), 322n.

40. King, *Land Reform*, 84n.

41. Grigory Ioffe, Tatyana Nefedova, and Ilya Zaslavsky, *The End of Peasantry? The Disintegration of Rural Russia* (Pittsburgh, PA: University of Pittsburgh Press, 2006), 28; and Zvi Lerman, Csaba Csaki, and Gershon Feder, *Agriculture in Transition: Land Policies and Evolving Farm Structures in Post-Soviet Countries* (Lanham, MD: Lexington Books, 2004), 216.

42. Shalmali Guttal et al., "Preface: A History and Overview of the Land Research Action Network," in Peter Rosset, Raj Patel, and Michael Courville, eds., *Promised Land: Competing Visions of Agrarian Reform*, xiii–xv (Oakland, CA: Food First Books, 2006), xiii.

43. Gabriel Ondetti, *Land, Protest, and Politics: The Landless Movement and the Struggle for Agrarian Reform in Brazil* (University Park: Pennsylvania State University Press, 2008).

44. Sam Moyo and Paris Yeros, eds., *Reclaiming the Land: The Resurgence of Rural Movements in Africa, Asia and Latin America* (London: Zed Books, 2005).

45. Ian Scoones et al., *Zimbabwe's Land Reform: Myths and Realities* (Woodbridge, UK: James Currey, 2005), 21.

46. See Nick Cullather, *The Hungry World: America's Cold War Battle Against Poverty in Asia* (Cambridge, MA: Harvard University Press, 2010); John H. Perkins, *Geopolitics and the Green Revolution: Wheat, Genes, and the Cold War* (New York: Oxford University Press, 1997); and Jonathan Harwood, *Europe's Green Revolution and Others Since: The Rise and Fall of Peasant-Friendly Plant Breeding* (London: Routledge, 2012).

47. Rodolfo Stavenhagen, "Indigenous Peoples: Land, Territory, Autonomy, and Self-Determination, Brazil," in Rosset, Patel, and Courville, *Promised Land*, 208–17.

48. Quoted in Kevin Gray and Susan Francis Gray, "The Idea of Property in Land," in Susan Bright and John Dewar, eds., *Land Law: Themes and Perspectives*, 15–51 (Oxford: Oxford University Press, 1998), 26.

49. R. Douglas Hurt, *Documents of the Dust Bowl* (Santa Barbara, CA: ABC-CLIO, 2019), 145.

50. See p. 205.

51. See Fred Pearce, *The Land Grabbers: The New Fight over Who Owns the Earth* (Boston: Beacon Press, 2012); Stefano Liberti, *Land Grabbing: Journeys in the New Colonialism* (London: Verso, 2013); and Mayke Kaag and Annelies Zoomers, eds., *The Global Land Grab: Beyond the Hype* (London: Zed Books, 2014).

52. Stuart Banner, *How the Indians Lost Their Land: Law and Power on the Frontier* (Cambridge, MA: Belknap Press of Harvard University Press, 2005), 11n.

7: BREADFRUIT

1. David Watts, *The West Indies: Patterns of Development, Culture and Environmental Change since 1492* (Cambridge: Cambridge University Press, 1987), 299.

2. David Mackay, "Banks, Bligh and Breadfruit," *New Zealand Journal of History* 8 (1974): 61–77; here, 65.

3. See Sherry Johnson, *Climate and Catastrophe in Cuba and the Atlantic World in the Age of Revolution* (Chapel Hill: University of North Carolina Press, 2011).

4. Jennifer Newell, *Trading Nature: Tahitians, Europeans, and Ecological Change* (Honolulu: University of Hawai'i Press, 2010), 144n.

5. John Cawte Beaglehole, ed., *The Endeavour Journal of Joseph Banks 1768–1771*, vol. 1 (Sydney: Angus & Robertson, 1962), 341.

6. Mackay, "Banks," 63. For cultural context, see Tim Fulford, "Romanticism, the South Seas and the Caribbean: The Fruits of Empire," *European Romantic Review* 11 (2000): 408–34; and Emma Spary and Paul White, "Food of Paradise: Tahitian Breadfruit and the Autocritique of European Consumption," *Endeavour* 28 (2004): 75–80.

7. Henry Trueman Wood, *A History of the Royal Society of Arts* (Cambridge: Cambridge University Press, 2011 [1913]), 95.

8. Richard B. Sheridan, "Captain Bligh, the Breadfruit, and the Botanic Gardens of Jamaica," *Journal of Caribbean History* 23 (1989): 28–50; here, 33.

9. Sheridan, "Captain Bligh," 36.

10. John G. T. Anderson, *Deep Things out of Darkness: A History of Natural History* (Berkeley: University of California Press, 2013), 194.

11. Newell, *Trading Nature*, 153–55, 158.

12. For a scholarly presentation that straddles the line between history and historical novel, see Greg Dening, *Mr. Bligh's Bad Language: Passion, Power and Theatre on the Bounty* (Cambridge: Cambridge University Press, 1992). For the historic account from the second secretary of the Admiralty, see John Barrow, *The Mutiny and Piratical Seizure of H.M.S. Bounty* (New York: Cambridge University Press, 2011 [1831]).

13. Nancy Davis Lewis, "The Pacific Islands," in Kenneth F. Kiple and Kriemhild Coneè Ornelas, eds., *The Cambridge World History of Food*, vol. 2, 1351–66 (Cambridge: Cambridge University Press, 2000), 1355. See also Glenn Petersen, "Micronesia's Breadfruit Revolution and the Evolution of a Culture Area," *Archeology in Oceania* 41 (2006): 82–92. See also Nyree J. C. Zerega, Diane Ragone, and Timothy J. Motley, "Complex Origins of Breadfruit (*Artocarpus Altilis*, Moraceae): Implications for Human Migrations in Oceania," *American Journal of Botany* 91 (2004): 760–66.

14. For popular accounts of the mutiny, see Caroline Alexander, *The Bounty: The True Story of the Mutiny on the Bounty* (London: HarperCollins, 2003); and Richard Hough, *Captain Bligh and Mr Christian: The Men and the Mutiny* (London: Cresset Library, 1988).

15. Sheridan, "Captain Bligh," 40–44; and Newell, *Trading Nature*, 160–67.

16. Andrew David, "Bligh's Successful Breadfruit Voyage," *RSA Journal* 141 (1993): 821–24; here, 824.

17. Richard Drayton, *Nature's Government: Science, Imperial Britain, and the 'Improvement' of the World* (Hyderabad: Orient Longman, 2005), 113.

18. James E. McClellan III, *Colonialism and Science: Saint Domingue in the Old Regime* (Chicago: University of Chicago Press, 2010), 159.

19. Sheridan, "Captain Bligh," 41.

20. McClellan, *Colonialism*, 158n.

21. Jeffrey M. Pilcher, "The Caribbean from 1492 to the Present," in Kiple and Ornelas, *Cambridge World History of Food*, 2:1278–88; here, 1281.

22. M. A. Emanuel and N. Benkeblia, "Ackee Fruit (*Blighia sapida* Konig)," in Elhadi M. Yahia, ed., *Postharvest Biology and Technology of Tropical and Subtropical Fruits*, vol. 2: *Açai to Citrus*, 54–64 (Oxford: Woodhead, 2011), 54.

23. B. W. Higman, *A Concise History of the Caribbean* (New York: Cambridge University Press, 2011), 166. See also p. 52–53.

24. Sheridan, "Captain Bligh," 45n; and Watts, *West Indies*, 505. Newell suggests that breadfruit "did well in the Caribbean colonies," but she does not engage with contrary assertions in the literature. (Newell, *Trading Nature*, 168.)

25. Drayton, *Nature's Government*, 114.

26. [Ernst Engel], "Die vorherrschenden Gewerbszweige in den Gerichtsämtern mit Beziehung auf die Productions- und Consumptionsverhältnisse des Königreichs Sachsen," *Zeitschrift des Statistischen Bureaus des Königlich Sächsischen Ministeriums des Innern* 3, no. 8–9 (November 22, 1857): 153–82.

27. [Engel], "Die vorherrschenden Gewerbszweige," 169. Translated in George J. Stigler, "The Early History of Empirical Studies of Consumer Behavior," *Journal of Political Economy* 62 (1954): 95–113; here, 98. See also Ernst Engel, "Die Lebenskosten Belgischer Arbeiter-Familien Früher und Jetzt," *Bulletin de L'Institut International de Statistique* 9 (1895): 1–124.

28. Engel, "Die vorherrschenden Gewerbszweige," 169.

29. Martin Browning, "Engel's Law," in Steven N. Durlauf and Lawrence E. Blume, eds., *The New Palgrave Dictionary of Economics*, vol. 2, 2nd ed., 850–51 (Basingstoke: Palgrave Macmillan, 2008), 850.

30. Giovanni Federico, *Feeding the World: An Economic History of Agriculture, 1800–2000* (Princeton, NJ: Princeton University Press, 2009), 19.

31. Engel, "Die vorherrschenden Gewerbszweige," 168; and Lisa C. Smith and Ali Subandoro, *Measuring Food Security Using Household Expenditure Surveys* (Washington, DC: International Food Policy Research Institute, 2007), 82.

32. Theodore M. Porter, *Trust in Numbers: The Pursuit of Objectivity in Science and Public Life* (Princeton, NJ: Princeton University Press, 1995).

33. Engel, "Die vorherrschenden Gewerbszweige," 169. Translations are mine.

34. Browning, "Engel's Law," 851.

35. James L. Hargrove, "History of the Calorie in Nutrition," *Journal of Nutrition* 136 (2006): 2957–61.

36. Nick Cullather, *The Hungry World: America's Cold War Battle against Poverty in Asia* (Cambridge, MA: Harvard University Press, 2010), 20.

37. Carey D. Miller, *Food Values of Breadfruit, Taro Leaves, Coconut, and Sugar Cane* (New York: Kraus Reprint, 1971 [originally published Honolulu: Bernice P. Bishop Museum, 1929]), 3, 22 (quotation).

38. See Heiko Stoff, *Wirkstoffe: Eine Wissenschaftsgeschichte der Hormone, Vitamine and Enzyme, 1920–1970* (Stuttgart: Steiner, 2012).

39. Can with 360 grams net drained weight, bought at the Morrisons store at Five Ways, Birmingham, August 12, 2017. Offer price £1.

40. On the mysteries of human tastes, see Marvin Harris, *Good to Eat: Riddles of Food and Culture* (New York: Simon and Schuster, 1985); Frederick J. Simoons, *Eat Not This Flesh: Food Avoidance from Prehistory to the Present* (Madison: University of Wisconsin Press, 1961); and Stephen Mennell, *All Manners of Food: Eating and Taste in England and France from the Middle Ages to the Present* (Oxford: Basil Blackwell, 1985).

41. Jun Morikawa, *Whaling in Japan: Power, Politics and Diplomacy* (London: Hurst, 2009), 30.

42. See Pierre Bourdieu, *Distinction: A Social Critique of the Judgement of Taste* (New York: Routledge, 2009 [French, 1979]), esp. 177–200.

43. John H. Parry, "Plantation and Provision Ground: An Historical Sketch of the Introduction of Food Crops into Jamaica," *Revista de Historia de América* 39 (June 1955): 1–20; here, 19.

44. For an overview, see Neela Badrie and Jacklyn Broomes, "Beneficial Uses of Breadfruit (*Artocarpus altilis*): Nutritional, Medicinal and Other Uses," in Ronald Ross Watson and Victor R. Preedy, eds., *Bioactive Foods in Promoting Health: Fruits and Vegetables*, 491–505 (Amsterdam: Elsevier, 2010).

45. Bonnie Thomas, *Breadfruit or Chestnut? Gender Construction in the French Caribbean Novel* (Lanham, MD: Lexington Books, 2006), 35n.

46. C. M. S. Carrington, R. Maharaj, and C. K. Sankat, "Breadfruit (*Artocarpus altilis* [Parkinson] Fosberg)," inYahia, *Postharvest Biology*, 251–71; here, 252; and Badrie and Broomes, "Beneficial Uses," 492, 501.

47. Fulford, "Romanticism," 412; Emma C. Spary, *Utopia's Garden: French Natural History from Old Regime to Revolution* (Chicago: University of Chicago Press, 2000), 130; and Ray Desmond, *The History of the Royal Botanic Gardens, Kew*, 2nd ed. (London: Royal Botanic Gardens, Kew, 2007), 322.

48. Carrington, Maharaj, and Sankat, "Breadfruit," 252n.

49. Jamaica Information Service, "MOH dislcoses [sic] 194 cases of ackee poisoning," February 25, 2011, accessed November 3, 2021, http://jis.gov.jm/moh-dislcoses-194-cases-of-ackee-poisoning/.

50. Laura R. Roberts-Nkrumah and George Legall, "Breadfruit (*Artocarpus altilis*, Moraceae) and Chataigne (*A. camansi*) for Food Security and Income Generation: The Case of Trinidad and Tobago," *Economic Botany* 67 (2013): 324–34; here, 329.

51. Mary B. Taylor and Valerie Saena Tuia, "Breadfruit in the Pacific Region," *Acta Horticulturae* 757 (2007): 43–50; here, 44.

52. Elizabeth Gibbons and Richard Garfield, "The Impact of Economic Sanctions on Health and Human Rights in Haiti, 1991–1994," *American Journal of Public Health* 89 (1999): 1499–504; here, 1500.

53. See Charles Nordhoff and James Norman Hall, *The Bounty Triology: Comprising the Three Volumes Mutiny on the Bounty, Men Against the Sea & Pitcairn's Island* (Boston: Little, Brown, 1940), 120.

54. Beaglehole, *The Endeavour Journal*, 341. On the religious overtones of the breadfruit discourse, see Fulford, "Romanticism"; and Sujit Sivasundaram, "Natural History Spiritualized: Civilizing Islanders, Cultivating Breadfruit, and Collecting Souls," *History of Science* 39 (2001): 417–43.

55. Beaglehole, *The Endeavour Journal*, 341.

56. Mackay, "Banks," 63.

57. Peter Griggs, "Improving Agricultural Practices: Science and the Australian Sugarcane Grower, 1864–1915," *Agricultural History* 78 (2004): 1–33; here, 6n.

58. Frank Uekötter, "Recollections of Rubber," in Dominik Geppert and Frank Lorenz Müller, eds., *Sites of Imperial Memory: Commemorating Colonial Rule in the Nineteenth and Twentieth Centuries*, 243–65 (Manchester: Manchester University Press, 2015), 251.

59. Elizabeth DeLoughrey, "Globalizing the Routes of Breadfruit and other Bounties," *Journal of Colonialism and Colonial History* 8, no. 3 (2007), https://muse.jhu.edu/.

60. Mennell, *All Manners of Food*, 266, 270 (quotation).

61. A. Maxwell P. Jones et al., "Nutritional and Morphological Diversity of Breadfruit (*Artocarpus*, Moraceae): Identification of Elite Cultivars for Food Security," *Journal of Food Composition and Analysis* 24 (2011): 1091–102; here, 1101. Similarly, Roberts-Nkrumah and Legall, "Breadfruit," 330–32.

62. Christina E. Turi et al., "Breadfruit (*Artocarpus altilis* and hybrids): A Traditional

Crop with the Potential to Prevent Hunger and Mitigate Diabetes in Oceania," *Trends in Food Science and Technology* 45 (2015): 264–72; here, 265. See also https://ntbg.org/breadfruit, accessed November 3, 2021.

63. Marcia H. Magnus, "The Top 25 Superfoods," *Cajanus* 30, no. 1 (1997): 3–13; here, 9.

64. I dedicate this sentence to my wife Simona. She was well familiar with the many risks of marrying a historian when I came to this chapter, but she had yet to learn that they include canned fruit from Morrisons.

8: GUANO

1. *Allgemeines Deutsches Commersbuch: Unter musikalischer Redaktion von Fr. Silcher und Fr. Erck*, 16th ed. (Straßburg: Moritz Schauenburg, 1873), 469.

2. See Heinz Dieter Kittsteiner, "Deutscher Idealismus," in Etienne François and Hagen Schulze, eds., *Deutsche Erinnerungsorte I*, 170–86 (Munich: Beck, 2001), 175.

3. Robert Somerville von Haddington, Christian August Wichmann, and Alexander Nicolaus Scherer, *Vollständige Uebersicht der gewöhnlichen, und mehrerer bisher minder bekannten Dünge-Mittel und deren Würksamkeit: Nach den Berichteten praktischer Landwirthe dem Britischen Landwirtschafts-Rath vorgelegt* (Leipzig: Breitkopf und Härtel, 1800).

4. Conrad Totman, *A History of Japan*, 2nd ed. (Malden, MA: Blackwell, 2005), 258.

5. Yong Xue, "'Treasure Nightsoil As If It Were Gold': Economic and Ecological Links between Urban and Rural Areas in Late Imperial Jiangnan," *Late Imperial China* 26 (2005): 41–71; and Dean T. Ferguson, "Nightsoil and the 'Great Divergence': Human Waste, the Urban Economy, and Economic Productivity, 1500–1900," *Journal of Global History* 9 (2014): 379–402.

6. Catalina Vizcarra, "Guano, Credible Commitments, and Sovereign Debt Repayment in Nineteenth-Century Peru," *Journal of Economic History* 69 (2009): 358–87. Scholars have offered conflicting assessments over the terms of trade and specifically the role of the British merchant house Antony Gibbs and Sons. See W. M. Mathew, *The House of Gibbs and the Peruvian Guano Monopoly* (London: Royal Historical Society, 1981); and Jonathan V. Levin, *The Export Economies: Their Pattern of Development in Historical Perspective* (Cambridge, MA: Harvard University Press, 1960), chap. 2. In summarizing this debate, Rory Miller and Robert Greenhill have noted, "It is difficult to see that the alternatives [to the chosen institutional solution] would have provided significantly better rewards for Peru and Chile." (Rory Miller and Robert Greenhill, "The Fertilizer Commodity Chains: Guano and Nitrate, 1840–1930," in Steven Topik, Carlos Marichal, and Zephyr Frank, eds., *From Silver to Cocaine: Latin American Commodity Chains and the Building of the World Economy, 1500–2000*, 228–70 [Durham, NC: Duke University Press, 2006], 261.)

7. Quoted in Celia Cordle, "The Guano Voyages," *Rural History* 18 (2007): 119–133; here, 122.

8. Jimmy M. Skaggs, *The Great Guano Rush: Entrepreneurs and American Overseas Expansion* (London: Macmillan, 1994), 86, 132.

9. Christina Duffy Burnett, "The Edges of Empire and the Limits of Sovereignty: American Guano Islands," *American Quarterly* 57 (2005): 779–803; here, 779n, 787.

10. Quoted in Burnett, "Edges of Empire," 786.

11. Edward D. Melillo, "The First Green Revolution: Debt Peonage and the Making of the Nitrogen Fertilizer Trade, 1840–1930," *American Historical Review* 117 (2012): 1028–60; here, 1039.

12. Steven Stoll, *Larding the Lean Earth: Soil and Society in Nineteenth-Century America* (New York: Hill and Wang, 2002), 188, 194.

13. Germany's guano imports totaled 85,233 tons in 1869, making it Europe's third

largest consumer after England and France. (Hermann Bielecke, *Die Geschichte der künstlichen Düngung und der Kunstdüngerversorgung* [Quakenbrück: Robert Kleinert, 1934], 35.)

14. Gregory T. Cushman, *Guano and the Opening of the Pacific World: A Global Ecological History* (New York: Cambridge University Press, 2013), 51.

15. Lise Sedrez, "Environmental History of Modern Latin America," in Thomas H. Holloway, ed., *A Companion to Latin American History*, 443–60 (Malden, MA: Wiley-Blackwell, 2011), 445.

16. Stoll, *Larding the Lean Earth*, 190.

17. Cushman, *Guano*, 77n. See also Brett Clark and John Bellamy Foster, "Ecological Imperialism and the Global Metabolic Rift: Unequal Exchange and the Guano/Nitrates Trade," *International Journal of Comparative Sociology* 50 (2009): 311–34.

18. Miller and Greenhill, "Fertilizer Commodity Chains," 239.

19. W. M. Mathew, "Peru and the British Guano Market, 1840–1870," *Economic History Review* 23 (1970): 112–28; here, 119.

20. William H. Brock, *Justus von Liebig: The Chemical Gatekeeper* (Cambridge: Cambridge University Press, 1997), 123.

21. Brock, *Justus von Liebig*, 128.

22. Stoll, *Larding the Lean Earth*, 49.

23. Jacques Hadamard, "Newton and the Infinitesimal Calculus," in The Royal Society, ed., *Newton Tercentenary Celebrations 15–19 July 1946*, 35–42 (Cambridge: Cambridge University Press, 1947), 35.

24. Brock, *Justus von Liebig*, 155.

25. See Ursula Schling-Brodersen, *Entwicklung und Institutionalisierung der Agrikulturchemie im 19. Jahrhundert: Liebig und die landwirtschaftlichen Versuchsstationen* (Stuttgart: Deutscher Apothekerverlag, 1989); and Frank Uekötter, *Die Wahrheit ist auf dem Feld: Eine Wissensgeschichte der deutschen Landwirtschaft* (Göttingen: Vandenhoeck & Ruprecht, 2010).

26. Paul S. Giller, "The Diversity of Soil Communities, the 'Poor Man's Tropical Rainforest,'" *Biodiversity and Conservation* 5 (1996): 135–68.

27. Max Gerlach, "Die Bestimmung des Düngerbedürfnisses der Böden," *Landwirtschaftliche Jahrbücher* 63 (1926): 339–68; here, 368.

28. Brock, *Justus von Liebig*, 128.

29. Badische Anilin- und Soda-Fabrik, *Die Landwirtschaftliche Versuchsstation Limburgerhof 1914–1964: 50 Jahre landwirtschaftliche Forschung in der BASF* (Ludwigshafen: BASF, 1965). See also p. 282.

30. Peter Longerich, *Heinrich Himmler: Biographie* (Munich: Siedler, 2008), 72n; and Helmut Heiber, ed., *Reichsführer! . . . Briefe an und von Himmler* (Munich: dtv, 1970), 110.

31. Cordle, "Guano Voyages," 119.

32. Paul Wagner, *Bericht über Arbeiten der landw. Versuchs- und Auskunfts-Station für das Großherzogthum Hessen zu Darmstadt* (Darmstadt: Hch. Kichler, 1874), 9.

33. Susanne Reichrath, *Entstehung, Entwicklung und Stand der Agrarwissenschaften in Deutschland und Frankreich* (Frankfurt: Lang, 1991), 118.

34. Vandana Shiva, *The Violence of the Green Revolution: Third World Agriculture, Ecology and Politics* (Penang: Third World Network, 1991), 104.

35. See Nick Cullather, *The Hungry World: America's Cold War Battle against Poverty in Asia* (Cambridge, MA: Harvard University Press, 2010); John H. Perkins, *Geopolitics and the Green Revolution: Wheat, Genes, and the Cold War* (New York: Oxford University Press, 1997); and Jonathan Harwood, *Europe's Green Revolution and Others Since: The Rise and Fall of Peasant-Friendly Plant Breeding* (London: Routledge, 2012).

36. Dominic Merriott, "Factors Associated with the Farmer Suicide Crisis in India," *Journal of Epidemiology and Global Health* 6 (2017): 217-27.

37. Michel Bernat, Michel Loubet, and Alain Baumer, "Sur l'origine des phosphates de l'atoll corallien de Nauru," *Oceanologica Acta* 14 (1991): 325-31.

38. Robert Craig, "The African Guano Trade," *Mariner's Mirror* 50 (1964): 25-55; here, 50; and John Robert Victor Prescott and Gillian D. Triggs, *International Frontiers and Boundaries: Law, Politics and Geography* (Leiden: Brill, 2008), 123.

39. Cushman, *Guano*, 81.

40. Skaggs, *Great Guano Rush*, 199.

41. Ben Daley and Peter Griggs, "Mining the Reefs and Cays: Coral, Guano and Rock Phosphate Extraction in the Great Barrier Reef, Australia, 1844-1940," *Environment and History* 12 (2006): 395-433; here, 410n.

42. Carl N. McDaniel and John M. Gowdy, *Paradise for Sale: A Parable of Nature* (Berkeley: University of California Press, 2000), 39-41.

43. Heinrich Schnee, ed., *Deutsches Kolonial-Lexikon* (Leipzig: Quelle & Meyer, 1920), 2:621n.

44. Paul Hambruch, *Nauru. Erster Halbband* (Hamburg: Friederichsen, 1914), 58.

45. Even an Independent Commission of Inquiry appointed by the government of Nauru was unable to produce evidence of concerns about rehabilitation. However, it did note that German mining law gave landowners the right to compensation for reduced value, which it read as implying a responsibility to rehabilitate the land. (Christopher Weeramantry, *Nauru: Environmental Damage under International Trusteeship* [Melbourne: Oxford University Press, 1992], 188-90.) The validity of this claim remains untested because Nauru settled its lawsuit against Australia with the International Court of Justice in The Hague out of court in 1993. (McDaniel and Gowdy, *Paradise for Sale*, 46.)

46. Burnett, "Edges of Empire," 780.

47. Cushman, *Guano*, 44.

48. Food and Agriculture Organization of the United Nations, *World Fertilizer Trends and Outlook to 2020: Summary Report* (Rome: FAO, 2017), 3.

49. Vizcarra, "Guano," 359, 384.

50. Nils Jacobsen, "Ausländische Wirtschaftsinteressen und der Konflikt zwischen Zentralismus und Regionalismus in Peru 1850-1930," *Geschichte und Gesellschaft* 14 (1988): 178-92; here, 181; and Vizcarra, "Guano," 361, 382.

51. Shane J. Hunt, "Growth and Guano in Nineteenth-Century Peru," in Roberto Cortés Conde, and Shane J. Hunt, eds., *The Latin American Economies: Growth and the Export Sector 1880-1930* (New York: Holmes and Meier, 1985), 255-318; and Leticia Arroyo Abad, "Failure to Launch: Cost of Living and Living Standards in Peru during the 19th Century," *Revista de Historia Económica* 32 (2014): 47-76. See also Paul Gootenberg, *Between Silver and Guano: Commercial Policy and the State in Postindependence Peru* (Princeton, NJ: Princeton University Press, 1989); and Paul Gootenberg, *Imagining Development: Economic Ideas in Peru's "Fictitious Prosperity" of Guano, 1840-1880* (Berkeley: University of California Press, 1993).

52. Gregory T. Cushman, "'The Most Valuable Birds in the World': International Conservation Science and the Revival of Peru's Guano Industry, 1900-1965," *Environmental History* 10 (2005): 477-509.

53. Fredrick B. Pike, *The Modern History of Peru* (London: Weidenfeld and Nicolson, 1967), 298.

54. Cushman, *Guano*, 289, 303, 325.

55. Cushman, *Guano*, 337.

56. Michael H. Glantz, "Science, Politics and Economics of the Peruvian Anchoveta Fishery," *Marine Policy* 3 (1979): 201–10; here, 201.

57. Sandra M. Calhuin et al., "Climatic Regimes and the Recruitment Rate of Anchoveta, *Engraulis ringens*, off Peru," *Estuarine, Coastal and Shelf Science* 84 (2009): 591–97; here, 592.

58. John Connell, "Nauru: The First Failed Pacific State?" *Round Table* 95 (2006): 47–63. For popular presentations of Nauru's history, see McDaniel and Gowdy, *Paradise for Sale*; and Luc Folliet, *Nauru, l'île dévastée: Comment la civilisation capitaliste a détruit le pays le plus riche du monde* (Paris: La Découverte, 2009).

9: WHALING

1. Wolfgang Radtke, *Die Preussische Seehandlung zwischen Staat und Wirtschaft in der Frühphase der Industrialisierung* (Berlin: Colloquium-Verlag, 1981); Bärbel Holtz, "Rother, Christian v.," in Historische Kommission bei der Bayerischen Akademie der Wissenschaften, ed., *Neue Deutsche Biographie*, vol. 22 (Berlin: Duncker & Humblot, 2005), 121–22; and W. O. Henderson, "Christian von Rother als Beamter, Finanzmann und Unternehmer im Dienste des preußischen Staates 1810–1848," *Zeitschrift für die gesamte Staatswissenschaft* 112 (1956): 523–50.

2. John F. Richards, *The Unending Frontier: An Environmental History of the Early Modern World* (Berkeley: University of California Press, 2005), 585.

3. Gordon Jackson, *The British Whaling Trade* (London: Adam & Charles Black, 1978), 55.

4. Adam Smith, *An Inquiry into the Nature and Causes of the Wealth of Nations*, ed. R. H. Campbell and A. S. Skinner, vol. 1 (Oxford: Clarendon Press, 1976), 518.

5. Geheimes Staatsarchiv Berlin III. HA MdA II no. 1443, Minister-Resident Rönne to Ministerium der Auswärtigen Angelegenheiten, October 23, 1838. Numbers should be taken as indications of scale rather than precise information. A comprehensive quantitative study has noted that contemporary employment stood at 16,600 seamen and that whaling agents directed 672 vessels valued at $21 million in 1880 prices. (Lance E. Davis, Robert E. Gallman, and Karin Gleiter, *In Pursuit of Leviathan: Technology, Institutions, Productivity, and Profits in American Whaling, 1816–1906* [Chicago: University of Chicago Press, 1997], 513.)

6. Jackson, *British Whaling Trade*, 85.

7. Philip Hoare, *Leviathan, or, The Whale* (London: Fourth Estate, 2009), 273.

8. The history of whaling produced a large literature long before environmental history came of age. As Gordon Jackson wrote in a book of 1978, "It is doubtful if any trade, save that in human beings, has attracted so much attention as whaling." (Jackson, *British Whaling Trade*, xi.) For some of the defining academic books, see J. N. Tønnessen and Arne Odd Johnsen, *The History of Modern Whaling* (London: Hurst, 1982); Briton Cooper Busch, *"Whaling Will Never Do for Me": The American Whaleman in the Nineteenth Century* (Lexington: University Press of Kentucky, 1994); Lisa Norling, *Captain Ahab Had a Wife: New England Women and the Whalefishery, 1720–1870* (Chapel Hill: University of North Carolina Press, 2000); Nancy Shoemaker, *Native American Whalemen and the World: Indigenous Encounters and the Contingency of Race* (Chapel Hill: University of North Carolina Press, 2015); Kurkpatrick Dorsey, *Whales and Nations: Environmental Diplomacy on the High Seas* (Seattle: University of Washington Press, 2013); Davis, Gallman, and Gleiter, *In Pursuit of Leviathan*; Jackson, *British Whaling Trade*; Charlotte Epstein, *The Power of Words in International Relations: Birth of an Anti-Whaling Discourse* (Cambridge, MA: MIT Press, 2008); Eric Jay Dolin, *Leviathan: The History of Whaling in America* (New York: Norton, 2007); and Richard Ellis, *Men and Whales* (London: Robert Hale, 1992).

9. Richards, *Unending Frontier*, 593; and Sayre A. Swarztrauber, *The Three-Mile Limit of Territorial Seas* (Annapolis: Naval Institute Press, 1972), 18–20, 24n, 35.

10. On the changing place of women in whaling, see Norling, *Captain Ahab*.

11. Davis, Gallman, and Gleiter, *In Pursuit of Leviathan*, 15.

12. See Hershel Parker, *Herman Melville: A Biography*, vol. 1 (Baltimore: Johns Hopkins University Press, 1996).

13. William S. McFeely, *Frederick Douglass* (New York: Norton, 1991), 77. For a discussion of race relations on whaling ships, see Busch, "Whaling," 32–50.

14. Geheimes Staatsarchiv Berlin III. HA MdA II no. 1443, Geheimer Staatsminister Rother to Ministerium der Auswärtigen Angelegenheiten, February 4, 1839.

15. Geheimes Staatsarchiv Berlin III. HA MdA II no. 1443, Geheimer Staatsminister Rother to Ministerium der Auswärtigen Angelegenheiten, February 4, 1839.

16. Geheimes Staatsarchiv Berlin III. HA MdA II no. 1443, Gesandter in Hamburg to Ministerium der Auswärtigen Angelegenheiten, March 13, 1839, p. 1.

17. Tønnessen and Johnsen, *History*, 25–32.

18. Jackson, *British Whaling Trade*, 182.

19. Tønnessen and Johnsen, *History*, 66.

20. See Charles Wilson, *The History of Unilever: A Study in Economic Growth and Social Change*, 2 vols. (London: Cassell, 1954).

21. Ingo Heidbrink, "A Second Industrial Revolution in the Distant-Water Fisheries? Factory-Freezer Trawlers in the 1950s and 1960s," *International Journal of Maritime History* 23 (2011): 179–92.

22. Robert C. Rocha Jr., Phillip J. Clapham, and Yulia V. Ivashchenko, "Emptying the Oceans: A Summary of Industrial Whaling Catches in the 20th Century," *Marine Fisheries Review* 76, no. 4 (2015): 37–48; here, 42–45.

23. Tønnessen and Johnsen, *History*, 356.

24. William M. Tsutsui, "The Pelagic Empire: Reconsidering Japanese Expansion," in Ian Jared Miller, Julia Adeney Thomas, and Brett L. Walker, eds., *Japan at Nature's Edge: The Environmental Context of a Global Power*, 21–38 (Honolulu: University of Hawai'i Press, 2013), 25; and Nancy Shoemaker, "Whale Meat in American History," *Environmental History* 10 (2005): 269–94; here, 274.

25. Ole Sparenberg, "Perception and Use of Marine Biological Resources under National Socialist Autarky Policy," in Frank Uekötter and Uwe Lübken, eds., *Managing the Unknown: Essays on Environmental Ignorance* (New York: Berghahn Books, 2014), 91–121.

26. Rocha, Clapham, and Ivashchenko, *Emptying the Oceans*, 44.

27. See the remarks in Jean-Pierre Proulx, *Whaling in the North Atlantic: From Earliest Times to the Mid-19th Century* (Ottowa: National Historic Parks and Sites Branch, Parks Canada, Environment Canada, 1986), 17.

28. Lance E. Davis, Robert E. Gallman, and Teresa D. Hutchins, "The Decline of U.S. Whaling: Was the Stock of Whales Running Out?" *Business History Review* 62 (1988): 569–95; here, 594.

29. Rocha, Clapham, and Ivashchenko, *Emptying the Oceans*, 43, 45.

30. Carmel Finley, *All the Fish in the Sea: Maximum Sustainable Yield and the Failure of Fisheries Management* (Chicago: University of Chicago Press, 2011).

31. Sparenberg, "Perception," 100.

32. Paul Josephson, "The Ocean's Hot Dog: The Development of the Fish Stick," *Technology and Culture* 49 (2008): 41–61.

33. Jean-Baptiste Malet, *Das Tomatenimperium: Ein Lieblingsprodukt erklärt den globalen Kapitalismus* (Cologne: Eichborn, 2018), 81, 223.

34. Dolin, *Leviathan*, 370–73; and Hiroyuki Watanabe, *Japan's Whaling: The Politics of Culture in Historical Perspective* (Melbourne: Trans Pacific Press, 2009), 155.
35. Rocha, Clapham, and Ivashchenko, *Emptying the Oceans*, 42–45.
36. Dolin, *Leviathan*, p. 13.
37. Genesis 1:21.
38. Jules Michelet, *The Sea* (London: T. Nelson & Sons, 1875), 255, 256.
39. Carl Schmitt, *Land and Sea*, trans. and with a Foreword by Simona Draghici (Corvallis, OR: Plutarch Press, 1997), 15. For the German original, see Carl Schmitt, *Land und Meer: Eine weltgeschichtliche Betrachtung* (Cologne: Hohenheim, 1981 [earlier editions published in Leipzig in 1942 and in Stuttgart in 1954]), 32.
40. Schmitt, *Land and Sea*, 15.
41. Schmitt, *Land and Sea*, 15.
42. Michelet, *Sea*, 248, 255.
43. See Anna-Katharina Wöbse, *Weltnaturschutz: Umweltdiplomatie in Völkerbund und Vereinten Nationen 1920–1950* (Frankfurt: Campus, 2012), 171–245.
44. Maglosia Fitzmaurice, *Whaling and International Law* (Cambridge: Cambridge University Press, 2015), 9.
45. Frank Zelko, *Make It A Green Peace! The Rise of Countercultural Environmentalism* (New York: Oxford University Press, 2013), 3; and Dorsey, *Whales and Nations*, 207n.
46. Fitzmaurice, *Whaling*, 7.
47. Dorsey, *Whales and Nations*, 92.
48. Tønnessen and Johnsen, *History*, 534.
49. Yulia V. Ivashchenko and Phillip J. Clapham, "Too Much Is Never Enough: The Cautionary Tale of Soviet Illegal Whaling," *Marine Fisheries Review* 76 (2014): 1–21; here, 18.
50. Dorsey, *Whales and Nations*, xx.
51. Zelko, *Make It a Green Peace!* 210; Rocha, Clapham, and Ivashchenko, *Emptying the Oceans*, 43, 45.
52. Dorsey, *Whales and Nations*, 208. The IWC had banned the hunting of blue whales in the 1960s (220).
53. NASA, "What Are the Contents of the Golden Record?" accessed November 3, 2021, https://voyager.jpl.nasa.gov/golden-record/whats-on-the-record/.
54. Zelko, *Make It a Green Peace!* 213.
55. Fitzmaurice, *Whaling*, 67.
56. Poul Johansen et al., "Human Exposure to Contaminants in the Traditional Greenland Diet," *Science of the Total Environment* 331 (2004): 189–206.
57. Jennifer Steinhauer, "Keeping Whale Off Sushi Plates Is Oscar Winners' Next Mission," *New York Times*, March 9, 2010.
58. Michelet, *Sea*, 254.
59. See W. Jeffrey Bolster, *The Mortal Sea: Fishing the Atlantic in the Age of Sail* (Cambridge, MA: Belknap Press of Harvard University Press, 2012).
60. Finley, *All the Fish*, 6n, 48.
61. Mark Kurlansky, *Cod: A Biography of the Fish That Changed the World* (New York: Penguin, 1998), 162–71; and Lawrence Juda, *International Law and Ocean Use Management: The Evolution of Ocean Governance* (London: Routledge, 1996): 171–80, 229.
62. See Dean Bavington, *Managed Annihilation: An Unnatural History of the Newfoundland Cod Collapse* (Vancouver: UBC Press, 2010); Alan Christopher Finlayson, *Fishing for Truth: A Sociological Analysis of Northern Cod Stock Assessments from 1977 to 1990* (St. John's: Institute of Social and Economic Research, 1994); and Michael Harris, *Lament for*

an Ocean: The Collapse of the Atlantic Cod Fishery; A True Crime Story (Toronto: McClelland and Stewart, 1999).

63. The role of fisheries policy in the making of Somali piracy is contested on several fronts. For one account, see U. Rashid Sumaila and Mahamudu Bawumai, "Fisheries, Ecosystem Justice and Piracy: A Case Study of Somalia," *Fisheries Research* 157 (September 2014): 154–63.

64. Charles Clover, *The End of the Line: How Overfishing Is Changing the World and What We Eat* (London: Ebury Press, 2005), 39.

65. Clover, *End of the Line*, 118.

10: UNITED FRUIT

1. Monica A. Rankin, *The History of Costa Rica* (Santa Barbara, CA: ABC-CLIO, 2012), 74–76; and Jason M. Colby, *The Business of Empire: United Fruit, Race, and U.S. Expansion in Central America* (Ithaca, NY: Cornell University Press, 2011), 36–39, 66–69.

2. Stacy May and Galo Plaza, *The United Fruit Company in Latin America* (Washington, DC: National Planning Association, 1958), 4–6.

3. Thomas L. Karnes, *Tropical Enterprise: The Standard Fruit and Steamship Company in Latin America* (Baton Rouge: Louisiana State University Press, 1978), 4.

4. Adam Smith, *An Inquiry into the Nature and Causes of the Wealth of Nations*, ed. R. H. Campbell and A. S. Skinner, vol. 1 (Oxford: Clarendon Press, 1976), 145.

5. Marcelo Bucheli, *Bananas and Business: The United Fruit Company in Colombia, 1899–2000* (New York: New York University Press, 2005), 8.

6. Alfred D. Chandler Jr., *The Visible Hand: The Managerial Revolution in American Business* (Cambridge, MA: Belknap Press of Harvard University Press, 1977), 4.

7. See Robert C. McMath Jr., *American Populism: A Social History 1877–1898* (New York: Noonday Press, 1993); and Daniel T. Rodgers, "In Search of Progressivism," *Reviews in American History* 10, no. 4 (December 1982): 113–32.

8. Paul J. Dosal, *Doing Business with the Dictators: A Political History of United Fruit in Guatemala 1899–1944* (Wilmington, DE: Scholarly Resources, 1993), 6.

9. Peter N. Davies, *Fyffes and the Banana: Musa Sapientum; A Centenary History 1888–1988* (London: Athlone Press, 1990), 122n.

10. Bucheli, *Bananas and Business*, 24.

11. Sarah C. Chambers, "New Nations and New Citizens: Political Culture in Nineteenth-Century Mexico, Peru, and Argentina," in Thomas H. Holloway, ed., *A Companion to Latin American History*, 215–29 (Malden, MA: Wiley-Blackwell, 2011), 215.

12. Lowell Gudmundson, "Mora Porrás, Juan Rafael," in Barbara A. Tenenbaum, ed., *Encyclopedia of Latin American History and Culture*, vol. 4, 111–12 (London: Simon and Schuster, 1996), 112.

13. Jeffrey M. Paige, *Coffee and Power: Revolution and the Rise of Democracy in Central America* (Cambridge, MA: Harvard University Press, 1997), 14.

14. Charles David Kepner Jr. and Jay Henry Soothill, *The Banana Empire: A Case Study of Economic Imperialism* (New York: Vanguard Press, 1935), 341.

15. Robert Fitzgerald, *The Rise of the Global Company: Multinationals and the Making of the Modern World* (Cambridge: Cambridge University Press, 2015), 89.

16. Pablo Neruda, *Selected Poems*, ed. and trans. Ben Belitt (New York: Grove Press, 1961), 149.

17. Dosal, *Doing Business*, 2.

18. Bucheli, *Bananas and Business*, 187. For Buheli's criticism of the approach, see 186–89.

19. Paul J. Dorsal, "United Fruit," in Akira Iriye and Pierre-Yves Saunier, eds., *The Pal-

grave Dictionary of Transnational History, 30–31 (Basingstoke: Palgrave Macmillan, 2009), 31.

20. Quoted in Douglas Southgate and Lois Roberts, *Globalized Fruit, Local Entrepreneurs: How One Banana-Exporting Country Achieved Worldwide Reach* (Philadelphia: University of Pennsylvania Press, 2016), 110.

21. See Dosal, *Doing Business*.

22. Mark Moberg and Steve Striffler, "Introduction," in Striffler and Moberg, eds., *Banana Wars: Power, Production, and History in the Americas*, 1–19 (Durham, NC: Duke University Press, 2003), 2.

23. Colby, *Business of Empire*, 119.

24. Bucheli, *Bananas and Business*, 1–3, 118–20, 131–33.

25. John Soluri, *Banana Cultures: Agriculture, Consumption, and Environmental Change in Honduras and the United States* (Austin: University of Texas Press, 2005), 53, 105.

26. Stephen Schlesinger and Stephen Kinzer, *Bitter Fruit: The Story of the American Coup in Guatemala* (Cambridge, MA: Harvard University Press, 1999), 49n, 55; Richard H. Immerman, *The CIA in Guatemala: The Foreign Policy of Intervention* (Austin: University of Texas Press, 1982), 80; and Piero Gleijeses, *Shattered Hope: The Guatemalan Revolution and the United States, 1944–1954* (Princeton, NJ: Princeton University Press, 1991).

27. Schlesinger and Kinzer, *Bitter Fruit*, 76.

28. Dan Koeppel, *Banana: The Fate of the Fruit That Changed the World* (New York: Penguin, 2009), 128.

29. Quoted in Schlesinger and Kinzer, *Bitter Fruit*, 227.

30. Lucía Álvarez de Toledo, *The Story of Che Guevara* (London: Quercus, 2010), 142–48, 155, 158; and Jon Lee Anderson, *Che Guevara: A Revolutionary Life* (London: Bantam Press, 1997), 121–61, 172–75.

31. Dosal, *Doing Business*, 229n.

32. Odd Arne Westad, *The Global Cold War: Third World Interventions and the Making of Our Times* (New York: Cambridge University Press, 2007), 148.

33. Soluri, *Banana Cultures*, 181.

34. Soluri, *Banana Cultures*, 161, 184.

35. Karnes, *Tropical Enterprise*, 284.

36. John Soluri, "Banana Cultures: Linking the Production and Consumption of Export Bananas, 1800–1980," in Striffler and Moberg, *Banana Wars*, 48–79; here, 76.

37. Bucheli, *Bananas and Business*, 51–53.

38. Colombian banana planters were practicing contract farming already in the first half of the twentieth century; see Marcelo Bucheli, "Enforcing Business Contracts in South America: The United Fruit Company and Colombian Banana Planters in the Twentieth Century," *Business History Review* 78, no. 2 (Summer 2004): 181–212.

39. See Chandler, *Visible Hand*.

40. Michael Evans, "'Para-politics' Goes Bananas," *The Nation*, April 16, 2007.

41. Thomas P. McCann, *An American Company: The Tragedy of United Fruit* (New York: Crown, 1976), 1–4, 118, 198.

42. Davies, *Fyffes and the Banana*, 229.

43. Bucheli, *Bananas and Business*, 76, 83.

44. James Wiley, *The Banana: Empires, Trade Wars, and Globalization* (Lincoln: University of Nebraska Press, 2008), 241.

45. Peter Clegg, *The Caribbean Banana Trade: From Colonialism to Globalization* (Basingstoke: Palgrave Macmillan, 2002), 182.

46. Mike Peed, "We Have No Bananas: Can Scientists Defeat a Devastating Blight?"

New Yorker, January 10, 2011, 28–34; and "Yes, We Have No Bananas," *The Economist*, March 1, 2014, 65.

47. Naomi Klein has made a similar argument about neoliberalism. See Naomi Klein, *The Shock Doctrine: The Rise of Disaster Capitalism* (London: Penguin, 2008).

48. Laura T. Raynolds, "The Global Banana Trade," in Striffler and Moberg, *Banana Wars*, 23–47; here, 43n.

49. Rebecca Smithers and Dominic Rushe, "Fyffes to Merge with Chiquita and Create World's Biggest Banana Company," *The Guardian*, March 10, 2014, accessed November 3, 2021, available online at https://www.theguardian.com/business/2014/mar/10/fyffes-merge-chiquita-banana-co-chiquitafyffes.

PART III
IRREVERSIBLE

1. Reinhart Koselleck, "Fortschritt," in Koselleck, Otto Brunner, and Werner Conze, eds., *Geschichtliche Grundbegriffe: Historisches Lexikon zur politisch-sozialen Sprache in Deutschland*, vol. 2 (Stuttgart: Klett-Cotta, 1975), 351–53, 363–423.

2. Joseph Needham, *The Grand Titration: Science and Society in East and West* (London: Routledge, 2005 [1969]), 301.

3. See John P. Parkes and F. Dane Panetta, "Eradication of Invasive Species: Progress and Emerging Issues in the 21st Century," in Mick N. Clout and Peter A. Williams, eds., *Invasive Species Management: A Handbook of Principles and Techniques*, 47–60 (New York: Oxford University Press, 2009).

4. Shane J. Hunt, "Growth and Guano in Nineteenth-Century Peru," in Roberto Cortés Conde and Shane J. Hunt, eds., *The Latin American Economies: Growth and the Export Sector 1880–1930*, 255–318 (New York: Holmes and Meier, 1985), 288.

5. Amitav Ghosh, "Petrofiction: The Oil Encounter and the Novel," in Imre Szemann and Dominic Boyer, eds., *Energy Humanities: An Anthology*, 431–40 (Baltimore: Johns Hopkins University Press, 2017), 431.

11: THE DODO

1. Sydney Selvon, *A New Comprehensive History of Mauritius*, vol. 1: *From Ancient Times to the Birth of Parliament*, 2nd ed. (n.p. [Port Louis]: M.D.S. Editions, 2017), 24n.

2. Jolyon C. Parish, *The Dodo and the Solitaire: A Natural History* (Bloomington: Indiana University Press, 2013), 3.

3. Perry J. Moree, *A Concise History of Dutch Mauritius, 1598–1710: A Fruitful and Healthy Land* (London: Kegan Paul International, 1998), 9n.

4. Quoted in Parish, *Dodo*, 5.

5. Staffan Müller-Wille, "Plant Taxonomy: The Love of Plants," *Nature* 446 (2007): 268; and Lisbet Koerner, *Linnaeus: Nature and Nation* (Cambridge, MA: Harvard University Press, 2000), 42.

6. Londa Schiebinger and Claudia Swan, "Introduction," in Schiebinger and Swan, eds., *Colonial Botany: Science, Commerce, and Politics in the Early Modern World*, 1–16 (Philadelphia: University of Pennsylvania Press, 2005), 8.

7. Samuel T. Turvey and Anthony S. Cheke, "Dead as a Dodo: The Fortuitous Rise to Fame of an Extinction Icon," *Historical Biology* 20 (2008): 149–63; here, 158. Linnaeus is known for "loading some of his names with an onerous burden of description." (Harriet Ritvo, *The Platypus and the Mermaid and Other Figments of the Classifying Imagination* [Cambridge, MA: Harvard University Press, 1997], 57.)

8. Julian P. Hume, "The History of the Dodo *Raphus cucullatus* and the Penguin of Mauritius," *Historical Biology* 18 (2006): 69–93; here, 70.

9. Natalie Lawrence, "Assembling the Dodo in Early Modern Natural History," *British Journal for the History of Science* 48 (2015): 387–408; here, 407.

10. Errol Fuller, *Dodo: From Extinction to Icon* (London: HarperCollins, 2002), 25. Similarly, Parish, *Dodo*, xi, and Hume, "History," 90.

11. Hume, "History," 78n, 79 (quotation).

12. David S. Landes, *The Wealth and Poverty of Nations: Why Some Are So Rich and Some So Poor* (New York: Norton, 1999), 146.

13. Lawrence, "Assembling the Dodo," 395.

14. Jo Elwyn Jones and J. Francis Gladstone, *The* Alice *Companion: A Guide to Lewis Carroll's* Alice *Books* (Basingstoke: Macmillan, 1998), 71; and Robert Douglas-Fairhurst, *The Story of Alice: Lewis Carroll and the Secret History of Wonderland* (Cambridge, MA: Belknap Press of Harvard University Press, 2015), 125.

15. Hume, "History," 80; and Turvey and Cheke, "Dead as a Dodo," 151.

16. Maglosia B. Nowak-Kemp and Julian P. Hume, "The Oxford Dodo, part 2: From Curiosity to Icon and Its Role in Displays, Education and Research," *Historical Biology* 29 (2017): 296–307; here, 297.

17. Hugo Edwin Strickland and Alexander Gordon Melville, *The Dodo and Its Kindred; Or The History, Affinities, and Osteology of the Dodo, Solitaire, and Other Extinct Birds of the Islands Mauritius, Rodriguez, and Bourbon* (Cambridge: Cambridge University Press, 2015 [1848]), 45.

18. Hume, "History," 89; and Parish, *Dodo*, 336.

19. Parish, *Dodo*, 236–44; Maglosia B. Nowak-Kemp and Julian P. Hume, "The Oxford Dodo, part 1: The Museum History of the Tradescant Dodo: Ownership, Displays and Audience," *Historical Biology* 29 (2017): 234–47; here, 234.

20. Beth Shapiro et al., "Flight of the Dodo," *Science* 295 (2002): 1683.

21. Delphine Angst et al., "Bone Histology Sheds New Light on the Ecology of the Dodo (*Raphus cucullatus*, Aves, Columbiformes)," *Scientific Reports* 7, no. 7993 (August 24, 2017).

22. Anthony Cheke and Julian Hume, *Lost Land of the Dodo: An Ecological History of Mauritius, Réunion and Rodrigues* (London: T & AD Poyser, 2008), 80.

23. Cheke and Hume, *Lost Land*, 117.

24. Steven R. Ewing et al., "Inbreeding and Loss of Genetic Variation in a Reintroduced Population of Mauritius Kestrel," *Conservation Biology* 22 (2008): 395–404; here, 401.

25. Parish, *Dodo*, 37.

26. Lawrence, "Assembling the Dodo," 406.

27. Parish, *Dodo*, 138; and Hume, "History," 84.

28. David Quammen, *The Song of the Dodo: Island Biogeography in an Age of Extinctions* (London: Pimlico, 1997), 262.

29. Jochen Dierschke et al., *Die Vogelwelt der Insel Helgoland* (Helgoland: OAG Helgoland, 2011), 25; and Erwin Stresemann, "Vor- und Frühgeschichte der Vogelforschung auf Helgoland," *Journal für Ornithologie* 108 (1967): 377–429; here, 377n.

30. Friedrich von der Decken, *Philosophisch-historisch-geographische Untersuchungen über die Insel Helgoland oder Heiligeland und ihre Bewohner* (Hanover: Hahnsche Hof-Buchhandlung, 1826), 159n.

31. John F. Richards, *The Unending Frontier: An Environmental History of the Early Modern World* (Berkeley: University of California Press, 2005), 619.

32. Cheke and Hume, *Lost Land*, 87.

33. Decken, *Philosophisch-historisch-geographische Untersuchungen*, 160. Later books doubted that service was ever suspended due to an approaching flock of birds, perhaps

because it reflected badly on the islanders' character. (Karl Reinhardt, *Von Hamburg nach Helgoland: Skizzenbuch* [Leipzig: Verlagsbuchhandlung von J. J. Weber, 1956], 120.)

34. Hume, "History," 86.

35. See Andrew Jackson, "Added Credence for a Late Dodo Extinction Date," *Historical Biology* 26 (2014): 699–701; and David L. Roberts and Andrew R. Solow, "Flightless Birds: When Did the Dodo Become Extinct?" *Nature* 426 (2003): 245.

36. See Martin J. S. Rudwick, *Bursting the Limits of Time: The Reconstruction of Geohistory in the Age of Revolution* (Chicago: University of Chicago Press, 2005), vi.

37. On paradigms in the history of science, see Thomas S. Kuhn, *The Structure of Scientific Revolutions* (Chicago: University of Chicago Press, 1962).

38. See Keith Thomson, *Jefferson's Shadow: The Story of His Science* (New Haven, CT: Yale University Press, 2012), 51–61, 89–92, 225.

39. Michael J. Benton and David A. T. Harper, *Introduction to Paleobiology and the Fossil Record* (Chichester: Wiley-Blackwell, 2009), 164, 179.

40. Edward O. Wilson, *The Future of Life* (London: Abacus, 2003), xxiii. See also Elizabeth Kolbert, *The Sixth Extinction: An Unnatural History* (London: Bloomsbury, 2014).

41. The argument of a sixth extinction remains contested. For a passionate defense of the concept, see Gerardo Ceballos and Paul R. Ehrlich, "The Misunderstood Sixth Mass Extinction," *Science* 360, no. 6393 (June 8, 2018): 1080–81.

42. Bernhard Gissibl, *The Nature of German Imperialism: Conservation and the Politics of Wildlife in Colonial East Africa* (New York: Berghahn Books, 2016), 154, 158.

43. At least that is the reading in Julio A. Baisre, "Shifting Baselines and the Extinction of the Caribbean Monk Seal," *Conservation Biology* 27 (2013): 927–35.

44. Mark V. Barrow Jr., *Nature's Ghosts: Confronting Extinction from the Age of Jefferson to the Age of Ecology* (Chicago: University of Chicago Press, 2009), 6–9.

45. E. O. Wilson, *The Diversity of Life* (London: Penguin, 2001), 335.

46. See Lewis Carroll, *The Annotated Alice: Alice's Adventures in Wonderland and through the Looking-Glass*, with an Introduction and notes by Martin Gardner (New York: Bramhall House, 1955), 47–50.

47. Ursula K. Heise, *Imagining Extinction: The Cultural Meanings of Endangered Species* (Chicago: University of Chicago Press, 2016), 37.

48. William M. Adams, *Against Extinction: The Story of Conservation* (London: Earthscan, 2004), 20.

49. Peter Cotgreave, "The Historian and the Dodo," *History Today* 47, no. 1 (January 1997): 7–8; here, 7.

50. Turvey and Cheke, "Dead as a Dodo," 150.

51. Wilson, *Diversity of Life*, xiii.

52. Adams, *Against Extinction*, 127.

53. See Thomas M. Bohn, Aliaksandr Dalhouski, and Markus Krzoska, *Wisent-Wildnis und Welterbe: Geschichte des polnisch-weißrussischen Nationalparks von Białowieża* (Cologne: Böhlau, 2017).

54. Douglas Brinkley, *The Wilderness Warrior: Theodore Roosevelt and the Crusade for America* (New York: Harper Perennial, 2010), 201n.

55. Richard Fitter and Sir Peter Scott, *The Penitent Butchers: The Fauna Preservation Society 1903–1978* (London: Fauna Preservation Society, 1978), 8.

56. Andrew C. Isenberg, *The Destruction of the Bison: An Environmental History, 1750–1920* (Cambridge: Cambridge University Press, 2000), p. 164; and David Hancocks, *A Different Nature: The Paradoxical World of Zoos and Their Uncertain Future* (Berkeley: University of California Press, 2001), 155–59.

57. Barrow, *Nature's Ghosts*, 4.

58. David A. Keith et al., "The IUCN Red List of Ecosystems: Motivations, Challenges, and Applications," *Conservation Letters* 8 (2015): 214–26. See also https://www.iucn.org/theme/ecosystem-management/our-work/red-list-ecosystems, accessed November 3, 2021.

59. Sir Peter Scott, John A. Burton, and Richard Fitter, "Red Data Books: The Historical Background," in Richard Fitter and Maisie Fitter, eds., *The Road to Extinction: Problems of Categorizing the Status of Taxa Threatened with Extinction: Proceedings of a Symposium held by the Species Survival Commission, Madrid, 7 and 9 November 1984*, 1–5 (Gland: International Union for the Conservation of Nature and Natural Resources, 1987), 3.

60. See https://www.iucnredlist.org/about/barometer-of-life, accessed November 3, 2021.

61. See Paul Munton, "Concepts of Threat to the Survival of Species Used in Red Data Books and Similar Compilations," in Fitter and Fitter, *Road to Extinction*, 71–111, here, 75, 78.

62. Scott, Burton, and Fitter, "Red Data Books," 2.

63. See http://cmsdocs.s3.amazonaws.com/summarystats/2017-2_Summary_Stats_Page_Documents/2017_2_RL_Stats_Table_4a.pdf, accessed November 3, 2021.

64. Ewing et al., "Inbreeding," 395.

65. Daniel Pauly, "Anecdotes and the Shifting Baseline Syndrome of Fisheries," *Trends in Ecology and Evolution* 10 (1995): 430.

66. See Parish, *Dodo*, and Hume, "History."

67. Tiago Saraiva, "Breeding Europe: Crop Diversity, Gene Banks, and Commoners," in Nil Disco and Eda Kranakis, eds., *Cosmopolitan Commons: Sharing Resources and Risks across Borders*, 185–211 (Cambridge, MA: MIT Press, 2013), 185.

68. Nicole C. Karafyllis, "Biofacts, Bioprospecting, Biobanks: A Reality Check of Seed Banks," in Sabine Maasen, Sascha Dickel, and Christoph Schneider, eds., *TechnoScienceSociety: Technological Reconfigurations of Science and Society*, 131–56 (Cham: Springer, 2020), 133.

69. Elisabeth Rosenthal, "Russia Defers Razing of Seed Repository," *New York Times*, September 10, 2010, accessed November 3, 2021, https://green.blogs.nytimes.com/2010/09/10/russia-defers-razing-of-seed-repository/?ref=science.

70. Kolbert, *Sixth Extinction*, 262.

71. Jane Carruthers, *National Park Science: A Century of Research in South Africa* (Cambridge: Cambridge University Press, 2017), 280n.

72. Antonie Marinus Harthoorn, *The Flying Syringe: Ten Years of Immobilising Wild Animals in Africa* (London: Geoffrey Bles, 1970).

73. Hancocks, *Different Nature*, 150.

74. Wilson, *Diversity of Life*, 319.

75. Wilson, *Future of Life*, 163.

76. Gregory Benford, "Saving the 'Library of Life,'" *Proceedings of the National Academy of Sciences of the United States of America* 89 (1992): 11098–101.

77. Kolbert, *Sixth Extinction*, 259.

78. Ben J. Novak, "The Great Comeback: Bringing a Species Back from Extinction," *Futurist* 47, no. 5 (September–October 2013): 40–44. For a critical commentary, see Ronald Sandler, "Techno-Conservation in the Anthropocene: What Does It Mean to Save a Species?" in Ursula K. Heise, Jon Christensen, and Michelle Niemann, eds., *The Routledge Companion to the Environmental Humanities*, 72–81 (London: Routledge, 2017).

79. Carroll, *Annotated Alice*, 49.

12: THE BOLL WEEVIL

1. R. N. Foster, "Boll Weevil," in Vincent H. Resh and Ring T. Cardé, eds., *Encyclopedia of Insects*, 2nd ed., 116–17 (Burlington, MA: Elsevier, 2009); Harry Bates Brown and Jacob Osborn Ware, *Cotton*, 3rd ed. (New York: McGraw-Hill, 1958), 204.

2. James C. Giesen, *Boll Weevil Blues: Cotton, Myth, and Power in the American South* (Chicago: University of Chicago Press, 2011), ix.

3. R. Douglas Hurt, *American Agriculture: A Brief History*, rev. ed. (West Lafayette, IN: Purdue University Press, 2002), 41–46.

4. Jack Temple Kirby, *Mockingbird Song: Ecological Landscapes of the South* (Chapel Hill: University of North Carolina Press, 2006), 95.

5. Willard W. Cochrane, *The Development of American Agriculture: A Historical Analysis*, 2nd ed. (Minneapolis: University of Minnesota Press, 1993), 268.

6. Sven Beckert, *Empire of Cotton: A New History of Global Capitalism* (London: Allen Lane, 2014), 256n; and C. Wayne Smith, "Production Statistics," in Smith and J. Tom Cothren, eds., *Cotton: Origin, History, Technology, and Production*, 435–49 (New York: John Wiley, 1999), 436. One bale weighs 480 lbs.

7. Giesen, *Boll Weevil Blues*, vii.

8. Alan L. Olmstead and Paul W. Rhode, *Creating Abundance: Biological Innovation and American Agricultural Development* (Cambridge: Cambridge University Press, 2008), 152n.

9. John Soluri, *Banana Cultures: Agriculture, Consumption, and Environmental Change in Honduras and the United States* (Austin: University of Texas Press, 2005), 52n.

10. Stuart McCook, "Global Rust Belt: *Hemileia vastatrix* and the Ecological Integration of World Coffee Production since 1850," *Journal of Global History* 1 (2006): 177–95.

11. George Gale, *Dying on the Vine: How Phylloxera Transformed Wine* (Berkeley: University of California Press, 2011).

12. Giesen, *Boll Weevil Blues*, 5.

13. Douglas Helms, "Technological Methods for Boll Weevil Control," *Agricultural History* 53 (January 1979): 286–99; here, 289.

14. James C. Giesen, "'The Truth about the Boll Weevil': The Nature of Planter Power in the Mississippi Delta," *Environmental History* 14 (October 2009): 683–704; here, 694.

15. Helms, "Technological Methods," 288.

16. Douglas Helms, "Revision and Reversion: Changing Cultural Control Practices for the Cotton Boll Weevil," *Agricultural History* 54 (January 1980): 108–25; here, 116n.

17. Giesen, "Truth," 684.

18. Helms, "Revision and Reversion," 108n.

19. Giesen, *Boll Weevil Blues*, 124–27.

20. Helms, "Revision and Reversion," 109, 111; and Fabian Lange, Alan L. Olmstead, and Paul W. Rhode, "The Impact of the Boll Weevil, 1892–1932," *Journal of Economic History* 69 (September 2009): 685–718; here, 692.

21. McCook, "Global Rust Belt," 185n.

22. See p. 161.

23. Helms, "Technological Methods," 291.

24. Helms, "Revision and Reversion," 118n.

25. Giesen, "Truth," 693n, 697.

26. Mart Stewart, "If John Muir Had Been an Agrarian: American Environmental History West and South," *Environment and History* 11 (2005): 139–62.

27. Giesen, *Boll Weevil Blues*, p. 95.

28. For a scholarly reflection of this view, see Roger L. Ransom and Richard Sutch, *One*

Kind of Freedom: The Economic Consequences of Emancipation (Cambridge: Cambridge University Press, 1977), 172–74.

29. Frank Uekötter, "Rise, Fall, and Permanence: Issues in the Environmental History of the Global Plantation," in Uekötter, ed., *Comparing Apples, Oranges, and Cotton: Environmental Histories of the Plantation*, 7–26 (Frankfurt: Campus, 2014), 8.

30. John Fraser Hart, "The Demise of King Cotton," *Annals of the Association of American Geographers* 67 (September 1977): 307–22; here, 307.

31. Giesen, *Boll Weevil Blues*, xii.

32. Arvarh E. Strickland, "The Strange Affair of the Boll Weevil: The Pest as Liberator," *Agricultural History* 68 (Spring 1994): 157–68; here, 167.

33. Lange, Olmstead, and Rhode, "Impact of the Boll Weevil," 696.

34. Lange, Olmstead, and Rhode, "Impact of the Boll Weevil," 687, 713.

35. Ricardo Carrere and Larry Lohmann, *Pulping the South: Industrial Tree Plantations in the World Paper Economy* (London: Zed Books, 1996), 16.

36. Kirby, *Mockingbird Song*, 320n.

37. Idus A. Newby, *The South: A History* (New York: Holt, Rinehart and Winston, 1978), 474n.

38. Helms, "Revision and Reversion," 123.

39. Eldon W. Downs and George F. Lemmer, "Origins of Aerial Crop Dusting," *Agricultural History* 39 (July 1965): 123–35; here, 129, 131; and W. David Lewis and Wesley Phillips Newton, *Delta: The History of an Airline* (Athens: University of Georgia Press, 1979), 11.

40. Kent Osband, "The Boll Weevil Versus 'King Cotton,'" *Journal of Economic History* 45 (September 1985): 627–43; here, 628, 640.

41. Helms, "Technological Methods," 295.

42. Helms, "Revision and Reversion," 124; and Helms, "Technological Methods," 297. See also p. 548.

43. Luther J. Carter, "Eradicating the Boll Weevil: Would It Be a No-Win War?" *Science* 183, no. 4124 (February 8, 1974): 494–99; here, 494.

44. Helms, "Revision and Reversion," 120.

45. Carter, "Eradicating the Boll Weevil," 495n, 499, 495 (quotation). On the fire ants campaign, see p. 575.

46. Jan Suszkiw and Jim De Quattro, "Evicting the Boll Weevil," *Agricultural Research* 42, no. 3 (March 1994): 4–10.

47. James Coppedge and Robert M. Faust, "Winning the Weevil War: Beating a $22 Billion Bug," *Agricultural Research* 51, no. 2 (February 2003): 2.

48. Suszkiw and De Quattro, "Evicting the Boll Weevil," 7.

49. Edmund Russell, *War and Nature: Fighting Humans and Insects with Chemicals from World War I to* Silent Spring (Cambridge: Cambridge University Press, 2001).

50. Coppedge and Faust, "Winning the Weevil War."

51. Smith, "Production Statistics," 437. For a comprehensive documentation of the campaign, see Willard A. Dickerson et al., *Boll Weevil Eradication in the United States through 1999* (Memphis, TN: Cotton Foundation, 2001).

52. Dennis O'Brien, "Mystery Solved: Detecting the Source of a Boll Weevil Outbreak," *Agricultural Research* 59, no. 1 (January 2011): 20–21.

53. See Nancy Leys Stepan, *Eradication: Ridding the World of Diseases Forever?* (Ithaca, NY: Cornell University Press, 2011).

54. Terrence J. Centner and Susana Ferreira, "Ability of Governments to Take Actions to Confront Incursions of Diseases—a Case Study: Citrus Canker in Florida," *Plant Pathology* 62 (2012): 821–28; here, 823.

55. Erik Stokstad, "New Disease Endangers Florida's Already-Suffering Citrus Trees,"

Science 312 (2006): 523–24; Darryl Fears, "The End of Florida Orange Juice? A Lethal Disease Is Devastating the State's Citrus Industry," *Washington Post*, November 9, 2019, accessed November 3, 2021, https://www.washingtonpost.com/climate-environment/2019/11/09/end-florida-orange-juice-lethal-disease-is-decimating-its-citrus-industry/.

56. Alison Abbott, "Thousands Turn Out to Support Science in Italy's Stricken Olive Region," *Nature*, January 22, 2019, accessed November 3, 2021, https://www.nature.com/articles/d41586-019-00224-8.

57. O'Brien, "Mystery Solved," 20.

58. Blake Layton, "The Boll Weevil in Mississippi: Gone, But Not Forgotten," Mississippi State University Extension Service Publication 2294 (2002): 1.

13: THE LITTLE GRAND CANYON

1. Georgia State Archives RG 37-8-35, Legislature, Commissions and Committees, Summary Committee Reports 1969–1972, folder "1969," Report of the Providence Canyons Study Committee (Senate Resolution No. 37), December 1969, p. 6.

2. Georgia State Archives RG 37-8-35, Legislature, Commissions and Committees, Summary Committee Reports 1969–1972, folder "1969," Report of the Providence Canyons Study Committee (Senate Resolution No. 37), December, 1969, p. 7.

3. Georgia State Archives RG 30-5-32, Parks and Historic Sites, Recreation Services Section, Land/Water Conservation Grant Project/Audit Files, folders "1300097 Acq. Providence Canyon" and "13-00196 Dev. Providence Canyon State Park."

4. See John C. Weaver, *The Great Land Rush and the Making of the Modern World, 1650–1900* (Montreal: McGill-Queen's University Press, 2003).

5. Jürgen Osterhammel, *The Transformation of the World: A Global History of the Nineteenth Century* (Princeton, NJ: Princeton University Press, 2014), 323.

6. Stuart McCook, *States of Nature: Science, Agriculture, and Environment in the Spanish Caribbean, 1760–1940* (Austin: University of Texas Press, 2002), 21, 83.

7. David Moon, *The Plough That Broke the Steppes: Agriculture and Environment on Russia's Grasslands, 1700–1914* (Oxford: Oxford University Press, 2013), 17.

8. Richard White, *The Middle Ground: Indians, Empires, and Republics in the Great Lakes Region, 1650–1815* (Cambridge: Cambridge University Press, 1991), x.

9. Kenneth Sylvester and Geoff Cunfer, "An Unremembered Diversity: Mixed Husbandry and the American Grasslands," *Agricultural History* 83 (2009): 352–83; here, 375n.

10. See Judith A. Carney, *Black Rice: The African Origins of Rice Cultivation in the Americas* (Cambridge, MA: Harvard University Press, 2001); David Eltis, Philip Morgan, and David Richardson, "Agency and Diaspora in Atlantic History: Reassessing the African Contribution to 'Rice Cultivation in the Americas," *American Historical Review* 112 (2007): 1329–58; and AHR Exchange: The Question of 'Black Rice,'" *American Historical Review* 115 (2010): 123–71.

11. Moon, *Plough*, 19.

12. Sven Beckert, "Emancipation and Empire: Reconstructing the Worldwide Web of Cotton Production in the Age of the American Civil War," *American Historical Review* 109 (2004): 1405–38.

13. See Steven Stoll, *Larding the Lean Earth: Soil and Society in Nineteenth-Century America* (New York: Hill and Wang, 2002).

14. Avery Craven, *Soil Exhaustion as a Factor in the Agricultural History of Virginia and Maryland, 1606–1860* (Columbia: University of South Carolina Press, 2006).

15. Diana K. Davis, *Resurrecting the Granary of Rome: Environmental History and French Colonial Expansion in North Africa* (Athens: Ohio University Press, 2007).

16. See, for instance, Jack Temple Kirby, *Mockingbird Song: Ecological Landscapes of the South* (Chapel Hill: University of North Carolina Press, 2006), 89; and Moon, *Plough*, 13.

17. Paul S. Sutter, *Let Us Now Praise Famous Gullies: Providence Canyon and the Soils of the South* (Athens: University of Georgia Press, 2015), 20.

18. Paul S. Sutter, "What Gullies Mean: Georgia's 'Little Grand Canyon' and Southern Environmental History," *Journal of Southern History* 76 (2010): 579–616; here, 611.

19. Georgia State Archives RG 50-2-033, Mines, Mining and Geology, Geological Field Agents, State Geologist Photographs and Negative Films, County Photographs, ca. 1870–1955, folder "Stewart," Negative Album # 7, Negative # 630.

20. See Manfred Hildermeier, *Geschichte der Sowjetunion 1917–1991: Entstehung und Niedergang des ersten sozialistischen Staates*, 2nd ed. (Munich: Beck, 2017), 828, 839; Marc Elie, "The Soviet Dust Bowl and the Canadian Erosion Experience in the New Lands of Kazakhstan, 1950s–1960s," *Global Environment* 8 (2015): 259–92; and Aaron T. Hale-Dorrell, *Corn Crusade: Khrushchev's Farming Revolution in the Post-Stalin Soviet Union* (New York: Oxford University Press, 2019).

21. Willard Sunderland, *Taming the Wild Field: Colonization and Empire on the Russian Steppe* (Ithaca, NY: Cornell University Press, 2004), 117.

22. Mary W. M. Hargreaves, *Dry Farming in the Northern Great Plains: Years of Readjustment, 1920–1990* (Lawrence: University Press of Kansas, 1993), 3; and Mary W. M. Hargreaves, "The Dry-Farming Movement in Retrospect," *Agricultural History* 51 (1997): 149–65; here, 152–56.

23. Moon, *Plough*, 243–45, 249n.

24. For a flattering biography, Wellington Brink, *Big Hugh: The Father of Soil Conservation* (New York: Macmillan, 1951).

25. National Archives of the United States RG 114, entry 1040, box 1, Report of the Chief of the Soil Conservation Service for the Fiscal Year Ending June 30, 1937, October 14, 1937, p. 49.

26. D. Harper Simms, *The Soil Conservation Service* (New York: Praeger, 1970), 71; and Willard W. Cochrane, *The Development of American Agriculture: A Historical Analysis*, 2nd ed. (Minneapolis: University of Minnesota Press, 1993), 292.

27. Robert J. Morgan, *Governing Soil Conservation: Thirty Years of the New Decentralization* (Baltimore: Resources for the Future, 1965), 10, 20–22.

28. See Wayne D. Rasmussen, *Taking the University to the People: Seventy-Five Years of Cooperative Extension* (Ames: Iowa State University Press, 1989).

29. Morgan, *Governing Soil Conservation*, 41; and R. Brunell Held and Marion Clawson, *Soil Conservation in Perspective* (Baltimore: Johns Hopkins University Press, 1965), 87.

30. Iowa State University Library, Special Collections Department, MS-198. box 3, folder 1, Oral History Interview with Donald A. Williams, June 2, 1981, p. 11.

31. National Archives of the United States RG 114, entry 1, box 1, folder "March–April 1934," Bennett, Memorandum to the Secretary of the Interior, April 13, 1934, p. 3.

32. Arthur M. Schlesinger Jr., *The Coming of the New Deal* (Boston: Houghton Mifflin, 1988), 342.

33. Iowa State University Library, Special Collections Department, MS-164, box 10, folder 27, This Is Your Land. Speech of H. H. Bennett, Broadcast over a Nation-Wide Network of the National Broadcasting Company, August 14, 1939, p. 4.

34. Russell Lord, *Behold Our Land* (Boston: Houghton Mifflin, 1938); Vernon Gill Carter and Tom Dale, *Topsoil and Civilization* (Norman: University of Oklahoma Press, 1955); John Seymour and Herbert Girardet, *Far from Paradise: The Story of Man's Impact on the Environment* (London: British Broadcasting Corporation, 1986); and David R. Montgomery, *Dirt: The Erosion of Civilizations* (Berkeley: University of California Press, 2007).

35. Randal S. Beeman and James A. Pritchard, *A Green and Permanent Land: Ecology and Agriculture in the Twentieth Century* (Lawrence: University Press of Kansas, 2001), 13.

36. National Archives of the United States RG 114, entry 1, box 1, folder "February 1934," H. H. Bennett, Director, Soil Erosion Service, Memorandum for the Secretary of the Interior, January 27, 1934; and folder "March–April 1934," Bennett, Memorandum to the Secretary of the Interior, April 13, 1934, p. 4.

37. Iowa State University Library, Special Collections Department, MS-164, box 10, folder 6, Statement presented by H. H. Bennett, Director, Soil Erosion Service, Department of the Interior, before Subcommittee of House Committee on Public Lands, March 20, 1935, p. 16.

38. Sutter, *Let Us Now Praise*, 83.

39. Hugh Hammond Bennett, *Soil Conservation* (New York: McGraw-Hill, 1939), 4. For an overview of literary references, see Sutter, *Let Us Now Praise*.

40. National Archives of the United States, Atlanta Branch, RG 114, no. 6107056, Soil Conservation Service, Southeastern Region, Tifton Area Office, Reports and Correspondence, 1938–1941, box 1, Project Monograph GA-2, Americus, Georgia, p. 31.

41. National Archives of the United States, Atlanta Branch, RG 114, no. 6107056, Soil Conservation Service, Southeastern Region, Tifton Area Office, Reports and Correspondence, 1938–1941, box 1, Project Monograph GA-2, Americus, Georgia, pp. 38, 72.

42. National Archives of the United States, Atlanta Branch, RG 114, no. 6107056, Soil Conservation Service, Southeastern Region, Tifton Area Office, Reports and Correspondence, 1938–1941, box 1, Project Monograph, Agate GA-3, Rome, Georgia, Southeastern Region, p. 123.

43. National Archives of the United States, Atlanta Branch, RG 114, no. 6107056, Soil Conservation Service, Southeastern Region, Tifton Area Office, Reports and Correspondence, 1938–1941, box 1, Project Monograph, Agate GA-3, Rome, Georgia, Southeastern Region, p. 42.

44. National Archives of the United States, Atlanta Branch, RG 114, no. 6106991, Soil Conservation Service, Southeastern Region, Reports and Correspondence, 1938–1941, General Plans and Procedures to be followed in conducting the demonstrational area in soil conservation at St. Matthews, South Carolina, May 24, 1937, p. 1.

45. National Archives of the United States RG 114, entry 1039, box 2, folder "Speeches—Salter," The Job Ahead. A Talk by Dr. Robt. M. Salter, Chief, Soil Conservation Service, US Department of Agriculture, at the sixth annual meeting of the National Association of Soil Conservation Districts at Cleveland, Ohio, on February 28, 1952, p. 2.

46. Hugh H. Bennett, "They're Wrecking Soil Conservation," *Country Gentleman* 72, no. 12 (December 1952): 21, 52, 56.

47. National Archives of the United States RG 114, Accession no. NN3-114-03-001, box 1, Walter C. Lowdermilk, The Eleventh Commandment. Written at Jerusalem, June 22, 1939, p. 1.

48. Sarah T. Phillips, "Lessons from the Dust Bowl: Dryland Agriculture and Soil Erosion in the United States and South Africa, 1900–1950," *Environmental History* 4 (1999): 245–66; here, 260.

49. *Soil Science Society of America Proceedings* 17 (1953): 86.

50. Donald Worster, *Under Western Skies: Nature and History in the American West* (New York: Oxford University Press, 1992), 102.

51. Moon, *Plough*, 146.

52. David B. Danbom, *Born in the Country: A History of Rural America* (Baltimore: Johns Hopkins University Press, 1995), 226.

53. For a comprehensive discussion of this point, see Frank Uekötter, "The Meaning of

Moving Sand: Towards a Dust Bowl Mythology," *Global Environment* 8 (2015): 349–79; and Uekötter, "In Search of a Dust Bowl Narrative for the Twenty-First Century," *Great Plains Quarterly* 40 (2020): 161–68.

54. See George M. Frederickson, *The Comparative Imagination: On the History of Racism, Nationalism, and Social Movements* (Berkeley: University of California Press, 1997), 37–46.

55. For more on the ambiguities of pictures, see p. 585–88.

56. Jan Arend, *Russlands Bodenkunde in der Welt: Eine ost-westliche Transfergeschichte 1880–1945* (Göttingen: Vandenhoeck & Ruprecht, 2017), 92.

57. Shane Hamilton, "Agriculture," in James Ciment, ed., *Postwar America: An Encyclopedia of Social, Political, Cultural, and Economic History*, vol. 1, 22–26 (Armonk, NY: M. E. Sharpe, 2007), 25.

58. "Wind Erosion on Cropland, by Region and Year," accessed November 3, 2021, http://www.nrcs.usda.gov/wps/portal/nrcs/detailfull/soils/survey/geo/?cid=nrcs143_013655.

59. "Providence Canyon State Park," accessed November 3, 2021, http://gastateparks.org/ProvidenceCanyon. Observations on Providence Canyon State Park are based on a visit in June 2015.

60. Georgia State Archives RG 37-8-35, Legislature, Commissions and Committees, Summary Committee Reports 1969–1972, folder "1969," Report of the Providence Canyons Study Committee (Senate Resolution No. 37), December, 1969, p. 6. For the long and ongoing effort to make sense of Providence Canyon, see Sutter, *Let Us Now Praise*.

61. Georgia State Archives RG 1-1-3, Governor, Executive Department, Executive Department Minutes—1998, October–December, folder "Providence Canyon," Executive Order of the Governor of Georgia, Zell Miller, November 6, 1998.

62. "Las Médulas," accessed November 3, 2021, http://whc.unesco.org/en/list/803.

14: CANE TOADS

1. Ross Fitzgerald, *From 1915 to the Early 1980s: A History of Queensland* (Brisbane: University of Queensland Press, 1984), 176.

2. Peter D. Griggs, *Global Industry, Local Innovation: The History of Cane Sugar Production in Australia, 1820–1995* (Bern: Peter Lang, 2011), 83.

3. William J. Lines, *Taming the Great South Land: A History of the Conquest of Nature in Australia* (Athens: University of Georgia Press, 1999), 143.

4. For a comprehensive study of Australia's sugar industry, see Griggs, *Global Industry*.

5. Peter Griggs, "'Rust' Disease Outbreaks and Their Impact on the Queensland Sugar Industry, 1870–1880," *Agricultural History* 69 (1995): 413–37; here, 434n.

6. Griggs, *Global Industry*, 521, 525.

7. See Edward H. Smith and George G. Kennedy, "History of Entomology," in Ring T. Cardé and Vincent H. Resh, eds., *Encyclopedia of Insects*, 2nd ed., 449–58 (Burlington, MA: Academic Press, 2009).

8. Griggs, *Global Industry*, 521.

9. Ian Tyrrell, *True Gardens of the Gods: Californian-Australian Environmental Reform, 1860–1930* (Berkeley: University of California Press, 1999), 181, 203; Nigel Turvey, *Cane Toads: A Tale of Sugar, Politics and Flawed Science* (Sydney: Sydney University Press, 2013), 55–73; and O. H. Swezey, "Biographical Sketch of the Work of Albert Koebele in Hawaii," *Hawaiian Planters' Record* 29 (1925): 364–68.

10. Turvey, *Cane Toads*, 49. On the global spread of toads, which were omnipresent except in Australia, Madagascar, and Antarctica, see Jennifer B. Pramuk et al., "Around the World in 10 Million Years: Biogeography of the Nearly Cosmopolitan True Toads (Anura: Bufonidae)," *Global Ecology and Biogeography* 17 (2008): 72–83.

11. Turvey, *Cane Toads*, 29, 109-12, 119-21, 135. For an overview of cane toad populations worldwide, see Christopher Lever, *The Cane Toad: The History and Ecology of a Successful Colonist* (Otley: Westbury Academic and Scientific, 2001), 34-140.

12. Turvey, *Cane Toads*, 106.

13. Rick Shine, *Cane Toad Wars* (Oakland: University of California Press, 2018), 7.

14. C. R. J. Boland, "Introduced Cane Toads *Bufo marinus* Are Active Nest Predators and Competitors of Rainbow Bee-Eaters *Merops ornatus*: Observational and Experimental Evidence," *Biological Conservation* 120 (2004): 53-62; here, 58.

15. Georgia Ward-Fear, Matthew J. Greenlees, and Richard Shine, "Toads on Lava: Spatial Ecology and Habitat Use of Invasive Cane Toads (*Rhinella marina*) in Hawai'i," *PLoS ONE* 11, no. 3 (2016): 1-16; here, 10.

16. Shine, *Cane Toad Wars*, 24.

17. Griggs, *Global Industry*, 536, 541n.

18. Turvey, *Cane Toads*, 157.

19. Eric Worrell, "The Unpopular Ones," in A. J. Marshall, ed., *The Great Extermination: A Guide to Anglo-Australian Cupidity, Wickedness and Waste*, 75-94 (London: Heinemann, 1966), 93.

20. Sarah Lowe et al., *100 of the World's Worst Invasive Alien Species: A Selection from the Global Invasive Species Database* (Auckland: Hollands Printing, 2000; rev. 2004), 6n.

21. Lowe et al., *100 of the World's Worst*, 3, 11.

22. See p. 529.

23. Leda Huta and Greg Kuether, "Dangerous Strangers: How Rogue Species Threaten Our Endangered Species," in Karl Weber, ed., *Cane Toads and Other Rogue Species*, 57-72 (New York: PublicAffairs, 2010), 60. On the consequences, see John D. Willson, "Indirect Effects of Invasive Burmese Pythons on Ecosystems in Southern Florida," *Journal of Applied Ecology* 54 (2017): 1251-58.

24. William Kremer, "Pablo Escobar's Hippos: A Growing Problem," accessed November 3, 2021, http://www.bbc.co.uk/news/magazine-27905743.

25. Turvey, *Cane Toads*, 113n, 138, 143-56.

26. Turvey, *Cane Toads*, 110n. On the scientific work of the Hawaiian Sugar Planters' Association, see Carol A. MacLennan, *Sovereign Sugar: Industry and Environment in Hawai'i* (Honolulu: University of Hawai'i Press, 2014), 241-48.

27. Swezey, "Biographical Sketch," 366n.

28. Richard Wassersug, "On the Comparative Palatability of Some Dry-Season Tadpoles from Costa Rica," *American Midland Naturalist* 86 (1971): 101-9; here, 101.

29. Steve Nadis, "Ig Nobel Glory for Levitating Frogs and Collapsing Toilets," *Nature* 407 (October 12, 2000): 665.

30. Mark Lewis, "The Making—and the Meaning—of *Cane Toads: The Conquest*," in Weber, *Cane Toads*, 19-29. For reviews, see Julian Thomas, "Film Reviews—Cane Toads: An Unnatural History," *American Historical Review* 96 (1991): 1118-20; Verina Glaessner, "Cane Toads—An Unnatural History," *Monthly Film Bulletin* 55, no. 659 (December 1, 1988): 361; and Michael Taussig, "Cane Toads: An Unnatural History," *American Anthropologist* 92 (1990): 1110-11.

31. Tim Low, *Feral Future: The Untold Story of Australia's Exotic Invaders*, 2nd ed. (Chicago: University of Chicago Press, 2002), 273.

32. Peter Darryl Griggs, "Too Much Water: Drainage Schemes and Landscape Change in the Sugar-Producing Areas of Queensland, 1920-90," *Australian Geographer* 49 (2018): 81-105; here, 100.

33. Lowe et al., *100 of the World's Worst*, 6n.

34. Shine, *Cane Toad Wars*, 161. They perished due to the harsh environmental conditions over the following two decades.

35. University of Queensland, "Bell Memorial Medal 2018," accessed November 3, 2021, https://agriculture.uq.edu.au/files/7150/Bell%20medal%202018.pdf.

36. Christa Beckmann and Richard Shine, "Impact of Invasive Cane Toads on Australian Birds," *Conservation Biology* 23 (2009): 1544–49.

37. Andres Taylor et al., "Impact of Cane Toads on a Community of Australian Native Frogs, Determined by 10 Years of Automated Identification and Logging of Calling Behaviour," *Journal of Applied Ecology* 54 (2017): 2000–2010.

38. Ruchira Somaweera and Rick Shine, "The (Non) Impact of Invasive Cane Toads on Freshwater Crocodiles at Lake Argyle in Tropical Australia," *Animal Conservation* 15 (2012): 152–63.

39. Edna Gonzáles-Bernal et al., "Toads in the Backyard: Why Do Invasive Cane Toads (*Rhinella marina*) Prefer Buildings to Bushland?" *Population Ecology* 58 (2016): 293–302.

40. Shine, *Cane Toad Wars*, 60.

41. Livia Albeck-Ripka, "Are We an Invasive Species?" *New York Times*, December 6, 2017, https://www.nytimes.com/2017/12/06/climate/climate-fwd-humans-invasive-species.html.

42. Geoffrey Bolton, *Spoils and Spoilers: Australians Make Their Environment 1788–1980* (Sydney: George Allen and Unwin, 1981), 138; and Tyrrell, *True Gardens*, 211.

43. Hannah M. Burgess and Ian Mohr, "Evolutionary Clash between Myxoma Virus and Rabbit PKR in Australia," *Proceedings of the National Academy of Sciences of the United States of America* 113, no. 15 (April 12, 2016): 3912–14.

44. Shine, *Cane Toad Wars*, 162n.

45. Cameron M. Hudson et al., "Constructing an Invasion Machine: The Rapid Evolution of a Dispersal-Enhancing Phenotype During the Cane Toad Invasion of Australia," *PLoS ONE* 11, no. 9 (2016): 1–12.

46. Matthew J. Greenlees, Benjamin L. Phillips, and Richard Shine, "Adjusting to a Toxic Invader: Native Australian Frogs Learn Not to Prey on Cane Toads," *Behavioral Ecology* 21 (2010): 966–71.

47. Ella Kelly and Ben L. Phillips, "Get Smart: Native Mammal Develops Toad-Smart Behavior in Response to a Toxic Invader," *Behavioral Ecology* 28 (2017): 854–58.

48. Ben L. Phillips and Richard Shine, "An Invasive Species Induces Rapid Adaptive Change in a Native Predator: Cane Toads and Black Snakes in Australia," *Proceedings of the Royal Society* B 273 (2006): 1545–50; Ben L. Phillips and Richard Shine, "Adapting to an Invasive Species: Toxic Cane Toads Induce Morphological Change in Australian Snakes," *Proceedings of the National Academy of Science of the United States of America* 101 (2004): 17150–55.

49. Damian C. Lettoof et al., "Do Invasive Cane Toads Affect the Parasite Burdens of Native Australian Frogs?" *International Journal for Parasitology: Parasites and Wildlife* 2 (2013): 155–64; here, 159.

50. Shine, *Cane Toad Wars*, 4.

51. Libby Robin, "Domestication in a Post-Industrial World," in Ursula K. Heise, Jon Christensen, and Michelle Niemann, eds., *The Routledge Companion to the Environmental Humanities*, 46–55 (London: Routledge, 2017), 51n.

52. Shine, *Cane Toad Wars*, 138.

53. See p. 405.

54. Mick N. Clout and Peter A. Williams, "Introduction," in Clout and Williams, eds., *Invasive Species Management: A Handbook of Principles and Techniques*, v–x (New York: Oxford University Press, 2009), vii.

55. Shine, *Cane Toad Wars*, 14.

56. Christopher J. Jolly, Richard Shine, and Matthew J. Greenlees, "The Impact of Invasive Cane Toads on Native Wildlife in Southern Australia," *Ecology and Evolution* 5 (2015): 3879-94; here, 3890. A similar point is made regarding Australia's general attitude toward invasive species in Low, *Feral Future*, viii.

15: SAUDI ARABIA

1. Daniel Yergin, *The Prize: The Epic Quest for Oil, Money and Power* (New York: Simon and Schuster, 1992), 403n.

2. Bernard Haykel, Thomas Hegghammer, and Stéphane Lacroix, "Introduction," in Haykel, Hegghammer, and Lacroix, eds., *Saudi Arabia in Transition: Insights on Social, Political, Economic and Religious Change*, 1-10 (New York: Cambridge University Press, 2015), 1.

3. Irvine H. Anderson, *Aramco, the United States, and Saudi Arabia: A Study of the Dynamics of Foreign Oil Policy, 1933-1950* (Princeton, NJ: Princeton University Press, 1981), 32.

4. For context, see Megan Black, *The Global Interior: Mineral Frontiers and American Power* (Cambridge, MA: Harvard University Press, 2018), 151.

5. Yergin, *Prize*, 300, 393, 395 (quotation).

6. Eduardo Galeano, *Open Veins of Latin America: Five Centuries of the Pillage of a Continent* (New York: Monthly Review Press, 1997 [1971]), 32.

7. Ricardo Soares de Oliveira, *Oil and Politics in the Gulf of Guinea* (London: Hurst, 2007), 338.

8. Charles David Kepner Jr. and Jay Henry Soothill, *The Banana Empire: A Case Study of Economic Imperialism* (New York: Vanguard Press, 1935), 38-40.

9. Yergin, *Prize*, 404.

10. See the classic Ida M. Tarbell, *The History of the Standard Oil Company*, 2 vols. (London: Heineman, 1905).

11. Timothy Mitchell, *Carbon Democracy: Political Power in the Age of Oil* (London: Verso, 2013), 166.

12. See Yergin, *Prize*.

13. Adam Smith, *An Inquiry into the Nature and Causes of the Wealth of Nations*, ed. R. H. Campbell and A. S. Skinner, vol. 1 (Oxford: Clarendon Press, 1976), 145.

14. Yergin, *Prize*, 187, 202.

15. Ronald W. Ferrier, *The History of the British Petroleum Company*, vol. 1: *The Developing Years 1901-1932* (Cambridge: Cambridge University Press, 1982), 176, 197.

16. Irena Cristalis, *East Timor: A Nation's Bitter Dawn*, 2nd ed. (London: Zed Books, 2009), 287.

17. Leslie McLoughlin, *Ibn Saud: Founder of a Kingdom* (Basingstoke: Macmillan, 1993), 22; and Stephen Kotkin, *Stalin*, vol. 1: *Paradoxes of Power, 1878-1928* (London: Allen Lane, 2014), 51n.

18. Steffen Hertog, *Princes, Brokers, and Bureaucrats: Oil and the State in Saudi Arabia* (Ithaca, NY: Cornell University Press, 2010), 42n.

19. On Dubai's career as a commercial hub, see Christopher M. Davidson, *Dubai: The Vulnerability of Success* (New York: Columbia University Press, 2008).

20. See Chad H. Parker, *Making the Desert Modern: Americans, Arabs, and Oil on the Saudi Frontier, 1933-1973* (Amherst: University of Massachusetts Press, 2015).

21. See Robert Vitalis, *America's Kingdom: Mythmaking on the Saudi Oil Frontier* (Stanford, CA: Stanford University Press, 2007).

22. See Soares de Oliveira, *Oil and Politics*.

23. Robert Fitzgerald, *The Rise of the Global Company: Multinationals and the Making of the Modern World* (Cambridge: Cambridge University Press, 2015), 242.

24. Parker, *Making the Desert Modern*, 11.

25. Hertog, *Princes*, 57.

26. Rachel Bronson, *Thicker than Oil: America's Uneasy Partnership with Saudi Arabia* (New York: Oxford University Press, 2006), 81–83, 112.

27. Simon Bromley, "The United States and the Control of World Oil," *Government and Opposition* 40 (2005): 225–55; here, 252.

28. "Crude Thinking: The Oil Price," *The Economist* 426, no. 9075 (January 20, 2018): 65–67; here, 65.

29. James P. Evans, *Environmental Governance* (London: Routledge, 2012), 148.

30. Giacomo Luciani, "From Price Taker to Price Maker? Saudi Arabia and the World Oil Market," in Haykel, Hegghammer, and Lacroix, *Saudi Arabia in Transition*, pp. 71–96; here, 72.

31. John V. Mitchell, "Petroleum Reserves in Question," Chatham House Sustainable Development Programme Briefing Paper 04/03, October 2004, p. 2. For a reflection on the multiple layers of complication in the making of petroknowledge, see Rüdiger Graf, *Öl und Souveränität: Petroknowledge und Energiepolitik in den USA und Westeuropa in den 1970er Jahren* (Berlin: De Gruyter, 2014).

32. Yergin, *Prize*, 762.

33. Haykel, Hegghammer, and Lacroix, "Introduction," 10.

34. Larry Rohter, "Chile Copper Windfall Forces Hard Choices on Spending," *New York Times*, January 7, 2007, accessed November 3, 2021, https://www.nytimes.com/2007/01/07/world/americas/07chile.html.

35. See Jeffrey Frankel, "How Can Commodity Exporters Make Fiscal and Monetary Policy Less Procyclical?" in Rabah Arezki, Thorvaldur Gylfason, and Amadou Sy, eds., *Beyond the Curse: Policies to Harness the Power of Natural Resources*, 167–92 (Washington: International Monetary Fund, 2011), 175–77.

36. Avinash K. Dixit, "The Cone of Uncertainty of the Twenty-First Century's Economic Hurricane," in Ignacio Palacios-Huerta, ed., *In 100 Years: Leading Economists Predict the Future*, 49–56 (Cambridge, MA: MIT Press, 2013), 52.

37. Orlando Figes, *Revolutionary Russia, 1891–1991* (London: Penguin, 2014), 375, 389.

38. See Sweder van Wijnbergen, "The 'Dutch Disease': A Disease After All?" *Economic Journal* 94 (March 1984): 41–55; and Warner Max Corden, "Booming Sector and Dutch Disease Economics: Survey and Consolidation," *Oxford Economic Papers* 36 (1984): 359–80.

39. Fernando Coronil, *The Magical State: Nature, Money, and Modernity in Venezuela* (Chicago: University of Chicago Press, 1997), 7.

40. Coronil, *Magical State*, 2.

41. See Fernando Coronil, "Magical Illusions or Revolutionary Magic? Chávez in Historical Context," *NACLA Report on the Americas* 33, no. 6 (2000): 34–42.

42. Hertog, *Princes*, 4 (quotation), 85.

43. Hertog, *Princes*, 51, 97 (quotation).

44. Hertog, *Princes*, 74, 88.

45. Anthony Ham, Martha Brekhus Shams, and Andrew Madden, *Saudi Arabia* (Footscray: Lonely Planet Publications, 2004), 52.

46. Ham, Shams, and Madden, *Saudi Arabia*, 18.

47. Ham, Shams, and Madden, *Saudi Arabia*, 3.

48. Rosalie Rayburn, *Living and Working in Saudi Arabia: Your Guide to a Successful Short or Long-Term Stay* (Oxford: How To Books, 2001), 46.

49. Karen Elliott House, *On Saudi Arabia: Its People, Past, Religion, Fault Lines—and Future* (New York: Vintage Books, 2013), 3.

50. Amitav Ghosh, "Petrofiction: The Oil Encounter and the Novel," in Imre Szemann

and Dominic Boyer, eds., *Energy Humanities: An Anthology*, 431–40 (Baltimore: Johns Hopkins University Press, 2017), 431.

51. Peter A. Shulman, "The Making of a Tax Break: The Oil Depletion Allowance, Scientific Taxation, and Natural Resources Policy in the Early Twentieth Century," *Journal of Policy History* 23 (2011): 281–322; here, 282.

52. Yergin, *Prize*, 494, 557.

53. House, *On Saudi Arabia*, 125.

54. Donald L. Losman, "The Rentier State and National Oil Companies: An Economic and Political Perspective," *Middle East Journal* 64 (2010): 427–45; here, 441.

55. Saudi Aramco, *Annual Review 2016*, accessed August 5, 2022, https://www.aramco.com/-/media/publications/corporate-reports/2016-annualreview-full-en.pdf.

56. OCP S.A., *2016 Annual Report*, accessed May 12, 2022, https://ocpsiteprodsa.blob.core.windows.net/media/2021-06/RA_OCP_2016_VUK.pdf, p. 19.

57. Mitchell, *Carbon Democracy*, p. 29.

58. See p. 518.

59. Losman, Rentier State, p. 428.

60. Michael L. Ross, "What Have We Learned about the Resource Curse?" *Annual Review of Political Science* 18 (2015): 239–59; here, 252.

61. Steve A. Yetiv, *Crude Awakenings: Global Oil Security and American Foreign Policy* (Ithaca, NY: Cornell University Press, 2004), 1.

62. Quoted in Terry Lynn Karl, *The Paradox of Plenty: Oil Booms and Petro-States* (Berkeley: University of California Press, 1997), 4, 187.

63. Elisabeth Bumiller and Adam Nagourney, "Bush: 'America Is Addicted to Oil,'" *New York Times*, February 1, 2006, accessed November 3, 2021, https://www.nytimes.com/2006/02/01/world/americas/01iht-state.html.

64. "A Second Crack of the Whip: Chile's Presidential Primaries," *The Economist*, July 1, 2013, accessed November 3, 2021, https://www.economist.com/blogs/americasview/2013/07/chile-s-presidential-primaries.

65. Bill McKibben, "Global Warming's Terrifying New Math," *Rolling Stone*, July 19, 2012, accessed November 3, 2021, https://www.rollingstone.com/politics/politics-news/global-warmings-terrifying-new-math-188550/.

66. Yergin, *Prize*, 762.

67. Kejal Vjas, "Guyana Dreams of Oil Riches," *Wall Street Journal*, June 22, 2018, p. A8.

PART IV
TECHNOLOGY TAKES COMMAND

1. Johan Hendrik Jacob van der Pot, *Encyclopedia of Technological Progress: Systematic Overview of Theories and Opinions*, 2 vols. (Delft: Eburon, 2004).

2. See Michael Adas, *Machines as the Measures of Men: Science, Technology, and Ideologies of Western Dominance* (Ithaca, NY: Cornell University Press, 1989).

3. See Edward Tenner, *Why Things Bite Back: Technology and the Revenge of Unintended Consequences* (New York: Knopf, 1996).

4. Melvin Kranzberg, "Technology and History: 'Kranzberg's Laws,'" *Technology and Culture* 27 (1986): 544–60; here, 549.

5. Kranzberg, "Technology and History," 548.

6. On the evolutionary advantages of sweating, see John R. McNeill, "The First Hundred Thousand Years," in Frank Uekoetter, ed., *The Turning Points of Environmental History*, 13–28 (Pittsburgh, PA: University of Pittsburgh Press, 2010), 14.

7. Adebayo Adedeji, "Comparative Strategies of Economic Decolonization in Africa,"

in Ali A. Mazrui, ed., *General History of Africa*, vol. 8: *Africa since 1935*, 393–431 (Paris: UNESCO, 1999), 424.

16: LONDON SMOG

1. For a cultural history of London smog, see Christine L. Corton, *London Fog: The Biography* (Cambridge, MA: Belknap Press of Harvard University Press, 2015). Corton discusses Monet's stay at the Savoy Hotel (182) and the terminology of smog (16–27).

2. William M. Cavert, *The Smoke of London: Energy and Environment in the Early Modern City* (Cambridge: Cambridge University Press, 2016), 21–23.

3. See Jürgen Osterhammel, *The Transformation of the World: A Global History of the Nineteenth Century* (Princeton, NJ: Princeton University Press, 2014), 651–58.

4. Gregory T. Cushman, *Guano and the Opening of the Pacific World: A Global Ecological History* (New York: Cambridge University Press, 2013), 121.

5. See Edmund Burke III, "The Big Story: Human History, Energy Regimes, and the Environment," in Burke and Kenneth Pomeranz, eds., *The Environment and World History*, 33–53 (Berkeley: University of California Press, 2009); Robert B. Marks, *The Origins of the Modern World: A Global and Ecological Narrative from the Fifteenth to the Twenty-First Century*, 3rd ed. (Lanham, MD: Rowman and Littlefield, 2015), 97; Rolf Peter Sieferle, *The Subterranean Forest: Energy Systems and the Industrial Revolution* (Cambridge: White Horse Press, 2001); and Rolf Peter Sieferle et al., *Das Ende der Fläche: Zum gesellschaftlichen Stoffwechsel der Industrialisierung* (Cologne: Böhlau, 2006).

6. See Joachim Radkau, *Technik in Deutschland: Vom 18. Jahrhundert bis heute* (Frankfurt: Campus, 2008), 29–38, among many others.

7. Cavert, *Smoke of London*, 26.

8. Vaclav Smil, *Energy Transitions: History, Requirements, Prospects* (Santa Barbara, CA: Praeger, 2010), 27.

9. Peter Brimblecombe, *The Big Smoke: A History of Air Pollution in London since Medieval Times* (London: Routledge, 1987), 7.

10. Adam Smith, *An Inquiry into the Nature and Causes of the Wealth of Nations*, ed. R. H. Campbell and A. S. Skinner, vol. 2 (Oxford: Clarendon Press, 1976), 371.

11. E. A. Wrigley, *Energy and the English Industrial Revolution* (Cambridge: Cambridge University Press, 2010), 10.

12. Paul Warde, "Commentary: W. S. Jevons, *The Coal Question* (1865)," in Libby Robin, Sverker Sörlin, and Paul Warde, eds., *The Future of Nature: Documents of Global Change*, 85–88 (New Haven, CT: Yale University Press, 2013), 85.

13. Quoted in Shellen Xiao Wu, *Empires of Coal: Fueling China's Entry into the Modern World Order, 1860–1920* (Stanford, CA: Stanford University Press, 2015), 28.

14. I have stressed the importance of real estate interests for the American anti-smoke movement in Frank Uekoetter, *The Age of Smoke: Environmental Policy in Germany and the United States, 1880–1970* (Pittsburgh, PA: University of Pittsburgh Press, 2009), 23.

15. Bill Luckin, "'The Heart and Home of Horror': The Great London Fogs of the Late Nineteenth Century," *Social History* 28 (2003): 31–48; here, 38.

16. Stephen S. Lim et al., "A Comparative Risk Assessment of Burden of Disease and Injury Attributable to 67 Risk Factors and Risk Factor Clusters in 21 Regions, 1990–2010: A Systematic Analysis for the Global Burden of Disease Study 2010," *Lancet* 380 (2012/2013): 2224–60; here, 2238.

17. Adam W. Rome, "Coming to Terms with Pollution: The Language of Environmental Reform, 1865–1915," *Environmental History* 1, no. 3 (1996): 6–28; here, 16.

18. Alain Corbin, *The Foul and the Fragrant: Odor and the French Social Imagination* (Cam-

bridge, MA: Harvard University Press, 1986), 66; and Peter Thorsheim, *Inventing Pollution: Coal, Smoke, and Culture in Britain since 1800* (Athens: Ohio University Press, 2006), p. 17.

19. Brimblecombe, *Big Smoke*, 8n.

20. William M. Cavert, "The Environmental Policy of Charles I: Coal Smoke and the English Monarchy, 1624–40," *Journal of British Studies* 53 (2014): 310–33; here, 311.

21. John Evelyn, *Fumifugium: or The Inconveniencie of the Aer and Smoak of London Dissipated. Together With some Remedies Humbly Proposed* (London: W. Godbid, 1661).

22. Thorsheim, *Inventing Pollution*, 30.

23. Christine Meisner Rosen, "Businessmen against Pollution in Late Nineteenth Century Chicago," *Business History Review* 69 (1995): 351–97.

24. For a full account of this story, see Frank Uekötter, "Die Kommunikation zwischen technischen und juristischen Experten als Schlüsselproblem der Umweltgeschichte: Die preußische Regierung und die Berliner Rauchplage," *Technikgeschichte* 66 (1999): 1–31; here, 3–8.

25. Geheimes Staatsarchiv Berlin I. HA Rep. 120 BB II a 2 Nr. 28 Bd. 4, p. 210r.

26. See Uekötter, "Kommunikation," 5–8.

27. Charles H. Benjamin, "Smoke and Its Abatement," *Transactions of the American Society of Mechanical Engineers* 26 (1905): 713–27; here, 716.

28. Joel A. Tarr and Carl Zimring, "The Struggle for Smoke Control in St. Louis: Achievement and Emulation," in Andrew Hurley, ed., *Common Fields: An Environmental History of St. Louis*, 199–220 (Saint Louis: Missouri Historical Society Press, 1997), 202.

29. Staatsarchiv Munich RA 58425, report of January 18, 1894; Roy Lubove, *Twentieth-Century Pittsburgh: Government, Business, and Environmental Change* (Pittsburgh, PA: University of Pittsburgh Press, 1995), 115; and E. Willard Miller, "Pittsburgh: An Urban Region in Transition," in Miller, ed., *A Geography of Pennsylvania*, 374–95 (University Park: Pennsylvania State University Press, 1995), 379.

30. Bernard Kettlewell, *The Evolution of Melanism: The Study of a Recurring Necessity* (Oxford: Clarendon Press, 1973), 60. For a discussion, see David W. Rudge, "The Beauty of Kettlewell's Classic Experimental Demonstration of Natural Selection," *BioScience* 55 (2005): 369–75. For an ongoing discussion of the precise mechanism, see L. M. Cook and I. J. Saccheri, "The Peppered Moth and Industrial Melanism: Evolution of a Natural Selection Case Study," *Heredity* 110 (2013): 207–12.

31. Stephen Mosley, "'A Network of Trust': Measuring and Monitoring Air Pollution in British Cities, 1912–1960," *Environment and History* 15 (2009): 273–302; here, 290.

32. The National Archives of the United Kingdom MH 58/398, Macleod to Fyfe, November 18, 1953, p. 2.

33. The National Archives of the United Kingdom CAB 129/64/22, "Smog." Memorandum by the Ministry of Housing and Local Government, November 18, 1953, p. 1. Macmillan's remarks have been cited as evidence of "breathtaking cynicism" (Devra Davis, *When Smoke Ran Like Water: Tales of Environmental Deception and the Battle against Pollution* [New York: Basic Books, 2002], 45; see also Thorsheim, *Inventing Pollution*, 167), but that ignores Macleod's even more lackluster stance as well as the wider context of the state's long search for its proper role. Macmillan made his remark at a time when the government was engaged in hectic discussions over whether it should purchase protective masks for sick people in anticipation of another smog episode—no wonder that Macmillan was keenly aware of the limits of government's options!

34. Erich Ashby and Mary Anderson, *The Politics of Clean Air* (Oxford: Clarendon Press, 1981), 104–19.

35. Anthony Seldon, *10 Downing Street: The Illustrated History* (London: HarperCollins,

1999), 16, 34; and Sir Anthony Seldon, *History: 10 Downing Street*, accessed November 3, 2021, https://www.gov.uk/government/history/10-downing-street.

36. Michelle L. Bell, Devra L. Davis, and Tony Fletcher, "A Retrospective Assessment of Mortality from the London Smog Episode of 1952: The Role of Influenza and Pollution," *Environmental Health Perspectives* 112 (2004): 6–8. Higher estimates of the death toll were already circulating in the aftermath of the disaster, and the government felt rather helpless in dealing with them. When the Ministry of Health discussed its response to an interim report of the Beaver Committee in November 1954, a memorandum indicated concern about inflated numbers but advised against making that an issue: "Perhaps this would not be a suitable opportunity to correct the mistaken impression spread by Professor Wilkins that the death toll was 12,000 and not 4,000?" (The National Archives of the United Kingdom MH 58/398, memorandum for the Secretary, November 25, 1953.)

37. A similar argument can be made about the American campaigns against domestic smoke a decade earlier. See Uekoetter, *Age of Smoke*, 83.

38. Alex Kemp, *The Official History of North Sea Oil and Gas*, vol. 1: *The Growing Dominance of the State* (London: Routledge, 2012), 138.

39. Christof Mauch, *Slow Hope: Rethinking Ecologies of Crisis and Fear* (Munich: Rachel Carson Center for Environment and Society, 2019), 23.

40. Committee on Air Pollution. *Report presented to Parliament by the Minister of Housing and Local Government, the Secretary of State for Scotland and the Minister of Fuel and Power by Command of Her Majesty* (London: Her Majesty's Stationery Office, November 1954), 34.

41. Chris Rose, *The Dirty Man of Europe: The Great British Pollution Scandal* (London: Simon and Schuster, 1990), 142.

42. Lim et al., "Comparative Risk Assessment," 2224, 2249.

43. Corton, *London Fog*, 329n.

44. See, for instance, Dongyong Zhang, Junjuan Liu, and Bingjun Li, "Tackling Air Pollution in China: What Do We Learn from the Great Smog of 1950s in London," *Sustainability* 6 (2014): 5322–38.

45. Adam Vaughan, "China Tops WHO List for Deadly Outdoor Air Pollution," *The Guardian*, September 27, 2016, accessed November 3, 2021, https://www.theguardian.com/environment/2016/sep/27/more-than-million-died-due-air-pollution-china-one-year.

46. Luke Howard, *The Climate of London Deduced from Meteorological Observations, Made at Different Places in the Neighbourhood of the Metropolis*, 2 vols. (London: W. Phillips, 1818 and 1820). For an assessment, see Gerald Mills, "Luke Howard and *The Climate of London*," *Weather* 63, no. 6 (2008): 153–57.

17: THE WATER CLOSET

1. Fernand Braudel, *Capitalism and Material Life, 1400–1800* (London: Weidenfeld and Nicolson, 1973), 210.

2. Kai Strittmatter, *Gebrauchsanweisung für China* (Munich: Piper, 2004), 57.

3. Dengchuan Cai and Manlai You, "An Ergonomic Approach to Public Squatting-Type Toilet Design," *Applied Ergonomics* 29 (1998): 147–53; here, 148.

4. United Nations General Assembly Resolution 67/291, adopted on July 24, 2013.

5. Kathleen Meyer, *How to Shit in the Woods: An Environmentally Sound Approach to a Lost Art* (Berkeley, CA: Ten Speed Press, 1989).

6. Lawrence Wright, *Clean and Decent: The Fascinating History of the Bathroom and the Water-Closet* (London: Penguin, 2000 [1960]), 7. Similarly, George Rosen, *A History of Public Health*, expanded ed. (Baltimore: Johns Hopkins University Press, 1993 [1959]), 1–3.

7. Rita P. Wright, *The Ancient Indus: Urbanism, Economy, and Society* (New York: Cambridge University Press, 2010), 122, 237n.

8. Richard Neudecker, *Die Pracht der Latrine: Zum Wandel öffentlicher Bedürfnisanstalten in der kaiserzeitlichen Stadt* (Munich: Pfeil, 1994), 8n.

9. Wright, *Clean and Decent*, 106.

10. Wright, *Clean and Decent*, 201. See also Maureen Ogle, *All the Modern Conveniences: American Household Plumbing, 1840–1890* (Baltimore: Johns Hopkins University Press, 1996).

11. Mary Douglas, *Purity and Danger: An Analysis of the Concepts of Pollution and Taboo* (London: Routledge, 1994 [1966]).

12. Norbert Elias, *The Civilizing Process: The History of Manners and State Formation and Civilization*, trans. Edmund Jephcott (Oxford: Blackwell, 1994), 111. For introductions to Elias's sociology, see Stephen Mennell, *Norbert Elias: Civilization and the Human Self-Image* (Oxford: Blackwell, 1989); Steven Loyal and Stephen Quilley, eds., *The Sociology of Norbert Elias* (Cambridge: Cambridge University Press, 2004); and Hermann Korte, "Norbert Elias and the Theory of Civilisation," in Bernhard Schäfers, ed., *Sociology in Germany: Development–Institutionalization–Theoretical Disputes* (Opladen: Leske + Budrich, 1994), 164–76.

13. Hans Peter Duerr, *Der Mythos vom Zivilisationsprozeß*, 5 vols. (Frankfurt: Suhrkamp, 1988–2002).

14. Joel A. Tarr, "The Separate vs. Combined Sewer Problem: A Case Study in Urban Technology Design Choice," *Journal of Urban History* 5 (1979): 308–39; here, 336.

15. Anonymous [August Julius Langbehn], *Rembrandt als Erzieher: Von einem Deutschen* (Leipzig: Hirschfeld, 1925 [1890], 312.

16. C. Vann Woodward, *The Strange Career of Jim Crow* (New York: Oxford University Press, 1974), 98, 116; and Daniel J. Walther, "Race, Space and Toilets: 'Civilization' and 'Dirt' in the German Colonial Order," *German History* 35 (2017): 551–67; here, 563.

17. Milena Angelova, "'The Model Village': The Modernization Project of the Villages in Bulgaria (1937–1944)," *Martor. The Museum of the Romanian Peasant Anthropology Review* 19 (2014): 89–96; here, 92.

18. David R. Boyd, "No Taps, No Toilets: First Nations and the Constitutional Right to Water in Canada," *McGill Law Journal* 57 (2011): 81–134.

19. "Build Toilets First and Temples Later, Narenda Modi Says," *Times of India*, October 2, 2013, accessed November 3, 2021, http://timesofindia.indiatimes.com/india/Build-toilets-first-and-temples-later-Narendra-Modi-says/articleshow/23422631.cms.

20. See Ayona Datta, "Another Rape? The Persistence of Public/Private Divides in Sexual Violence Debates in India," *Dialogues in Human Geography* 6 (2016): 173–77.

21. Jamie Benidickson, *The Culture of Flushing: A Social and Legal History of Sewage* (Vancouver: UBC Press, 2007), 93.

22. Heinrich Bauer, "Cloaca, Cloaca Maxima," in Eva Margareta Steinby, ed., *Lexicon Topographicum Urbis Romae*, vol. 1, A–C, 288–90 (Rome: Edizioni Quasar, 1993).

23. Ted Steinberg, *Gotham Unbound: The Ecological History of Greater New York* (New York: Simon and Schuster, 2014), 219n.

24. For the most comprehensive discussion of urban sanitation, see Martin V. Melosi, *The Sanitary City: Urban Infrastructure in America from Colonial Times to the Present* (Baltimore: Johns Hopkins University Press, 2000).

25. Pliny, *Natural History*, vol. 1: Books 1–2, trans. Harris Rackham (Cambridge, MA: Harvard University Press, 1938), 83.

26. See Stephen Halliday, *The Great Stink of London: Sir Joseph Bazalgette and the Cleansing of the Victorian Capital* (Stroud: Sutton, 1999), 71–76.

27. Karl Baedeker, *London and Its Environs: Handbook for Travellers*, 11th ed. (Leipsic [Leipzig]: Karl Baedeker, 1898), 96.

28. Quoted in Kalala J. Ngalamulume, "Coping with Disease in the French Empire: The

Provision of Waterworks in Saint-Louis-du-Senegal, 1860–1914," in Petri S. Juuti, Tapio S. Katko, and Heikki S. Vuorinen, eds., *Environmental History of Water: Global Views on Community Water Supply and Sanitation*, 147–63 (London: IWA, 2007), 152.

29. John Sheail, "Town Wastes, Agricultural Sustainability and Victorian Sewage," *Urban History* 23 (1996): 189–210; here, 194.

30. Ngalamulume, "Coping with Disease," 152.

31. Mike Davis, *City of Quartz: Excavating the Future in Los Angeles* (New York: Vintage Books, 1992), 196.

32. William Deverell and Tim Sitton, *Water and Los Angeles: A Tale of Three Rivers, 1900–1941* (Oakland: University of California Press, 2017), 12–14.

33. Harriett Ritvo, *The Dawn of Green: Manchester, Thirlmere, and Modern Environmentalism* (Chicago: University of Chicago Press, 2009); Peter Münch, *Stadthygiene im 19. und 20. Jahrhundert: Die Wasserversorgung, Abwasser- und Abfallbeseitigung unter besonderer Berücksichtigung Münchens* (Göttingen: Vandenhoeck & Ruprecht, 1993), 186–93; and David Soll, *Empire of Water: An Environmental and Political History of the New York City Water Supply* (Ithaca, NY: Cornell University Press, 2013).

34. F. J. J. Henry, comp., *The Water Supply and Sewerage of Sydney. Being an Account of the Development and History of the Water Supply and Sewerage Systems of Sydney and the South Coast from their Inception to the End of the First 50 Years of Control by a Board Specially Constituted for the Purpose* (Sydney: Halstead Press, 1939), 63.

35. See p. 394.

36. Soll, *Empire of Water*, 161–64.

37. Daniel Schneider, *Hybrid Nature: Sewage Treatment and the Contradictions of the Industrial Ecosystem* (Cambridge, MA: MIT Press, 2011), xxi.

38. Sherry H. Olson, *Baltimore: The Building of an American City* (Baltimore: Johns Hopkins University Press, 1980), 136n, 249.

39. See Christopher Hamlin, "Edwin Chadwick and the Engineers, 1842–1854: Systems and Antisystems in the Pipe-and-Brick Sewers War," *Technology and Culture* 33 (1992): 680–709.

40. Kenneth T. Jackson, *Crabgrass Frontier: The Suburbanization of the United States* (New York: Oxford University Press, 1987), 132.

41. See Melosi, *Sanitary City*, 287n.

42. See, for instance, Christoph Bernhardt, *Im Spiegel des Wassers: Eine transnationale Umweltgeschichte des Oberrheins (1800–2000)* (Cologne: Böhlau, 2016), 356n.

43. See Tarr, "Separate vs. Combined Sewer Problem."

44. Annabel Cooper et al., "Rooms of Their Own: Public Toilets and Gendered Citizens in a New Zealand City, 1860–1940," *Gender, Place and Culture* 7 (2000): 417–33; and Clara H. Greed, "Public Toilet Provision for Women in Britain: An Investigation of Discrimination against Urination," *Women's Studies International Forum* 18 (1995): 573–84.

45. Cai and You, "Ergonomic Approach"; and Allen Chun, "Flushing in the Future: The Supermodern Japanese Toilet in a Changing Domestic Culture," *Postcolonial Studies* 5 (2002): 153–70.

46. Arata Ichikawa, "Japan's Sewerage System," *International Journal of Water Resources Development* 4 (1988): 35–39; here, 35.

47. Houston Faust Mount II, *Oilfield Revolutionary: The Career of Everette Lee DeGoyler* (College Station: Texas A&M University Press, 2014), 40n.

48. M. A. Mammadova and S. I. Pashayeva, "Demands on, Condition, and Environmental Problems of the Baku Municipal Water Supply," in John H. Tellam, Michael O. Rivett, and Rauf G. Israfilov, eds., *Urban Groundwater Management and Sustainability*, 29–38

(Dordrecht: Springer, 2006), 32; and Farideh Heyat, *Azeri Women in Transition: Women in Soviet and Post-Soviet Azerbaijan* (Baku: Chashioglu, 2005), 65, 87.

49. Jürgen Büschenfeld, *Flüsse und Kloaken: Umweltfragen im Zeitalter der Industrialisierung (1870-1918)* (Stuttgart: Klett-Cotta, 1997), 266.

50. Clemens Zimmermann, *Die Zeit der Metropolen: Urbanisierung und Großstadtentwicklung* (Frankfurt: Fischer, 1996), 101.

51. Noyan Dinçkal, *Istanbul und das Wasser: Zur Geschichte der Wasserversorgung und Abwasserentsorgung von der Mitte des 19. Jahrhunderts bis 1966* (Munich: Oldenbourg, 2004), 87–89, 112, 125.

52. Dinçkal, *Istanbul und das Wasser*, 150, 195.

53. John Broich, *London: Water and the Making of the Modern City* (Pittsburgh, PA: University of Pittsburgh Press, 2013), 143.

54. Shahrooz Mohajeri, *100 Jahre Berliner Wasserversorgung und Abwasserentsorgung 1840-1940* (Stuttgart: Steiner, 2005), 142n.

55. Patrick Joyce, "What Is the Social in Social History," *Past and Present* 206 (2010): 213–48; here, 230.

56. Paul H. Garner, *British Lions and Mexican Eagles: Business, Politics, and Empire in the Career of Weetman Pearson in Mexico, 1889–1919* (Stanford, CA: Stanford University Press, 2011), 81.

57. Gerhard Meißl, "Gebirgswasser in Wien: Die Wasserversorgung der Großstadt im 19. und 20. Jahrhundert," in Karl Brunner and Petra Schneider, eds., *Umwelt Stadt: Geschichte des Natur- und Lebensraumes Wien*, 195–203 (Vienna: Böhlau, 2005), 198.

58. David P. Jordan, *Transforming Paris: The Life and Labors of Baron Haussmann* (New York: Free Press, 1995), 276.

59. Committee on Air Pollution. *Report presented to Parliament by the Minister of Housing and Local Government, the Secretary of State for Scotland and the Minister of Fuel and Power by Command of Her Majesty* (London: Her Majesty's Stationery Office, November 1954), 6.

60. Donald L. Miller, *City of the Century: The Epic of Chicago and the Making of America* (New York: Simon and Schuster, 1997), 427–29.

61. Harold L. Platt, *Shock Cities: The Environmental Transformation and Reform of Manchester and Chicago* (Chicago: University of Chicago Press, 2005), 421.

62. Baron von Liebig, *Letters on Modern Agriculture*, ed. John Blyth (London: Walton and Maberly, 1859), 221.

63. On the nightsoil trade in East Asia, see p. 130.

64. Baron von Liebig, *Letters*, 249.

65. Victor Hugo, *Les Misérables*, trans. and with an introduction by Norman Denny (London: Penguin, 1982), 1061.

66. Nicholas Goddard, "'A Mine of Wealth'? The Victorians and the Agricultural Value of Sewage," *Journal of Historical Geography* 22 (1996): 274–90; here, 275.

67. See Frank Uekötter, "City Meets Country: Recycling Ideas and Realities on German Sewage Farms," *Journal for the History of Environment and Society* 1 (2016): 89–107.

68. Melosi, *Sanitary City*, 167.

69. Alon Tal, "Rethinking the Sustainability of Israel's Irrigation Practices in the Drylands," *Water Research* 90 (2016): 387–94; here, 388n.

70. Henry, *Water Supply and Sewerage*, 172–74. For other historic events at Botany Bay, see p. 406.

71. See Joel A. Tarr, "The Search for the Ultimate Sink: Urban Air, Land, and Water Pollution in Historical Perspective," in Tarr, ed., *The Search for the Ultimate Sink: Urban Pollution in Historical Perspective*, 7–35 (Akron: University of Akron Press, 1996).

72. Olson, *Baltimore*, 249n.

73. J. H. J. Ensink et al., "Waste Stabilization Pond Performance in Pakistan and Its Implications for Wastewater Use in Agriculture," *Urban Water Journal* 4 (2007): 261–67.

74. Dana Cordell, Jan-Olof Drangert, and Stuart White, "The Story of Phosphorus: Global Food Security and Food for Thought," *Global Environmental Change* 19 (2009): 292–305; here, 296.

75. Donald Reid, *Paris Sewers and Sewermen: Realities and Representations* (Cambridge, MA: Harvard University Press, 1993), 154n.

76. Sarah K. Dickin et al., "A Review of Health Risks and Pathways for Exposure to Wastewater Use in Agriculture," *Environmental Health Perspectives* 124 (2016): 900–909.

77. See William Peysson and Emmanuelle Vulliet, "Determination of 136 Pharmaceuticals and Hormones in Sewage Sludge Using Quick, Easy, Cheap, Effective, Rugged and Safe Extraction Followed by Analysis with Liquid Chromatography–Time-of-Flight-Mass Spectrometry," *Journal of Chromatography A* 1290 (2013): 46–61.

78. Titus Livius, *The History of Rome*, books 1–8, trans. Daniel Spillan (Gloucester: Dodo Press, 2009), 67.

79. Susan B. Hanley, *Everyday Things in Premodern Japan: The Hidden Legacy of Material Culture* (Berkeley: University of California Press, 1997), 104n.

18: CHICAGO'S SLAUGHTERHOUSES

1. Upton Sinclair, *The Jungle* (London: T. Werner Laurie, 1946), 40, 41.

2. Louise Carroll Wade, *Chicago's Pride: The Stockyards, Packingtown, and Environs in the Nineteenth Century* (Urbana: University of Illinois Press, 1987), xiv, 370.

3. David E. Nye, *America's Assembly Line* (Cambridge, MA: MIT Press, 2013), 14.

4. Andrew J. Diamond, *Chicago on the Make: Power and Inequality in a Modern City* (Oakland: University of California Press, 2017), 27.

5. James R. Barrett, *Work and Community in the Jungle: Chicago's Packinghouse Workers, 1894–1922* (Urbana: University of Illinois Press, 1987), 57.

6. Gregory T. Cushman, *Guano and the Opening of the Pacific World: A Global Ecological History* (New York: Cambridge University Press, 2013), 100.

7. As befits a topic with so many connections, Chicago's slaughterhouses have been studied from multiple angles. Jimmy Skaggs wrote a brilliant overview of the American meat business over almost four centuries (Jimmy M. Skaggs, *Prime Cut: Livestock Raising and Meatpacking in the United States, 1607–1983* [College Station: Texas A&M University Press, 1986]), and William Cronon linked Chicago's meat business (as well as grain and lumber) with America's advancing frontier (William Cronon, *Nature's Metropolis: Chicago and the Great West* [New York: Norton, 1991]). Mary Yeager, a student of Alfred D. Chandler, looked at corporate development and the antitrust struggle (Mary Yeager, *Competition and Regulation: The Development of Oligopoly in the Meat Packing Industry* [Greenwich, CT: JAI Press, 1981]). Robert Slayton, Thomas Jablonsky, and Louise Carroll Wade discussed community development around the stockyards (Robert A. Slayton, *Back of the Yards: The Making of a Local Democracy* [Chicago: University of Chicago Press, 1986]; Thomas J. Jablonsky, *Pride in the Jungle: Community and Everday Life in Back of the Yards Chicago* [Baltimore: Johns Hopkins University Press, 1993]; and Wade, *Chicago's Pride*).

8. Margaret Walsh, *The Rise of the Midwestern Meat Packing Industry* (Lexington: University Press of Kentucky, 1982), 20n.

9. Jablonsky, *Pride in the Jungle*, 4.

10. Walsh, *Rise*, 92.

11. Skaggs, *Prime Cut*, 98n.

12. Cronon, *Nature's Metropolis*.

13. Jonathan Rees, *Refrigeration Nation: A History of Ice, Appliances, and Enterprise in America* (Baltimore: Johns Hopkins University Press, 2013), 90.

14. Roger Horowitz, Jeffrey M. Pilcher, and Sydney Watts, "Meat for the Multitudes: Market Culture in Paris, New York City, and Mexico City over the Long Nineteenth Century," *American Historical Review* 109 (2004): 1055–83; here, 1064, 1073, 1075n.

15. Jeffrey M. Pilcher, *The Sausage Rebellion: Public Health, Private Enterprise, and Meat in Mexico City, 1890-1917* (Albuquerque: University of New Mexico Press, 2006).

16. Anna Mazanik, "'Shiny Shoes' for the City: The Public Abattoir and the Reform of Meat Supply in Imperial Moscow," *Urban History* 45 (2018): 214–32.

17. The Progressive Era has received a great amount of scholarly scrutiny, including sharply diverging opinions on whether the name should stick. See Daniel T. Rodgers, "In Search of Progressivism," *Reviews in American History* 10, no. 4 (December 1982): 113–32, and Arthur S. Link and Richard L. McCormick, *Progressivism* (Arlington Heights, IL: Harlan Davidson, 1983), 21–25, as a response to Peter G. Filene, "An Obituary for 'The Progressive Movement,'" *American Quarterly* 22 (1970): 20–34.

18. Skaggs, *Prime Cut*, 129. See also Yeager, *Competition and Regulation*.

19. Upton Sinclair, *The Autobiography of Upton Sinclair* (London: W. H. Allen, 1963), 135.

20. Rick Halpern, *Down on the Killing Floor: Black and White Workers in Chicago's Packinghouses, 1904-54* (Urbana: University of Illinois Press, 1997); and Roger Horowitz, *"Negro and White, Unite and Fight!" A Social History of Industrial Unionism in Meatpacking, 1930-1990* (Urbana: University of Illinois Press, 1997).

21. Wade, *Chicago's Pride*, 90.

22. Sinclair, *Jungle*, 41.

23. Sam White, "From Globalized Pig Breeds to Capitalist Pigs: A Study in Animal Cultures and Evolutionary History," *Environmental History* 16 (2011): 94–120.

24. Keith Thomas, *Man and the Natural World: Changing Attitudes in England 1500-1800* (London: Penguin, 1984).

25. Karl Polanyi, *The Great Transformation* (Boston: Beacon Press, 1957).

26. P. W. Gerbens-Leenes, S. Nonhebel, and M. S. Krol, "Food Consumption Patterns and Economic Growth: Increasing Affluence and the Use of Natural Resources," *Appetite* 55 (2010): 597–608; here, 602.

27. John F. Love, *McDonald's: Behind the Arches* (New York: Bantam Press, 1987), 2, 12.

28. Garry L. Nall, "The Cattle-Feeding Industry on the Texas High Plains," in Henry C. Dethloff and Irvin M. May Jr., eds., *Southwestern Agriculture, Pre-Columbian to Modern*, 106–15 (College Station: Texas A&M University Press, 1982), 108.

29. John A. Jakle and Keith A. Sculle, *Fast Food: Roadside Restaurants in the Automobile Age* (Baltimore: Johns Hopkins University, 1999), 38; and David Gerard Hogan, *Selling 'em by the Sack: White Castle and the Creation of American Food* (New York: New York University Press, 1997), 24n.

30. James W. Whitaker, *Feedlot Empire: Beef Cattle Feeding in Illinois and Iowa, 1840-1900* (Ames: Iowa State University Press, 1975), 86; and Rudolf Alexander Clemen, *The American Livestock and Meat Industry* (New York: Ronald Press, 1923), 439.

31. Skaggs, *Prime Cut*, 177–81.

32. R. Douglas Hurt, *The Big Empty: The Great Plains in the Twentieth Century* (Tucson: University of Arizona Press, 2011), 190n, 204–6, 209.

33. Nall, "Cattle-Feeding Industry," 107.

34. Maxime Schwartz, *How the Cows Turned Mad: Unlocking the Mysteries of Mad Cow Disease* (Berkeley: University of California Press, 2004).

35. Skaggs, *Prime Cut*, 97, 190 (quotation).

36. Deborah Fink, *Cutting Into the Meatpacking Line: Workers and Change in the Rural Midwest* (Chapel Hill: University of North Carolina Press, 1998), 2.

37. Dominic A. Pacyga, *Slaughterhouse: Chicago's Union Stock Yard and the World It Made* (Chicago: University of Chicago Press, 2015), 196n.

38. Karl Ludwig Schweisfurth, *Wenn's um die Wurst geht: Gedanken über die Würde von Mensch und Tier* (Coburg: Riemann Verlag, 1999). For the Chicago experience, see p. 98.

39. Pilcher, *Sausage Rebellion*, 181–85.

40. Harvey Levenstein, *Paradox of Plenty: A Social History of Eating in Modern America* (Berkeley: University of California Press, 2003), 180.

41. For an idea of the range of options, see Marion Nestle, *Food Politics: How the Food Industry Influences Nutrition and Health* (Berkeley: University of California Press, 2003), and Morgan Spurlock and Jeremy Barlow, *Supersized: Strange Tales from a Fast-Food Culture* (Milwaukee, WI: Dark Horse Books, 2011).

42. Temple Grandin, "The Importance of Measurement to Improve the Welfare of Livestock, Poultry and Fish," in Grandin, ed., *Improving Animal Welfare: A Practical Approach*, 1–20 (Wallingford: CABI, 2010), 1.

43. Rachel Carson, "Foreword," in Ruth Harrison, *Animal Machines: The New Factory Farming Industry*, vii–viii (London: Vincent Stuart, 1964), viii.

44. Schweisfurth, *Wenn's um die Wurst geht*, 181–83, 187n, 223.

45. James C. Whorton, *Crusaders for Fitness: The History of American Health Reformers* (Princeton, NJ: Princeton University Press, 1982), 214.

19: SYNTHETIC NITROGEN

1. William Crookes, *The Wheat Problem: Based on Remarks Made in the Presidential Address to the British Association at Bristol in 1898* (London: John Murray, 1899), 2, 3.

2. Crookes, *Wheat Problem*, 37.

3. Crookes, *Wheat Problem*, 44, 46.

4. Crookes, *Wheat Problem*, 3. For the broader context of the wheat speech, see William H. Brock, *William Crookes (1832–1919) and the Commercialization of Science* (Aldershot: Ashgate, 2008), 367–88.

5. H. A. Bernthsen, "Synthetic Ammonia," in *Transactions and Organization: Eighth International Congress of Applied Chemistry, Washington and New York, September 4 to 13, 1912*, vol. 28, 182–201 (Concord, NH: The Rumford Press, 1912), 186.

6. Margit Szöllösi-Janze, *Fritz Haber 1868–1934: Eine Biographie* (Munich: Beck, 1998), 176, 181.

7. Vaclav Smil, *Enriching the Earth: Fritz Haber, Carl Bosch, and the Transformation of World Food Production* (Cambridge, MA: MIT Press, 2001), 159.

8. Kenneth Pomeranz, "Advanced Agriculture," in Jerry H. Bentley, ed., *The Oxford Handbook of World History*, 246–66 (Oxford: Oxford University Press, 2011), 254n.

9. Jeffrey Allan Johnson, "Die Macht der Synthese (1900–1925)," in Werner Abelshauser, ed., *Die BASF: Eine Unternehmensgeschichte*, 117–219 (Munich: Beck, 2002), 162–64.

10. Bernthsen, "Synthetic Ammonia," 201.

11. For a balanced assessment of Haber's biography, see Szöllösi-Janze, *Fritz Haber*.

12. Johnson, "Macht der Synthese," 180, 204.

13. Thomas Parke Hughes, "Technological Momentum in History: Hydrogenation in Germany 1898–1933," *Past and Present* 44 (1969): 106–32; here, 110.

14. Thomas P. Hughes, "Technological Momentum," in Merritt Roe Smith and Leo Marx, eds., *Does Technology Drive History? The Dilemma of Technological Determinism*, 101–13 (Cambridge, MA: MIT Press, 1994), 113.

15. Thomas P. Hughes, *Networks of Power: Electrification in Western Society, 1880–1930* (Baltimore: Johns Hopkins University Press, 1983).

16. Hughes, "Technological Momentum."

17. Szöllösi-Janze, *Fritz Haber*, 172.

18. J. L. Van Zanden, "The First Green Revolution: The Growth of Production and Productivity in European Agriculture, 1870–1914," *Economic History Review* 44 (1991): 215–39; here, 230–32.

19. Richard A. Wines, *Fertilizer in America: From Waste Recycling to Resource Exploitation* (Philadelphia: Temple University Press, 1985), 142.

20. Bundesarchiv R 3602 no. 606, Denkschrift des preußischen Landwirtschaftsministers zur Frage der Volksernährung, November 1, 1920, p. 1.

21. I have discussed the German soil fertility crisis extensively in Frank Uekötter, *Die Wahrheit ist auf dem Feld: Eine Wissensgeschichte der deutschen Landwirtschaft* (Göttingen: Vandenhoeck & Ruprecht, 2010), 183–275.

22. Helmut Zander, *Anthroposophie in Deutschland: Theosophische Weltanschauung und gesellschaftliche Praxis 1884–1945*, vol. 2 (Göttingen: Vandenhoeck & Ruprecht, 2007), 1579–607.

23. Sir Albert Howard, *An Agricultural Testament* (N.p.: Oxford City Press, 2010), 22.

24. Gunter Vogt, *Entstehung und Entwicklung des ökologischen Landbaus im deutschsprachigen Raum* (Bad Dürkheim: Stiftung Ökologie und Landbau, 2000), 201.

25. Uekötter, *Wahrheit*, 233.

26. Helmut Heiber, ed., *Reichsführer! . . . Briefe an und von Himmler* (Munich: dtv, 1970), 110.

27. See Robert Sigel, "Heilkräuterkulturen im KZ: Die Plantage in Dachau," *Dachauer Hefte* 4 (1988): 164–73.

28. Zander, *Anthroposophie*, 1600–604. For discussions of the link between biodynamic farming and National Socialism that betray a political agenda, see Anna Bramwell, *Blood and Soil: Walther Darré and Hitler's Green Party* (Abbotsbrook: Kensal Press, 1985), and Peter Staudenmaier, "Organic Farming in Nazi Germany: The Politics of Biodynamic Agriculture, 1933–1945," *Environmental History* 18 (2013): 383–411.

29. See Mechtild Rössler and Sabine Schleiermacher, eds., *Der "Generalplan Ost": Hauptlinien der nationalsozialistischen Planungs- und Vernichtungspolitik* (Berlin: Akademie-Verlag, 1993).

30. See Gregory A. Barton, "Albert Howard and the Decolonization of Science: From the Raj to Organic Farming," in Brett M. Bennett and Joseph M. Hodge, eds., *Science and Empire: Knowledge and Networks of Science across the British Empire, 1800–1970*, 163–86 (Basingstoke: Palgrave Macmillan, 2011).

31. Paul Wagner, *Stickstoffdüngung und Reingewinn* (Berlin: Paul Parey, 1906), 24.

32. See Theodore M. Porter, *Trust in Numbers: The Pursuit of Objectivity in Science and Public Life* (Princeton, NJ: Princeton University Press, 1995).

33. BASF AG, ed., *50 Jahre Nitrophoska* (Ludwigshafen: BASF Aktiengesellschaft, 1977).

34. Compo GmbH, https://www.compo.de/produkte/duenger-blattpflege/zimmerpflanzen/compo-gruenpflanzen-und-palmen-duengestaebchen-mit-guano, accessed November 3, 202).

35. Smil, *Enriching the Earth*, 245.

36. Hugh S. Gorman, *The Story of N: A Social History of the Nitrogen Cycle and the Challenge of Sustainability* (New Brunswick, NJ: Rutgers University Press, 2013), 5.

37. Smil, *Enriching the Earth*, 127.

38. Randal S. Beeman and James A. Pritchard, *A Green and Permanent Land: Ecology and Agriculture in the Twentieth Century* (Lawrence: University Press of Kansas, 2001), 78.

39. Michael Pollan, *The Omnivore's Dilemma: The Search for a Perfect Meal in a Fast-Food World* (London: Bloomsbury, 2007), 46.

40. Gorman, *Story of N*, 86–90.

41. Gregory T. Cushman, *Guano and the Opening of the Pacific World: A Global Ecological History* (New York: Cambridge University Press, 2013), 286n, 319n.

42. Gorman, *Story of N*, 95.

43. John Waterbury, *The Egypt of Nasser and Sadat: The Political Economy of Two Regimes* (Princeton, NJ: Princeton University Press, 1983), 65.

44. Smil, *Enriching the Earth*, 245.

45. David Tilman et al., "Agricultural Sustainability and Intensive Production Practices," *Nature* 418 (2002): 671–77; here, 673.

46. Paul J. Crutzen, "Geology of Mankind: The Anthropocene," *Nature* 415 (2002): 23. See also Will Steffen, Paul J. Crutzen, and John R. McNeill, "The Anthropocene: Are Humans Now Overwhelming the Great Forces of Nature?" *Ambix* 36 (2007): 614–21.

47. Pollan, *Omnivore's Dilemma*, 46.

48. Robert J. Diaz and Rutger Rosenberg, "Spreading Dead Zones and Consequences for Marine Ecosystems," *Science* 321 (2008): 926–29; here, 926, 928.

49. Anna M. Michalak et al., "Record-Setting Algal Bloom in Lake Erie Caused by Agricultural and Meteorological Trends Consistent with Expected Future Conditions," *Proceedings of the National Academy of Sciences* 110 (2013): 6448–52; here, 6449.

50. Commission of the European Communities, Report from the Commission: Implementation of Council Directive 91/676/EEC Concerning the Protection of Waters Against Pollution Caused by Nitrates from Agricultural Sources COM(2002) 407 final (Brussels, July 17, 2002), 30.

51. Keith S. Porter et al., *Nitrogen and Phosphorus: Food Production, Waste and the Environment* (Ann Arbor: Ann Arbor Science, 1975), 339.

52. Donella H. Meadows et al., *The Limits to Growth: A Report for the Club of Rome's Project on the Predicament of Mankind* (London: Earth Island, 1972).

53. Crookes, *Wheat Problem*, 51–133.

54. Brock, *William Crookes*, 377, 384.

55. See Paul Sabin, *The Bet: Paul Ehrlich, Julian Simon, and Our Gamble over Earth's Future* (New Haven, CT: Yale University Press, 2013).

56. Patrick Déry and Bart Anderson, "Peak Phosphorus," *Energy Bulletin* (August 13, 2007).

57. R. H. E. M. Koppelaar and H. P. Weikard, "Assessing Phosphate Rock Depletion and Phosphorus Recycling Options," *Global Environmental Change* 23 (2013): 1454–66; Roland W. Scholz and Friedrich-Wilhelm Wellmer, "Approaching a Dynamic View on the Availability of Mineral Resources: What We May Learn from the Case of Phosphorus?" *Global Environmental Change* 23 (2013): 11–27; and Dana Cordell, Jan-Olof Drangert, and Stuart White, "The Story of Phosphorus: Global Food Security and Food for Thought," *Global Environmental Change* 19 (2009): 292–305.

58. William Stanley Jevons, "The Coal Question," in Libby Robin, Sverker Sörlin, and Paul Warde, eds., *The Future of Nature: Documents on Global Change*, 78–84 (New Haven, CT: Yale University Press, 2013). For an overview of the long history of concerns about phosphorus scarcity, see Andrea E. Ulrich and Emmanuel Frossard, "On the History of a Reoccurring Concept: Phosphorus Scarcity," *Science of the Total Environment* 490 (2014): 694–707.

59. Uekötter, *Wahrheit*, 188.

60. Landesarchiv Schleswig-Holstein Abt. 422.5 no. 9, minutes of the assembly of the Wagrischer Landwirtschaftlicher Verein zu Lensahn on March 19, 1921, p. 6.

20: AIR-CONDITIONING

1. Allan Peskin, *Garfield: A Biography* (Kent, OH: Kent State University Press, 1978), 596–607.

2. Candice Millard, *Destiny of the Republic: A Tale of Madness, Medicine, and the Murder of a President* (New York: Anchor Books, 2011), 268, 293.

3. Dane Keith Kennedy, *The Magic Mountains: Hill Stations and the British Raj* (Berkeley: University of California Press, 1996), 14.

4. Alain Corbin, *The Lure of the Sea: The Discovery of the Seaside in the Western World, 1750–1840* (Berkeley: University of California Press, 1994), 270–72, 276.

5. Peskin, *Garfield*, 605.

6. See Salvatore Basile, *Cool: How Air Conditioning Changed Everything* (New York: Fordham University Press, 2014), 89–93.

7. Gail Cooper, *Air-Conditioning America: Engineers and the Controlled Environment, 1900–1960* (Baltimore: Johns Hopkins University Press, 1998), 7, 12, 29, 111n, 137.

8. Raymond Arsenault, "The End of the Long Hot Summer: The Air Conditioner and Southern Culture," *Journal of Southern History* 50 (1984): 597–628; here, 613. To the best of this author's knowledge, there is no article on another region of the world that had a similar impact, though that may say more about the extent to which the historical literature on air-conditioning centers on the United States.

9. Arsenault, "End."

10. Michael F. Logan, *Desert Cities: The Environmental History of Phoenix and Tucson* (Pittsburgh, PA: University of Pittsburgh Press, 2006), 144–46.

11. Arsenault, "End," 628.

12. Marsha E. Ackermann, *Cool Comfort: America's Romance with Air-Conditioning* (Washington, DC: Smithsonian Books, 2010), 70.

13. Jonah Engle, "Bush's Visit Shuts Down Senegal," *The Nation*, July 24, 2003, accessed November 3, 2021, https://www.thenation.com/article/bushs-visit-shuts-down-senegal/. Quotation from Remarks by the President on Goree Island, For Immediate Release, July 8, 2003, accessed November 3, 2021, https://georgewbush-whitehouse.archives.gov/news/releases/2003/07/20030708-1.html.

14. "Among the Costs of War: Billions a Year in A.C.?" National Public Radio, June 25, 2011, accessed November 3, 2021, http://www.npr.org/2011/06/25/137414737/among-the-costs-of-war-20b-in-air-conditioning.

15. Arsenault, "End," 607.

16. Cooper, *Air-Conditioning America*, 157.

17. Stan Cox, *Losing Our Cool: Uncomfortable Truths About Our Air-Conditioned World (and Finding New Ways to Get Through the Summer)* (New York: New Press, 2010), 63.

18. See Elizabeth Shove, *Comfort, Cleanliness and Convenience: The Social Organisation of Normality* (Oxford: Berg, 2003).

19. Jack Temple Kirby, *Mockingbird Song: Ecological Landscapes of the South* (Chapel Hill: University of North Carolina Press, 2006), 324.

20. Cooper, *Air-Conditioning America*, 163.

21. Cooper, *Air-Conditioning America*, 161n.

22. Dean Hawkes, *The Environmental Tradition: Studies in the Architecture of Environment* (London: E. & F. N. Spon, 1996), 20.

23. Hawkes, *Environmental Tradition*, 162.

24. Jeff Biddle "Explaining the Spread of Residential Air Conditioning, 1955–1980," *Explorations in Economic History* 45 (2008): 402–23; here, 420.

25. Cox, *Losing Our Cool*, 32.

26. Cox, *Losing Our Cool*, 38.

27. Cooper, *Air-Conditioning America*, 126.

28. Gordon Thomas and Max Morgan-Witts, *Trauma: The Search for the Cause of Legionnaires' Disease* (London: Hamish Hamilton, 1981), 425.

29. See G. J. Raw, *Sick Building Syndrome: A Review of the Evidence on Causes and Solutions* (London: HMSO, 1992).

30. Hawkes, *Environmental Tradition*, 99.

31. Hawkes, *Environmental Tradition*, 99. Similarly, Ackermann, *Cool Comfort*, 184. See also Joseph M. Siry, *Air-Conditioning in Modern American Architecture, 1890–1970* (University Park: Pennsylvania State University Press, 2021).

32. Nancy Y. Reynolds, "City of the High Dam: Aswan and the Promise of Postcolonialism in Egypt," *City & Society* 29 (2017): 213–35; here, 221.

33. Graham Parkhurst and Richard Parnaby, "Growth in Mobile Air-Conditioning: A Socio-Technical Research Agenda," *Building Research and Information* 36, no. 4 (2008): 351–62; here, 354.

34. K. Yammine and H. Al Adham, "The Status of Serum Vitamin D in the Population of the United Arab Emirates," *Eastern Mediterranean Health Journal* 22 (2016): 682–86.

35. Christopher M. Davidson, *Dubai: The Vulnerability of Success* (New York: Columbia University Press, 2008), 111.

36. Agis M. Papadopoulos, "The Influence of Street Canyons on the Cooling Loads of Buildings and the Performance of Air Conditioning Systems," *Energy and Buildings* 33 (2001): 601–7.

37. Cox, *Losing Our Cool*, 136.

38. Cox, *Losing Our Cool*, 132.

39. Lucas W. Davis and Paul J. Gertler, "Contribution of Air Conditioning Adoption to Future Energy Use under Global Warming," *Proceedings of the National Academy of Sciences of the United States of America* 112 (2015): 5962–67; here, 5966.

40. E. Krüger, P. Drach, and P. Bröde, "Implications of Air-Conditioning Use on Thermal Perception in Open Spaces: A Field Study in Downtown Rio de Janeiro," *Building and Environment* 94 (2015): 417–25.

41. This remark was inspired by Thomas P. Hughes, *American Genesis: A Century of Invention and Technological Enthusiasm 1870–1970* (New York: Viking, 1989).

42. Geoffrey J. Martin, *Ellsworth Huntington: His Life and Thought* (Hamden, CT: Archon Books, 1973), 64, 85, 90, 159n.

43. Ellsworth Huntington, *Mainsprings of Civilization* (New York: John Wiley, 1945), 275.

44. Huntington, *Mainsprings of Civilization*, 405.

45. Ackermann, *Cool Comfort*, 21, 23, 144.

46. Martin, *Ellsworth Huntington*, 175.

47. David S. Landes, *The Wealth and Poverty of Nations: Why Some Are So Rich and Some So Poor* (New York: Norton, 1999), 3.

48. See Geoffrey Parker, *Global Crisis: War, Climate Change and Catastrophe in the Seventeenth Century* (New Haven, CT: Yale University Press, 2013), and Wolfgang Behringer, *A Cultural History of Climate* (Cambridge: Polity, 2010).

49. Gore Vidal, *At Home: Essays 1982–1988* (New York: Random House, 1988), 6.

50. Ackermann, *Cool Comfort*, 62–76.

51. Gwyn Prins, "On Condis and Coolth," *Energy and Buildings* 18 (1992): 251-58; here, 251.

52. For an overview on the cultural critique of air-conditioning, see Ackermann, *Cool Comfort*, 138-57.

53. James Rodger Fleming, *Fixing the Sky: The Checkered History of Weather and Climate Control* (New York: Columbia University Press, 2010).

54. See Ronald E. Hester and Roy M. Harrison, eds., *Geoengineering of the Climate System* (Cambridge: Royal Society of Chemistry, 2014).

INTERLUDE
OPIUM

1. Howard Padwa, *Social Poison: The Culture and Politics of Opiate Control in Britain and France, 1821-1926* (Baltimore: Johns Hopkins University Press, 2012), 15 (quotation), 18.

2. Thomas Darnstädt, "Die Enthauptete Republik," *Der Spiegel* 47 (2003): 38-44, 47-49; here, 47.

3. See p. 42.

4. Richard Davenport-Hines, *The Pursuit of Oblivion: A Global History of Narcotics* (New York: Norton, 2004), 219.

5. David T. Courtwright, *Forces of Habit: Drugs and the Making of the Modern World* (Cambridge, MA: Harvard University Press, 2001), 79.

6. Bob Batchelor, "E-Commerce," in George R. Goethals, Georgia J. Sorenson, and James MacGregor Burns, eds., *Encyclopedia of Leadership*, vol. 1: *A-E*, 377-81 (Thousand Oaks, CA: Sage, 2004), 378.

7. Hunter S. Thompson, *Fear and Loathing in Las Vegas: A Savage Journey to the Heart of the American Dream* (London: Harper Perennial, 2005). I wish to record that the only stimulants used in the making of this book were caffeine and chocolate. I do not have firsthand knowledge of opium either and did not seek it for the purpose of this chapter. I was not in the mood for another gamble after the breadfruit fiasco.

8. Jan de Vries, *The Industrious Revolution: Consumer Behavior and the Household Economy, 1650 to the Present* (New York: Cambridge University Press, 2008), 165.

9. Thomas Dormandy, *Opium: Reality's Dark Dream* (New Haven, CT: Yale University Press, 2012), 7n, 32 (quotation).

10. Karl Marx, "A Contribution to the Critique of Hegel's Philosophy of Right: Introduction," in J. O. Malley, ed., *Marx: Early Political Writings*, 57-70 (Cambridge: Cambridge University Press, 1994), 57.

11. Carl A. Trocki, *Opium, Empire and the Global Political Economy: A Study of the Asian Opium Trade 1750-1950* (London: Routledge, 1999), 58.

12. Andrew B. Liu, "The Birth of a Noble Tea Country: On the Geography of Colonial Capital and the Origins of Indian Tea," *Journal of Historical Sociology* 23 (2010): 73-100; here, 78.

13. Dormandy, *Opium*, 195.

14. Paul Gootenberg, "Cocaine in Chains: The Rise and Demise of a Global Commodity, 1860-1950," in Steven Topik, Carlos Marichal, and Zephyr Frank, eds., *From Silver to Cocaine: Latin American Commodity Chains and the Building of the World Economy, 1500-2000*, 321-51 (Durham, NC: Duke University Press, 2006), 327.

15. Trocki, *Opium*, 58; Timothy Brook and Bob Tadashi Wakabayashi, "Introduction: Opium's History in China," in Brook and Wakabayashi, eds., *Opium Regimes: China, Britain, and Japan, 1839-1952*, 1-27 (Berkeley: University of California Press, 2000), 4.

16. See Motohiro Kobayashi, "An Opium Tug-of-War: Japan versus the Wang Jingwei Regime," in Brook and Wakabayashi, *Opium Regimes*, 344-59.

17. Adam Smith, *An Inquiry into the Nature and Causes of the Wealth of Nations*, ed. R. H. Campbell and A. S. Skinner, vol. 2 (Oxford: Clarendon Press, 1976), 936.

18. Orlando Figes, *Revolutionary Russia, 1891–1991* (London: Penguin, 2014), 391.

19. Victoria Berridge, *Opium and the People: Opiate Use and Drug Control Policy in Nineteenth and Early Twentieth Century England* (London: Free Association Books, 1987), xxx.

20. See Louise Foxcroft, *The Making of Addiction: The "Use and Abuse" of Opium in Nineteenth-Century Britain* (Aldershot: Ashgate, 2007).

21. Paul Gootenberg and Isaac Campos, "Towards a New Drug History of Latin America: A Research Frontier at the Center of Debates," *Hispanic American Historical Review* 95 (2015): 1–35; here, 9.

22. Gregory Blue, "Opium for China: The British Connection," in Brook and Wakabayashi, *Opium Regimes*, 31–54; here, 38.

23. See J. B. Brown, "Politics of the Poppy: The Society for the Suppression of the Opium Trade, 1874–1916," *Journal of Contemporary History* 8, no. 3 (1973): 97–111.

24. Brook and Wakabayashi, "Introduction," 13.

25. See James Windle, "A Very Gradual Suppression: A History of Turkish Opium Controls, 1933–1974," *European Journal of Criminology* 11 (2014): 195–212.

26. See Lucien Bianco, "The Responses of Opium Growers to Eradication Campaigns and the Poppy Tax, 1907–1949," in Brook and Wakabayashi, *Opium Regimes*, 292–319. See also Zheng Yangwen, *The Social Life of Opium in China* (Cambridge: Cambridge University Press, 2005), 190.

27. Davenport-Hines, *Pursuit of Oblivion*, 11.

28. Christian Parenti, "Flower of War: An Environmental History of Opium Poppy in Afghanistan," *SAIS Review* 35 (2015): 183–200; here, 183.

29. Davenport-Hines, *Pursuit of Oblivion*, 15.

30. Jill Jonnes, *Hep-Cats, Narcs, and Pipe Dreams: A History of America's Romance with Illegal Drugs* (Baltimore: Johns Hopkins University Press, 1999), 269.

31. Dan Baum, "Legalize It All: How to Win the War on Drugs," *Harper's Magazine*, April 2016, accessed November 3, 2021, https://harpers.org/archive/2016/04/legalize-it-all/.

32. National Public Radio, "Why Is the Opioid Epidemic Overwhelmingly White?" November 4, 2017, accessed November 3, 2021, https://www.npr.org/2017/11/04/562137082/why-is-the-opioid-epidemic-overwhelmingly-white?t=1531652560909; National Center for Health Statistics, "Drug Overdose Deaths in the U.S. Top 100,000 Annually," accessed July 24, 2022, https://www.cdc.gov/nchs/pressroom/nchs_press_releases/2021/20211117.htm.

33. See "Lawsuits Alone Will Not Fix the US Opioid Overdose Crisis," *Lancet* 394 (September 7, 2019), 805.

34. Paul Clammer, *Afghanistan* (Footscray: Lonely Planet Publications, 2007), 73.

35. See Gootenberg and Campos, "Towards a New Drug History."

PART V
RUPTURES

1. Donald Rumsfeld, *Known and Unknown: A Memoir* (London: Penguin, 2011), 476.

2. Christoph Gradmann, *Krankheit im Labor: Robert Koch und die medizinische Bakteriologie* (Göttingen: Wallstein, 2005), 290.

3. Geheimes Staatsarchiv Berlin I. HA Rep. 81 Gesandtschaft Dresden nach 1807 no. 232, Solms to Bismarck, June 19, 1885, p. 3.

4. Adam Smith, *An Inquiry into the Nature and Causes of the Wealth of Nations*, ed. R. H. Campbell and A. S. Skinner, vol. 2 (Oxford: Clarendon Press, 1976), 773.

5. See Ramachandra Guha, *India After Gandhi: The History of the World's Largest Democracy* (New York: Harper Perennial, 2008), 485.

6. Bidyut Chakrabarty, *Mahatma Gandhi: A Historical Biography* (New Delhi: Roli Books, 2007), 1.

7. Donald Anthony Low, *Britain and Indian Nationalism: The Imprint of Ambiguity 1929–1942* (Cambridge: Cambridge University Press, 1997), 31.

8. Judith M. Brown, "Introduction," in Brown and Anthony Parel, eds., *The Cambridge Companion to Gandhi*, 1–8 (New York: Cambridge University Press, 2011), 4.

9. Sean Scalmer, *Gandhi in the West: The Mahatma and the Rise of Radical Protest* (Cambridge: Cambridge University Press, 2011), 50.

10. Chris Johnston, "New Gandhi Statue Unveiled in London's Parliament Square," *The Guardian*, March 14, 2015, accessed November 3, 2021, https://www.theguardian.com/world/2015/mar/14/new-gandhi-statue-unveiled-in-londons-parliament-square.

11. Keishi Shiono et al., "Lessons Learned from Reconstruction Following a Disaster: Enhancement of Regional Seismic Safety Attained after the 1976 Tangshan, China Earthquake," *Doboku Gakkai Ronbunshu* 513 (1995): 9–15; here, 14.

12. Anthony Oliver-Smith, "Peru's Five-Hundred-Year Earthquake: Vulnerability in Historical Context," in Anthony Oliver-Smith and Susanna M. Hoffman. eds., *The Angry Earth: Disaster in Anthropological Perspective*, 74–88 (New York: Routledge, 1999).

21: CHOLERA

1. See Thomas Mann, *Death in Venice and Other Stories*, trans. and with an Introduction by David Luke (London: Vintage Books, 1998).

2. Numerous scholars have studied the history of cholera, and some of the best books focus on epidemics in specific places or countries (albeit usually in the West). See Charles E. Rosenberg, *The Cholera Years: The United States in 1832, 1849, and 1866* (Chicago: University of Chicago Press, 1987 [1962]); François Delaporte, *Disease and Civilization: The Cholera in Paris, 1832* (Cambridge, MA: MIT Press, 1986); Richard J. Evans, *Death in Hamburg: Society and Politics in the Cholera Years, 1830–1910* (Oxford: Clarendon Press, 1987); Frank M. Snowden, *Naples in the Time of Cholera, 1884–1911* (Cambridge: Cambridge University Press, 1995); Olaf Briese, *Angst in den Zeiten der Cholera*, vol. 1: *Über kulturelle Ursprünge des Bakteriums* (Berlin: Akademie Verlag, 2003); and Michael Zeheter, *Epidemics, Empire, and Environments: Cholera in Madras and Quebec City, 1818–1910* (Pittsburgh, PA: University of Pittsburgh Press, 2015).

3. Johann Wolfgang von Goethe, *Sämtliche Werke*, vol. 43 (Berlin: Propyläen-Verlag, 1930), 272; Mann, *Death in Venice*, 256.

4. K. David Patterson, "Cholera Diffusion in Russia, 1823–1923," *Social Science and Medicine* 38 (1994): 1171–91; here, 1189.

5. Christopher Hamlin, *Cholera: The Biography* (Oxford: Oxford University Press, 2009), 32.

6. Myron Echenberg, *Africa in the Time of the Cholera: A History of Pandemics from 1817 to the Present* (New York: Cambridge University Press, 2011), 174.

7. Goethe, *Sämtliche Werke*, 43:246, 262, 272n, 281, 293.

8. Alfred Pick, "Briefe des Feldmarschalls Grafen Neithardt v. Gneisenau an seinen Schwiegersohn Wilhelm v. Scharnhorst," *Historische Zeitschrift* 77 (1896): 448–60; here, 450.

9. Roger Parkinson, *Clausewitz: A Biography* (New York: Cooper Square Press, 2002), 327, 329.

10. Carolin Philipps, *Therese von Bayern: Eine Königin zwischen Liebe, Pflicht und Widerstand* (Munich: Piper, 2015), 371; Karl Heinz Götze, "Der absolute Geist, die Cholera und

die Himmelfahrt des Philosophen: Hegels Tod und Bestattung (1831)," *Merkur* 74, no. 853 (June 2020): 76-85; and K. Jack Bauer, *Zachary Taylor: Soldier, Planter, Statesman of the Old Southwest* (Baton Rouge: Louisiana State University Press, 1985), 314-16.

11. Heinrich Heine, *Historisch-kritische Gesamtausgabe der Werke*, ed. Manfred Windfuhr, vol. 12/1 (Hamburg: Hoffmann und Campe, 1980), 132, 134. For context, see Catherine J. Kudlick, *Cholera in Post-Revolutionary Paris: A Cultural History* (Berkeley: University of California Press, 1996).

12. Newberry Library, Chicago, Illinois Central Railroad Company Archives, box 92, folder 302, Johnson to Osborn, July 17, 1855, p. 1.

13. Dean C. Worcester, *A History of Asiatic Cholera in the Philippine Islands* (Manila: Bureau of Printing, 1908), 3. For context, see Warwick Anderson, *Colonial Pathologies: American Tropical Medicine, Race, and Hygiene in the Philippines* (Durham, NC: Duke University Press, 2006), 61-69, 72n.

14. Donald B. Cooper, "The New 'Black Death': Cholera in Brazil, 1855-1856," *Social Science History* 10 (1986): 467-88; here, 468.

15. Sean Burrell and Geoffrey Gill, "The Liverpool Cholera Epidemic of 1832 and Anatomical Dissection: Medical Distrust and Civil Unrest," *Journal of the History of Medicine and Allied Sciences* 60 (2005): 478-98; and Jeff Sahadeo, "Epidemic and Empire: Ethnicity, Class, and 'Civilization' in the 1892 Tashkent Cholera Riot," *Slavic Review* 64 (2005): 117-39. For an overview on cholera riots, see Samuel Kline Cohn Jr., "Cholera Revolts: A Class Struggle We May Not Like," *Social History* 42 (2017): 162-80.

16. Lowell Gudmundson, "Mora Porrás, Juan Rafael," in Barbara A. Tenenbaum, ed., *Encyclopedia of Latin American History and Culture*, vol. 4, 111-12 (New York: Simon and Schuster Macmillan, 1996), 111.

17. Andreas Weigl, "Tod und (Über-)Leben: Krankheiten und Lebenserwartung in Wien," in Karl Brunner and Petra Schneider, eds., *Umwelt Stadt: Geschichte des Natur- und Lebensraumes Wien*, 250-61 (Vienna: Böhlau, 2005), 254.

18. Newberry Library, Chicago, Illinois Central Railroad Company Archives, box 92, folder 303, Johnson to Osborn, August 15, 1855, p. 2.

19. Charles E. Rosenberg, "Cholera in Nineteenth-Century Europe: A Tool for Social and Economic Analysis," *Comparative Studies in Society and History* 8 (1966): 452-63; here, 452.

20. Margaret Pelling, *Cholera, Fever and English Medicine, 1825-1865* (Oxford: Oxford University Press, 1978), 4.

21. Manfred Vasold, *Grippe, Pest und Cholera: Eine Geschichte der Seuchen in Europa* (Stuttgart: Steiner, 2008), 100.

22. George Rosen, *A History of Public Health*, expanded ed. (Baltimore: Johns Hopkins University Press, 1993 [1959]), 251.

23. Hamlin, *Cholera: The Biography*, 10. The concept of a disease biography is open to critique on methodological grounds due to its pseudoanthropomorphic nature, but it provides an alternative to the hegemonic case studies. For an earlier "biography" of cholera, see Norman Longmate, *King Cholera: The Biography of a Disease* (London: Hamish Hamilton, 1966).

24. Mann, *Death in Venice*, 258.

25. Newberry Library, Chicago, Rodgers Family Papers, box 5, folder 212, Remedies for Asiatic Cholera Recommended by the Board of Health for the United Kingdom (England) in the case of humonitory diarrhea (undated note); and Terrence James Reed, *Thomas Mann: Der Tod in Venedig; Text, Materialien, Kommentar mit den bisher unveröffentlichten Arbeitsnotizen Thomas Manns* (Munich: Hanser, 1987), 112.

26. Geheimes Staatsarchiv Berlin I. HA Rep. 81 Madrid II no. 78, Volmar to von Pfuel, September 13, 1865.

27. Newberry Library, Chicago, Illinois Central Railroad Company Archives, box 92, folder 303, Johnson to Osborn, August 15, 1855, p. 3.

28. Ivan Turgenev, *Fathers and Sons*, trans. Richard Freeborn (Oxford: Oxford University Press, 1991), 54.

29. Burrell and Gill, "Liverpool Cholera Epidemic," 485.

30. Rosenberg, *Cholera Years*, 43, 47.

31. Geheimes Staatsarchiv Berlin I. HA Rep. 81 Gesandtschaft Dresden nach 1807 no. 269, Bülow to Caprivi, July 31, 1892, p. 1.

32. For an overview of the range of interpretations, see Ellis Shookman, *Thomas Mann's Death in Venice: A Novella and Its Critics* (Rochester, NY: Camden House, 2003).

33. Olaf Briese, *Angst in den Zeiten der Cholera*, vol. 4: *Das schlechte Gedicht. Strategien literarischer Immunisierung* (Berlin: Akademie Verlag, 2003).

34. Erwin H. Ackerknecht, "Anticontagionism between 1821 and 1867," *Bulletin of the History of Medicine* 22 (1948): 562–93; here, 565.

35. Brock, *Justus von Liebig*, 206n.

36. Nancy Tomes, *The Gospel of Germs: Men, Women, and the Microbe in American Life* (Cambridge, MA: Harvard University Press, 1998), 27.

37. Tomes, *Gospel of Germs*, 28.

38. See Thomas S. Kuhn, *The Structure of Scientific Revolutions* (Chicago: University of Chicago Press, 1962).

39. See Karin Knorr Cetina, "Objectual Practice," in Theodore R. Schatzki, Karin Knorr Cetina, and Eike von Savigny, eds., *The Practice Turn in Contemporary Theory*, 184–97 (London: Routledge, 2001).

40. Gerald M. Oppenheimer and Ezra Susser, "The Context and Challenge of von Pettenkofer's Contribution to Epidemiology," *American Journal of Epidemiology* 166 (2007): 1239–41; here, 1240.

41. Zeheter, *Epidemics*, 248. See also David S. Barnes, *The Great Stink of Paris and the Nineteenth-Century Struggle against Filth and Germs* (Baltimore: Johns Hopkins University Press, 2006), 11.

42. Tom Koch, "John Snow, Hero of Cholera: RIP," *Canadian Medical Association Journal* 178 (2008): 1736. See also Stephanie J. Snow, "The Art of Medicine: John Snow: The Making of a Hero?" *Lancet* 372 (2008): 22–23.

43. Norman Howard-Jones, "Choleranomalies: The Unhistory of Medicine as Exemplified by Cholera," *Perspectives in Biology and Medicine* 15 (1972): 422–34; here, 422n; Hamlin, *Cholera: The Biography*, 13.

44. Karl Kisskalt, *Max von Pettenkofer* (Stuttgart: Wissenschaftliche Verlagsgesellschaft, 1948), 116; and Harald Breyer, *Max von Pettenkofer: Arzt im Vorfeld der Krankheit* (Leipzig: Hirzel, 1981), 207.

45. Joel J. Epstein, *Francis Bacon: A Political Biography* (Athens: Ohio University Press, 1977), 175.

46. Kisskalt, *Max von Pettenkofer*, 117; and Breyer, *Max von Pettenkofer*, 208.

47. Ackerknecht, "Anticontagionism," 567.

48. Geheimes Staatsarchiv Berlin I. HA Rep. 81 Gesandtschaft Dresden nach 1807 no. 233, report from La Granja of August 5, 1885, p. 4.

49. Geheimes Staatsarchiv Berlin I. HA Rep. 81 Madrid II no. 78, letter to Konsul G. A. Lübbers, December 27, 1865.

50. Geheimes Staatsarchiv Berlin I. HA Rep. 81 Florenz/T no. 25, report of July 28, 1854, p. 3.

51. Geheimes Staatsarchiv Berlin I. HA Rep. 81 München (nach 1807) no. 2145, Mordtmann to von Winckler, March 18, 1890, p. 3.
52. Rosen, *History of Public Health*, 268, 457.
53. Snowden, *Naples*, 302n.
54. Mark Mazower, *Governing the World: The History of an Idea* (London: Penguin, 2013), 111.
55. Christoph Gradmann, *Krankheit im Labor: Robert Koch und die medizinische Bakteriologie* (Göttingen: Wallstein, 2005), 268–97.
56. Goethe, *Sämtliche Werke*, 43:293.
57. Asa Briggs, "Cholera and Society in the Nineteenth Century," *Past and Present* 19 (1961): 76–96; here, 85.
58. Flurin Condrau, "Demokratische Bewegung, Choleraepidemie und die Reform des öffentlichen Gesundheitswesens im Kanton Zürich (1867)," *Sudhoffs Archiv* 80 (1996): 205–19.
59. Christopher Hamlin, "'Cholera Forcing': The Myth of the Good Epidemic and the Coming of Good Water," *American Journal of Public Health* 99 (2009): 1946–54.
60. Richard J. Evans, "Epidemics and Revolutions: Cholera in Nineteenth-Century Europe," in Terence Ranger and Paul Slack, eds., *Epidemics and Ideas: Essays on the Historical Perception of Pestilence*, 194–73 (Cambridge: Cambridge University Press, 1995), 171.
61. Zeheter, *Epidemics*, 225.
62. Sahadeo, "Epidemic," 121.
63. Alexander James Hutchinson Russell, *A Memorandum on the Epidemiology of Cholera* (Geneva: League of Nations, 1925), 42.
64. Jean-Yves Tadié, *Marcel Proust* (New York: Viking, 2000), 30n.
65. Snowden, *Naples*, p. 2.
66. Thomas Rütten, "Cholera in Thomas Mann's *Death in Venice*," *Gesnerus* 66 (2009): 256–87.
67. Snowden, *Naples*, 332.
68. See Alfred W. Crosby, *America's Forgotten Pandemic: The Influenza of 1918* (Cambridge: Cambridge University Press, 1989).
69. Gordon Thomas and Max Morgan-Witts, *Trauma: The Search for the Cause of Legionnaires' Disease* (London: Hamish Hamilton, 1981), 428.
70. Gabriel García Márquez, *Love in the Time of Cholera* (New York: Alfred A. Knopf, 1988).
71. Robert Pollitzer, *Cholera* (Geneva: World Health Organization, 1959).
72. See Mohammad Ali et al., "The Global Burden of Cholera," *Bulletin of the World Health Organization* 90 (2012): 209–218A.
73. S. N. De, *Cholera: Its Pathology and Pathogenesis* (Edinburgh: Oliver and Boyd, 1961), 41. See also Debjani Bhattacharyya, *Empire and Ecology in the Bengal Delta: The Making of Calcutta* (Cambridge: Cambridge University Press, 2018), 97, 127.
74. Sheldon Watts, *Epidemics and History: Disease, Power and Imperialism* (New Haven, CT: Yale University Press, 1999), 167.
75. "Report of the Health Officer of Calcutta for 1893," *Lancet* 144 (1894): 1303.
76. Watts, *Epidemics*, 167.
77. Russell, *Memorandum*, 41.
78. Echenberg, *Africa*, 88.
79. Stephen W. Lacey, "Cholera: Calamitous Past, Ominous Future," *Clinical Infectious Diseases* 20 (1995): 1409–19; here, 1413.
80. Christopher Troeger et al., "Cholera Outbreak in Grande Comore: 1998–1999," *American Journal of Tropical Medicine and Hygiene* 94 (2016): 76–81.

81. "UN Admits for First Time That Peacekeepers Brought Cholera to Haiti," *The Guardian*, December 1, 2016, accessed November 3, 2021, https://www.theguardian.com/global-development/2016/dec/01/haiti-cholera-outbreak-stain-on-reputation-un-says.

82. Firdausi Qadri, Taufiqul Islam, and John D. Clemens, "Cholera in Yemen: An Old Foe Rearing Its Ugly Head," *New England Journal of Medicine* 377, no. 21 (November 23, 2017): 2005-7.

83. J. Eberhart-Phillips et al., "An Outbreak of Cholera from Food Served on an International Aircraft," *Epidemiology and Infection* 116 (1996): 9-13.

84. Pollitzer, *Cholera*, 7.

85. David A. Sack, "A New Era in the History of Cholera: The Road to Elimination," *International Journal of Epidemiology* 42 (2013): 1537-40; here, 1539.

86. Global Task Force on Cholera Control, Declaration to Ending Cholera, 4 October 2017, Annecy, France, accessed November 3, 2021, http://www.who.int/cholera/task_force/declaration-ending-cholera.pdf?ua=1.

87. Qadri, Islam, and Clements, "Cholera in Yemen," 2005n.

88. Alan L. Olmstead and Paul W. Rhode, *Arresting Contagion: Science, Policy, and Conflicts over Animal Disease Control* (Cambridge, MA: Harvard University Press, 2015), 1. On the eradication of the latter, see Amanda Kay McVety, *The Rinderpest Campaigns: A Virus, Its Vaccines, and Global Development in the Twentieth Century* (New York: Cambridge University Press, 2018).

89. Jacques Pauw, "The Politics of Underdevelopment: Metered to Death—How a Water Experiment Caused Riots and a Cholera Epidemic," *International Journal of Health Services* 33 (2003): 819-30.

90. Mohsin Hamid, *How to Get Filthy Rich in Rising Asia* (London: Penguin, 2014).

22: BAEDEKER

1. A. P. Herbert and A. Davies-Adams, *La Vie Parisienne: A Comic Opera in Three Acts* (Very remotely related to the Offenbach opera with the above title) (London: Ernest Benn, 1929), 39.

2. Jules Verne, *Clovis Dardentor* (London: Sampson Low, Marston & Company, 1897), 107.

3. Jules Verne, *Clovis Dardentor* (Vienna: Hartleben, 1897), 83; and Jules Verne, *Clovis Dardentor* (Berlin, 1984), 78.

4. Thomas Martyn, *A Tour through Italy: Containing Full Directions for Travelling in that Interesting Country* (London: C. and G. Kearsley, 1791), iii. See also James Buzard, "The Grand Tour and after (1660-1840)," in Peter Hulme and Tim Youngs, eds., *The Cambridge Companion to Travel Writing*, 37-52 (Cambridge: Cambridge University Press, 2002).

5. Benedikt Bock, *Baedeker & Cook—Tourismus am Mittelrhein 1756 bis ca. 1914* (Frankfurt: Peter Lang, 2010), 192n.

6. Johann August Klein, *Rheinreise von Mainz bis Köln: Historisch, Topographisch, Malerisch* (Koblenz: Fr. Röhling, 1828), iv.

7. Rudy Koshar, "'What Ought to Be Seen': Tourists' Guidebooks and National Identities in Modern Germany and Europe," *Journal of Contemporary History* 33 (1998): 323-40; here, 324.

8. Johann Wolfgang von Goethe, *Italienische Reise:* Herausgegeben und kommentiert von Herbert von Einem (Munich: Beck, 1980), 113, 125-27.

9. On his travel experience, see Norbert Miller, *Der Wanderer: Goethe in Italien* (Munich: Hanser, 2002).

10. Karl Baedeker, *Berlin and Its Environs: Handbook for Travellers*, 4th ed. (Leipzig: Karl Baedeker, 1910), 42.

11. Goethe, *Italienische Reise*, 127.

12. Bock, *Baedeker & Cook*, 278–80.

13. Peter H. Baumgarten and Monika I. Baumgarten, *Baedeker: Ein Name wird zur Weltmarke* (Ostfildern: Karl Baedeker, 1998), 23–26, 39, 72.

14. Alex Hinrichsen, *Baedeker's Reisehandbücher 1828–1945: Vollständiges Verzeichnis der deutschen, englischen und französischen Ausgaben* (Holzminden: Ursula Hinrichsen Verlag, 1979), 11, 35, 51.

15. Hinrichsen, *Baedeker's Reisehandbücher*, 20, 22, 24, 27n. All years refer to the German editions.

16. Arminia Kapusta and Robert Wiluś, "Geography of Tourism in Croatia," in Krzysztof Widawski and Jerzy Wyrzykowski, eds., *The Geography of Tourism of Central and Eastern European Countries*, 2nd ed., 109–47 (Cham: Springer, 2017), 142.

17. Frank Fonda Taylor, *To Hell with Paradise: A History of the Jamaican Tourist Industry* (Pittsburgh, PA: University of Pittsburgh Press, 1993), 4.

18. Hinrichsen, *Baedeker's Reisehandbücher*, 31n.

19. Baedeker, *Berlin and Its Environs*, 43.

20. Karl Baedeker, *Germany: A Handbook for Railway Travellers and Motorists* (Leipzig: Karl Baedeker, 1936), xlii.

21. K. Baedeker, *Italy: Handbook for Travellers*, part 3: *Southern Italy and Sicily*, 8th ed. (Leipzig: Karl Baedeker, 1883), xxiv.

22. K. Baedeker, *Palestine and Syria: Handbook for Travellers* (Leipzig: Karl Baedeker, 1876), 28.

23. Baedeker, *Palestine and Syria*, 9, 26, 37.

24. Buzard, "Grand Tour," 48n.

25. Goethe, *Italienische Reise*, 113.

26. Mark Twain, *The Innocents Abroad* (New York: Oxford University Press, 1996), 486.

27. Twain, *Innocents Abroad*, 486.

28. Karl Baedeker, *The Paris Exhibition of 1889: A Supplement to Paris and Environs* (Leipzig: Karl Baedeker, 1889), 8.

29. Karl Baedeker, *Berlin und Potsdam* (Leipzig: Karl Baedeker, 1936), 66.

30. Jules Verne, *Around the World in Eighty Days*, trans. with notes by Michael Glencross (London: Penguin, 2004), 19.

31. Susanne Müller, *Die Welt des Baedeker: Eine Medienkulturgeschichte des Reiseführers 1830–1945* (Frankfurt: Campus, 2012), 47; and E.A.G., "Griechenland," *Journal of Hellenic Studies* 9 (1888): 391-94; here, 391.

32. K. Baedeker, *The Rhine from Rotterdam to Constance: Handbook for Travellers*, 5th ed. (Koblenz: Karl Baedeker, 1873), v.

33. Baumgarten and Baumgarten, *Baedeker*, 92.

34. Edward Morgan Forster, *A Room with a View*, Introduction and notes, Malcolm Bradbury (New York: Penguin, 2000), 15.

35. Karl Baedeker, *Greece: Handbook for Travellers* (Leipzig: Karl Baedeker, 1889), v.

36. See John Urry, *The Tourist Gaze: Leisure and Travel in Contemporary Societies* (London: Sage, 1990). See also John Urry, "*The Tourist Gaze* 'Revisited,'" *American Behavioral Scientist* 36 (1992): 172–86.

37. Herbert and Davies-Adams, *La Vie Parisienne*, 37.

38. Baedeker, *Germany*, v.

39. See *Pschyrembel Weblog*, accessed November 3, 2021, https://diesteinlaus.wordpress.com/loriot-uber-die-steinlaus/.

40. P. S. Mstislavskii, "Hunger," in A. M. Prokhorov, ed., *Great Soviet Encyclopedia: A Translation of the Third Edition*, vol. 7, 555–56 (New York: Macmillan, 1975), 556.

41. *Catechism of the Catholic Church* (London: Geoffrey Chapman, 1994), 4.

42. T. F. Chipp, *The Forest Officers' Handbook of the Gold Coast, Ashanti and the Northern Territories* (London: Waterlow and Sons Limited, 1922), vi (quotation), 49.

43. Peter Hulme and Tim Youngs, "Introduction," in Hulme and Youngs, *Cambridge Companion*, 1–13; here, 1.

44. Evgenii V. Anisimov, Alexandra Bekasova, and Ekaterina Kalemeneva, "Books That Link Worlds: Travel Guides, the Development of Transportation Infrastructure, and the Emergence of the Tourism Industry in Imperial Russia, Nineteenth-Early Twentieth Centuries," *Journal of Tourism History* 8 (2016): 184–204; here 191; and Müller, *Welt des Baedeker*, 41.

45. Stephen Mennell, *All Manners of Food: Eating and Taste in England and France from the Middle Ages to the Present* (Oxford: Basil Blackwell, 1985), 282n.

46. My own desk reference was *Der Grosse Ploetz: Die Enzyklopädie der Weltgeschichte*, 35th ed. (Göttingen: Vandenhoeck & Ruprecht, 2008).

47. Christian Thorau, "Guides for Wagnerites: Leitmotifs and Wagnerian Listening," in Thomas S. Grey, ed., *Richard Wagner and His World*, 133–50 (Princeton, NJ: Princeton University Press, 2009), 133, 136, 142 (quotation).

48. M. K. Gandhi, *An Autobiography or The Story of My Experiments with Truth* (Ahmedabad: Navajivan Publishing House, 1927), 170.

49. Ernst Braches, *Der Tod in Venedig: Thomas Mann. Arbeitsnotizen* (Overveen: Braches, 2008), 67.

50. See Andrew G. Kirk, *Counterculture Green: The Whole Earth Catalog and American Environmentalism* (Lawrence: University Press of Kansas, 2007).

51. Rupert H. Wheldon, *No Animal Food* (London: C. W. Daniel, 1910), 143.

52. Müller, *Welt des Baedeker*, 52.

53. Karl Baedeker, *Das Generalgouvernement: Reisehandbuch* (Leipzig: Karl Baedeker, 1943).

54. Müller, *Welt des Baedeker*, 53.

55. Baedeker, *Rhine from Rotterdam to Constance*, 88. I have refrained from the popular juxtaposition of tourists and travelers in my narrative. With a Baedeker in hand, the purported open-mindedness of the traveler (as opposed to the simpleton tourist) is a matter of degrees.

56. Koshar, "'What Ought to Be Seen,'" 336n, 277 (quotation). Both guides are discussed extensively in Rudy Koshar, *German Travel Cultures* (Oxford: Berg, 2000).

57. Kathleen Meyer, *How to Shit in the Woods: An Environmentally Sound Approach to a Lost Art*, 3rd ed. (Emeryville, CA: Ten Speed Press, 2011). Characteristically, the first edition of 1989 was roughly a third shorter. See p. 000.

58. Müller, *Welt des Baedeker*, 232n, 257, 265n.

59. Tim Low, *Feral Future: The Untold Story of Australia's Exotic Invaders*, 2nd ed. (Chicago: University of Chicago Press, 2002), 235.

60. Andrew Holden, "Tourism and Natural Resources," in Tazim Jamal and Mike Robinson, eds., *The SAGE Handbook of Tourism Studies*, 203–14 (London: Sage, 2012), 211, 212 (quotation).

61. Quoted in Turgut Var and John Ap, "Tourism and World Peace," in William F. Theobald, ed., *Global Tourism*, 2nd ed., 44–57 (Oxford: Butterworth-Heinemann, 1998), 45.

62. Pertti Hämäläinen, *Yemen: A Travel Survival Kit*, 3rd ed. (Hawthorn: Lonely Planet, 1996), 87.

63. Stephan Löwenstein, "Wo sich Kulturen begegnen," *Frankfurter Allgemeine*, September 2, 2014, accessed November 3, 2021, http://www.faz.net/aktuell/politik

/ausland/europa/arabische-touristen-in-oesterreich-in-zell-am-see-begegnen-sich-kulturen-13131912.html.

64. Robert Reid and Michael Grosberg, *Myanmar (Burma)*, 9th ed. (Victoria: Lonely Planet, 2005), 337.

65. Paul H. Lewis, *Latin Fascist Elites: The Mussolini, Franco, and Salazar Regimes* (Westport, CT: Praeger, 2002), 137.

66. Filipe Ribeiro de Meneses, *Salazar: A Political Biography* (New York: Enigma Books, 2009), 600.

67. Paul Clammer, *Afghanistan* (Footscray: Lonely Planet, 2007), 40.

68. Baedeker, *Greece*, xi.

69. Patrick Young, *Enacting Brittany: Tourism and Culture in Provincial France, 1871–1939* (London: Routledge, 2016); and Katherine Haldane Grenier, *Tourism and Identity in Scotland, 1770–1914: Creating Caledonia* (Aldershot: Ashgate, 2005).

70. Hal K. Rothman, *Devil's Bargain: Tourism in the Twentieth-Century American West* (Lawrence: University Press of Kansas, 1998), 11.

71. Taylor, *To Hell with Paradise*, 37.

72. Sasha D. Pack, *Tourism and Dictatorship: Europe's Peaceful Invasion of Franco's Spain* (New York: Palgrave Macmillan, 2006), 2, 11.

73. Simon Richmond et al., *Korea*, 8th ed. (Footscray: Lonely Planet, 2010), 5.

74. Rothman, *Devil's Bargain*, 10.

75. P. J. O'Rourke, *Holidays in Hell* (London: Grove Press UK, 1988), 12, 17. In the absence of up-to-date guidebooks, O'Rourke relied on a twenty-year-old Hachette guide and an "1876 Baedeker I found in a New England thrift shop" (26).

76. Hämäläinen, *Yemen*, 3

77. Hämäläinen, *Yemen*, 87.

78. Serhil Plokhy, *Chernobyl: History of a Tragedy* (London: Allen Lane, 2018), 345.

79. Johann Wolfgang von Goethe, *Faust: Eine Tragödie*. Textkritisch durchgesehen und mit Anmerkungen versehen von Erich Trunz (Goethes Werke, Hamburger Ausgabe, vol. 3: Dramatische Dichtungen, vol. 1 [Hamburg: Christian Wegner, 1949]), 218.

80. Thorau, *Guides*, 138n.

81. Forster, *Room with a View*.

82. Markus Lenzen et al., "The Carbon Footprint of Global Tourism," *Nature Climate Change* 8 (2018): 522–28.

83. Santo Cilauro, Tom Gleisner, and Rob Sitch, *Molvanîa: A Land Still Untouched by Modern Dentistry* (Melbourne: Hardie Grant Books, 2013), 41; Santo Cilauro, Tom Gleisner, and Rob Sitch, *Phaic Tăn: Sunstroke on a Shoestring* (London: Quadrille, 2005), 75; and Santo Cilauro, Tom Gleisner, and Rob Sitch, *San Sombrèro: A Land of Carnivals, Cocktails and Coups* (London: Quadrille, 2006), 184.

23: GANDHI'S SALT

1. The literature on Gandhi defies comprehensive annotation, and it is not always driven by scholarly motives. For some of the more academic overviews, see Judith M. Brown, *Gandhi: Prisoner of Hope* (New Haven, CT: Yale University Press, 1989); B. R. Nanda, *Mahatma Gandhi: A Biography* (Delhi: Oxford University Press, 1982 [1958]); Dietmar Rothermund, *Mahatma Gandhi: An Essay in Political Biography* (New Delhi: Manohar, 1991); Arvind Sharma, *Gandhi: A Spiritual Biography* (New Haven, CT: Yale University Press, 2013); Shahid Amin, "Gandhi as Mahatma: Gorakhpur District, Eastern UP, 1921-2," in Ranajit Guha and Gayatri Chakravorty Spivak, eds., *Selected Subaltern Studies*, 288–348 (New York: Oxford University Press, 1988); and Ramachandra Guha, *Gandhi: The Years That Changed the World, 1914–1948* (London: Penguin, 2018).

2. B. R. Nanda, *In Search of Gandhi: Essays and Reflections* (New Delhi: Oxford University Press, 2008), 78–80, 78 (quotation).

3. Joachim Radkau, *The Age of Ecology: A Global History* (Cambridge: Polity Press, 2013).

4. Maria Misra, *Vishnu's Crowded Temple: India since the Great Rebellion* (London: Allen Lane, 2007), 194.

5. Quoted in Louis Fischer, *The Life of Mahatma Gandhi* (London: Vintage, [2015?]), 279.

6. Suchitra, "What Moves Masses: Dandi March as Communication Strategy," *Economic and Political Weekly* 30, no. 14 (April 8, 1995): 743–46; here, 744.

7. See Fischer, *Life*. The book had a chapter "My Week with Gandhi" (366–82.)

8. Sean Scalmer, *Gandhi in the West: The Mahatma and the Rise of Radical Protest* (Cambridge: Cambridge University Press, 2011), 39 (quotation), 44.

9. Scalmer, *Gandhi*, 40, 45.

10. Dennis Dalton, *Mahatma Gandhi: Nonviolent Power in Action* (New York: Columbia University Press, 1993), 107n.

11. See Scalmer, *Gandhi*, 50–60, 52 (quotation).

12. Nanda, *Mahatma Gandhi*, 316; and Neil Elkes, "New Future for Hotel Building Where Mahatma Gandhi Ate," *Birmingham Mail*, October 18, 2017, accessed June 13, 2020, https://www.birminghammail.co.uk/news/midlands-news/new-future-hotel-building-mahatma-13775084.

13. Fischer, *Life*, 283.

14. Fischer, *Life*, 10n.

15. Nanda, *In Search of Gandhi*, 79.

16. Kathryn Tidrick, *Gandhi: A Political and Spiritual Life* (London: I. B. Tauris, 2006), 225.

17. Nanda, *Mahatma Gandhi*, 294–97.

18. See p. 443–44.

19. See p. 456–57.

20. M. K. Gandhi, *An Autobiography or The Story of My Experiments with Truth* (Ahmedabad: Navajivan, 1927), ix (quotations), 18.

21. Gandhi, *Autobiography*,174. See also p. 339.

22. Brown, *Gandhi*, 31.

23. David Hardiman, *Gandhi in His Time and Ours: The Global Legacy of His Ideas* (London: Hurst, 2003), 39–59.

24. James R. Andrews and David Zarefsky, *Contemporary American Voices: Significant Speeches in American History, 1945–Present* (White Plains, NY: Longman, 1992), 80. See also Bidyut Chakrabarty, *Confluence of Thought: Mahatma Gandhi and Martin Luther King, Jr.* (New York: Oxford University Press, 2013).

25. See Hardiman, *Gandhi*, 255–93.

26. Brown, *Gandhi*, 167; and Rothermund, *Mahatma Gandhi*, 59.

27. See Theodore Sands and Chester Penn Higby, "France and the Salt Tax," *Historian* 11, no. 2 (March 1949): 145–65.

28. Nanda, *In Search of Gandhi*, 81.

29. Brown, *Gandhi*, 228.

30. Ramachandra Guha and Juan Martinez-Alier, *Varieties of Environmentalism: Essays North and South* (London: Earthscan, 1997), 155.

31. E. F. Schumacher, *Small Is Beautiful: A Study of Economics as if People Mattered* (London: Blond & Briggs, 1973), 29.

32. Knowledge in Civil Society, *Knowledge Swaraj: An Indian Manifesto on Science and Technology* (Tarnaka: Centre for World Solidarity, December 2009), p. 5, accessed

November 3, 2021, https://steps-centre.org/anewmanifesto/manifesto_2010/clusters/cluster5/Indian_Manifesto.pdf.

33. Rob Nixon, *Slow Violence and the Environmentalism of the Poor* (Cambridge, MA: Harvard University Press, 2011), 28. On the environmentalism of the poor, see p. 533.

34. Robert S. Emmett and David E. Nye, *The Environmental Humanities: A Critical Introduction* (Cambridge, MA: MIT Press, 2017), 11.

35. Ursula K. Heise, *Imagining Extinction: The Cultural Meanings of Endangered Species* (Chicago: University of Chicago Press, 2016), 117n.

36. Miriam Tola, "Between Pachamama and Mother Earth: Gender, Political Ontology and the Rights of Nature in Contemporary Bolivia," *Feminist Review* 118 (2018): 25–40; here, 26n.

37. Shashi Tharoor, *Inglorious Empire: What the British Did to India* (London: Penguin, 2017), 242.

38. Nelson Mandela, *Long Walk to Freedom* (London: Abacus, 2011 [1994]), 147.

39. Melinda Henneberger, "Nader Sees a Bright Side to a Bush Victory," *New York Times*, November 1, 2000, p. A29.

40. Hardiman, *Gandhi*, 70.

41. Hardiman, *Gandhi*, 71.

42. Rothermund, *Mahatma Gandhi*, 3.

43. Misra, *Vishnu's Crowded Temple*, 190–92.

44. Rothermund, *Mahatma Gandhi*, 63.

45. Paul Krugman, "The Theory of Interstellar Trade," *Economic Inquiry* 48 (2010): 1119–23.

46. See "All Prizes in Economic Sciences," accessed November 3, 2021, https://www.nobelprize.org/prizes/uncategorized/all-prizes-in-economic-sciences/.

47. Elinor Ostrom, *Governing the Commons: The Evolution of Institutions for Collective Action* (Cambridge: Cambridge University Press, 2015 [1990]), 26.

48. For Hardin's landmark essay, see Garrett Hardin, "The Tragedy of the Commons," *Science* 162 (1968): 1243–48.

49. Ostrom, *Governing the Commons*, 21n.

50. Ostrom, *Governing the Commons*, 214.

51. Lee Anne Fennell, "Ostrom's Law: Property Rights in the Commons," *International Journal of the Commons* 5 (2011): 9–27; here, 10.

52. Brown, *Gandhi*, 42n.

53. Anne Feuchter-Schawelka, "Siedlungs- und Landkommunebewegung," in Diethart Kerbs and Jürgen Reulecke, eds., *Handbuch der deutschen Reformbewegungen 1880–1933*, 227–44 (Wuppertal: Hammer, 1998), 237.

54. Joshua Clark Davis, *From Head Shops to Whole Foods: The Rise and Fall of Activist Entrepreneurs* (New York: Columbia University Press, 2017), 176–223; and Helga Willer, ed., *Ökologischer Landbau in Europa* (Holm: Deukalion, 1998).

55. James C. Whorton, *Crusaders for Fitness: The History of American Health Reformers* (Princeton, NJ: Princeton University Press, 1982), 201, 204 (quotation).

56. See p. 164.

57. Gandhi, *Autobiography*.

58. Sharma, *Gandhi*, 127.

59. Adam D. Shprintzen, *The Vegetarian Crusade: The Rise of an American Reform Movement, 1817–1921* (Chapel Hill: University of North Carolina Press, 2013), 162.

60. Marc Cluet, "Vorwort," in Cluet and Catherine Repussard, eds., *"Lebensreform": Die soziale Dynamik der politischen Ohnmacht*, 11–48 (Tübingen: Francke, 2013), 22.

61. Misra, *Vishnu's Crowded Temple*, 153.

62. Uwe Müller et al., "Agrarismus und Agrareliten im östlichen Mitteleuropa: Forschungsstand, Kontextualisierung, Thesen," in Eduard Kubů et al., eds., *Agrarismus und Agrareliten in Ostmitteleuropa*, 15–116 (Prague: Dokořán, 2013), 66.

63. Quoted in Fischer, *Gandhi*, 125.

64. Krugman, "Theory," 1119.

65. This is arguably a pathetic way to conclude, but please note that this is the last sentence in the final chapter that I wrote. If you ever spend five years of your life looking the full horror of modern history in the face, you will understand why I wrote this.

24: THE 1970 TOKYO RESOLUTION

1. Andrew Gordon, *A Modern History of Japan: From Tokugawa Times to the Present* (New York: Oxford University Press, 2003), 246, 265; Tertius Chandler and Gerald Fox, *3000 Years of Urban Growth* (New York: Academic Press, 1974), 341.

2. *The New Official Guide: Japan*, comp. Japan National Tourist Organization, pub. Japan Travel Bureau (Tokyo: Japan National Tourist Organization, 1966), 300.

3. Shigeto Tsuru, ed., *Proceedings of International Symposium: Environmental Disruption, March 1970, Tokyo* (Tokyo: Asahi Evening News, 1970), xviii–xxi, 319.

4. See Frank Uekötter, "Earth Day," in Edward J. Blum, ed., *America in the World, 1776 to the Present: A Supplement to the Dictionary of American History*, vol. 1, A–L, 305–6 (Farmington Hills, MI: Charles Scribner's, 2016).

5. John McCormick, *Reclaiming Paradise: The Global Environmental Movement* (Bloomington: Indiana University Press, 1991), 91.

6. Shigeto Tsuru, *The Political Economy of the Environment: The Case of Japan* (London: Athlone Press, 1999), 67; Shigeto Tsuru, "Foreword," in Tsuru, *Proceedings*, xiii–xiv; here, xiii.

7. Tsuru, *Proceedings*, 319n.

8. Declaration of the United Nations Conference on the Human Environment, adopted June 1972, accessed November 3, 2021, http://www.un-documents.net/unchedec.htm. Tsuru was one of the corresponding consultants for the unofficial conference report. (Barbara Ward and René Dubos, *Only One Earth: The Care and Maintenance of a Small Planet* [London: Andre Deutsch, 1972], 21.)

9. See p. 327.

10. See Jan-Henrik Meyer, "From Nature to Environment: International Organizations and Environmental Protection before Stockholm," in Wolfram Kaiser and Jan-Henrik Meyer, eds., *International Organizations and Environmental Protection: Conservation and Globalization in the Twentieth Century*, 31–73 (New York: Berghahn Books, 2017), 39–41.

11. Badisches Generallandesarchiv Karlsruhe Abt. 233 no. 3029, Kaiserlich Deutsche Botschaft in Frankreich to Reichskanzler von Bethmann Hollweg, November 11, 1909, p. 2.

12. See Frank Uekötter, "League of Nations," in Kathleen A. Brosnan, ed., *Encyclopedia of American Environmental History*, vol. 3, 834–35 (New York: Facts on File, 2011).

13. Tsuru, *Proceedings*, vii, xv–xvii. The hegemony of Western academia was also on display in the committee behind the report commissioned by the secretary-general of the United Nations Conference on the Human Environment. Of 152 individuals involved, 20 came from the United States and 53 from Western Europe while 44 came from countries of the Global South. (Ward and Dubos, *Only One Earth*, 13–22.)

14. Samy Friedman, "Facing Man and Society: The Challenge," in Tsuru, *Proceedings*, 32–37, here 37.

15. Tsuru, *Proceedings*, 319, 320.

16. Shigeto Tsuru, *Japan's Capitalism: Creative Defeat and Beyond* (Cambridge: Cambridge University Press, 1994), 138.

17. Joseph L. Sax, "The Public Trust Doctrine in Natural Resource Law: Effective Judicial Intervention," *Michigan Law Review* 68, no. 3 (January 1970), 471–566.

18. Tsuru, *Japan's Capitalism*, 138.

19. Joseph L. Sax, "Legal Redress of Environmental Disruption in the United States: The Role of Courts," in Tsuru, *Proceedings*, 223–32; here, 231.

20. See Brett L. Walker, *Toxic Archipelago: A History of Industrial Disease in Japan* (Seattle: University of Washington Press, 2010), 122–26, 138–40, 147–50, 208–10.

21. Simon Avenell, *Transnational Japan in the Global Environmental Movement* (Honolulu: University of Hawai'i Press, 2017), 36.

22. Tsuru, *Proceedings*, xxi.

23. Avenell, *Transnational Japan*, 9.

24. Tsuru, *Proceedings*, 93.

25. Avenell, *Transnational Japan*, 35.

26. Walker, *Toxic Archipelago*, 218.

27. Tsuru, *Japan's Capitalism*, 129.

28. Tsuru, *Political Economy*, 68.

29. Tsuru, *Proceedings*, 52.

30. See Paul Warde, Libby Robin, and Sverker Sörlin, *The Environment: A History of the Idea* (Baltimore: Johns Hopkins University Press, 2018).

31. Akira Iriye, *Global Community: The Role of International Organizations in the Making of the Contemporary World* (Berkeley: University of California Press, 2002), 144.

32. Tsuru, *Proceedings*, 52.

33. Indira Gandhi, *Of Man and His Environment* (New Delhi: Abhinav, 1992), 10. See also Stephen J. Macekura, *Of Limits and Growth: The Rise of Global Sustainable Development in the Twentieth Century* (New York: Cambridge University Press, 2015), 124n.

34. Martinez-Alier, *The Environmentalism of the Poor: A Study of Ecological Conflicts and Valuation* (Cheltenham: Edward Elgar, 2002).

35. World Commission on Environment and Development, *Our Common Future: Development and International Economic Co-operation; Environment* (N.p.: United Nations, 1987), 54.

36. Elke Seefried, "Rethinking Progress: On the Origins of the Modern Sustainability Discourse, 1970–2000," *Journal of Modern European History* 13 (2015): 377–99; here, 387–89.

37. For my own attempts to deliver, see Frank Uekötter, "Ein Haus auf schwankendem Boden: Überlegungen zur Begriffsgeschichte der Nachhaltigkeit," *Aus Politik und Zeitgeschichte* 64, no. 31 (July 28, 2014): 9–15, and "Wie bildet man für Nachhaltigkeit, wenn niemand mehr weiß, was Nachhaltigkeit ist? Eine historisch-politische Spurensuche," *Hessische Blätter für Volksbildung* 68 (2018): 111–18.

38. Stanley P. Johnson, *The Earth Summit: The United Nations Conference on Environment and Development (UNCED)* (London: Graham & Trotman, 1993), 4n.

39. Ingrid Boas, "Earth Summit," in Helmut K. Anheier and Mark Juergensmeyer, eds., *Encyclopedia of Global Studies*, 439–40 (Thousand Oaks, CA: Sage, 2012).

40. John Hemming, *Tree of Rivers: The Story of the Amazon* (London: Thames & Hudson, 2008), 307.

41. Armin Grunwald and Jürgen Kopfmüller, *Nachhaltigkeit: Eine Einführung* (Frankfurt: Campus, 2012), 23.

42. Ian Angus, *Facing the Anthropocene: Fossil Capitalism and the Crisis of the Earth System* (New York: Monthly Review Press, 2016), 27.

43. A. J. Tebble, *F. A. Hayek* (New York: Bloomsbury, 2013), 1.

44. Tsuru, *Proceedings*, 268.

45. Sax, "Legal Redress," 226n.

46. Joseph L. Sax, *Defending the Environment: A Strategy for Citizen Action* (New York: Alfred A. Knopf, 1971), 53.

47. Douglas Martin, "Joseph Sax, Who Pioneered Environmental Law, Dies at 78," *New York Times*, March 10, 2014, accessed November 3, 2021, https://www.nytimes.com/2014/03/11/us/joseph-l-sax-who-pioneered-legal-protections-for-natural-resources-dies-at-78.html?_r=0.

48. See pp. 182, 394.

49. John Lanchbery, "The Convention on International Trade in Endangered Species of Wild Fauna and Flora (CITES): Responding to Calls for Action from Other Nature Conservation Regimes," in Sebastian Oberthür and Thomas Gehring, eds., *International Interaction in Global Environmental Governance: Synergy and Conflict among International and EU Policies*, 157–79 (Cambridge, MA: MIT Press, 2006), 159.

50. See pp. 552–53.

51. See pp. 311, 329.

52. Isao Sakaguchi, "The Roles of Activist NGOs in the Development and Transformation of IWC Regime: The Interaction of Norms and Power," *Journal of Environmental Studies and Sciences* 3 (2013): 194–208; here, 199.

53. See European Commission, "Multilateral Environmental Agreements," accessed November 3, 2021, http://ec.europa.eu/environment/international_issues/agreements_en.htm.

54. John Vogler, "Environmental Issues," in John Baylis, Steve Smith, and Patricia Owens, eds., *The Globalization of World Politics: An Introduction to International Relations*, 7th ed., 385–401 (New York: Oxford University Press, 2017), 393.

55. See pp. 582–83.

56. See p. 95.

57. McCormick, *Reclaiming Paradise*, 98.

58. Avenell, *Transnational Japan*, 199.

59. Mark Mazower, *Governing the World: The History of an Idea* (London: Penguin, 2013), 331–42.

60. See p. 589. For the former, see Richard Elliot Benedick, *Ozone Diplomacy: New Directions in Safeguarding the Planet* (Cambridge, MA: Harvard University Press, 1998).

61. Thorsten Schulz-Walden, *Anfänge globaler Umweltpolitik: Umweltsicherheit in der internationalen Politik (1969–1975)* (Munich: Oldenbourg, 2013), 183–86.

62. Tsuru, "Foreword," xiii.

25: THE 1976 TANGSHAN EARTHQUAKE

1. James Palmer, *The Death of Mao: The Tangshan Earthquake and the Birth of the New China* (London: Faber and Faber, 2013), 44n.

2. Rhett Butler, Gordon S. Stewart, and Hiroo Kanamori, "The July 27, 1976 Tangshan, China Earthquake: A Complex Sequence of Intraplate Events," *Bulletin of the Seismological Society of America* 69 (1979): 207–20; here, 207. For information from the Earthquake Hazard Program of the US Geological Service (USGS), see https://earthquake.usgs.gov/earthquakes/browse/m7-world.php?year=1976, accessed November 3, 2021. The Richter scale number follows the opinion of the USGS, but it should be noted that Richter scale assessments diverge to a certain extent, and contemporary assessments of the Tangshan earthquake ranged from 7.5 to 8.2. See Chen Yong et al., eds., *The Great Tangshan Earthquake of 1976: An Anatomy of Disaster* (Oxford: Pergamon Press, 1988), 97n.

3. Yong et al., *Great Tangshan Earthquake*, 7n.

4. Palmer, *Death of Mao*, 236. Similar Butler, Stewart, and Kanamori, "July 27, 1976 Tangshan," 207.

5. Christiane Eifert, "Das Erdbeben von Lissabon 1755: Zur Historizität einer Naturkatastrophe," *Historische Zeitschrift* 274 (2002): 633–64; here, 644.

6. Andrew Gordon, *A Modern History of Japan: From Tokugawa Times to the Present* (New York: Oxford University Press, 2003), 140.

7. See Mitsuo Yamakawa and Daisaku Yamamoto, *Unravelling the Fukushima Disaster* (London: Routledge, 2017).

8. Palmer, *Death of Mao*, 140n.

9. Conrad Totman, *A History of Japan*, 2nd ed. (Malden, MA: Blackwell, 2005), 162.

10. Arno Borst, "Das Erdbeben von 1348: Ein historischer Beitrag zur Katastrophenforschung," *Historische Zeitschrift* 233 (1981): 529–69; here, 532.

11. See Hayden White, *Metahistory: The Historical Imagination in Nineteenth-Century Europe* (Baltimore: Johns Hopkins University Press, 1973).

12. Mark Elvin, *The Retreat of the Elephants: An Environmental History of China* (New Haven, CT: Yale University Press, 2004), 432.

13. See Manfred Jakubowski-Tiessen and Hartmut Lehmann, eds., *Um Himmels Willen: Religion in Katastrophenzeiten* (Göttingen: Vandenhoeck & Ruprecht, 2003).

14. Mark D. Anderson, *Disaster Writing: The Cultural Politics of Catastrophe in Latin America* (Charlottesville: University of Virginia Press, 2011), 9.

15. Christian Pfister, "Strategien zur Bewältigung von Naturkatastrophen seit 1500," in Pfister, ed., *Am Tag Danach: Zur Bewältigung von Naturkatastrophen in der Schweiz 1500–2000*, 209–55 (Bern: Haupt, 2002), 214. Similarly, François Walter, *Katastrophen: Eine Kulturgeschichte vom 16. bis ins 21. Jahrhundert* (Stuttgart: Reclam, 2010), 12.

16. Matthew 27:51.

17. Daniel Pick, *Rome or Death: The Obsession of General Garibaldi* (London: Jonathan Cape, 2005), 58–62. For context, see Oliver Logan, "The Clericals and Disaster: Polemic and Solidarism in Liberal Italy," in John Dickie, John Foot, and Frank M. Snowden, eds., *Disastro! Disasters in Italy since 1860: Culture, Politics, Society*, 98–112 (New York: Palgrave, 2002).

18. Leopold von Ranke, *Die Römischen Päpste in den letzten vier Jahrhunderten*, 7th ed. (Leipzig: Duncker & Humblot, 1878), 754.

19. Judith Shapiro, *Mao's War against Nature: Politics and the Environment in Revolutionary China* (New York: Cambridge University Press, 2001), 49.

20. J. A. G. Roberts, *A History of China*, 2nd ed. (Basingstoke: Palgrave Macmillan, 2006), 284.

21. Palmer, *Death of Mao*, 147. Authorities compiled a triumphalist booklet for international readers before the end of the year: *After the Tangshan Earthquake: How the Chinese People Overcame a Major Natural Disaster* (Peking: Foreign Language Press, 1976).

22. Shengrong Chen and Honggang Xu, "From Fighting against Death to Commemorating the Dead at Tangshan Earthquake Heritage Sites," *Journal of Tourism and Cultural Change* (August 5, 2017): 1–22; here, 9.

23. Palmer, *Death of Mao*, 172.

24. Roberts, *History of China*, 285.

25. David Birmingham, *A Concise History of Portugal*, 2nd ed. (Cambridge: Cambridge University Press, 2003), 75; and Harsh K. Gupta and Vineet K. Gahalaut, *Three Great Tsunamis: Lisbon (1755), Sumatra-Andaman (2004) and Japan (2011)* (Dordrecht: Springer, 2013), 28.

26. Geheimes Staatsarchiv Berlin I. HA Rep. 81 Gesandtschaft Dresden nach 1807 no. 232, Solms to Bismarck, June 20, 1885, p. 4.

27. Robin Harris, *Dubrovnik: A History* (London: Saqi Books, 2006), 322, 330.
28. Totman, *History of Japan*, 399.
29. Palmer, *Death of Mao*, 149.
30. Palmer, *Death of Mao*, 158.
31. Donald L. Miller, *City of the Century: The Epic of Chicago and the Making of America* (New York: Simon and Schuster, 1997), 168.
32. Greg Bankoff, *Cultures of Disaster: Society and Natural Hazards in the Philippines* (London: RoutledgeCurzon, 2003), 182.
33. David Alexander, "Messina, Italy, Tsunami (1908)," in K. Bradley Penuel and Matt Statler, eds., *Encyclopedia of Disaster Relief*, vol. 1, 414–16 (Thousand Oaks, CA: Sage, 2011), 415.
34. See Wooyeal Paik, "Authoritarianism and Humanitarian Aid: Regime Stability and External Relief in China and Myanmar," *Pacific Review* 24 (2011): 439–62.
35. Paik, "Authoritarianism," 451.
36. Joan Hoff Wilson, *Herbert Hoover: Forgotten Progressive* (Boston: Little, Brown, 1975), 114–17; Kendrick A. Clements, *Hoover, Conservation, and Consumerism: Engineering the Good Life* (Lawrence: University Press of Kansas, 2006), 111; and Felix Mauch, *Erinnerungsfluten: Das Sturmhochwasser von 1962 im Gedächtnis der Stadt Hamburg* (Munich: Dölling und Galitz, 2015), 62.
37. Edgar Wolfrum, *Rot-Grün an der Macht: Deutschland 1998–2005* (Munich: Beck, 2013), 488.
38. John Gooch, *The Unification of Italy* (London: Routledge, 2001), 37.
39. Julie A. Charlip, "Central America in Upheaval," in Thomas H. Holloway, ed., *A Companion to Latin American History*, 406–23 (Malden, MA: Wiley-Blackwell, 2011), 410.
40. John Dickie and John Foot, "Introduction," in Dickie, Foot, and Snowden, *Disastro!* 3–57; here, 44.
41. Naomi Klein, *The Shock Doctrine: The Rise of Disaster Capitalism* (London: Penguin, 2008), 4–6; 6 (quotation).
42. Alicia Dujovne Ortiz, *Eva Perón* (New York: St. Martin's Press, 1995), 55–62.
43. Pfister, "Strategien," 230n.
44. Richard M. Mizelle Jr., *Backwater Blues: The Mississippi Flood of 1927 in the African American Imagination* (Minneapolis: University of Minnesota Press, 2014), 136–39; Pete Daniel, *Deep'n as It Come: The 1927 Mississippi River Flood* (New York: Oxford University Press, 1977), 139–41; and John M. Barry, *Rising Tide: The Great Mississippi Flood of 1927 and How It Changed America* (New York: Touchstone, 1998), 323, 394n, 414.
45. Ted Steinberg, *Acts of God: The Unnatural History of Natural Disaster in America* (New York: Oxford University Press, 2000).
46. Gary Rivlin, *Katrina: After the Flood* (New York: Simon and Schuster, 2015), 387.
47. Wiebe E. Bijker, "The Oosterschelde Storm Surge Barrier: A Test Case for Dutch Water Technology, Management, and Politics," *Technology and Culture* 43 (2002): 569–84.
48. Pierre Milza, *Garibaldi* (Paris: Fayard, 2012), 604. For more details, see Pick, *Rome or Death*, 5–22, 185–91, 194–200, 218. Garibaldi was by no means the first to have that idea; see Maria Margarita Segarra Lagunes, *Il Tevere e Roma: Storia di Una Simbiosi* (Rome: Gangemi, 2004), 117–29.
49. Yang Zhang et al., "Planning and Recovery Following the Great 1976 Tangshan Earthquake," *Journal of Planning History* 14 (2015): 224–43; Beatrice Chen, "'Resist the Earthquake and Rescue Ourselves': The Reconstruction of Tangshan after the 1976 Earthquake," in Lawrence J. Vale and Thomas J. Campanella, eds., *The Resilient City: How Modern Cities Recover from Disaster*, 235–53 (New York: Oxford University Press, 2005), 248.

50. Yuval Noah Harari, *Sapiens: A Brief History of Humankind* (London: Vintage Books, 2011), 393.

51. Palmer, *Death of Mao*, 129. For photographs that confirm this observation, see *The Mammoth Tangshan Earthquake of 1976 Building Damage Photo Album*, comp. China Academy of Building Research (Beijing: China Aademic, 1986).

52. Gregory Clancey, *Earthquake Nation: The Cultural Politics of Japanese Seismicity, 1868-1930* (Berkeley: University of California Press, 2006), 2.

53. For the latter interpretation, see Richard C. Keller, *Fatal Isolation: The Devastating Paris Heat Wave of 2003* (Chicago: University of Chicago Press, 2015).

54. Micah S. Muscolino, *The Ecology of War in China: Henan Province, the Yellow River, and Beyond, 1938-1950* (New York: Cambridge University Press, 2015), 2.

55. Zhang et al., "Planning," 229, 231, 233n.

56. Zhang et al., "Planning," 228.

57. Keishi Shiono et al., "Lessons Learned from Reconstruction Following a Disaster: Enhancement of Regional Seismic Safety Attained after the 1976 Tangshan, China Earthquake," *Doboku Gakkai Ronbunshu* 513 (1995): 9-15; here, 14.

58. Zhang et al., "Planning," 232.

59. Harry Yeh, Shinji Sato and Yoshimitsu Tajima, "The 11 March 2011 East Japan Earthquake and Tsunami: Tsunami Effects on Coastal Infrastructure and Buildings," *Pure and Applied Geophysics* 170 (2013): 1019-31; here, 1021.

60. Hou-Can Zhang and Yi-Zhong Zhang, "Psychological Consequences of Earthquake Disaster Survivors," *International Journal of Psychology* 26 (1991): 613-21; here, 616.

61. Chen and Xu, "From Fighting against Death," 9-13.

62. Linda K. Richter, "The Politics of Heritage Tourism Development: Emerging Issues for the New Millennium," in Gerard Corsane, ed., *Heritage, Museums and Galleries: An Introductory Reader*, 257-71 (London: Routledge, 2005), 266. For the online journal, still a stub when the book went to press, see https://www.dark-tourism.org.uk, accessed November 3, 2021.

PART VI
THE FINAL RESERVES

1. Jean Brunhes, *Les Limites de notre Cage: Discours prononcé à l'occasion de l'inauguration solennelle des cours universitaires le 15 Novembre 1909* (Fribourg: Imprimerie de l'Œuvre de Saint-Paul, 1911), 5. All translations by the author.

2. Brunhes cited Peary's ongoing exploration as a reason for delaying publication of his lecture for more than a year (Brunhes, *Limites*, vi), but it is equally plausible that the delay was due to the notorious attitude of academics toward deadlines.

3. Brunhes, *Limites*, 3, 5 (quotation).

4. Sabine Höhler, *Spaceship Earth in the Environmental Age, 1960-1990* (London: Pickering & Chatto, 2015), 5. However, Höhler mistakenly dates Brunhes's lecture to 1911.

5. See Exodus 2:3.

6. Brunhes, *Limites*, 38.

26: KRUGER NATIONAL PARK

1. Leonard Thompson, *A History of South Africa*, 3rd ed. (New Haven, CT: Yale University Press, 2001), 221.

2. S. C. J. Joubert, "The Kruger National Park: An Introduction," *Koedoe* 29 (1986): 1-11; here, 9.

3. William Beinart and Peter Coates, *Environment and History: The Taming of Nature in the USA and South Africa* (London: Routledge, 1995), 77.

4. Jane Carruthers, *The Kruger National Park: A Social and Political History* (Pietermaritzburg: University of Natal Press, 1995), 48.

5. Jane Carruthers, "Dissecting the Myth: Paul Kruger and the Kruger National Parks," *Journal of Southern African Studies* 20 (1994): 263–83; here, 263.

6. Alfred Runte, *National Parks: The American Experience*, 2nd ed. (Lincoln: University of Nebraska Press, 1987), 17–22.

7. Melissa Harper and Richard White, "How National Were the First National Parks? Comparative Perspectives from the British Settler Societies," in Bernhard Gissibl, Sabine Höhler, and Patrick Kupper, eds., *Civilizing Nature: National Parks in Global Historical Perspective*, 50–67 (New York: Berghahn Books, 2012), 52, 58; and Theodore Catton, "A Short History of the New Zealand National Park System," in Adrian Howkins, Jared Orsi, and Mark Fiege, eds., *National Parks beyond the Nation: Global Perspectives on "America's Best Idea"*, 68–90 (Norman: University of Oklahoma Press, 2016), 72n.

8. Emily Wakild, *Revolutionary Parks: Conservation, Social Justice, and Mexico's National Parks, 1910–1940* (Tucson: University of Arizona Press, 2011).

9. David Thom, *Heritage: The Parks of the People* (Auckland: Lansdowne Press, 1987), 97.

10. Irena Cristalis, *East Timor: A Nation's Bitter Dawn*, 2nd ed. (London: Zed Books, 2009), 272; see also http://datazone.birdlife.org/site/factsheet/tasitolu-iba-timor-leste/text, accessed November 3, 2021.

11. Carolin Firouzeh Roeder, "Slovenia's Triglav National Park: From Imperial Borderland to National Ethnoscape," in Gissibl, Höhler, and Kupper, *Civilizing Nature*, 240–55.

12. Christopher Bennett, *Yugoslavia's Bloody Collapse: Causes, Course and Consequences* (London: Hurst, 1995), 150.

13. James Bryce, "Should Cars Be Admitted in Yosemite?" in David Harmon, ed., *Mirror of America: Literary Encounters with the National Parks*, 123–27 (Boulder, CO: Roberts Rinehart, 1989), 124.

14. For a genealogy of the remark, see Alan MacEachern, "Canada's Best Idea? The Canadian and American National Park Services in the 1910s," in Howkins, Orsi, and Fiege, *National Parks*, 51–67; here, 51.

15. Marine Deguignet et al., *2014 United Nations List of Protected Areas* (Cambridge: UNEP World Conservation Monitoring Centre, 2014), 2.

16. Sterling Evans, *The Green Republic: A Conservation History of Costa Rica* (Austin: University of Texas Press, 1999), 7; and Samuel Bridgewater, *A Natural History of Belize: Inside the Maya Forest* (Austin: University of Texas Press, 2012), 9, 205.

17. Martine Valo, "Gabon: Protecting Vital Forests, and Communities," *The Guardian*, August 27, 2015, accessed November 3, 2021, https://www.theguardian.com/world/2015/aug/27/gabon-forests-protect-communities-biodiversity-climate-change.

18. Deguignet et al., *2014 United Nations List*, 19.

19. José Drummond, "From Randomness to Planning: The 1979 Plan for Brazilian National Parks," in Howkins, Orsi, and Fiege, *National Parks*, 210–34; here, 213.

20. James Sievert, *The Origins of Nature Conservation in Italy* (Bern: Peter Lang, 2000), 200.

21. See Karen B. Wiley and Steven L. Rhodes, "From Weapons to Wildlife: The Transformation of the Rocky Mountain Arsenal," *Environment* 40, no. 5 (1998): 4–11, 28–35.

22. Kim Jeong-su, "DMZ Not Recognized as a Biosphere Reserve," *Hankyoreh*, July 14, 2012, accessed November 3, 2021, http://english.hani.co.kr/arti/english_edition/e_international/542513.html. See also Julia Adeney Thomas, "The Exquisite Corpses of Nature and History: The Case of the Korean DMZ," in Chris Pearson, Peter Coates, and Tim Cole, eds., *Militarized Landscapes: From Gettysburg to Salisbury Plain*, 151–68 (London: Continuum, 2010), and Ke Chung Kim, "Preserving Korea's Demilitarized Corridor for

Conservation: A Green Approach to Conflict Resolution," in Saleem H. Ali, ed., *Peace Parks: Conservation and Conflict Resolution*, 239–59 (Cambridge, MA: MIT Press, 2007); and Ko Dong-hwan, "South Korean Border Now UNESCO Biosphere Reserves," *Korea Times*, June 20, 2019, accessed November 3, 2021, https://www.koreatimes.co.kr/www/nation/2019/06/371_270949.html.

23. Zemaitija National Park, "The Cold War Exposition," accessed November 3, 2021, http://zemaitijosnp.lt/en/veikla/places-to-visit/cold-war-exposition/.

24. David Lawrence, *Kakadu: The Making of a National Park* (Victoria: Melbourne University Press, 2000).

25. Bundesarchiv B 245/137, p. 150.

26. Tenth General Assembly of I.U.C.N., New Delhi, December 1st, 1969, "Resolution 1: National Park Definition," in Richard van Osten, ed., *World National Parks: Progress and Opportunities*, 5 (Brussels: Hayez, 1972).

27. Corey Ross, *Ecology and Power in the Age of Empire: Europe and the Transformation of the Tropical World* (Oxford: Oxford University Press, 2017), 245.

28. See Maglosia B. Nowak-Kemp, Julian P. Hume, "The Oxford Dodo. Part 1: The Museum History of the Tradescant Dodo: Ownership, Displays and Audience," *Historical Biology* 29 (2017), 234–47; and Maglosia B. Nowak-Kemp, Julian P. Hume, "The Oxford Dodo. Part 2: From Curiosity to Icon and Its Role in Displays, Education and Research," *Historical Biology* 29 (2017): 296–307. For the general context, see Eric Baratay and Elisabeth Hardouin-Fugier, *Zoo: A History of Zoological Gardens in the West* (London: Reaktion Books, 2002), and Susanne Köstering, *Natur zum Anschauen: Das Naturkundemuseum des deutschen Kaiserreichs 1871–1914* (Cologne: Böhlau, 2003).

29. Kevin Y. L. Tan, *Of Whales and Dinosaurs: The Story of Singapore's Natural History Museum* (Singapore: NUS Press, 2015), 1.

30. Frank Uekoetter, *The Green and the Brown: A History of Conservation in Nazi Germany* (New York: Cambridge University Press, 2006), 49.

31. Alfred Runte, *National Parks: The American Experience*, 2nd ed. (Lincoln: University of Nebraska Press, 1987), 108. See also Jack E. Davis, *An Everglades Providence: Marjory Stoneman Douglas and the American Environmental Century* (Athens: University of Georgia Press, 2009), 366–72.

32. Carruthers, *Kruger National Park*, 63.

33. See Charles S. Maier, "Consigning the Twentieth Century to History: Alternative Narratives for the Modern Era," *American Historical Review* 105 (2000): 807–31. See also Charles S. Maier, "Leviathan 2.0: Inventing Modern Statehood," in Emily S. Rosenberg, ed., *A World Connecting, 1879–1845*, 27–282 (Cambridge, MA: Belknap Press of Harvard University Press, 2012).

34. Runte, *National Parks*, 47.

35. Carruthers, *Kruger National Park*, 19, 33–36, 53. See also Jane Carruthers, *Wildlife and Warfare: The Life of James Stevenson-Hamilton* (Pietermaritzburg: University of Natal Press, 2001).

36. Tom Mels, *Wild Landscapes: The Cultural Nature of Swedish National Parks* (Lund: Lund University Press, 1999), 94.

37. William M. Adams, *Against Extinction: The Story of Conservation* (London: Earthscan, 2004), 93n, 96.

38. Jeyamalar Kathirithamby-Wells, "From Colonial Imposition to National Icon: Malaysia's Taman Negara National Park," in Gissibl, Höhler, and Kupper, *Civilizing Nature*, 84–101; here, 95.

39. See Merlin Waterson, *The National Trust: The First Hundred Years* (London: National Trust, 1997).

40. See Mark David Spence, *Dispossessing the Wilderness: Indian Removal and the Making of the National Parks* (New York: Oxford University Press, 1999).

41. Teijo Rytteri and Riikka Puhakka, "Formation of Finland's National Parks as a Political Issue," *Ethics, Place and Environment* 12 (2009): 91–106; here, 95n.

42. Kevin McNamee, "From Wild Places to Endangered Spaces: A History of Canada's National Parks," in Philip Dearden and Rick Rollins, eds., *Parks and Protected Areas in Canada: Planning and Management*, 17–44 (Toronto: Oxford University Press, 1993), 32.

43. Carruthers, "Dissecting the Myth," 271.

44. For critical perspectives, see C. Michael Hall and Warwick Frost, "National Parks and the 'Worthless Land Hypothesis' Revisited," in Hall and Frost, eds., *Tourism and National Parks: International Perspectives on Development, Histories and Change*, 45–62 (London: Routledge, 2009), and Richard W. Sellars, "National Parks: Worthless Lands or Competing Land Values?" *Journal of Forest History* 27, no. 3 (1983): 130–34.

45. See Robert W. Righter, *The Battle over Hetch Hetchy: America's Most Controversial Dam and the Birth of Modern Environmentalism* (New York: Oxford University Press, 2005).

46. Runte, *National Parks*, 58, 129.

47. Carruthers, *Kruger National Park*, 23.

48. See Mark Dowie, *Conservation Refugees: The Hundred-Year Conflict between Global Conservation and Native Peoples* (Cambridge, MA: MIT Press, 2009); and Karl Jacoby, *Crimes against Nature: Squatters, Poachers, Thieves, and the Hidden History of American Conservation* (Berkeley: University of California Press, 2014 [2001]).

49. Deguignet et al., *2014 United Nations List*, 14.

50. See https://www.cbd.int/sp/targets/, accessed November 3, 2021.

51. Peter A. Lindsey et al., "Wildlife Viewing Preferences of Visitors to Protected Areas in South Africa: Implications for the Role of Ecotourism in Conservation," *Journal of Ecotourism* 6 (2007): 19–33; here, 20.

52. Sanette Ferreira and Alet Harmse, "Kruger National Park: Tourism Development and Issues around the Management of Large Numbers of Tourists," *Journal of Ecotourism* 13 (2014): 16–34; here, 22.

53. Tom Turner, *David Brower: The Making of the Environmental Movement* (Oakland: University of California Press, 2015), 140.

54. Jean Chrétien, *Straight from the Heart* (Toronto: Key Porter Books, 1994), 68.

55. See Kevin McNamee, "From Wild Places to Endangered Spaces: A History of Canada's National Parks," Dearden and Rollins, *Parks and Protected Areas*, 17–44; here, 33n.

56. L. Zhou and C.E.P. Seethal, "Tourism Policy, Biodiversity Conservation and Management: A Case of the Kruger National Park, South Africa," *International Journal of Sustainable Development and World Ecology* 18 (2011): 393–403; here, 393.

57. Jeff Schauer, "The Elephant Problem: Science, Bureaucracy, and Kenya's National Parks, 1955 to 1975," *African Studies Review* 58 (2015): 177–98.

58. Maier, "Consigning the Twentieth Century," 814.

59. Michael Slezak, "Conservation Report Reinforces Fears over 'Paper Parks,'" *New Scientist*, November 13, 2014, accessed November 3, 2021, https://www.newscientist.com/article/dn26552-conservation-report-reinforces-fears-over-paper-parks/.

60. Hans Schwenkel, *Taschenbuch des Naturschutzes* (Salach: Kaißer, 1941), 6, 46–51.

61. SANParks Annual Report 2015/16, pp. 6, 47, accessed November 3, 2021, https://www.sanparks.org/assets/docs/general/annual-report-2016.pdf.

62. Elizabeth Lunstrum, "Conservation Meets Militarisation in Kruger National Park: Historical Encounters and Complex Legacies," *Conservation and Society* 13 (2015): 356–69; here, 361.

63. Elizabeth Lunstrum, "Articulated Sovereignty: Extending Mozambican State

Power through the Great Limpopo Transfrontier Park," *Political Geography* 36 (2013): 1–11; here, 3n. On the failure of earlier attempts to create a transnational park, see Clapperton Mavhunga and Marja Spierenburg, "Transfrontier Talk, Cordon Politics: The Early History of the Great Limpopo Transfrontier Park in Southern Africa, 1925–1940," *Journal of Southern African Studies* 35 (2009): 715–35.

64. Patrick Kupper, *Creating Wilderness: A Transnational History of the Swiss National Park* (New York: Berghahn Books, 2014), 184 (quotation), 192.

65. Carruthers, *Kruger National Park*, 79.

66. Linda Flint McClelland, *Building the National Parks: Historic Landscape Design and Construction* (Baltimore: Johns Hopkins University Press, 1998); and Runte, *National Parks*, 82, 94.

67. Bryce, "Should Cars Be Admitted?" 126.

68. See Wilfried Huismann, *Schwarzbuch WWF: Dunkle Geschäfte im Zeichen des Panda* (Gütersloh: Gütersloher Verlagshaus, 2012), 22–28.

69. Lindsey, "Wildlife Viewing Preferences," 29.

70. Angela Gaylard, Norman Owen-Smith, and Jessica Redfern, "Surface Water Availability: Implications for Heterogeneity and Ecosystem Processes," in Johan T. du Toit, Kevin H. Rogers, and Harry C. Biggs, eds., *The Kruger Experience: Ecology and Management of Savanna Heterogeneity*, 171–88 (Washington, DC: Island Press, 2003), 171, 176.

71. Ian J. Whyte, Rudi J. van Aarde, and Stuart L. Pimm, "Kruger's Elephant Population: Its Size and Consequences for Ecosystem Heterogeneity," in du Toit, Rogers, and Biggs, *Kruger Experience*, 332–48; here, 338.

72. Llewellyn C. Foxcroft, David M. Richardson, and John R. U. Wilson, "Ornamental Plants as Invasive Aliens: Problems and Solutions in Kruger National Park, South Africa," *Environmental Management* 41 (2008): 32–51.

73. Michael G. L. Mills et al., "Reflections on the Kruger Experience and Reaching Forward," in du Toit, Rogers, and Biggs, *Kruger Experience*, 488–501; here, 491.

74. Julian Rademeyer, *Killing for Profit: Exposing the Illegal Rhino Horn Trade* (Cape Town: Zebra Press, 2012).

75. Antonio Cederna, *La Distruzione della Natura in Italia* (Torino: Einaudi, 1975), 196.

76. Paige West, James Igoe, and Dan Brockington, "Parks and Peoples: The Social Impact of Protected Areas," *Annual Review of Anthropology* 35 (2006): 251–77; here, 255.

77. Adams, *Against Extinction*, 120, 207n.

78. Hana Sakata and Bruce Prideaux, "An Alternative Approach to Community-Based Ecotourism: A Bottom-up Locally Initiated Non-Monetised Project in Papua New Guinea," *Journal of Sustainable Tourism* 21 (2013): 880–99.

79. See Peter Lundgreen, *Standardization—Testing—Regulation: Studies in the History of the Science-Based Regulatory State (Germany and the U.S.A., 19th and 20th Centuries)* (Bielefeld: Kleine, 1986).

80. David Turton, "The Mursi and National Park Development in the Lower Omo Valley," in David Anderson and Richard Grove, eds., *Conservation in Africa: People, Policies and Practice*, 169–86 (Cambridge: Cambridge University Press, 1987), 173, 180.

81. James Fairhead and Melissa Leach, "Practicing 'Biodiversity' in Guinea: Nature, Nation and an International Convention," *Oxford Development Studies* 31 (2003): 427–39.

82. Carruthers, "Dissecting the Myth," 264.

83. Rademeyer, *Killing for Profit*, 301.

84. Ferreira and Harmse, "Kruger National Park," 23.

85. Salomon Joubert, *The Kruger National Park: A History*, vol. 1 (Johannesburg: High Branching, 2007), 122.

86. For an influential critique of the wilderness trope, see William Cronon, "The

Trouble with Wilderness; or, Getting Back to the Wrong Nature," in Cronon, ed., *Uncommon Ground: Rethinking the Human Place in Nature*, 69–90 (New York: Norton, 1995). However, it is worth noting that a German scholar published a best-selling book on how landscapes were culturally constructed during the same year without triggering a similar controversy. (Hansjörg Küster, *Geschichte der Landschaft in Mitteleuropa: Von der Eiszeit bis zur Gegenwart* [Munich: Beck, 1995].)

87. Jane Carruthers, "Pilanesberg National Park, North West Province, South Africa: Uniting Economic Development with Ecological Design; A History, 1960s to 1984," *Koedoe* 53 (2011): 1–10; here, 8.

88. Ferreira and Harmse, "Kruger National Park," 19, 22.

27: EUCALYPTUS

1. Sandra M. Sufian, *Healing the Land and the Nation: Malaria and the Zionist Project in Palestine, 1920–1947* (Chicago: University of Chicago Press, 2007), 41.

2. Alon Tal, *Pollution in a Promised Land: An Environmental History of Israel* (Berkeley: University of California Press, 2002), 77.

3. Aaron Kalman, "Crikey! Eucalyptus Selected Most 'Israeli' Tree," *Times of Israel*, February 5, 2012, accessed November 3, 2021, http://www.timesofisrael.com/crikey-eucalyptus-elected-most-israeli-tree/.

4. A. R. Penfold and J. L. Willis, *The Eucalypts: Botany, Cultivation, Chemistry, and Utilization* (London: Leonard Hill, 1961), xix.

5. Tom Griffiths, *Forests of Ash: An Environmental History* (Cambridge: Cambridge University Press, 2001), 1.

6. Geoffrey Bolton, *Spoils and Spoilers: Australians Make their Environment 1788–1980* (Sydney: George Allen and Unwin, 1981), 45.

7. Robin W. Doughty, *The Eucalyptus: A Natural and Commercial History of the Gum Tree* (Baltimore: Johns Hopkins University Press, 2000), 34, 36.

8. See Daniel R. Headrick, *The Tools of Empire: Technology and European Imperialism in the Nineteenth Century* (New York: Oxford University Press, 1981).

9. Bolton, *Spoils and Spoilers*, 41; and Peter Abbott and Tegan Abbott, *Eucalyptus Oil: Australia's Natural Wonder* (Oakleigh: Felton Grimwade & Bickford, 2005).

10. Gregory Allen Barton, *Empire Forestry and the Origins of Environmentalism* (Cambridge: Cambridge University Press, 2002), 1n; and William Beinart, *The Rise of Conservation in South Africa: Settlers, Livestock, and the Environment 1770–1950* (New York: Oxford University Press, 2003), 77–88, 96n.

11. Brett M. Bennett, "The El Dorado of Forestry: The Eucalyptus in India, South Africa, and Thailand, 1850–2000," *International Review of Social History* 55 (S18) (2010): 27–50; here, 42.

12. Bolton, *Spoils and Spoilers*, 46.

13. Richard Pankhurst, *Economic History of Ethiopia 1800–1935* (Addis Ababa: Haile Sellassie I University Press, 1968), 246.

14. Andrew Hill Clark, *The Invasion of New Zealand by People, Plants and Animals: The South Island* (Westport, CT: Greenwood Press, 1970 [1949]), 367.

15. Ian Tyrrell, *True Gardens of the Gods: Californian-Australian Environmental Reform, 1860–1930* (Berkeley: University of California Press, 1999), 25. See also Jared Farmer, *Trees in Paradise: A California History* (New York: Norton, 2013), 109–220.

16. Karen Brown, "The Conservation and Utilisation of the Natural World: Silviculture in the Cape Colony, c. 1902–1910," *Environment and History* 7 (2001): 427–47; here, 434.

17. James E. McClellan III and François Regourd, "The Colonial Machine: French Sci-

ence and Colonization in the Ancien Régime," in Roy MacLeod, ed., *Nature and Empire: Science and the Colonial Enterprise*, 31–50 (Chicago: University of Chicago Press, 2000), 32.

18. Jayeeta Sharma, "British Science, Chinese Skill and Assam Tea: Making Empire's Garden," *Indian Economic and Social History Review* 43 (2006): 429–55; here, 430n.

19. Harald Witt, "The Emergence of Privately Grown Industrial Tree Plantations," in Stephen Dovers, Ruth Edgecombe, and Bill Guest, eds., *South Africa's Environmental History: Cases and Comparisons*, 90–111 (Athens: Ohio University Press, 2003), 92.

20. Ann Lindsay Mitchell and Syd House, *David Douglas: Explorer and Botanist* (London: Aurum Press, 1999), 52, 172.

21. Tyrrell, *True Gardens*, 26, 60.

22. Fa-ti Fan, *British Naturalists in Qing China: Science, Empire, and Cultural Encounter* (Cambridge, MA: Harvard University Press, 2004), 84, 152.

23. Richard H. Grove, *Green Imperialism: Colonial Expansion, Tropical Island Edens and the Origins of Environmentalism, 1600–1860* (Cambridge: Cambridge University Press, 1995), 168, 187, 191, 338.

24. Ray Desmond, *The History of the Royal Botanic Gardens, Kew*, 2nd ed. (London: Royal Botanic Gardens, Kew, 2007), 233.

25. Margaret Flanders Darby, "*Un*natural History: Ward's Glass Cases," *Victorian Literature and Culture* 35 (2007): 635–47; here, 645.

26. Tyrrell, *True Gardens*, 26.

27. Ray Desmond, *The European Discovery of the Indian Flora* (Oxford: Oxford University Press, 1992), 81. See also David Arnold, "Plant Capitalism and Company Science: The Indian Career of Nathaniel Wallich," *Modern Asian Studies* 42 (2008): 899–928; here, 925.

28. Desmond, *History*, 234.

29. Londa Schiebinger, *Plants and Empire: Colonial Bioprospecting in the Atlantic World* (Cambridge, MA: Harvard University Press, 2004), 5.

30. See Grove, *Green Imperialism*.

31. Richard Drayton, *Nature's Government: Science, Imperial Britain, and the "Improvement" of the World* (Hyderabad: Orient Longman, 2005), 180.

32. Drayton, *Nature's Government*, 3–25.

33. Grove, *Green Imperialism*, 184.

34. Katja Kaiser, "Exploration and Exploitation: German Colonial Botany at the Botanic Garden and Botanical Museum Berlin," in Dominik Geppert and Franz Lorenz Müller, eds., *Sites of Imperial Memory: Commemorating Colonial Rule in the Nineteenth and Twentieth Centuries*, 225–42 (Manchester: Manchester University Press, 2015), 226.

35. Jim Endersby, "A Garden Enclosed: Botanical Barter in Sydney, 1818–39," *British Journal for the History of Science* 33 (2000): 313–34; here, 314.

36. Doughty, *Eucalyptus*, 97, 101.

37. Karl Escherich, *Leben und Forschen: Kampf um eine Wissenschaft*, 2nd ed. (Stuttgart: Wissenschaftliche Verlagsgesellschaft, 1949), 208.

38. Tyrrell, *True Gardens*, 28n.

39. Drayton, *Nature's Government*, 210.

40. Stuart McCook, *States of Nature: Science, Agriculture, and Environment in the Spanish Caribbean, 1760–1940* (Austin: University of Texas Press, 2002), 83.

41. Doughty, *Eucalyptus*, 33.

42. Michael R. Dove, *The Banana Tree at the Gate: A History of Marginal Peoples and Global Markets in Borneo* (New Haven, CT: Yale University Press, 2011), 116.

43. See p. 119–21.

44. Tal, *Pollution*, 78.

45. Drayton, *Nature's Government*, 211, 265.

46. Drayton, *Nature's Government*, 252.

47. Alfred W. Crosby, *Ecological Imperialism: The Biological Expansion of Europe, 900-1900*, 2nd ed. (New York: Cambridge University Press, 2004).

48. Characteristically, Crosby mentions eucalyptus only in a conciliatory remark that points to the tree's enduring dominance in Australia as proof that "the triumph of Old World organisms was not total." (Alfred W. Crosby, *Germs, Seeds, and Animals: Studies in Ecological History* [Armonk, NY: M. E. Sharpe, 1994], 69.)

49. Royal Botanic Gardens, Kew, *Bulletin of Miscellaneous Information* no. 1 (1903): 1.

50. Harald Witt, "The Emergence of Privately Grown Industrial Tree Plantations," in Dovers, Edgecombe, and Guest, *South Africa's Environmental History*, 90-111; here, 93.

51. Pankhurst, *Economic History*, 247; James C. McCann, *Green Land, Brown Land, Black Land: An Environmental History of Africa, 1800-1990* (Portsmouth, NH: Heinemann, 1999), 14.

52. Tal, *Pollution*, 78n.

53. Doughty, *Eucalyptus*, 96.

54. Bennett, "El Dorado," 37.

55. Kate B. Showers, *Imperial Gullies: Soil Erosion and Conservation in Lesotho* (Athens: Ohio University Press, 2005), 60n.

56. Escherich, *Leben und Forschen*, 209.

57. Barry Gardiner and John Moore, "Creating the Wood Supply of the Future," in Trevor Fenning, ed., *Challenges and Opportunities for the World's Forests in the 21st Century*, 677-704 (Dordrecht: Springer, 2014), 679n.

58. Penfold and Willis, *Eucalypts*, 300.

59. Doughty, *Eucalyptus*, 176.

60. Bennett, "El Dorado," 44.

61. Doughty, *Eucalyptus*, ix.

62. Larry Lohmann, "Visitors to the Commons: Approaching Thailand's 'Environmental' Struggles from a Western Starting Point," in Bron Raymond Taylor, ed., *Ecological Resistance Movements: The Global Emergence of Radical and Popular Environmentalism*, 107-26 (Albany: State University of New York Press, 1995), 110, 117.

63. Tellingly, Shiva's anti-eucalyptus activism is not mentioned in Joachim Radkau, *The Age of Ecology: A Global History* (Cambridge: Polity Press, 2013), 225-30.

64. Lohmann, "Visitors," 112, 117, 121.

65. Vandana Shiva and J. Bandyopadhyay, *Ecological Audit of Eucalyptus Cultivation* (Dehradun: English Book Depot, 1987), 72.

66. J. L. M. Gonçalves et al., "Assessing the Effects of Early Silvicultural Management on Long-Term Site Productivity of Fast-Growing Eucalypt Plantations: The Brazilian Experience," *Southern Forests* 70 (2008): 105-18; here, 105.

67. Gabriel Dehon S. P. Rezende, Marcos Deon V. de Resende, and Teotônio F. de Assis, "*Eucalyptus* Breeding for Clonal Forestry," in Fenning, *Challenges and Opportunities*, 393-424; here, 394, 400 (quotation).

68. Kevan M. A. Gartland and Jill S. Gartland, "Forest Biotechnology Futures," in Fenning, *Challenges and Opportunities*, 549-65; here, 555n.

69. Alan Weisman, *The World Without Us* (New York: St. Martin's Press, 2007), 75.

70. Gary Kerr, "A Review of the Growth, Yield and Biomass Distribution of Species Planted in the English Network Trials of Short Rotation Forestry," in Helen McKay, ed., *Short Rotation Forestry: Review of Growth and Environmental Impacts*, 135-60 (Farnham: Forest Research, 2011). For earlier experiments, see Penfold and Willis, *Eucalypts*, 124n.

28: HYBRID CORN

1. John C. Culver and John Hyde, *American Dreamer: The Life and Times of Henry A. Wallace* (New York: Norton, 2000), 37, 54, 107, 130 (quotation), 251n.

2. Edward L. Schapsmeier and Frederick H. Schapsmeier, *Henry A. Wallace of Iowa: The Agrarian Years, 1910–1940* (Ames: Iowa State University Press, 1968), 20, 27.

3. Schapsmeier and Schapsmeier, *Henry A. Wallace*, 21.

4. See Jan Sapp, *Genesis: The Evolution of Biology* (New York: Oxford University Press, 2003), 117–24.

5. Jack Ralph Kloppenburg Jr., *First the Seed: The Political Economy of Plant Biotechnology, 1492–2000*, 2nd ed. (Madison: University of Wisconsin Press, 2004), 104.

6. Quoted in Thomas Wieland, *"Wir beherrschen den pflanzlichen Organismus besser,..." Wissenschaftliche Pflanzenzüchtung in Deutschland, 1889–1945* (Munich: Deutsches Museum, 2004), 68.

7. Deborah Fitzgerald, *The Business of Breeding: Hybrid Corn in Illinois, 1890–1940* (Ithaca, NY: Cornell University, 1990), 7; and Joseph Leslie Anderson, *Industrializing the Corn Belt: Agriculture, Technology, and Environment, 1945–1972* (DeKalb: Northern Illinois University Press, 2009), 7 (quotation), 172.

8. Zvi Griliches, "Hybrid Corn: An Exploration in the Economics of Technological Change," *Econometrica* 25 (1957): 501–22.

9. Howard S. Reed, *A Short History of the Plant Sciences* (Waltham, MA: Chronica Botanica, 1942), 97.

10. Fitzgerald, *Business of Breeding*, 220.

11. Jenny Leigh Smith, *Works in Progress: Plans and Realities on Soviet Farms, 1930–1963* (New Haven, CT: Yale University Press, 2014), 132.

12. William deJong-Lambert, *The Cold War Politics of Genetic Research: An Introduction to the Lysenko Affair* (Dordrecht: Springer, 2012), 148. See also Aaron T. Hale-Dorrell, *Corn Crusade: Khrushchev's Farming Revolution in the Post-Stalin Soviet Union* (New York: Oxford University Press, 2019).

13. Kloppenburg, *First the Seed*, 5.

14. W. Arthur Lewis, "The Export Stimulus," in Lewis, ed., *Tropical Development 1880–1913: Studies in Economic Progress*, 13–45 (London: George Allen and Unwin, 1970), 19.

15. Colin Barlow, *The Natural Rubber Industry: Its Development, Technology, and Economy in Malaysia* (Kuala Lumpur: Oxford University Press, 1978), 115–17.

16. James C. McCann, *Maize and Grace: Africa's Encounter with a New World Crop, 1500–2000* (Cambridge, MA: Harvard University Press, 2005), 167n. The three countries had made up the ill-fated Central African Federation in the late stages of British colonial rule.

17. McCann, *Maize and Grace*, 7.

18. Nick Cullather, *The Hungry World: America's Cold War Battle against Poverty in Asia* (Cambridge, MA: Harvard University Press, 2010), 44, 59.

19. See John H. Perkins, *Geopolitics and the Green Revolution: Wheat, Genes, and the Cold War* (New York: Oxford University Press, 1997).

20. Jonathan Swift, *Gulliver's Travels: Complete, Authoritative Text with Biographical and Historical Contexts, Critical History, and Essays from Five Contemporary Critical Perspectives*, ed. Christopher Fox (Boston: Bedford Books, 1995), 135.

21. For invocations of Swift, see Thomas Swann Harding, *Two Blades of Grass: A History of Scientific Development in the United States Department of Agriculture* (Norman: University of Oklahoma Press, 1947), and Sir Kenneth Lyon Blaxter and Noel Robertson, *From Dearth to Plenty: The Modern Revolution in Food Production* (Cambridge: Cambridge University Press, 1995), xi.

22. Cullather, *Hungry World*, 205.

23. For a concise overview, see Jonathan Harwood, *Europe's Green Revolution and Others Since: The Rise and Fall of Peasant-Friendly Plant Breeding* (London: Routledge, 2012), 118–23.

24. Cullather, *Hungry World*, 68.

25. Culver and Hyde, *American Dreamer*, 150.

26. Courtney Fullilove, *The Profit of the Earth: The Global Seeds of American Agriculture* (Chicago: University of Chicago Press, 2017), 48–50.

27. Philip J. Pauly, *Fruits and Plains: The Horticultural Transformation of America* (Cambridge, MA: Harvard University Press, 2007), 106; and Fullilove, *Profit*, 50 (quotation).

28. Timothy J. Farnham, *Saving Nature's Legacy: Origins of the Idea of Biological Diversity* (New Haven, CT: Yale University Press, 2007), 104.

29. Pauly, *Fruits*, 104; and Fullilove, *Profit*, 48.

30. Fitzgerald, *Business of Breeding*, 73.

31. Reed, *Short History*, 97, 99.

32. Perkins, *Geopolitics*, 218.

33. John Merson, "Bio-prospecting or Bio-piracy: Intellectual Property Rights and Biodiversity in a Colonial and Postcolonial Context," in MacLeod, *Nature and Empire*, 282–96; here, 284n.

34. Michael Astor, "Biopiracy Fears Hampering Research in Brazilian Amazon," Associated Press, October 30, 2005, accessed November 3, 2021, http://news.mongabay.com/2005/1030-ap.html. Similarly, Warren Dean, *Brazil and the Struggle for Rubber: A Study in Environmental History* (Cambridge: Cambridge University Press, 1987), 166.

35. Dean, *Brazil*, 44.

36. Dean, *Brazil*, 166. Similarly, Zephyr Frank and Aldo Musacchio, "Brazil in the International Rubber Trade, 1870–1930," in Steven Topik, Carlos Marichal, and Zephyr Frank, eds., *From Silver to Cocaine: Latin American Commodity Chains and the Building of the World Economy, 1500–2000*, 271–99 (Durham, NC: Duke University Press, 2006), 277.

37. Joe Jackson, *The Thief at the End of the World: Rubber, Empire and the Obsessions of Henry Wickham* (London: Duckworth Overlook, 2009), 191.

38. Michael R. Dove, *The Banana Tree at the Gate: A History of Marginal Peoples and Global Markets in Borneo* (New Haven, CT: Yale University Press, 2011), 36.

39. Jackson, *Thief*, 10.

40. David M. Kennedy, *Freedom from Fear: The American People in Depression and War, 1929–1945* (New York: Oxford University Press, 1999), 457, 788, 792.

41. Culver and Hyde, *American Dreamer*, 464, 477, 501.

42. Schapsmeier and Schapsmeier, *Henry A. Wallace*, 28.

43. Culver and Hyde, *American Dreamer*, 149.

44. Fitzgerald, *Business of Breeding*, 223.

45. George Gunset, "Dupont To Buy Pioneer Hi-bred as Agribusiness Mergers Heat Up," *Chicago Tribune*, March 16, 1999, accessed November 3, 2021, http://articles.chicagotribune.com/1999-03-16/business/9903160199_1_seed-corn-dekalb-genetics-pioneer-hi-bred-international.

46. Allan B. Bogue, "Changes in Mechanical and Plant Technology: The Corn Belt, 1910–1940," *Journal of Economic History* 43 (1983): 1–25; here, 11, 19.

47. Wayne D. Rasmussen, "Advances in American Agriculture: The Mechanical Tomato Harvester as a Case Study," *Technology and Culture* 9 (1958), 531–43; here, 540.

48. Jim Hightower, *Hard Tomatoes, Hard Times: A Report of the Agribusiness Accountability Project on the Failure of America's Land Grant College Complex* (Cambridge, MA: Schenkman, 1973), esp. 46–48.

49. Douglas Robinson and Nina Medlock, "Diamond v. Chakrabarty: A Retrospective

on 25 Years of Biotech Patents," *Intellectual Property and Technology Law Journal* 17, no. 10 (October 2005): 12-15.

50. United States Patent and Trademark Office, Performance and Accountability Report, Fiscal Year 2016, p. 180, accessed November 3, 2021, https://www.uspto.gov/sites/default/files/documents/USPTOFY16PAR.pdf.

51. Frederick H. Buttel, "Ever Since Hightower: The Politics of Agricultural Research Activism in the Molecular Age," *Agriculture and Human Values* 22 (2005), 275-83.

52. Marie-Monique Robin, *The World According to Monsanto: Pollution, Corruption, and the Control of the World's Food Supply* (New York: New Press, 2010), 3. See also Bartow J. Elmore, *Seed Money: Monsanto's Past and Our Food Future* (New York: Norton, 2021).

53. Emily Marden, "The Neem Tree Patent: International Conflict over the Commodification of Life," *Boston College International and Comparative Law Review* 22 (1999), 279-95; here, 293.

54. Marden, "Neem Tree Patent," 293.

55. Ted Genoways, "Corn Wars: The Farm-by-Farm Fight between China and the United States to Dominate the Global Food Supply," *New Republic* 246, no. 9/10 (September 2015): 26-37.

56. I am grateful to Andreas Bettray for pointing me to this information.

57. Jack Kloppenburg, "Impeding Dispossession, Enabling Repossession: Biological Open Source and the Recovery of Seed Sovereignty," *Journal of Agrarian Change* 10 (2010): 367-88.

58. Shalini Randeria, "Rechtspluralismus und überlappende Souveränitäten: Globalisierung und der 'listige Staat' in Indien," *Soziale Welt* 57 (2006): 229-58; here, 237n.

59. Carl M. Cannon, Lou Dubose, and Jan Reid, *Boy Genius: Karl Rove, the Architect of George W. Bush's Remarkable Political Triumphs* (New York: PublicAffairs, 2005), 41-47.

60. Norman E. Borlaug, "The Green Revolution Revisited and the Road Ahead," 19, 17, accessed June 20, 2017, https://www.nobelprize.org/nobel_prizes/peace/laureates/1970/borlaug-lecture.pdf.

61. Christopher Williams, "The Frankenstein Merger: How Bayer's Bid for Monsanto Could Create a Monster," *The Telegraph*, May 28, 2016, accessed November 3, 2021, http://www.telegraph.co.uk/business/2016/05/28/the-frankenstein-merger-how-bayers-bid-for-monsanto-could-create/.

62. Gary Toenniessen, Akinwumi Adesina, and Joseph DeVries, "Building an Alliance for a Green Revolution in Africa," *Annals of the New York Academy of Sciences* 1136 (2008): 233-42; here, 241.

63. P. E. Rajasekharan, "Gene Banking for *Ex Situ* Conservation of Plant Genetic Resources," in Bir Bahadur et al., eds., *Plant Biology and Biotechnology*, vol. 2: *Plant Genomics and Biotechnology*, 445-59 (New Delhi: Springer India, 2015), 446.

64. See Harwood, *Europe's Green Revolution*.

65. L. A. Tatum, "The Southern Corn Leaf Blight Epidemic," *Science* 171, no. 3976 (March 19, 1971): 1113-16.

66. Kloppenburg, *First the Seed*, 93. See also Frank Kutka, "Open-Pollinated vs. Hybrid Maize Cultivars," *Sustainability* 3 (2011): 1531-54.

29: ASWAN DAM

1. World Bank Archives, box 1376407, folder "Administration II," A. H. Steenbergen N.V., Summary of Evaluation Factors for Tile Laying Machines, presented to His Excellency, the Minister of Irrigation, Cairo, Egypt, June 1972, pp. 3, 2.

2. World Bank Archives, "Summary," 2 (quotation), 5.

3. World Bank Archives, "Summary," 4, 6 (quotations).

4. See Yoram Meital, "The Aswan High Dam and Revolutionary Symbolism in Egypt," in Haggai Erlich and Israel Gershoni, eds., *The Nile: Histories, Cultures, Myths*, 219–26 (Boulder, CO: Lynne Rienner, 2000).

5. Nancy Y. Reynolds, "Building the Past: Rockscapes and the Aswan High Dam in Egypt," in Alan Mikhail, ed., *Water on Sand: Environmental Histories of the Middle East and North Africa*, 181–205 (New York: Oxford University Press, 2013), 181.

6. Reynolds, "Building the Past," 185.

7. See Richard White, *The Organic Machine: The Remaking of the Columbia River* (New York: Hill and Wang, 1995).

8. Timothy Mitchell, *Rule of Experts: Egypt, Techno-Politics, Modernity* (Berkeley: University of California Press, 2002), 34.

9. D. A. Brading and Harry E. Cross, "Colonial Silver Mining: Mexico and Peru," *Hispanic American Historical Review* 52 (1972): 545–79, here, 554; and Kendall W. Brown, *A History of Mining in Latin America: From the Colonial Era to the Present* (Albuquerque: University of New Mexico Press, 2012), 21.

10. Sven Beckert, *Empire of Cotton: A New History of Global Capitalism* (London: Allen Lane, 2014), 132.

11. See David E. Nye, *American Technological Sublime* (Cambridge, MA: MIT Press, 1994).

12. Terje Tvedt, *The River Nile in the Age of the British: Political Ecology and the Quest for Economic Power* (London: I. B. Tauris, 2004), 81n.

13. Philipp Nicolas Lehmann, "Infinite Power to Change the World: Hydroelectricity and Engineered Climate Change in the Atlantropa Project," *American Historical Review* 121 (2016): 70–100; here, 90n.

14. Alexander Gall, *Das Atlantropa-Projekt: Die Geschichte einer gescheiterten Vision. Herman Sörgel und die Absenkung des Mittelmeers* (Frankfurt: Campus, 1998), 33.

15. Donald C. Jackson, "Engineering in the Progressive Era: A New Look at Frederick Haynes Newell and the U.S. Reclamation Service," *Technology and Culture* 34 (1993): 539–74; here, 556, 559 (quotation).

16. M. A. Abu-Zeid, and F. Z. El-Shibini, "Egypt's High Aswan Dam," *International Journal of Water Resources Development* 13 (1997): 209–17; here, 215.

17. See Harsh K. Gupta and B. K. Rastogi, *Dams and Earthquakes* (Amsterdam: Elsevier Scientific, 1976).

18. Mitchell, *Rule of Experts*, 19–31.

19. Claire Cookson-Hills, "The Aswan Dam and Egyptian Water Control Policy, 1882–1902," *Radical History Review* 116 (Spring 2013): 59–85; here, 68.

20. Righter, *Battle*, 4, 8, 59n.

21. Julia Tischler, "Cementing Uneven Development: The Central African Federation and the Kariba Dam Scheme," *Journal of Southern African Studies* 40 (2014): 1047–64; here, 1048.

22. Rick Shine, *Cane Toad Wars* (Oakland: University of California Press, 2018), 37n.

23. The commission defines a large dam as "a dam with a height of 15 metres or greater from lowest foundation to crest or a dam between 5 metres and 15 metres impounding more than 3 million cubic metres." (http://www.icold-cigb.net/GB/world_register/general_synthesis.asp, accessed November 3, 2021.)

24. Karl August Wittfogel, *Oriental Despotism: A Comparative Study of Total Power* (New York: Vintage Books, 1981 [1957]). For intellectual context, see Rolando Minuti, "Oriental Despotism," *European History Online*, published by the Leibniz Institute of European History, Mainz, May 3, 2012, accessed November 3, 2021, http://www.ieg-ego.eu/minutir-2012-en.

25. Donald Worster, *Rivers of Empire: Water, Aridity, and the Growth of the American West* (New York: Oxford University Press, 1985), 266.

26. American Heritage Center, University of Wyoming, Laramie, Wyoming, USA, Collection 2129-81-5-26, Floyd E. Dominy Papers, box 30, folder "Professional File 1969—August Asia Trip," The Prospect of Water Resource Development in the Han River. Interview with Floyd E. Dominy, Commissioner of the Bureau of Reclamation, the United States, *Seoul Economic Daily Press*, September 5, 1969, p. 1 (quotations).

27. Brett Hansen, "Conquering the Arizona Desert: The Theodore Roosevelt Dam," *Civil Engineering* 78, no. 8 (August 2008): 44–45; here, 45.

28. Tiago Saraiva, "Fascist Modernist Landscapes: Wheat, Dams, Forests, and the Making of the Portuguese New State," *Environmental History* 21 (2016): 54–75; here, 64.

29. See Donald J. Pisani, *Water and American Government: The Reclamation Bureau, National Water Policy, and the West, 1902–1935* (Berkeley: University of California Press, 2002), 23–29.

30. David Blackbourn, *The Conquest of Nature: Water, Landscape, and the Making of Modern Germany* (New York: Norton, 2006), 207. See also Wolfgang König, "Der Ingenieur als Politiker: Otto Intze, Staudammbau und Hochwasserschutz im Einzugsbereich der Oder," *Technikgeschichte* 73 (2006): 27–46.

31. Chandra Mukerji, "The New Rome: Infrastructure and National Identity on the Canal du Midi," *Osiris* 24 (2009): 15–32; here, 26.

32. Quoted in James Lawrence Powell, *Dead Pool: Lake Powell, Global Warming, and the Future of Water in the West* (Berkeley: University of California Press, 2008), 144.

33. See Exodus 17:6.

34. Matthew 14:25.

35. Jackson, "Engineering," 547, 552, 562.

36. William D. Rowley, *The Bureau of Reclamation: Origins and Growth to 1945*, vol. 1 (Denver: Bureau of Reclamation, 2006), 35.

37. See Ewald Blocher, *Der Wasserbau-Staat: Die Transformation des Nils und das moderne Ägypten 1882–1971* (Paderborn: Schöningh, 2016), 68–80.

38. Cookson-Hills, "Aswan Dam," 74.

39. Blocher, *Wasserbau-Staat*, 218–31.

40. Jennifer L. Derr, "Drafting a Map of Colonial Egypt: The 1902 Aswan Dam, Historical Imagination, and the Production of Agricultural Geography," in Diana K. Davis and Edmund Burke III, eds., *Environmental Imaginaries of the Middle East and North Africa*, 136–57 (Athens: Ohio University Press, 2011), 141n.

41. William U. Chandler, *The Myth of TVA: Conservation and Development in the Tennessee Valley, 1933–1983* (Cambridge, MA: Ballinger, 1984), 34. See also Erwin C. Hargrove, *Prisoners of Myth: The Leadership of the Tennessee Valley Authority, 1933–1990* (Princeton, NJ: Princeton University Press, 1994), 19–41.

42. R. Douglas Hurt, *The Big Empty: The Great Plains in the Twentieth Century* (Tucson: University of Arizona Press, 2011), 181–83.

43. Asit K. Biswas and Cecilia Tortajada, "Development and Large Dams: A Global Perspective," *Water Resources Development* 17 (2001): 9–21; here, 10.

44. David Ekbladh, *The Great American Mission: Modernization and the Construction of an American World Order* (Princeton, NJ: Princeton University Press, 2010), 85. See also Christopher Sneddon, *Concrete Revolution: Large Dams, Cold War Geopolitics, and the US Bureau of Reclamation* (Chicago: University of Chicago Press, 2015); and Kiran Klaus Patel, *The New Deal: A Global History* (Princeton, NJ: Princeton University Press, 2016), 97–103.

45. Stephan F. Miescher, "'Nkrumah's Baby': The Akosombo Dam and the Dream of Development in Ghana, 1952–1966," *Water History* 6 (2014): 341–66.

46. Herodotus, *The Histories. A New Translation by Robin Waterfield* (Oxford: Oxford University Press, 1998), 98.

47. Gupta and Rastogi, *Dams and Earthquakes*, 73.

48. Erik Swyngedouw, *Liquid Power: Contested Hydro-Modernities in Twentieth-Century Spain, 1898–2010* (Cambridge, MA: MIT Press, 2015), 14; and International Commission on Large Dams, "Number of Dams by Country Members," accessed November 3, 2021, http://www.icold-cigb.net/article/GB/world_register/general_synthesis/number-of-dams-by-country-members.

49. Nick Cullather, "Damming Afghanistan: Modernization in a Buffer State," *Journal of American History* 89 (2002): 512–37; here, 512, 529 (quotation).

50. Quoted in Nancy Y. Reynolds, "City of the High Dam: Aswan and the Promise of Postcolonialism in Egypt," *City & Society* 29 (2017): 213–35, here, 220.

51. Ekbladh, *Great American Mission*, 209.

52. American Heritage Center, University of Wyoming, Laramie, Wyoming, USA, Collection 2129-81-5-26, Floyd E. Dominy Papers, box 15, folder "Professional File, 1961–1969—Middle East Water Problems," memorandum, Under Secretary, Department of the Interior to The Secretary, July 10, 1967.

53. Robert Parnero, *A South American "Great Lakes" System* (Croton-on-Hudson, NY: Hudson Institute, 1967).

54. See Klaus Gestwa, *Die Stalinschen Großbauten des Kommunismus: Sowjetische Technik- und Umweltgeschichte, 1948–1967* (Munich: Oldenbourg, 2010).

55. American Heritage Center, University of Wyoming, Laramie, Wyoming, USA, Collection 2129-81-5-26, Floyd E. Dominy Papers, box 22, folder "Professional File, 1959—Afghanistan File—Travel," Dominy to Crabb, December 22, 1959.

56. American Heritage Center, University of Wyoming, Laramie, Wyoming, USA, Collection 2129-81-5-26, Floyd E. Dominy Papers, box 22, folder "Professional File, 1959—Afghanistan File—Travel," Dominy to Loren, December 22, 1959.

57. American Heritage Center, University of Wyoming, Laramie, Wyoming, USA, Collection 2129-81-5-26, Floyd E. Dominy Papers, box 5, folder "Correspondence, 1961 1962—January," Commissioner of Reclamation to Secretary of the Interior, November 30, 1961, p. 1.

58. Martin Meredith, *The Fate of Africa: From the Hopes of Freedom to the Heart of Despair; A History of Fifty Years of Independence* (New York: PublicAffairs, 2005), 40–43. On the domestic background, see Silvia Borzutzky and David Berger, "Dammed If You Do, Dammed If You Don't: The Eisenhower Administration and the Aswan Dam," *Middle East Journal* 64 (2010): 84–102.

59. World Bank Archives, box 1376407, folder "Administration I," office memorandum, Arab Republic of Egypt—Credit 181-UAR, Nile Delta Drainage Project, Full Supervision Report, Haynes to Sicely, March 29, 1972, p. 2; and folder "Administration II," Richardson to Da Costa, July 26, 1972, p. 1.

60. David Gilmartin, *Blood and Water: The Indus River Basin in Modern History* (Oakland: University of California Press, 2015), 205.

61. Marc Reisner, *Cadillac Desert: The American West and Its Disappearing Water* (New York: Penguin, 1993), 229.

62. American Heritage Center, University of Wyoming, Laramie, Wyoming, USA, Collection 2129-81-5-26, Floyd E. Dominy Papers, box 30, folder "Professional File 1969—August Asia Trip," The Prospect of Water Resource Development in the Han River. Interview with Floyd E. Dominy, Commissioner of the Bureau of Reclamation, the United States, *Seoul Economic Daily Press*, September 5, 1969, p. 1. It is crucial to recognize that dam builders had learned a lot by the postwar years, as recent scholarship tends to depict

the frenzy of dam construction as an act of technological naiveté. In light of scathing critiques that "the idea that one solution fit everywhere was seriously mistaken," one should note that builders like Dominy were long past these truisms. (Robert S. Emmett and David E. Nye, *The Environmental Humanities: A Critical Introduction* [Cambridge, MA: MIT Press, 2017], 119.)

63. American Heritage Center, University of Wyoming, Laramie, Wyoming, USA, Collection 2129-81-5-26, Floyd E. Dominy Papers, box 15, folder "Professional File, 1961-1969—Middle East Water Problems," Potential for Reclamation Assistance in Middle East Water Problems, p. 2.

64. American Heritage Center, University of Wyoming, Laramie, Wyoming, USA, Collection 2129-81-5-26, Floyd E. Dominy Papers, box 30, folder "Professional File 1969—August Asia Trip," Floyd E. Dominy, Considerations Prerequisite to Implementation of the Pa Mong Project with Emphasis on Operation and Management, July 18, 1969, p. 7.

65. American Heritage Center, University of Wyoming, Laramie, Wyoming, USA, Collection 2129-81-5-26, Floyd E. Dominy Papers, box 30, folder "Professional File 1969—August Asia Trip," Floyd E. Dominy, Considerations Prerequisite to Implementation of the Pa Mong Project with Emphasis on Operation and Management, July 18, 1969, p. 6. See also Sneddon, *Concrete Revolution*, 102-24.

66. Ceri Peach and Richard Gale, "Muslims, Hindus, and Sikhs in the New Religious Landscape of England," *Geographical Review* 93 (2003): 469-90; here, 474.

67. Worster, *Rivers of Empire*, 201.

68. American Heritage Center, University of Wyoming, Laramie, Wyoming, USA, Collection 2129-81-5-26, Floyd E. Dominy Papers, box 29, folder "Professional File, 1968—June—Travel," Irrigation Moves into the 21st Century. Address by Commissioner of Reclamation Floyd E. Dominy, Department of the Interior, before the Symposium on "Space Age Irrigation," Huron, South Dakota, June 13, 1968, p. 7.

69. American Heritage Center, University of Wyoming, Laramie, Wyoming, USA, Collection 2129-81-5-26, Floyd E. Dominy Papers, box 5, folder "Correspondence, 1962—Feb.," Remarks Scheduled by Commissioner of Reclamation Floyd E. Dominy, United States Department of the Interior, before the 43rd Annual Convention, Associated General Contractors of America, Los Angeles, CA, March 1, 1962, p. 2.

70. Donald J. Pisani, "A Tale of Two Commissioners: Frederick Newell and Floyd Dominy," in *The Bureau of Reclamation: History Essays from the Centennial Symposium*, vol. 2, 634-50 (Denver: Bureau of Reclamation, 2008), 637.

71. John McPhee, *Encounters with the Archdruid* (New York: Farrar, Straus and Giroux, 1971), 170.

72. Ali A. Mazrui, "Towards the Year 2000," in Mazrui, ed., *General History of Africa*, vol. 8: *Africa since 1935*, 905-34 (Paris: UNESCO, 1999), 909.

73. John Darwin, *Unfinished Empire: The Global Expansion of Britain* (London: Allen Lane, 2012), 362.

74. Tom Little, *High Dam at Aswan: The Subjugation of the Nile* (London: Methuen, 1965), 83.

75. Ewald Blocher, "Pyramiden der Lebenden: Der Assuan-Hochdamm als Erinnerungsort im Zeitalter technischer Großplanung," in Frank Uekötter, ed., *Ökologische Erinnerungsorte*, 252-72 (Göttingen: Vandenhoeck & Ruprecht, 2014), 263-67.

76. Alia Mossallam, "'We Are the Ones Who Made This Dam "High"!' A Builders' History of the Aswan High Dam," *Water History* 6 (2014): 297-314; here, 305.

77. Mossallam, "'We Are the Ones,'" 308.

78. Christine Folch, "Surveillance and State Violence in Stroessner's Paraguay: Itaipú

Hydroelectric Dam, Archive of Terror," *American Anthropologist* 115 (2013): 44–57; here, 45n.

79. Leonardo Mazzei and Gianmarco Scuppa, "The Role of Communication in Large Infrastructure: The Bumbuna Hydroelectric Project in Post-Conflict Sierra Leone," World Bank Working Paper no. 84 (Washington, DC: World Bank, 2006), 4n; "Upgrade of Bumbuna Hydropower Plant Inch," *EIU ViewsWire*, November 26, 2013.

80. American Heritage Center, University of Wyoming, Laramie, Wyoming, USA, Collection 2129-81-5-26, Floyd E. Dominy Papers, box 18, folder "Professional File, 1963–1969—U.S. Committee on Large Dams," T. W. Mermel, Report on Participation in 30th Executive Committee Meeting, International Commission on Large Dams, February 3 to 6, 1963, Cairo, U.A.R., and Study Tour in the United Arab Republic, February 7 to 15, 1963, p. 32.

81. Hal K. Rothman, *The Greening of a Nation? Environmentalism in the United States since 1945* (Fort Worth, TX: Hartcourt Brace College, 1998), 33–48, 73–79.

82. For the concept of brute force technology, more effective rhetorically than analytically, see Paul R. Josephson, *Industrialized Nature: Brute Force Technology and the Transformation of the Natural World* (Washington, DC: Island Press, 2002). For Josephson's take on Aswan, see p. 17.

83. Joachim Radkau, *Nature and Power: A Global History of the Environment* (New York: Cambridge University Press, 2008), 95.

84. Mahmoud Abu-Zeid and M. B. A. Saad, "The Aswan High Dam, 25 Years On," *UNESCO Courier* 46 (May 1993): 37.

85. See Benjamin Brendel, "Moderne—Macht—Morbid: Dammbau, Gesundheitshilfe und die Konstruktion von Macht im Kontext der Bilharziosebekämpfung im Ägypten der 1960er und frühen 1970er Jahre," *NTM: Zeitschrift für Geschichte der Wissenschaften, Technik und Medizin* 25 (2017): 349–82.

86. Hesham Abd-El Monseff, Scot E. Smith, and Kamal Darwish, "Impacts of the Aswan High Dam after 50 Years," *Water Resources Management* 29 (2015): 1873–85; here, 1880, 1882n.

87. Reynolds, "Building the Past," 197.

88. Nicholas S. Hopkins, "Irrigation in Contemporary Egypt," in Alan K. Bowman and Eugene Rogan, eds., *Agriculture in Egypt: From Pharaonic to Modern Times (Proceedings of the British Academy 96*, 367–85 (Oxford: Oxford University Press, 1999), 367.

89. Blackbourn, *Conquest*, 247.

90. Bruce Cumings, *Korea's Place in the Sun: A Modern History* (New York: Norton, 2005), 296; and Karl Schlögel, *Das Sowjetische Jahrhundert: Archäologie einer untergegangenen Welt* (Munich: Beck, 2017), 116.

91. Abu-Zeid and El-Shibini, "Egypt's High Aswan Dam," 211n.

92. Blocher, *Wasserbau-Staat*, 339.

93. American Heritage Center, University of Wyoming, Laramie, Wyoming, USA, Collection 2129-81-5-26, Floyd E. Dominy Papers, box 13, folder "Professional File, 1961–1969—Foreign Visitors," memorandum, Chief, Division of Foreign Activities to Commissioner, July 9, 1965, p. 1.

94. American Heritage Center, University of Wyoming, Laramie, Wyoming, USA, Collection 2129-81-5-26, Floyd E. Dominy Papers, box 18, folder "Professional File, 1963–1969—U.S. Committee on Large Dams," T. W. Mermel, Report on Participation in 30th Executive Committee Meeting, International Commission on Large Dams, February 3 to 6, 1963, Cairo, U.A.R., and Study Tour in the United Arab Republic, February 7 to 15, 1963, p. 11.

95. Scott S. Smith, "A Revised Estimate of the Life Span of Lake Nasser," *Environmental Geology and Water Science* 15 (1990): 123–29; here, 127n.

96. Reisner, *Cadillac Desert*, 490.

97. See http://shdegypt.com/inner.php?type=About&id=6, accessed November 3, 2021.

98. Flavius Josephus, *Translation and Commentary*, ed. Steve Mason, vol. 1B: *Judean War 2* (Leiden: Brill, 2008), 301. It should be noted that Flavius Josephus's estimate is probably rather generous, as other accounts suggest a population of only three million.

99. Coleen A. Fox, Francis J. Magilligan, and Christopher S. Sneddon, "'You Kill the Dam, You Are Killing a Part of Me': Dam Removal and the Environmental Politics of River Restoration," *Geoforum* 70 (2016): 93–104; here, 93.

100. Fox, Magilligan, and Sneddon, "'You Kill the Dam,'" 94.

101. Richard Stone, "Three Gorges Dam: Into the Unknown," *Science* 321, no. 5889 (August 1, 2008): 628–32; here, 629.

102. Christopher L. Pallas, *Transnational Civil Society and the World Bank: Investigating Civil Society's Potential to Democratize Global Governance* (Basingstoke: Palgrave Macmillan, 2013), 137.

103. Tom Phillips, "Avatar Director James Cameron Joins Amazon Tribe's Fight to Halt Giant Dam," *The Guardian*, April 18, 2010, accessed November 3, 2021, https://www.theguardian.com/world/2010/apr/18/avatar-james-cameron-brazil-dam.

104. Rowley, *Bureau of Reclamation*, 41.

105. Arun—III Hydropower Project, accessed November 3, 2021, https://www.power-technology.com/projects/arun-iii-hydropower-project/.

106. Pallas, *Transnational Civil Society*, 133, 138.

107. Paul Breeze, *Power Generation Technologies*, 2nd ed. (Amsterdam: Elsevier, 2014), 175.

108. Adebayo Adedeji, "Comparative Strategies of Economic Decolonization in Africa," in Mazrui, *General History of Africa*, 8:393–431, here, 428.

109. See Amanda E. Wooden, "Kyrgyzstan's Dark Ages: Framing and the 2010 Hydroelectric Revolution," *Central Asian Survey* 33 (2014): 463–81.

110. Folch, "Surveillance," 53.

111. Jacob Blanc, *Before the Flood: The Itaipu Dam and the Visibility of Rural Brazil* (Durham, NC: Duke University Press, 2019).

112. See Maurits W. Ertsen, *Improvising Planned Development on the Gezira Plain, Sudan, 1900–1980* (Basingstoke: Palgrave Macmillan, 2016).

113. Blocher, *Wasserbau-Staat*, 337, 341.

114. Rawia Tawfik, "Reconsidering Counter-Hegemonic Dam Projects: The Case of the Grand Ethiopian Renaissance Dam," *Water Policy* 18 (2016): 1033–52; here, 1047.

115. See Aibek Zhupankhan, Kamshat Tussupova, and Ronny Berndtsson, "Could Changing Power Relationships Lead to Better Water Sharing in Central Asia?" *Water* 9, no. 2 (2017): 139. For historical background see Maya K. Peterson, *Pipe Dreams: Water and Empire in Central Asia's Aral Sea Basin* (New York: Cambridge University Press, 2019).

116. "Rusumo Power Project Construction a Step Closer," *Energy Monitor Worldwide*, November 12, 2016.

117. Adegboyega Adeniran and Katherine Daniell, "The Attempt to Replenish Lake Chad's Water May Fail Again: Here's Why," *The Conversation*, October 5, 2017, accessed November 3, 2021, http://theconversation.com/the-attempt-to-replenish-lake-chads-water-may-fail-again-heres-why-84653.

118. Kenneth Rapoza, "In Russia, The World's Largest Lake Takes On the World Bank and Mongolian Power Build-Up," *Forbes*, April 7, 2017, accessed November 3, 2021, https://

www.forbes.com/sites/kenrapoza/2017/04/07/in-russia-the-worlds-largest-lake-takes-on-the-world-bank-and-mongolian-power-build-up/#2a5de4634df5.

119. Cullather, "Damming Afghanistan," 536.

120. Peter Reina, "Convoy Delivers Hydroelectric Turbine to Southern Afghanistan," *Engineering News-Record* 261, no. 8 (September 15, 2008): 16; and Alastair Leithead, "UK Troops in Huge Turbine Mission," *BBC News*, September 2, 2008, accessed November 3, 2021, http://news.bbc.co.uk/1/hi/uk/7593901.stm.

121. Mark Urban, "What Went Wrong with Afghanistan Kajaki Power Project?" *BBC News*, June 28, 2011, accessed November 3, 2021, http://www.bbc.co.uk/news/13925886.

30: THE RICE-EATING RUBBER TREE

1. My discussion of the rice-eating rubber dream is based on Michael R. Dove, "Rice-Eating Rubber and People-Eating Governments: Peasant versus State Critiques of Rubber Development in Colonial Borneo," *Ethnohistory* 43, no. 1 (Winter 1996): 33–63. For an updated version of his argument, see Dove, *Banana Tree*, 122–44.

2. John Loadman, *Tears of the Tree: The Story of Rubber—A Modern Marvel* (New York: Oxford University Press, 2005), 32n.

3. Corey Ross, *Ecology and Power in the Age of Empire: Europe and the Transformation of the Tropical World* (Oxford: Oxford University Press, 2017), 104–6.

4. Boris Fausto, *A Concise History of Brazil* (New York: Cambridge University Press, 2006), 176.

5. Seth Garfield, *In Search of the Amazon: Brazil, the United States, and the Nature of a Region* (Durham, NC: Duke University Press, 2013), 18.

6. Ross, *Ecology and Power*, 120.

7. Clifford Geertz, *Agricultural Involution: The Process of Ecological Change in Indonesia* (Berkeley: University of California Press, 1963), 60.

8. Ross, *Ecology and Power*, 122.

9. Vinson H. Sutlive Jr., *The Iban of Sarawak* (Arlington Heights, IL: AHM, 1978), 128, 129.

10. John H. Drabble, *Malayan Rubber: The Interwar Years* (London: Macmillan, 1991), 119.

11. John H. Drabble, *An Economic History of Malaysia, c. 1800–1990: The Transition to Modern Economic Growth* (Basingstoke: Macmillan, 2000), 108.

12. P. T. Bauer, *The Rubber Industry: A Study in Competition and Monopoly* (London: Longmans, 1948), 209.

13. Dove, *Banana Tree*, 133.

14. Dove, *Banana Tree*, 141.

15. Dove, "Rice-Eating Rubber," 35.

16. Vinson H. Sutlive Jr., *Tun Jugah of Sarawak: Colonialism and Iban Response* (Kuala Lumpur: Penerbit Fajar Bakti, 1992), 33, 86.

17. Dove, "Rice-Eating Rubber," 35.

18. Michael T. Taussig, *The Devil and Commodity Fetishism in South America* (Chapel Hill: University of North Carolina Press, 1980), 94.

19. Eduardo Galeano, *Open Veins of Latin America: Five Centuries of the Pillage of a Continent* (New York: Monthly Review Press, 1997 [1971]).

20. Robert M. Levine and John J. Crocitti, eds., *The Brazil Reader: History, Culture, Politics* (Durham, NC: Duke University Press, 1999), 224.

21. Louis A. Pérez Jr., "Dependency," in Michael J. Hogan and Thomas G. Paterson, eds., *Explaining the History of American Foreign Relations*, 2nd ed., 162–75. New York: Cambridge University Press, 2004), 164.

22. See Mary Nolan, "Where Was the Economy in the Global Sixties?" in Chen Jian et al., eds., *The Routledge Handbook of the Global Sixties: Between Protest and Nation-Building*, 315-27 (London: Routledge, 2018).

23. Ricardo Bielschowsky, "Sixty Years of ECLAC: Structuralism and Neo-Structuralism," *CEPAL Review* 97 (2009): 171-92; here, 174n.

24. Immanuel Wallerstein, *World-Systems Analysis: An Introduction* (Durham, NC: Duke University Press, 2004), 11.

25. See, for instance, Oscar Guardiola-Rivera, *What If Latin America Ruled the World? How the South Will Take the North into the 22nd Century* (London: Bloomsbury, 2011).

26. Andrés Velasco, "Dependency Theory," *Foreign Policy*, no. 133 (November 2002): 44-45; here, 44. For Velasco's subsequent career, see pp. 232, 236.

27. See Ilan Kapoor, "Capitalism, Culture, Agency: Dependency Theory versus Postcolonial Theory," *Third World Quarterly* 23 (2002): 647-64.

28. Pérez, "Dependency," 167.

29. James L. Dietz, "Dependency Theory: A Review Article," *Journal of Economic Issues* 14 (1980): 751-58; here, 755.

30. Velasco, "Dependency Theory," 44.

31. Dietz, "Dependency Theory," 751.

32. Andre Gunder Frank, *Capitalism and Underdevelopment in Latin America: Historical Studies of Chile and Brazil* (New York: Monthly Review Press, 1969), xi.

33. André Gunder Frank, "Dependence Is Dead, Long Live Dependence and the Class Struggle: An Answer to Critics," *Latin American Perspectives* 1 (1974): 87-106; here, 102.

34. Fernando Henrique Cardoso and Enzo Faletto, *Dependency and Development in Latin America* (Berkeley: University of Berkeley Press, 1979), 176.

35. Cardoso and Faletto, *Dependency*, xxiii.

36. Peter Evans, "From Situations of Dependency to Globalized Social Democracy," *Studies in Comparative International Development* 44 (2009), 318-36; here, 321.

37. For an overview of his biography, see Ted G. Goertzel, *Fernando Henrique Cardoso: Reinventing Democracy in Brazil* (Boulder, CO: Lynne Rienner, 1999).

38. Patrick Heller, Dietrich Rueschemeyer, and Richard Snyder, "Dependency and Development in a Globalized World: Looking Back and Forward," *Studies in Comparative International Development* 44 (2009): 287-95; here, 291.

39. Fernando Henrique Cardoso, "Problems of Social Change, Again?" *International Sociology* 2 (1987): 177-87; here, 185.

40. Cardoso, "Problems of Social Change," 186.

41. Goertzel, *Fernando Henrique Cardoso*, 157.

42. Edwin Williamson, *The Penguin History of Latin America*, rev. ed. (London: Penguin, 2009), 580.

43. Velasco, "Dependency Theory," 44.

44. Fernando Henrique Cardoso, "The Consumption of Dependency Theory in the United States," *Latin American Research Review* 12 (1977): 7-24.

45. Geertz, *Agricultural Involution*, 123.

46. Ross, *Ecology and Power*, 124-26, 129 (quotation).

47. Dove, *Banana Tree*, 143.

48. Nancy Lee Peluso, "Rubber Erasures, Rubber Producing Rights: Making Racialized Territories in West Kalimantan, Indonesia," *Development and Change* 40 (2009): 47-80.

49. Dove, "Rice-Eating Rubber," 37.

50. Ross, *Ecology and Power*, 133.

51. Dove, *Banana Tree*, 6.

52. Ross, *Ecology and Power*, 84.

53. Douglas Southgate and Lois Roberts, *Globalized Fruit, Local Entrepreneurs: How One Banana-Exporting Country Achieved Worldwide Reach* (Philadelphia: University of Pennsylvania Press, 2016), 2.

54. Nicola Vetter, *Ludwig Roselius: Ein Pionier der deutschen Öffentlichkeitsarbeit* (Bremen: Hauschild, 2002), 66.

55. I have discussed these events more extensively in Frank Uekötter, "Recollections of Rubber," in Dominik Geppert and Frank Lorenz Müller, eds., *Sites of Imperial Memory: Commemorating Colonial Rule in the Nineteenth and Twentieth Century*, 243-65 (Manchester: Manchester University Press, 2015).

56. Karl Fischer, *Blutgummi: Roman eines Rohstoffes* (Berlin: Büchergilde Gutenberg, 1938), 243.

57. See Kevin Niebauer, "The Endangered Amazon Rain Forest in the Age of Ecological Crisis," in Frank Uekötter, ed., *Exploring Apocalyptica: Coming to Terms with Environmental Alarmism*, 107-28 (Pittsburgh, PA: University of Pittsburgh Press, 2018). See also Kevin Niebauer, *Regenwald und ökologische Krise: Die Globalisierung Amazoniens im 20. Jahrhundert* (Frankfurt: Campus, 2021).

58. Andrew Revkin, *The Burning Season: The Murder of Chico Mendes and the Fight for the Amazon Rain Forest* (Washington, DC: Island Press, 2004), 203.

59. Garfield, *In Search of the Amazon*, 225.

60. David S. Salisbury, "Extractive Reserves," in Barney Warf, ed., *Encyclopedia of Geography*, vol. 2, 1072-73 (Thousand Oaks, CA: Sage, 2010), 1072.

61. Andrew Rivkin, *The Burning Season: The Murder of Chico Mendes and the Fight for the Amazon Rain Forest* (Washington, DC: Island Press, 2004), 10, 14, 280-88.

62. Quoted in Rivkin, *Burning Season*, 261.

63. See Greg Grandin, *Fordlandia: The Rise and Fall of Henry Ford's Forgotten Jungle City* (New York: Henry Holt, 2009).

64. Garfield, *In Search of the Amazon*, 171, 174, 187.

65. Dove, "Rice-Eating Rubber," 34.

PART VII
THE AGE OF CATASTROPHE

1. See Eric J. Hobsbawm, *The Age of Extremes: A History of the World, 1914-1991* (New York: Vintage Books, 1996).

2. Eric J. Hobsbawm, *How to Change the World: Marx and Marxism, 1840-2011* (London: Little, Brown, 2011).

3. Hobsbawm, *Age of Extremes*, 261.

4. Rudolf Bahro, *The Alternative in Eastern Europe*, trans. David Fernbach (London: NLB, 1978), 11.

5. Amartya Sen, *Development as Freedom* (Oxford: Oxford University Press, 2001), 16.

6. On the construction of obesity, see Nicolas Rasmussen, *Fat in the Fifties: America's First Obesity Crisis* (Baltimore: Johns Hopkins University Press, 2019).

7. For a more elaborate discussion of this point, see Frank Uekötter, "Memories in Mud: The Environmental Legacy of the Great War," in Richard P. Tucker et al., eds., *Environmental Histories of the First World War*, 278-95 (Cambridge: Cambridge University Press, 2018).

8. Charles S. Maier, "Consigning the Twentieth Century to History: Alternative Narratives for the Modern Era," *American Historical Review* 105 (2000): 807-31; here, 829.

31: HOLODOMOR

1. Stephen Kotkin, *Stalin*, vol. 2: *Waiting for Hitler, 1928-1941* (London: Allen Lane, 2017), 115-17.

2. Stéphane Courtois et al., *The Black Book of Communism: Crimes, Terror, Repression* (Cambridge, MA: Harvard University Press, 1999), 167. The French original was published in 1997. Quantifying famine deaths is an endeavor fraught with uncertainties, but the number likely represents the upper limit of serious estimates. Timothy Snyder suggested 3.3 million victims in Soviet Ukraine (of which 3 million were ethnic Ukrainians) and 5.5 million in the entire Soviet Union. (Timothy Snyder, *Bloodlands: Europe Between Hitler and Stalin* [London: Vintage, 2015], 53.)

3. S. A. Smith, *Russia in Revolution: An Empire in Crisis, 1890 to 1928* (Oxford: Oxford University Press, 2017), 232.

4. Alec Nove, *An Economic History of the USSR 1917-1991*, 3rd ed. (London: Penguin, 1992), 53-55.

5. Manfred Hildermeier, *Geschichte der Sowjetunion 1917-1991: Entstehung und Niedergang des ersten sozialistischen Staates*, 2nd ed. (Munich: Beck, 2017), 167, 245.

6. Orlando Figes, *Revolutionary Russia, 1891-1991* (London: Penguin, 2014), 207.

7. Tiago Saraiva, "Breeding Europe: Crop Diversity, Gene Banks, and Commoners," in Nil Disco and Eda Kranakis, eds., *Cosmopolitan Commons: Sharing Resources and Risks across Borders*, 185-211 (Cambridge, MA: MIT Press, 2013), 192.

8. R. W. Davies and Stephen G. Wheatcroft, *The Years of Hunger: Soviet Agriculture, 1931-1933* (Basingstoke: Palgrave Macmillan, 2004), 437-39.

9. Kotkin, *Stalin*, 122.

10. Gerd Koenen, *Die Farbe Rot: Ursprünge und Geschichte des Kommunismus* (Munich: Beck, 2017), 905.

11. Koenen, *Farbe Rot*, 906.

12. Figes, *Revolutionary Russia*, 373, 375.

13. Stephan Merl, *Bauern unter Stalin: Die Formierung des sowjetischen Kolchossystems 1930-1941* (Berlin: Duncker & Humblot, 1990), 261.

14. Nove, *Economic History*, 397.

15. Robert Service, *Comrades. Communism: A World History* (London: Pan Macmillan, 2008), 6.

16. R. W. Davies and Stephen G. Wheatcroft, *The Years of Hunger: Soviet Agriculture, 1931-1933* (Basingstoke: Palgrave Macmillan, 2004), 436.

17. Nove, *Economic History*, 397.

18. Merl, *Bauern unter Stalin*, 470

19. Service, *Comrades*, p. 296.

20. André Steiner, *Von Plan zu Plan: Eine Wirtschaftsgeschichte der DDR* (Berlin: Aufbau, 2007), 131n.

21. Tehila Sasson, "Ethiopia, 1983-1985: Famine and the Paradoxes of Humanitarian Aid," accessed November 3, 2021, http://wiki.ieg-mainz.de/ghra/articles/sasson-ethiopia.

22. Frank Dikötter, *Mao's Great Famine: The History of China's Most Devastating Catastrophe, 1958-1962* (New York: Walker, 2010), x. For a less personalistic interpretation, see Felix Wemheuer, *Famine Politics in Maoist China and the Soviet Union* (New Haven, CT: Yale University Press, 2014).

23. Martyn Rady, *Romania in Turmoil: A Contemporary History* (London: I. B. Tauris, 1992), 68, 119.

24. Norman Davies, *Heart of Europe: The Past in Poland's Present* (Oxford: Oxford University Press, 2001), 8, 11.

25. Kotkin, *Stalin*, 97.

26. Karl Marx and Friedrich Engels, *The Communist Manifesto*, with an introduction and notes by Gareth Stedman Jones (London: Penguin, 2004), 224.

27. Karl Schlögel, *Das Sowjetische Jahrhundert: Archäologie einer untergegangenen Welt* (Munich: Beck, 2017), 474.

28. See World Bank Report No. 25920-UA: Project Appraisal Document on a Proposed Loan in the Amount of US$193.15 Million to Ukraine for a Rural Land Titling and Cadastre Development Project, May 30, 2003, p. 2, accessed November 3, 2021, http://documents.worldbank.org/curated/en/286051468779357257/pdf/259201UA1Rural1tre0Dev1r20031011611.pdf.

29. Georgii Kas'ianov, "The Holodomor and the Building of a Nation," *Russian Social Science Review* 52, no.3 (2011): 71–93; here, 77.

30. Christopher W. Morris, "Introduction," in Morris, ed., *Amartya Sen*, 1–12 (New York: Cambridge University Press, 2010), 1n. The death toll of the Bengal famine is from Maria Misra, *Vishnu's Crowded Temple: India since the Great Rebellion* (London: Allen Lane, 2007), 216.

31. Amartya Sen, *Poverty and Famines: An Essay on Entitlement and Deprivation* (Oxford: Clarendon Press, 1988 [1981]), 58, 80 (quotation).

32. Sen, *Poverty and Famines*, 45.

33. Stephen Devereux, "Sen's Entitlement Approach: Critiques and Counter-Critiques," in Devereux, ed., *The New Famines: Why Famines Persist in an Era of Globalization*, 66–89 (London: Routledge, 2007), 67.

34. Fernand Braudel, *The Structures of Everyday Life: The Limits of the Possible* (vol. 1 of *Civilization and Capitalism*) (London: Collins, 1981), 73.

35. Braudel, *Structures of Everyday Life*, 77.

36. David Gilmour, *The Pursuit of Italy: A History of a Land, Its Regions and Their Peoples* (London: Penguin, 2012), 20.

37. Dagomar Degroot, *The Frigid Golden Age: Climate Change, the Little Ice Age, and the Dutch Republic, 1560–1720* (New York: Cambridge University Press, 2018), 126.

38. Genesis 42:1–8.

39. Fred Singleton, *A Short History of Finland*, 2nd ed. (Cambridge: Cambridge University Press, 2005), 86.

40. Cormac Ó Gráda, *Famine: A Short History* (Princeton, NJ: Princeton University Press, 2009), 36.

41. Geheimes Staatsarchiv Berlin I. HA Rep. 81 Dresden no. 341, report of April 24, 1897.

42. Alexander James Hutchinson Russell, *A Memorandum on the Epidemiology of Cholera* (Geneva: League of Nations, 1925), 42.

43. Robert Ross, *A Concise History of South Africa*, 2nd ed. (Cape Town: Cambridge University Press, 2008), 155.

44. See Sasson, "Ethiopia."

45. Cormad Ó Gráda, "Making Famine History," *Journal of Economic Literature* 45 (2007): 5–38; here, 18.

46. Geheimes Staatsarchiv Berlin I. HA Rep. 81 Gesandtschaft Dresden nach 1807 no. 269, report of February 7, 1892, pp. 1, 2.

47. Sen, *Poverty and Famines*, p. 162.

48. Amartya Sen, *Development as Freedom* (Oxford: Oxford University Press, 2001), 16.

49. Ó Gráda, *Famine*, 11.

50. Nove, *Economic History*, 176.

51. Kendrick A. Clements, *Hoover, Conservation, and Consumerism: Engineering the Good*

Life (Lawrence: University Press of Kansas, 2006), 29; and Johan den Hertog, "The Commission for Relief in Belgium and the Political Diplomatic History of the First World War," *Diplomacy and Statecraft* 21 (2010): 593-613.

52. Daniel R. Maul, "Appell an das Gewissen—Fridtjof Nansen und die Russische Hungerhilfe 1921-23," *Themenportal Europäische Geschichte*, 2011, accessed November 3, 2021, www.europa.clio-online.de/essay/id/artikel-3604.

53. Davis and Wheatcroft, *Years of Hunger*, 440.

54. Elena Zubkova, *Russia after the War: Hopes, Illusions, Disappointments, 1945-1957* (Armonk, NY: M. E. Sharpe, 1998), 40.

55. Frances Moore Lappé, Joseph Collins, and Peter Rosset, *World Hunger: Twelve Myths*, 2nd ed. (New York: Grove Press, 1998), 2.

56. FAO Regional Office for Europe and Central Asia, "Interview: Agriculture in Ukraine—What Does the Future Hold?" accessed November 3, 2021, http://www.fao.org/europe/news/detail-news/en/c/447159/.

57. P. S. Mstislavskii, "Hunger," in A. M. Prokhorov, ed., *Great Soviet Encyclopedia: A Translation of the Third Edition*, vol. 7, 555-56 (New York: Macmillan, 1975), 556.

58. David R. Marples, *Heroes and Villains: Creating National History in Contemporary Ukraine* (Budapest: Central European University Press, 2007), 36.

59. Anne Applebaum, *Red Famine: Stalin's War on Ukraine* (London: Penguin, 2018), 347n.

60. Robert Conquest, *The Harvest of Sorrow: Soviet Collectivization and the Terror Famine* (London: Hutchinson, 1986); and Olga Andriewsky, "Towards a Decentred History: The Study of the Holodomor and Ukrainian Historiography," *East/West: Journal of Ukrainian Studies* 2, no. 1 (2015): pp. 18-52; here, 25n. On the reception of Conquest's book, see Frank Sysyn, "Thirty Years of Research on the Holodomor: A Balance Sheet," *East/West: Journal of Ukrainan Studies* 2, no. 1 (2015): 3-16; here, 5n.

61. Marples, *Heroes*, 40.

62. David R. Marples, "Ethnic Issues in the Famine of 1932-1933 in Ukraine," *Europe-Asia Studies* 61 (2009): 505-18.

63. See Anna Kaminsky, ed., *Erinnerungsorte an den Holodomor 1932/33 in der Ukraine* (Leipzig: Bundesstiftung zur Aufarbeitung der SED-Diktatur, 2008).

64. Geert Mak, *Amsterdam: A Brief Life of the City* (London: Harvill Press, 1999), 284.

65. Andrew G. Newby and Timo Myllyntaus, "'The Terrible Visitation': Famine in Finland and Ireland, 1845 to 1868," in Declan Curran, Lubomyr Luciuk, and Andrew G. Newby, eds., *Famines in European Economic History: The Last Great European Famines Reconsidered*, 145-65 (London: Routledge, 2015), 158.

66. See Graham Auman Pitts, "'Make Them Hated in All of the Arab Countries': France, Famine, and the Creation of Lebanon," in Richard P. Tucker et al., eds., *Environmental Histories of the First World War*, 175-90 (New York: Cambridge University Press, 2018), and Zachary J. Foster, "Why Are Modern Famines so Deadly? The First World War in Syria and Palestine," in Tucker et al., *Environmental Histories*, 191-207.

67. See Christian Noack, Lindsay Janssen, and Vincent Comerford, eds., *Holodomor and Gorta Mór: Histories, Memories and Representations of Famine in Ukraine and Ireland* (London: Anthem Press, 2012).

68. Andriewsky, "Towards a Decentred History," 21. Similarly, Kas'ianov, "Holodomor," 82.

69. Conquest, *Harvest*, 3.

70. John-Paul Himka, "Encumbered Memory: The Ukrainian Famine of 1932-33," *Kritika: Explorations in Russian and Eurasian History* 14 (2013): 411-36; here, 420, 426;

Kas'ianov, "Holodomor," 86; and Jeremy Smith, *Red Nations: The Nationalities Experience in and after the USSR* (New York: Cambridge University Press, 2013), 110.

71. Mike Davis, *Late Victorian Holocausts: El Niño Famines and the Making of the Third World* (London: Verso, 2001). For his rationale, see p. 22.

72. Snyder, *Bloodlands*, 51.

73. Justin Trudeau, "Statement by the Prime Minister of Canada on Holocaust Memorial Day, Ottawa," November 25, 2017, accessed November 3, 2021, https://pm.gc.ca/eng/news/2017/11/25/statement-prime-minister-canada-holodomor-memorial-day.

74. Figes, *Revolutionary Russia*, 216.

75. Himka, "Encumbered Memory," 413, 427 (quotation). For a recent assessment of the debate, see the roundtable on Soviet famines in *Contemporary European History* 27 (2018): 432–81.

76. Kas'ianov, "Holodomor," 87; and Applebaum, *Red Famine*, 359.

77. Robert Kindler, *Stalins Nomaden: Herrschaft und Hunger in Kasachstan* (Hamburg: Hamburger Edition, 2014), 11. See also Sarah Cameron, *The Hungry Steppe: Famine, Mass Violence and the Making of Soviet Kazakhstan* (Ithaca, NY: Cornell University Press, 2018).

78. See Abdulaziz H. Al Fahad, "Rootless Trees: Genealogical Politics in Saudi Arabia," in Bernard Haykel, Thomas Hegghammer, and Stéphane Lacroix, eds., *Saudi Arabia in Transition: Insights on Social, Political, Economic and Religious Change*, 263–91 (New York: Cambridge University Press, 2015), 275–78.

79. Stanislav Kul'chyts'kyi, "The Holodomor of 1932–33: How and Why?" *East/West: Journal of Ukrainian Studies* 2, no. 1 (2015): 93–116; here, 93.

80. For a discussion of commemorative politics in Kazakhstan, see Kindler, *Stalins Nomaden*, 338–48.

32: THE PONTINE MARSHES

1. Jeffrey T. Schnapp, *A Primer of Italian Fascism* (Lincoln: University of Nebraska Press, 2000), 18.

2. Quoted in Philip Morgan, *Italian Fascism 1919–1945* (Basingstoke: Macmillan, 1995), 97.

3. John Whittam, *Fascist Italy* (Manchester: Manchester University Press, 1995), 60n.

4. See Steen Bo Frandsen, "'The War That We Prefer': The Reclamation of the Pontine Marshes and Fascist Expansion," *Totalitarian Movements and Political Religions* 2, no. 3 (2001): 69–82; here, 75.

5. Cesare Longobardi, *Land-Reclamation in Italy: Rural Revival in the Building of a Nation* (London: P. S. King & Son, 1936), 140–44.

6. Carl Ipsen, *Dictating Demography: The Problem of Population in Fascist Italy* (Cambridge: Cambridge University Press, 1996), 111.

7. Quoted in Guiseppe Tassinari, *Ten Years of Integral Land-Reclamation under the Mussolini Act* (Faenza: Fratelli Lega, 1939), 37. Littoria Province was formally created in 1934 (39.)

8. Federico D'Onofrio, "The Microfoundations of Italian Agrarianism: Italian Agricultural Economists and Fascism," *Agricultural History* 91 (2017): 369–96; here, 380.

9. Alexander Nützenadel, *Landwirtschaft, Staat und Autarkie: Agrarpolitik im faschistischen Italien (1922–1943)* (Tübingen: Niemeyer, 1997), 213–23; Adrian Lyttelton, *The Seizure of Power: Fascism in Italy, 1919–1929* (London: Weidenfeld and Nicolson, 1973), 193–95; and Anthony L. Cardoza, *Agrarian Elites and Italian Fascism: The Province of Bologna, 1901–1926* (Princeton, NJ: Princeton University Press, 1982).

10. Longobardi, *Land-Reclamation*, 3.

11. Ipsen, *Dictating Demography*, 73n, 114.

12. Quoted in Tassinari, *Ten Years*, 29.

13. Ipsen, *Dictating Demography*, 113.

14. Quoted in Tassinari, *Ten Years*, 39n.

15. Gaetano Salvemini, *Under the Axe of Fascism* (London: Victor Gollancz, 1936), 296. On the cinematic outfall of the Pontine Marshes project, see Frederico Caprotti and Maria Kaïka, "Producing the Ideal Fascist Landscape: Nature, Materiality and the Cinematic Representation of Land Reclamation in the Pontine Marshes," *Social and Cultural Geography* 9 (2008): 613–34.

16. Longobardi, *Land-Reclamation*, 141; and Frandsen, "'War That We Prefer,'" 78.

17. Ruth Sterling Frost, "The Reclamation of the Pontine Marshes," *Geographical Review* 24 (1934): 584–95; here, 595.

18. Sir E. J. Russell, "Agricultural Colonization in the Pontine Marshes and Libya," *Geographical Journal* 94 (1939): 273–89; here, 274. See also Roberta Pergher, *Mussolini's Nation-Empire: Sovereignty and Settlement in Italy's Borderlands, 1922–1943* (Cambridge: Cambridge University Press, 2018), 96.

19. Quoted in Tassinari, *Ten Years*, 36.

20. Robert Sallares, *Malaria and Rome: A History of Malaria in Ancient Italy* (Oxford: Oxford University Press, 2002), 181.

21. Sallares, *Malaria and Rome*, 185–90.

22. Frandsen, "'War That We Prefer,'" 75.

23. Frandsen, "'War That We Prefer,'" 75.

24. R. J. B. Bosworth, *Mussolini's Italy: Life under the Dictatorship 1915–1945* (New York: Penguin, 2006), 382.

25. See Federico Caprotti, "Scipio Africanus: Film, Internal Colonization and Empire," *Cultural Geographies* 16 (2009): 381–401.

26. Russell, "Agricultural Colonization."

27. Jürgen Osterhammel, *The Transformation of the World: A Global History of the Nineteenth Century* (Princeton, NJ: Princeton University Press, 2014), 368.

28. Osterhammel, *Transformation*, 369.

29. Russell, "Agricultural Colonization," 289. For views from Germany, see Patrick Bernhard, "Borrowing from Mussolini: Nazi Germany's Colonial Aspirations in the Shadow of Italian Expansionism," *Journal of Imperial and Commonwealth History* 41 (2013): 617–43.

30. Liesbeth van de Grift, "On New Land a New Society: Internal Colonisation in the Netherlands, 1918–1940," *Contemporary European History* 22 (2013): 609–26.

31. David Blackbourn, *The Conquest of Nature: Water, Landscape, and the Making of Modern Germany* (New York: Norton, 2006), 281–84; and Lars Armenda, "'Volk ohne Raum schafft Raum': Rassenpolitik und Propaganda im nationalsozialistischen Landgewinnungsprojekt an der schleswig-holsteinischen Westküste," *Informationen zur Schleswig-Holsteinischen Zeitgeschichte* 45 (2005): 4–31; here, 7, 9.

32. See Dan Diner, "Knowledge of Expansion: On the Geopolitics of Karl Haushofer," *Geopolitics* 4, no. 3 (1999): 161–88; and Hans-Adolf Jacobsen, "'Kampf um Lebensraum': Zur Rolle des Geopolitikers Karl Haushofer im Dritten Reich," *German Studies Review* 4 (1981): 79–104.

33. Ian Kershaw, *Hitler 1889–1936: Hubris* (London: Allen Lane, 1998), 248n.

34. Timothy Snyder, *Black Earth: The Holocaust as History and Warning* (London: Bodley Head, 2015), 9.

35. Nützenadel, *Landwirtschaft*, 30, 32, 63, 252.

36. Isabel Heinemann, "Wissenschaft und Homogenisierungsplanungen für Osteuropa: Konrad Meyer, der 'Generalplan Ost' und die Deutsche Forschungsgemeinschaft," in Heinemann and Patrick Wagner, eds., *Wissenschaft—Planung—Vertreibung: Neuordnungskonzepte und Umsiedlungspolitik im 20. Jahrhundert*, 45–72 (Stuttgart: Steiner, 2006).

37. Martin Broszat, "Soziale Motivation und Führer-Bindung des Nationalsozialismus," *Vierteljahreshefte für Zeitgeschichte* 18 (1970): 392–409; here, 405.

38. Patrick Bernhard, "Hitler's Africa in the East: Italian Colonialism as a Model for German Planning in Eastern Europe," *Journal of Contemporary History* 51 (2016): 61–90; here, 75, 81, 83. According to a stone tablet on the bell tower of Latina, Mussolini saw these towers as symbols of Fascist power and a sign showing settlers where they could turn to for help and justice. The tablet was still there when I visited Latina in April 2018. There was no commentary except for the flags of the Italian republic and the European Union that adorn government buildings in Italy. For a more sophisticated interpretation, see Mia Fuller, "Tradition as a Means to the End of Tradition: Farmers' Houses in Italy's Fascist-Era New Towns," in Nezar AlSayyad, ed., *The End of Tradition?* 171–86 (London: Routledge, 2004), 175–77.

39. R. Louis Gentilcore, "Reclamation in the Agro Pontino, Italy," *Geographical Review* 60 (1970): 301–27; here, 304.

40. Friedrich Ratzel, *Politische Geographie*, 3rd ed. (Munich: Oldenbourg, 1925), 35.

41. Gentilcore, "Reclamation," 312.

42. Marco Armiero and Wilko Graf von Hardenberg, "Green Rhetoric in Blackshirts: Italian Fascism and the Environment," *Environment and History* 19 (2013): 283–311; here, 306.

43. See Mark Mazower, *Hitler's Empire: How the Nazis Ruled Europe* (New York: Penguin, 2008), 223–93; Rüdiger Hachtmann and Winfried Süß, "Kommissare im NS-Herrschaftssystem: Probleme und Perspektiven der Forschung," in Hachtmann and Süß, eds., *Hitlers Kommissare: Sondergewalten in der nationalsozialistischen Diktatur*, 9–27 (Göttingen: Wallstein, 2006); and Hans Umbreit, "Die deutsche Herrschaft in den besetzten Gebieten 1942–1945," in Militärgeschichtliches Forschungsamt, ed., *Das Deutsche Reich und der Zweite Weltkrieg*, vol. 5: *Organisation und Mobilisierung des deutschen Machtbereichs*; part 2: *Kriegsverwaltung, Wirtschaft und personelle Ressourcen, 1942–1944/45*, 3–272 (Stuttgart: Deutsche Verlags-Anstalt, 1999).

44. Frank Snowden, "Latina Province, 1944–1950," *Journal of Contemporary History* 43 (2008): 509–26; here, 513–15. Snowden argued that flooding the Pontine Marshes, which increased the risk of malaria, was "the only known case of bioterror in twentieth-century Europe" (515), but that argument has been disputed. For a response, see Erhard Geissler and Jeanne Guillemin, "German Flooding of the Pontine Marshes in World War II: Biological Warfare or Total War Tactic?" *Politics and the Life Sciences* 29 (2010): 2–23.

45. Snowden, "Latina Province," 519. For an account of Italy's campaign against malaria, see Frank M. Snowden, *The Conquest of Malaria: Italy, 1900–1962* (New Haven, CT: Yale University Press, 2006).

46. Snowden, "Latina Province," 519n.

47. Correlli Barnett, *The Swordbearers: Studies in Supreme Command in the First World War* (London: Eyre & Spottiswoode, 1963), 35.

48. Ipsen, *Dictating Demography*, 115.

49. Nützenadel, *Landwirtschaft*, 232; and Snowden, *Conquest of Malaria*, 172.

50. Ingo Skoneczny, *Regionalplanung im faschistischen Italien: Die Besiedlung der pontinischen Sümpfe* (Berlin: Institut für Stadt- und Regionalplanung, 1983), 128–34.

51. James C. Scott, *Seeing Like a State: How Certain Schemes to Improve the Human Condition Have Failed* (New Haven, CT: Yale University Press, 1998), 93.

52. Skoneczny, *Regionalplanung*, 132.

53. Tiago Saraiva, "Fascist Modernist Landscapes: Wheat, Dams, Forests, and the Making of the Portuguese New State," *Environmental History* 21 (2016): 54–75; here, 59n, 65.

54. American Heritage Center, University of Wyoming, Laramie, Wyoming, USA, Collection 2129-81-5-26, Floyd E. Dominy Papers, box 24. folder "Professional File, 1962—USSR Trip," Luce to The Secretary, November 2, 1962, p. 7.

55. Joseph Morgan Hodge, "Writing the History of Development (Part 1: The First Wave)," *Humanity: An International Journal of Human Rights, Humanitarianism, and Development* 6 (2015): 429–63; here, 430.

56. James Ferguson, *The Anti-Politics Machine: "Development," Depoliticization, and Bureaucratic Power in Lesotho* (Minneapolis: University of Minnesota Press, 1994), xv.

57. See Hodge, "Writing the History (Part 1)," 451–55, and Joseph Morgan Hodge, "Writing the History of Development (Part 2: Longer, Deeper, Wider)," *Humanity: An International Journal of Human Rights, Humanitarianism, and Development* 7 (2016): 125–74.

58. Stephen Brain, *Song of the Forest: Russian Forestry and Stalinist Environmentalism, 1905–1953* (Pittsburgh, PA: University of Pittsburgh Press, 2011), 9.

59. Matteo Rizzo, "What Was Left of the Groundnut Scheme? Development Disaster and Labour Market in Southern Tanganyika, 1946–52," *Journal of Agrarian Change* 6 (2006): 205–38; here, 236. See also Stefan Esselborn, "Environment, Memory, and the Groundnut Scheme: Britain's Largest Colonial Agricultural Development Project and Its Global Legacy," *Global Environment* 11 (2013): 58–93.

60. Edward R. Tannenbaum, *The Fascist Experience: Italian Society and Culture, 1922–1945* (New York: Basic Books, 1972), 97; and Federico Caprotti, "Destructive Creation: Fascist Urban Planning, Architecture and New Towns in the Pontine Marshes," *Journal of Historical Geography* 33 (2007): 651–79; here, 665.

61. Nützenadel, *Landwirtschaft*, 240, 247n.

62. Salvemini, *Under the Axe*, 296.

63. Peter Aldhous, "Borneo Is Burning," *Nature* 432 (2004): 144–46; here, 144.

64. Hodge, "Writing the History (Part 1)," 429.

65. One such place is the University of Birmingham, which asked all scholars in my department to draft personal five-year plans during the writing of this book. Like most of my colleagues, I found this a pointless exercise, though I wish to record that it took me four years and eleven months to complete this manuscript.

66. David Gilmour, *The Pursuit of Italy: A History of a Land, Its Regions and Their Peoples* (London: Penguin, 2012), 8.

67. See Silvia Marchetti, "Mussolini's Latina Remains a Living Monument of Fascist Nostalgia," *Newsweek*, April 16, 2015, accessed November 3, 2021, http://www.newsweek.com/2015/04/24/mussolinis-latina-town-remains-living-monument-fascist-nostalgia-322777.html.

68. These observations are based on my visit in April 2018. For an insightful review of a local museum, see Suzanne Stewart-Steinberg, "Grounds for Reclamation: Fascism and Postfascism in the Pontine Marshes," *differences: A Journal of Feminist Cultural Studies* 27 (2016): 94–142.

69. Marco Armiero, *A Rugged Nation: Mountains and the Making of Modern Italy: Nineteenth and Twentieth Centuries* (Cambridge: White Horse Press, 2011), 116.

70. When I inquired about the name with receptionists during my stay at the Oasi di Kufra, responses varied. One did not know whether the Kufra oasis was imagined or real, another told me that the Kufra oasis was in Tunisia, and a third declared that the oasis was on an island. None of this is correct, though the last response has philosophical quality.

33: THE CHEMURGY MOVEMENT

1. Willard W. Cochrane, *The Development of American Agriculture: A Historical Analysis*, 2nd ed. (Minneapolis: University of Minnesota Press, 1993), 111.

2. William J. Hale, "Farming Must Become a Chemical Industry: Development of Co-Products Will Solve Present Agriculture Problem," *Dearborn Independent* 26, no. 50 (October 2, 1926): 4-5, 24-26; 4, 5, 24 (quotations).

3. Don Whitehead, *The Dow Story: The History of the Dow Chemical Company* (New York: McGraw-Hill, 1968), 95.

4. William J. Hale, "The Farm Chemurgic Movement," in Williams Haynes, ed., *American Chemical Industry*, vol. 5: *Decade of New Products*, appendix x, 486-90 (New York: Van Nostrand, 1954), 488.

5. Michigan State University Archives and Historical Collections, East Lansing, USA, LC 176 (William J. Hale Papers), box 2, folder 51, William J. Hale, "Chemurgy or Chaos." Address before Goodwyn Institute, Memphis, TN, November 15, 1949, pp. 1, 3.

6. Edwin T. Layton Jr., *The Revolt of the Engineers: Social Responsibility and the American Engineering Profession* (Cleveland: Press of Case Western Reserve University, 1971), 226-28; William E. Akin, *Technocracy and the American Dream: The Technocrat Movement, 1900-1941* (Berkeley: University of California Press, 1977); and Stefan Willeke, *Die Technokratiebewegung in Nordamerika und Deutschland zwischen den Weltkriegen: Eine vergleichende Analyse* (Frankfurt: Peter Lang, 1995).

7. For an overview, see Diane B. Paul and Marius Turda, "Eugenics, History of," in James D. Wright, ed., *International Encyclopedia of the Social and Behavioral Sciences*, 2nd ed., vol. 8, 253-57 (Amsterdam: Elsevier, 2015).

8. William J. Hale, *The Farm Chemurgic: Farmward the Star of Destiny Lights Our Way* (Boston: Stratford Company, 1934), 15.

9. Hale, *Farm Chemurgic*, 14.

10. Mark R. Finlay, *Growing American Rubber: Strategic Plants and the Politics of National Security* (New Brunswick, NJ: Rutgers University Press, 2009), 101.

11. Steven Watts, *The People's Tycoon: Henry Ford and the American Century* (New York: Alfred A. Knopf, 2005), 483.

12. Wayne K. Olson, *Chemurgy and Agriculture: 1934-1940* (Beltsville, MD: National Agricultural Library, 1993), 2.

13. Carroll W. Pursell Jr., "The Farm Chemurgic Council and the United States Department of Agriculture, 1935-1939," *Isis* 60 (1969): 307-17; here, 309.

14. Mark R. Finlay, "Old Efforts at New Uses: A Brief History of Chemurgy and the American Search for Biobased Materials," *Journal of Industrial Ecology* 7 (2004): 33-46; here, 40; Michigan State University Archives and Historical Collections LC 176, box 2, folder 12, letter of January 14, 1944.

15. Michigan State University Archives and Historical Collections LC 177 (Farm Chemurgic Council Records), box 32, folder 20, Farm Chemurgic Council, Minutes of Semi-Annual Meeting, Board of Governors and Committe Members, January 22, 1937, pp. 2, 4, 14, 21-22.

16. Ironically, George Washington Carver's image as "the peanut man" was itself a retrospective construction. (Mark D. Hersey, *My Work Is That of Conservation: An Environmental Biography of George Washington Carver* [Athens: University of Georgia Press, 2011], 164, 169-73.)

17. Finlay, "Old Efforts," 36.

18. Michigan State University Archives and Historical Collections LC 176, box 3, folder 52, Agrol Supreme. A Prospectus. Undated attachment to a letter from the National Agrol Company, July 9, 1941, p. 8.

19. Michigan State University Archives and Historical Collections LC 174 (Leo M. Christensen Papers), box 16, folder 49, Leo M. Christensen, The Agrol Opportunity, June 26, 1939, p. 1.

20. Michigan State University Archives and Historical Collections LC 176, box 3, folder 52, Agrol Supreme. A Prospectus, p. 9.

21. Michigan State University Archives and Historical Collections LC 176, box 3, folder 52, Agrol Supreme. A Prospectus, p. 8.

22. Michigan State University Archives and Historical Collections LC 176, box 3, folder 52, Agrol Supreme. A Prospectus, p. 8.

23. Bruce L. Gardner, *American Agriculture in the Twentieth Century: How It Flourished and What It Cost* (Cambridge, MA: Harvard University Press, 2002), 216.

24. David E. Wright, "Alcohol Wrecks a Marriage: The Farm Chemurgic Movement and the USDA in the Alcohol Fuels Campaign in the Spring of 1933," *Agricultural History* 67 (1993): 36–66; here, 57.

25. Hal Bernton, William Kovarik, and Scott Sklar, *The Forbidden Fuel: A History of Power Alcohol* (Lincoln: University of Nebraska Press, 2010), 17–21; and Michael S. Carolan, "A Sociological Look at Biofuels: Ethanol in the Early Decades of the Twentieth Century and Lessons for Today," *Rural Sociology* 74 (2009): 86–112; here, 96n.

26. Christian Warren, *Brush with Death: A Social History of Lead Poisoning* (Baltimore: Johns Hopkins University Press, 2000), 116–33. For the author's take on that critical threshold, see Frank Uekoetter, "The Merits of the Precautionary Principle: Controlling Automobile Exhausts in Germany and the United States before 1945," in E. Melanie DuPuis, ed., *Smoke and Mirrors: The Politics and Culture of Air Pollution*, 119–53 (New York: New York University Press, 2004).

27. Wright, "Alcohol," 61.

28. C. C. Furnas, "The Farm Problem—Chemurgy to the Rescue," *American Scholar* 10 (Winter 1940/41): 26–40; here, 27.

29. Wheeler McMillen, *New Riches from the Soil: The Progress of Chemurgy* (New York: Van Nostrand, 1946), 37.

30. William J. Hale, "Epistemocracy," *Review of Reviews and World's Work* 87 (February 1933): 27–29; here, 29.

31. Haynes, *American Chemical Industry*, 5:143; and Bernton, Kovarik, and Sklar, *Forbidden Fuel*, 25.

32. Anne B. W. Effland, "'New Riches from the Soil': The Chemurgic Ideas of Wheeler McMillen," *Agricultural History* 69 (Spring 1995): 288–97.

33. Mark R. Finlay, "The Failure of Chemurgy in the Depression-Era South: The Case of Jesse F. Jackson and the Central of Georgia Railroad," *Georgia Historical Quarterly* 81 (Spring 1997): 78–102; here, 99.

34. Michigan State University Archives and Historical Collections LC 174, box 16, folder 49, Leo M. Christensen, The Agrol Opportunity, June 26, 1939, p. 22.

35. Michigan State University Archives and Historical Collections LC 176, box 3, folder 52, Agrol Supreme: A Prospectus, p. 3.

36. Michigan State University Archives and Historical Collections LC 176, box 2, folder 12, Excerpts from remarks by Wheeler McMillen, President, National Farm Chemurgic Council, at the annual dinner, March 27, 1941, pp. 1–2. See also David E. Wright, "Agricultural Editors Wheeler McMillen and Clifford V. Gregory and the Farm Chemurgic Movement," *Agricultural History* 69 (Spring, 1995): 272–87; here, 275.

37. See p. 58.

38. Haynes, *American Chemical Industry*, 5:142. For the debate about an Iowa state law, see Jeffrey T. Manuel, "Iowa's Original Ethanol Debate: The Power Alcohol Movement of 1933-1934," *Annals of Iowa* 76 (2018): 41–78.

39. Finlay, *Growing American Rubber*, 193.

40. Use of Alcohol from Farm Products in Motor Fuel. Hearing before a Subcommittee

of the Committee on Finance, United States Senate, Seventy-Sixth Congress, First Session (Washington, DC: US Government Printing Office, 1939), 131.

41. Michigan State University Archives and Historical Collections LC 176, box 2, folder 53, William J. Hale, Out of Chaos into Chemurgy. Address before Renewable Resources Symposium, Philadelphia Engineers' Club, November 21, 1950, p. 14.

42. Michigan State University Archives and Historical Collections LC 176, box 1, folder 45, Hale to Earl Ubell, March 23, 1955; and folder 38, Hale to J. Earl Cooke, October 15, 1954.

43. Robert Lacey, *Ford: The Men and the Machine* (London: Heinemann, 1986), 546.

44. Hale, *Farm Chemurgic*, 163n.

45. William J. Hale, "When Agriculture Enters the Chemical Industry," *Industrial and Engineering Chemistry* 22 (1930): 1311–15; here, 1312.

46. Hale, *Farm Chemurgic*, 168.

47. Paul R. Ehrlich, *The Population Bomb* (New York: Ballantine Books, 1968).

48. See Thomas Robertson, *The Malthusian Moment: Global Population Growth and the Birth of American Environmentalism* (New Brunswick, NJ: Rutgers University Press, 2012).

49. R. Douglas Hurt, *Problems of Plenty: The American Farmer in the Twentieth Century* (Chicago: Ivan R. Dee, 2002), 77.

50. Érica Geraldes Castanheira, Helmer Acevedo, and Fausto Freire, "Greenhouse Gas Intensity of Palm Oil Produced in Colombia Addressing Alternative Land Use Change and Fertilization Scenarios," *Applied Energy* 114 (2014): 958–67.

51. Edward M. W. Smeets et al., "A Bottom-up Assessment and Review of Global Bio-Energy Potentials to 2050," *Progress in Energy and Combustion Science* 33 (2007): 56–106; here, 102.

52. Vaclav Smil, *Energy Transitions: History, Requirements, Prospects* (Santa Barbara, CA: Praeger, 2010), 116.

53. George Rapsomanikis, *The 2007–2008 Food Price Swing: Impact and Policies in Eastern and Southern Africa* (Rome: Food and Agriculture Organization of the United Nations, 2009), 10.

54. Frank Rosillo-Calle and Luis A. B. Cortez, "Towards ProAlcool II: A Review of the Brazilian Bioethanol Programme," *Biomass and Bioenergy* 14, no. 2 (1998): 115–24; here, 115. On subsequent developments, see Márcia Azanha Ferraz Dias de Moraes and David Zilberman, *Production of Ethanol from Sugarcane in Brazil: From State Intervention to a Free Market* (Cham: Springer, 2014).

55. Jennifer Eaglin, "The Demise of the Brazilian Ethanol Program: Environmental and Economic Shocks, 1985–1990," *Environmental History* 24 (2019): 104–29.

56. Lester R. Brown, *Outgrowing the Earth: The Food Security Challenge in an Age of Falling Water Tables and Rising Temperatures* (New York: Norton, 2004), 157.

57. Robert William Fogel, *Without Consent or Contract: The Rise and Fall of American Slavery* (New York: Norton, 1991), 18.

58. Eric J. Hobsbawm, *The Age of Extremes: A History of the World, 1914–1991* (New York: Vintage Books, 1996), 577.

59. Jennifer Eaglin, *Sweet Fuel: A Political and Environmental History of Brazilian Ethanol* (New York: Oxford University Press, 2022), 4.

60. Wendy Wolford, *This Land Is Ours Now: Social Mobilization and the Meanings of Land in Brazil* (Durham, NC: Duke University Press, 2010), 125–34.

61. Quoted in Thomas D. Rogers, *The Deepest Wounds: A Labor and Environmental History of Sugar in Northeast Brazil* (Chapel Hill: University of North Carolina Press, 2010), 210.

62. James N. Galloway et al., "Transformation of the Nitrogen Cycle: Recent Trends, Questions, and Potential Solutions," *Science* 320 (2008): 889–92, here, 891.

63. Manuel, "Iowa's Original Ethanol Debate," 42.

64. David Buchan, *Energy and Climate Change: Europe at the Crossroads* (Oxford: Oxford University Press, 2009), 154n.

65. Richard Doornbosch and Ronald Steenblik, *Biofuels: Is the Cure Worse Than the Disease?* SG/SD/RT [2007]3/REV1 (Paris: Organisation for Economic Co-operation and Development, 2007), 36.

66. I am stressing this legacy because of a lingering myth in green technology circles that there is "almost no institutional memory of what happened before the energy crisis of the '70s." (Alexis Madrigal, *Powering the Dream: The History and Promise of Green Technology* [Cambridge, MA: Da Capo Press, 2011], 1.) Institutional memory does not always take the shape of explicit commemoration!

34: AUTOBAHN

1. See Ulrich Kubisch and Gert Rietner, *Die Avus im Rückspiegel: Rennen, Rekorde, Rückstaus* (Berlin: Elefanten Press, 1987); and Richard Kitschigin, *Mythos Avus: Automobilsport in Berlin* (Berlin: Ullstein, 1995).

2. Maxwell G. Lay, *Ways of the World: A History of the World's Roads and of the Vehicles That Used Them* (New Brunswick, NJ: Rutgers University Press, 1992), 315.

3. Gijs Mom, *Atlantic Automobilism: Emergence and Persistence of the Car, 1895–1940* (New York: Berghahn Books, 2015), 59. See also Christophe Studeny, *L'invention de la vitesse: France, XVIIIe–XXe siècle* (Paris: Gallimard, 1995), and Joachim Radkau, *Das Zeitalter der Nervosität: Deutschland zwischen Bismarck und Hitler* (Munich: Hanser, 1998), 206–8, who showed that the automobile figured as therapy in the contemporary discourse on neurasthenia.

4. Wolfgang König, *Wilhelm II. und die Moderne: Der Kaiser und die technisch-industrielle Welt* (Paderborn: Schöningh, 2007), 205, 209, 217.

5. Christoph Maria Merki, *Der holprige Siegeszug des Automobils 1895–1930: Zur Motorisierung des Straßenverkehrs in Frankreich, Deutschland und der Schweiz* (Cologne: Böhlau, 2002), 181. See also Michael L. Berger, *The Devil Wagon in God's Country: The Automobile and Social Change in Rural America, 1893–1929* (Hamden, CT: Archon Books, 1979): 24–31, and Angela Jain and Massimo Moraglio, "Struggling for the Use of Urban Streets: Preliminary (Historical) Comparison between European and Indian Cities," *International Journal of the Commons* 8 (2014): 513–30.

6. Clive Emsley, "'Mother, What *Did* Policemen Do When There Weren't Any Motors?' The Law, the Police and the Regulation of Motor Traffic in England, 1900–1939," *Historical Journal* 36 (1993): 357–81; here, 358n.

7. Quoted in Tom McCarthy, *Auto Mania: Cars, Consumers, and the Environment* (New Haven, CT: Yale University Press, 2007), 12.

8. Jean Brunhes, *Les Limites de notre Cage. Discours prononcé à l'occasion de l'inauguration solennelle des cours universitaires le 15 Novembre 1909* (Fribourg: Imprimerie de l'Œuvre de Saint-Paul, 1911), 4. See also pp. 381–82.

9. Brunhes, *Limites* 5.

10. Barbara Haubner, *Nervenkitzel und Freizeitvergnügen: Automobilismus in Deutschland 1886–1914* (Göttingen: Vandenhoeck & Ruprecht, 1998), 164.

11. Emsley, "'Mother,'" 359.

12. Merki, *Holprige Siegeszug*, 147–55. See also Rudy Koshar, "Cars and Nations: Anglo-German Perspectives on Automobility between the World Wars," *Theory, Culture and Society* 21, no. 4/5 (2004): 121–44; here, 136.

13. Merki, *Holprige Siegeszug*, 451. See also Friedrich Lenger, *Werner Sombart 1863–1941: Eine Biographie* (Munich: Beck, 1994), esp. 162–70.

14. Friedrich Kittler, "Auto Bahnen / Free Ways," *Cultural Politics* 11 (2015): 376–83; here 378–80.

15. Gerald D. Feldman, *Hugo Stinnes: Biographie eines Industriellen, 1870–1924* (Munich: Beck, 1998), 660.

16. Andrea Greco and Giorgio Ragazzi, "History and Regulation of Italian Highways Concessionaires," *Research in Transportation Economics* 15 (2005): 121–33; here, 121. For a comprehensive study of Italy's autostrada projects before World War II, see Massimo Moraglio, *Driving Modernity: Technology, Experts, Politics, and Fascist Motorways, 1922–1943* (New York: Berghahn Books, 2017).

17. Timothy Davis, "The Rise and Decline of the American Parkway," in Christof Mauch, and Thomas Zeller, eds., *The World Beyond the Windshield: Roads and Landscapes in the United States and Europe*, 35–58 (Athens: Ohio University Press, 2008), 35.

18. Kubisch and Rietner, *Avus im Rückspiegel*, 79.

19. Thomas Zeller, "'Ganz Deutschland sein Garten': Alwin Seifert und die Landschaft des Nationalsozialismus," in Joachim Radkau and Frank Uekötter, eds., *Naturschutz und Nationalsozialismus*, 273–307 (Frankfurt: Campus, 2003), 306.

20. For a comprehensive discussion of the autobahn project in Nazi Germany, see Erhard Schütz and Eckhard Gruber, *Mythos Reichsautobahn: Bau und Inszenierung der "Straßen des Führers" 1933–1941* (Berlin: Ch. Links Verlag, 1996). Other publications of note include Thomas Zeller, *Driving Germany: The Landscape of the German Autobahn, 1930–1970* (New York: Berghahn Books, 2007); Charlotte Reitsam, *Reichsautobahn-Landschaften im Spannungsfeld von Natur und Technik: Transatlantische und interdisziplinäre Verflechtungen* (Saarbrücken: VDM Verlag Dr. Müller, 2009); and Richard Vahrenkamp, *The German Autobahn 1920–1945: Hafraba Visions and Mega Projects* (Lohmar: Josef Eul, 2010).

21. Schütz and Gruber, *Mythos Reichsautobahn*, 11n.

22. Quoted in Eduard Schönleben, *Fritz Todt: Der Mensch, der Ingenieur, der Nationalsozialist* (Oldenburg: Verlag Gerhard Stalling, 1943), 111. For more on Todt, see Frank Uekötter, "Techniker an der Macht: Der Ingenieur-Politiker im 20. Jahrhundert," *Historische Zeitschrift* 306 (2018): 396–423; here 411–16.

23. Schütz and Gruber, *Mythos Reichsautobahn*, 10.

24. John Christopher, *Organisation Todt: From Autobahns to the Atlantic Wall* (Stroud: Amberley, 2014), 11.

25. Schütz and Gruber, *Mythos Reichsautobahn*, 11.

26. See Friedrich Hartmannsgruber, "'. . . ungeachtet der noch ungeklärten Finanzierung': Finanzplanung und Kapitalbeschaffung für den Bau der Reichsautobahnen 1933–1945," *Historische Zeitschrift* 278 (2004): 625–81.

27. Alwin Seifert, *Ein Leben für die Landschaft* (Düsseldorf: Eugen Diederichs Verlag, 1962), 41.

28. Frank Uekoetter, *The Green and the Brown: A History of Conservation in Nazi Germany* (New York: Cambridge University Press, 2006), 167.

29. Christopher Kopper, "Germany's National Socialist Transport Policy and the Claim of Modernity: Reality or Fake?" *Journal of Transport History* 34 (2013): 162–76; here, 164.

30. Schütz and Gruber, *Mythos Reichsautobahn*, 12, 27n.

31. See Melanie Schiller, "'Fun Fun Fun on the Autobahn': Kraftwerk Challenges Germanness," *Popular Music and Society* 37 (2014): 618–37.

32. Alexander Gall, *"Gute Straßen bis ins kleinste Dorf!" Verkehrspolitik in Bayern zwischen Wiederaufbau und Ölkrise* (Frankfurt: Campus, 2005).

33. See Cotten Seiler, *Republic of Drivers: A Cultural History of Automobility in America* (Chicago: University of Chicago Press, 2008).

34. Richard Vahrenkamp, *Die Logistische Revolution: Der Aufstieg der Logistik in der Massenkonsumgesellschaft* (Frankfurt: Campus, 2011), 106.

35. Reitsam, *Reichsautobahn-Landschaften*, 204.

36. Scott L. Bottles, *Los Angeles and the Automobile: The Making of the Modern City* (Berkeley: University of California Press, 1987), 2–4.

37. Robert A. Caro, *The Power Broker: Robert Moses and the Fall of New York* (New York: Vintage Books, 1975), 546.

38. Kenneth T. Jackson, "Robert Moses and the Planned Environment: A Re-Evaluation," in Joann P. Krieg, ed., *Robert Moses: Single-Minded Genius*, 21–30 (Interlaken, NY: Heart of the Lakes, 1989), 26.

39. Christophe Bonneuil and Jean-Baptiste Fressoz, *The Shock of the Anthropocene: The Earth, History and Us* (London: Verso, 2016), 138. As Thomas Zeller notes on the military uses of the autobahn, "It is now generally accepted in the scholarly literature that this view is one of the myths surrounding the roadways." (Zeller, *Driving Germany*, 56.)

40. Alfred Gottwaldt, *Julius Dorpmüller, die Reichsbahn und die Autobahn: Verkehrspolitik und Leben des Verkehrsministers bis 1945* (Berlin: Argon, 1995), 45.

41. Public Papers of the Presidents of the United States: Dwight D. Eisenhower 1955. Containing the Public Messages, Speeches, and Statements of the President, January 1 to December 31, 1955 (Washington, DC: US Government Printing Office, 1960), 276.

42. Bruce E. Seely, "'Push' and 'Pull' Factors in Technology Transfer: Moving American-Style Highway Engineering to Europe, 1945–1965," *Comparative Technology Transfer and Society* 2 (2004): 229–46; here, 236.

43. Pascal Menoret, *Joyriding in Riyadh: Oil, Urbanism, and Road Revolt* (New York: Cambridge University Press, 2014), 70.

44. Menoret, *Joyriding in Riyadh*, 79.

45. See Frank Uekötter, *Der deutsche Kanal: Eine Mythologie der alten Bundesrepublik* (Stuttgart: Steiner, 2020), 151–53, 200.

46. Geheimes Staatsarchiv Berlin I. HA Rep. 77 Tit. 1328 no. 24 vol. 2, memorandum of Geheimes Civil Cabinet, March 26, 1910, and AVUS GmbH to Oberbürgermeister Berlin, September 15, 1910, p. 2.

47. Stephen E. Ambrose, *Eisenhower*, vol. 2: *The President* (London: George Allen and Unwin, 1984), 547.

48. Quoted in Dietmar Klenke, *Bundesdeutsche Verkehrspolitik und Motorisierung: Konfliktträchtige Weichenstellungen in den Jahren des Wiederaufstiegs* (Stuttgart: Steiner, 1993), 164.

49. M. Luísa Sousa and Rafael Marques, "Political Transitions, Value Change and Motorisation in 1970s Portugal," *Journal of Transport History* 34 (2013): 1–21; here, 3.

50. Lewis H. Siegelbaum, *Cars for Comrades: The Life of the Soviet Automobile* (Ithaca, NY: Cornell University Press, 2011), 224–27.

51. See Luminita Gatejel, *Warten, hoffen und endlich fahren: Auto und Sozialismus in der Sowjetunion, in Rumänien und der DDR (1956–1989/91)* (Frankfurt: Campus, 2014); and Lewis H. Siegelbaum, ed., *The Socialist Car: Automobility in the Eastern Bloc* (Ithaca, NY: Cornell University Press, 2011).

52. Gijs Mom, "Roads without Rails: European Highway-Network Building and the Desire for Long-Range Motorized Mobility," *Technology and Culture* 46 (2005): 745–72; here, 752.

53. James J. Flink, *The Automobile Age* (Cambridge, MA: MIT Press, 1990), 374; and Irving Brinton Holley Jr., *The Highway Revolution, 1895–1925: How the United States Got Out of the Mud* (Durham, NC: Carolina Academic Press, 2008), 153.

54. Stephen B. Goddard, *Getting There: The Epic Struggle between Road and Rail in the American Century* (New York: Basic Books, 1994), 179, 192.

55. G. de Clercq, "Fifteen Years of Road Lighting in Belgium," *International Lighting Review* 36 (1985): 2–7; here, 3.

56. Richard Kitschigin, *Rennen, Reifen und Rekorde: Die Avus Story* (Stuttgart: Motorbuch Verlag, 1972), 63, 67, 70.

57. See Mark Kahn, *Death Race: Le Mans 1955* (London: Barrie & Jenkins, 1976), esp. 102–29.

58. Statistisches Bundesamt, *Verkehrsunfälle: Zeitreihen 2015* (Wiesbaden: Statistisches Bundesamt, 2016), 18.

59. See Klenke, *Bundesdeutsche Verkehrspolitik*, 83–91.

60. Statistisches Bundesamt, *Verkehrsunfälle*, 201.

61. Flink, *Automobile Age*, 383.

62. Björn Kluge, Moritz Werkenthin, and Gerd Wessolek, "Metal Leaching in a Highway Embankment on Field and Laboratory Scale," *Science of the Total Environment* 493 (2014): 495–504; here, 503.

63. Tom Lewis, *Divided Highways: Building the Interstate Highways, Transforming American Life* (New York: Penguin, 1999), 179–210.

64. Quoted in Flink, *Automobile Age*, 374.

65. For information on new personal car registrations or sales, see http://www.oica.net/wp-content/uploads//pc-sales-20151.pdf, accessed November 3, 2021.

66. Calculations based on World Health Organization, *Global Status Report on Road Safety 2015* (Geneva: World Health Organization, 2015), 95, 120, 141, 232n, 246.

67. Emma Rothschild, *Paradise Lost: The Decline of the Auto-Industrial Age* (London: Allen Lane, 1974), 5–7; Sousa and Marques, "Political Transitions," 2.

68. Rothschild, *Paradise Lost*, 9.

69. Paul Virilio, *City of Panic* (Oxford: Berg, 2005), 3.

70. Landesarchiv Schleswig-Holstein Abt. 539 no. 1–6.

71. Welsley Johnston, *The Belfast Urban Motorway: Engineering Ambition and Social Conflict* (Newtownards: Colourpoint Books, 2014), 101n.

72. Ueli Haefeli, "Gas geben oder das Steuer herumreissen? Verkehrspolitik und Verkehrsplanung in Bielefeld nach dem Zweiten Weltkrieg," *Jahresbericht des Historischen Vereins für die Grafschaft Ravensberg* 85 (1999): 239–62.

73. Quoted in Zack Furness, *One Less Car: Bicycling and the Politics of Automobility* (Philadelphia: Temple University Press, 2010), 211. See also Barbara Schmucki, *Der Traum vom Verkehrsfluss: Städtische Verkehrsplanung seit 1945 im deutsch-deutschen Vergleich* (Frankfurt: Campus, 2001), and Michael Hascher, *Politikberatung durch Experten: Das Beispiel der deutschen Verkehrspolitik im 19. und 20. Jahrhundert* (Frankfurt: Campus, 2006).

74. Uekoetter, *The Green and the Brown*, 167.

75. Marc Augé, *Non-Places: An Introduction to Supermodernity* (London: Verso, 1995), 79, 97.

76. Menoret, *Joyriding in Riyadh*, 67.

35: THE PINE ROOTS CAMPAIGN

1. Basil Henry Liddell Hart, *History of the Second World War* (London: Cassell, 1970), 682n, 691.

2. Jerome B. Cohen, *Japan's Economy in War and Reconstruction* (London: Routledge, 2000), 142–44.

3. Euan Graham, *Japan's Sea Lane Security, 1950–2004: A Matter of Life and Death?* (London: Routledge, 2005), 82.

4. U.S. Naval Technical Mission to Japan, *Miscellaneous Targets: Japanese Fuels and Lubricants. Article 6: Research on Diesel and Boiler Fuel at the First Naval Fuel Depot, OFUNA* (Tokyo: U.S. Naval Technical Mission to Japan Report X-38(N)-6, February 1946), 1.

5. Cohen, *Japan's Economy*, 146.

6. C. J. Argyle, *Japan at War, 1937-45* (London: Arthur Barker, 1976), 182.

7. Satoshi Hirose, "Taiheiyōsensō Makki no Nihon no Kōkūnenryō: Daiyōnenryō to shite no Shōkon'yu" [Japan's aviation fuel in the last stage of the Pacific War: focusing on pine root oil as a substitute], in *Sensō to Heiwa: Ōsaka Kokusai Heiwa Kenkyūjo Kiyō* [War and peace: proceedings of Osaka International Peace Research Institute] 13 (2004): 15-34. I wish to thank Kazuki Okauchi for making this article available to me.

8. Cohen, *Japan's Economy*, 146n.

9. Daniel Yergin, *The Prize: The Epic Quest for Oil, Money and Power* (New York: Simon and Schuster, 1992), 362.

10. David Edgerton, "Controlling Resources: Coal, Iron Ore and Oil in the Second World War," in Michael Geyer and Adam Tooze, eds., *The Cambridge History of the Second World War*, vol. 3: *Total War: Economy, Society and Culture*, 122-48 (Cambridge: Cambridge University Press, 2015), 143; Francis Pike, *Hirohito's War: The Pacific War, 1941-1945* (London: Bloomsbury, 2016), 330; and Yergin, *Prize*, 363.

11. Hauptstaatsarchiv Dresden 13530 Forstamt Bärenfels mit Revieren no. 2801, Der Reichskanzler to sämtliche Bundesregierungen, December 17, 1916, and Kriegsausschuß für pflanzliche und tierische Oele und Fette, Die Fichtenharzgewinnung, undated (ca. July 1917).

12. For the administrative record of Lipinsky's project, see Bundesarchiv NS 19/758. Memorandum of February 7, 1945 (quotation).

13. See Kumkum Chatterjee, *Merchants, Politics and Society in Early Modern India: Bihar, 1733-1820* (Leiden: Brill, 1996), 79.

14. Ernst Jünger, *In Stahlgewittern* (Stuttgart: Klett-Cotta, 2014 [1920]).

15. Mark R. Finlay, *Growing American Rubber: Strategic Plants and the Politics of National Security* (New Brunswick, NJ: Rutgers University Press, 2009), 1.

16. A. D. Harvey, *Collision of Empires: Britain in Three World Wars, 1793-1945* (London: Phoenix, 1994), 584. For an assertive version of this argument, see Roland H. Worth Jr., *No Choice But War: The United States Embargo against Japan and the Eruption of War in the Pacific* (Jefferson, NC: McFarland, 1995).

17. Mats Ingulstad, Espen Storli, and Robin S. Gendron, "Introduction: Opening Pandora's Bauxite; A Raw Materials Perspective on Globalization Processes in the Twentieth Century," in Gendron, Ingulstad, and Storli, eds., *Aluminum Ore: The Political Economy of the Global Bauxite Industry*, 1-23 (Vancouver: UBC Press, 2013), 7. Similarly, Mimi Sheller, *Aluminum Dreams: The Making of Light Modernity* (Cambridge, MA: MIT Press, 2014), 62.

18. Luitgard Marschall, *Aluminium: Metall der Moderne* (Munich: oekom, 2008), 166, 171.

19. Brian C. Black, *Crude Reality: Petroleum in World History* (Lanham, MD: Rowman and Littlefield, 2012), 80.

20. William M. Tsutsui, "Landscapes in the Dark Valley: Toward an Environmental History of Wartime Japan," *Environmental History* 8 (2003): 294-311; here, 299n.

21. Tessa Morris-Suzuki, *Showa: An Inside History of Hirohito's Japan* (London: Athlone Press, 1984), 148. For other memories of the pine roots campaign, see Gail Lee Bernstein, *Isami's House: Three Centuries of a Japanese Family* (Berkeley: University of California Press, 2005), 155, and Joseph Coleman, *Unfinished Work: The Struggle to Build an Ageing American Workforce* (New York: Oxford University Press, 2015), 135.

22. Eugene Soviak, ed., *A Diary of Darkness: The Wartime Diary of Kiyosawa Kiyoshi* (Princeton, NJ: Princeton University Press, 1999), 333.

23. Marius B. Jansen, "Foreword," in Eugene Soviak, ed., *A Diary of Darkness: The Wartime Diary of Kiyosawa Kiyoshi*, vii–viii (Princeton, NJ: Princeton University Press, 1999), vii.

24. For this tradition, see Conrad Totman, *The Green Archipelago: Forestry in Preindustrial Japan* (Berkeley: University of California Press, 1989).

25. Cohen, *Japan's Economy*, 147. According to a postwar intelligence report of the US Naval Technical Mission to Japan, experimental use of pine root oil as a diesel surrogate in a Japanese Navy laboratory led to a similar conclusion: "Pine root oil was undesirable because the resinous matters gummed the fuel injection system." The report also noted that "the supply of bunker fuel from this source was too small to be of significance." Test results were so unambiguous that "only two or three experiments were made." (U.S. Naval Technical Mission to Japan, *Miscellaneous Targets*, 13, 9, 76.)

26. Cohen, *Japan's Economy*, 147.

27. Thomas Arthur Bisson, *Japan's War Economy* (New York: Macmillan, 1945), 164.

28. Cohen, *Japan's Economy*, 137–40; 140 (quotation).

29. Ulrich Herbert, *Geschichte Deutschlands im 20. Jahrhundert* (Munich: Beck, 2014), 143.

30. Peter Koch, *Die Erfindungen des Dr. Konrad Adenauer* (Reinbek: Wunderlich, 1986), 91–104; 97 (quotation). Ironically, Adenauer received a sausage patent in Great Britain in 1918 (104).

31. "Wo es das Brot von Konrad Adenauer gibt," accessed November 3, 2021, http://www.koeln.de/koeln/unfertig_adenauerbrot_428837.html.

32. Ulrich Wengenroth, "Die Flucht in den Käfig: Wissenschafts- und Innovationskultur in Deutschland 1900–1960," in Rüdiger vom Bruch and Brigitte Kaderas, eds., *Wissenschaften und Wissenschaftspolitik: Bestandsaufnahmen zu Formationen, Brüchen und Kontinuitäten im Deutschland des 20. Jahrhunderts*, 52–59 (Stuttgart: Steiner, 2002), 53.

33. See Günther Luxbacher, *Ersatzstoffe und neue Werkstoffe: Metalle, Technik und Forschungspolitik in Deutschland im 20. Jahrhundert* (Stuttgart: Steiner, 2020).

34. Frank Uekötter, "Mutmaßungen über Mais: Anmerkungen zu Westfalens erfolgreichstem Neophyten," *Westfälische Forschungen* 57 (2007): 151–71; Sarah Waltenberger, *Deutschlands Ölfelder: Eine Stoffgeschichte zur Kulturpflanze Raps (1897–2017)* (Paderborn: Schöningh, 2020).

35. Tsutsui, "Landscapes," 304n.

36. Susanne Heim, *Kalorien, Kautschuk, Karrieren: Pflanzenzüchtung und landwirtschaftliche Forschung in Kaiser-Wilhelm-Instituten 1933–1945* (Göttingen: Wallstein, 2003), 152–72.

37. See Edmund Russell, *War and Nature: Fighting Humans and Insects with Chemicals from World War I to Silent Spring* (Cambridge: Cambridge University Press, 2001).

38. See Jacob Darwin Hamblin, *Arming Mother Nature: The Birth of Catastrophic Environmentalism* (New York: Oxford University Press, 2013); and Spencer R. Weart, *The Discovery of Global Warming* (Cambridge, MA: Harvard University Press, 2008).

39. Ching-Sang Chiu et al., "Detection and Censusing of Blue Whale Vocalizations along the Central California Coast Using a Former SOSUS Array," *Journal of the Acoustical Society of America* 106 (1999): 2163.

40. Konrad Rubner, "Das durch Artilleriegeschosse verursachte Fichtensterben," *Mitteilungen der Bayerischen Botanischen Gesellschaft zur Erforschung der heimischen Flora* 3, no. 13 (January 1, 1916): 273–76.

41. Joseph P. Hupy, "The Environmental Footprint of War," *Environment and History* 14 (2008): 405–21; here, 413.

42. Peter Coates et al., "Defending Nation, Defending Nature? Militarized Landscapes and Military Environmentalism in Britain, France, and the United States," *Environmental History* 16 (2011): 456–91; here, 466.

43. Edward A. Ackerman, *Japan's Natural Resources and their Relation to Japan's Economic Future* (Chicago: University of Chicago Press, 1953), 333, 335, 349n.

44. Ackerman, *Japan's Natural Resources*, 342.

45. American Heritage Center, University of Wyoming, Collection 5363 Edward A. Ackerman Papers, box 4, folder "Japan—Forests," Rash Wartime Deforestation behind Timber Dearth Crisis. Article from the *Nippon Times*, January 8, 1947.

46. Heinz Ellenberg and Christoph Leuschner, *Vegetation Mitteleuropas mit den Alpen in ökologischer, dynamischer und historischer Sicht*, 6th ed. (Stuttgart: Ulmer, 2010), 1071n.

47. Tom Warwick, "The Colonization of Bomb-Crater Ponds at Marlow, Buckinghamshire," *Journal of Animal Ecology* 18 (1949): 137–41.

48. Warwick, "Colonization," 137.

49. Simon Richmond et al., *Korea*, 8th ed. (Footscray: Lonely Planet, 2010), 153.

50. Jeffrey A. McNeely, "Conserving Forest Biodiversity in Times of Violent Conflict," *Oryx* 37, no. 2 (2003): 142–52; here, 148.

51. Thomas Wrbka, Katharina Zmelik, and Franz Michael Grünweis, eds., *The European Green Belt: Borders. Wilderness. Future* (Weitra: Verlag Bibliothek der Provinz, 2009).

52. See Saleem H. Ali, ed., *Peace Parks: Conservation and Conflict Resolution* (Cambridge, MA: MIT Press, 2007).

53. Chris Pearson, *Mobilizing Nature: The Environmental History of War and Militarization in Modern France* (Manchester: Manchester University Press, 2012), 136.

54. Brian Black, "The Nature of Preservation: The Rise of Authenticity at Gettysburg," *Civil War History* 58 (2012): 348–73; here, 353n.

55. Steven D. Warren et al., "Biodiversity and the Heterogeneous Disturbance Regime on Military Training Lands," *Restoration Ecology* 15 (2007): 606–12.

56. See Chris Pearson, Peter Coates, and Tim Cole, eds., *Militarized Landscapes: From Gettysburg to Salisbury Plain* (London: Continuum, 2010).

57. See Edwin A. Martini, *Agent Orange: History, Science, and the Politics of Uncertainty* (Amherst: University of Massachusetts Press, 2012); and David Zierler, *The Invention of Ecocide: Agent Orange, Vietnam, and the Scientists Who Changed the Way We Think about the Environment* (Athens: University of Georgia Press, 2011).

58. Robert F. Durant, *The Greening of the U.S. Military: Environmental Policy, National Security, and Organizational Change* (Washington, DC: Georgetown University Press, 2007), 5.

59. Judith A. Bennett, "Pests and Disease in the Pacific War: Crossing the Line," in Richard P. Tucker and Edmund Russell, eds., *Natural Enemy, Natural Ally: Towards an Environmental History of War*, 217–52 (Corvallis: Oregon State University Press, 2004), 230.

60. Bryan L. Stuart and Peter Davidson, "Use of Bomb Crater Ponds by Frogs in Laos," *Herpetological Review* 30 (1999): 72–73; and Csaba F. Vad et al., "Wartime Scars or Reservoirs of Biodiversity? The Value of Bomb Crater Ponds in Aquatic Conservation," *Biological Conservation* 209 (2017): 253–62.

61. Douglas V. Prose and Howard G. Wilshire, *The Lasting Effects of Tank Maneuvers on Desert Soils and Intershrub Flora*, Open-File Report OF 00–512 (Washington, DC: US Geological Survey, 2000), 15.

62. For the defining source on the pine roots campaign, see a few lines in a book by Jerome Cohen, who studied the Japanese war economy attached to the US Strategic Bombing Survey during the Occupation. First published in 1949, his monograph was

a doctoral dissertation at Columbia University's faculty of political science that went through reprints in 1973 and 2000. (Cohen, *Japan's Economy*.)

63. Economics of Defense Policy: Adm. H. G. Rickover. Hearing before the Joint Economic Committee, Congress of the United States, Ninety-Seventh Congress, Second Session, Part 1 (January 28, 1982). Printed for the use of the Joint Economic Committee (Washington, DC: Government Printing Office, 1982), 61. For context, see Francis Duncan, *Rickover: The Struggle for Excellence* (Annapolis, MD: Naval Institute Press, 2001), 277–94.

64. "Spanbroekmolen Mine Crater Memorial: The Pool of Peace, Belgium," accessed November 3, 2021, http://www.greatwar.co.uk/ypres-salient/memorial-spanbroekmolen-pool-of-peace.htm.

65. Martin Gilbert, *First World War* (London: Weidenfeld and Nicolson, 1994), 336. On the European tradition of the *lieux de mémoire*, see Pierre Nora, ed., *Les lieux de mémoire*, 7 vols. (Paris: Gallimard, 1984–1992); Etienne François and Hagen Schulze, eds., *Deutsche Erinnerungsorte*, 3 vols. (Munich: Beck, 2001); Jan Bank et al., eds., *Plaatsen van Herinnering*, 4 vols. (Amsterdam: Bakker, 2005–2007); and Mario Isnenghi, ed., *I Luoghi della Memoria*, 3 vols. (Rome: Laterza, 1996–1997).

PART VIII
THE GREAT ENTRENCHMENT

1. Nigel Fisher, *Harold Macmillan: A Biography* (London: Weidenfeld and Nicolson, 1982), 192, 220n.

2. Eric J. Hobsbawm, *The Age of Extremes: A History of the World, 1914–1991* (New York: Vintage Books, 1996), 6.

3. See Christian Pfister, ed., *Das 1950er Syndrom: Der Weg in die Konsumgesellschaft* (Bern: Haupt, 1995). For a more recent presentation of his argument, see Christian Pfister, "The '1950s Syndrome' and the Transition from a Slow-Growing to a Rapid Loss of Global Sustainability," in Frank Uekoetter, ed., *The Turning Points of Environmental History*, 90–118 (Pittsburgh, PA: University of Pittsburgh Press, 2010).

4. John R. McNeill and Peter Engelke, *The Great Acceleration: An Environmental History of the Anthropocene since 1945* (Cambridge, MA: Belknap Press of Harvard University Press, 2014).

5. Will Steffen et al., "The Trajectory of the Anthropocene: The Great Acceleration," *Anthropocene Review* 2 (2015): 81–98; here, 82.

6. Joachim Radkau, *The Age of Ecology: A Global History* (Cambridge: Polity Press, 2013), 10.

7. Joan Martinez-Alier, "The Environmentalism of the Poor: Its Origin and Spread," in John R. McNeill and Erin Stewart Mauldin, eds., *A Companion to Global Environmental History*, 513–29 (Malden, MA: Wiley-Blackwell, 2012). See also http://ejatlas.org, accessed November 3, 2021.

8. Christophe Bonneuil and Jean-Baptiste Fressoz, *The Shock of the Anthropocene: The Earth, History and Us* (London: Verso, 2016), 66.

9. Klaus Damme and Ralf-Achim Hildebrand, *Geflügelhaltung: Legehennen, Hähnchen, Puten, Management, Tierschutz, Umwelt, Ökonomie* (Stuttgart: Ulmer, 2002), 15.

10. See Martin V. Melosi, *Atomic Age America* (Boston: Pearson, 2013), 158–60.

11. Vannevar Bush, *Science: The Endless Frontier; A Report to the President* (Washington, DC: US Government Printing Office, 1945).

12. The National Archives of the United Kingdom MAF 293/169, J. A. Barrah, "Animal Machines," April 25, 1964.

36: BATTERY CHICKEN

1. Gerald Frost, *Antony Fisher: Champion of Liberty* (London: Profile Books, 2002), 2, 14, 39-41.

2. Antony Fisher, *The Case for Freedom* (London: Runnymede Press, 1949).

3. Daniel Stedman Jones, *Masters of the Universe: Hayek, Friedman, and the Birth of Neoliberal Politics* (Princeton, NJ: Princeton University Press, 2012), 156.

4. Norman McCord, "Fisher, Sir Antony George Anson (1915-1988)," in *The Oxford Dictionary of National Biography*, vol. 19, 663-64 (Oxford: Oxford University Press, 2004).

5. Roger Horowitz, *Putting Meat on the American Table: Taste, Technology, Transformation* (Baltimore: Johns Hopkins University Press, 2006), 103, 108.

6. Annie Potts, *Chicken* (London: Reaktion Books, 2012), 141; and Edith H. Whetham, *The Agrarian History of England and Wales*, vol. 8: *1914-39* (Cambridge: Cambridge University Press, 1978), 14.

7. Andres Godley and Bridget Williams, "Democratizing Luxury and the Contentious 'Invention of the Technological Chicken' in Britain," *Business History Review* 83 (2009): 267-90; here, 273.

8. William Boyd, "Making Meat: Science, Technology, and American Poultry Production," *Technology and Culture* 42 (2001): 631-64; here, 645.

9. See, for instance, Mark Skousen, *The Making of Modern Economics: The Lives and Ideas of the Great Thinkers* (Armonk, NY: M. E. Sharpe, 2001), 444; Kenneth R. Hoover, *Economics as Ideology: Keynes, Laski, Hayek, and the Creation of Contemporary Politics* (Lanham, MD: Rowman and Littlefield, 2003), 203; David Miller, "How Neoliberalism Got Where It Is: Elite Planning, Corporate Lobbying and the Release of the Free Market," in Kean Birch and Vlad Mykhnenko, eds., *The Rise and Fall of Neoliberalism: The Collapse of an Economic Order?* 23-41 (London: Zed Books, 2010), 27; and Andrew Marr, *A History of Modern Britain* (London: Macmillan, 2007), 163.

10. Frost, *Antony Fisher*, 47.

11. Alessandra Tessari and Andrew Godley, "Made in Italy: Made in Britain; Quality, Brands and Innovation in the European Poultry Market, 1950-80," *Business History* 56 (2014): 1057-83; here, 1060.

12. Tessari and Godley, "Made in Italy," 1070.

13. Patrice G. Poutrus, "Lebensmittelkonsum, Versorgungskrisen und die Entscheidung für den 'Goldbroiler': Problemlagen und Lösungsversuche der Agrar- und Konsumpolitik in der DDR 1958-1965," *Archiv für Sozialgeschichte* 39 (1999): 391-421.

14. Eric Schlosser, *Fast Food Nation: What the All-American Meal Is Doing to the World* (London: Penguin, 2002), 140.

15. Steve Striffler, *Chicken: The Dangerous Transformation of America's Favorite Food* (New Haven, CT: Yale University Press, 2005), 27.

16. Hans-Wilhelm Windhorst, "Die sozialgeographische Analyse raum-zeitlicher Diffusionsprozesse auf der Basis der Adoptorkategorien von Innovationen: Die Ausbreitung der Käfighaltung von Hühnern in Südoldenburg," *Zeitschrift für Agrargeschichte und Agrarsoziologie* 27 (1979): 244-66; here, 256. See also Hans-Wilhelm Windhorst, *Spezialisierte Agrarwirtschaft in Südoldenburg: Eine agrargeographische Untersuchung* (Leer: Schuster, 1975).

17. Monica R. Gisolfi, *The Takeover: Chicken Farming and the Roots of American Agribusiness* (Athens: University of Georgia Press, 2017), 3n.

18. Carl Weinberg, "Big Dixie Chicken Goes Global: Exports and the Rise of the North Georgia Poultry Industry," *Business and Economic History* 1 (2003): 1-32; here, 18.

19. Kiran Klaus Patel, *Europäisierung wider Willen: Die Bundesrepublik Deutschland in der Agrarintegration der EWG, 1955-1975* (Munich: Oldenbourg, 2009), 242n; and Ross B.

Talbot, *The Chicken War: An International Trade Conflict between the United States and the European Economic Community, 1961-64* (Ames: Iowa State University Press, 1978).

20. Andrew C. Godley and Bridget Williams, "The Chicken, the Factory Farm, and the Supermarket: The Emergence of the Modern Poultry Industry in Britain," in Warren Belasco and Roger Horowitz, eds., *Food Chains: From Farmyard to Shopping Cart*, 47-61 (Philadelphia: University of Pennsylvania Press, 2009), 59.

21. William Boyd and Michael Watts, "Agro-Industrial Just-in-Time: The Chicken Industry and Postwar American Capitalism," in David Goodman and Michael Watts, ed., *Globalising Food: Agrarian Questions and Global Restructuring*, 192-225 (London: Routledge, 1997), 199.

22. Frost, *Antony Fisher*, 54-56, 115-17.

23. Archie Carr, "Caribbean Green Turtle: Imperiled Gift of the Sea," *National Geographic* 131, no. 6 (1967): 876-90.

24. James J. Parsons, "Sea Turtles and Their Eggs," in Kenneth F. Kiple and Kriemhild Coneè Ornelas, eds., *The Cambridge World History of Food*, vol. 1, 567-74 (Cambridge: Cambridge University Press, 2000), 569. See also Sharika D. Crawford, *The Last Turtlemen of the Caribbean: Waterscapes of Labor, Conservation, and Boundary Making* (Chapel Hill: University of North Carolina Press, 2020).

25. Alison Rieser, *The Case of the Green Turtle: An Uncensored History of a Conservation Icon* (Baltimore: Johns Hopkins University Press, 2012), 103, 148; Parsons, "Sea Turtles," 570; and Frederick Rowe Davis, *The Man Who Saved Sea Turtles: Archie Carr and the Origins of Conservation Biology* (New York: Oxford University Press, 2007), 180.

26. Carr, "Caribbean Green Turtle," 879.

27. See Archie Carr, *Handbook of Turtles: The Turtles of the United States, Canada, and Baja California* (Ithaca, NY: Cornell University Press, 1952); and Archie Carr and Coleman J. Goin, *Guide to the Reptiles, Amphibians and Fresh-Water Fishes of Florida* (Gainesville: University of Florida Press, 1959). For his contributions to popular science, see Archie Carr and the editors of *Life*, *The Reptiles* (Amsterdam: Time-Life International Nederland, 1964 [1963]), and Archie Carr, *So Excellent a Fishe: A Natural History of Sea Turtles* (Gainesville: University Press of Florida, 2011 [1965]).

28. Carr, *So Excellent a Fishe*, 12.

29. Frost, *Antony Fisher*, 119n.

30. Davis, *Man Who Saved Sea Turtles*, 192.

31. Neil D'Cruze, Rachel Alcock, and Marydele Donnelly, "The Cayman Turtle Farm: Why We Can't Have Our Green Turtle and Eat It Too," *Journal of Agricultural and Environmental Ethics* 28 (2015)L 57-66; here, 58.

32. Rieser, *Case of the Green Turtle*, 108.

33. Rieser, *Case of the Green Turtle*, 194.

34. Staatsarchiv Oldenburg Re 400 Akz. 226 B no. 486, Der Regierungsveterinärarzt des Landkreises Vechta to Landeskriminalpolizeistelle Oldenburg, March 21, 1962.

35. Louis Galambos with Jane Eliot Sewell, *Networks of Innovation: Vaccine Development at Merck, Sharp & Dohme, and Mulford, 1895-1995* (Cambridge: Cambridge University Press, 1995), 127.

36. T. A. B. Corley and Andrew Godley, "The Veterinary Medicine Industry in Britain in the Twentieth Century," *Economic History Review* 64 (2011): 832-54; here, 838.

37. Godley and Williams, "Democratizing Luxury," 271.

38. Robert Bud, *Penicillin: Triumph and Tragedy* (Oxford: Oxford University Press, 2007), 163-69; and Claas Kirchhelle, "Toxic Confusion: The Dilemma of Antibiotic Regulation in West German Food Production (1951-1990)," *Endeavour* 40 (2016): 114-27.

39. H. Faller, "Antibiotika im Preisvergleich," *Deutsche Landwirtschaftliche Presse* 88 (1965): 331–32; here, 331.

40. Joseph L. Anderson, *Industrializing the Corn Belt: Agriculture, Technology, and Environment, 1945–1972* (DeKalb: Northern Illinois University Press, 2009), 97.

41. Abigail Woods, "Science, Disease and Dairy Production in Britain, c. 1927 to 1980," *Agricultural History Review* 62 (2014): 294–314.

42. Staatsarchiv Oldenburg Re 400 Akz. 226 B no. 486.

43. Harald Ebbell, *Eier-Erzeugung in Legekäfigen und -batterien*, Geflügelzucht-Bücherei 12, 2nd ed. (Stuttgart: Ulmer, 1967), 8, 75, 77.

44. Geoffrey Sykes, *Poultry: A Modern Agribusiness* (London: Crosby Lockwood, 1963), 231. On the concept of agribusiness, see Shane Hamilton, "Agribusiness, the Family Farm, and the Politics of Technological Determinism in the Post-World War II United States," *Technology and Culture* 55 (2014): 560–90.

45. Meredith McKittrick, "Industrial Agriculture," in John R. McNeill and Erin Stewart Mauldin, eds., *A Companion to Global Environmental History*, 411–32 (Malden, MA: Wiley-Blackwell, 2012), 424.

46. Charles J. Price and Josephine E. Reed, *Poultry Husbandry II* (Rome: Food and Agriculture Organization of the United Nations, 1971), 79.

47. Weinberg, "Big Dixie Chicken," 28.

48. Sam White, "From Globalized Pig Breeds to Capitalist Pigs: A Study in Animal Cultures and Evolutionary History," *Environmental History* 16 (2011): 94–120.

49. Boyd, "Making Meat," 656–658.

50. Potts, *Chicken*, 143.

51. Klaus Damme and Ralf-Achim Hildebrand, *Geflügelhaltung: Legehennen, Hähnchen, Puten, Management, Tierschutz, Umwelt, Ökonomie* (Stuttgart: Ulmer, 2002), 14; and Potts, *Chicken*, 155.

52. Godley and Williams, "Democratizing Luxury," 271.

53. Arend Laurens Hagedoorn and Geoffrey Sykes, *Poultry Breeding: Theory and Practice* (London: Crosby Lockwood, 1953), 67.

54. Margaret E. Derry, *Masterminding Nature: The Breeding of Animals 1750–2010* (Toronto: University of Toronto Press, 2015), 166n.

55. Carr and the editors of *Life*, *The Reptiles*, 174n.

56. Carr and the editors of *Life*, *The Reptiles*, 175. See also Carr, *So Excellent a Fishe*, 234–38.

57. Rieser, *Case of the Green Turtle*, 8, 200, 205.

58. David W. Ehrenfeld, "Conserving the Edible Sea Turtle: Can Mariculture Help? Commercial Husbandry Does Not Necessarily Protect Endangered Species," *American Scientist* 62, no. 1 (1974): 23–31; here, 29.

59. Archie Carr, "Great Reptiles, Great Enigmas," *Audubon* 74 (March 1972): 24–35; here, 34.

60. Rieser, *Case of the Green Turtle*, 4.

61. Frost puts Fisher's personal loss at more than $600,000. (Frost, *Antony Fisher*, 130n.) However, a coffee-table book on the turtle farm project records the sum as the *total* loss of all the directors. (Peggy D. Fosdick and Sam Fosdick, *Last Chance Lost? Can and Should Farming Save the Green Sea Turtle? The Story of Mariculture, Ltd.,—Cayman Turtle Farm* [York, PA: I. S. Naylor, 1994], 181.)

62. Fosdick and Fosdick, *Last Chance Lost?* 209.

63. Rieser, *Case of the Green Turtle*, 9; and Davis, *Man Who Saved Sea Turtles*, 194.

64. Rieser, *Case of the Green Turtle*, 237, 260.

65. Lisa M. Campbell, "Science and Sustainable Use: Views of Marine Turtle Conservation Experts," *Ecological Applications* 12 (2002): 1229–46; here, 1232.

66. Campbell, "Science," 1240.

67. Ebbell, *Eier-Erzeugung*, 6n.

68. Landesarchiv Nordrhein-Westfalen NW 831 Paket 255, memorandum of October 8, 1952. See also William Percy Blount, *Hen Batteries* (London: Baillière, Tindall & Cox, 1951), 245–48.

69. *Spectator* 211, no. 7052 (August 23, 1963): 222.

70. Ruth Harrison, *Animal Machines: The New Factory Farming Industry* (London: Vincent Stuart, 1964), 3. On Harrison, see Claas Kirchhelle, *Bearing Witness: Ruth Harrison and British Farm Animal Welfare (1920–2000)* (Basingstoke: Palgrave Macmillan, 2021).

71. Joy A. Mench, "Thirty Years after Brambell: Whither Animal Welfare Science?" *Journal of Applied Animal Welfare Science* 1, no. 2 (1998): 91–102; here, 91.

72. *Report of the Technical Committee to Enquire into the Welfare of Animals Kept under Intensive Livestock Husbandry Systems*. Presented to Parliament by the Secretary of State for Scotland and the Minister of Agriculture, Fisheries and Food by Command of Her Majesty, December, 1965 (London: Her Majesty's Stationary Office, 1965), 13.

73. Landesarchiv Nordrhein-Westfalen NW 131 no. 23 4R (brochure of The Harvey Laying Battery Co. Ltd., Colchester, Essex, ca. 1950).

74. The National Archives of the United Kingdom MAF 293/172, The British Chicken Association to Ministry of Agriculture, Fisheries and Food, April 23, 1964.

75. The National Archives of the United Kingdom MAF 293/169, E. S. Virgo, "Animal Machines," March 12, 1963.

76. Siegfried Scholtyssek, ed., *Käfighaltung von Hühnern: Referate einer fachwissenschaftlichen Vortrags- und Diskussionstagung in Stuttgart-Hohenheim* (Stuttgart: Ulmer, 1971), 145. See also Christian Teubner, Sybil Gräfin Schönfeldt, and Siegfried Scholtyssek, *The Chicken and Poultry Bible: The Definitive Sourcebook* (New York: Penguin Studio, 1997).

77. For evidence on a gender gap in opinions on battery chicken, see Ulrich Planck, *Situation der Landjugend: Die ländliche Jugend unter besonderer Berücksichtigung des landwirtschaftlichen Nachwuchses* (Münster-Hiltrup: Landwirtschaftsverlag, 1982), 175n.

78. Horowitz, *Putting Meat on the American Table*, 120.

79. The National Archives of the United Kingdom MAF 293/169, News from the National Farmers Union, "Animal Machines": N.F.U. Comment on New Book, March 6, 1964.

80. Frost, *Antony Fisher*, 56. Fisher was not shy about defending his turtle farm project either. When the air was getting thin for the endeavor in 1974, he declared unabashedly that "the worst attacks came from people who had taken no trouble to find out what we actually were doing." (Antony G. A. Fisher, "Conservation," in Fosdick and Fosdick, *Last Chance Lost?* A61–A76; here, A61.)

81. The National Archives of the United Kingdom MAF 369/75, National Farmers' Union, Farm Animal Welfare—Proposed Legislation, N.F.U. Comment, November 16, 1966, 4. Poultry interests also criticized the Brambell Committee because it "expressly excludes Northern Ireland and Eire" and thus opened the door for "unfair discrimination." (The National Archives of the United Kingdom MAF 369/41, National Association of Poultry Packers Limited to Ministry of Agriculture, Fisheries, and Food, February 28, 1966, p. 2.) The National Farmers' Union felt that Northern Ireland was "the most acute case of the whole import question." (The National Archives of the United Kingdom MAF 369/77, The National Farmers' Union to Mr. Fawcett, September 15, 1967, p. 1.)

82. The National Archives of the United Kingdom MAF 369/78, Fawcett to Hewer, October 3, 1967. For a more extensive discussion of the political consequences, see Abigail

Woods, "From Cruelty to Welfare: The Emergence of Farm Animal Welfare in Britain, 1964-71," *Endeavor* 36, no. 1 (2012): 14-22.

83. See Landesarchiv Nordrhein-Westfalen NW 831 Paket 127 and 320.

84. For anti-cage statements of animal welfare organizations, see MAF 369/75, Humane Farming Campaign, November 17, 1966, p. 5, and Royal Society for the Prevention of Cruelty to Animals, October 31, 1966, p. 2; and MAF 369/76, Council of Justice to Animals and Humane Slaughter Association, December 2, 1966. On the history of mad cow disease, see Maxime Schwartz, *How the Cows Turned Mad: Unlocking the Mysteries of Mad Cow Disease* (Berkeley: University of California Press, 2004).

85. Council Directive 1999/74/EC of July 19, 1999.

86. Godley and Williams, "Democratizing Luxury," 286; and Horowitz, *Putting Meat on the American Table*, 125.

87. Damme and Hildebrand, *Geflügelhaltung*, 14.

88. Damme and Hildebrand, *Geflügelhaltung*, 21.

89. R. Emmett Tyrrell Jr., *After the Hangover: The Conservatives' Road to Recovery* (Nashville, TN: Thomas Nelson, 2010), 169 (quotation); and Frost, *Antony Fisher*, 135n, 143n.

90. Jamie Peck, *Constructions of Neoliberal Reason* (Oxford: Oxford University Press, 2010), 171.

91. Tessari and Godley, *Made in Italy*, 1060; and Frost, *Antony Fisher*, 49.

92. D'Cruze, Alcock, and Donnelly, "Cayman Turtle Farm," 59.

93. Fosdick and Fosdick, *Last Chance Lost?* 297.

94. See https://www.turtle.ky/about-us/about-cayman-turtle-centre/, accessed November 3, 2021.

95. Ryan Ver Berkmoes et al., *Caribbean Islands*, 6th ed. (London: Lonely Planet, 2011), 278; and https://www.turtle.ky/explore/dining/, accessed November 3, 2021.

37: LUCKY DRAGON NO. 5

1. Robert A. Divine, *Blowing in the Wind: The Nuclear Test Ban Debate 1954-1960* (New York: Oxford University Press, 1978), 4-7.

2. John Swenson-Wright, *Unequal Allies? United States Security and Alliance Policy Toward Japan, 1945-1960* (Stanford, CA: Stanford University Press, 2005), 152-54.

3. Aya Homei, "The Contentious Death of Mr Kuboyama: Science as Politics in the 1954 Lucky Dragon Incident," *Japan Forum* 25 (2013): 212-32; here, 215.

4. Donald F. Glut, *Classic Movie Monsters* (Metuchen, NJ: Scarecrow Press, 1978), 376-78.

5. Divine, *Blowing in the Wind*, 11, 20n.

6. David E. Lilienthal, *The Journals of David E. Lilienthal*, vol. 3: *Venturesome Years, 1950-1955* (New York: Harper and Row, 1966), 497.

7. Catherine Caufield, *Multiple Exposures: Chronicles of the Radiation Age* (London: Secker & Warburg, 1989), 3, 9, 11, 13.

8. As of 2009, there were more English-language children's books about Marie Curie than about Albert Einstein. (Trevor Owens, "Going to School with Madame Curie and Mr. Einstein: Gender Roles in Children's Science Biographies," *Cultural Studies of Science Education* 4 [2009]: 929-43; here, 930.)

9. Hermann Vogel, "Das Ehrenmal der Radiologie in Hamburg: Ein Beitrag zur Geschichte der Röntgenstrahlen," *RöFo. Fortschritte auf dem Gebiet der Röntgenstrahlen und der bildgebenden Verfahren* 178 (2006): 753-56. After two amendments in 1940 and 1960, the memorial now bears 359 names. See also Hermann Holthusen, Hans Meyer, and Werner Molineus, eds., *Ehrenbuch der Röntgenologen und Radiologen aller Nationen*, 2nd ed. (Munich: Urban & Schwarzenberg, 1959), ix.

10. Quoted in Robert Jungk, *Brighter Than a Thousand Suns: A Personal History of the Atomic Scientists* (Harmondsworth: Penguin Books, 1982), 183.

11. Quoted in Richard Rhodes, *The Making of the Atomic Bomb* (New York: Simon and Schuster, 1986), 675.

12. Andrew Gordon, *A Modern History of Japan: From Tokugawa Times to the Present* (New York: Oxford University Press, 2003), 274.

13. Allan M. Winkler, *Life under a Cloud: American Anxiety about the Atom* (Urbana: University of Illinois Press, 1999), 96.

14. Ralph H. Lutts, "Chemical Fallout: Rachel Carson's *Silent Spring*, Radioactive Fallout, and the Environmental Movement," *Environmental Review* 9 (1985): 210–25.

15. Divine, *Blowing in the Wind*, 239, 317.

16. Vitaly Fedchenko, "Nuclear Explosions, 1945–2013," in Stockholm International Peace Research Institute, ed., *SIPRI Yearbook 2014: Armaments, Disarmament and International Security*, 346–51 (Oxford: Oxford University Press, 2014), 350n.

17. J. Samuel Walker, "The Atomic Energy Commission and the Politics of Radiation Protection, 1967–1971," *Isis* 85 (1994): 57–78; here, 59.

18. Glut, *Classic Movie Monsters*, 383, 387–91.

19. Shoichiro Kawasaki, *Daigo Fukuryu Maru: Present-Day Meaning of the Bikini Incident* (Tokyo: Daigo Fukuryu Maru Foundation, 2008), 56–58.

20. See http://en.japantravel.com/view/lucky-dragon-no-5 and http://www.diveadventures.com/pages/destinations/Micronesia/bikini_lagoon.htm, accessed November 3, 2021.

21. Homei, "Contentious Death," 223. See also Swenson-Wright, *Unequal Allies*, 182.

22. Ran Zwigenberg, "'The Coming of a Second Sun': The 1956 Atoms for Peace Exhibit in Hiroshima and Japan's Embrace of Nuclear Power," *Asia-Pacific Journal* 10, issue 6, no. 1 (February 6, 2012). See also Simon Avenell, "From Fearsome Pollution to Fukushima: Environmental Activism and the Nuclear Blind Spot in Contemporary Japan," *Environmental History* 17 (2012): 244–76; here, 269.

23. John Krige, "Atoms for Peace, Scientific Internationalism, and Scientific Intelligence," *Osiris* 21 (2006): 161–81; here, 180. See also Kenneth Osgood, *Total Cold War: Eisenhower's Secret Propaganda Battle at Home and Abroad* (Lawrence: University Press of Kansas, 2006), 155.

24. Krige, "Atoms for Peace," 174.

25. Paul R. Josephson, *Red Atom: Russia's Nuclear Power Program from Stalin to Today* (Pittsburgh, PA: University of Pittsburgh Press, 2005), 174.

26. Krige, "Atoms for Peace," 174n.

27. Conrad Totman, *A History of Japan*, 2nd ed. (Malden, MA: Blackwell, 2005), 472.

28. Brian Balogh, *Chain Reaction: Expert Debate and Public Participation in American Commercial Nuclear Power, 1945–1975* (Cambridge: Cambridge University Press, 1991), 90, 102, 177.

29. Vaclav Smil, *Energy Myths and Realities: Bringing Science to the Energy Policy Debate* (Washington, DC: AEI Press, 2010), 31.

30. Rhodes, *Making of the Atomic Bomb*, 438. I dedicate this sentence to the research assistants of my Munich years. To the best of my knowledge, they all came out of the job physically unharmed.

31. David Okrent, *Nuclear Reactor Safety: On the History of the Regulatory Process* (Madison: University of Wisconsin Press, 1981), 295n.

32. On the Fukushima Daiichi accident, see International Atomic Energy Agency, *The Fukushima Daiichi Accident: Report by the Director General* (Vienna: International Atomic Energy Agency, 2015), and National Research Council, *Lessons Learned from the Fukushima*

Nuclear Accident for Improving Safety of U.S. Nuclear Plants (Washington, DC: National Academies Press, 2014). On the international reception, see Jens Wolling and Dorothee Arlt, eds., *Fukushima und die Folgen: Medienberichterstattung, Öffentliche Meinung, Politische Konsequenzen* (Ilmenau: Universitätsverlag Ilmenau, 2014).

33. Joachim Radkau, "GAU: Nuclear Reactors and the 'Maximum Credible Accident,'" *Global Environment* 11 (2013): 42–56; 51.

34. Martin V. Melosi, *Atomic Age America* (Boston: Pearson, 2013), 222n.

35. United States Nuclear Regulatory Commission, IE Bulletin No. 75–04A: Cable Fire at Browns Ferry Nuclear Plant, April 3, 1975.

36. Melosi, *Atomic Age America*, 223.

37. Winkler, *Life under a Cloud*, 163.

38. Gabrielle Hecht, *The Radiance of France: Nuclear Power and National Identity after World War II* (Cambridge, MA: MIT Press, 1998), 285n, 297.

39. Masa Takubo, "Closing Japan's Monju Fast Breeder Reactor: The Possible Implications," *Bulletin of the Atomic Scientists* 73 (2017): 182–87; "Monju Reactor project Failed to Pay Off after Swallowing ¥1.13 Trillion of Taxpayers' Money: Auditors," *Japan Times*, May 11, 2018, accessed November 3, 2021, https://www.japantimes.co.jp/news/2018/05/11/national/monju-reactor-project-failed-pay-off-swallowing-¥1-13-trillion-taxpayers-money-auditors/#.W31zqy2X_OQ.

40. Helmut Lackner, "Von Seibersdorf bis Zwentendorf: Die 'friedliche Nutzung der Atomenergie' als Leitbild der Energiepolitik in Österreich," *Blätter für Technikgeschichte* 62 (2000): 201–26; here, 223n.

41. See http://www.world-nuclear.org/info/Country-Profiles/Countries-A-F/Brazil/, accessed November 3, 2021.

42. Frank Uekötter, *Am Ende der Gewissheiten: Die ökologische Frage im 21. Jahrhundert* (Frankfurt: Campus, 2011), 160.

43. Chris Timlinson, "Nuclear Power As We Know It Is Finished," *Houston Chronicle*, August 3, 2017, accessed November 3, 2021, https://www.houstonchronicle.com/business/columnists/tomlinson/article/Nuclear-power-as-we-know-it-is-finished-11727465.php.

44. David E. Lilienthal, *Change, Hope, and the Bomb* (Princeton, NJ: Princeton University Press, 1963), 132.

45. See, for instance, *Zeitschrift für Semiotik* 6, no. 3 (1984).

46. United States Government Accountability Office, Commercial Nuclear Waste: Effects of a Termination of the Yucca Mountain Repository Program and Lessons Learned (GAO-11-229, Washington, April 2011), 2–3, 27.

47. Steven L. Simon and André Bouville, "Health Effects of Nuclear Weapons Testing," *Lancet* 386 (2015): 407–9.

48. Simon and Bouville, "Health Effects," 408. To their credit, the authors hastened to add that the effect was "very substantial to those individuals affected by it" (408).

49. "Fukushima Disaster: Nuclear Executives Found Not Guilty," *BBC News*, September 19, 2019, accessed November 3, 2021, https://www.bbc.co.uk/news/world-asia-49750180.

50. Gerald Jacob, *Site Unseen: The Politics of Siting a Nuclear Waste Repository* (Pittsburgh, PA: University of Pittsburgh Press, 1990), 121.

51. See, for instance, Jacob Darwin Hamblin, "'A Dispassionate and Objective Effort': Negotiating the First Study on the Biological Effects of Atomic Radiation," *Journal of the History of Biology* 40 (2007): 147–77.

52. Office of Nuclear Energy, "7 Things The Simpsons Got Wrong about Nuclear," https://www.energy.gov/ne/articles/7-things-simpsons-got-wrong-about-nuclear, accessed November 3, 2021.

53. See Joachim Radkau, *Aufstieg und Krise der deutschen Atomwirtschaft 1945–1975: Ver-*

drängte Alternativen in der Kerntechnik und der Ursprung der nuklearen Kontroverse (Reinbek: Rowohlt, 1983), and Balogh, *Chain Reaction*.

54. Jacob Darwin Hamblin, "Fukushima and the Motifs of Nuclear History," *Environmental History* 17 (2012): 285-99; here, 292. See also Spencer R. Weart, *The Rise of Nuclear Fear* (Cambridge, MA: Harvard University Press, 2012).

55. Information from Internet Movie Database, http://www.imdb.com/title/tt1921116/, accessed November 3, 2021.

56. Machiko Osawa, Myoung Jung Kim, and Jeff Kingston, "Precarious Work in Japan," *American Behavioral Scientist* 57 (2013): 309-34; here, 326-28.

57. Josephson, *Red Atom*, 264.

58. See Melanie Arndt, *Tschernobylkinder: Die transnationale Geschichte einer nuklearen Katastrophe* (Göttingen: Vandenhoeck & Ruprecht, 2020).

59. Ulrich Beck, *Risk Society: Towards a New Modernity* (London: Sage, 1998), 23.

60. Yukuo Sasamoto, "The Bikini Incident and Radiation Surveys," in Shigeru Nakayama with Kunio Goto and Hitoshi Yoshioka, eds., *A Social History of Science and Technology in Contemporary Japan*, vol. 2: *Road to Self-Reliance 1952-1959*, 125-43 (Melbourne: Trans Pacific Press, 2005), 137.

38: DDT

1. Lukas Straumann, *Nützliche Schädlinge: Angewandte Entomologie, chemische Industrie und Landwirtschaftspolitik in der Schweiz 1874-1952* (Zurich: Chronos, 2005), 203-5.

2. David Kinkela, *DDT and the American Century: Global Health, Environmental Politics, and the Pesticide That Changed the World* (Chapel Hill: University of North Carolina Press, 2011), 29.

3. Nicholas P. Cheremisinoff and Paul E. Rosenfeld, *Handbook of Pollution Prevention and Cleaner Production*, vol. 3: *Best Practices in the Agrochemical Industry* (Amsterdam: William Andrew, 2011), 253.

4. Kinkela, *DDT*, 15.

5. Thomas R. Dunlap, "Introduction," in Dunlap, ed., *DDT, Silent Spring, and the Rise of Environmentalism*, 3-10 (Seattle: University of Washington Press, 2008), 4.

6. James L. A. Webb Jr., *The Long Struggle against Malaria in Tropical Africa* (New York: Cambridge University Press, 2014), 69, 79 (quotation), 83. See also Thomas Zimmer, *Welt ohne Krankheit: Geschichte der internationalen Gesundheitspolitik, 1940-1970* (Göttingen: Wallstein, 2017), 219-53; Marcos Cueto, Theodore M. Brown, and Elizabeth Fee, *The World Health Organization: A History* (Cambridge: Cambridge University Press, 2019), esp. 96-114.

7. Joseph L. Anderson, *Industrializing the Corn Belt: Agriculture, Technology, and Environment, 1945-1972* (DeKalb: Northern Illinois University Press, 2009), 42n.

8. Kinkela, *DDT*, p. 42.

9. Edmund Russell, *War and Nature: Fighting Humans and Insects with Chemicals from World War I to Silent Spring* (Cambridge: Cambridge University Press, 2001), 124n.

10. Thomas R. Dunlap, *DDT: Scientists, Citizens, and Public Policy* (Princeton, NJ: Princeton University Press, 1981), 93; and Gordon Patterson, *The Mosquito Crusades: A History of the American Anti-Mosquito Movement from the Reed Commission to the First Earth Day* (New Brunswick, NJ: Rutgers University Press, 2009), 158-60.

11. See Russell, *War and Nature*.

12. Brigadier General James Stevens Simmons, "How Magic Is DDT?" in Dunlap, *DDT, Silent Spring*, 31-38; here, 38.

13. Straumann, *Nützliche Schädlinge*, 262.

14. Anderson, *Industrializing the Corn Belt*, p. 21.

15. Frank Uekötter, "Why Panaceas Work: Recasting Science, Knowledge, and Fertilizer Interests in German Agriculture," *Agricultural History* 88 (2014): 68–86.

16. Zimmer, *Welt ohne Krankheit*, 246–49.

17. Joshua Blu Buhs, *The Fire Ant Wars: Nature, Science, and Public Policy in Twentieth-Century America* (Chicago: University of Chicago Press, 2004), 1n.

18. I have discussed this process extensively in Frank Uekötter, *Die Wahrheit ist auf dem Feld: Eine Wissensgeschichte der deutschen Landwirtschaft* (Göttingen: Vandenhoeck & Ruprecht, 2010).

19. Linda Lear, *Rachel Carson: Witness for Nature* (New York: Henry Holt, 1997), 319.

20. Lear, *Rachel Carson*, 74, 118n, 228, 233.

21. Priscilla Coit Murphy, *What a Book Can Do: The Publication and Reception of Silent Spring* (Amherst: University of Massachusetts Press, 2005), 4; and Frank Graham Jr., *Since Silent Spring* (London: Hamish Hamilton, 1970), 48.

22. Quoted in Lear, *Rachel Carson*, 335.

23. James Whorton, *Before Silent Spring: Pesticides and Public Health in Pre-DDT America* (Princeton, NJ: Princeton University Press, 1974), 253.

24. Michael Smith, "'Silence, Miss Carson!' Science, Gender, and the Reception of *Silent Spring*," in Lisa H. Sideris and Kathleen Dean Moore, eds., *Rachel Carson: Legacy and Challenge*, 168–87 (Albany: State University of New York Press, 2008).

25. Lear, *Rachel Carson*, 450–55, 471n, 480; and Mark Hamilton Lytle, *The Gentle Subversive: Rachel Carson, Silent Spring, and the Rise of the Environmental Movement* (New York: Oxford University Press, 2007), 188n.

26. Those who feel that this is an overstatement should read Lawrence Culver, Christof Mauch, and Katie Ritson, eds., "Rachel Carson's *Silent Spring*: Encounters and Legacies," *RCC Perspectives*, no. 7 (2012). See also Lytle, *Gentle Subversive*, 219, 256.

27. Rachel Carson, *Silent Spring*, 25th anniversary ed. (Boston: Houghton Mifflin, 1987), 278.

28. Ted Nordhaus and Michael Shellenberger, *Break Through: Why We Can't Leave Saving the Planet to Environmentalists* (Boston: Houghton Mifflin Harcourt, 2009), 130; and Paul B. Thompson, *The Spirit of the Soil: Agriculture and Environmental Ethics* (London: Routledge, 1995), 1.

29. Carson, *Silent Spring*, 297.

30. Christopher J. Bosso, *Pesticides and Politics: The Life Cycle of a Public Issue* (Pittsburgh, PA: University of Pittsburgh Press, 1987), 121.

31. Buhs, *Fire Ant Wars*, 143.

32. Craig Osteen, "Pesticide Use Trends and Issues in the United States," in David Pimentel and Hugh Lehman, eds., *The Pesticide Question: Environment, Economics, and Ethics*, 307–36 (New York: Chapman and Hall, 1993), 308.

33. J. Brooks Flippen, *Nixon and the Environment* (Albuquerque: University of New Mexico Press, 2000), p. 172; Cheremisinoff and Rosenfeld, *Handbook of Pollution Prevention*, 249n; and Dunlap, *DDT* (1981), 5, 209, 234.

34. See Adam Rome, *The Genius of Earth Day: How a 1970 Teach-In Unexpectedly Made the First Green Generation* (New York: Hill and Wang, 2013); Scott Hamilton Dewey, *Don't Breathe the Air: Air Pollution and U.S. Environmental Politics, 1945–1970* (College Station: Texas A&M University Press, 2000); and Frank Uekoetter, *The Age of Smoke: Environmental Policy in Germany and the United States, 1880–1970* (Pittsburgh, PA: University of Pittsburgh Press, 2009).

35. See Jack Doyle, *Taken for a Ride: Detroit's Big Three and the Politics of Pollution* (New York: Four Walls Eight Windows, 2000).

36. Apoorva Mandavilli, "DDT Returns," *Nature Medicine* 12, no. 8 (August 2006): 870–71; here, 870.

37. Cheremisinoff and Rosenfeld, *Handbook of Pollution Prevention*, 249.

38. Bosso, *Pesticides and Politics*, 141. See also Frederick Rowe Davis, *Banned: A History of Pesticides and the Science of Toxicology* (New Haven, CT: Yale University Press, 2014).

39. James L. A. Webb Jr., *Humanity's Burden: A Global History of Malaria* (New York: Cambridge University Press, 2009), 166–72; Webb, *Long Struggle*, 93–95; Charles C. Hughes and John M. Hunter, "The Role of Technological Development in Promoting Disease in Africa," in M. Taghi Farvar and John P. Milton, eds., *The Careless Technology: Ecology and International Development*, 69–101 (London: Tom Stacey, 1973), 82n.

40. Buhs, *Fire Ant Wars*, 184n.

41. Michael Crichton, *State of Fear: A Novel* (New York: HarperCollins, 2004), 487.

42. Frank M. Snowden, *The Conquest of Malaria: Italy, 1900–1962* (New Haven, CT: Yale University Press, 2006), 201.

43. Frank Snowden, "Latina Province, 1944–1950," *Journal of Contemporary History* 43 (2008): 509–26; here, 522.

44. Snowden, *Conquest of Malaria*, 201.

45. Snowden, "Latina Province," 524.

46. Richard Tren and Roger Bate, *Malaria and the DDT Story* (London: Institute of Economic Affairs, 2001), 56, 58.

47. Apoorva Mandavilli, "Health Agency Backs Use of DDT against Malaria," *Nature* 443 (2006): 250–51.

48. Kinkela, *DDT*, 179n; and U.S. Department of Health and Human Services, Public Health Service, Agency for Toxic Substances and Disease Registry, Toxicological Profile for DDT, DDE, and DDD, Atlanta, GA, September 2002, accessed January 6, 2014, http://www.atsdr.cdc.gov/toxprofiles/tp35.pdf, p. 222.

49. David Leonard Downie, Jonathan Krueger, and Henrik Selin, "Global Policy for Hazardous Chemicals," in Regina S. Axelrod, David Leonard Downie, and Norman J. Vig, eds., *The Global Environment: Institutions, Law, and Policy*, 2nd ed., 125–45 (Washington: CQ Press, 2005); Philippe Sands et al., *Principles of International Environmental Law*, 3rd ed. (Cambridge: Cambridge University Press, 2012), 523–26; and Patricia Birnie, Alan Boyle, and Catherine Redgwell, *International Law and the Environment*, 3rd ed. (Oxford: Oxford University Press, 2009), 448–51.

50. See Bosso, *Pesticides and Politics*; Pimentel and Lehman, *Pesticide Question*.

51. Lytle, *Gentle Subversive*, 226.

39: TORREY CANYON

1. For the government's account of the course of events, see Cabinet Office, *The* Torrey Canyon: *Report of the Committee of Scientists on the Scientific and Technological Aspects of the* Torrey Canyon *Disaster* (London: Her Majesty's Stationery Office, 1967), 5–9. For two accounts from journalists, see Edward Cowan, *Oil and Water: The* Torrey Canyon *Disaster* (Philadelphia: Lippincott, 1968), and Richard Petrow, *The Black Tide: In the Wake of* Torrey Canyon (London: Hodder and Stoughton, 1968). For an overview that concentrates on political responses, see John Sheail, "*Torrey Canyon*: The Political Dimension," *Journal of Contemporary History* 42 (2007): 485–504.

2. Charles Ryle Fay, *Great Britain from Adam Smith to the Present Day: An Economic and Social Survey* (London: Longmans, 1953), 45.

3. Katja Patzel-Mattern, "Von der Unmöglichkeit nicht zu kommunizieren: Unternehmerische Kommunikation nach industriellen Unfällen, BASF 1921 und Hoffmann-La Roche 1976," *Jahrbuch für Wirtschaftsgeschichte* 57 (2016): 423–53; here, 426.

4. Hugh W. Stephens, *The Texas City Disaster, 1947* (Austin: University of Texas Press, 1997), pp. 6, 100.

5. Daniel Yergin, *The Prize: The Epic Quest for Oil, Money and Power* (New York: Simon and Schuster, 1992), 352.

6. Hans Peter Bläuer, "Der Bergsturz von Elm am 11. September 1881: Ursache und gesellschaftliche Bewältigung einer menschengemachten Naturkatastrophe," in Christian Pfister, ed., *Am Tag Danach: Zur Bewältigung von Naturkatastrophen in der Schweiz 1500–2000*, 113–31 (Bern: Haupt, 2002), 118n.

7. Hans Joachim Ritzau, *Schatten der Eisenbahngeschichte: Ein Vergleich britischer, US- und deutscher Bahnen*, vol. 1: *Von den Anfängen bis 1945* (Pürgen: Ritzau, Verlag Zeit und Eisenbahn, 1987), 163.

8. On the reception of the Lisbon earthquake, see Christoph Daniel Weber, *Vom Gottesgericht zur verhängnisvollen Natur: Darstellung und Bewältigung von Naturkatastrophen im 18. Jahrhundert* (Hamburg: Meiner, 2015).

9. Susan Sontag, *Regarding the Pain of Others* (London: Penguin Books, 2003), 16.

10. Kathryn Morse, "There Will Be Birds: Images of Oil Disasters in the Nineteenth and Twentieth Centuries," *Journal of American History* 99 (2012): 124–34; here, 129.

11. Tony Tran, Aida Yazdanparast, and Eric A. Suess, "Effect of Oil Spill on Birds: A Graphical Assay of the Deepwater Horizon Oil Spill's Impact on Birds," *Computational Statistics* 29 (2014): 133–40; here, 139.

12. Sontag, *Regarding the Pain of Others*, 94.

13. Cabinet Office, *Torrey Canyon*, 44. A global survey listed the *Torrey Canyon* as number 18 among the worst oil spills up to January 2010. According to this list, it was by far the worst disaster in 1967. (Dagmar Schmidt-Etken, "Spill Occurrences: A World Overview," in Mervin Fingas, ed., *Oil Spill Science and Technology: Prevention, Response, and Cleanup*, 7–48 [Amsterdam: Elsevier, 2011], 13–15.)

14. Cabinet Office, *Torrey Canyon*, 44. See also Petrow, *Black Tide*, 218–25.

15. See Timothy Cooper and Anna Green, "The *Torrey Canyon* Disaster, Everyday Life, and the 'Greening' of Britain," *Environmental History* 22 (2017): 101–26; here, 106–10.

16. British authorities spread no less than 10,000 tons of detergent fluids to cope with some 14,000 tons of oil that had washed up on Cornish beaches. (J. E. Smith, *"Torrey Canyon": Pollution and Marine Life; A Report by the Plymouth Laboratory of the Marine Biological Association of the United Kingdom* [Cambridge: Cambridge University Press, 1968], 11.)

17. Cabinet Office, *Torrey Canyon*, 6.

18. Smith, "Torrey Canyon," 7.

19. Cowan, *Oil and Water*, 176n, 184–86.

20. Cabinet Office, *Torrey Canyon*.

21. Gaetano Librando, "Influence of the *Torrey Canyon* Incident on the Liability and Compensation Regimes Developed under the Auspices of the IMO," in Norman A. Martínez Gutiérrez, ed., *Serving the Rule of International Maritime Law: Essays in Honour of Professor David Joseph Attard*, 315–27 (London: Routledge, 2010), 316.

22. Commandant L. Oudet, *In the Wake of the* Torrey Canyon: *Reflections on a Disaster* (London: Royal Institute of Navigation, 1972), 10n.

23. L. R. Beynon, *The* Torrey Canyon *Incident: A Review of Events* ([London?]: British Petroleum Limited, 1967), 1.

24. Cowan, *Oil and Water*, 195.

25. Paul Burrows, Charles Rowley, and David Owen, "The Economics of Accidental Oil Pollution by Tankers in Coastal Waters," *Journal of Public Economics* 3 (1974): 251–68; here, 260.

26. Librando, "Influence."

27. Stephen Haycox, "'Fetched Up': Unlearned Lessons from the *Exxon Valdez*," *Journal of American History* 99 (2012): 219–28; here, 223.

28. Zoran Perunovic and Jelena Vidic-Perunovic, "Environmental Regulation and Innovation Dynamics in the Oil Tanker Industry," *California Management Review* 55, no. 1 (2012): 130–48; here, 135–37.

29. See Cowan, *Oil and Water*, 36–45; and Petrow, *Black Tide*, 11–18, 33–49.

30. Charles Perrow, *Normal Accidents: Living with High-Risk Technologies* (New York: Basic Books, 1984). For his account of Three Mile Island, see pp. 15–31.

31. Charles Perrow, *The Next Catastrophe: Reducing Our Vulnerabilities to Natural, Industrial, and Terrorist Disasters* (Princeton, NJ: Princeton University Press, 2007), 142.

32. Perrow, *Normal Accidents*, 16; and Cowan, *Oil and Water*, 10, 32.

33. Perrow, *Next Catastrophe*, 142.

34. Paul Shrivastava, *Bhopal: Anatomy of a Crisis*, 2nd ed. (London: Paul Chapman, 1992), 55.

35. Kim Fortun, *Advocacy after Bhopal: Environmentalism, Disaster, New Global Orders* (Chicago: University of Chicago Press, 2001), 27n.

36. Jack Healy and Christine Hauser, "Inside Southwest Flight 1380, 20 Minutes of Chaos and Terror," *New York Times*, April 18, 2018, accessed November 3, 2021, https://www.nytimes.com/2018/04/18/us/southwest-plane-engine-failure.html.

37. See Robert D. Bullard, *Dumping in Dixie: Race, Class, and Environmental Quality* (Boulder, CO: Westview, 1994); and Ellen Griffith Spears, *Baptized in PCBs: Race, Pollution, and Justice in an All-American Town* (Chapel Hill: University of North Carolina Press, 2014).

38. Cabinet Office, *Torrey Canyon*, 3n.

39. Mervin Fingas, "Introduction," in Fingas, ed., *Oil Spill Science and Technology: Prevention, Response, and Cleanup*, 3–5 (Amsterdam: Elsevier, 2011), 5.

40. Phia Steyn, "Oil, Ethnic Minority Groups and Environmental Struggles against Multinational Oil Companies and the Federal Government in the Nigerian Niger Delta since the 1990s," in Marco Armiero and Lise Sedrez, eds., *A History of Environmentalism: Local Struggles, Global Histories*, 57–81 (London: Bloomsbury, 2014).

41. "A Nation Unprepared," *Spectator* 218, no. 7240 (March 31, 1967): 357–58.

42. Cabinet Office, *Torrey Canyon*, 44.

43. See Fortun, *Advocacy after Bhopal*.

44. Petrow, *Black Tide*, 127, 135; and Crispin Gill, Frank Booker, and Tony Soper, *The Wreck of the Torrey Canyon* (Newton Abbot: David & Charles, 1967), 88–103, 123–28.

45. Cabinet Office, *Torrey Canyon*, 42.

46. Petrow, *Black Tide*, 136n.

47. Joachim Fest, *Inside Hitler's Bunker: The Last Days of the Third Reich* (London: Macmillan, 2004), 76.

48. Manfred P. Fleischer, "John Maynard—Dichtung und Wahrheit," *Zeitschrift für Religions- und Geistesgeschichte* 16, no. 2 (1964): 168–73.

40: PLASTIC BAGS

1. "Fünf rätselhafte Objekte begleiten 'Atlantis,'" *Spiegel*, September 20, 2006, accessed November 3, 2021, http://www.spiegel.de/wissenschaft/weltall/space-shuttle-fuenf-raetselhafte-objekte-begleiten-atlantis-a-438188.html. See also Gene Stansbery, "Space Waste," in Trevor M. Letcher and Daniel A. Vallero, eds., *Waste: A Handbook for Management*, 377–91 (Amsterdam: Elsevier, 2011).

2. This comment focuses on the immediate cause in technical terms. For insights into the wider context of NASA's safety culture, see Diane Vaughan, *The Challenger Launch Decision: Risky Technology, Culture, and Deviance at NASA* (Chicago: University of Chicago Press,

1996), and Edward R. Tufte, *The Cognitive Style of PowerPoint: Pitching Out Corrupts Within* (Cheshire, CT: Graphic Press, 2006).

3. Jeffrey L. Meikle, *American Plastics: A Cultural History* (New Brunswick, NJ: Rutgers University Press, 1997), 30.

4. Alison J. Clarke, *Tupperware: The Promise of Plastics in 1950s America* (Washington, DC: Smithsonian Institution Press, 1999), 2.

5. Susannah Handley, *Nylon: The Story of a Fashion Revolution* (Baltimore: Johns Hopkins University Press, 1999).

6. Peter Dauvergne, *The Shadows of Consumption: Consequences for the Global Environment* (Cambridge, MA: MIT Press, 2008), 49.

7. Neil MacGregor, *A History of the World in 100 Objects* (London: Penguin, 2010), 647–51.

8. Meikle, *American Plastics*, 8.

9. See Katja Böhme and Andreas Ludwig, eds., *Alles aus Plaste: Versprechen und Gebrauch in der DDR* (Cologne: Böhlau, 2012).

10. Roland Barthes, *Mythologies*, selected and translated from the French by Annette Lavers (London: Jonathan Cape, 1972), 97.

11. Barthes, *Mythologies*, 98, 99.

12. For an overview of the development of plastic shopping bags, see Heinz Schmidt-Bachem, *Tüten, Beutel, Tragetaschen: Zur Geschichte der Pappe, Papier und Folien verarbeitenden Industrie in Deutschland* (Münster: Waxmann, 2001), 222–40.

13. Susan Mossman, "Perspectives on the History and Technology of Plastics," in Mossman, ed., *Early Plastics: Perspectives, 1850–1950*, 15–71 (London: Leister University Press in association with Science Museum, 1997), 54–57.

14. Meikle, *American Plastics*, 1.

15. Anne Sudrow, *Der Schuh im Nationalsozialismus: Eine Produktgeschichte im deutsch-britisch-amerikanischen Vergleich* (Göttingen: Wallstein, 2010), 526, 531, 536.

16. Vance Packard, *The Waste Makers* (London: Longmans, 1961), 107. For Packard's view of consumer society, see Daniel Horowitz, *The Anxieties of Affluence: Critiques of American Consumer Culture, 1939–1979* (Amherst: University of Massachusetts Press, 2004), 108–20.

17. Robert Begiebing, "Twelth Round: An Interview with Norman Mailer," in J. Michael Lennon, ed., *Conversations with Norman Mailer*, 306–29 (Jackson: University Press of Mississippi, 1988), 321.

18. Begiebing, "Twelth Round," 322.

19. Meikle, *American Plastics*, 249–52.

20. Meikle, *American Plastics*, 252n. On Heckman, see also Sarah A. Vogel, *Is It Safe? BPA and the Struggle to Define the Safety of Chemicals* (Berkeley: University of California Press, 2013), 15–19.

21. "Jerome H. Heckman," accessed November 3, 2021, https://www.plasticshof.org/members/jerome-h-heckamn. See also his obituary in the *Washington Post*, January 23/24, 2013, accessed November 3, 2021, http://www.legacy.com/obituaries/washingtonpost/obituary.aspx?pid=162584255.

22. Maggie Murdoch, "The Road to Zero Waste: A Study of the Seattle Green Fee on Disposable Bags," *Environmental Practice* 12 (2010): 66–75; here, 69.

23. Susan Strasser, *Waste and Want: A Social History of Trash* (New York: Henry Holt, 1999), 211–13.

24. See Geheimes Staatsarchiv Berlin I. HA Rep. 87 B no. 16293.

25. Schmidt-Bachem, *Tüten*, 234–47.

26. Rudolf Strahm, "Der aktionserprobte Achtundsechziger im Team der EvB 1974–

1978," in Anne-Marie Holenstein, Regula Renschler, and Rudolf Strahm, eeds., *Entwicklung heisst Befreiung: Erinnerungen an die Pionierzeit der Erklärung von Bern (1968-1985)*, 113-66 (Zurich: Chronos, 2008), 134-40.

27. Heinz Schmidt-Bachem, *Aus Papier: Eine Kultur- und Wirtschaftsgeschichte der Papier verarbeitenden Industrie in Deutschland* (Berlin: De Gruyter, 2011), 818n.

28. Chris Edwards and Jonna Meyhoff Fry, *Life Cycle Assessment of Supermarket Carrier Bags: A Review of the Bags Available in 2006* (Bristol: Environment Agency, 2011).

29. Jamie Furniss, "Alternative Framings of Transnational Waste Flows: Reflections Based on the Egypt-China PET Plastic Trade," *Area* 47 (2015): 24-30; here, 25; Yvonne A. Braun and Assitan Sylla Traore, "Plastic Bags, Pollution, and Identity: Women and the Gendering of Globalization and Environmental Responsibility in Mali," *Gender and Society* 29 (2015): 863-87; here, 864; and Andy Fourie, "Balancing Environmental and Social Considerations in a Developing Country: The Great South African Plastic Bag Debate," *Waste Management* 24 (2004): 531-640; here, 531.

30. Edwards and Fry, *Life Cycle Assessment*, 61.

31. Martina Häßler, "Wegwerfen: Zum Wandel des Umgangs mit Dingen," *Zeitschrift für Erziehungswissenschaft* 16 (2013), supplement 2: 253-66; here, 260.

32. Émilie Clavel, "Think You Can't Live Without Plastic Bags? Consider This: Rwanda Did It," *The Guardian*, February 15, 2014, accessed November 3, 2021, https://www.theguardian.com/commentisfree/2014/feb/15/rwanda-banned-plastic-bags-so-can-we.

33. "Mauritania Bans Plastic Bag Use," *BBC News*, January 2, 2013, accessed November 3, 2021, https://www.bbc.co.uk/news/world-africa-20891539.

34. Frank Convery, Simon McDonnell, and Susana Ferreira, "The Most Popular Tax in Europe? Lessons from the Irish Plastic Bags Levy," *Environmental and Resource Economics* 38 (2007): 1-11; here, 9.

35. Convery, McDonnell, and Ferreira, "Most Popular Tax?" 3, 6n.

36. Tariq Malik, "Atlantis Landing Delayed After Mystery Object Spotted," accessed November 3, 2021, http://www.space.com/2915-atlantis-landing-delayed-mystery-object-spotted.html.

37. Jens Soentgen, "Dissipation," in Kijan Espahangizi and Barbara Orland, eds., *Stoffe in Bewegung: Beiträge zu einer Wissensgeschichte der materiellen Welt*, 275-83 (Zurich: diaphanes, 2014), 282.

38. Dauvergne, *Shadows of Consumption*, 49.

39. John R. McNeill and Peter Engelke, *The Great Acceleration: An Environmental History of the Anthropocene since 1945* (Cambridge, MA: Belknap Press of Harvard University Press, 2014), 138.

40. W. R. P. Bourne, "Marine Debris," *Marine Pollution Bulletin* 20 (1989): 579-80.

41. David W. Laist, "Overview of the Biological Effects of Lost and Discarded Plastic Debris in the Marine Environment," *Marine Pollution Bulletin* 18 (1987), 319-26; here, 320.

42. See Charles Moore with Cassandra Phillips, *Plastic Ocean: How a Sea Captain's Chance Discovery Launched a Determined Quest to Save the Oceans* (New York: Avery, 2012).

43. Alan Weisman, *The World Without Us* (New York: St. Martin's Press, 2007), 114.

44. See Arjun Makhijani and Kevin R. Gurney, *Mending the Ozone Hole: Science, Technology, and Policy* (Cambridge, MA: MIT Press, 1995).

45. Weisman, *World Without Us*, 124-28.

46. Daniel R. Headrick, *Humans versus Nature: A Global Environmental History* (New York: Oxford University Press, 2020), 4. Similarly, Daniel R. Headrick, *Power over Peoples: Technology, Environments, and Western Imperialism, 1400 to the Present* (Princeton, NJ: Princeton University Press, 2010), 370.

CODA
THE PANDEMIC

1. See Peter Alagona et al., "Reflections: Environmental History in the Era of COVID-19," *Environmental History* 25 (2020): 595-686; "COVID-19 & Environmental History," *Journal for the History of Environment and Society* 5 (2020); and *World History Bulletin* 36, no. 1 (2020).

THE MESS WE'RE IN
AN INCONCLUSIVE CONCLUSION

1. For more on this topic, see Frank Uekötter, ed., *Exploring Apocalyptica: Coming to Terms with Environmental Alarmism* (Pittsburgh, PA: University of Pittsburgh Press, 2018).

2. See Frank Uekötter, "Confronting the Pitfalls of Current Environmental History: An Argument for an Organisational Approach," *Environment and History* 4 (1998): 31-52; Frank Uekötter, "Wie Neu sind die Neuen Sozialen Bewegungen? Revisionistische Bemerkungen vor dem Hintergrund der umwelthistorischen Forschung," *Mitteilungsblatt des Instituts für soziale Bewegungen* 31 (2004): 109-31; Frank Uekötter, "Native Plants: A Nazi Obsession?" *Landscape Research* 32 (2007): 379-83; Frank Uekötter, "Consigning Environmentalism to History? Remarks on the Place of the Environmental Movement in Modern History," *RCC Perspectives* 7 (2011): 3-35; Frank Uekötter, "Eine ökologische Ära? Perspektiven einer neuen Geschichte der Umweltbewegungen," *Zeithistorische Forschungen / Studies in Contemporary History* 9 (2012): 108-14; Frank Uekötter, "Myths, Big Myths and Global Environmentalism," in Stefan Berger and Holger Nehring, eds., *The History of Social Movements in Global Perspective: A Survey*, 419-47 (London: Palgrave Macmillan, 2017); and Frank Uekötter, "Umweltgeschichte in Umbruchszeiten: Von der Entgrenzung eines Forschungsfeldes," *Geschichte für heute* 14 (2021): 7-22.

3. For a classic in this vein, see Samuel P. Hays with Barbara D. Hays, *Beauty, Health, and Permanence: Environmental Politics in the United States, 1955-1985* (New York: Cambridge University Press, 1987).

4. See Frank Uekötter, convener, "Roundtable: Should Agricultural Historians Care about the New Materialism?" *Agricultural History* 96 (2022): 223-70. For some key readings on the "new materialism," see Timothy J. LeCain, *The Matter of History: How Things Create the Past* (Cambridge: Cambridge University Press, 2017); John Bellamy Foster, *Marx's Ecology: Materialism and Nature* (New York: Monthly Review Press, 2000); Marina Fischer-Kowalski and Helmut Haberl, eds., *Socioecological Transitions and Global Change: Trajectories of Social Metabolism and Land Use* (Cheltenham: Edward Elgar, 2007); Jane Bennett, *Vibrant Matter: A Political Ecology of Things* (Durham, NC: Duke University Press, 2010); Dipesh Chakrabarty, "The Climate of History: Four Theses," *Critical Inquiry* 35 (2009): 197-222; and Bruno Latour, "Can We Get Our Materialism Back, Please?" *Isis* 98 (2007): 138-42.

5. For a recent attempt, see Frank Trentmann, *Empire of Things: How We Became a World of Consumers, from the Fifteenth Century to the Twenty-First* (London: Allen Lane, 2016).

6. For my understanding of jurisdiction, see Andrew Abbott, *The System of Professions: An Essay on the Division of Expert Labor* (Chicago: University of Chicago Press, 1988).

7. Jason W. Moore, *Capitalism in the Web of Life: Ecology and the Accumulation of Capital* (London: Verso, 2015), 53. More recently, Moore offered a diluted version of this argument that offers more wooly categories: "cheap nature," "cheap money," "cheap work," "cheap care," and "cheap lives." (Raj Patel and Jason W. Moore, *A History of the World in Seven Cheap Things: A Guide to Capitalism, Nature, and the Future of the Planet* [Oakland:

University of California Press, 2017]). Unlike its sequel, the original argument focuses on factors of industrial production and offers perspectives beyond moral condemnation.

8. Karl William Kapp, *The Social Costs of Private Enterprise* (Cambridge, MA: Harvard University Press, 1950).

9. It should be noted that, as categories go, the Global South is one of the more difficult ones. First, it comprises an extraordinary diversity. It includes authoritarian regimes and failed states, megacities and rural regions, tropical rain forests and the penguins of Argentina, and so forth. Second, inclusion ultimately hinges on things that the Global South did *not* have: the resources that wealthy industrialized countries had at their disposal when they confronted environmental challenges. The Global South has faced most of the challenges within a few years that Western countries dealt with over the course of a century and more, and the Global South had command over less money, expertise, and experience. It is this common predicament that the term seeks to highlight, knowing that the ways in which countries in the Global South dealt with this predicament differed considerably.

10. Paul S. Sutter, "The World with Us: The State of American Environmental History," *Journal of American History* 100 (2013): 94–119; here, 104.

11. See John R. McNeill and Peter Engelke, *The Great Acceleration: An Environmental History of the Anthropocene since 1945* (Cambridge, MA: Belknap Press of Harvard University Press, 2014).

12. See, for instance, Saleem H. Ali, *Treasures of the Earth: Need, Greed, and a Sustainable Future* (New Haven, CT: Yale University Press, 2009), 236.

13. See p. 302.

14. Oswald Spengler, *Man and Technics: A Contribution to a Philosophy of Life* (New York: Alfred A. Knopf, 1932), 104.

15. Theodor W. Adorno, *Negative Dialectics*, trans. E. B. Ashton (London: Routledge & Kegan Paul, 1973), p. 3.

16. See p. 357. An eminent historian has pointed out that this statement evokes a Ralph Waldo Emerson quote that "doesn't fit Frank's meaning, because Emerson was arguing one has to change one's views when facts and circumstances change, rather than cling to them when overtaken by events." I am not sure that this is philologically correct, but the comment does support my general point.

17. Paul S. Sutter, "Nature Is History," *Journal of American History* 100 (2013): 145–48; here, 146.

APPENDIX
MAKING CHOICES

1. See Jan Assmann, *Cultural Memory and Early Civilization: Writing, Remembrance, and Political Imagination* (New York: Cambridge University Press, 2011).

2. Michael Bess, *The Light-Green Society: Ecology and Technological Modernity in France, 1960–2000* (Chicago: University of Chicago Press, 2003), 4.

3. I have certainly argued in this vein: see Frank Uekötter, "Zeit für einen Plan B: Der gescheiterte Klimagipfel von Kopenhagen ist auch eine Chance für die Umweltpolitik," *Umwelt aktuell* 3 (March 2010): 3–5.

4. Paul S. Sutter, *Let Us Now Praise Famous Gullies: Providence Canyon and the Soils of the South* (Athens: University of Georgia Press, 2015), 2.

5. For more on this point, see Frank Uekötter, "The Meaning of Moving Sand: Towards a Dust Bowl Mythology," *Global Environment* 8 (2015): 349–79; and Frank Uekötter, "In Search of a Dust Bowl Narrative for the Twenty-First Century," *Great Plains Quarterly* 40 (2020): 161–68.

6. This point is an outgrowth of my attempt to deal with a diffuse commodity: see Frank Uekötter, "Recollections of Rubber," in Dominik Geppert and Frank Lorenz Müller, eds., *Sites of Imperial Memory: Commemorating Colonial Rule in the Nineteenth and Twentieth Century*, 243–65 (Manchester: Manchester University Press, 2015).

7. Fernand Braudel, *Capitalism and Material Life, 1400–1800* (London: Weidenfeld and Nicolson, 1973), 183. This is one reason that opium, originally a chapter topic, was demoted to an "interlude."

8. For the latter topic, see Stefan Esselborn, "Environment, Memory, and the Groundnut Scheme: Britain's Largest Colonial Agricultural Development Project and Its Global Legacy," *Global Environment* 11 (2013): 58–93.

INDEX

Places are listed under the country's current name. Where historical names differ, that name is listed as well. Chapter topics are printed in **bold**.

Aboriginal people, 116
absolutism, 35, 65–66, 230
Abu Dhabi, 229
Abu Ghraib, 207
Académie des Sciences, 65
Acapulco, 42, 130
Achard, Franz Karl, 56
Achelous River, 436
acid rain, 83, 255
ackee fruit, 121, 125
Addams, Jane, 273
addiction, 171, 235–36, 302, 305–6, 622
Addis Ababa, 407
Adelaide, 212
Adenauer, Konrad, 509, 513, 515, 524–25
Adorno, Theodor W., 636
aerial dusting, 195
Afghanistan, 298, 306–7, 342, 436–37, 445
African National Congress, 403
Afrikaners, 208, 392
Agde, 69
Agency for International Development (USA), 107
Agenda 21, 365
Agent Orange, 529
Agricola, Georg, 40
Agricultural Path, the, 22
agriculture. *See* Agricultural Path, the
agrochemistry, 134–37. *See also* fertilizer; synthetic nitrogen
agroforestry, 84–85, 127, 447–50
Ahmedabad, 346–47, 357
air pollution, 239, 245–57, 266, 271, 315, 331–32, 359, 361–64, 402
air travel, 70–71, 195, 331, 591
air-conditioning, 238, 242–43, 294–302, 329, 341, 377, 401, 444, 515, 600, 603, 617, 619, 622, 627–28, 648

al-Banna, Hassan, 71–72
al-Qantara, 72
Alabama, 189, 216, 288–89, 575
Alamo Mission, 296
Alaska, 589
Albania, 79
Albert, Prince, 176
alcohol, 232, 303–5, 411, 465, 497–505, 520
Alfonso XII (Spain), 373
Algeria, 201, 407
Alice in Wonderland, 166, 176, 180, 185, 633
Allende, Salvador, 451
Alliance for a Green Revolution in Africa, 428
alligators, 398
Almadén, 42
almonds, 60
Alps, 262, 266, 335, 368, 401
alternative projects, 15, 314, 355–58
aluminum, 522, 647
Amazon rain forest, 46, 127, 134, 350, 365, 388, 409, 443, 448, 455–57, 641
Amazon River, 43
American Academy of Arts and Letters, 577–78
American Chemistry Council, 599
American Civic Association, 393
American Farm Bureau Federation, 496
American Geographical Society, 484, 577
American Humane Society, 274
American Medical Association, 598
American Relief Administration, 477
American Society of Civil Engineers, 266
Americus, Georgia, 205
Amnesty International, 457
Amsterdam, 475, 478
Amundsen, Roald, 381
Anaconda, 158
anchoveta, 103, 140–41
Andean condor, 140

Andersen, Hans Christian, 597
Andes, 33, 39, 42, 48, 227, 527
Anglo-Iranian Oil Company, 47
Anglo-Oriental Society for the Suppression of the Opium Trade, 306
Anglo-Persian Oil Company, 228
Angola, 69, 391
animal welfare, 209, 241, 274–75, 278–79, 285, 347, 534–35, 538, 540, 553–56, 628, 632
animals. *See* Path of the Animals, the
Antarctica, 146–47, 381, 646
antelope, 401
Anthropocene, 5, 10, 15–18, 290, 366, 532–33, 615–16, 630–31
anthroposophy, 285
Anti-Trust Company, 155–56
antibiotics, 548–49
Antigua, 51, 215, 217
antisemitism, 467, 494
antitrust proceedings, 157, 161, 241, 272–73
Antony Gibbs and Sons, 675n6
Anzio, 488
apartheid, 352, 391, 401, 403, 640
apocalyptic thinking, 97, 199, 204, 211, 612, 616
Apollo, 645
Appalachia, 47, 139
appeasement, 60
applied entomology. *See* entomology
aquaculture, 551
aqueducts, 263–64, 431, 641
Arab Agricultural Revolution, 51
Arab Revolt, 340
Arab seafaring, 173
Arab Spring, 226
Arabian-American Oil Company. *See* Saudi Aramco
arabica coffee, 191
Arbenz Guzmán, Jacobo, 159–61
archaeology, 91, 259, 309–10, 435, 439, 517, 603, 620
Arctic, 381
Arctic Sea, 143, 152
Argentina, 46, 57, 150, 152, 160, 200, 375, 407
Aristotle, 134
Arizona, 296, 434
Armour, 270, 272, 277
Armstrong, Neil, 595
Army Corps of Engineers, 67
arsenic, 47–48, 191, 573–74, 627. *See also* calcium arsenate
Arun III project, 443–44
Arzáns de Orsúa y Vela, Bartolomé, 39, 43–46

asbestos, x, 95, 97, 647
Ascension crake, 180
Ashurbeyli, Nabat, 265
Aspen, 43
Assmann, Jan, 641
Asturias, Miguel Ángel, 158
Aswan Dam, 113, 289–90, 300, 386–87, 429–37, 439–45, 627, 633, 635
Atchison, Kansas, 499–500
Atlacamani, 372
Atlacoya, 372
Atlanta, 70, 194
Atlantis, 595–96, 602–3
Atlantropa, 431
Atlas Economic Research Foundation, 556
atomic energy. *See* nuclear power
Atomic Energy Commission, 558, 563, 565, 568
Attenborough, Richard, 348
Audubon Society, 150
Augé, Marc, 518
Augustus, 65
Australia, 44, 47, 53, 116, 121, 127, 138–40, 169, 171, 179, 201, 211–24, 228, 263, 267, 327, 341, 383–84, 392, 394, 400, 405–6, 408–10, 412, 415, 433, 529, 620, 628, 634, 677n45
Austria, 40, 92, 251, 265–66, 321, 329, 342, 361, 366, 545, 567
autarky, 57, 83, 147, 150, 388, 463, 525, 597, 625
autobahn, 32, 67, 335, 460–61, 465, 509–12, 514–15, 517–18, 633, 644
Automobil-Verkehrs- und Übungsstraße (AVUS), 506, 509, 513, 515–16
automobilism, 16, 32, 228, 235, 277, 296, 300, 362, 401, 448, 465–66, 468, 497–518, 538, 596, 602–3, 619, 621–24, 627
Auyuittuq National Park, 399
Avengers: Age of Ultron, 97
avian malaria, 219
Avicenna, 304
Ayako, Tsutsui, 523
Azerbaijan, 265
Aztecs, 372, 418

Bacon, Francis, 326
bacteria, 181, 188, 197, 290, 299, 320, 329, 426
bacteriology, 324–25, 329
Baedeker, 262, 312–13, 333–42, 344, 640
Baedeker, Karl, 312, 333–34, 340
Baffin Island, 399
Baghdad, 207, 309–10, 327, 510, 513
Bahamas, 59–60, 335
Bahrain, 229

784

INDEX

Bahre, Louis, 145
Bahro, Rudolf, 462
Bakelite, 596–97
Bakiyev, Kurmanbek, 444
Baku, 265
Balikpapan, 519, 585
Baltic Sea, 67, 83
Baltimore, 56, 96, 263, 267
Banaba, 246
banana war, 162
bananas, 19, 53–54, 126, 154–64, 188, 191, 197, 339, 343, 356, 455
Bangladesh, 18, 32, 36–37, 87–90, 93–97, 474, 600, 627, 646
Bangladesh Environmental Lawyers Association, 96
Banks, Joseph, 102, 119, 121, 125–27, 383, 406, 415
Barbados, 51–52, 54, 60, 212, 215, 335, 411
barbed wire, 110, 648
Barcelona, 323, 330
Bardot, Brigitte, 347
Barracuda Tanker Company, 589
Barthes, Roland, 597
Basel Convention on the Control of Transboundary Movements of Hazardous Wastes, 95
BASF, 135, 281–83, 285, 288, 566, 585, 640
Basques, 103, 143
Basra, 327
Bath, 341
Batista, Fulgencio, 161
battery cages, 114, 278, 534–35, 549–50, 553–56, 627, 648. *See also* chicken
battle for grain, 464, 482, 490
Batumi, 228
bauxite, 46
Bavaria, 135, 304, 320, 327, 428, 526
Bay of Bengal, 87, 176
Bayer, 56, 305, 427
Bayreuth, 339
Beard, Daniel P., 444
beautification movement, 644
Beaver Committee, 253, 255, 266
Beck, Ulrich, 571
Bedford, 531
Bedouins, 72
beer, 44, 218, 304
Beethoven, Ludwig van, 5
Behrens, Peter, 431
Beijing, 255, 370
Beirut, 343
Belarus, 181
Belfast, 517
Belgium, 57, 122, 335, 344, 382, 401, 477, 492, 514, 529
Belize, 393, 411
Bell, Alexander Graham, 294
Bell, Arthur, 219
Belo Monte Dam, 350, 444
Benares, 305
Benford, Gregory, 185
Bengal, 408–9, 474
Benin, 516
Bennett, Hugh Hammond, 203–7, 489, 633
Bentham, Jeremy, 108
benzene hexachloride, 215
Bergen-Belsen, 480
Berlin, 56, 95, 250, 265, 327, 334–36, 340, 344, 410, 423, 472, 506, 509–10, 517, 594
Bermuda, 221, 545, 589, 646
Bernhardi, Theodor von, 111
Bernthsen, August, 281–82, 284
Béziers, 68–70
Bhopal, 586, 588, 591, 593
Bhutan, 4
Białowieża Forest, 181
bible, 75, 126, 149, 168, 204, 354, 371–72, 386, 431, 441, 462, 475, 554
Biden, Joe, 226, 642
Bielefeld, 517
Bihar, 305, 421, 522
Bikini Atoll, 558–59, 563
Bill and Melinda Gates Foundation, 428
Bin Laden, Mohammad, 233
Bin Laden, Osama, 233
Bingham Canyon Pit, 47
biodynamic farming, 285–86
biofuels, 465, 497–505, 625, 629
biological diversity, 4–5, 85, 166–67, 171, 182–85, 365, 398, 402–3, 427
biological pest control, 213–15, 217, 221
biological warfare, 753n44
biopiracy, 423–24, 427, 448
biosphere reserves, 367
birds, 20, 103, 129, 138, 140–41, 166–67, 171, 173–81, 183–85, 219, 277, 361, 369, 393, 395, 401, 538, 544–57, 577, 586, 593–94, 620, 633, 640. *See also* chicken
Birmingham (Great Britain), 60, 349
Birmingham, Alabama, 70
Bismarck, Otto von, 250–51
bison, 181
Black, Eli, 162

Black Country, 139, 252
Black Sea, 228
Bligh, William, 119–21
Boer War, 397
Bogotá, 517
Bohemia, 40, 393
Boko Haram, 85
Bolívar, Simón, 46
Bolivia, 39–48, 140, 353, 431
boll weevil, 19, 60, 81, 113, 163, 167, 186–98, 222, 236, 525, 608, 622, 625, 633–34, 640
Bond, James, 589
Bonn, 509, 513
Boone and Crockett Club, 181
Bophuthatswana, 403
Bordeaux, 92
Borlaug, Norman, 421, 427
Borneo, 387, 447, 449–50, 454–55, 457, 492, 585, 633
Bosch, Carl, 281
Bosnia and Herzegovina, 342
Boston (Great Britain), 59
Boston Fruit Company, 155
botanical exchange, 54, 62, 65, 80, 100, 102, 118–22, 127, 177, 213–14, 216, 383–85, 408–12, 423, 448, 625, 640–41, 643
botanical gardens, 62, 65, 102, 119, 121, 125, 127, 184, 384, 408–10, 448, 625
Bounty, 102, 119–20, 126, 411
Bourdieu, Pierre, 125
Brambell Report, 554–55
Brandis, Dietrich, 78
Brandt, Willy, 366
Branson, Richard, 344
Brasília, 456
Braudel, Fernand, 9–10, 258, 474, 609, 647
Brazil, 42–43, 46, 51, 58, 79, 115, 132, 188, 201, 226, 263, 300, 321, 350, 353, 365, 388, 394, 407, 410, 413–15, 423–24, 440, 444, 448, 450–53, 456–57, 504–5, 567, 646
breadfruit, 52, 100–2, 118–22, 124–28, 144, 213, 224, 411, 619, 625
Breslau, 512
Brest, 66
Brexit, 59, 607, 646
Brezhnev, Leonid, 232
Brighton, 295
Brisbane, 212, 216–17
Bristol, 43, 280, 291
British Association for the Advancement of Science, 176, 280
British Chicken Association, 554

British Honduras, 411
British Petroleum (BP), 47, 589
Brittany, 342, 585
Brockhaus, 339
Brotherhood of the Green Turtle, 546
Brower, David, 399
brown tree snake, 216, 529
Browns Ferry Nuclear Power Plant, 566
Brundtland report, 76, 365
Brunhes, Jean, 381–82, 388, 508
Brussels, 344
Brussels Sugar Convention, 58, 461
Bryce, James, 393, 401
Budapest, 71
Buenos Aires, 57, 437
Buffalo, 594
Building Professions, 26
Building the State, 26
Bukhansan National Park, 394
Bulgaria, 57, 71, 260, 472
Bumbuna Dam, 440
Bureau of Entomology (USA), 81
Bureau of Reclamation, 387, 432, 434–35, 437–39, 442, 444
Burkina Faso, 444
Burma, 42, 474
Burmese python, 216
Burundi, 445
Bush, George H. W., 529
Bush, George W., 236, 297, 354
Bush, Vannevar, 538
Butz, Earl, 208
Buxted Chicken, 544
buzzwords, 15, 76, 78, 362–66, 402

Cadbury, 455
cadmium, 47, 152, 363, 565
Cairo, 71, 429
calcium arsenate, 19, 191, 195, 525
Calcutta, 295, 311, 327, 330, 408–9, 474
California, 44, 58, 60, 184, 214, 276, 298, 407, 553
calories, 51, 124, 304, 421, 475
Calw, 423
Cambodia, 438
Cambridge, 474, 543
Cameron, James, 350, 443–44
Cameroon, 445, 448
campaign mode of political action, 203–4, 240, 254–56, 306, 347–48, 352, 464, 470, 476, 490, 520, 573–75, 579–83, 614
Campbell, Hardy Webster, 70, 203

INDEX

Canaan, 475
Canada, 44, 47–48, 50, 59, 116, 153, 260, 361, 397, 399, 480, 627
Canal du Midi, 21, 32, 64–74, 431, 434, 625, 633, 641
Canary Islands, 51
cancer, 536, 540, 559–61, 569, 578, 598
Cancun, 617
cane toads, 139, 166, 169, 171, 179, 211, 214–24, 433, 620, 622, 628, 640, 649
Cantillon, Richard, 408
Cape Horn, 119
Cape of Good Hope, 119
Cape Town, 121, 408
Cape Verde, 51
carbon dioxide, 239, 505, 540. *See also* climate change
carbon sequestration, 86
Carcassonne, 72
Cárdenas, Lárazo, 112, 392
Cardiff, 252
Cardoso, Fernando Henrique, 452–54
Caribbean, 34, 48, 50–54, 58, 60–62, 100–101, 118–19, 121, 125, 128, 154–55, 158, 162, 180, 200, 224, 226–27, 305, 335, 411, 450, 497, 546–47, 551, 618, 630
Caribbean Conservation Corporation, 551
Carlowitz, Hans Carl von, 36, 75–78, 81, 86, 626
Carnegie, Andrew, 81
Carnivorous Path, the, 27
Carr, Archie, 546, 551–52
Carroll, Lewis, 176, 633
Carson, Rachel, 12, 279, 347, 536–37, 540–41, 554, 561, 576–80, 582–83, 622, 629, 634, 637, 642, 647
Carter, Jimmy, 505, 582
Carver, George Washington, 497, 755n16
Castle Lager, 653n34
Castro, Fidel, 160
Catholics, 88, 338, 372, 517
Catskill Mountains, 262–63
cattle, 54, 276–77, 476, 529, 548–49, 551, 574, 601
cattle ticks, 529
Caucasus, 265
Caulerpa seaweed, 216
Cavendish bananas, 161, 163, 191
Cayman Islands, 367, 535, 546–47, 551–52, 557
Ceaușescu, Nicolae, 472
celebrities, 313–14, 347–50, 443
celibacy, 350–51
cellophane, 597
cellulose, 496

Centers for Disease Control and Prevention, 307
Central African Federation, 433
Central African Republic, 445
Central Intelligence Agency (CIA), 160–61, 598
Central Philippine University, 68
Cerro Rico, 33, 41–42, 44–46, 48, 431. *See also* Potosí
Ceylon, 79, 409, 573
Chad, 445
Chadwick, Edwin, 263, 267
Challenger, 596
Chamberlain, Joseph, 59–60
Chamberlain, Neville, 60
Chandler, Alfred, 157, 162
Chantico, 372
charcoal, 247
charisma, 203, 314, 347–8, 355, 387, 392, 399, 426, 438, 492
Charlemagne, 65
Charles I (England), 249
Charles II (England), 250
Charles V (Holy Roman Empire), 39–40
charros, 644
Chauri Chaura, 352
Chávez, César, 58
Chávez, Hugo, 233
Chemical Foundation, 496
chemical weapons, 282, 394, 498
chemistry. *See* Path of Chemistry, the
chemurgy movement, 460, 464–65, 468, 473, 494–505, 625, 628–29
Cheney, Richard, 529
Chernobyl, 232, 344, 478, 571
Chesapeake Bay, 267
Chiang Kai-shek, 377
Chicago, 12, 20, 59, 70, 143, 148, 155, 240–41, 250, 266, 269–75, 277–79, 284, 356, 374, 419, 461, 512, 534, 565, 568, 623, 625, 628, 644, 648
chicken, 18, 55, 59, 83, 114, 128, 141, 277–78, 326, 355, 386, 420, 425, 461, 512, 534–35, 540, 544–57, 573, 625, 627–29, 632, 644, 648
chicken war, 545
Chile, 47, 140, 152, 232, 236, 266, 281, 284, 288, 451
China, 32, 41–42, 51, 69, 71, 76, 78, 91, 96, 130–31, 148, 168, 248, 255, 258, 266–67, 305–6, 316–17, 330, 370–79, 408, 426, 440, 443–45, 454–55, 472–73, 475, 486, 550, 571, 592, 601, 605, 607, 624, 630, 641, 646
Chincha Islands, 102, 130–32, 139
Chinchona tree, 410

787

Chipko movement, 77
Chiquita, 161, 191, 339
Chiquita Brands International, 105, 162–64, 356. *See also* United Fruit
Chittagong, 32, 36–37, 87–90, 93–97, 646
chlorofluorocarbons, 299, 561, 603
cholera, 10, 73, 260, 262, 310–11, 316, 319–32, 339, 361, 367–68, 373, 475, 606, 633
Chrétien, Jean, 399
Crichton, Michael, 581, 629
Christian, Fletcher, 120
Chungju, 434
Churchill, Winston, 225, 228, 315, 348, 431, 444, 516, 546, 646
Cicero, 75
Ciénaga, 159
cigarettes, 304
cigars, 296, 304
Cincinnati, 162, 240, 270
cinnamon, 121
Circeo National Park, 394, 402, 488
cities. *See* Path Toward the City, the
citrus, 10, 54, 197, 214
civil disobedience, 313, 347–48, 352, 354
civil rights movement, 274, 352
Civil War (USA), 147, 187, 192, 201, 263, 528, 545
Clausewitz, Carl von, 320
Clean Air Act (Great Britain), 253–55
cleanliness, 238–40, 259–61, 352
climate change, 4–5, 9–10, 14, 17, 83, 86, 179, 221, 239, 255, 301–2, 313, 344, 347, 353, 365, 368, 416, 503, 505, 526, 539–41, 561, 567, 601, 614, 623–24, 635–36, 642, 648
climate control, 294–302
Clinton, Bill, 367
Clipperton Island, 130–31
clover, 54
Club of Rome, 291, 388
coal, 5, 44, 46–47, 83, 238–39, 246–48, 251–52, 255, 282–83, 293, 370, 377, 508
Coal Smoke Abatement Society (London), 250
coca leaves, 42, 44, 304
Coca-Cola, 58, 90, 158
cocaine, 304
cockroaches, 215
cocoa, 51, 54, 455
coconut, 139
cod, 152–53
cod wars, 152
coffee, 51, 53–54, 132, 154, 157, 188, 191, 304, 317, 448, 631, 647
coffee leaf rust, 188, 191

Coimbra, 410
Colbert, Jean Baptiste, 65, 67, 69, 73
Cold War, 101, 114–16, 160–61, 206, 229, 368, 420, 437, 459, 464, 490–91, 513, 525–27, 533, 562–63, 569, 625, 631
collectivization, 101, 112, 114, 461–62, 470–73, 477, 481
College Station, Tex., 197
Cologne, 334, 509, 524–25
Colombia, 46, 73, 92–93, 159, 162, 216, 450, 503, 517, 527
colonialism. *See* Colonial Path, the
Colonial Path, the, 25
Colonial Sugar Refining Company, 53, 213
Colorado (state), 43–44
Colorado River, 434
Columbia, 595
Commoner, Barry, 13
Common Agricultural Policy, 505
commons, 355–56, 382, 415, 635
Communism, 115, 160–61, 316, 340, 357–58, 372–75, 378, 397, 425, 462, 469–73, 477–81, 490–91, 513–14. *See also* Socialism
community-based conservation, 402–3
Comoros, 126, 331
comparative advantages, 55, 451
concrete, 647
condor, 184
conferencing, 315, 317, 359–69, 627, 645
Congo Free State, 455
Congo River, 445
Congress Party, 347–49
Conquest, Robert, 478, 480
conservation biology, 167, 182, 546, 552. *See also* Convention on Biological Diversity
conservation diplomacy, 181, 385, 547, 552–53
conspicuous consumption, 51, 93. *See also* inconspicuous consumption
Constantinople, 327, 335
contagionists, 311, 324, 326
Convention of Constantinople, 660n16
Convention on Biological Diversity, 181, 365, 398, 403, 425
Convention on International Trade in Endangered Species (CITES), 367, 552–53
Convention on the Law of the Sea, 152
Cook, James, 119, 126, 406
cooking oil, 505
cooperatives, 18, 45, 53, 285, 357, 600
Copenhagen, 409, 642
copper, 40, 46–47, 69, 232, 573–74
coral mining, 138
corn. *See* maize

Corn Laws, 132, 201
Cornell University, 544
Cornwall, 40, 585, 587–88, 642
coronavirus. *See* COVID-19
Corporación Minera de Bolivia (COMIBOL), 45
Corps des Ingénieurs des Ponts et Chaussées, 67
Costa Rica, 154, 157, 218, 224, 321, 393
Côte d'Ivoire, 444
cotton, 52, 54, 132, 140, 167, 186–89, 191, 193–98, 200–2, 296, 432, 496, 525, 539–40, 545, 599–601, 622, 625, 627, 629, 632, 634
Council of Europe, 479
Cousteau, Jacques, 347
COVID-19, 5, 13, 605–9, 611, 635–36, 642
cowboy, 644
Craven, Avery, 201
creative destruction, 538
creole ecology, 54
Crete, 259
cricket, 348
Croatia, 335, 393
crocodiles, 219, 557
Crookes, William, 280–81, 284, 291–92
crop rotations, 54–55, 62, 132, 205, 385, 471
Crosby, Albert, 412
Crutzen, Paul, 290, 366
Cuba, 52–54, 91–92, 112, 115, 160–61, 200, 451
Cudahy, 270
Cuernavaca, 366
Cultural Revolution, 372
culture of self-observation, 635–38
Curie, Marie, 559, 770n8
Cuvier, Georges, 178
cyclic thinking, 288–93
cypress trees, 405
Czech Republic, 40, 114, 472, 512
Czechoslovakia, 472, 512

Dachau concentration camp, 286
Dakota Badlands, 181
Dalí, Salvador, 152
Dallas, 81, 276, 341
Dallas, 234
Dally, Clarence, 559
dams, 6, 65, 113, 262, 350, 371, 386–87, 395–96, 398, 429–46, 495, 620, 624, 627, 633
Dandi, 313, 346, 348, 350
Dante Alighieri, 76
Danube, 71, 436
Dar es Salaam, 69
dark tourism, 379
Darwin, Charles, 179. *See also* Social Darwinism

Dauerwald. See mixed forestry
David against Goliath, 349, 354, 577
Davis, Mike, 480
DDT, 12, 19, 55, 137, 196, 209, 213, 215, 347, 368, 461, 525, 536–37, 555, 561, 572–83, 600, 622, 625, 629, 633, 636, 642, 647–48
dead zones, 290
Dearborn, 496
Deepwater Horizon, 586
Dehradun, 78
DeLay, Tom, 581
Delta Air Lines, 195
Democratic Republic of the Congo, 455, 516
Deng Xiaoping, 372–73
Denison, Iowa, 276
Denmark, 143, 251, 409, 553
Denver, 394
Department of Energy, 570
dependency, 43, 46, 387–88, 447–57, 622, 625, 633
dependency theory, 43, 46, 52, 140, 158, 161, 226, 388, 424, 450–55
Des Moines, 419
Des Plaines, Ill., 276
Des Voeux, Henry Antoine, 246, 250
desiccation discourse, 406
determinism, 12, 165, 201, 283, 301, 570
Detroit, 501, 596
development. *See* Developmentalist Path, the
Developmentalist Path, the, 27
devil, 450
Dhaka, 88, 474
diabetes, 61, 128
Díaz, Porfirio, 57, 264–65
Dickens, Charles, 246
Diderot, Denis, 339
dingo, 220
dinosaurs, 179, 185
Dion, Celine, 645
dioxin, 528
disaster capitalism, 375
disasters. *See* Path of Disasters, the
diseases, 9–10, 54, 61, 73, 91–92, 161, 171, 188, 191, 197, 213, 217, 229, 235, 239, 248–49, 255–56, 260, 262, 274–75, 277, 299, 310–11, 316, 319–32, 335, 339, 361–62, 367–68, 373–74, 384, 387, 396, 398, 405–6, 411–12, 428, 432, 441, 475, 483–84, 488, 491, 536, 540, 547–49, 554–55, 558–63, 568–69, 572–75, 578, 580–83, 590, 598, 601, 605–9, 633
Disneyland, 557

dissipation, 179, 540, 602–4
divided highways, 21, 32, 67, 71, 277, 335, 460–61, 465–66, 506–18, 621–22, 624, 627, 633, 644
Divina Commedia, 76
divine punishment, 371
Djerba, 80
Dnepr River, 441
Dodgson, Charles Lutwidge, 176
dodo, 20, 166–67, 171, 173–80, 183, 185, 395, 620, 633, 640
Dokuchaev, Vasilii, 203, 208
Domesday Book, 108
Dominica, 411
Dominican Republic, 52–54, 118, 121–22
Dominy, Floyd, 387, 434, 437–39
Donbass, 139, 247
Douglas, David, 408
Douglas, Mary, 259
Douglas fir, 408
Douglass, Frederick, 144
Dow Chemical, 495
Doyle, Conan, 246, 256
Dreyfus, Alfred, 73
droughts, 9, 84, 203, 207, 209, 263, 306, 372, 401, 442, 504
drugs, 56, 216, 303–7, 537
dry farming, 70, 203
Duarte, Eva. *See* Evita
Dubai, 90, 229, 300
Dublin, 247
Dubrovnik, 373–74
Duflo, Esther, 355
Dufour, Guillaume-Henri, 67
Dulles, Allen, 160
Dulles, John Foster, 160
DuPont, 425, 498, 598
Durban, 408
Düsseldorf, 553
Dust Bowl, 116, 166, 168, 203, 207–9, 644–45
Dutch disease, 232–36
Dutch East Indies, 53, 62, 173, 330, 409, 448
Dutch West Indies Company, 156

Earth Day, 5, 360
Earth Summit (Rio de Janeiro 1992), 365, 368, 541, 645
earth system science, 16–17
earthquakes, 310, 316–17, 331, 344, 370–79, 432, 436, 586, 607, 649
East India Company (British), 305, 408, 522
East India Company (Dutch). *See* Vereenigde Oostindische Compagnie

East Timor, 228, 342, 392–93
Eberswalde, 78
ebony, 177
econometrics, 69, 419
economics. *See* Economics 101
Economics 101, 27
ecotourism, 341, 402
Ecuador, 53, 367, 455
Edison, Thomas Alva, 155, 559
Edward VIII (England), 335
Edwards, George, 175
Egypt, 71–72, 113, 201, 225, 289–90, 300, 327, 335, 371, 387–88, 429–37, 439–45, 475, 601
Ehrenfeld, David, 552
Ehrlich, Paul, 292, 502
Ehrlichman, John, 307
Einstein, Albert, 770n8
Eisenhower, Dwight D., 160, 513–14, 563–64, 568–69
El Dorado, 644
El Niño, 103, 140, 480
El Salvador, 53, 157
Elba, 73
Elbe Lateral Canal, 513
Elbe River, 375
Elders and Fyffes, 157. *See also* Fyffes
Eleanor of Provence, 249
Électricité de France (EDF), 567
elephants, 178, 181, 392, 399–401
Elias, Norbert, 259–60
Ellsworth, Henry, 422
Elm, 585
Emerson, Ralph Waldo, 357
Empress Dowager Cixi, 475
Empty Quarter, 394
Encyclopædia Britannica, 339
endangered species. *See* extinction
energy. *See* Path of Energy, the
Engel, Ernst, 102, 122
Engel's law, 102, 122–24
Enterprise, Alabama, 189–90, 192, 194
entomology, 36, 79–82, 188–89, 196, 214, 620, 634
Environment Agency for England and Wales, 600–601
Environmental Defense Fund, 579
environmental justice, 95, 559, 571, 592, 619–20. *See also* racism
Environmental Protection Agency (USA), 47, 579, 581
environmentalism of the poor, 353, 365, 533
eradication campaigns, 169, 196–98, 217, 221, 306–7, 575, 579–80, 583

Erasmus of Rotterdam, 260
Erika, 589
Eritrea, 79
Erklärung von Bern, 600
erosion. *See* soil erosion
ersatz products, 524–25
Escherich, Karl, 36, 79–81, 410, 413, 634, 637, 647
Escobar, Pablo, 216
espresso machines, 304
Esser Mittag, Judith, 552
Estonia, 112
Estoril, 342
ethical consumption, 241
Ethiopia, 107–8, 111, 114–15, 402, 407, 413, 420, 437, 445, 472, 476, 485
Ethyl Gasoline Corporation, 498
Eton, 543
eucalyptus, 19, 48, 70, 121, 177, 222, 383–86, 405–8, 410–16, 420, 426, 488, 634, 640–41
eugenics, 495
Eunápolis, 414
European Community. *See* European Union
European Patent Office, 426–27
European Union, 59, 96, 116, 162–63, 197, 291, 368, 505, 545, 555, 591, 607
eutrophication, 242, 290–92
Evelyn, John, 250
Everglades, 216, 395–96, 398
Evita, 375
evolution, 179
Exeter, 341
experts. *See* Building Professions
externalization, 618–21
extinction, 4, 20, 165–67, 173–85, 216, 367, 395, 546, 614, 620, 624, 626, 640
extractive reserves, 456
Exxon Valdez, 369, 538, 589

factory farming, 276–79, 534–35, 537, 544–57, 618, 627, 629, 636, 643
Fahlberg, Constantin, 56
fair trade, 163–64, 356, 626
Faisalabad, 268
Faletto, Enzo, 452
Falklands War, 18, 557
fallout, 536, 558–63, 569, 571, 643
Falun, 40
famines. *See* hunger
Fanta, 90
fascism, 16, 202, 205, 354, 394, 414, 460–61, 463–65, 467, 482–93, 509, 514, 581, 625, 633

Federal Bureau of Investigations (FBI), 426, 598
feedlots, 276
Fermi, Enrico, 565, 568
Fernow, Bernhard, 78
fertile soil, 19, 54, 57, 75, 134, 201, 208, 267, 285, 288, 385. *See also* soil erosion
Fertilia, 483
fertilizers, 19, 54, 91, 100, 102–3, 105, 107, 116, 129–41, 161, 189, 209, 228, 240–41, 243, 266–68, 270, 276, 280–93, 415, 421, 425, 441–42, 463, 487, 496, 502–3, 522, 534, 550, 573, 575, 599–600, 631–32, 640, 643
feudalism, 108, 111–12, 451
Fiat, 516
figs, 121
Fiji, 53, 125, 213, 215, 408
Finland, 343, 397, 474–75, 478–79, 510, 567
fire ants, 196, 216, 536, 575
Firestone, 60, 455, 512
firs, 121
Fish and Wildlife Service, 552
fish stick, 148
Fisher, Antony, 355, 367, 535, 537, 540, 543–48, 555–57, 637, 647
fishing, 103, 140–41, 146, 148, 151–53, 183, 535, 558–59, 561, 563, 587, 590, 619, 629, 640
fishmeal, 103, 130, 140–41
Five Star Movement (Italy), 197
Flamanville Nuclear Power Plant, 567
Flanders, 175, 475
Flavius Josephus, 443
floods, 159, 261–62, 371–72, 375–76, 431, 527, 602, 627
Florence, 326–27, 334, 337, 475
Florida, 197, 216
Fogel, Robert, 69, 355
Fontane, Theodor, 594
food aid, 374, 476–77, 581
Food and Agriculture Organization, 153, 504, 550
Food and Drug Administration, 549, 574. *See also* Pure Food and Drug Act
food choices, 100–101, 122–28
food safety, 19, 241, 272
food security, 127
foot-and-mouth disease, 548
forced labor, 45, 53, 73, 131, 212–13, 525
Ford, 158, 196
Ford, Henry, 270, 457, 494, 496, 501
Ford Foundation, 421
Fordlândia, 457
Forest of Dean, 83

791

forestry, 32, 35–36, 48, 54, 65, 71, 75–86, 177, 194, 216, 365, 405–7, 410, 413–15, 520, 523, 526, 528, 548, 620, 625–26, 632, 634, 647
forests. *See* Wooden Path, the
Forillon National Park, 397
Forster, E. M., 337, 344
Fortune, Robert, 408
fossil fuels, 5, 10, 44, 46–47, 68, 83, 85, 89, 146, 155, 166, 170–71, 226–36, 238–39, 243, 246–48, 251–55, 265, 282–83, 293, 305–6, 361–62, 369–70, 377, 393, 426, 456, 461, 465, 497–505, 508, 514, 516, 518–20, 522, 538, 540, 550, 584–90, 592–94, 599–601, 627, 625, 640–41, 647
Foucault, Michel, 490
foxes, 554
Foyn, Svend, 146
Framework Convention on Climate Change, 365, 541
France, 35, 42, 50, 52, 56, 62, 64–74, 78, 85, 89–93, 100, 112, 118, 121–22, 125, 127, 130, 142–43, 149, 201, 251, 262, 265–66, 268, 278, 321, 327–28, 333, 335–36, 339, 342, 347, 352, 361, 377, 381, 407–8, 423, 431, 439, 448–49, 508–9, 515, 517–18, 526, 528–29, 531, 567, 569, 585, 587, 597, 613, 633, 642, 645–46
Franco, Francisco, 343, 436
François I (France), 65
Frank, Andre Gunder, 452
Frank, Hans, 340
Frankfurt, 509
Franz Joseph I of Austria, 266
free trade, 34, 55, 57–58, 140, 451, 456, 500–1, 504, 546–57, 556, 581. *See also* neoliberalism
French Guiana, 73
French Revolution, 52, 62, 112, 127, 336, 352
Freud, Sigmund, 266
Fribourg, 381
Friedman, Milton, 375
frogs, 169, 219, 221–22
frontiers, 144, 168, 181, 200, 208, 271, 411, 422, 435, 644
frozen food, 148, 271, 544–46
Fruili, 371
fuel additives, 465, 497–99
Fugger, 40
Fukushima, 369, 371, 378, 536, 566, 569–70, 586, 588
Fuller, Luther, 594
fungal diseases, 74, 104, 159, 163, 188, 191, 196–98, 213, 428, 457

Fyffes, 157, 162, 164

Gabon, 47, 393
Gainsbourg, Serge, 585, 645
Galeano, Eduardo, 43, 45, 226, 450
game reserves, 395–96, 402
Gandhi, Indira, 364
Gandhi, Mahatma, 15, 313–15, 339, 346–58, 450, 637, 647
Gandhi, Manilal, 354
Gang of Four, 372–73
Ganges, 320
Gansu Province, 248
garbage, 4, 36–37, 47–48, 87–97, 100, 153, 270, 342, 359, 362, 496, 539, 550, 561, 568, 601–3, 620, 626, 643
García Márquez, Gabriel, 158–59, 329–30
gardens, 34, 55, 62, 91, 122–23, 125, 219, 384, 408–9, 412, 471–72, 576. *See also* botanical gardens
Garfield, James, 294–96
Garibaldi, Giuseppe, 376
Garonne River, 64
Garst, Roswell, 420
Garvan, Francis, 499
gas warfare, 282
gauchos, 644
Gaulle, Charles de, 567
Gayer, Karl, 80
Geigy chemical company, 572
gender, 33, 44, 60, 67, 77, 144, 169, 218, 222, 228, 240, 248, 260, 274, 315, 342, 353, 355–56, 361, 387, 439, 476, 512, 518, 534, 549, 551, 554–55, 570, 577, 599–600, 632, 634, 644, 653n34
General Electric, 296, 426, 559, 566
General Land Office, 109
General Motors, 498, 512
Generalplan Ost, 286, 487–88
genetic conservation, 166, 171, 185
genetically modified organisms, 386, 415, 426–28, 590
genetics, 166, 171, 183–85, 276–77, 385–86, 415, 418–20, 423–28, 470, 495, 556, 590
Geneva, 337, 565
geoengineering, 17, 302, 635
geography (academic discipline), 381–82
Geological Society of London, 15
geology, 16, 134, 178, 247, 370, 532
George III (England), 119
George V (England), 349
Georgia (state), 168, 186, 197, 199–200, 202, 205–6, 209–10, 545, 645

INDEX

Gerlach, Max, 135

German Democratic Republic (GDR), 95, 462, 472, 544, 596–97

German East Africa, 179, 409, 448

Germany, 7, 36, 43–44, 55–57, 67, 69–70, 75–83, 86, 89–90, 95–96, 109–11, 122, 129–36, 139, 142–45, 147–48, 177–79, 214, 241–43, 247, 250–51, 258, 260, 265–66, 270, 277–78, 281–88, 295, 304–5, 311–12, 320–21, 324–29, 333–36, 338–41, 344, 347, 352, 357, 361, 366, 375, 394–95, 400, 409–10, 413, 418–19, 423, 427–28, 431, 434, 439, 441, 455–56, 460–67, 472, 475–77, 480, 486–88, 495, 506–18, 520, 522, 524–26, 529, 531, 544–45, 547–49, 552–54, 559–60, 567, 571, 585–86, 594, 596–98, 600, 634, 644–46

Gettysburg, 528

Gezira, 445

Ghana, 148, 338, 436, 444, 455

Ghosh, Amitav, 171, 234

Gibraltar, 431

Gilded Age, 155, 227

Giza, 430

gladiolas, 425

Gladstone, William, 248

Glasgow, 252

glass, 90, 298–99, 431, 602

Glass, Lisa, 653n34

global environmental governance, 95, 150–51, 181–82, 216, 219–20, 311, 316, 327, 365, 367–69, 385, 394, 398, 400, 403, 425, 461, 541, 547, 552–53, 589, 607, 627, 635, 642, 645

global warming. *See* climate change

Gneisenau, Neidhardt, 320

God, 39, 64, 158, 178, 357, 371–72, 434

Godzilla, 559, 563

Goethe, Johann Wolfgang von, 320, 327, 334, 336, 344

gold, 43–44, 48, 211, 627, 641

Gold Coast, 338, 455

Gold Kist, 545

gold rush, 33, 130, 643

Goldbroiler, 544

Gomułka, Władysław, 473

Goodyear, Charles, 448

Gorbachev, Mikhail, 305, 471–72

Gore, Al, 347

Gorée Island, 297

Göring, Hermann, 82, 486

Göttingen, 125

granaries, 201, 382

Grand Canal (China), 71, 641

Grand Canyon, 199

Grand Coulee Dam, 444

Grand Ethiopian Renaissance Dam, 445

Grand Tour, 81, 312, 334

Grandin, Temple, 278

grapes, 188, 492. *See also* wine

grasshoppers, 523

Graubünden, 508

Great Acceleration, 532, 540, 630–31

great auk, 180

Great Barrier Reef, 138

Great Britain, 5, 40, 42–44, 46, 50–51, 54, 59, 61, 66, 69–71, 82–84, 90, 96, 100, 103, 108–11, 113–14, 118–22, 124–27, 130–33, 136, 138–39, 143–44, 146–47, 152, 157, 162, 176–77, 181, 186–87, 201, 213, 217, 228, 243, 245–50, 252–57, 259, 261–63, 265, 268, 280–81, 283, 295, 298, 305, 313–15, 319, 321, 324–25, 330, 333–42, 344, 346–50, 354–56, 374, 393, 397, 406, 408, 410, 415–16, 423–24, 427, 431–32, 435–39, 445, 449–50, 474–75, 478–79, 484–86, 491, 507–8, 510, 514, 516–17, 526–27, 529–31, 535, 538, 543–46, 548–49, 553–57, 559, 563, 567–69, 584–85, 587–89, 593, 597, 600–601, 607, 619, 628, 633, 645–46

Great Depression, 58, 168, 205, 207–8, 217, 273, 296, 355, 459, 463–65, 502, 599

Great Leap Forward, 472

Great Migration, 193

Great Pacific Garbage Patch, 603

Great Soviet Encyclopedia, 338–39, 358, 478

Great Stink, 262

Greece, 87, 150, 259, 327, 335–36, 342, 436, 513, 545, 573

Green Belt, 527

Green International, 114

Green Party (Germany), 347

Green Party (USA), 347, 354

Green Revolution, 114, 116, 137, 289, 385, 421, 423, 428, 593, 625

Greenfield, Mass., 443

Greenland, 151–52, 393

Greenpeace, 104, 151–52, 368

Greenwood, Miss., 193

Grenada, 411

Griliches, Zvi, 419

Grimm's Fairy Tales, 76

Gros Michel bananas, 161, 163, 191

Grotius, Hugo, 144

Groundnut Scheme, 491, 647

growth hormones, 549

Guadeloupe, 50–51

793

Guam, 215–16, 520, 523, 529
Guanajuato, 44
guano, 19, 53, 102–3, 129–33, 135–41, 170, 201, 226–27, 232, 266, 281, 284, 286, 289, 293, 421, 575, 631–32, 644, 646
Guano Island Act, 131, 139
Guatemala, 53, 159–61
guavas, 121
Guevara, Che, 112, 160, 453
Guggenheim Exploration Company, 47
Guide Michelin, 339
Guinea, 402, 516
Gujarat, 346
Gulbenkian, Calouste, 227
Gulf of Guinea, 226–27, 229
Guyana, 236
gypsies, 74

Haber, Fritz, 281, 284
Haber-Bosch process, 241–42, 281–83, 285, 288–90, 496, 522, 566, 585
Habsburg, 39–40, 393
Haiti, 52–54, 118, 121–22, 125–26, 331
hajj, 72–73, 233, 320
Hale, William Jay, 473, 494–96, 499–503, 647
Hall, James Norman, 126
Hambruch, Paul, 139
Hamburg, 70, 145, 328, 375, 517, 559–60
Hamid, Mohsin, 311, 332
Han River, 434
handbooks, 127, 233, 259, 312, 337–40, 640
Handorf, 287
Hanekom, Derek, 403
Hannibal, 485
Hanover, 178
Hardin, Garrett, 355
Haridwar, 320
Harrisburg, 590
Harrison, Ruth, 279, 553–54
Hartig, Georg Ludwig, 82
Harvard University, 231–32, 236, 301, 339, 451, 491, 495
Haushofer, Karl, 487
Haussmann, Georges-Eugène, 278
Havana, 92
Hawaii, 124, 127, 214–15, 217, 407–8
Hawaiian Sugar Planters' Association, 217
Hayek, Friedrich August von, 355, 366, 369, 436, 543
Hayward, Elijah, 109
heart disease, 61
heat wave, 377
Heckman, Jerome H., 599

Hegel, Georg Wilhelm Friedrich, 130, 321
Heine, Heinrich, 129, 321
Heligoland, 177–78
Helmand Valley, 436–37, 445
Henan Province, 377
Henry III (England), 249
Henry IV (France), 65
Heraclitus, 18, 28
herbicides, 573. *See also* pesticides
Herodotus, 436, 441
heroin, 56, 305
Herrenchiemsee Abbey, 304
Hess, Rudolf, 487
Hetch Hetchy, 398, 432–33
Heyerdahl, Thor, 603
hibakusha, 564, 571
Hidalgo, Anne, 256
Hightower, Jim, 426–27
Hillsboro, Tex., 585
Himalaya, 295
Himmler, Heinrich, 136, 286, 466–67, 520
Hind Swaraj, 314, 353–54
Hindus, 320, 354–55, 561
Hinkley Point Nuclear Power Plant, 567
Hintze, Otto, 434
hippopotamus, 216
Hiroshima, 559, 561, 564, 569
Hiroshima Peace Memorial Museum, 564
Hitler, Adolf, 60, 79, 149, 460–61, 465, 473, 486–87, 509–10, 512, 518, 581, 594
Hobsbawm, Eric, 459–60, 504, 531
Hochtief, 439
Hoffman, Dustin, 596
holistic thinking, 36, 81, 242, 287–93, 537, 578–79
Hollywood, 97, 120, 151, 185, 234, 256, 262, 276, 313, 348, 350, 427, 443, 612, 645
Holocaust, 136, 463, 467, 553, 598
Holodomor, 112, 460–63, 469–71, 473, 477–81, 534
Honcamp, Franz, 293
Honda, 516
Honduras, 53, 546
honey, 32, 51
Hong Kong, 545
Hong Kong Convention for the Safe and Environmentally Sound Recycling of Ships, 95, 368
Honolulu, 603
Honshu, 378
Hooker, William, 410
Hoover, Herbert, 296, 370, 375–76, 477
Houghton Mifflin, 577

INDEX

Houston, 194, 581
Howard, Albert, 285
Howard, John, 228
Howard, Luke, 256
Hu Jintao, 375
Hu Yaobang, 373
Hua Guofeng, 373, 378
Huancavelica, 42
Hudson Bay Company, 156
Hudson Institute, 437
Hudson River, 297
Huff Daland Dusters, 195, 525
Hugo, Victor, 266
Huguenots, 74
Humboldt Current, 103, 140–41
humus, 134
Hungary, 71, 361, 529
hunger, 9, 61–62, 84, 112, 114, 123, 148, 280, 285, 293, 338, 355, 421, 460–63, 467, 469–81, 495, 505, 553, 622
hunting, 76, 151, 169, 177–81, 220, 222, 395–96, 546, 551, 647
Huntington, Ellsworth, 301
hurricanes, 119, 155, 371–72, 374–76, 607
Huskisson, William, 585
Huxley, Aldous, 262
Hwange National Park, 181
hybrid seeds. *See* seed improvement
hydrogen bomb, 558
hydrogenation technology, 103, 146, 241, 283
hydrological regimes, 67, 386–87, 429–46, 643

Iban people, 450
Ibiza, 617
Ibn Saud, 170, 225–28, 230, 234
Iceland, 152
Ichaboe Island, 138
Ickes, Harold, 204, 226
Idrija, 42
IG Farben, 283–84, 286
Ig Nobel award, 218
Ilha do Cirne, 173. *See also* Mauritius
illicit activities, 33, 37, 42, 55–56, 150, 153, 306, 350, 406, 471, 556
Illiers, 328
Illinois, 274, 276
Illinois Central Railroad, 321, 323
Illinois Humane Society, 274
Imperial Chemical Industries (ICI), 597
import substitution, 451, 457
Inca roads, 32
Incas, 43, 66

inconspicuous consumption, 298
India, 32, 41, 51, 53, 77–78, 90, 96, 130, 138, 171, 175, 201, 219, 234, 240, 255, 260, 285, 289, 295, 300, 305, 311, 313, 320, 327–28, 330, 335, 346–58, 361, 364, 401, 407–9, 414, 421, 426–27, 435, 437, 442, 444, 449, 474, 539, 559, 567, 586, 591, 593, 627, 646
Indigenous people, 5, 73–74, 100, 109, 116, 139, 143, 151–52, 200, 260, 353, 374, 384, 387, 392, 394, 397–99, 401–2, 406, 414, 424, 443–44, 447–50, 454–57, 546, 633
indigo, 54, 496
Indochina, 330, 448
Indonesia, 53, 62, 173, 330, 393, 409, 448, 454–55, 492, 519, 585, 646
Indore, 285
Indus civilization, 259
Indus River, 437, 442
influenza, 329
Infrastructure Path, the, 24
infrastructures. *See* Infrastructure Path, the
insider perspective, 366–67
Institute of Economic Affairs, 543
International Atomic Energy Agency, 567
International Commission on Large Dams, 433, 440, 442
International Commission on Stratigraphy, 16
International Conference on Peaceful Uses of Atomic Energy, 565
International Congress of Applied Chemistry, 281
International Convention for the Prevention of Pollution from Ships, 589
International Geosphere-Biosphere Program, 366, 532
International Maritime Organization, 95, 588–90
International Opium Convention, 306
International Ornithological Congress, 361, 366
International Rubber Regulation Agreement, 449, 633
International Social Science Council, 360, 364
International Society of Sugar Cane Technologists, 214, 217
International Sociological Association, 453
International Technical Conference on the Protection of Nature, 182
International Union for Conservation of Nature (IUCN), 182, 216, 219–20, 367–68, 394, 400, 547
International Whaling Commission, 150–51, 368

interstellar trade, 355, 357
Inuit, 116, 399
invasive species, 6, 165, 169, 171, 178, 211–24, 341, 400, 402, 415, 432–33, 536, 575, 622, 628, 640, 649
Iowa, 276, 287–88, 290, 417, 419–20, 425–26, 505, 549
Iowa Beef Packers (IBP), 276
Iran, 47, 226, 304, 567
Iraq, 207, 226, 298, 309–10, 320, 330, 510, 513, 550
Iraq War, 226, 309–10, 550
Ireland, 162, 247, 270, 475, 479, 601–2, 769n81
iron, 46–47, 125, 294, 661n39
irrigation, 213, 264, 267, 300, 306, 429–446, 490
Irschenberg, 511, 518
Ismailia, 71
Israel, 206, 222, 267, 335, 372, 400, 405–7, 411, 413, 436, 439, 441, 463, 475
Istanbul, 265
itai itai, 363
Itaipú Dam, 440, 444
Italy, 89, 147–48, 197, 202, 205, 311, 319–20, 325–27, 329–30, 334–36, 343–44, 367, 372, 374–76, 394, 402, 407, 460–61, 463–65, 475, 482–93, 509, 544, 572, 574, 586, 589, 591, 607, 633, 646

Jack the Ripper, 246, 256
Jackling, Daniel, 47
Jackson, Andrew, 109, 324, 422
Jahangir, 175
Jainism, 354
Jamaica, 52, 60–61, 118–19, 121–22, 124–25, 157, 215, 343, 408, 411, 546
Japan, 112, 115, 124, 130, 147–49, 152, 264, 268, 305, 315, 359–66, 368–69, 371, 374, 377–78, 402, 423, 434, 455, 457, 460–61, 466–67, 474, 486, 519–27, 529, 535–36, 558–59, 561–67, 570–71, 585–86, 619, 625, 640, 646
Java, 53, 62, 330
jeans, 44
Jefferson, Thomas, 111, 178, 435
Jerusalem, 168, 206, 372
Jerusalem artichoke, 496
Jesuits, 408
Jesus Christ, 88, 372, 434, 439
Jetlag Travel, 344–45
Jevons, William Stanley, 248, 293
Jewish National Fund, 405, 413
Jews, 89–90, 340, 405, 409, 413, 434, 544, 577
Jiang Qing, 372
Jinji, 444

Joachimsthal, 40
John Murray, 334
John Paul II, 338, 393
Johns Hopkins University, 56, 68, 436, 576
Johnson, Lyndon B., 436, 438
Johnson v. McIntosh, 109, 117
Jones, Donald, 422
Joplin, Scott, 585
Jordan, 513
José I (Portugal), 373
Judas Iscariot, 88
Julius Caesar, 485
Jünger, Ernst, 522
Jupiter, 12
jute, 600–601

Kahn, Herman, 437
Kajakai Dam, 445
Kakadu National Park, 394
kale, 59
Kalimantan, 455
kamikaze (divine wind), 371
kamikaze (military tactic), 519
Kandahar, 445
Kansas, 116, 276, 499
Kansas City, 277
Kanton, 258
Kapp, Karl William, 621
Karbala, 320
Kariba Dam, 432–33
Karlsruhe, 79, 281
Kashmir, 438
Katrina, 375–76
Kazakhstan, 203, 445, 463, 481
Keeling Curve, 539–40
Keith, Minor, 60, 154–55
Kellogg, John Harvey, 279, 356
Kelly, Petra, 347, 352
Kennedy, John F., 304, 341, 576–77
Kentucky Fried Chicken (KFC), 545
Kenya, 397, 399–400, 444, 547
kerosene, 228
kestrel, 177, 183
Keudell, Walter von, 83
Kew Gardens, 26, 119, 125, 127, 409–12, 423, 448, 641
Keynesianism, 232
keystone species, 181
Khadera, 407
Khan, Imran, 348
Khartoum, 445
Khashoggi, Adnan, 233
Khrushchev, Nikita, 203, 420, 514

Khuriya Muriya Islands, 138
Kiautschou Bay, 78
Kiel, 409
Kiel Canal, 67
Kikawada, Kazutaka, 369
Kimmel, Roy, 116
King, Martin Luther, 274, 352
King Kong, 563
Kiribati, 125, 246
Kiyosawa, Kiyoshi, 523
Klein, Naomi, 375
Knapp, Friedrich, 133
Kneese, Allen, 360
Knossos, 259
Koblenz, 333–34
Koch, Robert, 179, 311, 324–27
Koebele, Albert, 214, 217
kok-saghyz, 525
Kolkata. *See* Calcutta
kolkhoz, 112, 462, 470–73
Kölreuter, Joseph Gottlieb, 419, 423
Kompienga Dam, 444
Korea, 374. *See also* North Korea; South Korea
Korean War, 441
Koselleck, Reinhart, 165
Kraftwerk, 511
Kraków, 40
Kranzberg's laws of technology, 238
Kremasta Dam, 436
Kroc, Ray, 276
Kruger, Paul, 392
Kruger National Park, 20, 382–83, 391–92, 395–403, 640, 649
Krugman, Paul, 355, 357
Kuala Lumpur, 220
kudzu, 211
Kufra oasis, 485, 493
kulaks, 460, 462, 470
Kuwait, 226, 229, 545
Kyiv, 479
Kyrgyzstan, 444–45
Kyushu, 362

La Villette, 278
Labrador, 153
Ladejinsky, Wolf, 115
ladybird, 214
Lagardère Group, 334
Lahore, 347, 349
Lake Baikal, 445
Lake Chad, 445
Lake District National Park, 262
Lake Erie, 291, 594
Lake Michigan, 266
Lake Success, 182
Lake Tegernsee, 262
Lake Victoria, 216
land grabbing, 117
land reclamation, 92, 147, 394, 405, 461, 463, 482–93, 581, 633
land reform, 101, 111–16, 159–60, 192, 483, 487
landfills, 92–93, 95, 603, 620
Landless Workers Movement (Brazil), 58, 115, 414, 444, 504
landownership, 12, 19, 66, 100–101, 105, 107–17, 158, 160, 167, 386, 415, 433, 455, 473, 483, 500, 625, 632
landraces, 183–84
Langbehn, Julius, 260
Langmuir, Irving, 302
Languedoc, 64–65, 68–69, 73–74
Laos, 434, 438, 529
Lapland, 423
Lappé, Frances Moore, 278
large technological systems, 33, 59, 238–40, 261, 283–84, 386, 426–27, 430, 461, 566, 585, 590, 593, 633
Las Médulas, 211
Latin, 75, 121, 174, 506
Latina, 490, 493, 581. *See also* Littoria
Latvia, 112
Laughton, Charles, 120
Laur, Ernst, 483
law of the minimum, 134–35, 280, 284, 293
Lawrence, T. E., 340
Le Mans, 515
lead, 47, 516. *See also* tetraethyl lead
League of Nations, 60, 330, 361, 476–77
leather, 597
Lebanon, 61–62, 230, 343, 479
Lebensraum, 336, 463–64, 485–88, 492–93, 541
Legionnaires' Disease, 299, 329, 606
Leipzig, 423
Lenin, Vladimir Ilyich, 470
Leningrad, 184. *See also* Saint Petersburg
Leopold, Aldo, 83
Leopold II (Belgium), 455
Lesotho, 413, 491
Lesseps, Ferdinand de, 72–73
Lessepsian migration, 72, 74
Lessona, Alessandro, 485
leukemia, 561, 569
Lever Brothers, 455
Lewis, Mark, 218, 222
Lewis and Clark Expedition, 178
Liberal Party, 113

Liberia, 60, 455, 588–89, 646
Libya, 445, 485–86, 493
Liebig, Justus von, 133–36, 138, 266–67, 280–82, 284–85, 293, 325, 497
life reform movement, 356
light water reactors, 536, 566–67
lightbulb, 647
Lilienthal, David, 568
Limburgerhof, 135, 282
Limits to Growth, 291–92, 388
Lincolnshire, 59
Lindner, Carl, 162
Linnaeus, Carl, 174, 423. *See also* taxonomy
lions, 220, 392
liquidators, 571
Lisbon, 330, 371, 373, 586
Lithuania, 270, 394
Little Grand Canyon, 168, 199–200, 202, 205, 209–11, 633, 645
Little Ice Age, 475
Littoria, 482–85, 487–89, 493
Liverpool, 133, 135, 144, 321, 324, 585
Livorno, 327
Lloyd George, David, 113–14
Lloyd's Register, 584
locusts, 479
London, 5, 15, 20, 46, 50, 121, 139, 175, 239–40, 245–48, 250, 252–57, 261, 265–66, 268, 314, 325, 348–49, 356, 406, 409, 411, 474, 479, 536, 543, 561, 628, 632–33, 640–41
London smog disaster (1952), 239, 253–56
Long Island, 512, 576
Loreley, 129, 340
Loriot, 338
Los Alamos, 561
Los Angeles, 92, 152, 262, 298, 350, 444, 512, 589, 643
Louis XIV, 65, 69, 633, 641
Louisiana, 189, 196
Louvre, 333
Love Canal, 592
Lowdermilk, Walter, 168, 206
Lower Chattahoochee Valley Area Planning and Development Commission, 199, 210
Lubbock, 276
Lucky Dragon No. 5, 533, 535–36, 558–59, 562–63, 571, 619, 640, 643, 649
Ludlow massacre, 44
Ludwig I (Bavaria), 320
Ludwigshafen, 281, 585, 640
Lyon, 91–92
Lysenko, Trofim Denisovich, 420, 464, 491

MacArthur, Douglas, 112, 349
Macaulay, Thomas, 66
machine gun, 67
Macmillan, Harold, 253–54, 531, 699n33
mad cow disease, 277, 555
Madagascar, 47
Madeira, 51
Madras, 328
Madrid, 311, 642
Magnitogorsk, 644
Mailer, Norman, 598
Mainz, 334
MairDumont, 340
maize, 32, 54, 83, 137, 201–2, 213, 276, 385–86, 417–20, 422–23, 425, 428, 501, 525, 550, 619, 644
malaria, 73, 229, 368, 398, 405–6, 432, 483–84, 488, 491, 536, 572–75, 580–83, 601
Malawi, 420
Malaysia, 220, 330, 387–88, 397, 420, 447–50, 454–55, 457
Mali, 85, 601
Malthus, Thomas Robert, 123, 474–75, 477, 502
Managua, 375
Manaus, 43, 46, 226, 641
Manchester, 262, 585
Manchuria, 486
Mandela, Nelson, 354
Mangla Dam, 438
mango, 121
Manhattan Project, 283, 560–61
Manila, 42, 321
Mann, Thomas, 262, 310–11, 320, 322, 324, 326, 329, 339, 633
Mansur, Ustad, 175
Mao Zedong, 316, 372–73, 472
Maoism, 316, 378
Maori, 109, 392
Maradi region, 85
mare liberum, 144
margarine, 103, 146, 148
Marggraf, Andreas Sigismund, 56
Mariculture Limited, 546–47, 551–52
Marine Biological Association, 588
maritime resources, 72, 100, 103–4, 130, 140–53, 183, 347, 546–47, 551–53, 556–57, 603, 619, 640, 643
Maroon communities, 52
Marseille, 91
Marsh, George Perkins, 407
Marshall Islands, 558, 563
Martinique, 50–51, 215
Marx, Karl, 40, 79, 304, 451, 473

Maryland, 201
Mascarene Islands, 180
masculinity, 144, 169, 218, 222, 228, 387, 439, 476, 512, 518, 534, 554–55, 634, 644
mass destruction mining, 33, 47
Massachusetts, 443
Massachusetts Institute of Technology, 291
mastitis, 549
mastodon, 178
Mauna Loa Observatory, 539
Maurice of Nassau, 173
Mauritania, 601
Mauritius, 20, 53–54, 121, 166, 173–74, 176–80, 183, 408, 410, 640
Maynard, John, 594
McDonald's, 276, 518, 544–45, 623, 647
McDowell, Mary, 273
McKibben, Bill, 236
McMillen, Wheeler, 496, 499–500
McNeill, William H., 13
Meadows, Dennis, 291
measles, 322
meat consumption. *See* Carnivorous Path, the
Meat Inspection Act, 272, 279
Mecca, 72, 320
medical expertise, 56, 68, 125, 197, 249, 305–6, 311, 323–32, 338–39, 483, 488, 548, 568–69, 581, 598, 606–8
Mediterranean, 9–10, 32, 51, 64, 66, 70, 72, 216, 300, 431, 619
Mega Rice Project, 492
Mehta, Gaganvihari Lallubhai, 559
Meiggs, Henry, 154
Mekong, 436, 438
Melanesia, 53
Melbourne, 138, 140, 212, 408, 410, 634
Melville, Herman, 144
Mendel's laws of inheritance, 385, 418
Mendes, Chico, 350, 353, 388, 456–57
Mennonites, 203
Mercedes, 513, 515
Merck, 548
mercury, 33, 42, 48, 152, 362
Messina, 374
Messines, 529
methyl isocyanate, 591
Meuse, 66
Mexican Revolution, 112, 271
Mexico, 42, 57, 60, 66, 112, 116, 130–31, 145, 152, 157, 160, 187, 196, 264–65, 271, 278, 289, 296, 335, 366, 372, 392, 418, 420–22, 453, 617
Mexico City, 160, 265, 392

Miami, 71
miasma theory, 249, 324–26, 606
Michelet, Jules, 149, 152
Michigan, 356, 362, 496
Micronesia, 120
Migros, 285
Milford Haven, 584, 589
Military. *See* Path of War, the
Miller, Zell, 210
Minamata, 362–63
mining. *See* Mining Path, the
Mining Path, the, 23
Minoan civilization, 259
mirex, 579, 581–82
Miskito people, 546
Mississippi (state), 189, 193, 196, 198
Mississippi River, 193, 195, 266, 375–76, 516
Missouri River, 499
Missouri Valley Authority, 436
mita, 45
mixed forestry, 36, 80–81, 83, 413, 634
Mobile, Alab., 216, 575
mobility. *See* Mobility Path, the
Mobility Path, the, 28
Modi, Narenda, 260
Mojave Desert, 529
Moldova, 477
Moltke, Hellmuth von, 489
Molvanîa, 344
Monaco, 216
Monet, Claude, 245, 247–48, 257
Mongolia, 445, 473
Mongols, 371
mongoose, 219
Monju Breeder Reactor, 567
monk seal, 180
monoculture, 36, 54–55, 80–83, 104, 113, 132, 165, 167, 171, 186–98, 212, 413, 421, 580, 619, 622, 624, 632, 640, 664n31
monosodium glutamate, 647
Monsanto, 426–27
Monserrat, 51
monsoon, 262
Mont Pèlerin Society, 556
Monte Verità, 356
Montreal, 59
Montreal Protocol, 368
Moore, Charles, 603
Morales, Eva, 353
Moravia, 512
Morgan, Arthur E., 436
Morocco, 234, 381
Morris, 270

Moscow, 271, 276
Moses, 386, 434
Moses, Robert, 261, 512, 518
mosquitoes, 539, 573–74, 580, 601
Mother Teresa, 233
Moton, Robert Russa, 376
motorway. *See* divided highway
Moule, Henry, 262
Mozambique, 391, 400–401, 409
Muckalee Creek, 205
Mugabe, Robert, 115
Mughal Empire, 175, 177
Muhammad Ali Pasha, 431
Muir, John, 347
mulberry, 62
Müller, Ferdinand von, 408–10, 634
Müller, Hans, 285
Müller, Paul, 572–73, 575
Munich, 79–80, 136, 262, 286, 329, 410, 426, 511
Münster, 287, 395
Murcia, 373
Mursi people, 402
Muscle Shoals, 288–90
mushrooms, 76
Musk, Elon, 344
Muslim Brotherhood, 72
Muslims, 72–73, 85, 226, 234, 304, 320, 342, 354
Muspratt works, 133
Mussolini, Arnaldo, 493
Mussolini, Benito, 205, 394, 402, 461, 464, 467, 482–85, 489, 492–93, 509
Myanmar, 42, 342, 374, 474
mynah bird, 219

Nader, Ralph, 347, 354
Nagasaki, 559, 561, 569
Nagoya, 362, 377
Nairobi, 547
Najaf, 320
Namibia, 47, 138
Nancy, 78
Nansen, Fridtjof, 477
Naples, 330, 335, 572, 574
Napoleon, 56, 66, 70, 73, 142, 376
Napoleon III, 130
narcotics. *See* drugs
Nasser, Gamal Abdel, 226, 430, 434, 436, 439
Natal, 408
National Aeronautics and Space Administration (NASA), 595–96, 602–4
National Audubon Society, 577

National Cotton Council, 196, 599
National Farmers' Union, 555
National Grange, 496
National Institutes of Health, 574
National Marine Fisheries Services, 552, 603
National Park Service, 401, 528
national parks, definition of, 110, 367, 394. *See also* nature reserves
National Tropical Botanical Garden (Hawaii), 127
National Trust for England and Wales, 397
National Wildlife Federation, 150
National Wildlife Refuge, 394
natural disasters, 119, 155, 193–94, 203, 220, 236, 310–11, 316–17, 324, 331, 344, 370–79, 431, 566, 586, 594, 607, 636–37
natural gas, 232, 254
natural history, 166, 174–76, 180, 182, 395
Natural Resources Conservation Service, 203. *See also* Soil Conservation Service
nature reserves, 5, 20, 48–49, 60, 69, 110, 168, 181, 184–85, 199–200, 209–11, 262–63, 335, 344, 353, 361, 364, 367, 382–83, 391–404, 424, 456, 488, 527–28, 619–20, 628, 640, 644
Nauru, 138–41, 232, 292, 677n45
Nautilus, 536
Navarro de Andrade, Edmundo, 410, 413
Nazis, 9, 67, 82–83, 90, 92, 136, 147–49, 242, 284, 286, 336, 340–41, 455, 460–61, 463–67, 480, 486–87, 509–12, 515, 517, 525, 597–98, 633, 637
Nebraska, 500
nectarine trees, 121
neem trees, 426–27
Nehru, Jawaharlal, 349, 559
Nehru, Motilal, 349
Nelson, Horatio, 83
neoliberalism, 18, 58, 85–86, 115, 140, 163, 213, 278, 355, 369, 453, 500, 504, 537, 540, 543–44, 546–47, 556–57, 581, 593, 637, 683n47
Nepal, 331, 443–44
Neptun-Bratwurst, 148
Nero, 485
Neruda, Pablo, 158
Nestlé, 279
Netherlands, 42, 62, 66, 90, 143, 146, 166, 173–79, 203, 232, 237, 261, 335, 376, 423, 429, 442, 449, 455, 475, 478, 486, 553, 585
neurasthenia, 758n3
Nevada, 568
Nevis, 48, 51

New Bedford, 144
New Caledonia, 529
New Deal, 203–7, 297, 417, 422, 436, 460–61, 465, 498–99, 502, 625
New Delhi, 155, 474
New Economic Policy, 470
New Guinea, 213–15
New Jersey, 74, 295
new materialism, 6–7, 11–15, 615–16
New Mexico, 524, 560
New Orleans, 155, 375–76, 499, 516
New South Wales, 44, 212
New York (state), 182, 512, 576, 592
New York City, 71, 89, 91–92, 110, 150, 162, 261–63, 271, 281–82, 298, 425, 512, 544, 607
New Zealand, 69, 109, 392, 407
Newcastle, 248, 252
Newcomb, Simon, 294
Newell, Frederick Haynes, 432, 434–35
Newfoundland, 153
Newson, Dale, 196
Nicaragua, 53, 112–13, 375, 451, 453, 546
Nicobar Islands, 176
Niger, 47, 85
Niger River, 592
Nigeria, 85, 235, 353, 445, 538, 592–93, 647
Nile, 113, 386, 429–37, 439–45
Nile Perch, 216
1950s syndrome, 532
nitrates, 116, 242, 284, 288, 290–91
Nitrates Directive, 291
nitrogen, 133–35. *See also* synthetic nitrogen
Nitrophoska, 286
Nixon, Richard, 304, 306–7
Nkrumah, Kwame, 436, 444
Nobel awards, 69, 158, 281–82, 285, 302, 310, 347, 355, 421, 427, 474, 559, 561, 573
noise, 151, 249, 256, 513, 516, 619
nomads, 9, 72, 74, 200, 436, 463, 481, 486, 644
Norderney, 295
Nordhaus, Ted, 578
Nordhoff, Charles, 126
Norfolk four-course system, 54, 59
North Atlantic Treaty Organization (NATO), 152
North Carolina, 197
North Korea, 343, 394, 441, 527, 567, 571
North Pole, 381
North Sea, 67, 177, 248, 254, 376, 486
North-South Commission, 366
Northeast Greenland National Park, 393
Northern Ireland, 96, 517, 769n81

Norway, 83, 96, 103, 146–47, 235, 335, 427, 477, 579, 603
Nottingham, 249
Nouakchott, 601
Nowa Huta, 644
NPK mentality, 285
Nubians, 439
nuclear power, 47, 232, 344, 369, 371, 418, 427, 461, 478, 492, 524, 529, 533, 535–37, 540–41, 558–71, 590–91, 618, 620, 625–26, 635, 640, 649
nuclear semiotics, 568
nuclear waste, 568, 620
Nuclear Waste Policy Act, 568
nuclear weapons testing, 535–36, 558–63, 568–69, 640
nudists, 335
nylon, 596

O'Rourke, P. J., 343
Oahu Sugar Company, 217
oak trees, 83
Oakland, Calif., 92
Obama, Barack, 568
obesity, 61, 463, 477, 617
Odessa, 330
Offenbach, Jacques, 333
Office of Civilian Radioactive Waste Management, 568
Ogoni people, 592
Ohio, 229
oil. *See* petroleum
oil crisis (1973), 231, 504, 514–16, 600
Oil Pollution Act, 589
oil spills, 290, 369, 426, 538, 540, 584–90, 592–94, 642
oil states. *See* resource states
Okinawa, 520
Oktoberfest, 321
Oldenburg Münsterland, 545, 547–49
olive trees, 197, 405
Olkiluoto Nuclear Power Plant, 567
Olympic Games, 336
Oman, 138, 235
Onassis, Aristotle, 150, 368
OPC Group, 234
open seas, 143–44
opencast mining, 47
opera houses, 43, 46, 226, 641
Operation Ranch Hand, 528
opioid crisis, 307
opium, 303–7, 322, 608, 782n7
Opium Wars, 71, 305, 408

Oppau, 281
Oppenheimer, Robert, 561
Orange Revolution, 479
oranges, 10, 197
organic farming, 242, 279, 285–87, 356, 427, 492, 512, 537, 555–56, 576, 626
organic machine, 430
Organisation for Economic Co-operation and Development (OECD), 505
Organization for Tropical Studies, 218
Organization of Petroleum Exporting Countries (OPEC), 230, 235
Orinoco, 437
Osaka, 359
Ostrom, Elinor, 355–56
Otago, 407
Ottoman Empire, 340–41, 479
Ouro Preto, 44
overproduction, 57, 161, 492, 498, 505, 545
Owens Valley, 262
Oxford, 176, 395, 474
ozone hole, 299, 368–69, 561, 603

Pa Mong, 438
Pachamama, 353
Pacini, Filippo, 325–26
Packard, Vance, 93, 598
Pakistan, 90, 112, 230, 267–68, 310, 332, 367, 421, 436–38, 442, 567, 627, 646
paleontology, 178–79
Palestine, 335–36, 340, 405–7, 411, 413
palm oil, 455, 503
Pan Am building, 162
Panama, 71, 73, 150
Panama Canal, 71, 73, 266
Panama disease, 159, 188
Papal States, 336, 372
paper and pulp industry, 83, 384, 414–15
paper parks, 383, 400. *See also* nature reserves
Papua New Guinea, 213–15, 402
Paraguay, 440, 444
Paris, 93, 256, 266, 268, 278, 321, 327–28, 336–37, 360–61, 367–68, 377, 517, 642
passenger pigeon, 180, 185
Pasteur, Louis, 324
patents, 133, 135, 138, 259, 262, 323, 385, 422–23, 426–27, 525, 572
Path of Chemistry, the, 25
Path of Disasters, the, 27
Path of Energy, the, 24
Path of Pollution, the, 24
Path of the Animals, the, 23
Path of War, the, 28
Path Toward the City, the, 28
Pauling, Linus, 561
Paulista Railroad Company, 410
peace parks, 393, 528
peak phosphorus, 293
peanuts, 189, 497, 520
Pearl Harbor, 522
Pearson, Weetman, 265
Peary, Robert, 381
peas, 54
peasant parties, 114
Peenemünde, 283
Pennsylvania, 229, 565, 590
Peoria, Ill., 499
pepper, 121, 448
Perdue, Frank, 555
Pérez Alfonzo, Juan Pablo, 235
Perón, Juan, 375
Perrow, Charles, 590
Perry, Rick, 427
Persian Gulf, 90, 226, 229, 510, 584
persistent organic pollutants, 97, 215–16, 368, 537, 561, 582–83, 618, 625
Peru, 42, 46, 53, 102–3, 130–31, 136, 140–41, 152, 170, 187, 226, 232, 266, 289, 317, 330, 545, 646
pesticides, 19, 81, 116, 137, 189, 191, 195–97, 209, 213, 215, 421, 425, 427, 496, 525, 536–37, 572–83, 634
pests, 6, 19, 60, 79–81, 213–15, 428, 572–83, 622, 625, 632, 634
Pettenkofer, Max von, 326
petroleum, 10, 68, 89, 146, 155, 166, 170–71, 226–36, 265, 283, 305–6, 361–62, 369, 393, 426, 456, 461, 465, 497–505, 514, 516, 518–20, 522, 538, 540, 584–90, 592–94, 599–601, 627, 647
Pfeil, Wilhelm, 80, 84
Phaic Tan, 344
pharmaceutics, 548–50
Philadelphia, 155, 299, 329, 499
Philae Island, 435
Philip II (Spain), 9
Philippines, 42, 68, 215, 321, 330–31, 374, 421, 513, 519
phlogiston, 25
Phoenix, 296, 432
phosphates, 133, 135, 138–39, 141, 232, 234, 246, 284–85, 289, 293
phylloxera, 188
Picasso, Pablo, v

INDEX

Pietermaritzburg, 351–52, 408
pigeons, 176, 180, 185
pigs, 76, 92, 101, 122, 125, 216, 267, 269–70, 274, 276, 465, 502, 512, 550–51
Pilanesberg National Park, 403
Pinchot, Gifford, 78
pine roots campaign, 460–61, 466–67, 519–24, 526–27, 529, 625
pine trees, 194, 405
pineapples, 121
pink flamingos, 596
Pioneer Hi-Bred Corn, 425
piracy, 153
Pitcairn Island, 120
Pittsburgh, 361, 363, 436, 619
Pius XII, 349, 559
plague, 10
plant breeding. *See* seed improvement
plantain, 126
plastics, 83, 93, 150–51, 179, 228, 402, 461, 496, 538–40, 595–604, 620, 625, 627–29, 636, 645
Pliny the Elder, 261
Plitvice Lakes National Park, 393
plutonium, 540, 567, 602, 618
Plymouth, 588
Poivre, Pierre, 408
Poland, 40, 59, 63, 181, 319–20, 340, 394, 473, 475, 478, 644
Polanski, Roman, 262
Polanyi, Karl, 275
polders, 377–78
Pollan, Michael, 288, 290
pollution. *See* Path of Pollution, the
polychlorinated biphenyl (PCB), 152, 426, 582
polyethylene, 597–98, 601, 603
Polynesia, 120, 144
pomegranates, 121
Pompeii, 636
Pondoland, 476
Pontine Marshes, 147, 202, 394, 407, 460–61, 463–64, 467, 482–93, 581, 625, 633, 635
popes, 338, 349, 372, 393, 559. *See also* Papal States
population bomb, 502
Populist Revolt, 114, 157
Port Douglas, 223
Portugal, 42, 51, 173, 330, 342, 371, 373, 407, 409, 414, 434, 489, 513–14, 586
potash, 284–85, 293
potato famine, 133
potatoes, 32, 213

Potosí, 19–20, 32–33, 39–48, 139, 226, 247, 304–5, 424, 431, 497, 630, 641, 648
Potosí, La Paz and Peruvian Mining Association, 46
Potsdam, 334
Poubelle, Eugène, 93
Prague, 114
Prestige, 589
prickly pear cactus, 221
Prince Philip, 557
Prince William Sound, 589
professions. *See* Building Professions
Progressive movement, 157, 272, 278–79, 425
prohibition, 499
prostitution, 72, 304, 322
Proust, Adrien, 328
Proust, Marcel, 328
Providence Canyon. *See* Little Grand Canyon
Prussia, 78–79, 82, 142–45, 285, 311, 323, 326–27, 373, 434, 488, 513
Pschyrembel, 338
Public Eye, 600
public health, 91, 240, 260, 265, 272, 326–27, 396, 573–74
public trust doctrine, 315, 362, 367
Puerto Limón, 154
Puerto Rico, 53, 59, 214–15, 217, 219, 545
pulp production. *See* paper and pulp industry
Punjab, 300
Pure Food and Drug Act, 272
Puttkamer, Robert von, 250–51
Putumayo affair, 455–56
Pyongyang, 343
pyrethrum, 525

Qatar, 229
quagga, 180
Quaker, 553
quantification, 45, 69, 82, 101–2, 105, 122–24, 183, 256, 275, 286, 322, 384, 502, 551, 554, 621
quarantines, 10, 188, 311, 326–27, 383, 606
Quebec, 397
Queen Elizabeth, 557
Queensland, 53, 138, 169, 179, 212–15, 217–19, 223–24, 620
Quran, 431

rabbits, 221, 574
racism, 59, 95, 115, 158–59, 167, 187, 191–93, 207, 213, 229, 260, 270, 273–74, 281, 291, 296, 301, 306–7, 340, 347, 351–52, 376, 391,

803

racism (*cont.*): 397–98, 454–55, 512, 592, 634
radiation, 332, 371, 528, 536, 558–63, 569–71, 622, 640
radium, 559
Ragusa. *See* Dubrovnik
railroads, 21, 67–71, 154–56, 158, 203, 258, 265, 271–72, 274, 277, 294, 296, 321–23, 351–52, 355, 370, 381, 396, 406, 410, 512, 545, 585
rainmaking, 302
ranching, 456, 552
rape, 240, 260
rapeseed, 525
rats, 178–79, 272
Ratzel, Friedrich, 336, 487–88
Reagan, Ronald, 369, 505, 543, 582
recycling, 37, 87–97, 539, 602–3
red lists, 166, 181–82, 367
Red Sea, 72
refineries, 89, 520, 585
Reisner, Marc, 442
Remarque, Erich Maria, 526
renewable resources, 103, 150, 239, 246, 416, 501–4
rentier economy, 170. *See also* resource states
repair science, 36, 189
reserves, 62, 77, 208, 381–89
resource curse, 235
resource states, 130, 140–41, 165–66, 170–71, 225–36, 353, 393, 424, 518, 620, 628, 644, 647
Resources for the Future, 360
restoration of mining landscapes, 139
Réunion, 178, 180
Revolutionary Armed Forces of Colombia (FARC), 527
Rheinberg, 66
Rhine, 66, 282, 312, 333–35, 340, 586
rhinoceros, 184, 402–3
Rhodesia, 69, 397
Ricardo, David, 34, 55, 57, 451, 624
rice, 42, 58, 62, 186, 201, 387–88, 421, 433, 447–50, 454, 457, 492, 625, 633, 645
rice-eating rubber tree, 387, 447–50, 454, 457, 625, 633, 645
Richards, Mary, 576
Rickover, Hyman, 529
Ridley, Henry Nicholas, 127, 448
rinderpest, 197, 311, 332
Ring of the Nibelung, 339
Rio de Janeiro, 263, 300, 394. *See also* Earth Summit
Riquet, Pierre-Paul, 67–68, 70, 73–74, 434

Riyadh, 228, 513, 518
robusta coffee, 191
Rochefort, 66
Rockefeller, John D., 68, 227
Rockefeller Foundation, 421, 428
Rockefeller Institute for Medical Research, 68
Rocky Mountain Arsenal, 394
Rodrigues, 173, 180
Roentgen Society, 559–60
Roerich, Nicholas, 417
Roll Back Malaria Partnership, 582
Roman Empire, 66, 201, 259, 261, 268, 334, 431, 443, 485, 488, 509, 636, 641
Romania, 71, 79, 472
Rome, 261, 268, 334, 372, 375–76, 463, 483–85, 492, 509, 641
Rönne, Friedrich Ludwig von, 143
Röntgen, Wilhelm Conrad, 559
Roosevelt, Franklin D., 203, 225–28, 234, 297, 417, 425, 436, 498
Roosevelt, Theodore, 181, 187, 434
Rostock, 293
Rothamsted Experimental Station, 484
Rother, Christian von, 142–45, 153
Rothschild, 411
Rousseau, Jean-Jacques, 586
Rove, Karl, 427
Rowe, Phil, 163
Royal Agricultural Society, 136
Royal Geographical Society, 484–85
Royal National Park, 392
Royal Society, 406
Royal Society of Arts, 119, 121
rubber, 43, 46, 60, 127, 387–88, 409, 411, 420, 423–24, 447–50, 454–57, 522, 596–97, 625, 633, 645
Ruckelshaus, William, 579
Ruhr Area, 44, 247, 340, 361
rum, 54, 305
Rümker, Kurt von, 418
Rumsfeld, Donald, 309
Russell, E. John, 484–86
Russia, 3, 57, 69, 111–12, 115, 183–84, 200–201, 203, 207–8, 265, 271, 320, 323–24, 328, 330, 335, 344, 374, 381, 417, 423, 425, 445, 470, 473, 475–77, 479, 481, 490–91. *See also* Soviet Union
Russian Academy of Science, 111
Rusumo Falls Hydroelectric Project, 445
Rwanda, 342, 445, 601

Sabaudia, 493

SABMiller, 44
saccharine, 34, 55–56, 58, 61
Sachsenhausen Concentration Camp, 598
sacrificial death, 594
Saddam City, 513
Sadr City, 513
safari, 392, 434
Sahara, 84
Sahel, 35, 62, 84–86, 632
Saint Helena, 73, 121
Saint Kitts and Nevis, 48, 51, 54
Saint Louis, 266
Saint Lucia, 411
Saint Petersburg, 57, 112, 184, 265, 271, 423
Saint Vincent, 119, 121
Saint-Domingue, 52, 54, 118, 121–22
Saint-Étienne, 91–92
Saint-Ferréol, 65, 431
Saint-Louis (Senegal), 262
Salazar, António, 342, 434
Salman of Saudi Arabia, 226
salt, 40, 305, 314, 346, 352, 354–55
salt march 313–14, 346–50, 352, 354–55, 357, 450
Salt River Dam, 432, 434
Salter, Robert M., 206
saltpeter, 281, 522
Salvemini, Gaetano, 484, 491
Salzburg, 511
Sami people, 397
Samoa, 125
San Antonio, 296
San Bernardino, 276
San Diego, 185
San Francisco, 432, 499
San Francisco Bay, 92
San Luis Potosí, 44
San Sombrèro, 344
Sandinistas, 113, 451, 453
Sandoz, 586
sanitation, 5–6, 91, 100, 139, 238–40, 243, 248–49, 258–68, 272, 281, 284, 311, 324–28, 330, 352, 359, 387, 396, 461, 512, 561, 619–20, 622, 624, 641
Santa Barbara oil spill, 592
Santa Monica, 152, 262
Santander, 326
Santiago, 451
São Tomé, 51
Sarawak, 450
Sardinia, 483
Saro-Wiwa, Ken, 353, 592

satyagraha, 313, 352, 354
Saudi Arabia, 166, 170–71, 225–31, 233–36, 394, 481, 501, 513–14, 518, 620, 628, 644
Saudi Aramco, 170–71, 229–30, 234–35
Savery, Roelandt, 175
Sax, Joseph, 315, 362, 366–67
Saxony, 76, 78, 81, 122
Scharnhorst, Wilhelm von, 320
Scheffel, Victor von, 129–32
schistosomiasis, 441
Schleißheim, 136
Schlich, Wilhelm (William), 78
Schmidt, Helmut, 375
Schmitt, Carl, 149
Scholtyssek, Siegfried, 554
Schröder, Gerhard, 375
Schumacher, Ernst Friedrich, 353
Schumpeter, Joseph, 537
Schutzstaffel (SS), 467, 520
Schwarzschild & Sulzberger, 270
Schwaz, 40
Schweinitz, Hans Lothar von, 476
Schweisfurth, Karl Ludwig, 277, 279, 556, 636
Schweitzer, Albert, 559
Scilly Isles, 584
Scotland, 82–83, 342, 646
Scott, Robert, 381
Sea Island Cotton, 191
Sea Shepherd, 152
Sears catalog, 271
Seattle, 599
sedentarization, 481
seed companies, 385, 425, 427–28
seed improvement, 19, 55–56, 107, 114, 116, 137, 150, 183, 189, 195, 276, 385–86, 417–28, 463, 472, 487, 534, 550, 574, 619, 622, 625, 628, 632, 644
seedbanks, 166, 183–84
Sen, Amartya, 355, 463, 474, 476–77
Senegal, 85, 262, 297
Seoul, 394
Serampore, 409
Serpieri, Arrigo, 487, 491
Sète, 64
Seven Sisters, 227, 230
Seven Years' War, 50, 78, 118
Seveso, 586, 588, 591
Seveso Directives, 591
sewage farms, 267–68, 281, 619
sewage treatment, 5, 267–68, 359, 512, 619
sewers, 6, 91, 100, 139, 239–40, 243, 249, 260–68, 284, 328, 396, 619, 641

Seychelles, 368
Shakespeare, 262
Shanghai, 91, 306, 601
Shanxi Province, 248
sharecropping, 167, 191–93, 471, 545, 632
Sharjah, 90
Shat-el-Arab, 330
Shcherbytsky, Volodymyr, 478
sheep, 201, 476, 601
Shelbyville, Ill., 321, 323
Shell, 585, 592–93
Shellenberger, Michael, 578
Sherlock Holmes, 246, 256
Sherman, William Tecumseh, 296
shifting baseline syndrome, 183
Shimla, 295
Shine, Rick, 220, 222, 224, 433
shipbreaking, 32, 36–37, 87–90, 93–97, 368, 626–27
Shippingport Atomic Power Station, 529, 565
Shiva, Vandana, 137, 414–15, 426–27
shock doctrine, 36
shoe industry, 597–98
Siam, 449
Siberia, 381
Sichuan Province, 374–75, 475
sick building syndrome, 299
Sierra Club, 150, 399
Sierra Leone, 440, 573
Siesta, 295, 303
Sigatoka fungus, 159
Silesia, 142
silicosis, 45
silk, 62, 496, 523
silver, 19–20, 32–33, 35, 39–43, 45–46, 48, 66, 88, 226–27, 247, 305, 431, 648
Simon, Julian, 291
Simpson, Wallis, 335
Simpsons, 570
Sinai desert, 434
Sinclair, Upton, 269, 272–74, 277, 279
Singapore, 127, 395, 409, 448, 589
sisal, 59–60, 400
Sitka spruce, 82
skyscrapers, 298–99
slaughterhouses, 12, 20, 59, 70, 143, 148, 155, 239–41, 266, 269–75, 277–79, 284, 356, 461, 512, 534, 618–19, 623, 625, 627–28, 644, 648
slavery, 33–34, 43, 51–53, 58, 61–62, 101, 118–19, 121–22, 125, 127, 131, 144, 187, 191, 201, 297, 321, 450, 504, 630
Slovakia, 472

Slovenia, 42, 393
smallpox, 197, 311, 322, 332
Smith, Adam, 39, 143, 153, 156, 227, 248, 305, 312, 646
smog. *See* smoke
smoke, 5, 20, 239, 245–57, 266, 536, 561, 622–23, 625, 628, 632–33, 640–41
Smuts, Jan, 315, 413
snakes, 216, 222, 433, 529
Snow, John, 325
soap, 103, 146, 148
Social Darwinism, 487
Social Democratic Party (Germany), 340
Socialism, 35–36, 63, 79, 92, 112, 114–15, 146, 160–61, 233, 260, 272–73, 316, 338, 437, 451, 453, 461, 469–73, 477–81, 490–91, 507, 544, 562, 597, 644
Society for the Preservation of the Wild Fauna of the Empire, 181
Society for the Prevention of Smoke (Chicago), 250
Society of the Plastics Industry, 598–99
Sofia, 57
Soil Conservation Service (USA), 116, 168, 203–6, 209, 211, 287, 489, 633
soil erosion, 116, 124, 139, 165–67, 171, 199–211, 213, 216, 287–88, 406, 413, 486, 489, 527, 620, 628, 633, 643, 645
Solomon Islands, 212
Somalia, 153
Sombart, Werner, 508
Somoza, Anastasio, 316, 375, 546
Somoza family, 112
Sontag, Susan, 586
Sörgel, Herman, 431
South Africa, 44, 69, 77, 121, 184, 207–8, 315, 332, 347, 351–52, 354, 356, 368, 391–92, 395–403, 408, 412–13, 476, 601, 640
South Carolina, 191, 197
South Korea, 90, 394, 434, 438, 527, 627
South Pole, 381
South Sudan, 431
South Vietnam, 438
Southern corn leaf blight, 428
Southwest Airlines, 591
sovereign wealth funds, 235
Soviet Union, 101, 112, 114–15, 139, 150, 184, 202, 230, 232, 276, 300, 305, 338–39, 361, 368, 394, 420, 437, 439–41, 445, 460–63, 466, 469–73, 477–78, 480–81, 490–91, 514, 559, 563, 565, 571, 644, 646
Soyang Dam, 434
soybeans, 496, 501, 520, 525

space shuttle, 18, 595–96, 602–3
spaceflight, 18, 151, 256, 355, 357, 439, 539–40, 595–96, 602–4, 645
Spain, 40, 42–43, 51, 66, 102–3, 143, 211, 265, 311, 323, 326, 330, 342–43, 373, 407, 414, 434, 436, 617, 645–46
Spanbroekmolen, 529
Spanish-American War, 91
Spengler, Oswald, 431, 636
Spiderman, 427
Spielberg, Steven, 185
Spitsbergen, 143, 184, 428
Spock, Marjorie, 576
Sporn, Philip, 566
Spreda, 547
Sprengel, Carl, 134
springbok, 392
Springfield Nuclear Power Plant, 570
squatting, 109, 258–59, 264, 397
Sri Lanka, 79, 409, 427, 573
Stalin, Joseph, 112, 225, 228, 461, 469–73, 475, 477, 480, 514
Stalin Plan for the Transformation of Nature, 491
Standard Oil, 155, 227, 498, 512
Stanford Research Institute, 107, 115
Stanford University, 68
Staples Vegetables, 59
Star Trek, 256
Starbucks, 151
statehood. *See* Building the State, 26
Statens Pensjonsfond, 235
statistics, 122–23, 256, 322, 502. *See also* quantification
steam engines, 33, 46, 246–47, 585
steel production, 155–56, 284
Steenbergen, 429–30, 442–43
Steiner, Rudolf, 285
Steller's sea cow, 180
Stevenson-Hamilton, James, 396–97, 403
stilbestrol, 549
stimulants, 303–7
Stinnes, Hugo, 509
Stockholm Conference on the Human Environment, 315, 360, 364, 367–69, 579
Stockholm Convention on Persistent Organic Pollutants, 215–16, 368, 582–83, 618
Stone, Alfred, 189
stone louse, 338
Strauss, Levi, 44
Strauss, Lewis, 565
strawberries, 121, 425
strikes, 44

Stroessner, Alfredo, 440, 444
Struve, Gustav, 357
Stuttgart, 515
subsidies, 34, 57–58, 85, 143, 153, 168, 204, 206, 265, 411, 457, 503, 557, 566, 568
subsistence production, 62–63, 117, 123, 387, 449
suburbs, 194, 219, 264, 276, 296, 298, 501, 511, 561, 596
Sudan, 430, 435, 445
Suez Canal, 67, 71–73, 225, 439
Suez crisis, 437, 439
sugar, 32–35, 43, 50–58, 61–62, 128, 132, 148, 523, 599, 649
sugar beets, 34, 53, 56–58, 488
sugarcane, 32–34, 48, 51–54, 56, 61–62, 118–19, 127, 138, 140, 169, 177, 179, 200, 212–15, 217–18, 224, 226–27, 305, 335, 411, 420, 450, 497, 504–5, 618, 620, 630
Suharto, 492
Sulawesi, 330
sulfur dioxide, 239, 255, 362
Summers, Lawrence, 95
Sumner Nuclear Power Plant, 568
Sunbelt, 296
supermarkets, 59, 94, 124, 277, 540, 545, 556, 600–601, 647
supertanker, 369, 540, 584–90, 592–94
Surabaya, 519
Suriname, 60, 553
sustainable development, 365–66
sustainable forestry, 32, 35–36, 65, 75–86, 177, 365, 406, 523, 625–26, 647. *See also* forestry
Suzuku, Shinzo, 558
Svalbard Global Seed Vault, 184, 428
Swabia, 129, 132
swamps, 19, 384, 405–6, 411, 413, 431, 492
Swaziland, 396
Sweden, 40, 77, 82, 335, 360, 397, 423, 516, 579
swidden farming, 448
Swift, 270, 272, 277
Swift, Jonathan, 421
Switzerland, 55–56, 67, 285, 328, 335, 337, 356, 376, 381, 401, 483, 508, 516, 545, 565, 572, 574, 585, 600
Sydney, 212, 221, 263, 267, 392, 400, 410
Sykes, Geoffrey, 549
synthetic gasoline, 283–84, 511, 524, 573
synthetic nitrogen, 19, 55, 114, 135, 146, 195, 209, 238–39, 241–42, 268, 276, 281–93, 442, 496, 503, 505, 522, 534, 548, 550, 566, 575, 585, 620, 622–23, 626, 628, 640
Syria, 335, 340

tachinid flies, 213
Tahiti, 119–21, 411
Taiwan, 90, 215, 259, 627
Tajikistan, 445
Taliban, 342, 445
Taman Negara, 397
Tammany Hall, 91
tampons, 18, 552
Tanganyika, 397, 491
Tangshan, 310, 316–17, 344, 370–74, 376–79, 636, 649
Tanker Owners Agreement Concerning Liability for Oil Pollution, 589
Tanzania, 69, 179, 397, 409, 445, 448, 491
Tariki, Abdullah, 230
taro, 126
Tashkent, 321, 328
Tasi Tolu, 392
Tasmania, 121
taxonomy, 121, 174, 409
Taylor, Zachary, 321
tea, 51, 305, 408–9
technocracy, 495
technological disasters, 232, 426, 461, 538, 566, 584–94, 641
technological momentum, 283–84. *See also* large technological systems
Telgte, 70
10 Downing Street, 253
Tennessee Valley Authority (TVA), 289–90, 436, 438, 566
Tepeyollotl, 372
territoriality, 382, 395–400
tetraethyl lead, 465, 498–99
Texas, 81, 186, 188–89, 196–97, 234, 276, 287, 296, 354, 427, 581, 585
Texas City, 585
Texas Railroad Commission, 234
Thailand, 414–15, 434, 438, 449
Thames, 245–47, 261–62
Tharandt, 78–79
Thatcher, Margaret, 82, 543
Therese of Saxe-Hildburghausen, 320
Thirlmere Reservoir, 262
Thomas Cook, 312, 329
Thompson, Elihu, 559
Thompson, Hunter S., 304
Thoreau, Henry David, 352
Three Gorges Dam, 440, 443–44, 624
Three Mile Island Nuclear Power Plant, 590
Tianjin, 370
Tiber, 372, 376

Tibet, 375
Tiflis, 228
tiger, 220, 320, 401
Tijuca National Park, 263, 394
timber famine, 77–78
Timor, 120
tin, 40, 46
Titanic, 589, 645
Titus Livius, 268
toads, 169, 214. *See also* cane toads
tobacco, 52, 56, 186, 304–5
Todt, Fritz, 509–10, 512, 518
Togo, 201
Tokyo, 268, 315, 359–66, 369, 371, 562–63
Tokyo Electric Power Company (TEPCO), 369
Tokyo Resolution, 315, 359–63, 367, 369, 628, 645
Tolstoy, Leo, 352, 567
tomatoes, 425–26
Tongariro National Park, 392
Torrey Canyon, 426, 533, 538, 584–90, 592–94, 607, 642, 645–46
tortoises, 178
total history, 613
total war, 218, 460, 466–67, 519–30, 643
Toulon, 66
Toulouse, 64, 73
tourism, 69, 71, 74, 200, 211, 233, 266, 306–7, 311–13, 317, 327, 333–45, 359, 379, 383, 393–94, 399–403, 442, 492–93, 527–28, 557, 563, 587, 617, 624, 640
Toynbee, Arnold, 436, 445
tractors, 137, 196, 296, 469, 471, 573
Trans-Siberian Railroad, 69, 381
transhumance, 9
Transvaal, 392, 398
trash. *See* garbage
Treaty of Paris, 50
Triglav National Park, 393
Trinidad and Tobago, 125, 411
Trinitrotoluene (TNT), 647
Tropical Race Four, 163, 197
Tropical Trading and Transport Company, 154
Trudeau, Justin, 480
Truman, Harry S., 229, 425
Trump, Donald, 348, 357, 427, 607
trustbusting. *See* antitrust proceedings
Tsavo National Park, 399–400
tsetse fly, 179
tsunami, 369, 371, 378, 566
Tsuru, Shigeto, 360, 363–64, 723n8
tuberculosis, 322, 590

Tucson, 296
tuna, 559
tung oil, 496
Tunisia, 80, 201
Tupac Amaru Indian nationalist uprising, 45
Tupper, Earl Silas, 596
Tupperware, 596
Turgenev, Ivan, 323
Turkey, 96, 150, 187, 265, 306, 335–36, 513
turkeys, 548, 551
Turkish Petroleum Company, 227
Turkmenistan, 201, 445
Turner, Frederick Jackson, 208
Turner, J. M. W., 88
turnips, 54
turtles, 18, 367, 535, 540, 546–47, 551–53, 556–57
Tuskegee Institute, 376
Tutu, Desmond, 351
Tuvalu, 125
Twain, Mark, 336
typhoons, 371
typhus, 572
Tyrol, 40

Ubico, Jorge, 159
Udall, Stewart, 490
Uganda, 367, 397, 431, 444
Ui, Jun, 363
Ukraine, 247, 330, 344, 441, 460–63, 469–71, 473, 477–81, 534
Unilever, 146
Union Carbide, 591, 593
Union Oil Company, 589
unions, 44, 89, 241, 273–74, 277, 350, 456–57
United Arab Emirates, 90, 229, 300, 514
United Brands, 162
United Farm Workers, 58
United Fruit, 19, 53, 60, 100, 104–5, 154–64, 191, 227, 271, 343, 356, 455, 461, 631, 640
United Kingdom. *See* Great Britain
United Nations, 33, 48, 71, 153, 211, 259, 298, 315, 360, 364–65, 367, 393–94, 439, 451, 479, 504, 541, 550, 563–64, 579, 582
United Nations Children's Fund (UNICEF), 582
United Nations Education, Scientific and Cultural Organization (UNESCO), 33, 48, 71, 211, 367, 394, 439, 563
United Packinghouse Workers of America, 274
United Self-Defense Forces of Colombia, 162
United States of America, 3–4, 43–44, 47, 53, 58–59, 67–69, 73–74, 78, 81, 89, 91–96, 102–3, 107–10, 113–18, 120, 122, 127, 131, 138–39, 141, 143–44, 147–50, 152, 154–64, 166–68, 170, 178–79, 185–211, 214–17, 219, 225–27, 229–31, 240–41, 243, 260–61, 263, 266–67, 269–79, 281–84, 287–91, 294–302, 305–7, 309, 321–24, 329, 335–36, 340–43, 347–50, 352, 357, 360–63, 367–69, 375–76, 387, 392–401, 407–8, 417–23, 425–28, 432, 434–39, 441–45, 451–53, 455–57, 460–61, 464–65, 477–78, 484, 490, 494–502, 505, 507, 509, 512–16, 519–20, 522–29, 535–38, 543–45, 549–53, 555–56, 558–82, 585–86, 589–92, 595–99, 602–4, 607–8, 622, 625, 631–32, 642, 645–46
United States Steel, 155–56
University of Chicago, 68, 419, 565
University of Michigan, 362, 495
Uppsala, 423
uranium, 47, 394, 533, 565
urban growth. *See* Path Toward the City, the
Uruguay, 150, 450
Utah, 47
Uttarakhand, 77
Uzbekistan, 445

vaccines, 608
Valencia, 373
Vanuatu, 212
Vargas, Getúlio, 450–51
Vauban, 68, 74
Vavilov, Nikolai, 184, 470
Veblen, Thorstein, 51, 93
vegan diet, 279, 340
vegetarian diet, 279, 349–50, 356–57
Velasco, Andrés, 232, 236, 451–53
Velsicol Chemical Corporation, 577
Venezuela, 232–33, 235, 573
Venice, 310–11, 319–20, 322, 326, 329
Verdun, 508–9, 528
Vereenigde Oostindische Compagnie (VOC), 166, 176, 408, 522
Verne, Jules, 333, 336
Versailles, 66, 68, 633
vertical integration, 53, 88–89, 99, 104–5, 146, 154–57, 161–63, 545, 617, 625, 640
Vesuvius, 335, 636
Victor Emmanuel II (Italy), 375
Victoria (Australia), 44, 408
Vidal, Gore, 301
Vienna, 40, 79, 92, 265–66, 321, 361, 366, 512, 567
Vietnam, 330, 436, 438, 448, 528–29

Vietnam War, 196, 436, 438, 528–29
Virgin Lands campaign, 203
Virginia, 178, 196–97, 201
visual media, 151, 169, 181, 207–8, 218–19, 222, 256, 290, 538, 574, 585–88, 591–92, 607
vitamins, 124, 300, 476–77
Vlissingen, 261
vodka, 232, 305
Vogtle Nuclear Power Plant, 568
Voie Sacrée, 509
volcanoes, 335, 371–72, 377
Volga, 320, 470, 477, 486
Volkswagen, 511, 515, 547
Volta River Dam, 436, 444
Voltaire, 586
Voronezh, 324
Voyager spacecrafts, 151
vulcanization, 448

Wagner, Richard, 339, 344, 594
Waldkirch, 214, 217
Waldron, John, 217
Wales, 109, 111, 397, 584, 600
Walker, William, 321
Wallace, Henry A., 417–18, 421–22, 425
Wallerstein, Immanuel, 51–52, 59, 451, 457, 619. *See also* world-system
Wallich, Nathaniel, 409
Waluma, 402
war. *See* Path of War, the
War of the Pacific, 140
Ward, Nathaniel Bagshaw, 409
Wardian case, 409
Waring, George, 91–92
Washington DC, 143, 181, 188, 294, 296–97, 301, 321, 360, 422, 429, 513, 599
waste. *See* garbage
water. *See* Waterway, the
water buffalo, 492
water closet, 238, 240, 243, 258–63, 265, 267, 431, 626
water control. *See* hydrological regimes
water hyacinth, 216
water pollution, 262, 270–71, 282, 290–92, 315, 361, 364
water supply, 43, 261–66, 328, 394, 486
Waterway, the, 25
Wayne, John, 276
Weimar, 327
Wen Jiabao, 375
Wesel, 70
West India Company (Dutch), 66

West Pakistan, 112
Westinghouse, 566
Westinghouse, George, 155
Westphalia, 70, 395
whales, 103–4, 124, 142–53, 180, 220, 361, 368, 526, 626, 629, 640, 649
Whanganui National Park, 69
Whanganui River, 69
wheat, 10, 200, 276, 280–81, 284, 291, 306, 420–21, 423, 488, 490
Wheeler, Tony, 343
Wheldon, Rupert, 340
White, Alfred, 91
White, Hayden, 371
White Castle, 276
White House, 294–96. *See also individual presidents*
Whole Earth Catalog, 339–40
Wichita, 276
Wickham, Henry, 423–24
Wieliczka, 40
Wieringermeer, 486
Wilhelm II, 434, 507, 513
Willcocks, William, 435
Willoughby, Francis, 60
Wilson, E. O., 179–82, 184–85
Wilson, Woodrow, 507
wine, 69, 188, 304
Wise, Brownie, 596
Wittfogel, Karl August, 386, 433, 438
Witwatersrand, 643
wolves, 220
Wolzogen, Hans von, 339, 344
Wooden Path, the, 23
wood scarcity, 36
Worcester, Dean, 321
Works Progress Administration, 204
World Bank, 58, 95, 101, 115–16, 429, 437, 440, 442–43, 445, 582
World Health Organization, 256, 330–32, 516, 573, 575, 581–82, 605
World Toilet Day, 259
World Trade Organization, 162–63
World War I, 58, 62, 89, 124, 136, 228, 241, 282, 285, 288, 293, 354, 459–60, 479, 482–84, 494, 506, 508–9, 520, 522, 524–26, 528–30, 600, 626
World War II, 90, 112, 146, 149–50, 184, 208, 216, 226, 283, 286–87, 289, 305, 340–41, 354, 397, 420, 432, 441, 445, 455–57, 459–60, 466–67, 478, 487–88, 492, 500, 509, 519–27, 529, 531, 543, 553, 560–61, 572, 574, 581, 597

World Wildlife Fund, 401, 557
world's fairs, 155, 157, 250, 266, 270, 321, 336, 356, 359, 406
world-system, 42, 46, 51–52, 59, 451, 457, 619, 635
Wuhan, 605
Württemberg, 423
Würzburg, 559

Xylella fastidiosa, 197

Yale University, 301, 453, 590
Yalta, 225
Yamani, Ahmed Zaki, 230–31, 235–36
yams, 54
Yangtze River, 624
yellow fever, 73, 92, 262, 322, 572, 580
Yellow River, 377
Yellowknife mine, 47–48, 627
Yellowstone National Park, 392, 396, 644
Yemen, 331–32, 342–43, 513
Yokkaichi, 362–63
Yokkaichi asthma, 362
Yokohama, 371
York, 341
Yosemite National Park, 347, 392, 398, 401, 432, 644
Young, Arthur, 114
Yucca Mountain, 568
Yudkin, John, 61
Yugoslavia, 361, 373, 393
Yunnan Province, 42
Yushchenko, Victor, 479
Yushu Tibetan Autonomous Prefecture, 375

Zacatecas, 44
Zambesi, 432–33
Zambia, 69, 420, 432–33
Zapatista movement, 116, 453
Zell am See, 342
Žemaitija National Park, 394
Zemurray, Sam, 158–59
Zhonglin, Tan, 248
Zhou Enlai, 372–73
Zimbabwe, 69, 115–16, 181, 391, 401, 420, 432–33, 463
Zionists, 405–6
zoological gardens, 181, 184–5, 216, 395
Zuiderzee, 486
Zurich, 56, 328, 483
Zwentendorf, 567